Essentials of Electric Circuits

2nd edition

A RESTON BOOK
Prentice-Hall
Englewood Cliffs,
New Jersey

Essentials of Electric Circuits
2nd edition

JAMES H. HARTER
Professor, Electro-Mechanical Technology
Los Angeles Harbour College

PAUL Y. LIN
Watkins-Johnson Company
Vice President & Manager
Components Division

Marolyn Young, *Technical Illustrator*

Library of Congress Cataloging-in-Publication Data

HARTER, JAMES H.
    Essentials of electric circuits.

    Includes index.
    1. Electric circuits. I. Lin, Paul Y. II. Title.
TK454.H37   1986   621.319′2   85-14485
ISBN 0-8359-1787-8

© 1986 by James H. Harter and Paul Y. Lin

All rights reserved.
No part of this book may be reproduced
in any way, or by any means,
without permission in writing from the publisher.

10 9 8 7 6 5 4 3 2 1

Printed in the United States of America

CONTENTS

PREFACE 9

I Introduction/Review 1

1 Getting Started 3

1-1 STRUCTURE OF THE TEXT, 4
1-2 THE ELECTRIC CIRCUIT, 7

2 Electricity 11

2-1 WORK AND ENERGY, 12
2-2 FIELD FORCE, 14
2-3 ATOMIC STRUCTURE, 15
2-4 UNIT QUANTITY OF ELECTRIC CHARGE, 19
2-5 VOLTAGE, 19
2-6 ELECTRIC CURRENT, 21
2-7 SAFETY, 24

3 Measurement and Number Notation 29

3-1 SYSTEMS OF MEASUREMENT, 30
3-2 THE SI SYSTEM, 32
3-3 ENGINEERING NOTATION AND DECIMAL NOTATION, 38
3-4 FORMING DECIMAL MULTIPLES AND SUBMULTIPLES OF THE SI UNITS, 42
3-5 UNIT ANALYSIS AND CONVERSION WITHIN AND BETWEEN SYSTEMS OF UNITS, 44

II DC Circuits, Concepts, and Devices 53

4 Selected Properties of Electrical Materials 55

Part I: INSULATORS, SEMICONDUCTORS, AND CONDUCTORS, 56

4-1 ENERGY BANDS, 56
4-2 INSULATORS, 57
4-3 SEMICONDUCTORS, 62
4-4 CONDUCTORS, 63
4-5 RESISTANCE, 64

Part II: ROUND ELECTRIC CONDUCTORS, 77

- 4-6 RESISTANCE OF METALLIC WIRE, 78
- 4-7 EFFECT OF TEMPERATURE ON WIRE RESISTANCE, 79
- 4-8 AMERICAN WIRE GAUGE (AWG), 81
- 4-9 APPLICATIONS USING THE AWG AND FORMULAS, 85

5 Resistance and Resistors 91

- 5-1 OHM'S LAW, 92
- 5-2 FIXED LINEAR RESISTORS, 98
- 5-3 VARIABLE LINEAR RESISTORS, 103
- 5-4 NONLINEAR RESISTORS, 107
- 5-5 RESISTANCE DESIGNATION, 112

6 Power and Energy 125

- 6-1 COMPUTING AVERAGE POWER, 126
- 6-2 POWER CONSIDERATIONS IN LINEAR RESISTORS, 129
- 6-3 EFFICIENCY AND THE COST OF ENERGY, 134
- 6-4 CIRCUIT CONTROL AND PROTECTIVE DEVICES, 141

7 The Series Circuit 151

- 7-1 DEFINITION OF A SERIES CIRCUIT, 152
- 7-2 APPLICATION OF THE SERIES CIRCUIT RULES, 154
- 7-3 CURRENT DIRECTION AND THE POLARIZATION OF CIRCUIT COMPONENTS, 156
- 7-4 KIRCHHOFF'S VOLTAGE LAW, 158
- 7-5 VOLTAGE SOURCES IN SERIES, 161
- 7-6 FORMING AN EQUIVALENT SERIES CIRCUIT, 167
- 7-7 NOTATION AND REFERENCES, 170
- 7-8 VOLTAGE SOURCE CHARACTERISTICS, 174

8 The Parallel Circuit 187

- 8-1 DEFINITION OF A PARALLEL CIRCUIT, 188
- 8-2 APPLICATION OF THE PARALLEL CIRCUIT RULES, 190
- 8-3 KIRCHHOFF'S CURRENT LAW, 193
- 8-4 CURRENT SOURCES IN PARALLEL, 196
- 8-5 FORMING AN EQUIVALENT CIRCUIT, 198
- 8-6 CURRENT SOURCE CHARACTERISTICS, 202

9 The Series-Parallel Network 217

- 9-1 FORMING AN EQUIVALENT CIRCUIT FOR THE SERIES-PARALLEL NETWORK, 218
- 9-2 SOLVING SERIES-PARALLEL NETWORKS, 226
- 9-3 POWER DISSIPATION IN THE SERIES-PARALLEL NETWORK, 232
- 9-4 APPLICATIONS WITH THE SERIES-PARALLEL NETWORK, 236

10
Voltage and Current Dividers
251

10-1 VOLTAGE DIVIDERS, 252
10-2 VOLTAGE DIVIDER APPLICATIONS, 260
10-3 CURRENT DIVIDERS, 263
10-4 CURRENT DIVIDER APPLICATIONS, 269
10-5 LOADED AND UNLOADED CONDITIONS, 272
10-6 APPROXIMATIONS, 278

11
Equivalent Circuits and the Network Theorems
289

11-1 SUPERPOSITION THEOREM, 290
11-2 THEVENIN'S THEOREM, 301
11-3 NORTONS THEOREM, 311
11-4 CONVERSION OF THEVENIN'S AND NORTON'S EQUIVALENT CIRCUITS, 315
11-5 MAXIMUM POWER TRANSFER THEOREM, 318
11-6 MILLMAN'S THEOREM, 323
11-7 APPLYING THE NETWORK THEOREMS, 326

12
Loop and Node Analysis
343

12-1 LOOP ANALYSIS, 344
12-2 ESTABLISHING THE NUMBER OF LOOP EQUATIONS, 345
12-3 WRITING AND SOLVING LOOP EQUATIONS, 348
12-4 NODE ANALYSIS, 361
12-5 ESTABLISHING THE NUMBER OF NODE EQUATIONS, 362
12-6 WRITING AND SOLVING NODE EQUATIONS, 363
12-7 SELECTING NODE OR LOOP ANALYSIS, 365

13
Capacitance
379

13-1 ELECTROSTATICS, 380
13-2 CAPACITANCE, 385
13-3 CAPACITOR, 386
13-4 DIELECTRIC, 390
13-5 CONNECTING CAPACITORS IN SERIES AND PARALLEL, 396
13-6 CAPACITOR APPLICATIONS, CHARACTERISTICS, AND SPECIFICATIONS, 398

14
RC Transient Circuits
405

14-1 CHARGE AND DISCHARGE OF CAPACITORS, 406
14-2 RC TIME CONSTANT, 407
14-3 INITIAL AND FINAL CONDITIONS OF RC CIRCUITS, 409
14-4 INSTANTANEOUS VALUES OF RC CIRCUITS, 415
14-5 APPLICATIONS, 420

15
Magnetism
427

15-1 MAGNETS, 428
15-2 MAGNETIC FIELD CHARACTERISTICS, 429
15-3 SELECTED MAGNETIC QUANTITIES, 431
15-4 MAGNETIC PROPERTIES OF MATERIALS, 436

16 Electromagnetism and the Magnetic Circuit 443

16-1 ELECTROMAGNETISM, 444
16-2 SOLENOID, 447
16-3 SERIES MAGNETIC CIRCUITS, 450
16-4 AIR GAP IN THE MAGNETIC CIRCUIT, 455
16-5 SYMMETRICAL PARALLEL MAGNETIC CIRCUITS, 457
16-6 ADDITIONAL MAGNETIC UNITS, 459

17 Inductance 465

17-1 ELECTROMAGNETIC INDUCTION, 466
17-2 FLUX LINKAGE, 467
17-3 FARADAY'S LAW, 468
17-4 LENZ'S LAW, 468
17-5 SELF-INDUCTION, 471
17-6 INDUCTANCE OF A COIL, 472
17-7 ENERGY STORED BY INDUCTORS, 474
17-8 INDUCTORS IN SERIES AND PARALLEL, 475

18 RL Transient Circuits 479

18-1 CHARGE AND DISCHARGE OF INDUCTORS, 480
18-2 L/R TIME CONSTANT, 481
18-3 INITIAL AND FINAL CONDITIONS OF RL CONSTANTS, 484
18-4 INSTANTANEOUS VALUES OF RL CIRCUITS, 486
18-5 APPLICATIONS, 490

III AC Circuits, Concepts, and Devices 497

19 Alternating Current 499

19-1 ELEMENTARY GENERATOR AND AC TERMINOLOGY, 500
19-2 SINE WAVE, 502
19-3 PHASE RELATION OF R, L, AND C ELEMENTS, 505
19-4 EFFECTIVE AND AVERAGE VALUE OF CURRENT AND VOLTAGE, 513
19-5 AVERAGE POWER, 519

20 Mathematics of Phasors 527

20-1 COMPLEX NUMBERS, 528
20-2 TRANSFORMING COMPLEX NUMBER FORMS, 531
20-3 ARITHMETIC OPERATIONS WITH COMPLEX NUMBERS, 534
20-4 PHASORS, 538
20-5 OPERATING WITH PHASOR QUANTITIES, 544

21 Series and Parallel AC Circuits 551

21-1 VOLTAGE PHASOR FOR SERIES AC CIRCUITS, 552
21-2 IMPEDANCE OF SERIES AC CIRCUITS, 555
21-3 SOLVING SERIES AC CIRCUITS, 558
21-4 VOLTAGE DIVISION IN SERIES AC CIRCUITS, 561
21-5 CURRENT PHASOR FOR PARALLEL AC CIRCUITS, 563
21-6 ADMITTANCE OF PARALLEL AC CIRCUITS, 566
21-7 SOLVING PARALLEL AC CIRCUITS, 570

- 21-8 CURRENT DIVISION IN PARALLEL AC CIRCUITS, 575
- 21-9 FORMING SERIES AND PARALLEL AC EQUIVALENT CIRCUITS, 577

22 Series-Parallel AC Networks 589

- 22-1 SOLVING SERIES-PARALLEL AC NETWORKS, 590

23 Power in Alternating Current Circuits 601

- 23-1 INTRODUCTION TO AC POWER PARAMETERS, 602
- 23-2 AVERAGE POWER, 605
- 23-3 REACTIVE POWER, 607
- 23-4 POWER IN AC NETWORKS, 614
- 23-5 MAXIMUM POWER TRANSFER, 620

24 Transformers 627

- 24-1 TRANSFORMER BEHAVIOR, 628
- 24-2 TRANSFORMER VOLTAGE AND CURRENT RATIO, 630
- 24-3 TURNS RATIO, IMPEDANCE RATIO, AND REFLECTED IMPEDANCE, 632
- 24-4 TRANSFORMER CONFIGURATIONS, 637
- 24-5 TRANSFORMER OPERATING CONSTRAINTS, 639

25 Decibels 647

- 25-1 DECIBEL DEFINITION AND NOTATION, 648
- 25-2 APPROXIMATING DECIBEL GAIN AND LOSS, 650
- 25-3 COMPUTING ABSOLUTE POWER, 653
- 25-4 VOLTAGE AND CURRENT RATIOS AND DECIBELS, 656
- 25-5 APPLICATION OF DECIBELS, 659

26 Resonance 669

- 26-1 INTRODUCTION TO RESONANCE, 670
- 26-2 SERIES-RESONANT CIRCUIT, 673
- 26-3 PARALLEL-RESONANT CIRCUIT, 685
- 26-4 IMPEDANCE OF THE UNLOADED PARALLEL-RESONANT CIRCUIT, 686
- 26-5 LOADING THE PARALLEL-RESONANT CIRCUIT, 690

27 An Introduction to Three-Phase Power Systems 699

- 27-1 CHARACTERISTICS OF THREE-PHASE SYSTEM, 700
- 27-2 THREE-PHASE POWER GENERATION, 702
- 27-3 THREE-PHASE POWER TRANSMISSION AND DISTRIBUTION, 755
- 27-4 CONFIGURATIONS AND PARAMETERS OF THREE-PHASE LOAD 713
- 27-5 POWER AND POWER FACTOR OF THE THREE-PHASE LOAD 730

IV Frequency Domain and Filters 745

28 Complex Wave Forms and the Frequency Domain 747

28-1 SUPERPOSITION OF SINUSOIDS, 748
28-2 FOURIER THEOREM, 750
28-3 FREQUENCY DOMAIN VERSUS TIME DOMAIN, 751
28-4 SPECTRUM ANALYSIS, 751
28-5 HARMONIC DISTORTION, 753
28-6 INTERMODULATION DISTORTION, 755
28-7 PRACTICAL APPLICATIONS, 757

29 Introduction to Filters 761

29-1 LOW-PASS FILTERS, 762
29-2 HIGH-PASS FILTERS, 765
29-3 BODE PLOT, 768
29-4 BANDPASS FILTERS, 772
29-5 BAND-REJECT FILTERS, 774
29-6 TUNABLE FILTERS, 775
29-7 FILTER SPECIFICATIONS AND TERMINOLOGY, 776

APPENDIX Answers to Selected Problems 781

INDEX 801

PREFACE

This entry level text in electric circuits is structured so the reader, with pencil, paper, and calculator, is an active participant in the educational process. To expedite the learning process, each chapter has a list of Performance Objectives along with many detailed examples, figures, and tables. Additionally, an extensive number of related problems, categorized by chapter section, are provided so that the material may be mastered section by section.

Rather than burden the reader with an enormous number of topics presented in a superficial, encyclopedic manner, we have instead selected the *Essentials of Electric Circuits* and presented them in a complete, ordered, and interrelated manner. Thus, by taking an active role in the use of the text, the reader may insure a *real* understanding of the essential topics of electric circuits, thereby creating for himself the foundation for entry into the study of active devices and circuits.

For the purpose of description, the chapters of the text are arranged into several general categories, as noted in Table 1. The preparatory chapters (Chapters 1–3) have been written in an informative manner, without a detailed mathematical treatment. Depending on the reader's prior experience with electric circuits, this material may be either scanned or studied in depth.

TABLE 1 CHAPTERS ARRANGED BY GENERAL CATEGORIES

Category	*Chapters*	*Sections*
Preparatory	1–3	Introduction/Review
Pre-circuit	4–6	
Circuits and networks	7–12	DC Circuits, Devices and Concepts
Capacitance, inductance, and magnetics	13–18	
Introduction to AC, AC mathematics, and time domain	19–20	AC Circuits, Devices, and Concepts
AC circuits and networks	21–22	
AC principles, properties, and circuits	23–27	
Waveforms, filters, and frequency domain	28–29	Frequency Domain and Filters

A comprehensive coverage—both quantitative and qualitative—of the essentials of electric circuits begins with Chapter 4 and continues throughout the remaining chapters. Contained in that coverage are such topics as:

- ☐ Resistance and resistors
- ☐ Device characteristic curves
- ☐ Voltage and current sources—both practical and theoretical
- ☐ DC and AC circuits and networks
- ☐ Network theorems and network analysis
- ☐ Capacitance and capacitors
- ☐ Inductance and inductors
- ☐ Magnetism, magnetic induction, and magnetic circuits
- ☐ RC and RL transient behavior
- ☐ Impedance and admittance
- ☐ Resonance and filters
- ☐ Frequency domain and time domain
- ☐ Decibels and decibel systems
- ☐ Power and power triangle
- ☐ Transformation of current, voltage, and impedance
- ☐ Harmonic and intermodulation distortion
- ☐ Complex waveforms
- ☐ Three-phase power

These and many additional topics are included to provide a solid understanding of the essentials of electric circuits. The selection of topics and the depth of presentation were made to enable the reader to make an orderly transition into the study of active devices and active circuits.

The authors wish to acknowledge and thank Wallace Beitzel and Leonard Glover for their constructive comments, Carol Rankin and Roger Scheunemann for their diligence in checking the mathematics of the examples and solutions to the chapter problems, Marolyn Young for her creative talents in the preparation of the illustrations, Janet Billings for her expertise in typing the manuscript, and Patricia Rayner for her talent in the production of the second edition text.

James Harter

Paul Lin

I Introduction/Review

This section of the book serves both as an introduction to and a review of electricity. It is an introduction for those who have no background in electricity and, at the same time, it is a review for those who have had some experience with electricity.

This section introduces some of the electrical principles, components, units, and concepts, along with the electric circuit, current, voltage, and electrical safety.

1 Getting Started

1–1 STRUCTURE OF THE TEXT
1–2 THE ELECTRIC CIRCUIT

CHAPTER PREVIEW The first part of this chapter is intended to help you gain an understanding of the structure of the text so that you may effectively work with the material in it. Once this introduction is completed, we devote the remainder of this chapter to familiarizing you with the components of the electric circuit.

PERFORMANCE OBJECTIVES Once you have read each section of this chapter and worked each problem, you will be able to:

☐ Work effectively with the rest of the chapters in this text.
☐ Understand the need for a scientific calculator.
☐ Recall the components of an electric circuit.
☐ Draw several schematic symbols.
☐ Draw an elementary schematic diagram.

1–1 STRUCTURE OF THE TEXT We have selected essential topics from the fields of chemistry, physics, and electronic technology to give you a well-rounded educational experience so that you may go on to the advanced study of electronic circuits without any hindrance. We have selected only those topics essential to your understanding of the technology; hence the name of the text, *Essentials of Electric Circuits*.

Structure We have structured the book so that you can be successful in the course. We have provided you with hundreds of detailed examples to aid you in *"learning on your own,"* thereby becoming responsible for your own education. If your class is taught in a traditional *lecture/demonstration manner*, then the self-educating aspects of the text may be used to prepare for the next day's lesson and to assist in doing the assigned out-of-class work. In addition, if you have been absent, the material missed may be studied and mastered on an individual basis.

Each chapter is formed from the following parts:

Chapter Preview: presents the central theme(s) of the chapter.
Performance Objectives: focus on the main concepts that are to be learned.
Chapter Section Headings: serve to divide the chapter into learning modules (sections), each having one or two key ideas as indicated in the title of the section.
Chapter Subsection Headings: divide the sections into single concepts.
Selected Technical Terms: key technical words or phrases that will become part of your vocabulary.
Problems: at the end of the chapter, arranged by section.

Using the Structure of the Text

The quickest and most successful way to master the material in this text is to be an active participant in the educational process. That is, your success comes by doing, not by idly watching someone else.

We suggest that you work through this text by first reading the *Chapter Preview* and *Performance Objectives*. Then read the material in each section several times, making notes of the key ideas and working through each example with pencil, paper, and calculator. Once you feel you have a good understanding of the section material, go to the end of the chapter and work the problems for that section. Using the answers in the Appendix, check the answers to the odd-numbered problems and rework those found incorrect.

Repeat this procedure for every section in the chapter. Once the entire chapter has been mastered, return to the performance objectives and check yourself. Judge if you are able to meet each of the stated objectives and review those you feel unsure of.

Finally, look once again at the technical terms to see if you can define each in your own words; review those you have difficulty with.

Enhance Your Educational Experience

We have made some assumptions about you that we would like to share with you.

- ☐ We assume that you will make all calculations with a calculator as its use enhances the learning process. Table 1–1 lists the functions and operations used in this text.
- ☐ We assume that you have mastered the skills of arithmetic, including adding, subtracting, multiplying, dividing, fractions, and percentage.
- ☐ We assume that you are concurrently enrolled in a mathematics course and will cover such topics as algebra, exponential functions, logarithmic functions, trigonometric functions, circular functions, graphing, vectors, phasors, and phasor arithmetic.
- ☐ We assume that you are concurrently enrolled in an electronics laboratory course.
- ☐ We assume that you will work through each example, since a great deal of information has been included in the examples.
- ☐ We assume that you will make a list of technical terms so that you can learn both their meaning and spelling.
- ☐ We assume that you have access to supplemental materials including a technical dictionary and the owner's guide for your calculator.
- ☐ We assume that you have read these assumptions.

TABLE 1–1 SELECTED CALCULATOR STROKES USED IN THIS TEXT

Typical Key Symbol	Function/Operation	Comments	First Used in Chapter	Alternative Key Stroke
$+$	Add	Simple arithmetic operations	4	
$-$	Subtract		4	
$\times$	Multiply		3	
$\div$	Divide		3	
CHS	Change sign		3	$+/-$
y^x	Exponential	Raise number to power	5	x^y
$1/x$	Reciprocal	Calculate the reciprocal	8	
$\sqrt{x}$	Square root	Calculate square root	6	
x^2	Square x	Square a number	4	
FIX	Fix point notation	Display and rounding	3	
SCI	Scientific notation		3	
ENG	Engineering notation		3	
STO	Store	Memory		
RCL	Recall			
$x \rightleftarrows y$	Exchange x and y			EXC
EE	Enter exponent		3	EEX
log	Common logarithm	Logarithmic and exponential functions	25	
10^x	Common antilog		25	INV LOG
ln x	Natural logarithm		14	
e^x	Natural antilogarithm		14	INV LN
SIN	Sine	Trigonometric functions	19	
COS	Cosine		19	
TAN	Tangent		20	
SIN^{-1}	Arc sine	Inverse trigonometric functions	20	INV SIN
COS^{-1}	Arc cosine		20	INV COS
TAN^{-1}	Arc tangent		20	INV TAN
DEG	Degree	Angular mode selection	20	DRG
RAD	Radian		20	
$\rightarrow$P	Rectangular to polar	Polar/rectangular coordinate conversion	20	
$\rightarrow$R	Polar to rectangular		20	

General Information   In addition to the previous assumptions, you need to be aware of the following:

7
The Electric Circuit

☐ Most answers have been solved using the full capacity of the calculator and then rounded to three significant figures. This may cause some variation in the third digit of the answer when the problem is worked by first rounding each factor in the problem to three significant figures.

☐ Numbers are set off in groups of three without commas, as in 2 482 306 rather than 2,482,306.

☐ Selected answers to the problems are in the Appendix.

1-2 THE ELECTRIC CIRCUIT

An elementary knowledge of the electric circuit is needed as a starting point for the study of the *essentials of electric circuits.* The electric circuit in its simplest form is a complete path through which electricity travels.

Components of the Electric Circuit

At one time or another, most of us have held all components of an electric circuit in our hand. The flashlight has all four components of the electric circuit: a source of **electric energy** (the flashlight battery), a **conductive path** (case, contact springs, reflector, and lamp base, all made of metal), a way to **control** the electricity (slide switch), and a **load** (lamp).

☐ The *source of electric energy* may be as simple as a small battery in a child's toy or as complex as a regulated power supply used in a modern computer. Regardless of the complexity of the circuit, every electric circuit must have a source of energy.

☐ The *conductive path* may have many forms. The form may vary from the round copper wire in house wiring or the flat copper ribbons in electric circuit boards to the metal case of the flashlight or the metal frame of the automobile. Although the shape and material of the conductive path may vary, the need for a complete path for the electric circuit does not vary.

☐ The *control* of the electricity in the electric circuit is necessary if electricity is to be useful, convenient, and safe. The control device may be hand operated like the push button used with the doorbell circuit or the switch used to control the light in a room. The control may be automatic, as with a fuse or circuit breaker, both of which respond automatically to an electric circuit *overload*.

☐ The *load* is the point in the electric circuit where electric energy does useful work such as ringing a bell, spinning a motor, lighting a light, or heating water. The load might be a simple device such as a lamp or a complex electronic system like a hi-fi amplifier.

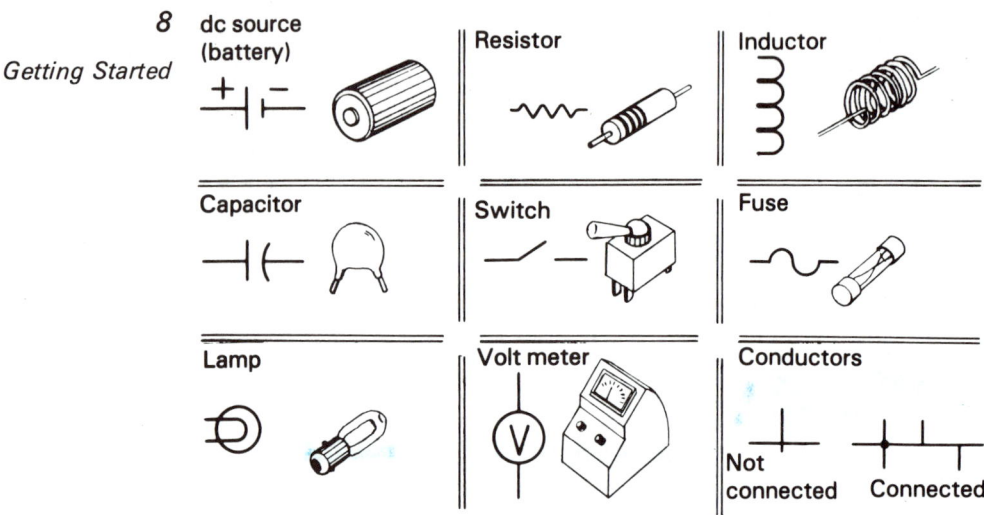

FIGURE 1-1 Selected schematic symbols with their corresponding circuit component.

Electronic Symbols and Components

Electronic symbols, shown in Figure 1-1, are used to diagram electric circuits. These diagrams, which are called *schematic* diagrams, are used to indicate how the components are connected to form the electric circuit. Figure 1-2 is a schematic diagram of a flashlight. Notice how the symbols of Figure 1-1 are used to indicate the four components of the electric circuit: the dc source (battery), the conductive path, the control device (switch), and the load (lamp).

The remaining chapters in this text will amplify your understanding of the electric circuit. Chapter 2 will aid in your understanding of electricity.

SELECTED TECHNICAL TERMS

Several of the technical terms used in this chapter are defined here to aid your understanding of them.

Conductive path: a carrier of electricity that is usually made of metal, such as copper.

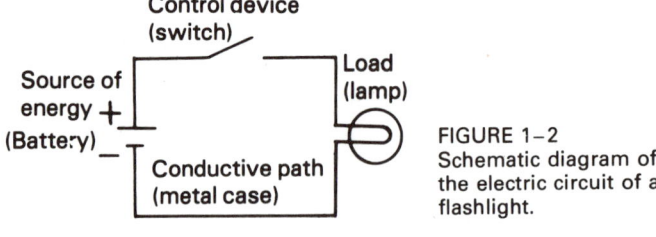

FIGURE 1-2
Schematic diagram of the electric circuit of a flashlight.

9

Problems

Control device: used to modify the movement of electricity in the electric current. The switch and the fuse are examples of control devices.

Electric energy source: causes electricity to move through the electric circuit. In the automobile, the storage battery and the alternator are examples of electric energy sources.

Electricity: a general label used to describe either a stationary or moving electric charge.

Load: the point in the electric circuit where work is done to change electric energy to another energy form, such as heat or light.

Overload: when the conductive path of an electric circuit carries a greater load of electric energy than it is designed to carry.

Schematic diagram: a diagram of the electric circuit showing the interconnection of the electric circuit components using standardized symbols to represent the parts.

PROBLEMS

Section 1-2

1-1 List two sources of electric energy.
1-2 Electric conductors are usually made from _____, a conductive metal.
1-3 A fuse is used to interrupt the flow of electricity when an _____ condition is present in the electric circuit.
1-4 Electric energy is converted by a lamp into _____ and _____ energy.
1-5 A flashlight is an example of an _____ _____.
1-6 Draw the schematic symbol for a switch.
1-7 Draw the schematic symbol for a lamp.
1-8 Draw the schematic symbol for a dc source.
1-9 Draw the schematic symbol for a resistor.
1-10 Draw the schematic symbol for a fuse.
1-11 Draw the schematic diagram for the electric circuit pictured in Figure 1-3.
1-12 Draw the schematic diagram for the electric circuit pictured in Figure 1-4.

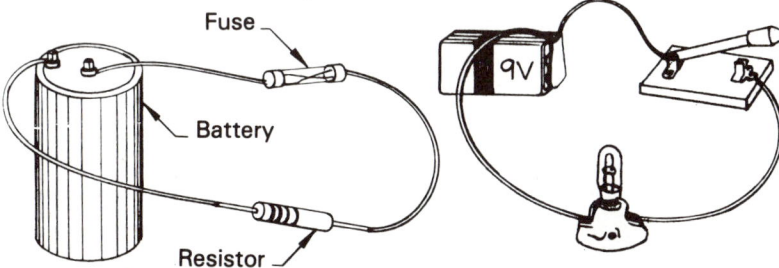

FIGURE 1-3 (left) Pictorial diagram of Problem 1-11.

FIGURE 1-4 (right) Pictorial diagram of Problem 1-12.

1-13 Redraw the schematic diagram of Figure 1–5(a) with a switch symbol in place of the fuse symbol.

1-14 Redraw the schematic diagram of Figure 1–5(b) with a lamp symbol in place of the resistor symbol.

1-15 Draw a schematic diagram of a battery connected to a resistor connected to a lamp connected back to the battery.

1-16 Draw a schematic diagram of a battery connected to a capacitor connected to a fuse connected to an inductor connected back to the battery.

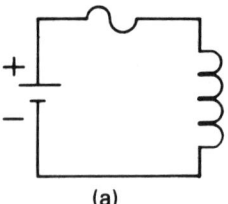
(a)

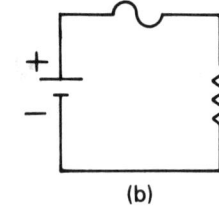
(b)

FIGURE 1–5
Pictorial diagram of Problems 1–13 and 1–14.

2 Electricity

2-1 WORK AND ENERGY
2-2 FIELD FORCE
2-3 ATOMIC STRUCTURE
2-4 UNIT QUANTITY OF ELECTRIC CHARGE
2-5 VOLTAGE
2-6 ELECTRIC CURRENT
2-7 SAFETY

CHAPTER PREVIEW The intention of this chapter is to introduce you to *electricity*. The word electricity is a general label used to indicate properties of both electric charge and electric charge in motion, that is, *electric current*. We will begin the study of electricity by exploring some of its properties. This study is not meant to be highly technical but, rather, a survey of the nature of electric charge at rest and in motion.

PERFORMANCE OBJECTIVES Once you have read each section, worked each example with pencil, paper, and calculator, and answered each problem for every section, you will be able to:

☐ Understand the nature of electricity and how electric current carries energy through conductors.
☐ Recognize and use the terms joule, coulomb, volt, and ampere.
☐ Have a mental picture of the atom and its component parts.
☐ Understand the meaning of voltage, electromotive force, and potential difference.
☐ Have an elementary understanding of a field force (action at a distance) and that electric charge has an electric field.
☐ Recall the names and characteristics of the three types of atomic bonding.
☐ Differentiate between ac and dc current and give an example of a source of each.
☐ Recall the physiological response of your body to varying amounts of electric current.
☐ Appreciate the need for a safe and healthy attitude and constant awareness when working with electricity.

2-1 WORK AND ENERGY

Work

One of the many characteristics of electricity is its ability to transport energy from one place to another. How remarkable that huge amounts of energy can be transmitted long distances between the source and the consumer. This property of electricity is the key to our modern industrialized society.

Even more remarkable is that electric energy is readily and conveniently converted into other useful energy forms, such as light, heat, and mechanical energy.

To convert energy from one form to another, *work* must be done. To convert the energy of falling water to electric energy, work must be done on the electric charge by a device called an electric generator. As the charge moves through the converting device, the energy of the moving charge is increased by the work done.

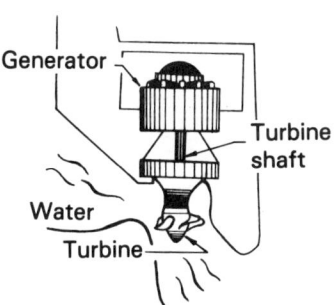

FIGURE 2-1
Force of the moving water strikes the turbine causing it to spin the generator which in turn converts mechanical work to electric energy.

The term *work* is used to describe the result of applying a force. When the turbine of Figure 2-1 turns, work is done by the force of the falling water. If the turbine remains stationary, no work is done. As a general statement, then:

> Work is done only when an object is moved from its original position.

Physical exertion is not an indication that work has been done. Only when an applied force results in an object being displaced has work been done.

Energy Energy is expended when an object is moved against an opposing force. When energy is expended and an object is moved, work is done. Energy must be present before work can be done. Thus:

> Energy is the ability to do work.

Whenever energy is converted from one form to another, work is performed. In the conversion of energy, the amount of work done to convert energy is exactly equal to the energy converted. That is, energy and work are the same physical quantity and are measured with the same unit, the *joule*. Thus:

> In a closed system, the total energy remains constant. That is, energy is neither created nor destroyed, but it may be converted from one form to another.

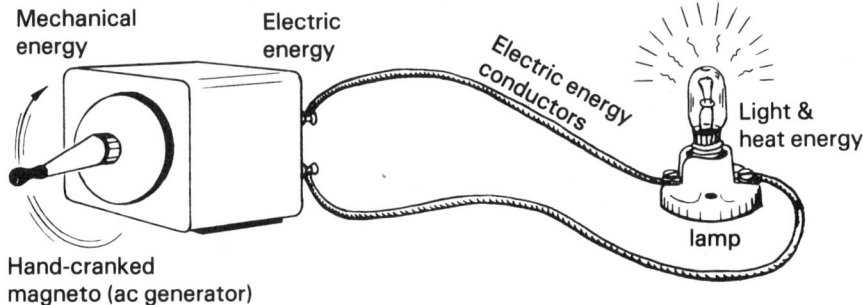

FIGURE 2-2 Mechanical work of turning the crank is converted to electric energy by the conversion device (magneto). Electric energy is carried by the flow of electric charge moving in the conductors (wires) to the lamp. Moving electric charge performs work in the lamp to convert electric energy to heat and light energy.

Therefore, whenever energy of one form disappears, an equivalent amount of energy of another form is produced. This principle is the **law of conservation of energy.** Figure 2-2 pictures the conversion of energy first from mechanical energy to electric energy, and then to heat and light energy.

2-2 FIELD FORCE

Contact Force

When opening a book, a force is applied to the book by your hand in contact with the book. Your hand must touch the book and a *contact force* must be applied to the cover and pages before the book will open.

In science fiction stories, we learn about an invisible mysterious force that captures the space traveler. This invisible beam (force) is coming from a point in space and has a wide *sphere of influence*, trapping and pulling unsuspecting space travelers into it with ever-increasing force. This science fiction *"tractor beam"* has all the characteristics of a *field force*, or action at a distance.

Field Force

The idea of an action at a distance is precisely what we experience living here on earth. We do not fly off into space because of a field force called *gravity*. Space travel has taught us that the gravitational force of the earth diminishes as we move away from the earth.

Two bodies having *mass* will have a gravitational force of attraction between them. An ocean's tides are influenced by the presence of the moon separated from the earth by several hundred thousand kilometers of empty space.

Electric Field

Electric charge has a field force (force at a distance) called an *electric field*. Figure 2-3(a) shows a quantity of electric charge

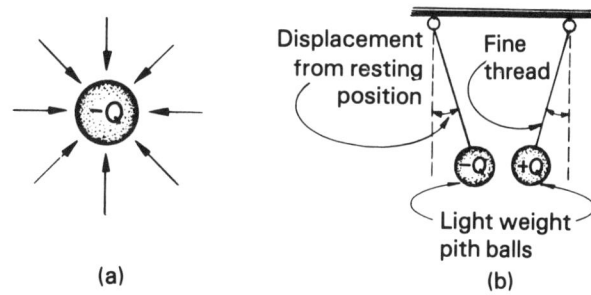

(a)   (b)

FIGURE 2-3 (a) Quantity of electric charge pictured as $-Q$ has a sphere of influence on an outside quantity of electric charge which enters the electric field. A force will be *felt* by the outside electric charge.
(b) Two pith balls having an equal quantity of opposite polarity electric charge (one positively charged, the other negatively charged) have an attractive force.

(Q) with lines of force, representing the electric field, depicted as arrows.

Electric charge may be thought of as having two states: one positive ($+Q$) and one negative ($-Q$). When a quantity of *positive charge* is brought into the sphere of influence of a quantity of *negative charge*, an attractive force is noted. This is pictured by Figure 2-3(b).

Notice that electric lines of force are conventionally pictured as starting from $+Q$ and finishing at $-Q$. From the action of Figure 2-4, we may state the following:

> Unlike charges attract; like charges repel.

In summary, an electric charge is pictured as being surrounded by an electric field that exerts an attractive or repulsive force on any other electric charge that comes into its sphere of influence.

2-3 ATOMIC STRUCTURE

The atom is an extremely small electrically neutral particle consisting of a core (center) called the *nucleus* and one or more *electrons* rotating about the nucleus.

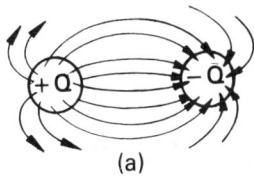

(a)

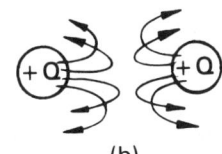

(b)

FIGURE 2-4 (a) Lines of force depict the electric field of two equal electric charges with opposite polarity. Unlike charges have an attractive force.
(b) Lines of force depict the electric field of two equal electric charges with the same polarity. Like charges have a repulsive force. Lines of force never cross.

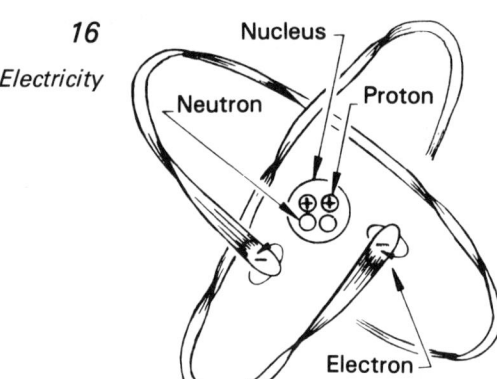

FIGURE 2-5
Artist's concept of a helium atom picturing the nucleus with two protons (+ charge) and two neutrons. Electrons (− charge) are seen orbiting around the nucleus as well as having spin. Modern research indicates the existence of several additional particles in the nucleus besides those shown here.

Electric Charge and Atomic Structure

The nucleus is said to consist of *protons* and *neutrons*. The proton has a positive electric charge that is exactly the same amount as the negative charge of the electron. The neutron has no electric charge.

The atom is electrically neutral because the positive charge of the nucleus is neutralized by the negative charge of the orbiting electrons. Figure 2-5 is an artist's conception of the parts of the atom discussed in this section. Notice that the number of protons is equal to the number of electrons. This is the state of a *neutral* atom.

Electron Distribution

Electrons orbit around the nucleus within well-defined *zones* at varying distances from the core of the atom. The electrons are attracted to the nucleus because of the electric field (unlike charges attract). This field force is called an *electrostatic force*.

The zones, called shells, represent varying levels of *potential* energy. The shells are lettered K, L, M, N, O, P, and Q. The K shell, the closest to the nucleus, may have up to 2 electrons, while the L shell may have up to 8, the M shell may have up to 18, and the N shell may have up to 32 electrons. The electron structure for germanium (Ge) is given in Figure 2-6.

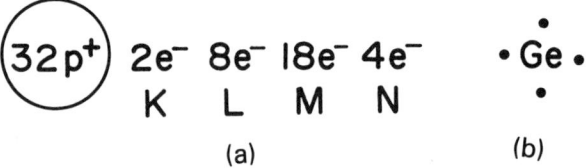

FIGURE 2-6 Diagrams of the electron structure for an atom of germanium (Ge): (a) Shown are 32 protons in the nucleus (32p$^+$) and the distribution of the 32 electrons (32e$^-$) in each of the main energy levels (shells).
(b) Abbreviated diagram showing the chemical symbol (Ge) surrounded by 4 dots representing the 4 electrons in the outer or valence shell.

Valence Electrons and Bonds

The electrons in the outermost shell are called the *valence* electrons. It is these electrons that *bond* atoms together to form compounds and solid metals. The valence electrons are lightly held to the nucleus as the electrostatic force is diminished at the outermost shell. These electrons are either shared by or transferred to other atoms to form a configuration of 8 electrons in the outer shell. Either 2 or 8 electrons in the outer shell create a very stable substance. The sharing or transferring of valence electrons bonds atoms together. This is called *bonding*. All matter is formed by bonds of one or more of three general types: *ionic*, *covalent*, and *metallic*. An example of each type of bond is given in the following series of examples.

EXAMPLE 2–1 Discuss the bonding of sodium (Na) and chlorine (Cl) to form table salt, an *ionic compound*.

SOLUTION In this case, an **ionic** bond is created when the sodium atom transfers its one valence electron to the chlorine atom, forming a sodium *ion* with a positive charge (positive ion) and a chlorine *ion* with a negative charge (negative ion). Each ion has a stable (8-electron) outer shell; however, each has an opposite charge. The two oppositely charged ions attract each other with an electrostatic force. The reaction between a sodium and a chlorine atom is diagrammed in Figure 2–7.

∴ Sodium and chlorine are combined into ordinary table salt by an electron transfer forming a very strong ionic bond.

EXAMPLE 2–2 Discuss the bonding of hydrogen (H) and oxygen (O) to form water, a *molecule*.

SOLUTION In this case, the one unpaired electron of 2 hydrogen atoms (2 electrons) is shared with 2 elec-

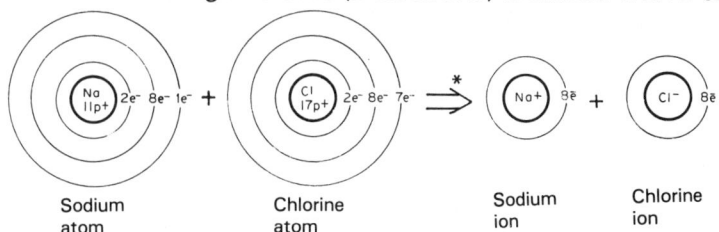

Sodium atom   Chlorine atom   Sodium ion   Chlorine ion

FIGURE 2–7 Sodium atom (K $2e^-$, L $8e^-$, M $1e^-$) transfers its valence electron to the chlorine atom (K $2e^-$, L $8e^-$, M $7e^-$) to yield sodium chloride (NaCl), a common salt. The sodium ion (K $2e^-$, L $8e^-$) forms an *ionic* bond with the chlorine ion (K $2e^-$, L $8e^-$, M $8e^-$), each ion has a complete outer shell of 8 electrons.

* This is a yield symbol and is read "*yields*".

2H· + ·Ö: ⇒ H:Ö:
                       H

FIGURE 2-8 Valence electrons (represented by dots) are shared by two hydrogen atoms (K 1e⁻) and one oxygen atom (K 2e⁻, L 8e⁻, M 6e⁻) to form water (H₂O). A single uncharged particle formed by *covalent* bonding is called a *molecule*.

trons of an oxygen atom (6 electrons) to form a stable (8-electron) covalent compound. A shared pair of electrons is called a **covalent** bond. The resulting particle is called a *molecule*. The covalent bonding of hydrogen and oxygen is diagrammed in Figure 2-8.

∴ A covalent bond results when electrons of two or more atoms are shared.

EXAMPLE 2-3 Discuss metallic bonding and how it relates to the physical properties of metals, such as copper, silver, and gold.

SOLUTION In metals, the atoms are thought to be present as ions distributed in a "sea" of valence electrons. The valence electrons do not belong to a particular atom but, instead, belong to the solid metal as a whole. It is believed that each valence electron is shared not by an adjoining atom as in a covalent bond, but by up to 12 of its nearest neighbors. The atoms within the metal are held in place by the **metallic** bond, which consists of several forces (including electrostatic) acting to line up the atoms into a regular pattern of cells called a *lattice*. Figure 2-9 is a representation of the structure of the lattice that forms many of the

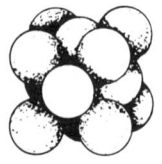

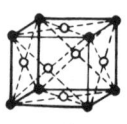
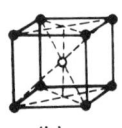

(a)   (b)

FIGURE 2-9 (a) *Face-centered cubic lattice* is formed when arrays of atoms are strongly bonded by *metallic bonds*. These lattices form crystals making up the solid metal. Metals including aluminum, copper, gold, iron (γ), lead, nickel, platinum and silver are formed in this manner.
(b) Additional metals are formed with the *body-centered cubic lattice* including molybdenum, tantalum, chromium and tungsten.

Voltage

conductive metals, including copper. Lattices are formed into *crystals*, and crystals form the solid metal.

It is the relatively free motion of the valence electrons among the ions that bonds the crystal together. It is the "randomness" of these *free electrons* that makes the metals electrically and thermally conductive.

∴ *Metallic bond* is a name given to the many forces acting to position the ions into a lattice and thus establish an equilibrium between all the forces involved. The valence electrons are free to carry electric energy through the solid metal. These carriers are called *free electrons*.

2-4 UNIT QUANTITY OF ELECTRIC CHARGE

In your study of electricity, you must learn the meaning of many technical words. It is important that new technical words be learned. An understanding of the new words will come with their use. The first of these words is *coulomb*, which will be explained in the following paragraphs.

Remember in our discussion of work and energy in Section 2-1 it was stated that a very important property of electricity is its ability to transport electric energy. The electric energy is moved through an electric conductor (a metal such as copper) by an electric charge transported by the *free electrons* of the metal.

The electric charge of one electron is very, very small (1.6×10^{-19} coulomb). Because of this small charge, it is not practical to use just the charge of one electron. Instead, the charge carried by a very, very large number of electrons (6.242×10^{18} electrons) is used. This *unit quantity of electric charge* is called the **coulomb**. Thus:

One coulomb equals the charge of 6.242×10^{18} electrons.

2-5 VOLTAGE

In this section, you will learn how electric charge has work done on it to increase its energy. The new technical words are *potential difference* and *electromotive force*.

Potential Difference

To get electric charge to carry energy through electric conductors (copper wire), the charge moves from a region of high energy to a region of low energy. The electric charge must acquire energy by having work done to it so that it can move.

When charge acquires energy, its potential to do work is raised. Remember the definition of energy: "the ability to do work."

When charge moves from a higher to a lower energy, its potential to do work is lessened because energy is expended in moving a unit of electric charge. The *difference* in the energy of the electric charge from the higher energy level to the lower energy level is the *potential energy difference* of the electric charge or, simply, the **potential difference**.

Potential difference is measured in units of energy per unit charge. Thus:

> Potential difference = joules/coulomb

The potential difference of the charge is usually measured in *volts* rather than joules per coulomb. That is:

> 1 volt = 1 joule/1 coulomb

Figure 2–10 pictures a light bulb. The lamp gives off light and heat when the electric charge moves through the tungsten filament (metal conductor). The electric energy is converted to heat

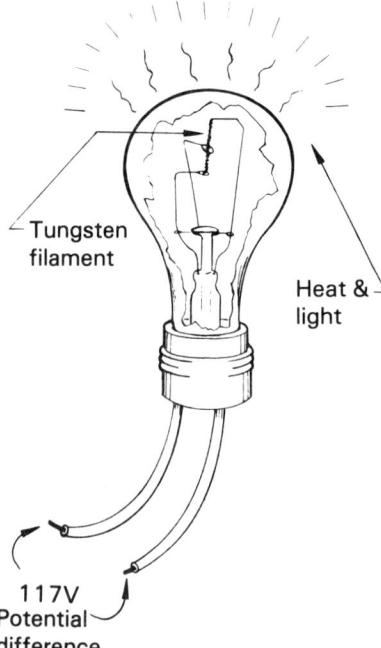

FIGURE 2–10
Potential difference of 117 volts causes the electric charge to do work as it passes through the tungsten filament. Electric energy of the moving charge is converted to heat and light energy.

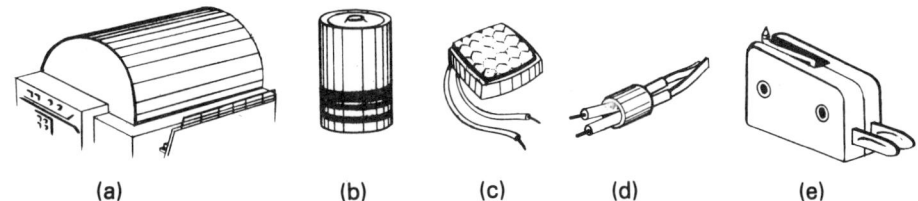

FIGURE 2-11 Devices used to convert various energy forms into electric energy. These devices are sources of *electromotive force* (emf). The work of the device in transferring a unit of electric charge from one terminal to the other terminal is called the electromotive force. The emf is measured in a unit called the **volt**.
(a) Mechanical energy is converted to electric energy in a *rotating machine* called a generator. Magnetism plays an important part in this machine.
(b) Chemical energy is converted to electric energy in a *dry cell* or a battery.
(c) Light energy is converted to electric energy in a *photovoltaic cell* (solar cell).
(d) Heat energy is converted to electric energy in a *thermocouple* formed by joining two different metals.
(e) Mechanical energy is converted to electric energy in a *crystal* of Rochelle salt mounted in the crystal *pickup cartridge* of a phonograph. Lateral motion of the stylus in the record groove mechanically stresses the crystal, generating a potential difference. This action is called the piezoelectric effect.

and light energy when a potential difference (117 volts) is created across the terminals of the lamp. **Electric charge will not move through an electric conductor unless there is a potential difference.**

Electromotive Force

To raise the energy level of the electric charge, work must be done on the charge by a device such as one of those pictured in Figure 2-11. A device that is used to convert mechanical energy, chemical energy, light energy, or heat energy to electric energy is called a *source of electromotive force* or emf. Electromotive force is the name given to the work done to each unit of charge that is moved through the source. The emf of a source is measured in units of joules per coulomb or volts.

Because both potential difference and emf are measured in volts, each is often referred to by the general word *voltage*. The term voltage, however, does not indicate whether a source of electromotive force (emf) or a potential difference is being indicated. In future chapters, we will develop a labeling system that will help to keep the two identified.

2-6 ELECTRIC CURRENT

Flow of Charge

Electric current is charge in motion. The charge is put in motion by a source of electromotive force. Consider the action in the elementary *circuit* of Figure 2-12. The dry cell, through chemical energy, creates an excess of electrons at the negative terminal and an excess of positive ions at the positive terminal. The excess charge at each terminal has a potential difference of 1.5 joules per coulomb (1.5 volts).

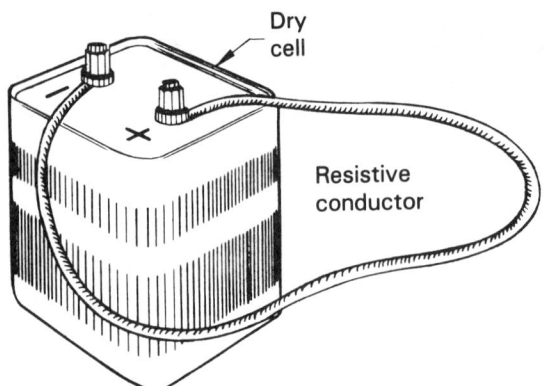

FIGURE 2-12
Elementary circuit. The dry cell is the source of emf. A potential difference of 1.5 volts is created between the positive (+) and negative (−) terminals. The resistive conductor is made from metal like that used in a toaster.

A metal conductor is connected between the negative and positive terminals of the cell. The free electrons in the metal immediately begin to *drift* (move) toward the positive terminal of the battery at a rate of approximately 1 centimeter per second ($\approx$ 0.4 inch per second). This slow movement of a single electron through the positive ions making up the metallic lattice is due to the random motion of the free electrons. It should be noted that the positive ions of the lattice making up the crystalline metal remain stationary, as do the positive ions at the positive terminal of the cell.

The free electrons move and combine with positive ions at the positive terminal of the cell to form neutral atoms. This action creates a net imbalance of electrons in the metal, which is *instantaneously* overcome by electrons entering the conductor from the negative terminal of the battery. Unlike the slow drift of electrons through the metal, the process of an electron entering the conductor from the negative terminal of the cell and an electron combining with the positive ion at the positive terminal of the cell takes place in the incredibly short period of time of about one billionth (1×10^{-9}) of a second. This is based on the assumption that the conductor is 30 centimeters (12 inches) long and the rate of the event is the speed of light, 3×10^{10} centimeters per second (9.8×10^{8} feet per second).

The Ampere of Electric Current

Electric current is the movement of a unit of electric charge per time. That is:

$$\text{Current} = \text{coulombs/second}$$

A current of 1 **ampere** is flowing when 1 coulomb of charge passes a point in the conductor in 1 second. Thus:

Electric Circuit

Electronics is the study of the control and application of electric current flow in electric circuits and related devices. Since the understanding of the electric circuit is vital to your understanding of electronics, the electric circuit will be the topic of much of the remainder of this book. Several important concepts of an electric circuit must be understood at this time. These are summarized in the following statements:

☐ A source of emf (voltage source) must be present if electric current is to carry energy to the *load* (see Figure 2–13).
☐ The electric current is the same throughout the circuit. That is, the current leaving the source is equal to the current returning to the source.
☐ Current flow has the potential for doing work.
☐ A potential difference (voltage) exists across the load when current flows through the load. This is the characteristic of a complete or closed circuit.
☐ An incomplete or *open* circuit has no current flow.
☐ Current moves from the source to the load and back to the source through a complete path (electric circuit) formed by connecting the load to the source with metallic conductors (copper wire).
☐ The electrons making up the current flow are called free electrons and are readily available in the metal conductors.

Direct and Alternating Current

Direct current (dc) is the name given to current that moves in only one direction. The direction of current flow is determined by the source of emf (source voltage). If the source is unidirec-

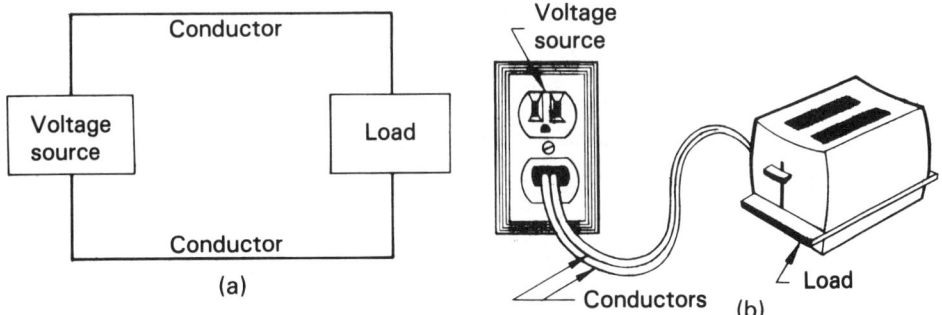

FIGURE 2–13 Electric circuit:
(a) Complete circuit consists of voltage source (source of emf), electric conductors (copper wire), and a load (to convert electric energy to some other energy form).
(b) Pictorial diagram showing voltage source, electric conductors and load.

tional (one direction), like a battery, the current will be unidirectional or direct current. In contrast to dc, *alternating current* changes direction. If the voltage source changes polarity (alternates polarity) periodically, then the current must also change direction. Current that alternates back and forth is called alternating current. Figure 2–14 pictures both a steady-state dc wave form, obtainable from a battery as shown in Figure 2–11(b), and an ac wave form, obtainable from a rotating ac generator as pictured in Figure 2–11(a).

2–7 SAFETY

Sources of Hazards

When you work as a technician, you will be faced with hazards of several types in the course of a day's work. The obvious hazard, electric shock, is well known; other hazards, like radiation and exposure to certain chemicals, are becoming known as their cumulative effect is understood. Hazards resulting from hand and power tools are ever present to technicians involved in manufacture, installation, and maintenance activities.

Certain job categories have hazards peculiar to activities associated with those jobs, such as climbing antenna towers or lifting heavy racks.

Typical safety hazards encountered by those working in all parts of the electronics industry include exposure to lasers; ultraviolet, microwave, and x-ray radiation; chemicals; hand and power tools; static and dynamic electricity; and high, intermediate, and low levels of voltage.

Safety is an attitude that you must develop if you are to work in the electronics industry and avoid the pain and physical damage of an injury. If you value yourself, you will stop to evaluate your work habits for unsafe practices. Before you can work safely, you must first learn to think safely.

Electric Shock

Electric shock is a potential source of injury to anyone working in the electronics industry. An electric shock will be experienced

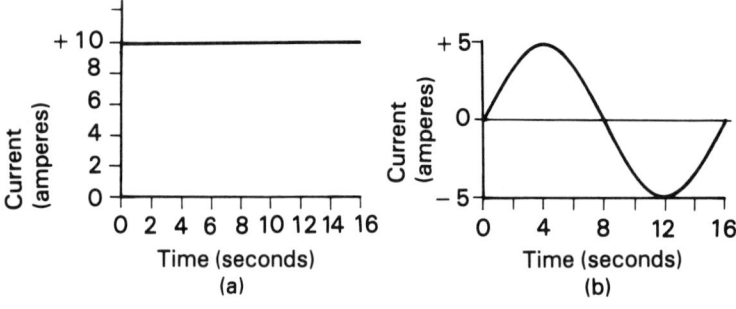

FIGURE 2–14
(a) The dc wave form resulting from a battery.
(b) An ac wave form resulting from a rotating ac generator.

25
Safety

when an electric conductor is touched while the electric circuit is energized. The resulting electric shock (due to current passing through the body) is accompanied by involuntary muscle contractions and pain. If a large enough current passes through the body, the electric shock and/or the resulting electric burn will cause very serious injury or death.

The effect that electric shock will have on your body will depend upon the amount of current (measured in milliamperes or $\frac{1}{1000}$ of an ampere) passing through your body, the amount of time the current passes through your body, and the path the current takes through your body. Table 2–1 summarizes the response of the human body to various amounts of current.

You must exercise **extreme caution** when working on *live* circuits, even circuits operating with low voltage. Some basic safety procedures are:

☐ Where possible, shut the circuit OFF by unplugging from the power source before working on equipment.
☐ Before working on a circuit, study the schematic diagram, read the operation manual, and pay particular attention to the references to safety hazards.
☐ Never assume a circuit is OFF, even when a switch is open (off). Measure to be certain.
☐ When making measurements on an energized circuit, keep yourself insulated from *grounded* objects.

Realize that safety procedures are ineffective and will not keep you from being electrically shocked if your work habits and attitude are poor and careless.

TABLE 2–1 ELECTRIC SHOCK

Amount of Current Flow through Body*	*Physiological Response*
1 milliampere (mA)†	Feel current; voluntarily let go of shock source
1 to 6 mA	Tingle to painful shock; not fatal
6 to 24 mA	Intense pain; unable to turn loose of shock source
25 to 50 mA	Excruciating pain; muscles are in violent contractions
50 to 200 mA	Ventricular fibrillation of the heart, heart stopped and/or breathing stopped; fatal

* Duration of current flow is greater than 25 milliseconds (25/1000 second).
† 1 milliampere = 1/1000 ampere.

SELECTED TECHNICAL TERMS

Several of the technical terms used in this chapter are defined here to aid your understanding of them.

Electric current: a flow of electric charge (electricity) along a conductor in a closed circuit. Current (as it is usually called) carries energy in conductors (wires) by means of free electrons. The energy carried by the current may do work by producing heat, light, mechanical force, or magnetic attraction or repulsion.

Electric field: a region in which a force is felt by an electric charge. An electric field is produced by an electric charge, and the electric charge is affected by the presence of an electric field.

Electromotive force (emf): the name given to work done on electric charge as it passes through a source. Electromotive force produces an electric current in a conductor by creating a difference in potential.

Energy: the capacity to do work by overcoming resistance.

Free electron: responsible for electric current in a conductor. The amount of electric current in a conductor depends upon the number of free electrons.

Ion: an atom that has gained or lost an electron. Negative ions have more electrons than a neutral atom, whereas positive ions have fewer electrons than a neutral atom.

Potential difference: voltage between two points in an electric circuit, resulting from current doing work in converting electric energy into heat or light or some other energy form.

Valence electrons: electrons in the atom that are responsible for bonding between chemical elements. In metals, such as electric conductors, the valence electrons are often free electrons.

Voltage: electromotive force (emf) or potential difference expressed in volts.

PROBLEMS

Section 2-1

2-1 In addition to chemical energy, list four other forms of energy.

2-2 Electric energy is transported from one place to another by _____ _____ moving in electric conductors.

2-3 Using your own words, state the law of conservation of energy.

2-4 Work and energy are measured in the same unit, the _____.

2-5 When electric energy is converted to heat and light in a lamp, the amount of _____ done is equal to the _____ converted.

2-6 Using the generator pictured in Figure 2-1 as an example, explain the law of conservation of energy.

Section 2–2 2–7 Using your own words, describe the difference between a contact force and a field force.

2–8 The field force associated with electric charge is called an _____ _____.

2–9 Describe how you could demonstrate that electric charge has a field force rather than a contact force.

2–10 The electric field of a positive charge will exert an attractive force on a _____ charge.

2–11 Like charges _____ and unlike charges _____, regardless of whether the electric charge is stationary or in motion.

Section 2–3 2–12 Briefly describe the structure of the atom.

2–13 Neon (Ne) is an inert gas having an atomic number of 10 (10 orbiting electrons). Using the concept of shells, diagram the structure of neon and discuss the likelihood of this element combining with another element to form a chemical compound.

2–14 Copper, a good conductor of electric energy, has an atomic number of 29 (29 orbiting electrons). Using the concept of shells, diagram the copper atom.

2–15 Using the total charge of the atom, distinguish between a neutral atom, a negative ion, and a positive ion.

2–16 Describe the role played by valence electrons in forming matter.

2–17 In a short statement, describe the formation of an ionic bond and contrast this with the formation of a covalent bond.

2–18 Using the shell structure of the copper atom developed in Problem 2–14 and your understanding of metallic bonding, describe how copper can conduct electric energy.

2–19 Differentiate between the terms lattice and crystal.

Section 2–4 2–20 The unit quantity of electric charge is the combined charge of _____ electrons.

2–21 Using your own words, define the term coulomb.

2–22 Free electrons serve to move _____ _____ through a metal.

Section 2–5 2–23 Energy per unit charge (joules/coulomb) is commonly called _____.

2–24 When an electric charge carries energy through an electric circuit, the charge moves from a region of _____ energy to a region of _____ energy.

2–25 When electric charge gains energy, its potential to do _____ is raised.

2-26 The potential energy difference of electric charge is measured in units of _____.
2-27 List three devices, not including a battery, that raise the energy level of electric charge.
2-28 A rise in the energy level of electric charge is carried out in a source of _____ force.
2-29 In a complete electric circuit such as a flashlight, electric charge moves from the source to the load and then back to the source carrying electric energy to the load. Using the words higher energy and lower energy, explain why you think energy is not transferred from the load back to the source.

Section 2-6
2-30 Charge in motion is called electric _____.
2-31 React to the appropriateness of using the word drift to describe the movement of free electrons in a metallic conductor.
2-32 Free electrons entering and leaving a voltage source (source of emf) do so at the speed of _____.
2-33 Electric current is defined as the movement of a unit of _____ _____ per _____.
2-34 Define the word ampere.
2-35 In your own words, describe the characteristics of an electric circuit.
2-36 Describe both direct current and alternating current. Include in your discussion similarities, differences, and an example of a source of each.

Section 2-7
2-37 List five sources of safety hazards that may be encountered by an electronics technician.
2-38 Electric shock is the result of electric _____ passing through the body.
2-39 Using your own words, list several safety procedures when working with electricity.
2-40 No safety procedure will keep you from being electrically shocked if your work habits and _____ are careless.

3 Measurement and Number Notation

3-1 SYSTEMS OF MEASUREMENT
3-2 THE SI SYSTEM
3-3 ENGINEERING NOTATION AND DECIMAL NOTATION
3-4 FORMING DECIMAL MULTIPLES AND SUBMULTIPLES OF THE SI UNITS
3-5 UNIT ANALYSIS AND CONVERSION WITHIN AND BETWEEN SYSTEMS OF UNITS

CHAPTER PREVIEW

In the study of electronics, you will have an opportunity to measure many physical quantities. These measurements will have numbers as well as units associated with them. It is important that you learn from the onset of your study that a measurement or a calculation without an appropriate unit is utterly meaningless. If you are told that a swimming pool is 20 long, you still don't know how long the pool is. Only when the unit of length is given does the statement have real meaning. Thus, 20 long is meaningless, but 20 meters long has meaning (assuming, of course, you have an understanding of length in meters).

In this chapter, you will be introduced to many units common to science, engineering, and electronic technology and how to write prefixed units. The chapter concludes with a demonstration of the techniques of unit analysis.

You will be using your calculator to multiply and divide as well as format numbers in *engineering notation*, but only if your calculator has the ENG stroke. If not, then this operation will be carried out by hand. We recommend that you keep your calculator owner's guide available with your calculator.

PERFORMANCE OBJECTIVES

Once you have read each section, worked each example with pencil, paper, and calculator, and answered each problem for every section, you will be able to:

☐ Distinguish between SI and English units of measurement.
☐ Recall selected units and symbols of the SI system.
☐ Convert from engineering notation to decimal notation, and conversely.
☐ Use prefixes and the associated symbols with SI units of measurement.
☐ Apply the technique of unit analysis to convert between or within systems of measurement.

3-1 SYSTEMS OF MEASUREMENT

In science and technology, problems are solved in a systematic manner using an investigative process called the *scientific method*. Measurement plays a very important part in this problem-solving process. A measured physical *quantity* has both a numerical value (how much) as well as a *unit* (of what). Throughout the centuries of scientific discovery, various systems of *standard units* have been created and used. Today, two systems are used in the United States for measuring quantities. These are the *English* system and the *metric* system.

Systems of Measurement

TABLE 3-1 ENGLISH SYSTEM FOR MECHANICS

Physical Quantity	Unit Name	Unit Symbol or Definition
BASE UNITS		
Length	foot	ft
Force	pound	lb
Time	second	s
DERIVED UNITS		
Mass	slug	lb·s^2/ft
Pressure	pound per square foot	lb/ft^2
Work	foot-pound	ft-lb
Power	horsepower	hp
Energy (heat)	British thermal unit	Btu

English System   The English system, originated in thirteenth-century England, has developed over many centuries to its present form. One difficulty in applying the English system is its use of units and subunits that are not systematically chosen. Thus, if the names foot, inch, and yard were not known to you, you would have no clue that each was related to the other. Furthermore, without prior knowledge of the size of each of them, you would not know that 12 inches equals 1 foot or 3 feet equals 1 yard. In the English system, each time a multiple or submultiple of a unit was needed, a new name was applied. Table 3-1 lists part of the English system.

Because of these shortcomings and other complexities, the English system has not been adopted as a standard for the various scientific and engineering fields. Instead, the International System of Units (SI) has been selected as the standard system. The abbreviation SI is rooted in the French Le Système International d'Unités.

SI Metric System   The SI system (metric system), like the English system, has undergone changes to get to its present form. The metric system in use today is called the *International System of Units,* or simply the SI system. The SI system uses unit prefixes to form smaller and larger units of a quantity. Thus, centi**meter**, **meter,** and kilo**meter** are obviously related to one another and are all units of length. The SI system keeps the same unit name for both multiples and submultiples. Because of this simplicity, the SI system is *"the standard of the industry"* and as such has been adopted by the Institute of Electrical and Electronic Engineers (IEEE).

3-2 THE SI SYSTEM

The International System of Units has seven *base* units, two *supplementary* units, several *derived* units with special names, and many derived units with compound names. In this section, we will introduce and categorize many of the units peculiar to electronic technology, as well as those in general use throughout science and technology.

Base Units

The seven *base* units and the two *supplementary* units listed in Table 3-2 are the building blocks from which the *derived* units are constructed. Each base unit has been defined in terms of a reproducible standard, as indicated in the following:

- Length: *meter*. Defined as 1 650 763.73 wavelengths of the orange-red spectral line (in a vacuum) of the krypton 86 isotope.
- Mass: *kilogram*. Defined in the United States by an exact duplicate of the standard mass made of a metal alloy, platinum-iridium. The original standard mass, a cylinder, is kept in Sèvres, France.
- Time: *second*. Defined as the duration of 9 192 631 770 periods of the radiation resulting from the cesium 133 isotope shifting between the two ground-state energy levels.
- Electric current: *ampere*. Defined as the amount of constant current required to produce 2×10^{-7} newtons of force per

TABLE 3-2 BUILDING BLOCKS OF THE SI SYSTEM

Physical Quantity	*Physical Symbol*	*Unit Name*	*Unit Symbol*
BASE UNITS			
Length	ℓ	meter	m
Mass	m	kilogram	kg
Time	t	second	s
Electric current	I	ampere	A
Temperature	T	kelvin	K
Luminous intensity	I	candela	cd
Amount of substance	M	mole	mol
SUPPLEMENTARY UNITS			
Plane angle	—	radian	rad
Solid angle	—	steradian	sr

The SI System

meter of length (N/m) in two very long parallel conductors in a vacuum separated from each other by 1 meter.

From these definitions, you may now have an understanding of the precision of the standards that are used by science and technology to define the base units. Of the seven base units and two supplementary units, you will become familiar with length, mass, time, electric current, temperature, and plane angle in your study of this text. These units, along with many of the derived units, will be explained as they are used. For now, become aware of the names of the units and symbols used to designate the units.

Derived Units The SI-derived units are formed from the combination of two or more SI base and supplementary units. Many derived units have been given convenient names and symbols in place of the complicated ones that result from the derivation from the base units. For example, the *watt* (W), a measure of both mechanical and electrical power, is such a unit. The base units of the watt include the meter, kilogram, and second. Without the special name watt and the symbol W, power would be expressed as $m^2 \cdot kg/s^3$, read "meter squared kilogram per second cubed." Several SI-derived units are listed in Table 3–3. These units, as well as some not mentioned in Table 3–3, will become familiar to you in your study of this text.

TABLE 3–3 SI-DERIVED UNITS WITH SPECIAL NAMES

Physical Quantity	*Physical Symbol**	*Unit Name*	*Unit Symbol†*
Capacitance	C	farad	F
Conductance	G	siemens	S
Electric charge	Q	coulomb	C
Electric resistance	R	ohm	Ω
Electromotive force	E	volt	V
Energy, Work	W	joule	J
Force	F	newton	N
Frequency	f	hertz	Hz
Inductance	L	henry	H
Magnetic flux	ϕ	weber	Wb
Potential difference	V	volt	V
Power	P	watt	W

* Standardized letters used in formulas or equations to represent the physical quantity in the calculation.
† An abbreviation used to indicate the units in the statement of physical measurement.

34

Measurement and Number Notation

Before moving on to further topics, a word of caution is needed. Great care must be used when writing the *unit symbol* (abbreviation) for a particular unit. The uppercase letters must be easily distinguished from the lowercase letters. In Table 3–2, you might have noticed that all the unit symbols are lowercase letters except the A for ampere and the K for kelvin. Because these two units are named for scientists, the symbols are uppercase. In Table 3–3, all the unit symbols are uppercase, which indicates that each unit is named for a person. Furthermore, two of the units listed have a two-letter symbol (hertz, Hz, and weber, Wb). When this is the case, the first letter is uppercase and the other letter is lowercase.

Selected SI Units

Chapter 2 introduced you to several SI units: energy, electric charge, electric current, electromotive force, as well as potential difference. Each of these SI units, along with length, mass, temperature, force, power, and electric resistance, will be summarized here to provide you with a *glossary of terms* to be used as a reference for future study.

Length (ℓ)   *Unit name:* meter (m)
Use: Length is the physical quantity that describes space and position. 1 meter = 39.37 inches (1 m = 39.37 in.).
See: Figure 3–1

EXAMPLE: The length of a room is 8 m, the width 5 m, and the height 3 m.

Mass (m)   *Unit name:* kilogram (kg)
Use: Mass is the physical quantity that describes the property of *matter* called *inertia*. Matter in motion resists a change in motion because of the inertia of the matter. The greater the mass of an object, the greater the inertia. 14.6 kilograms = 1 slug (14.6 kg = 1 lb·s²/ft).
See: Figure 3–2

EXAMPLE: The mass of a certain book is 4 kg.
NOTE: Mass is independent of location; that is,

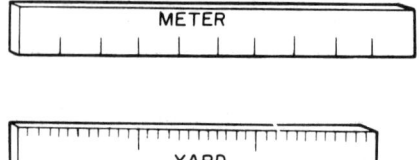

FIGURE 3–1
Comparison of units of length—meter, SI System, to yard, English System. The meter (m) is the base SI unit for length. 1 meter = 39.37 inches and 0.914 meter = 1 yard. As shown, the meter is slightly longer than the yard.

The SI System

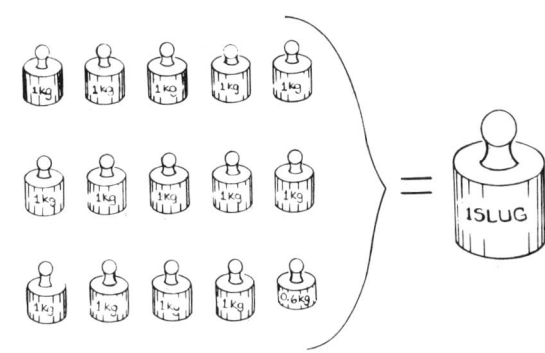

FIGURE 3-2 Comparison of units of mass—kilogram, SI System, to slug, English System. The kilogram (kg) is the base SI unit for mass. 14.6 kilograms = 1 slug.

an object has the same mass whether located on earth or in free space.

Temperature (*T*) *Unit name:* kelvin (K)
Use: Temperature is the physical quantity that describes the level of internal energy possessed by the individual atoms and molecules of an object. Temperature is also an indicator of the directions of flow of thermal energy. Heat flows from regions of higher temperature to regions of lower temperature in an isolated system.
See: Figure 3-3

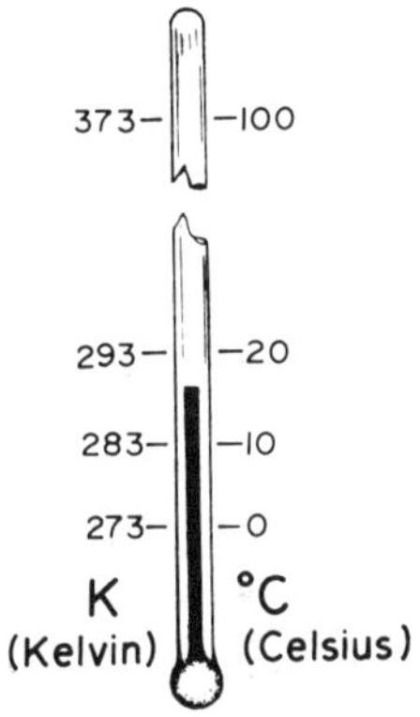

EXAMPLE: Water boils at 373 K.
NOTE: Temperature is commonly measured in degrees Celsius (°C). Kelvin and degrees Celsius are related in the following way: K = °C + 273.15.

FIGURE 3-3 Thermometer illustrating relationship between kelvin (K) and the common unit of temperature, the degree Celsius (°C). Degree Celsius is used for temperature measurement in electronics. Notice that a 10°C temperature change is the same size as a 10 K temperature change. The difference in the two scales is the starting point.

Force (*F*) *Unit name:* newton (N)
Use: Force is the physical quantity that describes the result of acting on a mass to change its state or position. The words exert, extend, compress, pull, or push are used to indicate the application of a force to the mass of an object. 4.45 newtons = 1 pound (4.45 N = 1 lb).
See: Figure 3-4

EXAMPLE: Due to the gravitational force (field force) of attraction between the mass of the earth and objects on the surface of the earth, a 1-kg mass of

FIGURE 3-4
Comparison of units of force—newton, SI System, to pound, English System. The newton (N) is the derived SI unit for force. 4.45 newtons = 1 pound.

potato will exert a force of 9.81 N on a spring scale.

Energy (W) *Unit name:* joule (J)
Use: Energy is the physical quantity that describes the ability to do work. Work is accomplished when a mass is moved against a force. Work then describes the effect of applying a force to a mass. 1 joule = 1 newton-meter (1 J = 1 N·m).
See: Figure 3-5

EXAMPLE: In sliding a heavy (150-kg) piece of electrical equipment 15 m across a floor, 800 J of work was done.

Power (P) *Unit name:* watt (W)
Use: Power is the physical quantity that describes the time rate of doing work. One watt of power is delivered when work is done at the rate of one joule per second. 1 watt = 1 joule/second (1 W = 1 J/s).
See: Figure 3-6

EXAMPLE: Ten watts of power is delivered in moving a 100-kg mass 10 m in 10 s when a force of 10 N is applied.

Electric Charge (Q) *Unit name:* coulomb (C)
Use: Electric charge refers to the unit quantity of charge, the coulomb, defined as the quantity of charge on 6.242×10^{18} electrons.

EXAMPLE: The free electrons of a certain copper wire have a combined electric charge of 4 C.

Electric Current (I) *Unit name:* ampere (A)
Use: Electric current is electric charge in motion. As such, it carries electric energy from the source to the load.
See: Figure 3-7(a)

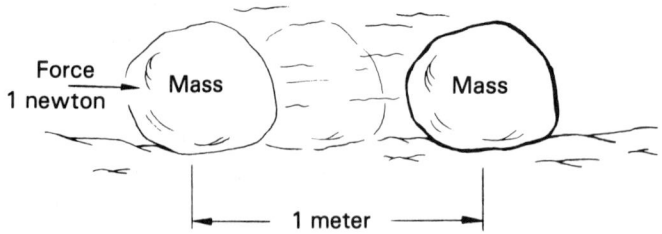

FIGURE 3-5
One joule of work is done when the mass is moved 1 meter when a force of 1 newton is applied. One joule = 1 newton-meter.

The SI System

FIGURE 3-6 Two vehicles with the same 1400 kg mass travel the same 10 000 m distance in different times. Assuming equal cornering abilities, coefficient of drag, gear ratio, etc., the vehicle that can do work at a greater rate will arrive first. One watt = 1 joule/second.

EXAMPLE: The measured current flow in a certain electric light is 1.2 A.

NOTE: Electric current is one of the seven base units. However, it is commonly referred to as the rate of electric charge passing a given point in an electric circuit in 1 second. 1 ampere = 1 coulomb/second (1 A = 1 C/s).

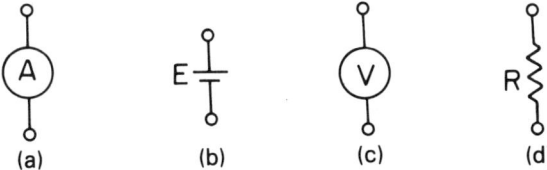

(a)　(b)　(c)　(d)

FIGURE 3-7 Symbols used in electrical circuit diagram called *schematic diagram*.
(a) Schematic symbol used to represent an ammeter. The ammeter is used to measure electric current.
(b) Schematic symbol used to represent dc voltage source. The longer line indicates the positive terminal and the letter E represents electromotive force (emf).
(c) Schematic symbol used to represent a voltmeter. The voltmeter is used to measure potential difference.
(d) Schematic symbol used to represent any kind of electric resistance. The letter R represents resistance.

	Electromotive force (E)	*Unit name:* volt (V) *Use:* Electromotive force is the quantity used to describe the work done on each unit of electric charge to raise the energy level of the charge. Emf causes current to flow. *See:* Figure 3–7(b)
EXAMPLE:		A certain source of electromotive force has raised the energy level on 5 C of charge by 100 V.
	Potential Difference (V)	*Unit name:* volt (V) *Use:* Potential difference describes the result of moving an electric charge from a higher energy level to a lower energy level. The difference in the two energy levels is the potential energy difference or, simply, the potential difference. *See:* Figure 3–7(c)
EXAMPLE:		When 2 A of electric current moves through a certain lamp, a potential difference of 117 V is measured. NOTE: Electromotive force and potential difference are both measured in units of volts (V). It is a common practice in electronics to refer to either as voltage.
	Electric Resistance (R)	*Unit name:* ohm (Ω) *Use:* Electric resistance describes the opposition to the movement of electric current. Energy of the electric current is transformed into heat energy when current moves through resistance. *See:* Figure 3–7(d)
EXAMPLE:		A current of 3 A moving through a resistance of 12 Ω creates a potential difference of 36 V.

3–3 ENGINEERING NOTATION AND DECIMAL NOTATION

Engineering Notation

The *base* and *derived* SI units of measurement are formed into multiple and submultiple units to form a system for measuring physical quantities. The vast majority of multiple and submultiple SI units used with electrical units have *exponents of ten*, which are given in multiples of 3.

Numbers written with an exponent of ten as a multiple of 3 are defined as numbers written in *engineering notation*. Table 3–4 shows several examples of engineering notation. You will notice that each number written in engineering notation has an exponent of ten that is a multiple of 3. Numbers are formatted in

Engineering Notation and Decimal Notation

TABLE 3-4 COMPARISON OF NOTATION

Decimal Notation	Powers of Ten Notation	Engineering Notation
8000	80.00×10^2	8.000×10^3
24 000	2.4000×10^4	24.000×10^3
0.0500	5.00×10^{-2}	50.0×10^{-3}
0.000 667	66.7×10^{-5}	667×10^{-6}
1 800 000	180.0000×10^4	$1.800\ 000 \times 10^6$

engineering notation so that multiples and submultiples of the SI units may be easily formed.

As shown in Table 3-4, engineering notation is a special form of powers of ten notation. Rule 3-1 summarizes the procedure for writing a decimal number in engineering notation, and Example 3-1 demonstrates the procedure outlined in the rule.

RULE 3-1. EXPRESSING DECIMAL NUMBERS IN ENGINEERING NOTATION

To express a decimal number in engineering notation:

1. Form the decimal coefficient so the exponent of 10 is a multiple of 3 by moving the decimal point 3, 6, 9, etc., places to the right or left.
2. Determine the exponent by counting the number of places the decimal point was moved.
3. Assign a sign to the exponent as:
   (a) A minus sign (−) if the decimal point is moved to the right.
   (b) A plus sign (+) if the decimal point is moved to the left.
4. Write the number in engineering notation as a product of the decimal coefficient and the power of ten.

EXAMPLE 3-1 Write 4700 in engineering notation.

SOLUTION Use Rule 3-1.

Step 1: Form the decimal coefficient:

$$4700 \Rightarrow 4.700$$

Steps 2-3: Count places and determine the sign of the exponent:

Measurement and Number Notation

$$4{,}\underbrace{700}_{3\ 2\ 1}{,} \quad \text{moved three places to the left } (+3)$$

Step 4: Form the number:

$$4.700 \times 10^3$$

$$\therefore \quad 4700 = 4.700 \times 10^3$$

EXAMPLE 3–2 Write (a) 0.03 and (b) 225 500 in engineering notation.

SOLUTION (a) Move the decimal point three places to the right (−3) and form the number:

$$0.03 = 0.\underbrace{030}_{1\ 2\ 3}{,} \Rightarrow 30 \times 10^{-3}$$

Notice that the zero was added to fill out the number.

$$\therefore \quad 0.03 = 30 \times 10^{-3}$$

(b) Move the decimal point three places to the left (+3) and form the number:

$$225\ 500 = 225\ \underbrace{500}_{3\ 2\ 1} = 225.500 \times 10^3$$

$$\therefore \quad 225\ 500 = 225.500 \times 10^3$$

ALTERNATE (b) Move the decimal point six places to the left (+6) and form the number:

$$225\ 500 = {.}\underbrace{225\ 500}_{6\ 5\ 4\ \ 3\ 2\ 1} = 0.225\ 500 \times 10^6$$

$$\therefore \quad 225\ 500 = 0.225\ 500 \times 10^6$$

Observation In the previous example, two solutions were given to part (b) of the problem. It is *conventional* to express the decimal coefficient as a number from 0.100 to 1000. Because of this convention, we have two solutions in Example 3–2(b).

Decimal Notation Numbers written without powers of ten are written in *decimal notation*. Thus 925, 0.0033, and 14.3 are all written in decimal notation. When working with electrical units, it is sometimes nec-

Engineering Notation and Decimal Notation

essary to change numbers in engineering notation to decimal notation. A general procedure for doing this is given in Rule 3-2. Example 3-3 shows how to apply the rule.

RULE 3-2. CONVERTING FROM ENGINEERING NOTATION TO DECIMAL NOTATION

To change from engineering notation to decimal notation, form the decimal number from the decimal coefficient and:

1. If the exponent is positive, move the decimal point to the right the number of places equal to the exponent. If necessary, add trailing zeros to hold the position of the decimal point.
2. If the exponent is negative, move the decimal point to the left the number of places equal to the exponent. If necessary, add leading zeros to hold the position of the decimal point. The number will be in decimal notation.

EXAMPLE 3-3 Write (a) 6.28×10^3 and (b) 530×10^{-6} in decimal notation. Use Rule 3-2.

SOLUTION (a) Move the decimal point 3 places to the right:

$$6.28 \times 10^3 = 6.280. = 6280.$$

Notice that the zero was added to fill out the number.

∴ $\quad 6.28 \times 10^3 = 6280$

(b) Move the decimal point 6 places to the left:

$$530 \times 10^{-6} = 0.000\ 530. = 0.000\ 530$$

Notice that zeros were added to fill out the number.

∴ $\quad 530 \times 10^{-6} = 0.000\ 530$

Table 3-5 contains more examples that illustrate the use of Rule 3-2.

Measurement and Number Notation

TABLE 3-5 APPLICATION OF RULE 3-2

Engineering Notation	Movement of Decimal Point		Decimal Notation
	Direction	No. of Places	
68.0×10^{-3}	Left	3	0.068
0.25×10^{6}	Right	6	250 000
500×10^{-9}	Left	9	0.000 000 500
320×10^{-6}	Left	6	0.000 320
4.70×10^{3}	Right	3	4700

3-4 FORMING DECIMAL MULTIPLES AND SUBMULTIPLES OF THE SI UNITS

When working with the SI units, it often becomes necessary to use larger (multiple) or smaller (submultiple) units of a particular quantity. The multiple and submultiple units of the SI quantities are formed by using the prefixes in Table 3-6. The name of the base or derived SI unit is modified by the addition of a prefix. Some of the *prefixed units* of electric current would be the milliampere (mA), microampere (μA), and nanoampere (nA).

As mentioned in Section 3-3, in forming numbers with dimensions, it is conventional to select a prefix so that the number values may be expressed as a number from 0.100 to 1000. Thus,

TABLE 3-6 PREFIXES AND SYMBOLS FOR MULTIPLES AND SUBMULTIPLES OF THE SI UNITS

Prefix	Symbol	Multiplying Factor	Engineering Notation
MULTIPLE PREFIXES			
kilo	k	1 000	10^{3}
mega	M	1 000 000	10^{6}
giga	G	1 000 000 000	10^{9}
tera	T	1 000 000 000 000	10^{12}
peta	P	1 000 000 000 000 000	10^{15}
exa	E	1 000 000 000 000 000 000	10^{18}
SUBMULTIPLE PREFIXES			
milli	m	0.001	10^{-3}
micro	μ	0.000 001	10^{-6}
nano	n	0.000 000 001	10^{-9}
pico	p	0.000 000 000 001	10^{-12}
femto	f	0.000 000 000 000 001	10^{-15}
atto	a	0.000 000 000 000 000 001	10^{-18}

43

Forming Decimal Multiples and Submultiples of the SI Units

an appropriate multiple or submultiple of the SI unit is chosen so that the number value lies in the conventional range. Table 3–7 has several examples of prefixes being applied to SI units to form prefixed units. Rule 3–3 outlines the procedure for forming prefixed units. The rule is followed by several examples demonstrating the use of the rule in forming prefixed SI units.

RULE 3–3. PROCEDURE FOR FORMING PREFIXED UNITS

In forming multiples and submultiples of SI units:

1. Write the number in engineering notation with a decimal coefficient expressed as a number from 0.100 to 1000.
2. Form the prefixed unit by writing the decimal coefficient with the engineering notation replaced with the appropriate prefix. Attach the unit symbol.

EXAMPLE 3–4 Write 6600 volts with prefixed units.

   SOLUTION Apply Rule 3–3.
   Step 1: Express 6600 V in engineering notation:

   $$6600 \text{ V} \Rightarrow 6.6 \times 10^3 \text{ V}$$

   Step 2: Replace engineering notation with prefixed unit:

   $$6.6 \times 10^3 \text{ V} \Rightarrow 6.6 \text{ kV}$$

   $$\therefore \quad 6600 \text{ V} = 6.6 \text{ kV}$$

EXAMPLE 3–5 Write 330 000 ohms using a prefix from Table 3–6.

   SOLUTION (a) $330\,000 \ \Omega \Rightarrow 330 \times 10^3 \ \Omega \Rightarrow 330 \text{ k}\Omega$
   (b) $330\,000 \ \Omega \Rightarrow 0.330 \times 10^6 \ \Omega \Rightarrow 0.330 \text{ M}\Omega$

TABLE 3–7 FORMING PREFIXED SI UNITS

Measured or Computed Value	Engineering Notation	Value Expressed as Prefixed SI Units
0.002 80 A	2.80×10^{-3} A	2.80 mA
8 470 000 Hz	8.47×10^6 Hz	8.47 MHz
47 000 Ω	$47 \times 10^3 \ \Omega$	47 kΩ
0.009 20 V	9.20×10^{-3} V	9.20 mV
0.000 000 053 s	53×10^{-9} s	53 ns
1390 W	1.39×10^3 W	1.39 kW

NOTE: Either answer is correct as it meets the conventional guidelines (0.100 to 1000) for expressing dimensional numbers in SI units.

$$330\ 000\ \Omega = 330\ k\Omega \quad \text{or} \quad 0.330\ M\Omega$$

EXAMPLE 3–6 Change 52 ns (nanoseconds) to the basic unit seconds.

SOLUTION Replace n with 10^{-9}.

$$52\ ns \Rightarrow 52 \times 10^{-9}\ s$$

Express in decimal notation:

$$52 \times 10^{-9}\ s \Rightarrow 0.000\ 000\ 052\ s$$

$$\therefore \quad 52\ ns = 0.000\ 000\ 052\ s$$

Combining Prefixed Units

When adding or subtracting dimensional prefixed numbers, first convert each number to its basic unit, then add or subtract. Write the answer with an appropriate prefixed unit. Example 3–7 demonstrates this concept.

EXAMPLE 3–7 Add 3.128 kJ and 72.0 J.

SOLUTION Convert 3.128 kJ to joules:

$$3.128\ kJ \Rightarrow 3128\ J$$

Add 3128 J and 72.0 J:

$$3128\ J + 72.0\ J = 3200\ J$$

Write 3200 J with prefixed units:

$$3200\ J \Rightarrow 3.2 \times 10^3\ J \Rightarrow 3.2\ kJ$$

$$\therefore \quad 3.128\ kJ + 72.0\ J = 3.2\ kJ$$

3–5 UNIT ANALYSIS AND CONVERSION WITHIN AND BETWEEN SYSTEMS OF UNITS

Unit Analysis

When converting from basic units to prefix units or converting between the English and SI systems, you may become confused in knowing when to divide and when to multiply. If you use a technique called *unit analysis*, you are less likely to have this problem. This technique uses the concept that a quantity may be multiplied by *one* without changing its value. Thus, the identity 100 centimeters = 1 meter may be expressed as a fraction; 100 cm/1 m = 1, or in its reciprocal form, 1 m/100 cm = 1. In either case, we see the ratio (fraction) is worth 1 and has no dimensions. The technique of unit analysis is summarized in Rule 3–4.

Unit Analysis and Conversion Within and Between Systems of Units

---

RULE 3–4. UNIT ANALYSIS TECHNIQUE

Unit analysis is generally applied by:

1. Expressing an identity (identical equation) that has the desired units.
2. Transforming the identity into a ratio having the desired units in the numerator.
3. Multiplying the unit to be acted on by the ratio.
4. Writing the resulting product with the desired units.
5. Repeating the preceding steps if more than one conversion is needed.

Conversion within the SI System

The following examples will demonstrate the use of unit analysis. Table 3–8 lists several identities useful in conversion within the SI system.

EXAMPLE 3–8 Convert 0.2 hours (h) to seconds (s).

SOLUTION Select identities from Table 3–8.
Use 1 h = 60 min to convert hours to minutes.
Express the identity as the ratio $\frac{60 \text{ min}}{1 \text{ h}}$:

$$0.2 \text{ h} \times \frac{60 \text{ min}}{1 \text{ h}} = 12 \text{ min}$$

Convert min to s (1 min = 60 s):

$$12 \text{ min} \times \frac{60 \text{ s}}{1 \text{ min}} = 720 \text{ s}$$

∴ 0.2 h = 720 s

TABLE 3–8 SELECTED SI IDENTITIES

Length: meter (m)		Time: second (s)	
Centimeter (cm)	100 cm = 1 m	Minute (min)	1 min = 60 s
Millimeter (mm)	1000 mm = 1 m	Hour (h)	1 h = 60 min
Mass: kilogram (kg)		Day (d)	1 d = 24 h
Gram (g)	1000 g = 1 kg	Energy: joule (J)	
Milligram (mg)	1000 mg = 1 g	Kilowatthour (kWh)	1 kWh = 3.6 MJ

EXAMPLE 3–9 Convert 380 centimeters (cm) to meters (m).

SOLUTION Use the identity 100 cm = 1 m expressed as the ratio 1 m/100 cm:

$$380 \text{ cm} \times \frac{1 \text{ m}}{100 \text{ cm}} = 3.80 \text{ m}$$

$$\therefore \quad 380 \text{ cm} = 3.80 \text{ m}$$

EXAMPLE 3–10 Convert 0.5 grams (g) to milligrams (mg).

SOLUTION Use the identity 1000 mg = 1 g expressed as the ratio 1000 mg/1 g:

$$0.5 \text{ g} \times \frac{1000 \text{ mg}}{1 \text{ g}} = 500 \text{ mg}$$

$$\therefore \quad 0.5 \text{ g} = 500 \text{ mg}$$

EXAMPLE 3–11 Convert 0.820 kilovolts (kV) to volts (V).

SOLUTION Use the identity 1000 V = 1 kV expressed as the ratio 1000 V/1 kV.

$$0.820 \text{ kV} \times \frac{1000 \text{ V}}{1 \text{ kV}} = 820 \text{ V}$$

$$\therefore \quad 0.820 \text{ kV} = 820 \text{ V}$$

As may be seen from the previous examples, unit analysis treats the units as factors in an algebraic expression. Thus, algebraic operations may be applied to units in the same way that they are applied to numbers and variables.

The common definition for the SI units listed in Table 3–9 may be used in conjunction with unit analysis to solve problems. The following examples will demonstrate this concept.

EXAMPLE 3–12 Determine the amount of energy (in joules) changed to heat and light when a 100-watt (W) lamp is operated for 5.0 hours (h).

SOLUTION Select an identity from Table 3–9 that relates joules, watts, and seconds: joule = watt · second (1 J = 1 W·s).
Since time is in seconds, convert 5.0 h to seconds:

$$5.0 \text{ h} \times \frac{60 \text{ min}}{1 \text{ h}} \times \frac{60 \text{ s}}{1 \text{ min}} = 18\,000 \text{ s}$$

TABLE 3-9 SI UNITS WITH COMMON DEFINITION AND DIMENSIONAL ORIGIN

Physical Quantity	*Unit Name*	*Common Definition*	*Expressed in SI Base Units*
Acceleration	meter per second squared	—	$\dfrac{m}{s^2}$
Force	newton	mass · acceleration	$\dfrac{kg \cdot m}{s^2}$
Work (energy)	joule	newton · meter watt · second	$\dfrac{kg \cdot m^2}{s^2}$
Power	watt	joule/second volt · ampere	$\dfrac{kg \cdot m^2}{s^3}$
Charge	coulomb	ampere · second	A·s
Current	ampere	coulombs/second volts/ohm	A
Voltage	volt	joules/coulomb ampere · ohm	$\dfrac{kg \cdot m^2}{s^3 \cdot A}$
Resistance	ohm	volts/ampere	$\dfrac{kg \cdot m^2}{s^3 \cdot A^2}$

Convert watt-seconds (W·s) to joules using the $\dfrac{1 \text{ J}}{1 \text{ W·s}}$:

$$100 \, \cancel{W} \times 18\,000 \, \cancel{s} \times \frac{1 \text{ J}}{1 \, \cancel{W \cdot s}} = 1\,800\,000 \text{ J}$$

Express the solution to two significant figures with prefixed units.

$$1\,800\,000 \text{ J} \Rightarrow 1.8 \times 10^6 \text{ J} \Rightarrow 1.8 \text{ MJ}$$

∴ $\quad 100 \text{ W} \times 5.0 \text{ h} = 1.8 \text{ MJ}$

EXAMPLE 3-13 Determine the potential difference (in volts) measured across a load when 48 joules (J) of work is done in moving 3.0 coulombs (C) of charge through the load.

SOLUTION Select the identity volts = joules/coulomb (1 V = 1 J/C).

Measurement and Number Notation

Convert joules per coulomb to volts using the ratio 1 V·C/1 J:

$$\frac{48 \text{ J}}{3.0 \text{ C}} \times \frac{1 \text{ V·C}}{1 \text{ J}} = 16 \text{ V}$$

∴ Potential difference is 16 V.

Conversion Between Systems of Units

Unit analysis may also be used when converting between systems. Table 3–10 has several conversion factors (identities) listed. These may be used as ratios to convert between the English and the SI systems. There are many intersystem conversion factors listed in engineering, physics, and chemical handbooks.

EXAMPLE 3–14 Convert 0.0930 inch (in.) to millimeters.

SOLUTION From Table 3–10, select the identity 1 mm = 0.0394 in.

Convert inches to millimeters using the ratio 1 mm/0.0394 in.:

$$0.0930 \text{ in.} \times \frac{1 \text{ mm}}{0.0394 \text{ in.}} = 2.36 \text{ mm}$$

∴ 0.0930 in. = 2.36 mm

EXAMPLE 3–15 Convert 50.0 foot-pounds of work to joules.

SOLUTION From Table 3–10, select the identity 1 ft-lb = 1.36 J.

Convert foot-pounds to joules using the ratio $\frac{1.36 \text{ J}}{1 \text{ ft-lb}}$:

$$50.0 \text{ ft-lb} \times \frac{1.36 \text{ J}}{1 \text{ ft-lb}} = 68.0 \text{ J}$$

∴ 50.0 ft-lb = 68.0 J

TABLE 3–10 CONVERSION FACTORS BETWEEN SYSTEMS

1 inch	= 2.540 centimeters	1 gallon	= 3.785 liters
1 meter	= 39.37 inches	1 cubic yard	= 0.7646 cubic meter
1 millimeter	= 0.0394 inch	1 Btu (heat)	= 1055 joules
1 pound	= 4.45 newtons	1 horsepower	= 746 watts
1 foot-pound	= 1.36 joules	1 slug	= 14.6 kilograms

EXAMPLE 3–16 Convert 320 pounds of force to newtons.

SOLUTION From Table 3–10, select the identity 1 lb = 4.45 N.

Convert pounds to newtons using the ratio $\frac{4.45 \text{ N}}{1 \text{ lb}}$:

$$320 \cancel{\text{lb}} \times \frac{4.45 \text{ N}}{1 \cancel{\text{lb}}} = 1424 \text{ N}$$

Express 1424 N to three significant figures with a prefixed unit.

$$1424 \text{ N} \Rightarrow 1.42 \times 10^3 \text{ N} \Rightarrow 1.42 \text{ kN}$$

∴ 320 lb = 1.42 kN to three significant figures

SELECTED TECHNICAL TERMS

Several of the technical terms used in this chapter are defined here to aid your understanding of them.

Ampere: the SI unit of electric current.
Coulomb: the SI unit of electric charge; equal to the electric charge transported through any point in a conductor in 1 second by 6.24×10^{18} electrons.
Degrees Celsius: unit commonly used for temperature measurement in electronic technology. Degrees Celsius (°C) is the same size as the SI base unit of temperature, the kelvin.
Engineering notation: a form of power of ten notation in which the decimal point is adjusted so that the power of ten is a multiple of 3. Also, one of the formatting functions on many scientific calculators.
Ohm: the SI unit of electrical resistance.
Volt: the SI unit of electric potential difference and electromotive force.
Watt: the SI unit of power.

PROBLEMS

Section 3–1

3–1 A physical quantity such as 10 meters (10 m) has both a _____ value and a _____.

3–2 List the two systems of measurement used in the United States.

3–3 What do the letters SI represent?

3–4 In your own words, express the strengths and weaknesses of both the English system and the SI system of measurement.

Section 3–2 3–5 Using the SI system, list the unit name and unit symbol for the following physical quantities: length, electric current, plane angle, time, and mass.

3–6 Because many units are named for scientists, the unit symbols for these names are _____.

3–7 Match one item from the list at the right to one item at the left. Some items in the right list may be used more than once.

(1) 9.81 N
(2) resistance
(3) meter
(4) J/s
(5) mass
(6) coulomb
(7) temperature
(8) English unit
(9) work
(10) SI base unit

(a) °C
(b) ampere-second
(c) watt
(d) K
(e) N·m
(f) force
(g) 39.37 inches
(h) slug
(i) ohm
(j) J/C

3–8 State why electromotive force is not actually a force.

Section 3–3 3–9 Express the following decimal numbers in engineering notation:
a. 230
b. 1750
c. 47 300
d. 2 220 000
e. 8240
f. 560 000

3–10 Express the following decimal numbers in engineering notation:
a. 0.000 68
b. 0.000 000 091
c. 0.0033
d. 0.000 020
e. 0.000 000 000 470
f. 0.000 384

3–11 Convert the following numbers written in engineering notation to decimal notation:
a. 52×10^{-3}
b. 1.25×10^{-6}
c. 520×10^{-9}
d. 43×10^{-12}
e. 2.5×10^{-3}
f. 820×10^{-6}

3–12 Convert the following numbers written in engineering notation to decimal notation:
a. 0.620×10^{6}
b. 92×10^{3}
c. 160×10^{12}
d. 0.39×10^{3}
e. 2.5×10^{9}
f. 75×10^{6}

Section 3–4 3–13 Express the following quantities in the named multiple or submultiple unit:

Problems

		Express in:		Express in:
a.	43 000 Ω	kΩ	e. 1 800 000 Ω	MΩ
b.	0.000 000 8 s	ns	f. 0.033 V	mV
c.	0.0045 C	mC	g. 0.000 25 A	μA
d.	9430 W	kW	h. 0.75 m	mm

3–14 Express the following quantities in the named multiple or submultiple unit:

		Express in:		Express in:
a.	1080 J	kJ	e. 0.064 rad	mrad
b.	0.094 N	mN	f. 520 000 g	Mg
c.	0.000 025 V	μV	g. 0.000 000 000 04 s	ps
d.	898 K	kK	h. 0.000 000 175 A	nA

3–15 Express the following quantities as a multiple or submultiple of the stated SI unit:
  a. 9240 V
  b. 0.078 300 C
  c. 510 000 Ω
  d. 0.285 A
  e. 0.000 003 s
  f. 82 300 W

3–16 Combine the following as indicated and state the answer with an appropriate prefixed SI unit:
  a. Add 120 μA and 2.08 mA
  b. Add 0.853 mrad and 2.18 rad
  c. Add 4 kΩ and 0.126 MΩ
  d. Subtract 0.528 mV from 1.08 V
  e. Subtract 430 kΩ from 2.7 MΩ
  f. Subtract 102 J from 0.708 kJ

Section 3–5
3–17 Convert 18 millihours (mh) to seconds (s).
3–18 Convert 32 centimeters (cm) to millimeters (mm).
3–19 Convert 0.2 gram (g) to milligrams (mg).
3–20 Convert 22 megajoules (MJ) to kilowatthours (kWh).
3–21 Determine the electromotive force (voltage) of a source if 100 joules (J) of work is done in moving 12 coulombs (C) of charge through the source.
3–22 Convert 0.320 watthour (Wh) to joules (J).
3–23 Convert 42 newtons (N) to pounds (lb).
3–24 Convert 6 horsepower (hp) to kilowatts (kW).
3–25 Convert 8.2 kilometers (km) per liter (ℓ) to miles (mi) per gallon (gal) where 5380 feet (ft) equals 1 mile (mi).
3–26 Convert 10 square feet (ft^2) to square meters (m^2).
3–27 Express 80 coulombs per hour (80 C/h) in amperes (A).
3–28 Express 4 mC/min in amperes.

II DC Circuits, Concepts, and Devices

This section of the book covers the important principles of conductors, resistance, Ohm's law, and Kirchhoff's laws. Circuits (both series and parallel) are covered in detail, along with voltage and current sources. Equivalent loads and sources are studied, along with network theorems and loop and node analysis. Magnetic circuits, inductance, capacitance, and transient concepts round out this section.

By mastering dc circuits, devices, and concepts, you will set the stage for the study of active devices.

4 Selected Properties of Electrical Materials

PART I INSULATORS, SEMICONDUCTORS, AND CONDUCTORS

4-1 ENERGY BANDS
4-2 INSULATORS
4-3 SEMICONDUCTORS
4-4 CONDUCTORS
4-5 RESISTANCE

CHAPTER PREVIEW: PART I

Solid crystalline materials, such as copper, silicon, and mica, used in electronic technology are grouped into one of three categories: *conductors*, such as copper, *semiconductors*, such as silicon, and *insulators*, such as mica.

The properties of these three categories of materials are determined by the bonds between the atoms, which align the atoms into regular patterns (lattice) forming the crystalline structure of the material. In this part of the chapter, you will learn about the properties of electric conductors, semiconductors, and insulators.

PERFORMANCE OBJECTIVES: PART I

Once you have read each section, worked each example with pencil, paper, and calculator, and answered each problem for every section, you will be able to:

☐ Understand the fundamental concept of energy levels within the atom.
☐ Identify the physical properties of conductors, semiconductors, and insulators.
☐ Understand the effect of temperature on resistive material.
☐ Compute the resistance of conductive materials.

4–1 ENERGY BANDS

Instead of having discrete energy levels like the single unbonded atom (as noted in Chapter 2), atoms that form solid crystalline materials have bands of permissible energy levels called *energy bands*. Within the energy bands are many energy levels that have resulted from the interaction between the electrons of the bonded atoms.

Figure 4–1 shows only the two outer energy bands of an atom. Materials may be identified as conductors, semiconductors, or insulators, depending on the position of these energy bands with respect to one another.

Energy-Band Diagrams

In an atom at *ground state* (lowest energy level of a stable substance), all the valence electrons occupy energy levels within the *valence band*. If an electron in the valence band is to move through the crystal, it must first be elevated from the valence band into the *conduction band* by the addition of a discrete amount of energy. If less than the required amount of work is done to elevate the electron, the electron does not move into the conduction band, nor does it move into the *energy gap* between the energy bands. Instead, the electron remains in the valence band. Electrons cannot reside in the energy gap between the energy bands. When an electron in the valence band acquires enough energy to move into the conduction band, it is a *free*

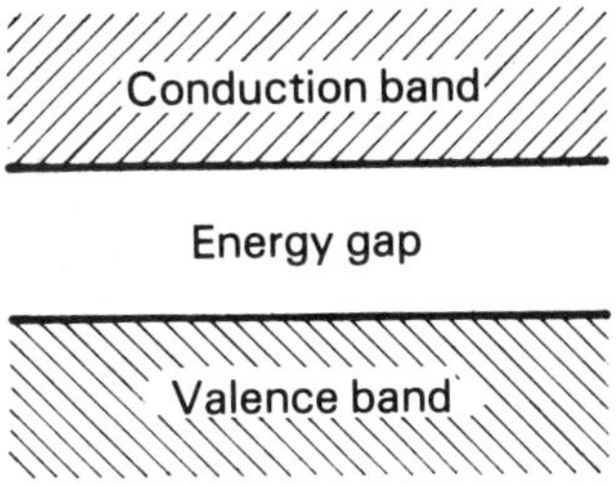

Insulators

FIGURE 4-1
Energy levels of outer shell electrons of an atom. Electrons in conduction band provide electric current in material when acted upon by external potential (voltage).

electron. The free electron can move through the crystal lattice, thus providing an electric current when a voltage source is connected across the material. The input of energy to *excite* (move) the electron from the valence band into the conduction band is usually measured in units of electron volts (eV).

$$1 \text{ eV} = 1.6 \times 10^{-19} \text{ J}$$

4-2 INSULATORS

Materials with energy-band diagrams having the conduction band separated from the valence band by a wide energy gap are called *insulators*. Figure 4-2 pictures the energy-band diagrams of an insulator, semiconductor, and conductor.

Insulating Material

Insulating materials are characterized by bonding structures that hold the valence electrons very tightly to the individual atoms, thus producing a very stable substance. Very few free electrons are present in an insulating material at room temperature (25°C); consequently, very large energies are required to free the valence electrons by moving them from the valence energy band into the conduction band. This large amount of energy corresponds to the wide energy gap of the insulator of Figure 4-2(a).

Insulating materials are manufactured from compounds of carbon and silicon. Both of these elements are group IV elements; that is, they both have four valence electrons. Bonds formed by atoms of these elements are usually covalent; in such

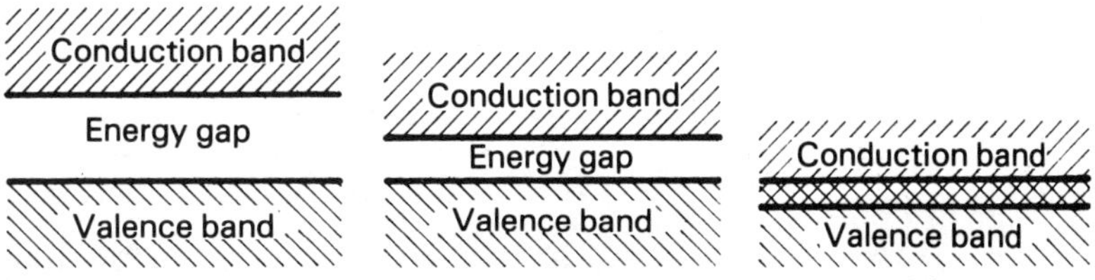

FIGURE 4-2
Energy levels in outer shell of several different classifications of materials.
(a) The *insulator* has a wide energy gap separating the conduction and valence bands.
(b) The *semiconductor* has a smaller energy gap between the conduction and valence bands.
(c) The *conductor* has an overlap of conduction and valence bands.

FIGURE 4–3
Plastic products used by the electronics industry.
(a) Cable clamp.
(b) Fuseholder.
(c) Integrated circuit (IC) socket.

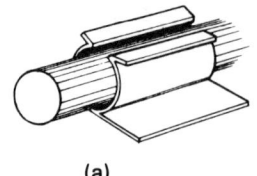
(a)

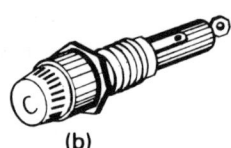
(b)

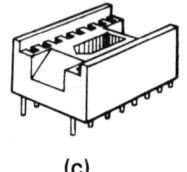
(c)

bonds, electrons are shared by the combined atoms to fill the valence band with eight electrons.

Carbon-Based Insulators Plastics are very important insulative materials for the electronics industry. Plastics are formed from molecules of hydrocarbons called *monomers* found in petroleum (oil). The separate molecules are linked to each other over and over again, forming a structure of several repeating parts. The linking of thousands and thousands of monomers is called *polymerization*. Polymers (plastics) are used to insulate and jacket conductors and cables, as well as to fasten, hold, and secure electric components and conductors. See Figure 4–3. Table 4–1 lists the chemical and common names of several polymers used by the electronics industry.

Silicon-Based Insulators Ceramic and glass insulators are formed naturally or are manufactured commercially from silicon dioxide (silica) and its derivatives, the *silicates*. About 85% of the materials in the earth's crust are made up of these materials. The silicates have very complex bonding structures that form very stable materials. Among the hundreds of natural mineral silicates, two are used in their natural state as insulators. These are *mica* and *asbestos*. Mica is found in its natural form as thin sheets, which are used between metal plates in forming electrical components. Asbestos has a fibrous structure that allows it to be wrapped around conductors to insulate them. Asbestos has very good insulative properties even at high temperatures and, unlike many of the plastic insulators, may be used well above 100°C.

Commercial ceramic insulating materials are made from materials containing silica (SiO_2), alumina (Al_2O_3), and water.

(a)

(b)

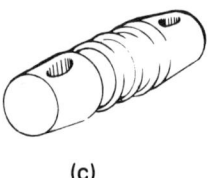
(c)

FIGURE 4–4
Ceramic insulators.
(a) Standoff.
(b) Plate cap.
(c) Antenna insulator.

Additionally, oxides of other elements may also be present. These materials are mixed and molded into the desired shapes. After drying and coating with a glazing compound, they are *fired* at high temperatures (900°C), forming the ceramic material. Ceramic materials include porcelain, steatite, and barium titanate. Figure 4–4 pictures several ceramic insulators.

TABLE 4–1 PLASTIC INSULATING MATERIALS FOR WIRE AND CABLE

Trade Name	Chemical Composition	Maximum Typical Operating Temperature	Use and Comments
Kynar*	Polyvinylidene	125°C (257°F)	Thin, durable insulation for conductors used in wire wrapping computer back planes; excellent resistance to chemicals and solvents
Polyethylene	Polyethylene	60°C (140°F)	Excellent high-frequency properties used in solid or foamed form in coaxial cable for commercial and CATV applications
PVC	Polyvinylchloride	80°C (176°F)	Usual insulator found on hook-up wire and multiconductor cables; good resistance to moisture, ozone, and most chemicals and solvents
Neoprene	Polychloroprene	60°C (140°F)	Used to jacket electric power cords; excellent resistance to ultraviolet, oil, ozone; recommended for outdoor use
FEP (Teflon)†	Fluorinated ethylene propylene	200°C (392°F)	Has excellent temperature characteristics; outstanding resistance to chemicals, solvents, and ozone, and is nonflammable
TFE (Teflon)†	Polytetrafluoroethylene	200°C (392°F)	

* Kynar is a registered trademark of Penwalt Chemical Co.
† Teflon is a registered trademark of Du Pont Co.

Insulation Breakdown

At room temperature (25°C), there are only a few free electrons available to carry current through the insulating material. As the insulators are elevated in temperature, still more electrons cross over the energy gap from the valence band to the conduction band. This is due to the increased heat energy.

When a small voltage (potential difference) is applied between two conductors, as in the cable of Figure 4–5, a very small electric current, called a *leakage* current, will pass through the insulative jacket surrounding the two metallic conductors. If the voltage of the source across the two conductors is increased, a point will be reached where the insulator will *break down* or *rupture*, and a very large current will move between the conductors and through the insulator. The current in the insulator is caused by countless electrons moving from the valence band into the conduction band owing to the added energy of the external voltage; once the insulation is ruptured, it no longer acts as an insulator and the cable must be replaced.

The voltage, which will cause the breakdown of an insulator, is given in units of volts per mil of thickness (English system,

FIGURE 4–5
Leakage current passes through insulation surrounding electrical conductors when voltage is applied between conductors. This is the case when a line cord is plugged into an ac outlet.

TABLE 4–2 TYPICAL BREAKDOWN VOLTAGE OF INSULATING MATERIALS AT 25°C

Material	*V/Mil*	*V/mm*
CERAMICS, SILICATES, AND GLASSES		
Glass	2000	80 000
Mica	5000	200 000
Porcelain	200	8 000
Steatite	300	12 000
Titanates (Ba, Sr)	100	4 000
PLASTICS		
Bakelite	300	12 000
Neoprene rubber	500	20 000
Nylon	400	16 000
Polyethylene	800	30 000
Polyvinylchloride (PVC)	800	30 000
Styrene	600	24 000
Teflon	1000	40 000
MISCELLANEOUS		
Air (dry)	75	3 000
Oil	400	16 000
Paper (waxed)	1200	50 000
Shellac	800	30 000

Insulators

where 1 mil equals 0.001 in.) or volts per millimeter (SI, where 1 mm equals 0.0394 in.). Table 4–2 lists the breakdown voltage of various materials. These are approximate values as they will vary with temperature. The following examples will amplify several points relevant to insulator properties.

EXAMPLE 4–1 Using a cable manufacturer's catalog, determine the type of insulative material and the leakage current in an electrical cable specifically designed for medical and dental test and analytical equipment.

SOLUTION A particular *low-leakage medical safety* cable is listed as having fluorinated ethylene propylene (FEP Teflon) insulation over the copper conductors, with polyethylene filler between the insulated conductors and a polyvinylchloride (PVC) jacket over the entire cable. Maximum leakage current at 120 V ac is listed as 15 μA through the jacket and 5 μA between conductors for a 3-m (10-ft) length of cable.

EXAMPLE 4–2 Using a *parts catalog*, determine the rated operating temperature and the breakdown voltage for Teflon *heat* shrinkable tubing.

SOLUTION The operating temperature is given as a range of $-275°F$ to $+500°F$ ($-170°C$ to $+260°C$); the breakdown voltage is given as 600 V/mil (23.6 kV/mm).

EXAMPLE 4–3 Determine the maximum operating voltage for a PVC (polyvinylchloride) covered conductor having a *wall thickness* of 0.016 in. See Figure 4–6 for an indication of how wall thickness is measured.

SOLUTION Convert 0.016 in. to mils using 1 mil = 0.001 in.:

$$0.016 \text{ in.} \times \frac{1 \text{ mil}}{0.001 \text{ in.}} = 16 \text{ mil}$$

Consult Table 4–2 for the breakdown voltage for PVC, which is 800 V/mil:

$$800 \frac{V}{\text{mil}} \times 16 \text{ mil} = 12.8 \text{ kV}$$

NOTE: In actual practice, this conductor would never be operated at this voltage. The insulation

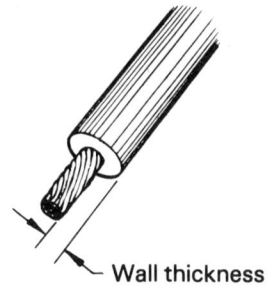

FIGURE 4–6
Wall thickness refers to thickness of insulative material around the electric conductor.

breakdown would be *derated* by a factor of 10 or more.

∴ 12.8 kV is the theoretical operating voltage, while less than 1.2 kV is the actual operating voltage.

4-3 SEMICONDUCTORS

In this section, we will be discussing only pure (*intrinsic*) semiconductor materials, such as silicon, used in the manufacture of semiconductor devices, some of which are shown in Figure 4-7. You will no doubt study more about *extrinsic* semiconductors in courses relating to *active* electronic devices. Extrinsic semiconductors have impurities added that appear to narrow the gap between energy bands. This process (called *doping*) is common in transistor manufacture.

Intrinsic Semiconductors

Silicon, a very commonly used semiconductor material, is *grown* into single crystals, with the chemical impurities removed by a process called *zone refining*. Because the silicon atom has four valence electrons, it will form covalent bonds with neighboring silicon atoms, as pictured in Figure 4-8. In looking at the bonding of the silicon atoms in Figure 4-8, it would appear that all the valence electrons are held and no free electrons are available for an electric current. This is the characteristic of an insulator. However, this is not the case at room temperature (25°C). Sufficient thermal energy is available to break some of the covalent bonds, creating free electrons, which will support an electric current through the crystal with the application of an emf. Because the material behaves as a poor conductor, it is called a *semiconductor*.

When a valence electron is elevated (by an addition of energy) into the conduction band, a *free hole* (an imaginary positive charge) is left in the valence band. In intrinsic (pure) semiconductive materials, the number of free electrons in the conductive band corresponds to the number of free holes in the valence band.

Charge Carriers

For a semiconductor to conduct electricity, energy must be added to break covalent bonds and elevate valence electrons into the conduction band. When this is done, a vacancy at the

  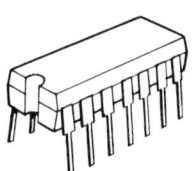

FIGURE 4-7 Transistors and integrated circuits are available in various styles of packaging.

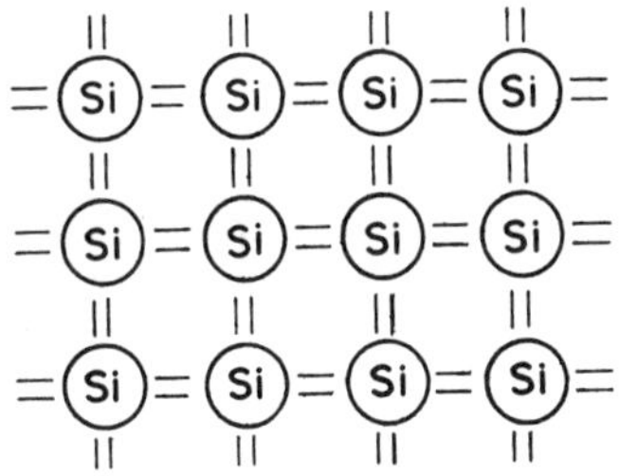

FIGURE 4-8 Silicon (Si) atoms are bonded to one another by covalent bonds. Each line represents one shared electron. Notice that each silicon atom has 8 electrons, which is a very stable state.

bond is created. The space vacated by an electron is called a *hole* and represents the absence of an electron. Since this leaves the atom with a positive charge, an electron from a neighboring atom may move into this vacancy, leaving a hole in its place. As free electrons move in one direction filling vacancies, the vacancies (holes) move in the opposite direction. Both electrons and holes in motion contribute to the electric current in a semiconductor. Thus, the amount of electric current in a semiconductor depends upon both the free electrons in the conduction band and the free holes in the valence band.

4-4 CONDUCTORS

Because of the properties exhibited by metals as a result of metallic bonds, metals are extremely good conductors. As was previously noted in Figure 4-2, the valence and conduction bands in the energy-band diagram overlap in conductors, indicating an abundance of free electrons. Hence, the application of only a small amount of voltage will result in a large amount of electric current.

The ability of a material to conduct an electric current is called its **conductance**. Each type of material has a different *conductivity*. Table 4-3 lists materials by their *relative conductivity*, with silver as the reference. Silver, the best conductor, is given an arbitrary value of 100; copper, the next best conductor, is given 95, and so forth throughout the table.

Materials that have poor conductive properties, such as carbon or nichrome (a metal used in heating elements), are referred to as being *resistive* rather than conductive.

Using the information of Table 4-3, you can appreciate the conductive properties of metals such as copper, gold, and silver when compared to semiconductors and insulators. Silver, the most conductive of the metals, is 30 billion times more conductive than intrinsic silicon, a semiconductor, and one billion billion times more conductive than polyvinylchloride (PVC), an insulator. Even nichrome, a very poor conductor when compared to copper or silver, is hundreds of millions times more conductive than intrinsic silicon, a semiconductor.

Selected Properties of Electrical Materials

TABLE 4–3 RELATIVE CONDUCTIVITY OF SELECTED MATERIALS*

Material	Relative Conductivity	
CONDUCTORS		
Silver	100	
Copper (annealed)	95	
Gold	67	
Aluminum	58	
Tungsten	30	
Tin-lead (Sn 60, Pb 40)	10	
Nichrome	1.6	
SEMICONDUCTORS		
Carbon	4×10^{-2}	(0.04)
Germanium (intrinsic)	3×10^{-6}	(0.000 003)
Silicon (intrinsic)	3×10^{-9}	(0.000 000 003)
INSULATORS		
Polyvinyl chloride (PVC)	1×10^{-16}	(0.000 000 000 000 000 1)
Mica	1×10^{-18}	(0.000 000 000 000 000 001)
Teflon	1×10^{-21}	(0.000 000 000 000 000 000 001)

* The actual conductivity of silver is 6.10×10^7 $(1/\Omega \cdot m)$ at 20°C.

4–5 RESISTANCE

The electric resistance, measured in units of ohms, may be determined for materials by applying Equation 4–1.

$$R = \rho \frac{\ell}{A} \quad (\Omega) \qquad (4\text{–}1)$$

where R = resistance in ohms (Ω)
  ρ = resistivity in ohm-meters ($\Omega \cdot m$)
  ℓ = length in meters (m)
  A = area in square meters (m²)

Resistivity The resistance, the ability of a material to oppose electric current, depends upon several factors. These include:

65
Resistance

☐ Length ☐ Temperature
☐ Cross-sectional area ☐ Kind of material

The resistance of a material is directly proportional to the length of the material. This is, $R \propto \ell$.* Furthermore, the resistance of a material is inversely proportional to the cross-sectional area of the material: $R \propto 1/A$.

Assuming a constant temperature, the resistance of a conductor may be stated mathematically as

$$R \propto \frac{\ell}{A}$$

The proportional statement of resistance becomes an equation when a *proportionality constant* is introduced.

$$R = \rho \frac{\ell}{A} \quad (\Omega) \tag{4-1}$$

The Greek letter rho (ρ) is used as the symbol for the proportionality constant called **resistivity**.

Resistivity is measured in units of ohm-meters ($\Omega \cdot$ m) at a specified temperature and is based on the resistance of a cube of material 1 meter on a side. Resistivity is the reciprocal of conductivity.

Table 4–4 lists the resistivity at 20°C of various materials from each of three categories: conductors, semiconductors, and insulators.

---

*The symbol $\propto$ is read *"proportional to."*

TABLE 4–4 RESISTIVITY OF SELECTED MATERIALS AT 20°C

Material	Resistivity ($\Omega \cdot$ m)	Material	Resistivity ($\Omega \cdot$ m)
CONDUCTORS		SEMICONDUCTORS	
Aluminum	2.83×10^{-8}	Germanium (intrinsic)	0.47
Constantan*	49.0×10^{-8}	Silicon (intrinsic)	640
Copper	1.72×10^{-8}		
Gold	2.44×10^{-8}	INSULATORS	
Nichrome*	100×10^{-8}	Polyvinylchloride (PVC)	$>10^{10}$
Silver	1.64×10^{-8}	Mica	$>10^{12}$
Tungsten	5.5×10^{-8}	Teflon	$>10^{15}$

*Metal alloy.

EXAMPLE 4-4 Using Equation 4-1 and the resistivities of Table 4-4, determine the resistance of a nichrome conductor at 20°C. The conductor is 0.5 m long with a cross-sectional area of 0.2 square millimeter (0.2 mm²).

SOLUTION Substitute into Equation 4-1. Use 0.5 m for length (ℓ), 100×10^{-8} Ω·m for resistivity (ρ), and convert 0.2 mm² of area (A) to square meters (m²). Use 1 m² = 10^6 mm².

$$A = 0.2 \text{ mm}^2 \times \frac{1}{10^6} \frac{\text{m}^2}{\text{mm}^2} = 0.2 \times 10^{-6} \text{ m}^2$$

Substitute:

$$R = \rho \frac{\ell}{A}$$

$$R = \frac{(100 \times 10^{-8})(0.5)}{0.2 \times 10^{-6}}$$

$$\therefore R = 2.5 \text{ Ω}$$

EXAMPLE 4-5 Determine the resistance from end to end of a cylinder of intrinsic silicon at 20°C if it is 2 cm long and 4 cm in diameter. 1 cm = 10^{-2} m.

SOLUTION Compute area:

$$\frac{\pi d^2}{4} = \frac{\pi (4 \times 10^{-2})^2}{4}$$

$$A = 1.26 \times 10^{-3} \text{ m}^2$$

Substitute into Equation 4-1:

$$R = \rho \frac{\ell}{A}$$

$$R = \frac{(640)(2 \times 10^{-2})}{1.26 \times 10^{-3}}$$

$$\therefore R = 10.2 \text{ kΩ}$$

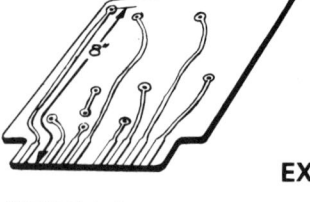

FIGURE 4-9
Printed wiring of printed circuit board is made of copper foil.

EXAMPLE 4-6 The copper foil conductor of the *printed circuit* board pictured in Figure 4-9 is 8 in. long, 0.030 in. wide, and 0.002 in. thick. Assuming the temperature is 20°C, determine the resistance of the copper foil conductor.

SOLUTION Convert from English units to the appropriate SI units.

Length 39.37 in. = 1 m:

$$8 \text{ in.} \times \frac{1}{39.37} \frac{m}{\text{in.}} = 0.203 \text{ m}$$

Width 0.0394 in. = 1 mm and 10^3 mm = 1 m:

$$0.030 \text{ in.} \times \frac{1}{0.0394} \frac{mm}{\text{in.}} = 0.761 \text{ mm}$$

$$0.761 \text{ mm} \times \frac{1}{10^3} \frac{m}{mm} = 0.761 \times 10^{-3} \text{ m}$$

Thickness 0.0394 in. = 1 mm and 10^3 mm = 1 m:

$$0.002 \text{ in.} \times \frac{1}{0.0394} \frac{mm}{\text{in.}} = 5.08 \times 10^{-2} \text{ mm}$$

$$5.08 \times 10^{-2} \text{ mm} \times \frac{1}{10^3} \frac{m}{mm} = 5.08 \times 10^{-5} \text{ m}$$

Area = width × thickness:

$$A = 0.761 \times 10^{-3} \times 5.08 \times 10^{-5}$$

$$A = 3.86 \times 10^{-8} \text{ m}^2$$

Compute resistance using Equation 4–1:

$$R = \rho \frac{\ell}{A}$$

$$R = \frac{(1.72 \times 10^{-8})(0.203)}{3.86 \times 10^{-8}}$$

$$R = 9.05 \times 10^{-2}$$

$$\therefore R = 90.5 \text{ m}\Omega$$

Effects of Temperature on Resistance

In the preceding examples, the resistance was determined for several different materials, all at a temperature of 20°C. As temperature varies, the resistivity of most materials will also vary.

Semiconductors and insulators have more covalent bonds broken as the temperature rises, resulting in more charge carriers. Because more free electrons and holes are available, the material is less resistive (more conductive). Semiconductors and insulators have less resistance at elevated temperatures and more resistance at lowered temperatures. These materials are subject to *thermal runaway* owing to a self-elevating temperature. When electric current passes through a semiconductor or an insulator, the temperature of the material is elevated, which is an increase in thermal energy that, in turn, provides for more

Selected Properties of Electrical Materials

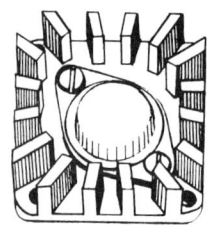

FIGURE 4-10
Power transistor is mounted on heat sink so that operating temperature is maintained within specified operating range.

charge carriers, thus providing more thermal energy and more current, and so it goes. When working with semiconductor devices (transistors, diodes, and integrated circuits), great care is used not to exceed the recommended maximum operating temperature of the devices (usually 200°C for silicon and 100°C for germanium). These devices are mounted on *heat sinks*, as shown in Figure 4-10, for safe operation.

Because of the possibility for thermal runaway, insulators are designed to operate well below their theoretical breakdown voltage. This then limits their leakage current to a value so small that no internal heating will result. The resistance of the material will therefore remain very high ($>10^{12}$ Ω) and the current will remain very small ($<10^{-6}$ A).

In conductors such as copper, silver, and tungsten, the resistance increases with an increase in temperature. As the temperature of the metal rises, the atoms (positive ions) making up the crystalline lattice vibrate owing to thermal energy. The moving atoms present an obstacle to the otherwise free electrons, resulting in an increase in the electric resistivity of the metal.

Temperature Coefficient of Resistivity

In unalloyed metals (pure metals, such as copper), the change in electric resistivity is directly proportional to the change in temperature. The resistivity of a metal (ρ_x) at any temperature (t_x) is stated as

$$\rho_x = \rho[1 + \alpha(t_x - t)] \quad (\Omega \cdot m) \qquad (4-2)$$

where ρ_x = resistivity at any temperature t_x in ohm-meters (Ω · m)

ρ = resistivity at t (usually 20°C) in ohm-meters (Ω·m)

α = temperature coefficient of resistivity in ohms per degree Celsius per ohm (Ω/°C/Ω) at 20°C

The *temperature coefficient of resistivity*, commonly referred to as simply the *temperature coefficient*, is an indication of the effect change in temperature has on the electrical resistance of a material. For a given material, each ohm of that material's resistance will change by a constant amount equal to α for each degree Celsius of change in temperature from a reference temperature. Thus, the temperature coefficient, α, is the change in resistance (Ω) for each degree Celsius of change in temperature (°C) for each ohm resistance (Ω) of the material, Ω/°C/Ω.

In general, metals have positive temperature coefficients (resistance increases with a rise in temperature), whereas insulators and semiconductors have negative temperature coefficients (resistance decreases with a rise in temperature). Certain alloyed metals, such as constantan, may have a temperature

Resistance as a Function of Temperature

The temperature coefficients of Table 4-5 are used to compute resistance at temperatures other than 20°C. The following examples will demonstrate the application of Equations 4-1 and 4-2 and Tables 4-4 and 4-5 to determine the resistance of metals at elevated temperatures.

EXAMPLE 4-7 Determine the resistivity of a tungsten lamp filament at 2250°C. Assume the temperature coefficient α is constant.

SOLUTION Determine the resistivity of tungsten at 2250°C using Equation 4-2, where

$\rho = 5.5 \times 10^{-8}$ (Table 4-4)

$\alpha = 4.5 \times 10^{-3}$ (Table 4-5)

$t_x = 2250°C$ and $t = 20°C$

Substituting results in

$\rho_x = \rho[1 + \alpha(t_x - t)]$

$\rho_x = 5.5 \times 10^{-8}[1 + 4.5 \times 10^{-3}(2250 - 20)]$

$\therefore \quad \rho_x = 60.7 \times 10^{-8}\ \Omega \cdot m$

EXAMPLE 4-8 Determine both the *cold* resistance (at 20°C) and the *hot* resistance (at 2250°C) of a tungsten lamp filament 8 cm long and 0.040 mm in diameter.

SOLUTION Compute the area in square meters (m²):

$$A = \frac{\pi d^2}{4} = \frac{(\pi)(40 \times 10^{-6})^2}{4} = 1.26 \times 10^{-9}\ m^2$$

TABLE 4-5 TEMPERATURE COEFFICIENT OF RESISTIVITY (α) OF SELECTED METALS AND ALLOYS AT 20°C

Materials	$\alpha\ (\Omega/°C/\Omega)$	*Materials*	$\alpha\ (\Omega/°C/\Omega)$
Aluminum	3.9×10^{-3}	Nichrome*	4×10^{-4}
Constantan*	4×10^{-6}	Silver	3.8×10^{-3}
Copper	3.93×10^{-3}	Tungsten	4.46×10^{-3}
Gold	3.4×10^{-3}		

* Metal alloy.

Determine R at 20°C:

$$R = \rho \frac{\ell}{A}$$

$$R = \frac{(5.5 \times 10^{-8})(8 \times 10^{-2})}{1.26 \times 10^{-9}}$$

$\therefore \quad R = 3.49 \ \Omega$

Determine R at 2250°C. Use ρ_x from Example 4-7:

$$R = \rho \frac{\ell}{A}$$

$$R = \frac{(60.7 \times 10^{-8})(8 \times 10^{-2})}{1.26 \times 10^{-9}}$$

$$R = 38.5 \ \Omega$$

$\therefore \quad$ The cold resistance is 3.49 Ω, and the hot resistance is 38.5 Ω.

Observation Because the *cold resistance* of the lamp filament is small compared to the hot resistance, a large current (called a *surge current*) passes through the filament when the lamp is first turned on, causing the filament to heat very rapidly. The resulting *thermal stress* produced by the rapid heating sometimes causes the filament to break. Thus, lights are most likely to *burn out* when first turned on.

From Example 4-8 we see that the resistance of the material increased with temperature. This was expected since the temperature coefficient for metals is positive. These observations may be formulated into an equation that relates the resistance of a metallic conductor to temperature change (ΔT) and the temperature coefficient of the material (α). Thus, the following proportions may be stated:

$$\frac{\rho_x}{\rho} = \frac{R_x}{R} \qquad (4-3)$$

where ρ = resistivity at 20°C in ohm-meters ($\Omega \cdot$m)
ρ_x = resistivity at operating temperature t_x in ohm-meters ($\Omega \cdot$ m)
R = resistance at 20°C in ohms (Ω)
R_x = resistance at operating temperature t_x in ohms (Ω)

Equation 4-2 may be restated as

$$\frac{\rho_x}{\rho} = [1 + \alpha(t_x - t)]$$

However, $\rho_x/\rho = R_x/R$ from Equation 4-3. Thus:

$$\frac{R_x}{R} = [1 + \alpha(t_x - t)]$$

and

$$R_x = R[1 + \alpha(t_x - t)]$$

Let $t_x - t = \Delta T$:

$$R_x = R(1 + \alpha \Delta T) \quad (\Omega) \tag{4-4}$$

For calculator use, distribute R. Thus:

$$R_x = R + \alpha R \Delta T \quad (\Omega) \tag{4-5}$$

where R_x = resistance at operating temperature t_x in ohms (Ω)
R = resistance at 20°C in ohms (Ω)
α = temperature coefficient of resistivity at 20°C ($\Omega/°C/\Omega$)
ΔT = difference between t_x, the operating temperature, and t, the reference temperature (20°C): $\Delta T = t_x - t$ (°C)

EXAMPLE 4-9 Determine the resistance of the tungsten filament of Example 4-8 at both (a) 800°C and (b) 2250°C, given 3.49 Ω as the resistance at 20°C.

SOLUTION Use Equation 4-5:

(a) $R_x = R + \alpha R \Delta T$

$R_x = 3.49 + (4.5 \times 10^{-3})(3.49)(780)$

$R_x = 15.7 \ \Omega$

∴ $R_x = 15.7 \ \Omega$ at 800°C

(b) $R_x = R + \alpha R \Delta T$

$R_x = 3.49 + (4.5 \times 10^{-3})(3.49)(2230)$

$R_x = 38.5 \ \Omega$

∴ $R_x = 38.5 \ \Omega$ at 2250°C

Equation 4-1 may now be restated as Equation 4-6 in order to relate the effects of temperature to the resistance of a metallic conductor. Restating Equation 4-1,

$$R = \rho \frac{\ell}{A} \quad (\Omega) \tag{4-1}$$

and Equation 4–4,

$$R_x = R(1 + \alpha \Delta T) \quad (\Omega) \tag{4-4}$$

Substituting Equation 4–1 into Equation 4–4 results in

$$R_x = \rho \frac{\ell}{A}(1 + \alpha \Delta T) \quad (\Omega) \tag{4-6}$$

where R_x = resistance at operating temperature t_x in ohms (Ω)
ρ = resistivity of the material at 20°C in ohm-meters ($\Omega \cdot$m)
ℓ = length in meters (m)
A = area in square meters (m²)
α = temperature coefficient of resistivity in ohms/°C/ohms (Ω/°C/Ω) at 20°C
ΔT = difference between the operating temperature and 20°C

EXAMPLE 4–10 Repeat Example 4–8 using Equation 4–6. Determine both the cold resistance at 20°C and the hot resistance at 2250°C of a tungsten lamp filament 8 cm long and 0.040 mm in diameter.

SOLUTION Compute the area in square meters (m²):

$$A = \frac{\pi d^2}{4} = \frac{\pi (40 \times 10^{-6})^2}{4} = 1.26 \times 10^{-9} \text{ m}^2$$

Substitute into Equation 4–6 and determine R_x at 20°C:

$$R_x = \rho \frac{\ell}{A}(1 + \alpha \Delta T)$$

$$R_x = \frac{(5.5 \times 10^{-8})(8 \times 10^{-2})}{1.26 \times 10^{-9}} (1 + 4.5 \times 10^{-3} \times 0)$$

$$R_x = \frac{(5.5 \times 10^{-8})(8 \times 10^{-2})}{1.26 \times 10^{-9}}$$

$$R = 3.49 \, \Omega$$

$$R_x = 3.49 \, \Omega \text{ at } 20°C$$

Substitute into Equation 4–6 and determine R_x at 2250°C:

$$R_x = \rho \frac{\ell}{A}(1 + \alpha \Delta T)$$

$$R_x = \frac{(5.5 \times 10^{-8})(8 \times 10^{-2})}{1.26 \times 10^{-9}} (1 + 4.5 \times 10^{-3} \times 2230)$$

$$R_x = 38.5 \ \Omega$$

$$\therefore \quad R_x = 38.5 \ \Omega \text{ at } 2250°C$$

Equation 4–6 is very important since it allows us to calculate for any one of six parameters of a conductor; thus resistance (R_x), temperature (ΔT), temperature coefficient (α), area (A), length (ℓ), and kind of material (ρ) may be determined when given the other five parameters.

SELECTED TECHNICAL TERMS: PART I

Several of the technical terms used in Part I of this chapter are defined here to aid your understanding of them.

Conductance: the measure of the ease with which a conductor carries an electric current; the reciprocal of resistance.
Conductivity: the reciprocal of resistivity; an indication of the ability of a material to pass an electric current.
Conductor: a material that offers a small resistance (opposition) to the passage of an electric current; a wire, cable, or bus used to carry current.
Hole: a vacancy in the valence band of a semiconductor due to the loss of an electron from the band.
Insulator: a material that offers a very high resistance to the passage of an electric current.
Resistivity: the resistance between opposite parallel faces of a 1-meter cube of a material; also called specific resistance. Resistivity ($\Omega \cdot$m) depends only on the properties of the material itself, whereas resistance (Ω) depends not only on the properties of the material but on its length and cross-sectional area.
Temperature coefficient: the change in resistance of a material per degree of change in temperature. As a general rule, conductors have a positive temperature coefficient, whereas semiconductors and insulators have a negative temperature coefficient.

PROBLEMS: PART I

Section 4–1

4–1 Crystalline materials have bands of permissible energy levels called _____ _____.

4–2 The two outer energy bands of an atom are the _____ band and the _____ band.

4–3 Briefly describe what is needed to elevate an electron from the valence band into the conduction band where it

Selected Properties of Electrical Materials

is a free electron.

4-4 The quantity of energy needed to excite the electron is measured in units of _____ _____.

4-5 Convert 72.6 × 10⁻¹⁴ J into electron volts (eV).

4-6 Express 2 keV as joules (J).

Section 4-2

4-7 Relate the classification of crystalline materials as conductors, semiconductors, or insulators by the relative size of the energy gap.

4-8 Insulating materials used by the electronics industry are made from _____ and _____, both having _____ valence electrons.

4-9 Explain how polymers are formed from monomers.

4-10 Another less technical name for polymer is _____.

4-11 Using an electronics parts catalog, locate and describe two items that use Teflon in their manufacture.

4-12 PVC is an abbreviation of _____.

4-13 An example of a silicate that is used in its natural state is _____.

4-14 Porcelain, steatite, and barium titanate are all _____ materials.

4-15 List the typical breakdown voltage in volts/mil for the following insulating materials: (a) paper, (b) styrene, (c) steatite, (d) Bakelite, (e) Teflon, and (f) mica.

4-16 Explain the meaning of the phrase "break down."

4-17 Reduce the information of Table 4-2 by a factor of 10 and determine the actual operating voltage that may be dropped across the plates of a capacitor that uses mica 0.002 in. thick as the insulator between the metal plates.

4-18 Determine a practical operating voltage for a Teflon-covered conductor having a wall thickness of 0.25 mm. Remember to reduce the breakdown voltage rating by a factor of 10 to arrive at a practical operating voltage.

Section 4-3

4-19 Silicon, a semiconductor material, has four valence electrons and forms _____ bonds.

4-20 In an intrinsic semiconductor material, when an electron is elevated in energy level into the conduction band, a _____ is created in the valence band.

4-21 The electric charge carriers making up electric current in a semiconductor are the _____ in the valence band and the _____ in the conduction band.

Section 4-4

4-22 Due to the metallic bonding of the atoms, metals have an abundance of _____ _____ to serve as charge carriers.

4-23 Working with the materials listed in Table 4-3, determine the actual conductivity of each in units of $1/\Omega \cdot m$ from the listed values of relative conductivity, where 6.10×10^7 $(1/\Omega \cdot m)$ is the actual conductance of silver at 20°C.

Section 4-5

4-24 Since resistivity $(\Omega \cdot m)$ is the reciprocal of conductivity $(1/\Omega \cdot m)$, determine the resistivity of iron if its conductivity is 8.33×10^6 $(1/\Omega \cdot m)$.

4-25 Repeat Problem 4-24 for nickel, which has a conductivity of 1.28×10^7 $(1/\Omega \cdot m)$.

4-26 Using Equation 4-1 and the resistivity of Table 4-4, determine the resistance of a round copper wire at 20°C. The wire is 50 m long and 1.63 mm in diameter.

4-27 Using Equation 4-1 and the resistivity of Table 4-4, determine the resistance at 20°C of a round aluminum wire 2.59 mm in diameter and 380 m long.

4-28 A copper foil conductor of a printed circuit board, operating at 20°C, is 35 cm long, 1.5 mm wide, and 63 μm thick; determine the resistance of the conductor.

4-29 A rectangular bar made of constantan, operating at 20°C, is 10 cm long, 1 cm wide, and 0.5 cm thick; determine the resistance of the bar.

4-30 Determine the resistance of a mica insulator 0.70 in. long, 0.30 in. wide, and 0.004 in. thick having a resistivity at 20°C of 2.72×10^{12} $\Omega \cdot m$.

4-31 A round electrical insulating *standoff* made from TFE Teflon is 1.25 in. long and 0.19 in. in diameter; determine the resistance of the standoff when operating at 20°C if the resistivity of the Teflon is 4.82×10^{15} $\Omega \cdot m$.

4-32 State the types of materials that are subject to thermal runaway; give an example of each type of material.

4-33 Explain why the conductivity of silicon and mica increases with temperature, whereas the conductivity of copper and aluminum decreases with temperature.

4-34 Determine the resistance of a tungsten wire operating at 500°C if the resistance of the wire at 20°C is 10 Ω.

4-35 Determine the resistance of a nichrome heating element operating at 1200°C if its resistance at 20°C is 7 Ω.

4-36 Determine the resistance at 100°C of a round copper wire 0.64 mm in diameter and 12 m in length.

4-37 Determine the resistance at 85°C of a round copper wire 0.020 in. in diameter and 5 ft in length.

4-38 Determine the resistance of a gold alloy wire used in the manufacture of transistors. The gold alloy's resistivity is

14×10^{-8} Ω·m, and its temperature coefficient is 5.8×10^{-4} Ω/°C/Ω, both of which are specified at 20°C. The wire, which is 3 mm long and 13 μm in diameter, is operating at 200°C.

4-39 Determine the resistance of a rectangular bar of aluminum 6 ft long, 1 in. wide, and 0.25 in. thick if its temperature is 75°C.

4-40 Determine the resistance of a ribbon of copper foil at 80°C if the foil is 18 in. long, 0.093 in. wide, and 0.003 in. thick.

4 Selected Properties of Electrical Materials

PART II ROUND ELECTRIC CONDUCTORS

4-6 RESISTANCE OF METALLIC WIRE
4-7 EFFECT OF TEMPERATURE ON WIRE RESISTANCE
4-8 AMERICAN WIRE GAUGE (AWG)
4-9 APPLICATIONS USING THE AWG AND FORMULAS

CHAPTER PREVIEW: PART II

In Part I of this chapter, it was learned that resistance of a material depends upon four parameters: length (ℓ), cross-sectional area (A), temperature (T), and kind of material (ρ). This is true for round metallic electric conductors (wire) as well.

This part of the chapter deals with round *wire*. Because wire length is usually specified by manufacturers in feet and wire diameter in mils (1 mil = 0.001 in.), the formulas used in conjunction with round wire are defined in English units.

PERFORMANCE OBJECTIVES: PART II

Once you have read each section, worked each example with pencil, paper, and calculator, and answered each problem for every section, you will be able to:

☐ Compute the resistance of round conductors.
☐ Understand how temperature affects the resistance of round conductors.
☐ Make approximations using the key relations of the AWG.
☐ Use the AWG to specify electric conductors.

4–6 RESISTANCE OF METALLIC WIRE

The resistance of any round conductor at a specified temperature may be computed using the formula

$$R = \rho \frac{\ell}{A} \quad (\Omega) \tag{4-7}$$

where R = resistance of the wire in ohms (Ω)
ρ = resistivity of the metal in ohm-circular mils per foot (Ω-cmil/ft)
ℓ = length in feet (ft)
A = cross section in circular mils (cmils); A equals diameter in mils squared ($A = d^2$).

In Equation 4–7, the cross section of the wire (A) is specified in units of circular mils, abbreviated cmil. This unit is defined as the diameter of the wire in mils squared (d^2). Thus, a wire with a diameter of 0.010 in. (10 mils) would have a cross section of 10^2 or 100 cmils. Furthermore, Equation 4–7 has resistivity defined in units of Ω-cmil/ft. This unit results from the need for a *standard conductor* to compare resistances of wires of different cross sections, lengths, and materials. The standard conductor is defined as a 1-ft conductor with a cross section of 1 cmil. Thus, resistivity is defined from Equation 4–7 as

$$\rho = \frac{RA}{\ell} \Rightarrow \Omega\text{-cmil/ft}$$

The resistivity for selected metals and alloys at 20°C is given in Table 4–6.

TABLE 4-6 TYPICAL RESISTIVITY (ρ) OF SELECTED METALS AND ALLOYS AT 20°C

Metals	Resistivity (Ω-cmil/ft)	Metals	Resistivity (Ω-cmil/ft)
Aluminum	17.0	Nichrome*	600
Constantan*	295	Silver	9.8
Copper	10.4	Tungsten	33.0
Gold	14.1		

* Metal alloy.

EXAMPLE 4-11 Using Equation 4-7, determine the resistance of 80 ft of aluminum wire 0.050 in. in diameter at a temperature of 20°C.

SOLUTION Determine cross section:

Diameter = 0.050 = 50 mils

$$A = d^2 = 50^2 = 2500 \text{ cmil}$$

Let ρ = 17.0 Ω-cmil/ft, ℓ = 80 ft, and A = 2500 cmil. Substitute into Equation 4-7:

$$R = \rho \frac{\ell}{A} = \frac{(17.0)(80)}{2500}$$

$$\therefore \quad R = 0.544 \ \Omega$$

4-7 EFFECT OF TEMPERATURE ON WIRE RESISTANCE

Because the resistance of a wire varies directly with temperature, the resistance of a conductor of fixed length at 20°C may have a substantial increase in resistance when used at a temperature of 80°C. The actual amount of resistance would depend upon the temperature coefficient of resistivity (α) of the metal. A very useful formula resulting from the relation of temperature, resistance, and type of material is stated as Equation 4-8.

$$\frac{R_2}{R_1} = \frac{|t_0| + T_2}{|t_0| + T_1} \quad (4\text{-}8)$$

where R_1 = resistance of wire in ohms (Ω) at temperature T_1
R_2 = resistance of wire in ohms (Ω) at temperature T_2
$|t_0|$ = absolute value of the temperature of the *inferred zero resistance point* of the material in degrees Celsius (°C)

Inferred Zero Point

The *inferred zero resistance point* is the theoretical temperature where a given material has zero resistance. In reality, the material has a finite resistance at this temperature owing to impurities in the metal as well as imperfections in the crystalline

Selected Properties of Electrical Materials

TABLE 4-7 ABSOLUTE VALUE OF THE INFERRED ZERO RESISTANCE TEMPERATURES

Metal	$|t_0|$ °C	Metal	$|t_0|$ °C
Aluminum	236	Nichrome	2480
Constantan	250 000	Silver	243
Copper	234.5	Tungsten	204
Gold	274		

structure. The temperature of the inferred zero resistance point for a particular material may be found by substituting the material's temperature coefficient into Equation 4-9.

$$|t_0| = \frac{1}{\alpha_t} - t \qquad (4-9)$$

where $|t_0|$ = absolute value of the temperature of the inferred zero resistance (°C)
α_t = temperature coefficient of the material in Ω/°C/Ω at temperature t (°C)
t = temperature of the material at which α_t is specified

EXAMPLE 4-12 Determine the absolute temperature of the inferred zero resistance point for the materials listed in Table 4-5 by substituting the listed temperature coefficients (α) into Equation 4-9.

SOLUTION A sample calculation for tungsten is presented here. However, all the absolute temperatures of the inferred zero resistances are listed in Table 4-7.

$$|t_0| = \frac{1}{\alpha_t} - t$$

$$|t_0| = \frac{1}{4.46 \times 10^{-3}} - 20$$

∴ $|t_0| = 204°C$, for tungsten

Observation The temperature of the inferred zero resistance is actually $-204°C$. Taking the absolute value yields 204°C. Thus:

$$|-204°C| \Rightarrow 204°C$$

EXAMPLE 4-13 Determine the resistance of a constantan wire at 80°C if the resistance of the constantan wire is 1.000 Ω at 20°C.

SOLUTION Using Equation 4–8 and the inferred zero resistance for constantan listed in Table 4–7, calculate the resistance at 80°C:

$$\frac{R_2}{R_1} = \frac{|t_0| + T_2}{|t_0| + T_1}$$

$$R_2 = R_1 \left(\frac{|t_0| + T_2}{|t_0| + T_1}\right)$$

Substitute:

$$R_2 = 1.000 \left(\frac{2.5 \times 10^5 + 80}{2.5 \times 10^5 + 20}\right)$$

∴ $R_2 = 1.0002 \; \Omega$ at 80°C

Observation Even though the temperature varied by 60°C (80°C − 20°C), the resistance varied by only 0.2 mΩ (1.0002 Ω − 1.000 Ω). Temperature has little effect on the resistance of constantan.

EXAMPLE 4–14 Determine the resistance of a tungsten wire at 1000°C if the resistance is 40 Ω at 20°C.

SOLUTION Use Equation 4–8 and Table 4–7:

$$R_2 = R_1 \left(\frac{|t_0| + T_2}{|t_0| + T_1}\right)$$

Substitute:

$$R_2 = 40 \left(\frac{204 + 1000}{204 + 20}\right)$$

∴ $R_2 = 215 \; \Omega$ at 1000°C

4–8 AMERICAN WIRE GAUGE (AWG)

Table 4–8 lists the gauge numbers, diameters, and cross sections of the American wire gauge (AWG), as well as the resistance per 1000 ft of annealed solid copper wire at 20°C. The AWG is used to specify both *conductive wire* made of copper and aluminum and *resistive wire* made of nichrome and constantan. Please note the following about Table 4–8:

☐ Gauge numbers range from 4/0 to 40 and the diameters get smaller as the gauge numbers get larger.
☐ Resistance per 1000 ft varies inversely with diameter and cross section. Thus, larger diameters and cross sections have smaller resistance per 1000 ft, and conversely.
☐ A decrease of three gauge numbers represents a doubling of

the cross section. For example, #40 has 10 cmils, #37 has 20 cmils, #34 has 40 cmils, #31 has 80 cmils, and #28 has 160 cmils.
☐ An increase of three gauge numbers results in the cross section being cut in half. Thus #10 has 10 400 cmils, #13 has 5200 cmils, #16 has 2600 cmils, and #19 has 1300 cmils.
☐ A change of 10 gauge numbers results in an approximate 10 to 1 change in cross section. For example, #1/0 has about 100 000 cmils, #10 has about 10 000 cmils, #20 has 1000 cmils, #30 has 100 cmils, and #40 has 10 cmils.
☐ A decrease of six gauge numbers represents a doubling of the diameter. For example, #38 has 4 mils diameter, #32 has 8 mils, #26 has 16 mils, #20 has 32 mils, and #14 has 64 mils.

TABLE 4-8 AMERICAN WIRE GAUGE

Gauge Number	Diameter (mil)*	Cross Section† (cmil)	$\Omega/1000$ ft‡	Gauge Number	Diameter (mil)	Cross Section (cmil)	$\Omega/1000$ ft
4/0	460	212 000	0.0490	19	36	1300	8.05
3/0	410	168 000	0.0618	20	32	1000	10.2
2/0	365	133 000	0.0779	21	28.5	810	12.8
1/0	325	106 000	0.0983	22	25.3	640	16.1
1	289	83 700	0.124	23	22.6	510	20.4
2	258	66 400	0.156	24	20.1	400	25.7
3	229	52 600	0.197	25	17.9	320	32.4
4	204	41 700	0.249	26	15.9	250	40.8
5	182	33 100	0.313	27	14.2	200	51.5
6	162	26 200	0.395	28	12.6	160	64.9
7	144	20 800	0.498	29	11.3	130	81.8
8	128	16 500	0.628	30	10.0	100	103
9	114	13 100	0.792	31	8.9	80	130
10	102	10 400	0.999	32	8.0	63	164
11	91	8200	1.26	33	7.1	50	207
12	81	6500	1.59	34	6.3	40	261
13	72	5200	2.00	35	5.6	32	329
14	64	4100	2.53	36	5.0	25	415
15	57	3300	3.18	37	4.5	20	523
16	51	2600	4.02	38	4.0	16	660
17	45	2000	5.06	39	3.5	12	832
18	40	1600	6.39	40	3.1	10	1049

* 1 mil = 0.001 in.
† Nominal values are rounded for ease of calculation.
‡ Annealed solid copper wire at 20°C.

American Wire Gauge (AWG)

- The resistance per 1000 ft of copper wire approximately doubles for an increase of three gauge numbers. Thus #10 has 1 Ω/1000 ft, #13 has 2 Ω/1000 ft, #16 has 4 Ω/1000 ft, and #19 has 8 Ω/1000 ft.
- The resistance per 1000 ft of copper wire increases by a factor of 10 for an increase of 10 gauge numbers. For example, #1/0 has 0.1 Ω/1000 ft, #10 has 1 Ω/1000 ft, #20 has 10 Ω/1000 ft, #30 has 100 Ω/1000 ft, and #40 has 1000 Ω/1000 ft.
- The conductance of a copper wire is cut in half for an increase of three gauge numbers. Thus #10 may conduct about 30 A, #13 about 15 A, and #16 about 7.5 A.
- The conductance of a copper wire is reduced by a factor of 10 for an increase of 10 gauge numbers. For example, #10 may conduct about 30 A, #20 about 3 A, and #30 about 0.3 A.
- In addition to the previous observations, memorizing the AWG specifications for #10 copper wire lets you approximate the dimensions for the other gauge numbers. Thus #10 gauge copper wire at 20°C is 100 mils in diameter, has a 10 000 cmils cross section, and is 1 Ω/1000 ft.

EXAMPLE 4-15 Using the previous observations and #10 AWG as a reference, approximate the resistance of the following solid copper conductors at 20°C: (a) 1000 ft of #26, (b) 100 ft of #14, and (c) 250 ft of #16.

SOLUTION Since #10 copper wire at 20°C is 1 Ω/1000 ft, then

(a) #10 is 1 Ω/1000 ft, #20 is 10 Ω/1000 ft, #23 is 20 Ω/1000 ft, and #26 is 40 Ω/1000 ft.

∴ 1000 feet of #26 wire is 40 Ω.

(b) #10 is 1 Ω/1000 ft, #20 is 10 Ω/1000 ft, #17 is 5 Ω/1000 ft, and #14 is 2.5 Ω/1000 ft. Since resistance is directly proportional to length,

$$\frac{R_1}{R_2} = \frac{\ell_1}{\ell_2} \qquad (4-10)$$

and the resistance of 100 feet of #14 is

$$\frac{2.5 \ \Omega}{R_2} = \frac{1000 \ \text{ft}}{100 \ \text{ft}}$$

$$R_2 = \frac{2.5 \times 100}{1000} = 0.25 \ \Omega$$

100 feet of #14 is 0.25 Ω.

Selected Properties of Electrical Materials

(c) #10 is 1 Ω/1000 ft, #13 is 2 Ω/1000 ft, and #16 is 4 Ω/1000 ft. Since resistance is directly proportional to length, then:

$$\frac{R_1}{R_2} = \frac{\ell_1}{\ell_2}$$

And the resistance of 250 feet of #16 is:

$$\frac{4}{R_2} = \frac{1000 \text{ ft}}{250 \text{ ft}}$$

$$R_2 = \frac{4 \times 250}{1000} = 1 \text{ Ω}$$

∴ 250 feet of #16 is 1 Ω.

The AWG is constructed so that an increase of one gauge number results in an approximate change in cross section of 25% from the preceding number in the series. Thus, Equation 4–11 may be used to determine the approximate cross section of any number in the AWG.

$$A_n = \frac{A}{(1.25)^{\Delta n}} \quad \text{(cmil)} \tag{4-11}$$

where A = the cross section of the reference conductor (i.e., #1/0, #10, #20, or #30) in cmils

Δn = absolute difference between the reference gauge # and the gauge # of the desired conductor. Thus:

$\Delta n = 7$ when the reference is #20 and the desired gauge is #27 (27 − 20 = 7)

A_n = cross section of desired conductor in cmils

The result of Equation 4–11 is used with the following proportion to determine the approximate resistance per 1000 ft for any desired gauge in the AWG. Because resistance is inversely proportional to cross section, the following is true:

$$\frac{R_1}{R_2} = \frac{A_2}{A_1} \tag{4-12}$$

EXAMPLE 4–16 Approximate the resistance of 400 ft of #24 solid copper wire at 20°C.

SOLUTION Use #20 as the reference:

#10 has 10 000 cmils and 1 Ω/1000 ft

#20 has 1000 cmils and 10 Ω/1000 ft

Using Equation 4–11, determine the cross section of #24:

$$A_{24} = \frac{A}{(1.25)^{\Delta n}} = \frac{1000}{(1.25)^4} = 410$$

$A_{24} = 410$ cmils

Using Equation 4–12, determine the resistance per 1000 ft of #24:

$$\frac{R_1}{R_2} = \frac{A_2}{A_1} \Rightarrow \frac{R_{20}}{R_{24}} = \frac{A_{24}}{A_{20}}$$

Substitute:

$$\frac{10}{R_{24}} = \frac{410}{1000}$$

$$R_{24} = \frac{(10)(1000)}{410} = 24.4$$

$R_{24} = 24.4 \;\Omega/1000$ ft

Use Equation 4–10 to determine the resistance of 400 ft of #24:

$$\frac{R_1}{R_2} = \frac{\ell_1}{\ell_2} \Rightarrow \frac{R_{1000}}{R_{400}} = \frac{\ell_{1000}}{\ell_{400}}$$

Substitute:

$$\frac{24.4}{R_{400}} = \frac{1000}{400}$$

$$R_{400} = \frac{(24.4)(400)}{1000} = 9.8 \;\Omega$$

∴ The approximate resistance of 400 feet of #24 is ≈ 10 Ω.

Observation The actual resistance is 10.3 Ω and the percentage of error is

$$\frac{9.8 - 10.3}{10.3} \times 100\% = -4.85\%$$

Therefore, the approximation is quite acceptable.

4–9 APPLICATIONS USING THE AWG AND FORMULAS

The following examples emphasize the use of the relations of the wire table and the formulas in Part II of this chapter to solve problems dealing with round conductors.

EXAMPLE 4–17 Determine the resistance of 100 ft of #16 aluminum wire at 80°C.

SOLUTION Determine the resistance at 20°C using Equation 4–7, the resistivity of aluminum from Table 4–6, and the cross section from the AWG of Table 4–8. Thus:

$$R = \rho \frac{\ell}{A}$$

where

$\rho = 17.0 \ \Omega - \text{cmil/ft}$

$\ell = 100 \ \text{ft}$

$A = 2600 \ \text{cmils}$

Substitute:

$$R = \frac{(17.0)(100)}{2600} = 0.654 \ \Omega$$

∴ The resistance is 0.654 Ω at 20°C.

Now adjust the resistance for the increase in temperature from 20° to 80°C using Equation 4–8 and the inferred zero resistance temperature from Table 4–7. Thus:

$$\frac{R_2}{R_1} = \frac{|t_0| + T_2}{|t_0| + T_1}$$

Substitute:

$$\frac{R_2}{0.654} = \frac{236 + 80}{236 + 20}$$

$$R_2 = \frac{(316)(0.654)}{256} = 0.807 \ \Omega$$

∴ The resistance of 100 ft of #16 aluminum wire at 80°C is 0.803 Ω.

EXAMPLE 4–18 A two-conductor 5-mile-long cable connecting a remote weather station to the central station has been *short circuited* during the installation of the cable. It is theorized that the short has occurred at one of the nine *junction boxes* where the cable was *spliced* (connected) together. The conductors in the cable are solid copper #24 wire at 15°C. Determine where the short circuit is located if the nine junction boxes are spaced at half-mile intervals.

SOLUTION Starting at one end of the cable, measure the resistance between the two conductors with an

Applications Using the AWG and Formulas

ohmmeter as shown in Figure 4–11(a). The resistance was measured and found to be 400 Ω. Because the tables are referenced to 20°C, this resistance is changed from 400 Ω at 15°C to resistance at 20°C. Use Equation 4–8 and the inferred zero resistance temperature from Table 4–7. Thus:

$$\frac{R_2}{R_1} = \frac{t_0 + T_2}{t_0 + T_1}$$

Substituting:

$$\frac{R_2}{400} = \frac{234.5 + 20}{234.5 + 15}$$

$$R_2 = \frac{(254.5)(400)}{249.5} = 408 \ \Omega$$

∴ The resistance between the two shorted conductors is 408 Ω at 20°C.

Determine the length represented by a resistance of 408 Ω. Use Equation 4–7 and solve for ℓ:

$$R = \rho \frac{\ell}{A}$$

$$\ell = \frac{RA}{\rho}$$

Substitute, using 10.4 Ω-cmil/ft from Table 4–6 for ρ and 400 cmils for A from Table 4–8:

$$\ell = \frac{(408)(400)}{10.4} = 15\ 692 \text{ ft}$$

∴ The length of the conductors from the end to the short circuit and back to the end is 15 700 ft.

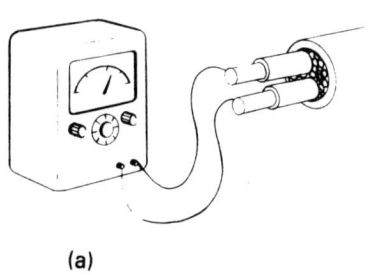

(a)

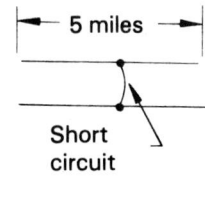

|← 5 miles →|

Short circuit

(b)

FIGURE 4–11
(a) Ohmmeter is used to measure resistance of a shorted cable.
(b) Schematic diagram showing two five mile long conductors with a short circuit connecting them together.

Divide the length by 2 to determine the length of the cable from the end to the short circuit. Thus:

$$\text{Cable length} = \frac{15\,700}{2} = 7850 \text{ ft}$$

Determine the distance in miles:

$$5280 \text{ ft} = 1 \text{ mi}$$

$$7850 \text{ ft} \times \frac{1 \text{ mi}}{5280 \text{ ft}} = 1.5 \text{ miles}$$

∴ Because the junction boxes are spaced at half-mile intervals, the short is located in the third junction box.

EXAMPLE 4–19 A motor used to circulate refrigerated air through a computer console is located 600 ft from the power source. It is determined that the aluminum wire connecting the motor cannot have more than 0.30-Ω resistance at 50°C. Specify the gauge number of the connecting wires.

SOLUTION Since two wires are needed from the power source to the motor, the conductor length is twice 600 ft, or 1200 ft. Because the tables are referenced to 20°C, determine the resistance at 20°C. Use Equation 4–8 and Table 4–7. Thus:

$$\frac{R_2}{R_1} = \frac{|t_0| + T_2}{|t_0| + T_1}$$

Substituting for aluminum:

$$\frac{R_2}{0.3} = \frac{236 + 20}{236 + 50}$$

$$R_2 = \frac{(256)(0.3)}{286} = 0.269 \text{ }\Omega$$

∴ The conductor resistance for 1200 ft at 20°C is 0.269 Ω.

Determine the cross section of the aluminum conductor using Equation 4–7.

$$R = \rho \frac{\ell}{A}$$

Solve for A:

$$A = \frac{\rho \ell}{R}$$

Substitute 17.0 for ρ from Table 4–6:

$$A = \frac{(17.0)(1200)}{0.269} = 75\,836 \text{ cmils}$$

Using the AWG of Table 4–8, determine the closest gauge number equal to or greater than 75 836 cmils in cross section.

∴ Number 1 is selected.

SELECTED TECHNICAL TERMS: PART II

Several of the technical terms used in Part II of this chapter are defined here to aid your understanding of them.

AWG: American wire gauge, a table of standard wire diameters indicated by gauge numbers.
Circular mil: the cross section of a round conductor equal to the diameter in mils squared.
Inferred zero point: in theory, the temperature at which the material of the conductor has zero resistance.
Splice: the joining together of the ends of two wires to form a conductive path.

PROBLEMS: PART II

Section 4–6

4–41 Using Equation 4–7 and the values of resistivity of Table 4–6, determine the resistance of 10 ft of copper wire 0.010 in. in diameter at a temperature of 20°C.

4–42 Using Equation 4–7 and the values of resistivity of Table 4–6, determine the resistance of 8 in. of constantan wire 0.005 in. in diameter at a temperature of 20°C.

Section 4–7

4–43 Determine the resistance of a tungsten wire at 2000°C if its resistance is 10 Ω at 60°C.

4–44 Determine the resistance of a nichrome wire at 600°C if its resistance is 100 Ω at 30°C.

4–45 Determine the resistance of an iron wire at 900°C if its resistance is 18 Ω at 40°C and the temperature coefficient, α, is 6.8×10^{-3} at 100°C.

4–46 Determine the resistance of a brass wire at 300°C if its resistance is 2.5 Ω at 25°C and the temperature coefficient, α, is 1.0×10^{-3} at 15°C.

Section 4–8

4–47 Using #10 AWG copper wire as a reference (1 Ω/1000 ft at 20°C), approximate the resistance of 1000 ft of the following copper wire at 20°C: (a) #20, (b) #30, and (c) #40.

4–48 Using #10 AWG copper wire as a reference (1 Ω/1000 ft at 20°C), approximate the resistance of 100 ft of the following copper wire at 20°C: (a) #13, (b) #23, and (c) #33.

4-49 Using #10 AWG copper wire as a reference, approximate the resistance of 500 ft of the following copper wire at 20°C: (a) #17, (b) #24, and (c) #37.

4-50 Assuming #10 AWG may conduct a maximum of 30 A of current, approximate the maximum current in each of the following conductors at 20°C: (a) #24, (b) #36, and (c) #14.

4-51 Using the information in the wire table, Table 4-8: (a) record the resistance of 300 ft of #24 annealed solid copper wire at 20°C, (b) indicate the cross section in cmil for #18, and (c) note the diameter in mils for #30.

4-52 Using the information in the wire table, Table 4-8: (a) record the resistance of 16.5 ft of #36 annealed solid copper wire at 20°C, (b) indicate the cross section in cmil for #26, and (c) note the diameter in mils for #12.

Section 4-9

4-53 Determine the resistance of 8.2 ft of #24 nichrome wire at 55°C.

4-54 Determine the resistance of 27 in. of #32 constantan wire at 72°C.

4-55 Determine the amount of copper wire (in feet) left on a large spool of #28 wire. The measured resistance between the two ends of the spooled wire is 292 Ω and the temperature is 27°C.

4-56 A cable connecting a computer terminal to the CPU (central processing unit) has been short circuited. The resistance measured between the two #20 copper wires indicates 328 mΩ at 25°C. Determine the distance, in feet, from the point of measurement to the short circuit.

4-57 A computer console is located 85 ft from the power panel. It is desired that the aluminum wire used to power the computer console not have more than 0.5-Ω resistance at 40°C. Specify the gauge number of the connecting wires.

5 Resistance and Resistors

5-1 OHM'S LAW
5-2 FIXED LINEAR RESISTORS
5-3 VARIABLE LINEAR RESISTORS
5-4 NONLINEAR RESISTORS
5-5 RESISTANCE DESIGNATION

CHAPTER PREVIEW In this chapter you will learn about resistance and conductance, and about a very important electrical device, the resistor. Additionally, various resistive devices, including the potentiometer, thermistor, varistor, and photoresistor, are introduced. Finally, various methods of indicating the resistance of a resistor are explored.

PERFORMANCE OBJECTIVES Once you have read each section, worked each example with pencil, paper, and calculator, and answered each problem for every section, you will be able to:

☐ Distinguish between linear and nonlinear resistors.
☐ Read different resistor coding systems.
☐ Understand the need for resistors of different power rating.
☐ Compute electric resistance and conductance.
☐ Calculate the resistive change due to temperature variation.
☐ Distinguish the identifying characteristics of the various types of linear resistors.

5–1 OHM'S LAW

Resistance

The electrical resistance of a material may be determined by placing a source of voltage across the material and then measuring the resulting electric current in the material. The amount of voltage applied to the material is determined by placing the leads of a voltmeter across the material, as shown in Figure 5–1. The current is measured by placing an ammeter into the path of the current so that the current passes through the ammeter as pictured in Figure 5–1. The *ratio* of the applied voltage to the current in the material is the electrical resistance of the material. Thus:

Resistance = volts per ampere

and

1 ohm = 1 volt/1 ampere

Example 5–1 demonstrates this concept.

EXAMPLE 5–1 Determine the resistance of an electric heating element when the element is connected to a 220-V source and 20 A of current flows.

SOLUTION Since resistance is defined as volts per ampere,

Ohm's Law

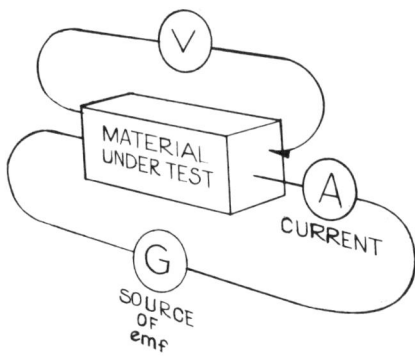

FIGURE 5–1
All materials offer opposition to movement of electric current. *Resistance* is the term used to describe opposition to flow of electric current in a material. Resistance is defined as the ratio between voltage across material to the current in the material. Thus: 1 ohm = 1 volt/1 ampere.

$$\text{Resistance} = \frac{220 \text{ V}}{20 \text{ A}}$$

∴   Resistance = 11 Ω

NOTE: Since heating elements are made from material with nearly a zero temperature coefficient, the 11-Ω resistance of this heating element is constant over a wide temperature range.

Because the electrical resistance (R) of a material is defined as the voltage across the material (V) divided by the resulting current (I) in the material, the following equation may be stated.

$$R = \frac{V}{I} \quad (\Omega) \tag{5-1}$$

A material has a resistance of 1 *ohm* when the application of 1 *volt* results in an electric current of 1 *ampere*.

EXAMPLE 5–2 Determine the resistance of a carbon rod that has 20 mA of current in the rod with a voltage of 100 V across the rod.

SOLUTION Use Equation 5–1.

$$R = \frac{V}{I}$$

Substitute 100 for V and 20×10^{-3} for I:

$$R = \frac{100}{20 \times 10^{-3}} = 5 \times 10^3$$

∴ $R = 5$ kΩ

Conductance Electric *conductance*, symbolized by the letter G, is defined as the reciprocal of resistance R. Thus:

$$G = \frac{1}{R} \quad \text{(S)} \tag{5-2}$$

where G = conductance in units of siemens (S)
R = resistance in units of ohms (Ω)

EXAMPLE 5-3 Determine the conductance of a gold conductor with a resistance of 10 mΩ.

SOLUTION Use Equation 5-2.

$$G = \frac{1}{R}$$

Substitute 10×10^{-3} for R:

$$G = \frac{1}{10 \times 10^{-3}} = 0.1 \times 10^3$$

$$\therefore \quad G = 100 \text{ S}$$

Because conductance is defined as the reciprocal of resistance, the conductance of a material (G) is also equal to the electric current (I) divided by the voltage (V). Thus:

$$G = \frac{I}{V} \quad \text{(S)} \tag{5-3}$$

EXAMPLE 5-4 A silicon semiconductor has a voltage across it of 18 V and a current in it of 60 μA. Determine the conductance of the semiconductor.

SOLUTION Use Equation 5-3.

$$G = \frac{I}{V}$$

Substitute 60×10^{-6} for I and 18 for V:

$$G = \frac{60 \times 10^{-6}}{18} = 3.3 \times 10^{-6}$$

$$\therefore \quad G = 3.3 \text{ }\mu\text{S}$$

Ohm's Law of Constant Proportionality Ohm's law, named for the nineteenth-century German physicist, Georg Simon Ohm, relates voltage to current in an electric circuit. Ohm learned through experimentation that the potential difference measured across a metallic conductor is related to the current in the metallic conductor by a numerical constant (k). Stated mathematically:

Ohm's Law

$$V = kI \quad (V) \tag{5-4}$$

where V = potential difference across the conductor in volts (V)
I = current through the conductor in amperes (A)
k = numerical constant in volts per ampere (V/A)

By applying Equation 5-4 to an electrical circuit, it is then possible to determine the circuit current when the numerical constant (k) and the applied voltage are known. Table 5-1 gives several examples of the application of Equation 5-4.

EXAMPLE 5-5 Determine the current in a metallic conductor when k = 10 V/A and the voltage of the source connected across the conductor is 50 V.

SOLUTION Substitute into Equation 5-4 and solve for I.

$$V = kI$$

$$50 = 10I$$

Solve for I:

$$I = \frac{50}{10}$$

$$\therefore \quad I = 5 \text{ A}$$

Since the numerical constant, k, is actually the resistance, R, of the metallic conductor, the symbol R may be used instead of k. Equation 5-5 is Ohm's law of constant proportionality.

$$\frac{V}{I} = k = R \quad \text{or simply} \quad R = \frac{V}{I} \quad (\Omega) \tag{5-5}$$

where R = resistance in ohms (Ω)
V = voltage across the resistance in volts (V)
I = current in the resistance in amperes (A)

TABLE 5-1 CURRENT IN AN ELECTRIC CIRCUIT WITH k = 10 V/A

V (volts)	k (volts/amperes)	I (amperes)
50	10	5
25	10	2.5
10	10	1
5	10	0.5
1	10	0.1

Equation 5–5 (Ohm's law) and Equation 5–1 (the definition of resistance) are the same equation; however, there is a subtle difference. Equation 5–1 defines the resistance of any material or device as the voltage across the material divided by the resulting electric current in the material. However, Equation 5–5 describes the operation of an electric circuit, where the ratio of voltage to current results in a numerical constant: resistance, R. The following examples will illustrate this difference.

EXAMPLE 5–6 Determine the voltage and current coordinates of points A and B of Figure 5–2 and then compute the resistance of the diode at each of these points.

SOLUTION Point A: 0.7 V, 100 mA
Point B: 0.9 V, 600 mA
Resistance point A:

$$R_{\text{point }A} = \frac{V}{I} = \frac{0.7}{0.1}$$

$$\therefore \quad R_{\text{point }A} = 7 \ \Omega$$

Resistance point B:

$$R_{\text{point }B} = \frac{V}{I} = \frac{0.9}{0.6}$$

$$\therefore \quad R_{\text{point }B} = 1.5 \ \Omega$$

From Example 5–6 it is seen that the resistance of the diode is not a constant. The diode is said to have a *nonlinear resistance*. Contrast the nonlinear resistance of the diode to the *linear resistance* (constant resistance) of the resistor in Example 5–7.

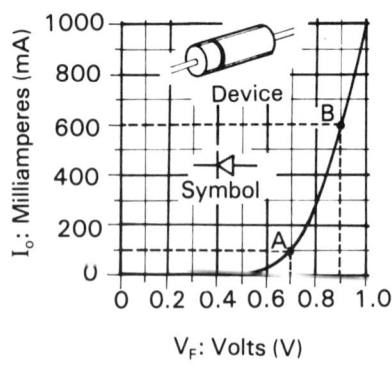 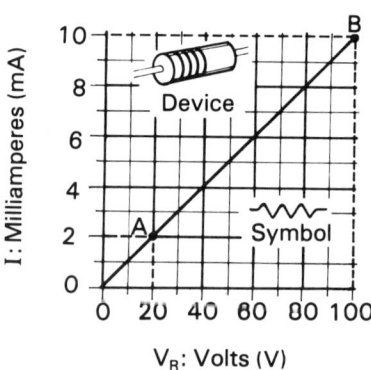

FIGURE 5–2 (left) Voltage-current characteristic of a silicon diode.
FIGURE 5–3 (right) Voltage-current characteristic of a carbon resistor.

EXAMPLE 5-7 Determine the voltage and current coordinates of points A and B of Figure 5-3 and then compute the resistance of the resistor at each of these points.

SOLUTION *Point A:* 20 V, 2 mA
Point B: 100 V, 10 mA
Resistance point A:

$$R_{\text{point }A} = \frac{V}{I} = \frac{20}{2 \times 10^{-3}}$$

$\therefore \quad R_{\text{point }A} = 10 \text{ k}\Omega$

Resistance point B:

$$R_{\text{point }B} = \frac{V}{I} = \frac{100}{10 \times 10^{-3}}$$

$\therefore \quad R_{\text{point }B} = 10 \text{ k}\Omega$

Because the voltage-current (*V-I*) characteristic of the resistor of Figure 5-3 is a straight line, the resistance of the device is a constant. Ohm's law may be applied to a circuit that has a load with a linear *V-I* characteristic curve, such as a carbon resistor, because the resistance is constant. However, Ohm's law may not be applied to a circuit that has a load with a nonlinear *V-I* characteristic curve, such as a diode, because the resistance of the device is not constant. Example 5-8 amplifies this point.

EXAMPLE 5-8 Apply Ohm's law to determine the current in a circuit that contains a *power diode* when 0.4 V is measured across the device. The resistance of the diode is 800 mΩ when 0.8 V is measured across the device.

SOLUTION To apply Ohm's law to a circuit, two conditions must be met. First, two parameters of the circuit must be known and, second, the load must have a linear voltage-current characteristic. Even though the diode resistance is specified at 0.8 V as 800 mA, it cannot be used to solve for current at 0.4 V because it is not constant owing to the nonlinear characteristic of the diode. Thus, this circuit cannot be solved by applying Ohm's law.

From Example 5-8 it is apparent that not all circuit conditions can be determined by the application of Ohm's law. Only after determining that the load is linear and that sufficient information is given can Ohm's law be successfully applied.

5-2 FIXED LINEAR RESISTORS

A resistor is an electrical component that has been manufactured with a specified amount of resistance. Resistors are used for controlling the flow of electric current and providing desired amounts of voltage in electronic circuits.

Resistors are classified as pictured in Figure 5-4. Both linear and nonlinear resistors are *bilateral* devices; that is, current passing in either direction through the device will experience the same electrical resistance. Because of the bilateral characteristics of resistors, resistors are *nonpolarized symmetrical* devices, which may be connected into an electric circuit without concern for *lead polarization*.

Linear resistors, which are available as both fixed and variable resistors, have the property of little or no change in the resistance value for variations in temperature, applied voltage, or light intensity. Therefore, linear resistors have, for all practical purposes, a constant resistance value.

Thermal Stability

The preceding statement is true for resistors made from metals; however, resistors made from carbon may have a significant change in resistance value owing to a change in operating temperature. The thermal stability of a particular resistor is indicated by the temperature coefficient specification. The temperature coefficient (TC) is usually expressed in parts per million per degree Celsius ($\pm$ppm/°C). The smaller the temperature coefficient, the less variations in temperature will affect the resistance of the resistor. Resistors with temperature coefficients as small as 1 ppm/°C are commercially available. However, typical tem-

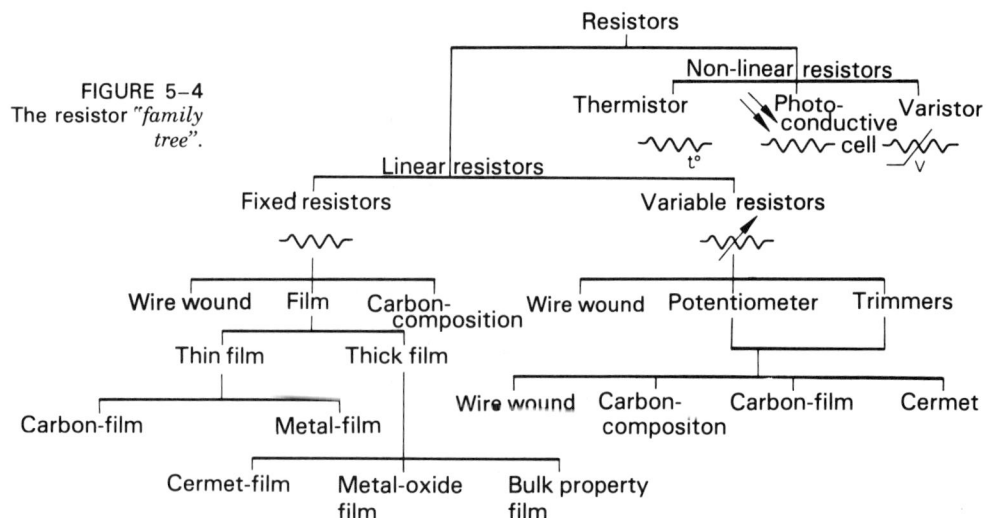

FIGURE 5-4
The resistor *"family tree"*.

perature coefficients for resistors using metal as the resistive material are 50 to 100 ppm/°C, while resistors using carbon as the resistive material are 300 ppm/°C or more.

EXAMPLE 5-9 Determine the change in resistance of a 1.0-megohm (1.0-MΩ) resistor constructed from carbon. The temperature coefficient is -500 ppm/°C.

SOLUTION The nominal (named) resistance of 1.0 MΩ is usually specified at an *ambient* temperature of 25°C. The temperature coefficient tells us that, for every degree Celsius above or below the ambient temperature, the resistance of the device will change by 500 Ω for each 1 million ohms in the resistor (500 parts per 1 million parts). Since the nominal resistance is 1.0 MΩ, then each degree Celsius of change will change the nominal resistance by 500 Ω. Therefore, an increase in operating temperature to 50°C would cause the nominal resistance to change by 12.5 kΩ, as calculated using the following equation:

$$\Delta R = \Delta T \times TC \times R \qquad (5-6)$$

where ΔR = the change in resistance in ohms (Ω) due to the ΔT
ΔT = the change in the device temperature from the specified ambient temperature in °C
TC = temperature coefficient in $\pm$ ppm/°C
R = nominal (named) resistance in **megohms** (MΩ) of the device at a specified ambient temperature

Substituting into Equation 5-6, where $\Delta T = 25$, $TC = -500$, and $R = 1$:

$$\Delta R = (25)(-500)(1) = -12\,500$$

$\therefore \quad \Delta R = -12.5 \text{ k}\Omega$

EXAMPLE 5-10 Determine the resistance of an 82-kΩ resistor (specified at 23°C) with a TC of $+80$ ppm/°C when the resistor is operating at a temperature of 100°C.

SOLUTION Using Equation 5-6, first convert 82 kΩ into megohms to meet the conditions of the equation. Thus:

$$82 \, k\Omega \times \frac{1 \, M\Omega}{1000 \, k\Omega} = 0.082 \, M\Omega$$

Substitute into Equation 5–6 where $\Delta T = 77$, $TC = +80$, and $R = 0.082$:

$$\Delta R = \Delta T \times TC \times R$$
$$= (77)(+80)(0.082)$$
$$= +505 \, \Omega$$

And the resistance is:

$$82\,000 + 505 = 82\,505 \, \Omega$$

∴ The resistance of the 82-kΩ resistor increases to 82.5 kΩ at an operating temperature of 100°C.

Fixed Resistors Fixed resistors are available in a variety of sizes and shapes with both *axial* and *radial* leads, as pictured in Figure 5–5. Additionally, fixed resistors are available with *lugs* for installation by soldering or mounting with machine screws or rivets.

As indicated in Figure 5–4, fixed linear resistors are available in three fabrication technologies: carbon-composition, film (both thin and thick film), and wire-wound. The properties and characteristics of each of these resistor types will be investigated in the following material.

Carbon-composition resistors are made by mixing carbon powder and insulative binders to produce the desired value of resistance. The resulting resistance values are within ±10% of the desired nominal (named) value; devices with ±5% tolerance

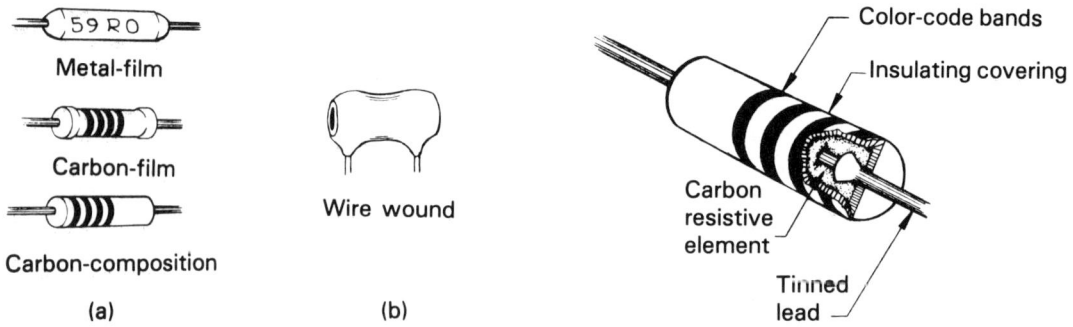

FIGURE 5–5 (left) Fixed resistors are available with both (a) axial leads and (b) radial leads.
FIGURE 5–6 (right) Cutaway of a carbon composition resistor.

Fixed Linear Resistors

are obtained by selective sorting of each run. Closer-tolerance devices are pointless since the stability of carbon-composition material is such that the nominal resistance value cannot be preserved over a long period of operating time to a close tolerance. Figure 5–6 pictures a cutaway of a carbon-composition resistor.

Film resistors are made by depositing a conductive material onto an insulative rod, tube, or plate made of ceramic or glass. Thin-film fixed resistors include both carbon and metal film resistors.

Carbon-film resistors cost less than carbon-composition resistors and have significantly better performance characteristics. These devices are characterized by low noise, good stability, and wide operating frequency range.

Metal-film resistors are made by depositing a very thin film of metal onto a ceramic or glass rod. The metal film is spiral cut to the desired resistance. Tolerances ranging from ±0.025% to ±2% of the nominal value are commercially available.

Thick film resistors include metal-oxide, bulk-property, and cermet-film resistors.

Metal-oxide film resistors are manufactured by oxidizing tin chloride on a heated glass substrate (base material). The resulting resistors exhibit excellent properties, including low noise and good temperature stability. Units with very high voltage ratings (8 kV/in., 3 kV/cm) and very high resistance (up to 10^6 MΩ) are available on special order.

Bulk-property film resistors are made of metal film that is photoetched to provide close resistance tolerances (±0.1% to ±1%). These devices are extremely low noise, virtually noninductive, high-frequency resistors with a very low temperature coefficient and extremely good long-term stability.

Cermet-film resistors are made by placing a coating of metal alloy along with insulating materials onto a ceramic substrate and firing it into a ceramic metal, or *cermet*. Cermet-film resistors are available as axial lead resistors and resistor networks, as shown in Figure 5–7. Cermet material provides the highest resistance to size of any resistive material. Resistor values to 500 MΩ with tolerances of ±0.5% to 2% are available.

Wire-wound resistors are made by winding resistive wire such as nichrome (nickel-chromium) on a ceramic form. The wire is then coated with an insulative *vitreous enamel* or silicon ceramic material. Although costly when compared to other resistor types, the wire-wound resistor has outstanding electrical properties, including very low noise, very good time stability,

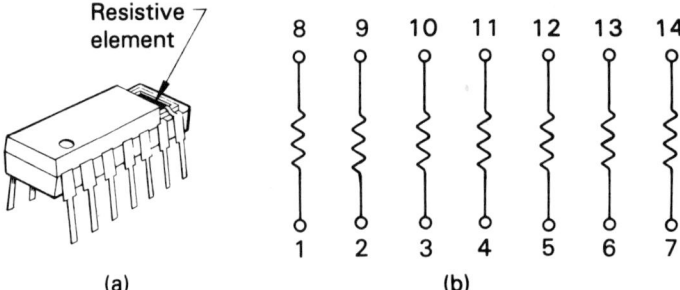

FIGURE 5-7
Cermet resistor network.
(a) Cutaway showing cermet-film resistive element.
(b) *Connection* diagram of resistive network.

and good overload (high current) characteristics. Because of the inductance and capacitance inherent in its method of manufacture, the wire-wound resistor is not used in high-frequency circuits.

Power wire-wound resistors are the least standardized of all the resistor types and are, therefore, available in a variety of sizes and shapes, as shown in Figure 5-8. Precision wire-wound resistors are available with tolerances between ±0.001% and ±1%, while ultraprecision wire-wound resistors have tolerances as small as ±0.0002% (±2 ppm) with temperature coefficients less than 1 ppm/°C.

Resistor Facts
- ☐ Table 5-2 summarizes the characteristics of the various fixed resistors.
- ☐ The voltage rating of a fixed resistor is specified by the manufacturer as the maximum continuous voltage at which the resistor may operate without failing.
- ☐ The voltage rating depends upon the dielectric strength of the insulating materials used in manufacturing the device.
- ☐ As the voltage is increased across the fixed resistor, both the chance for failure as well as *resistor noise* will increase.
- ☐ Most wire-wound resistors are limited to dc and audio-frequency applications owing to the large amounts of inductance and capacitance, which makes them unsuited for high-frequency applications.

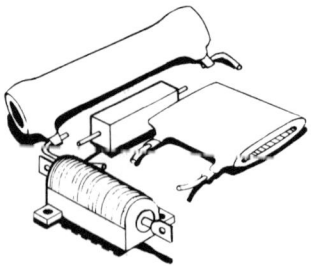

FIGURE 5-8
Various power wire-wound resistors. Notice variety of package styles.

TABLE 5-2 SUMMARY OF THE RANGE OF TYPICAL SPECIFICATIONS FOR SELECTED FIXED RESISTORS*

Resistor Type	Resistance Range	Power Range (W)	Tolerance (%)	Temp Coef. (ppm/°C)	Ambient Temperature Range (°C)
Carbon-composition	1 to 22 MΩ	$\frac{1}{8}$ to 2	±5 or ±10	±300	−55 to 100
Carbon-film	0.5 to 33 MΩ	$\frac{1}{10}$ to 1	±2 or ±5	±300	−55 to 150
Metal-film	0.5 to 100 MΩ	$\frac{1}{10}$ to $\frac{1}{4}$	±1	±100	−55 to 150
Cermet-film	10 to 10 MΩ	$\frac{1}{10}$ to $\frac{1}{2}$	±1, 2 or 5	±50	−55 to 165
Wire-wound					
Power	1 to 100 kΩ	3 to 225	±5	±30	−55 to 300
Precision	0.001 to 150 kΩ	$\frac{1}{2}$ to 25	0.01 to 1	±10	−55 to 250

* These specifications do not include *"special-order"* devices; also, the specifications vary depending on the manufacturer.

☐ Carbon-composition, carbon-film, and some product lines of metal-film resistors are *color coded* to designate both their resistance and resistance tolerance values.

☐ Cermet-film and most metal-film and wire-wound resistors are labeled with their resistance and tolerance values using one of two systems.

(a) Industrial/consumer types are labeled with standard numbers like 10 kΩ ± 5%.

(b) Military types have a MIL-spec designation of numbers and letters to indicate both resistance and resistance tolerance values.

5-3 VARIABLE LINEAR RESISTORS

Variable resistors, like fixed resistors, are used to control current flow and provide desired amounts of voltage. Figure 5-9(a) shows a variable resistor used to limit current in a lamp. This circuit dims the lamp by allowing the intensity of the lamp to be changed from full brightness to total darkness. Figure 5-9(b) shows a variable resistor used to regulate the gain of the amplifier. When a variable resistor is used in this manner it is called a *volume control*.

Variable resistors may be classified as variable wire-wound resistors, potentiometers, and trimmers. Each of these three categories will be investigated in the following material.

Variable Wire-wound Resistors

Variable wire-wound resistors are constructed of nichrome wire wound on a ceramic core and covered with an insulative coating. A *window*, as shown in Figure 5-10, is left in the insu-

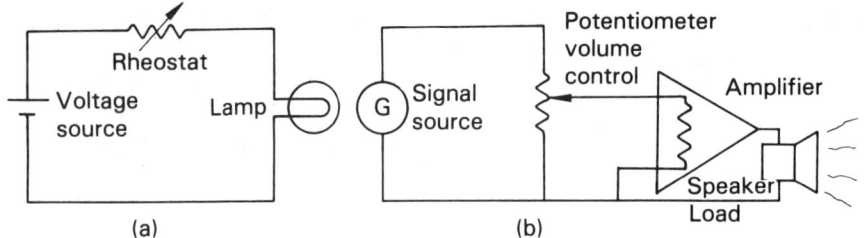

FIGURE 5-9 Variable resistors are used to control current and provide voltage division.
(a) *Rheostat* (variable resistor) is used to vary illumination of lamp by varying electric current in the lamp.
(b) Potentiometer (variable resistor) is used to vary *volume* of sound from speaker by varying voltage level across input to amplifier.

lating cover, exposing the resistive wire. An adjustable tap rides along the exposed wire making electrical contact. A nut and bolt with a screwdriver slot is used to secure the tap at the desired resistance. Like the fixed wire-wound resistors, these resistors find application in power supplies as voltage-divider resistors and in low-frequency ac circuits. Because of their inductive and capacitive properties, they are not suited for high-frequency application. They are available in resistive sizes from 1 Ω to 150 kΩ with ±5% and ±10% tolerances and power rating to 200 W.

Potentiometer Potentiometer is the name given to a three-terminal variable resistor constructed, as shown in Figure 5-11, so that a wiper moves over a resistive element when the control shaft is moved, providing a continuous variation in resistance between the center terminal and either outside terminal. The name potentiometer comes from the use of this device as a *potential meter*, that is, a voltage divider.

As noted in Figure 5-11, the potentiometer has three terminals; two are fixed, and the other is variable. When only two of the three terminals (one fixed and one variable) are used to connect the potentiometer into a circuit, the device is called a *rheostat*. An example of this circuit connection is shown in Figure 5-11(c).

The resistive element is made from any one of several materials, including carbon composition, carbon film, cermet, and

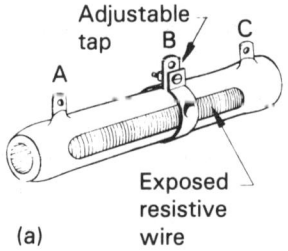

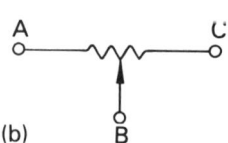

FIGURE 5-10
Variable wire-wound resistor.
(a) Pictorial showing exposed resistive wire.
(b) Schematic symbol.

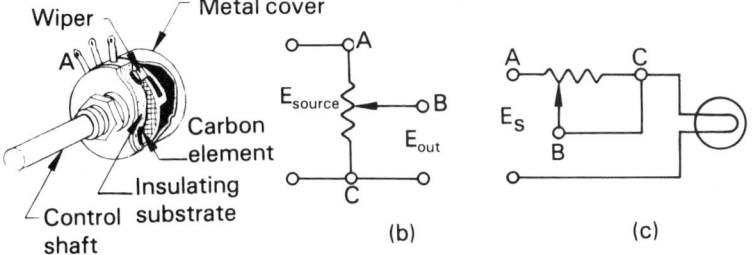

FIGURE 5-11
(a) Cutaway of carbon composition potentiometer.
(b) Schematic of potentiometer used as a voltage divider.
(c) Schematic of potentiometer connected as a rheostat to dim light.

wire. Wire-wound potentiometers are of three types, each specifically constructed for a particular application.

Power potentiometers and **rheostats** are made from ribbon or heavy-gauge resistance wire wound on a toroidal (donut shape) ceramic form. This type of potentiometer is designed for both large current flow and large power dissipation (to 250 W).

Low-power potentiometers are used for controlling small-signal voltages. They are made from fine wire closely wound on a plastic form, which is bent to conform to the housing.

High-resolution potentiometers are made with very fine wire and are machined to very close tolerances. Although this type of potentiometer is very expensive, it provides outstanding electrical and mechanical characteristics, including small *backlash*, very low noise, low tolerance (to ±0.1%), high resistance resolution, and very good linearity. Most high-resolution potentiometers are supplied as multiturn (up to 10 turns) devices.

Carbon-composition potentiometers are by far the most popular of the potentiometers. They are low cost and have long life, low noise, acceptable linearity, and very good reliability. Carbon-composition potentiometers may be used as volume controls, tone controls, and as fractional wattage rheostats.

Trimmer Trimmers are used in electronic circuits to *trim* the circuit to the required operating conditions by inserting a small screwdriver into a slot and turning one or more times. Because adjustments are made infrequently, materials and fabrication processes unsuited for potentiometer construction are used to advantage in trimmer construction. Small physical size (as noted in Figure 5-12), very fine wire, and high wiper pressure are found only in trimmers. Wire, carbon composition, carbon film, and cermet are materials commonly used in trimmer construction. Both single-turn and multiturn trimmers are available. Table 5-3 is a summary of selected specifications for variable resistors.

Resistance Taper Because low-power (carbon-composition) potentiometers are used for a variety of control applications, there are several dif-

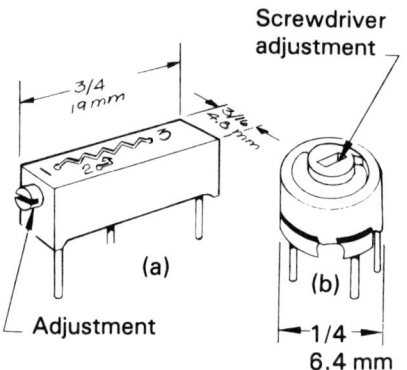

FIGURE 5-12
Trimmers.
(a) Multiturn rectangular trimmer.
(b) Single-turn round trimmer.

ferent types available that are designated by their *taper*. Taper refers to the relationship between the movement of the control shaft and the resistance between one of the outside terminals and the center terminal. Figure 5-13 shows three standard resistance taper curves.

Curve 1 is a linear taper used for general control applications, including TV picture adjustments, audio tone, and uniform voltage division. Notice that 50% of the resistance corresponds to 50% clockwise (cw) rotation.

Curve 2 is a modified left-hand log curve used for volume controls. Notice that 20% of the resistance corresponds to 50% cw rotation.

Curve 3 is a left-hand log taper commonly used for volume controls. Notice that 10% of the resistance corresponds to 50% of the cw rotation.

The type of taper for any potentiometer may be determined by first placing an ohmmeter (a meter designed to measure resistance) across the center and one outside terminal and then following Procedure 5-1.

TABLE 5-3 SUMMARY OF THE RANGE OF SELECTED SPECIFICATIONS FOR SEVERAL VARIABLE RESISTORS

Resistor Type	*Resistance*	*Power (W)*	*Rotation*
Variable wire-wound	1 to 150 kΩ	3 to 200	Slide
Potentiometers			
Power	0.5 to 10 kΩ	7 to 500	300°
Low power	50 to 5 MΩ	$\frac{1}{2}$ to 2	280°
High resolution	10 to 1 MΩ	$\frac{1}{2}$ to 1	Single turn Multiturn
Trimmers	50 to 5 MΩ	$\frac{1}{4}$ to $\frac{3}{4}$	Single turn Multiturn

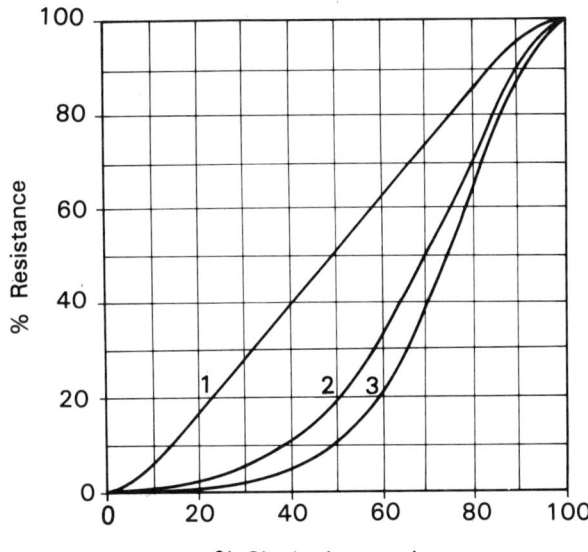

FIGURE 5-13
Basic *resistance taper curves* for control potentiometers.

PROCEDURE 5-1. DETERMINING POTENTIOMETER TAPER

With an ohmmeter connected to the center and one outside terminal:

1. Rotate the control shaft so that the total resistance of the potentiometer is read on the meter. Record this reading.
2. Rotate the control shaft so that the meter reads zero ohms.
3. Rotate the control shaft 50% of its travel and note the resistance. The taper is:
   (a) Linear if 50% of the total resistance is seen on the meter scale.
   (b) Logarithmic if 10% or 20% of the total resistance is seen on the meter scale.

5-4 NONLINEAR RESISTORS

Nonlinear resistors are made from semiconductive materials. The breaking of the covalent bonds in the semiconductive materials gives nonlinear resistors their unique properties. The source of energy for the creation of charge carriers in these semiconductor devices varies.

Thermistors achieve their properties through thermally generated charge carriers, **varistors** through voltage across the semiconductive material, and **photoconductive cells** (photoresistors) by light energy irradiating the semiconductor surface. Each of these nonlinear bilateral resistive devices (the thermistor, the varistor, and the photoconductive cell) will be discussed in this section.

Thermistor The thermistor is a thermally sensitive bilateral resistor used to detect very small changes in temperature. The variation in temperature is reflected through an appreciable variation of the resistance of the device. Figure 5-14 is a semilog plot of temperature and resistance of both the negative temperature coefficient (NTC) and the positive temperature coefficient (PTC) type of thermistor. The NTC thermistor devices are nonlinear resistors having a high negative temperature coefficient resulting from the metal oxides of cobalt (Co_2O_3), nickel (NiO), and manganese (Mn_2O_3) used to manufacture these devices.

Positive temperature coefficient (PTC) thermistors are made from *doped* barium titanate semiconductor material. This material is characterized by a very large change in resistance for a small change in temperature, as noted in Figure 5-14.

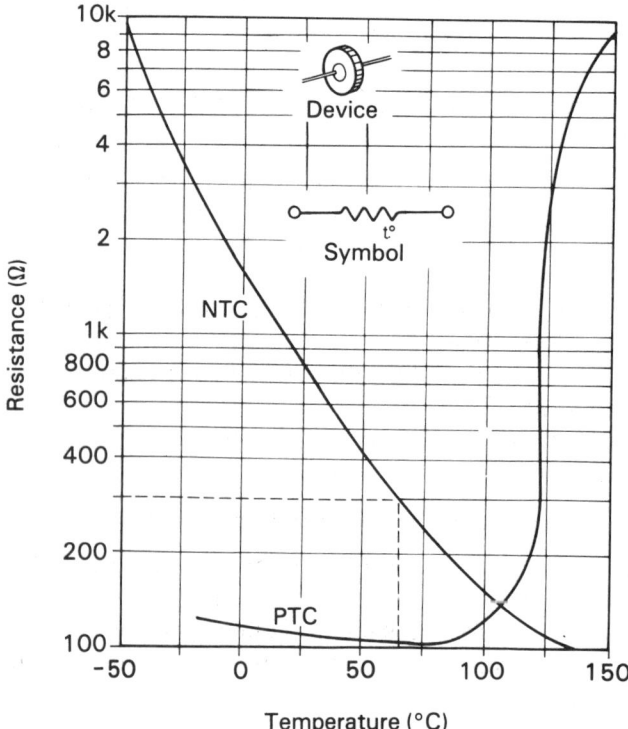

FIGURE 5-14
Thermistor characteristic curve. NTC is curve for negative temperature coefficient thermistor. PTC is curve for positive temperature coefficient thermistor.

TABLE 5-4 TYPICAL THERMISTOR TYPES AND CHARACTERISTICS

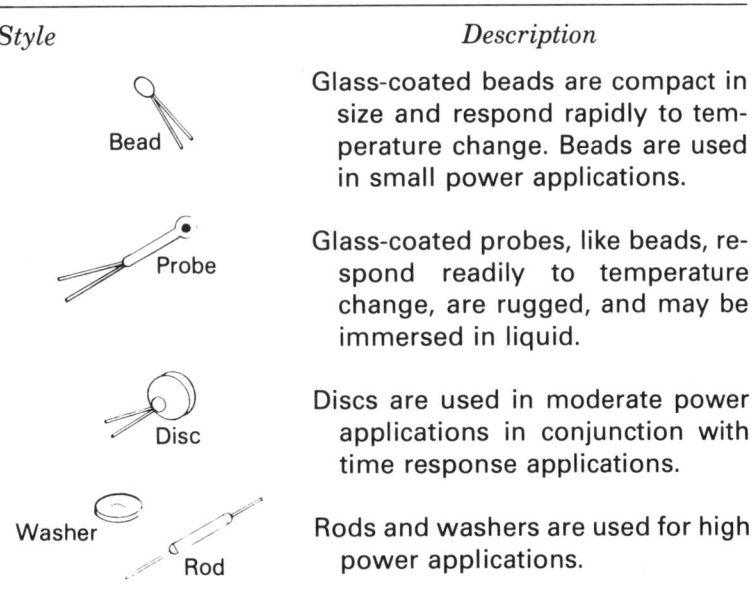

Style	Description
Bead	Glass-coated beads are compact in size and respond rapidly to temperature change. Beads are used in small power applications.
Probe	Glass-coated probes, like beads, respond readily to temperature change, are rugged, and may be immersed in liquid.
Disc	Discs are used in moderate power applications in conjunction with time response applications.
Washer / Rod	Rods and washers are used for high power applications.

The temperature where the resistance rises rapidly is referred to as the *switching point*. The switching point temperature depends upon the particular type of PTC thermistor. Typical values of the switching point temperature range from 30° to 140°C.

Thermistors are manufactured in various shapes, as noted in Table 5-4, through a process called *sintering*. This process uses metal oxides mixed with ceramic materials, which are then sintered into hard ceramic beads, rods, and discs by the application of high pressure and high temperature.

The thermal characteristic of precision thermistors is *curve matched* to a published resistance-temperature curve. The resulting close resistance tolerance (±1%) over a wide temperature span (−50° to 150°C) provides the precision thermistor with a definite repeatable response to temperature change. The following example explores the relationship between temperature and resistance.

EXAMPLE 5-11 Using the curve of the NTC (negative temperature coefficient) thermistor of Figure 5-14, determine the temperature of the thermistor for a resistance of 300 Ω.

SOLUTION Enter the curve of Figure 5-14 on the vertical axis at 300 Ω, move horizontally to the right until

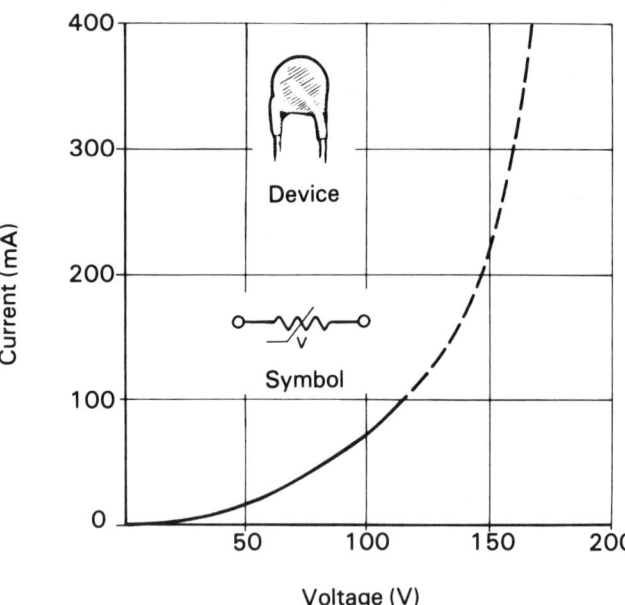

FIGURE 5–15
Typical characteristic curve of a *voltage dependent resistor* (VDR), commonly called a *varistor*. Solid curve represents region for continuous operation while dashed curve represents surge response. Notice that a 50% increase in voltage (120 V to 170 V) produces a 300% increase in current (100 mA to 400 mA).

the NTC curve is intersected, then project vertically down to the horizontal axis. Read 65°C.

∴ 300 Ω represents a temperature of 65°C.

In addition to temperature measurement and control, thermistors are used for temperature stabilization of semiconductor circuits, measurement of RF power, and in time-delay circuits.

Varistors Varistors, also referred to as metal-oxide varistors (MOVs), are voltage-dependent bilateral resistors used to protect circuitry from high-energy voltage transients by rapidly changing from a high *standby* resistance to a low *conducting* resistance. This action *clamps* the voltage to a safe level. Figure 5–15 shows a typical response curve of a varistor. Circuitry is protected from destructive energy levels caused by high-voltage transients by dissipating the energy in the varistor, thus protecting the circuitry from damage as a result of overvoltage conditions.

Applications include transient suppression in inductive and transformer-switching circuits, switch contact arc suppression, and the protection of circuit insulation.

Varistors are available in a variety of packages capable of instantaneous currents up to 2000 A with ac operating voltage ranging from 12 to 660 V at temperatures from −40° to +85°C.

Photoconductive Cells (Photoresistors)

Photoconductive cells are made from semiconductor materials that change resistivity when illuminated with light energy. Materials such as cadmium selenide (CdSe) and cadmium sulfide (CdS) have insulator properties when in the dark. When these materials are exposed to light, covalent bonds are broken, producing charge carriers (electrons in the conduction band and holes in the valence band). The amount of illumination on the surface determines the number of electron-hole pairs generated in the material and consequently the resistance of the photoconductive cells. Since the resistance of the cell depends solely on light energy (these materials have few thermally generated charge carriers), photoconductive cells are referred to as *photoresistors*.

Figure 5–16 is an example of the characteristic curve of resistance versus illumination for a typical CdS cell. This photoconductive cell is specified as having a maximum dark resistance of 100 kΩ 5 seconds after removing an illumination of 20 lux (2 footcandles).

The amount of light falling on a surface, the surface illumination (E), is defined as the incident luminous flux (F) per unit of surface area (A). Expressed as an equation:

$$E = \frac{F}{A} \quad \text{(lx)} \tag{5-7}$$

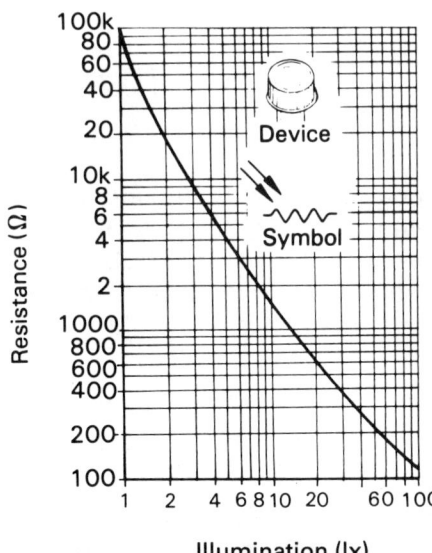

FIGURE 5–16
Typical characteristic curve of cadmium sulfide (CdS) *photoconductive cell* (also referred to as a photoresistor).

where E = surface illumination in lux (lx)
F = luminous flux in lumens (lm)
A = area in square meters (m²)

In the SI system of measurement, the lux (lx) is the unit of surface illumination (E), while the footcandle (fc) is the unit of surface illumination (E) in the English system. Table 5–5 may be used to gain a feel for units of illumination. Thus in the SI system:

1 lux = 1 lumen per square meter

1 lx = 1 lm/m²

And in the English system:

1 footcandle = 1 lumen per square foot

1 fc = 1 lm/ft²

The conversion factor between the two systems is often needed since both units are commonly used. Thus:

1 fc = 10.77 lx

1 lx = 0.0929 fc

Example 5–12 illustrates the use of the conversion factor for surface illumination.

EXAMPLE 5–12 Determine the illumination of 2 fc in SI units.

SOLUTION The SI unit of illumination is the lux (lx). 1 fc = 10.77 lx.

$$2 \text{ fc} \times \frac{10.77 \text{ lx}}{1 \text{ fc}} = 21.54 \text{ lx}$$

∴ 2 fc = 21.54 lx

As noted in Figure 5–16, photoconductors respond to a wide range of illumination (1 lux to 100 lux) with a wide range of resistive values (100 kΩ to 100 Ω). Because of this wide variation in resistance, photoconductors are used in many applications, including light meters and light-activated relay control circuits.

5–5 RESISTANCE DESIGNATION

Resistors are *branded;* that is, they have their resistance value marked on the body of the device. Several methods are used to brand the resistance value on the resistor. These include:

☐ The resistance value printed on the body of the device, such as 10 Ω, 2.2 kΩ, or 1.2 MΩ. Industrial/consumer resistor types are marked this way.

TABLE 5-5 APPROXIMATE SURFACE ILLUMINATION RESULTING FROM NATURAL EVENTS

Natural Event	Illumination in Units of Lux (lx)	Illumination in Units of Footcandles (fc)
Sunlight	100 000	10 000
Daylight sky	10 000	1 000
Overcast sky	1 000	100
Evening sky (twilight)	10	1

Resistance Designation

☐ The resistance value printed on the body of the device as a series of numbers or numbers and the letter R such as 123 or 4R70. MIL-spec resistors (military specification resistors) use this system.

☐ The resistance value printed on the body of the device as a code using colored bands. Both industrial-consumer resistors and MIL-spec resistors use *color code* to designate resistance value, as well as the percent of tolerance.

Examples of the latter two systems are pictured in Figure 5-17.

Preferred Number Series

Both MIL-spec and industrial-consumer fixed resistors may be specified by using any one of five *preferred number* series. These number series are used in *electronic parts catalogs* as the standard resistor, capacitor, and inductor values. These five preferred number series, which when first encountered seem to have no logical relationship, are actually formed by raising 10 to a fractional exponent, as noted in the following formula:

$$N_{x+1} = 10^{x/n} \quad (5-8)$$

where N_{x+1} = a value in the preferred number series
n = number of values in one of the five preferred number series ($n \neq 0$)
x = whole numbers ranging from 0 to $n - 1$ in value.

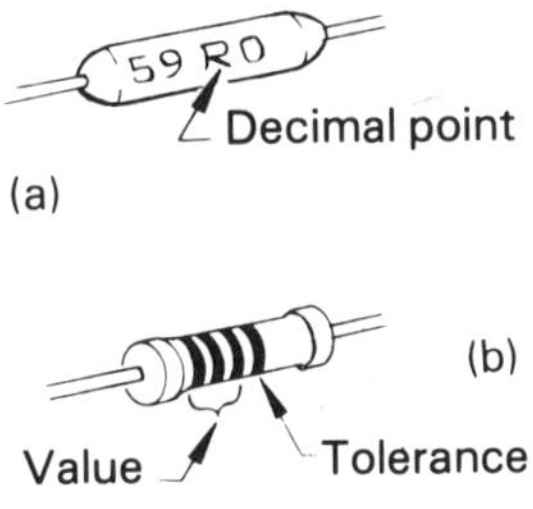

NOTE: Use either the $\boxed{y^x}$ key or the $\boxed{10^x}$ key to solve this equation.

As noted in Table 5-6, the E6 preferred number series is written with two significant figures in each number. This is also true for the E12 and E24 number series; however, the E96 and E192 have three significant figures in each number.

Manufacturers of resistors specify a *resistance range* for a given tolerance, listing the available range of standard resis-

FIGURE 5-17
Methods of designating resistance.
(a) Military (MIL-spec) resistor with 59.0 Ω resistance indicated as 59R0.
(b) Color code used to indicate resistor value and tolerance.

TABLE 5-6 PREFERRED NUMBER SERIES FOR STANDARD RESISTANCE VALUES FOR THE 1 TO 10 DECADE

E6 ±20%	E12 ±10% (*bold*) E24 (all) ±2% and ±5%	E96 ±1% (*bold*) and E192 ±0.1%, ±0.25%, and ±0.5% (all numbers)							
1.0	**1.0**	**1.00**	1.01	**1.02**	1.04	**1.05**	1.06	**1.07**	1.09
	1.1	**1.10**	1.11	**1.13**	1.14	**1.15**	1.17	**1.18**	1.20
	1.2	**1.21**	1.23	**1.24**	1.26	**1.27**	1.29	**1.30**	1.32
	1.3	**1.33**	1.35	**1.37**	1.38	**1.40**	1.42	**1.43**	1.45
1.5	**1.5**	**1.47**	1.49	**1.50**	1.52	**1.54**	1.56	**1.58**	1.60
	1.6	**1.62**	1.64	**1.65**	1.67	**1.69**	1.72	**1.74**	1.76
	1.8	**1.78**	1.80	**1.82**	1.84	**1.87**	1.89	**1.91**	1.93
	2.0	**1.96**	1.98	**2.00**	2.03	**2.05**	2.08	**2.10**	2.13
2.2	**2.2**	**2.15**	2.18	**2.21**	2.23	**2.26**	2.29	**2.32**	2.34
	2.4	**2.37**	2.40	**2.43**	2.46	**2.49**	2.52	**2.55**	2.58
	2.7	**2.61**	2.64	**2.67**	2.71	**2.74**	2.77	**2.80**	2.84
	3.0	**2.87**	2.91	**2.94**	2.98	**3.01**	3.05	**3.09**	3.12
3.3	**3.3**	**3.16**	3.20	**3.24**	3.28	**3.32**	3.36	**3.40**	3.44
	3.6	**3.48**	3.52	**3.57**	3.61	**3.65**	3.70	**3.74**	3.79
	3.9	**3.83**	3.88	**3.92**	3.97	**4.02**	4.07	**4.12**	4.17
	4.3	**4.22**	4.27	**4.32**	4.37	**4.42**	4.48	**4.53**	4.59
4.7	**4.7**	**4.64**	4.70	**4.75**	4.81	**4.87**	4.93	**4.99**	5.05
	5.1	**5.11**	5.17	**5.23**	5.30	**5.36**	5.42	**5.49**	5.56
	5.6	**5.62**	5.69	**5.76**	5.83	**5.90**	5.97	**6.04**	6.12
	6.2	**6.19**	6.26	**6.34**	6.42	**6.49**	6.57	**6.65**	6.73
6.8	**6.8**	**6.81**	6.90	**6.98**	7.06	**7.15**	7.23	**7.32**	7.41
	7.5	**7.50**	7.59	**7.68**	7.77	**7.87**	7.96	**8.06**	8.16
	8.2	**8.25**	8.35	**8.45**	8.56	**8.66**	8.76	**8.87**	8.98
	9.1	**9.09**	9.20	**9.31**	9.42	**9.53**	9.65	**9.76**	9.88

*Note: Once a particular number series is selected, then the standard values within a specified resistance range are determined from the 1 to 10 decade by multiplying by 0.1, 10, 100, 1000, etc.

tance values for a particular type of resistor. This concept is demonstrated in Example 5–13.

EXAMPLE 5-13 Specify an appropriate preferred number series for the following specified resistance ranges.

(a) Cermet-film resistor ±1% 10 Ω to 1 MΩ.
(b) Carbon-composition resistor ±10% 2.2 Ω to 2.2 MΩ.
(c) Wire-wound resistor ±5% 0.10 Ω to 15 kΩ.

Resistance Designation

(d) Metal-film resistor ±0.5% 49.9 Ω to 100 kΩ.
(e) Carbon-film resistor ±2% 1 Ω to 10 MΩ.

SOLUTION From Table 5–6:

(a) ±1%: Use E96 preferred number series. Typical values from this series for 10-Ω to 1-MΩ resistance range include 196 Ω, 1.65 kΩ, 43.2 kΩ, 536 kΩ, and 1.00 MΩ.

(b) ±10%: Use E12 preferred number series. Typical resistor values for this series for 2.2 Ω to 2.2 MΩ resistance range include 3.9 Ω, 0.68 kΩ, 15 kΩ, 0.27 MΩ, and 1.8 MΩ.

(c) ±5%: Use E24 preferred number series. Typical resistor values from this series for 0.10 Ω to 15 kΩ resistance range include 0.36 Ω, 47 Ω, 0.91 kΩ, 6.2 kΩ, and 11 kΩ.

(d) ±0.5%: Use E192 preferred number series. Typical resistor values from this series for 49.9 Ω to 100 kΩ resistance range include 64.2 Ω, 741 Ω, 8.06 kΩ, 13.3 kΩ, and 98.8 kΩ.

(e) ±2%: Use E24 preferred number series. Typical resistor values from this series for 1.0 Ω to 10 MΩ resistance range include 2.0 Ω, 27 Ω, 5.6 kΩ, 0.75 MΩ, and 9.1 MΩ.

MIL-Spec Resistance Designation

The code for specifying military types of resistors is explored in the following series of examples.

EXAMPLE 5–14 Describe the resistance designation for resistor tolerances of ±2% or larger (≥ ±2%) for:

(a) Resistors 10 Ω or greater ($R \geq 10\ \Omega$)
(b) Resistors less than 10 Ω but greater than or equal to 1 Ω ($1\ \Omega \leq R < 10\ \Omega$)
(c) Resistors less than 1 Ω ($R < 1\ \Omega$)

SOLUTION (a) MIL-spec resistors ≥10 Ω with tolerances ≥ ±2% are specified with a three-number system. The first two digits are significant figures in the resistance value, while the third digit indicates the number of zeros. Thus:

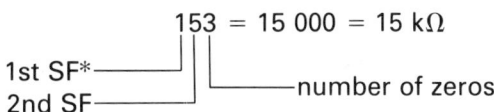

153 = 15 000 = 15 kΩ
1st SF*
2nd SF
number of zeros

*SF stands for significant figure.

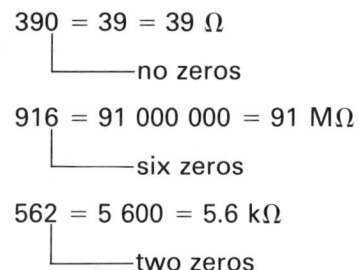

$390 = 39 = 39\ \Omega$ (no zeros)

$916 = 91\,000\,000 = 91\ M\Omega$ (six zeros)

$562 = 5\,600 = 5.6\ k\Omega$ (two zeros)

(b) MIL-spec resistors $<10\ \Omega$ but $\geq 1\ \Omega$ with tolerances $\geq \pm 2\%$ are specified with two digits, each being a significant figure and a decimal point indicated by the letter R. Thus:

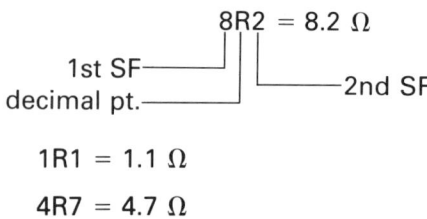

$8R2 = 8.2\ \Omega$ (1st SF, decimal pt., 2nd SF)

$1R1 = 1.1\ \Omega$

$4R7 = 4.7\ \Omega$

(c) MIL-spec resistors $<1\ \Omega$ (fractional values of an ohm) with tolerance $\geq \pm 2\%$ are specified with two digits, each being a significant figure and a decimal point indicated by the letter R. Thus:

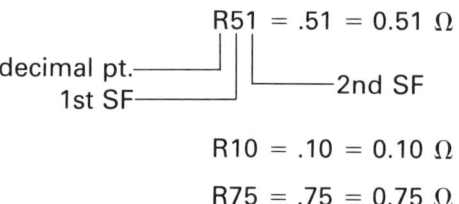

$R51 = .51 = 0.51\ \Omega$ (decimal pt., 1st SF, 2nd SF)

$R10 = .10 = 0.10\ \Omega$

$R75 = .75 = 0.75\ \Omega$

EXAMPLE 5-15 Describe the resistance designation for resistor tolerances of $\pm 1\%$ or smaller ($\leq \pm 1\%$) for:
(a) Resistors $100\ \Omega$ or greater ($R \geq 100\ \Omega$)
(b) Resistors less than $100\ \Omega$, including fractional values of resistance

SOLUTION (a) MIL-spec resistors $\geq 100\ \Omega$ with tolerances $\leq \pm 1\%$ are specified with a four-digit number system. The first three digits are significant figures in the resistance value, while the fourth digit indicates the number of zeros. Thus:

Resistance Designation

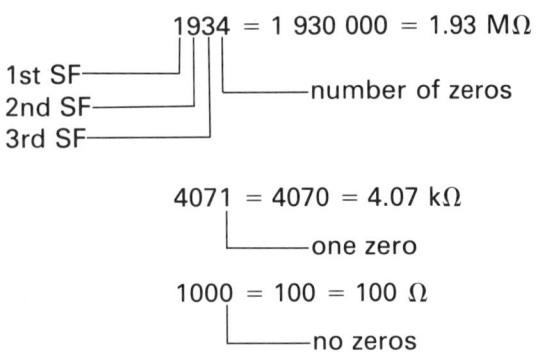

(b) MIL-spec resistors <100 Ω with tolerances ≤ ±1% are specified with three digits, each being a significant figure and a decimal point indicated by the letter R. Thus:

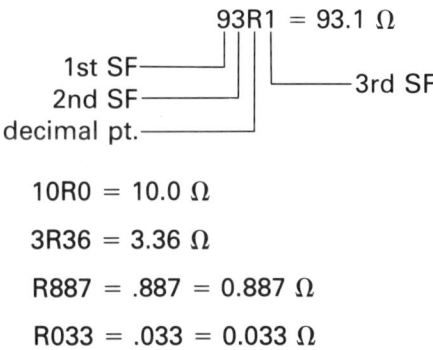

10R0 = 10.0 Ω

3R36 = 3.36 Ω

R887 = .887 = 0.887 Ω

R033 = .033 = 0.033 Ω

Color-Code Resistance Designation

Color code (bands of color) is used on carbon-composition, carbon-film, and some metal-film resistors to specify both the resistance value and the resistance tolerance of the fixed resistor. Two systems are in use: (1) a four-band system for the E6 (±20%), E12 (±10%), and the E24 (±5% and ±2%) preferred values and tolerances; and (2) a five-band system for the E96 (±1%) and E192 (±0.1%, etc.) preferred values and tolerances. The system is based on a series of valued colors, as noted in Table 5–7.

In the four-band system, the first two bands represent significant figures while the third band is the *multiplier* (essentially, the number of zeros). The fourth color band represents the resistor tolerance. The following examples will demonstrate the application of the color code.

TABLE 5-7 COLOR CODE FOR SPECIFYING PREFERRED RESISTOR VALUES AND TOLERANCES

Four-band Color Code for E6, E12, and E24 Series of Preferred Values

Color	1st Band	2nd Band	3rd Band	4th Band
Black	0	0	×1	—
Brown	1	1	×10	—
Red	2	2	×100	±2%(G)*
Orange	3	3	×1000	—
Yellow	4	4	×10 000	—
Green	5	5	×100 000	—
Blue	6	6	×1 000 000	—
Violet	7	7	—	—
Gray	8	8	—	—
White	9	9	—	—
Gold	—	—	×0.1	±5%(J)
Silver	—	—	×0.01	±10%(K)
Plain	—	—	—	±20%(M)

Five-band Color Code for E96 and E192 Series of Preferred Values

1st Band	2nd Band	3rd Band	4th Band	5th Band
0	0	0	×1	—
1	1	1	×10	±1%(F)*
2	2	2	×100	—
3	3	3	×1000	†
4	4	4	×10 000	—
5	5	5	×100 000	±0.5%(D)
6	6	6	×1 000 000	±0.25%(C)
7	7	7	—	±0.1%(B)
8	8	8	—	—
9	9	9	—	—
—	—	—	×0.1	—
—	—	—	×0.01	—
—	—	—	—	—

* Tolerance code letters for specifying MIL-spec resistor tolerance.
† A yellow 5th band is sometimes used to indicate the reliability level of MIL-spec carbon-composition resistors.

Resistance Designation

EXAMPLE 5–16 Using Table 5–7, specify the color code for the following resistor values:
(a) 10 Ω ± 10%,
(b) 4.3 kΩ ± 2%,
(c) 6.8 MΩ ± 20%,
(d) 2.7 Ω ± 5%,
(e) 3.6 Ω ± 10%,
(f) 0.2 Ω ± 5%.

SOLUTION From Table 5–7:

	1st SF	*2nd SF*	*Multiplier*	*Tolerance*	
(a)	Brown 1	Black 0	Black ×1	Silver ±10%	⇒ 10 Ω ± 10%
(b)	Yellow 4	Orange 3	Red ×100	Red ±2%	⇒ 4.3 kΩ ± 2%
(c)	Blue 6	Gray 8	Green ×100 000	Plain ±20%	⇒ 6.8 MΩ ± 20%
(d)	Red 2	Violet 7	Gold ×0.1	Gold ±5%	⇒ 2.7 Ω ± 5%
(e)	Orange 3	Blue 6	Gold ×0.1	Silver ±10%	⇒ 3.6 Ω ± 10%
(f)	Red 2	Black 0	Silver ×0.01	Gold ±5%	⇒ 0.2 Ω ± 5%

EXAMPLE 5–17 Using Table 5–7, specify the color code for the following resistor values:
(a) 51.1 Ω ± 1%,
(b) 2.64 Ω ± 0.5%,
(c) 16 kΩ ± 5%,
(d) 0.47 MΩ ± 20%,
(e) 732 Ω ± 1%,
(f) 397 kΩ ± 0.1%.

SOLUTION From Table 5–7:

(a) 51.1 Ω ± 1% ⇒ Green   Brown   Brown   Gold   Brown
                  5   1   1   ×0.1   ±1%

(b) 2.64 Ω ± 0.5% ⇒ Red   Blue   Yellow   Silver   Green
                    2   6   4   ×0.01   ±0.5%

(c) 16 kΩ ± 5% ⇒ Brown   Blue   Orange   Gold
                 1   6   ×1000   ±5%

(d) 0.47 MΩ ± 20% ⇒ Yellow   Violet   Yellow   Plain
                    4   7   ×10 000   ±20%

(e) 732 Ω ± 1% ⇒ Violet   Orange   Red   Black   Brown
                 7   3   2   ×1   ±1%

(f) 397 kΩ ± 0.1% ⇒ Orange   White   Violet   Orange   Violet
                        3   9   7   ×1000   ±0.1%

EXAMPLE 5–18 Using Table 5–7, record both the value and the tolerance of the resistors pictured in Figure 5–18.

SOLUTION (a) From Figure 5–18(a):

Green   Blue   Orange   Red   ⇒ 56 kΩ ± 2%
  5   6   ×1000   ±2%

(b) From Figure 5–18(b):

Yellow   Black   Red   Red   Brown
  4   0   2   ×100   ±1%   ⇒ 40.2 kΩ ± 1%

(c) From Figure 5–18(c):

Yellow   Orange   Brown   Gold
  4   3   ×10   ±5%   ⇒ 430 Ω ± 5%

SELECTED TECHNICAL TERMS

Several of the technical terms used in this chapter are listed here to aid in your understanding of them.

Ambient temperature: the temperature of the air (or other coolant) surrounding electric components or equipment.

Axial leads: leads that are connected on the axis (center line) of a component.

Bilateral: conducts current in both directions; conduction of current in one direction is termed *unilateral*.

Linear resistance: resistance that remains constant for variation in voltage, current, or temperature.

MIL-spec: abbreviation for military specification that indicates that the component's characteristics meet standards set down by the U.S. Department of Defense.

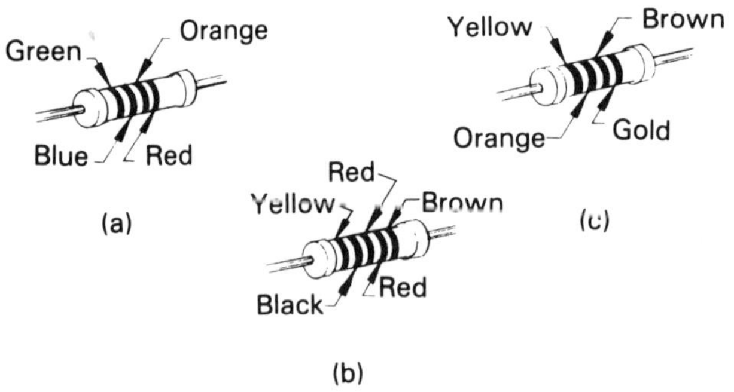

FIGURE 5–18
Color-coded resistors for Example 5–21.

Potentiometer: a variable resistor used to divide voltage. The fixed terminals are connected to either end of the resistive element and a third movable terminal is connected to the wiper contact. The input voltage across the two outside fixed terminals is divided as the wiper contact moves over the resistive element in the potentiometer circuit.

Radial leads: leads that are connected on the side of a component.

Rheostat: a variable resistor used to alter the current flowing through the circuit.

Tolerance: the allowed variation permitted in the electric properties or the physical dimensions of a component or device.

PROBLEMS

Section 5-1

5-1 Determine the resistance of an aluminum rod that has 6 A of current in the rod with 18 mV across the rod.

5-2 Determine the resistance of a stainless steel wire that has a measured current of 238 mA in the wire when 0.110 V is measured across the wire.

5-3 Determine the resistance of the neoprene insulating material on an underground cable when an applied test voltage of 600 V results in a leakage current of 12 μA.

5-4 Determine the resistance of a nylon standoff when the application of a 300-V source produces a measured current of 800 nA.

5-5 Determine the conductance of an 8-mΩ copper ribbon conductor.

5-6 Determine the conductance of a 12-Ω heating element.

5-7 Determine the conductance of a silicon wafer that has a current of 90 mA in it when a 36-V source is applied.

5-8 Determine the conductance of the material in a 12-V, 7-A automotive window defogger.

5-9 Determine the resistance of the electric circuit of a flashlight that has a voltage source of 3 V and a current in the lamp of 300 mA.

5-10 Determine the resistance of the electric circuit of a 120-V digital clock radio that requires an operating current of 120 mA.

5-11 Determine the reading of an ammeter used to measure the electric current in the circuit of a 117-V electric water heater if the resistance of the heater is 6.9 Ω.

5-12 Determine the current in the electric motor circuit of a 24-V, 18 Ω equipment cooling fan.

5-13 Determine the voltage of the source of an electric alarm circuit having 67 Ω of resistance with a current of 180 mA.

5-14 Determine the number of 1.5-V dry cells (batteries) used as the voltage source in a flashlight circuit if the light is rated at 500 mA and the circuit has a resistance of 15 Ω.

Section 5-2

5-15 A resistor is a manufactured circuit component that has a specified _____.

5-16 Resistors are used in electric circuits to _____ the current in the circuit.

5-17 Explain what is meant when a resistor is described as a nonpolarized, symmetrical, bilateral device.

5-18 Linear resistors have nearly _____ resistance.

5-19 The temperature coefficient of resistance is an indication of the _____ stability of a resistor.

5-20 Rank the resistors listed in Table 5-2 in order from the least to the most affected by variations in temperature.

5-21 Determine the change in resistance of a 100-kΩ cermet-film resistor due to a +60°C temperature variation. The TC of the resistor is specified as +50 ppm/°C.

5-22 A carbon-film resistor is specified as 2 MΩ at 20°C. Determine the resistance change due to temperature when the device is operated at 85°C if the TC is −250 ppm/°C.

5-23 Determine the resistance of a 180-kΩ (nominal) metal-film resistor specified at 25°C when it is operating at an ambient temperature of 72°C. The temperature coefficient is +80 ppm/°C.

5-24 A precision wire-wound resistor is specified as 22.3 kΩ at 20°C. Determine the value of the resistor when it is operated at 200°C if the temperature coefficient is +5 ppm/°C.

Section 5-3

5-25 Describe how to connect a potentiometer so that it functions as a rheostat.

5-26 From the first curve of Figure 5-13, determine the percentage of resistance that corresponds to an 80% cw rotation of the potentiometer control shaft.

5-27 From the modified left-hand log taper curve of Figure 5-13, determine the resistance between the wiper contact and an outside fixed terminal for a 70% cw rotation of the control shaft of a 500-Ω potentiometer.

5-28 Using the linear taper curve of Figure 5-13, determine the resistance between the wiper contact and an outside fixed terminal for a 40% cw rotation of the control shaft of a 1-kΩ potentiometer.

Section 5-4

5-29 From the PTC thermistor characteristic curve of Figure 5-14, approximate the thermistor resistance at (a) 75°C and (b) 125°C.

5-30 Using the NTC thermistor characteristic curve of Figure 5-14, approximate the thermistor resistance at (a) 25°C and (b) 75°C.

5-31 Using the NTC thermistor characteristic curve of Figure 5-14, approximate the temperature when the resistance is (a) 2 kΩ and (b) 200 Ω.

5-32 From the PTC thermistor characteristic curve of Figure 5-14, approximate the temperature when the resistance is (a) 100 Ω and (b) 1 kΩ.

5-33 Using the varistor characteristic curve of Figure 5-15, approximate the resistance of the device when 50 V is dropped across the device.

5-34 Using the varistor characteristic curve of Figure 5-15, approximate the resistance of the device when 165 V is dropped across the device.

5-35 From the photoconductive cell characteristic curve of Figure 5-16, approximate the cell's resistance for a surface illumination of 4 lux.

5-36 From the photoconductive cell characteristic curve of Figure 5-16, approximate the cell's resistance for a surface illumination of 20 lux.

5-37 Using the CdS "photocell" characteristic curve of Figure 5-16, approximate the surface illumination on the cell in footcandles (fc) when a cell resistance of 380 Ω is measured.

5-38 Using the CdS "photocell" characteristic curve of Figure 5-16, approximate the surface illumination on the cell in footcandles (fc) when a cell resistance of 9 kΩ is measured.

Section 5-5

5-39 State the resistance of a MIL-spec resistor when the following numbers are noted on the body of the resistor: (a) 910, (b) 7R5, (c) 3831, (d) 1R2, (e) 683, (f) R47, (g) 53R6, and (h) R20.

5-40 State the resistance of a MIL-spec resistor when the following numbers are noted on the body of the resistor: (a) R988, (b) 392, (c) 7594, (d) 10R4, (e) 5R1, (f) 14R3, (g) 185, and (h) R43.

5-41 Record both the value and the tolerance of the resistors with the following color codes.
(a) Brown, gray, yellow, silver

(b) Red, black, green, silver, brown
(c) Yellow, violet, orange
(d) Orange, orange, black, gold
(e) Yellow, white, orange, red, green
(f) Red, red, red, red

5-42 Record both the value and the tolerance of the resistors with the following color codes.
(a) Blue, violet, orange, gold, blue
(b) Yellow, orange, green, gold
(c) Brown, green, gold, silver
(d) Gray, yellow, green, red, brown
(e) Brown, black, blue, silver, violet
(f) Blue, red, black, silver

5-43 Specify the color code for the following resistor values:
(a) 510 Ω ± 5%, (b) 82 kΩ ± 10%, and (c) 0.24 MΩ ± 2%.

5-44 Specify the color code for the following resistor values:
(a) 3.16 kΩ ± 1%, (b) 2.2 MΩ ± 20%, and (c) 1.0 kΩ ± 10%.

5-45 Using Equation 5-8, verify the 10th entry ($N_{10} = 2.4$) in the E24 preferred number series (Table 5-6) by computing its value.

5-46 Using Equation 5-8, verify the 15th entry ($N_{15} = 1.40$) in the E96 preferred number series (Table 5-6) by computing its value.

5-47 Compute the maximum resistance of a carbon-film 100 Ω ± 10% resistor.

5-48 Compute the maximum resistance of a metal-film 100 Ω ± 1% resistor.

5-49 For the carbon-film resistor of Problem 5-47, compute its maximum resistance at 80°C if its nominal (named) value is specified at 20°C and the temperature coefficient is −300 ppm/°C.

5-50 For the metal-film resistor of Problem 5-48, compute its maximum resistance at 80°C if its nominal value is specified at 20°C and the temperature coefficient is +100 ppm/°C.

6 Power and Energy

6-1 COMPUTING AVERAGE POWER
6-2 POWER CONSIDERATIONS IN LINEAR RESISTORS
6-3 EFFICIENCY AND THE COST OF ENERGY
6-4 CIRCUIT CONTROL AND PROTECTIVE DEVICES

CHAPTER PREVIEW This chapter will introduce power dissipation in linear resistors, the concepts of efficiency in energy conversion, and the computation of energy cost. The chapter concludes with an introduction to switch, fuse, and circuit breaker application and terminology.

PERFORMANCE OBJECTIVES Once you have read each section, worked each example with pencil, paper, and calculator, and answered each problem for every section, you will be able to:

☐ Compute power dissipation in a resistive device.
☐ Understand the meaning of average power.
☐ Use a power derating curve to aid in selecting the power rating of a resistor.
☐ Calculate the efficiency of an electric circuit or system.
☐ Select a fuse type and value for a particular current.

6–1 COMPUTING AVERAGE POWER In Section 2–1 the concepts of work and energy were introduced. Additionally, in Chapter 2 we talked of electric charge in motion (current) as a carrier of electric energy. The effectiveness of electric current as an energy carrier is indicated by the *effective value* of current.

When current passes through a linear resistive load, the energy carried by the current does work on the load, causing the resistive device to gain energy by raising its temperature. The *average* amount of energy imparted by the *current* to the resistive load per second is measured in units of joules per second (J/s), also called watts (W). Thus:

---

1 watt = 1 joule/second

Power Equations Average power may be computed with one of several forms of the *power equation*.

$$P = I^2 R \quad \text{(W)} \tag{6–1}$$

where P = average power in watts
I = effective current (rms) in amperes
R = resistance of the linear resistance in ohms

Additional forms of the formula for power may be derived by substituting forms of Ohm's law ($I = E/R$ and $R = E/I$) into Equation 6–1. Thus, substitute E/R for I in $P = I^2R$:

Computing Average Power

$$P = (E/R)^2 R$$

$$P = \frac{E^2}{R} \quad \text{or} \quad P = \frac{V^2}{R} \quad \text{(W)} \tag{6-2}$$

Substitute E/I for R in $P = I^2 R$:

$$P = I^2(E/I)$$

$$P = IE \quad \text{or} \quad P = IV \quad \text{(W)} \tag{6-3}$$

where P = average power in watts
I = effective current in amperes
E = effective voltage of a voltage source in volts
V = effective voltage of a voltage drop in volts

Since power is the rate of producing or using energy as well as the rate of doing work,

$$P = \frac{W}{t} \quad \text{(W)} \tag{6-4}$$

where P = average power in watts
W = work (energy) in joules
t = time in seconds

The power equations are summarized in Table 6–1. The following series of examples will demonstrate the use of various forms of the power equations.

EXAMPLE 6-1 Determine the power delivered to the lamp of Figure 6–1 when the ammeter rads 0.5 A and the voltmeter indicates 6 V.

SOLUTION Select the equation from Table 6–1 that relates power to voltage and current:

$$P = IV$$

$$P = (0.5)(6) = 3 \text{ W}$$

∴ 3 W of power is delivered to the lamp.

Observation Electric energy is being converted in the lamp to heat and light at the rate of 3 J/s (3 W).

TABLE 6–1   POWER EQUATIONS

Formula	Equation Number
$P = I^2 R$	(6–1)
$P = E^2/R$ or $P = V^2/R$	(6–2)
$P = IE$ or $P = IV$	(6–3)
$P = W/t$	(6–4)

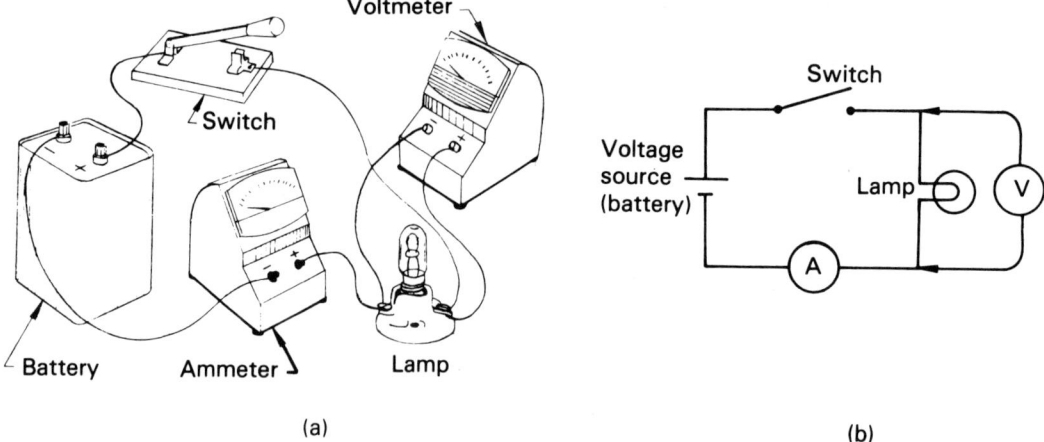

FIGURE 6-1 A light circuit made up of a voltage source (battery), a switch, and a light. It is assumed that the meters have no *loading* effect on the circuit. (a) Pictorial diagram. (b) Schematic diagram.

EXAMPLE 6-2 Determine the power rating of a microwave oven if it has a resistance of 10 Ω. The oven is attached to a 120-V ac source as shown in Figure 6-2.

SOLUTION Select an equation from Table 6-1 that relates power to voltage and resistance:

$$P = \frac{E^2}{R}$$

$$P = \frac{120^2}{10} = 1440$$

$$\therefore \quad P = 1.44 \text{ kW}$$

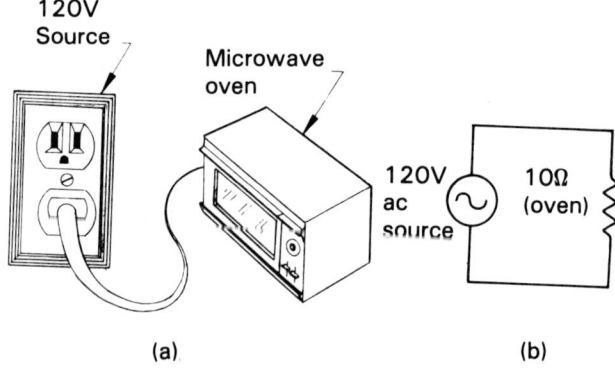

FIGURE 6-2
Circuit of Example 6-2.
(a) Pictorial diagram.
(b) Schematic diagram.

Power Considerations in Linear Resistors

EXAMPLE 6-3 Determine the rate at which an electric water heater converts electric energy to heat energy if 64 kJ of heat is produced in 40 s.

SOLUTION Select an equation from Table 6-1 that relates power to work (energy) and time.

$$P = \frac{W}{t}$$

$$P = \frac{64 \times 10^3}{40} = 1600$$

$$\therefore \quad P = 1.6 \text{ kW}$$

6-2 POWER CONSIDERATIONS IN LINEAR RESISTORS

Linear resistors are specified by type, resistance in ohms (Ω), tolerance, and *power dissipation*. Power dissipation is the rate of converting electric energy to heat energy measured in units of watts. Thus, a carbon-film resistor may have a rating of $\frac{1}{4}$ W, 5.1 kΩ ± 2% while a power wire-wound resistor might be rated as 3 W, 100 Ω ± 5%. Figure 6-3 pictures several carbon-composition resistors arranged by physical size. Notice that as size increases so does the wattage rating of the resistor. In general, the larger the physical size of the resistor, the greater will be its ability to dissipate power.

Resistor Power Derating Curve

The power rating for a resistor is the nominal (named) power, which indicates the highest power the resistor can dissipate when in *free air* at a specified air temperature (ambient temperature). For example, the power rating of carbon-film, carbon-composition, metal-film, cermet, and metal-oxide fixed axial lead resistors is usually specified at 70°C ambient temperature. However, cermet resistor networks have their power rating specified at 40°C ambient, and low-power axial lead wire-wound re-

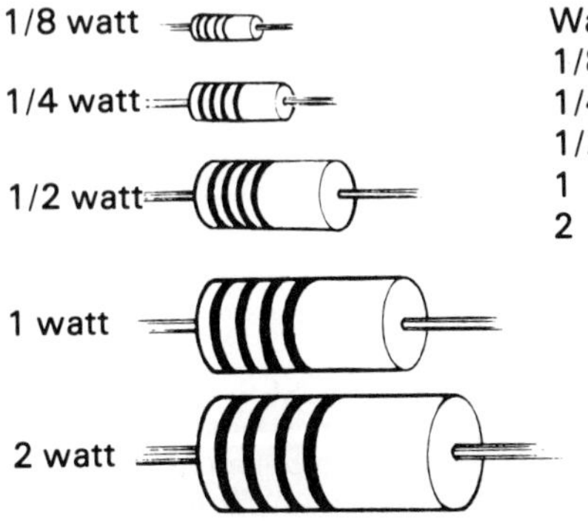

Watts	* Length	* Dia
1/8	.145/3.68	.062/1.59
1/4	.250/6.35	.090/2.29
1/2	.375/9.52	.140/3.56
1	.562/14.3	.225/5.72
2	.688/17.5	.312/7.94

* inch/mm

1/8 watt
1/4 watt
1/2 watt
1 watt
2 watt

FIGURE 6-3 Power rating of carbon-composition resistors is determined by their size.

sistors have their power rating specified at 125°C ambient. Except for high-power wire-wound resistors that are mounted on heat sinks, resistors may be operated up to their full power rating at the specified ambient temperature. For temperatures beyond the specified ambient temperature, the power is derated linearly to zero watts at the maximum specified ambient temperature. This concept is demonstrated in Example 6–4.

EXAMPLE 6–4 Construct the *power derating curve* for a $\frac{1}{4}$-W cermet-film fixed resistor with the following power and temperature specifications:

Power rating (P_{nom}) at 70°C = 0.25 W

Ambient temperature range = −55° to 165°C

SOLUTION The power derates linearly from the full rated power of 0.25 W ($\frac{1}{4}$ W) at 70°C to zero power (0 W) at 165°C. Using linear graph paper, construct a vertical and horizontal axis as shown in Figure 6–4. Divide the vertical axis into five divisions of 50 mW each. Divide the horizontal axis into six divisions of 30°C each. Graduate the axis as noted in Figure 6–4. Construct the power derating curve starting at (0°C, 250 mW) and moving horizontally to the right intersecting point (70°C, 250 mW); then move diagonally down at a constant slope intersecting the horizontal axis at (165°C, 0 W).

Observation From the given specifications and Figure 6–4, it may be seen that the resistor is capable of dissipating the full rated power of $\frac{1}{4}$ W (250 mW) from −55° to 70°C, and then the power derates linearly (constant slope) to zero power at 165°C.

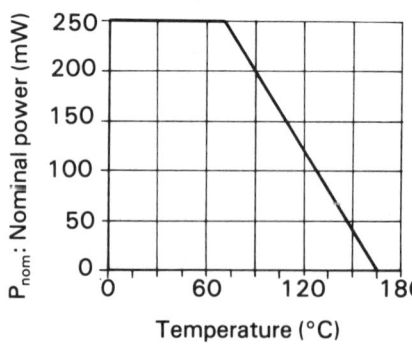

FIGURE 6–4
Power derating curve for cermet-film resistor of Example 6–4.

Power Considerations in Linear Resistors

EXAMPLE 6–5 Determine the power rating for the cermet resistor of the previous example operating at an ambient temperature of 90°C.

SOLUTION From Figure 6–4, the power rating is 200 mW at 90°C.

Relating Power and Voltage Specifications

In addition to the power rating at a specified temperature, the operation of a fixed resistor is limited to a specified dc or ac (rms) working voltage. The established working voltage is the highest voltage that the resistor can continuously operate at without failure. The specified working voltage (also called the maximum continuous rms voltage) various from power rating to power rating as indicated in Table 6–2.

Because voltage is related to power by the formula $P = E^2/R$, the *rated continuous working voltage* for any resistance (R) in ohms may be computed from the following equation.

$$RCWV = \sqrt{P_{nom} \times R} \quad \text{or the specified working voltage, whichever is less} \quad (6\text{–}5)$$

where $RCWV$ = rated continuous working voltage in volts
P_{nom} = nominal power rating in watts
R = nominal value of resistance in ohms

EXAMPLE 6–6 Determine the voltage rating for a $\frac{1}{2}$-W, 470-Ω carbon-film resistor operating at a 50°C ambient temperature.

SOLUTION Using Equation 6–5 and Table 6–2:

$$RCWV = \sqrt{P_{nom} \times R} \quad \text{or 350 V, whichever is less}$$

$$RCWV = \sqrt{0.5 \times 470} = 15.3 \text{ V}$$

∴ The rated continuous working voltage for a $\frac{1}{2}$-W, 470-Ω carbon-film resistor is 15.3 V, not 350 V.

TABLE 6–2 TYPICAL WORKING VOLTAGES FOR SELECTED RESISTORS

Power Rating at 70°C (W)	Carbon-Film (V)	Carbon-Composition (V)	Cermet-Film (V)	Metal-Film (V)
$\frac{1}{4}$	250	250	250	250
$\frac{1}{2}$	350	350	350	350
1	500	500	—	500

Observation If more than 15.3 V is dropped across the 470-Ω resistor, the power rating of ½ W is exceeded.

Not all values of resistance in a given power series may dissipate the rated power, as demonstrated in Example 6–7.

EXAMPLE 6–7 Determine the value of voltage needed to be applied to a ¼-W, 560-kΩ carbon-composition resistor so that the resistor will dissipate ¼ W.

SOLUTION Solve for E in:

$$P = \frac{E^2}{R}$$

$$E = \sqrt{PR}$$

Substitute:

$$E = \sqrt{0.25 \times 560 \times 10^3} = 374 \text{ V}$$

∴ An applied voltage of 374 V will result in a power dissipation of ¼ W in a 560-kΩ carbon-composition resistor.

Observation From Table 6–2, the working voltage for a ¼-W carbon-composition resistor is 250 V. The applied voltage of 374 V exceeds the maximum continuous voltage by 124 V.

Because both the power rating and the voltage rating must be complied with when using a resistor in a circuit application, both the power equation, Equation 6–2, as well as the rated continuous voltage equation, Equation 6–5, must be used in determining the acceptability of the selected resistor.

Table 6–3 lists the *critical resistance*, the value of resistance for a power series where both the maximum specified working voltage and the maximum specified power occur simultaneously. Below the critical resistance the specified voltage must be reduced to keep from exceeding the specified power rating.

TABLE 6–3 CRITICAL RESISTANCE FOR SEVERAL POWER SERIES

Power Rating at 70°C (W)	Working Voltage (V)	Critical Resistance (MΩ)
¼	250	0.25
½	350	0.25
1	500	0.25

Power Considerations in Linear Resistors

Above the critical resistance, the power rating cannot be reached without exceeding the specified working voltage.

EXAMPLE 6–8 A 100-kΩ carbon-composition resistor is to be used in a circuit operating at 100°C. Specify the power rating if 210 V is dropped across the resistor.

SOLUTION Obtain the following data for the carbon-composition resistor from a *parts catalog:*

Power Rating at 70°C (W)	Working Voltage (V)	Maximum Ambient Temperature (°C)
$\frac{1}{4}$	250	125
$\frac{1}{2}$	350	125
1	500	125
2	500	125

Construct a family of power derating curves for the carbon-composition resistors as shown in Figure 6–5. Compute the power dissipation of the 100-kΩ resistor.

$$P = \frac{V^2}{R}$$

$$P = \frac{210^2}{100 \times 10^3} = 441 \text{ mW}$$

Determine the derated nominal power rating by entering the derating curve of Figure 6–5 at 100°C and reading:

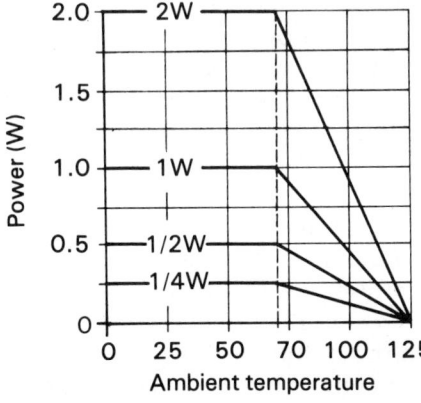

FIGURE 6–5
Family of power derating curves for carbon-composition resistor of Example 6–8.

$\frac{1}{4}$ W derates to approximately 0.1 W at 100°C

$\frac{1}{2}$ W derates to approximately 0.25 W at 100°C

1 W derates to approximately 0.45 W at 100°C

2 W derates to approximately 0.9 W at 100°C

Select power rating:

1 W derated to 0.45 W is greater than the computed power of 0.441 W

2 W derated to 0.9 W is greater than the computed power of 0.441 W

Check working voltage rating:

1 W is 500 V, which is greater than the 210 V dropped across the 100-kΩ resistor

2 W is 500 V, which is greater than the 210 V dropped across the 100-KΩ resistor

∴ Specify the carbon-composition resistor as 100 kΩ, 2 W.

Observation The selection of the 2-W rating is explained in the following paragraph.

Because heat is very detrimental to electronic components, it is very poor practice to operate resistors at 100% of their permitted nominal power. Instead, derate the nominal power by a factor of 2 or more; that is, if $\frac{1}{2}$-W rating is needed, use 1 W (minimum).

As a general rule, when selecting the resistor wattage rating, choose a wattage rating that is reasonable, affordable, and functional. Excess wattage rating will provide the following benefits.

☐ More resistance stability
☐ Quieter circuit operation
☐ Lower resistor temperature
☐ Smaller thermally derived resistance change
☐ Lower rate of component failure

6-3 EFFICIENCY AND THE COST OF ENERGY

Efficiency

Many devices, including the electric motor, electric light, and loudspeaker, have been designed specifically to convert electrical energy to other useful forms of energy. In the process of converting energy from one form to another, some input energy is lost to the output.

Energy conversion is governed by the *law of conservation of*

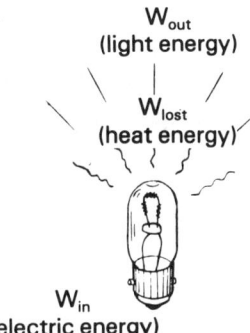

FIGURE 6–6
Conservation of energy requires that energy input equal energy output plus energy lost ($W_{in} = W_{out} + W_{lost}$).

energy, which states that energy cannot be created or destroyed, but may be changed from form to form. Figure 6–6 illustrates this law of physics. The electric lamp changes the input electric energy (W_{in}) to the desired output light energy (W_{out}), but in the process an unwanted heat energy (W_{lost}) is also developed, which is lost to the output.

The *efficiency* of an energy-converting device is expressed as a ratio that indicates how well the device performs as an energy converter. Equation 6–6 is used to compute the efficiency of an energy-converting device.

$$\eta = \frac{W_{out}}{W_{in}} \qquad (6\text{–}6)$$

where η = efficiency of the device
W_{out} = useful output energy
W_{in} = input energy

Since power is the rate of converting energy ($P = W/t$), W in Equation 6–6 may be replaced with P, as in Equation 6–7.

$$\eta = \frac{P_{out}}{P_{in}} \qquad (6\text{–}7)$$

Expressed as a percentage (%):

$$\text{Percent efficiency} = \frac{P_{out}}{P_{in}} \times 100 \qquad (6\text{–}8)$$

EXAMPLE 6–9 A stereo loudspeaker system develops 5 W of acoustical energy while operating at an efficiency of 8%. Determine the electric power into the speaker system from the amplifier.

SOLUTION Use Equation 6–8:

$$\text{Percent efficiency} = \frac{P_{out}}{P_{in}} \times 100$$

Solve for P_{in} and substitute:

$$P_{in} = \frac{P_{out} \times 100}{\%\eta}$$

$$P_{in} = \frac{5 \times 100}{8} = 62.5 \text{ W}$$

∴ The power into the loudspeaker system is 62.5 W.

Mechanical Power   In the SI system of units, the watt is the unit for both electrical and mechanical power. However, horsepower (hp) is still a commonly used unit for measuring the mechanical rate of doing work. To convert from horsepower to watts, multiply the horsepower by 746 W/hp. Thus:

$$1 \text{ hp} = 746 \text{ W}$$

EXAMPLE 6–10   A 120-V electric motor has an efficiency of 80% and delivers 0.5 hp. Determine the current in the line cord.

SOLUTION   Figure 6–7 pictures the input and output conditions for the motor. From Equation 6–8:

$$P_{in} = \frac{P_{out} \times 100}{\%\eta}$$

Substitute where $P_{out} = 0.5 \times 746 = 373$ W:

$$P_{in} = \frac{373 \times 100}{80} = 466 \text{ W}$$

Solve for I:

$$P = EI$$

$$I = \frac{P}{E}$$

Substitute:

$$I = \frac{466}{120} = 3.9 \text{ A}$$

∴   The current in the line cord is 3.9 A.

System Efficiency   When several energy-converting devices are connected one after another, they form a *cascaded* system. The efficiency of each component in a cascaded system contributes to the overall percent of efficiency, as described in Equation 6–9.

$$\%\eta_{ov} = \eta_1 \times \eta_2 \times \eta_3 \times \cdots \times \eta_n \times 100 \qquad (6\text{–}9)$$

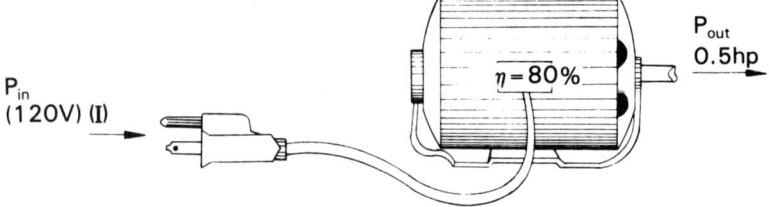

FIGURE 6–7
Input and output conditions for Example 6–10.

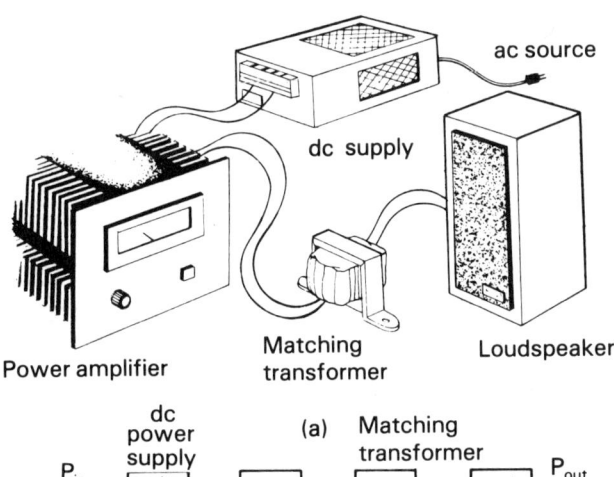
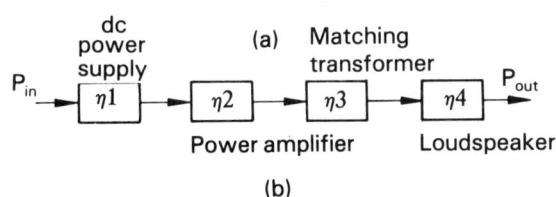

FIGURE 6–8
Amplifier system.
(a) Pictorial of the system.
(b) Cascaded block diagram of amplifying system.

where $\%\eta_{ov}$ = overall percent of efficiency of the system
η_n = efficiency of component n expressed as a decimal fraction

In Figure 6–8(a), the ac source is converted to dc by the power supply, which is then used to power the amplifier, which in turn drives the loudspeaker through the matching transformer. Finally, the loudspeaker converts electric energy to *acoustical* energy. The overall percent of efficiency of any system decreases each time an additional conversion device is added to the cascaded system.

EXAMPLE 6–11 Determine the overall percent of efficiency of the audio system of Figure 6–8(b) when $\eta_1 = 0.73$, $\eta_2 = 0.40$, $\eta_3 = 0.85$, and $\eta_4 = 0.05$.

SOLUTION Use Equation 6–9:

$$\%\eta_{ov} = \eta_1 \times \eta_2 \times \eta_3 \times \eta_4 \times 100$$

$$\%\eta_{ov} = 0.73 \times 0.40 \times 0.85 \times 0.05 \times 100$$

$$\therefore \quad \%\eta_{ov} = 1.24\%$$

Observation Notice how dependent $\%\eta_{ov}$ is on each efficiency in the cascaded system. From this example, it is learned that the overall efficiency of a system is always smaller than the smallest efficiency in the cascaded system.

Another Unit of Energy

Electrical energy is measured in units of joules (J). However, this unit is not practical for the sale of electric energy because it is very small. Instead, a larger, more practical unit, the kilowatt-hour (kWh) is used. Thus:

$$1.0 \text{ kWh} = 3.6 \text{ MJ}$$

Residential and commercial energy usage is measured by an instrument called a kilowatthour meter. This instrument records energy use on a series of dials, as pictured in Figure 6–9.

EXAMPLE 6–12 Determine the energy used (in kilowatthours) to operate a 100-W light for 6 h a day for a period of 1 year (365 days).

SOLUTION Use Equation 6–4 and solve for W. Thus:

$$P = \frac{W}{t}$$

and

$$W = Pt$$

where P is in kilowatts and t in hours. Express power in kilowatts:

$$P = 100 \text{ W} \times \frac{1 \text{ kW}}{1000 \text{ W}} = 0.100 \text{ kW}$$

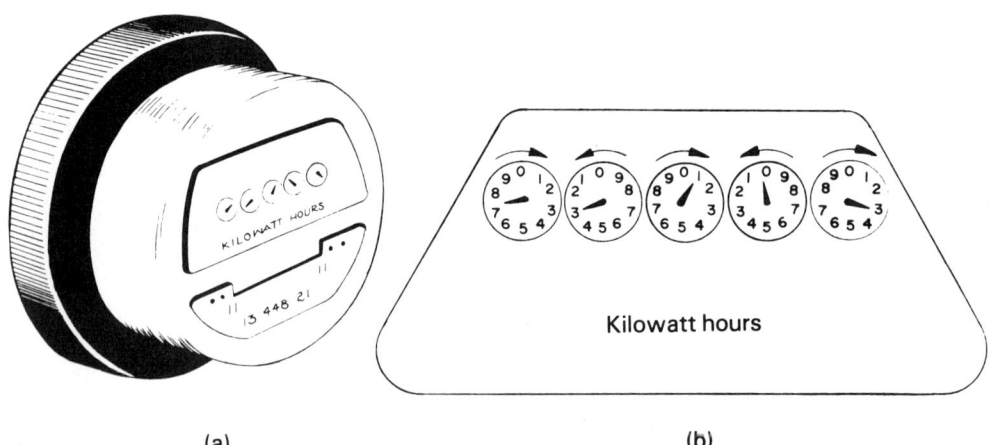

(a) (b)

FIGURE 6–9 (a) Kilowatt-hour meter is used to measure electric energy usage. (b) Dials, read left to right, record energy use. The reading is 73 103 kWh.

Efficiency and the Cost of Energy

Express time in hours:

$$t = 365 \text{ d} \times \frac{6 \text{ h}}{\text{d}} = 2190 \text{ h}$$

Substitute into $W = Pt$:

$$W = (0.100 \text{ kW})(2190 \text{ h}) = 219 \text{ kWh}$$

∴ The energy used for 1 year by a 100-W light operating 6 h a day is 219 kWh.

Computing Energy Cost

The cost of electric energy is determined from the rate charged by the utility company. The rate in cents per kilowatthour (¢/kWh) along with the amount of energy used in kilowatthours is substituted into the following equation.

$$\text{Cost} = \text{rate} \times \text{number of kilowatthours} \tag{6–10}$$

EXAMPLE 6–13 If the rate for electric energy is 8¢/kWh, compute the cost of operating the 100-W lamp of Example 6–12.

SOLUTION The energy used for 1 year of operation was 219 kWh. Use Equation 6–10 and substitute:

$$\text{Cost} = \text{rate} \times \text{number of kilowatthours}$$

$$\text{Cost} = \frac{8¢}{\text{kWh}} \times 219 \text{ kWh} = 1752¢$$

∴ The cost of operating a 100-W light 6 h/day for 1 year at 8¢/kWh is $17.52.

EXAMPLE 6–14 A 10-hp motor is operated continuously for 2 h and a kilowattmeter attached to the motor indicates a use of 17.9 kWh during the time of operation. Determine (a) the percent of efficiency of the motor; (b) the cost of operation at 10.5¢/kWh; and (c) the amount of lost energy in joules (J).

SOLUTION (a) Use Equation 6–8:

$$\% \eta = \frac{P_{\text{out}}}{P_{\text{in}}} \times 100$$

Express P_{out} in kilowatts:

$$P_{\text{out}} = 10 \text{ hp} \times \frac{746 \text{ W}}{1 \text{ hp}} = 7460 \text{ W} = 7.46 \text{ kW}$$

Use Equation 6–4 and express P_{in} in kilowatts:

$$P_{in} = \frac{W}{t}$$

Substitute:

$$P_{in} = \frac{17.9 \text{ kWh}}{2 \text{ h}} = 8.95 \text{ kW}$$

Substitute into Equation 6–8:

$$\%\eta = \frac{P_{out}}{P_{in}} \times 100 = \frac{7.46 \text{ kW} \times 100}{8.95 \text{ kW}} = 83.4\%$$

∴ The motor is 83.4% efficient in converting electric energy to mechanical energy.

(b) Use Equation 6–10:

$$\text{Cost} = \text{rate} \times \text{number of kilowatthours}$$

Substitute:

$$\text{Cost} = \frac{10.5 \text{ ¢}}{\text{kWh}} \times 17.9 \text{ kWh} = 188\text{¢}$$

∴ The cost of operating the motor for 2 h is $1.88.

(c) Apply the law of conservation of energy ($W_{in} = W_{out} + W_{lost}$). Solve for W_{lost}:

$$W_{lost} = W_{in} - W_{out}$$

Express W_{out} in kilowatthours:

$$W_{out} = P_{out}(t) = 10 \text{ hp} \times \frac{746 \text{ W}}{\text{hp}} \times 2 \text{ h}$$

$$W_{out} = 14\,920 \text{ Wh}$$

$$W_{out} = 14.92 \text{ kWh}$$

Substitute into $W_{lost} = W_{in} - W_{out}$:

$$W_{lost} = 17.9 \text{ kWh} - 14.92 \text{ kWh} = 2.98 \text{ kWh}$$

Determine the power in joules by applying the identity 1.0 kWh = 3.6 MJ:

$$2.98 \text{ kWh} \times \frac{3.6 \text{ MJ}}{1.0 \text{ kWh}} = 10.73 \text{ MJ}$$

∴ The energy lost (to heat) is 10.73 MJ.

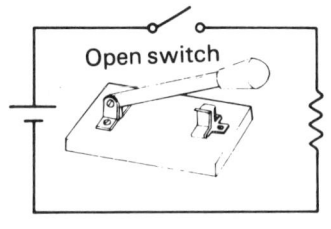 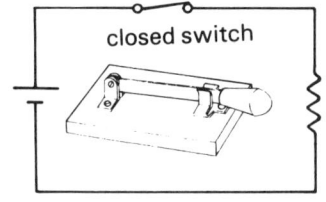

(a) (b)

FIGURE 6–10
A single-pole single-throw (SPST) *knife* switch shown in its:
(a) Open or OFF position, and
(b) Closed or ON position.

6–4 CIRCUIT CONTROL AND PROTECTIVE DEVICES

The control of electric current and the protection of electronic circuits and devices from excessive current is a very important consideration. Without provision for the safe control of electric current, equipment may be damaged and people injured. Switches, fuses, and circuit breakers have been developed to protect and control electrical equipment and circuits.

Switches

A *switch* is used in series with the load and source to switch the power ON and OFF to the load. When the switch in Figure 6–10 is in the closed position (also called the ON or *make* position), current flows in the load because the resistance of the switch contacts is only a few milliohms. However, when the switch is in the open position (also called the OFF or *break* position), the current to the load is interrupted by the high resistance between the switch contacts. Because the entire source voltage is across the open switch contacts, the insulating material and the *air gap* between the contacts must have a breakdown voltage rating very much greater than the voltage of the source.

When selecting a switch for ON/OFF application, both the load current and the source voltage must be considered. Because contact wear is accelerated by large currents, only switches specifically designed for ON/OFF application should be used for power control. The *toggle* switch of Figure 6–11(b) has been specifically designed with very low contact resistance for ON/OFF application in electronic circuits.

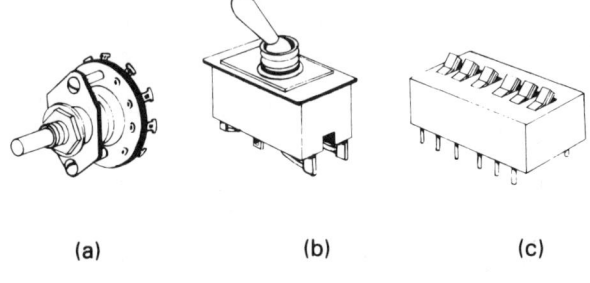

(a) (b) (c)

FIGURE 6–11
Assortment of switches.
(a) Rotary switch.
(b) Toggle switch.
(c) DIP switch.

The life of a switch is shortened when higher than rated currents are switched. Even under normal conditions, the contacts of a switch will gradually deteriorate (due to oxidation, pitting, and weld formation) until the contact resistance rises to an excessive amount (20 to 100 mΩ), signaling the end-of-life for the switch.

Thus, switch *end-of-life* is not a mechanical failure, but rather results when the contact resistance becomes excessive. Example 6–15 explores this concept.

EXAMPLE 6–15 Compare the power dissipated at the contacts of a new switch with a contact resistance of 2 mΩ and an old switch with a contact resistance of 100 mΩ when 15 A of current is switched.

SOLUTION Apply the power equation:

$$P = I^2 R$$

Substitute for 2-mΩ contact resistance:

$$P = 15^2(2 \times 10^{-3}) = 0.45 \text{ W}$$

∴ The new switch dissipates 0.45 W at the contacts.

Apply the power equation and substitute for 100-mΩ contact resistance:

$$P = 15^2(100 \times 10^{-3}) = 22.5 \text{ W}$$

∴ The *worn-out* switch dissipates 22.5 W at the contacts.

Observation In high-current circuits, the contact resistance must be very small; otherwise, excessive heating at the switch contacts will result in early failure of the switch.

It is good practice to operate a switch below its maximum current rating. This practice will ensure long life for the switch.

The voltage and current rating for a given switch may differ for ac and dc operation. The dc switch ratings are usually lower for a given switch as greater contact arcing and erosion result from switching dc.

EXAMPLE 6–16 List the current and volatage rating for a toggle switch.

SOLUTION Rating:

6 amperes   125 V ac

6 amperes   24 V dc

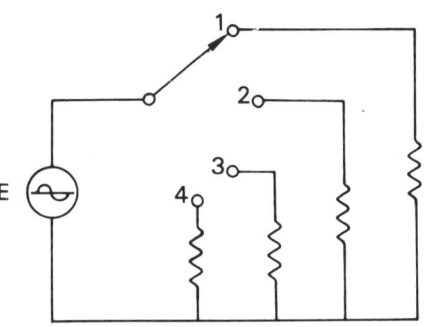

FIGURE 6–12
A one-pole, four-position rotary switch is used to direct signal currents from source E to various loads.

Observation   The power switched in ac operation is 6 A times 125 V or 750 W, while the power switched in dc operation is 6 A times 24 V or 144 W.

In addition to switching load current, switches are used to direct signals and set up operating conditions. The rotary switch pictured in Figure 6–11(a) may be used to direct small-signal currents from one circuit to another by changing the connections when the switch is turned. Figure 6–12 shows the schematic diagram of a one-pole, four-position rotary switch.

The DIP switch (dual in-line package switch) pictured in Figure 6–11(c) is a miniature switch that is designed for installation on printed circuit boards where it is used for individual circuit switching or for encoding a binary code in computer circuitry.

Switch contacts are arranged in well-defined patterns, which are referred to as the number of *poles* (pairs of mating contacts operated by the switch handle) and the number of *throws* (the number of positions in which the switch provides circuit closure). Table 6–4 lists several contact arrangements, and Figure 6–13 shows two single-pole double-throw switches (SPDT) used to control the operation of the lamp from two locations.

Fuses   Fuses are used in electrical circuits to protect the circuitry from excessive current that would, unless interrupted by the fuse, destroy the components of the circuit. Excessive current passing through the metallic fuse element heats the fine fuse wire until it melts, thus opening the circuit and halting the flow of current.

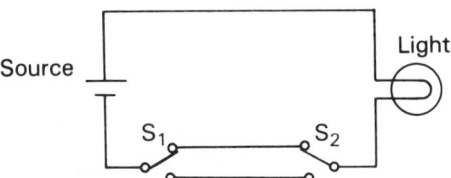

FIGURE 6–13
Switch S_1 and switch S_2 are both single-pole, double-throw (SPDT) switches. Operation of the light may be controlled from either of the two switches.

TABLE 6-4 CONTACT ARRANGEMENTS FOR SWITCHES

Contact Arrangement	Abbreviation	Schematic Symbol
Single pole Single throw	SPST	
Single pole Double throw	SPDT	
Double pole Single throw	DPST	
Double pole Double throw	DPDT	

Fuses are rated by current capacity, voltage, and fusing characteristics. Selecting a fuse for a particular application requires an understanding of each of these ratings. Figure 6-14 pictures a fusing characteristic curve for a glass-cartridge type of fuse, a popular type of fuse used in electronic equipment.

☐ The **current rating** of a fuse varies from a few milliamperes to many amperes of current. When a fuse *blows*, it should be replaced with another fuse with the same current, voltage, and

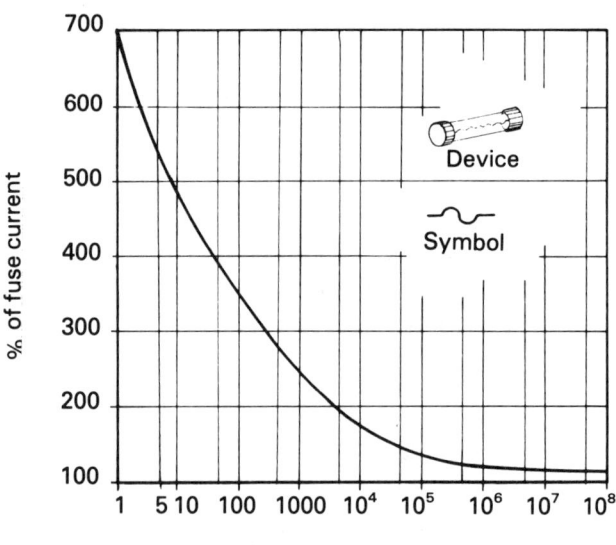

FIGURE 6-14
Fusing characteristic curve for medium action type AGC fuse.

fusing characteristic, but only after the reason for the failure of the fuse is determined.

☐ The **voltage rating** of a fuse becomes significant once the fuse opens and the entire source voltage is across the open fuse. If the voltage across the fuse exceeds the voltage rating of the fuse, it is then possible for an arc of sufficient length to form between the open fuse ends, resulting in a continuation of the flow of current in the load.

☐ The **fusing characteristic** of a fuse depends upon the material, design, construction, and operating temperature of the fuse.

Each of these factors combines to govern the time it takes for the fuse wire to open. As noted in Table 6–5, the fusing time versus current (fusing characteristic) may be divided into three categories: *fast action*, ranging from 1 to 500 ms; *medium action*, ranging from 10 ms to 5 s; and *delayed action*, ranging from 3 to 25 s.

EXAMPLE 6–17 Select a glass-cartridge fuse for the following circuit conditions. (a) High initial current of approximately five times the operating current of 1 A; the source voltage is 117 V and the operating temperature is 70°C. (b) Constant load current of 100 mA from a voltage source of 12 V operating at 25°C.

SOLUTION (a) Because of the high initial current, a delayed-action fuse (also called a slo-blo fuse) of the MDX or MDL type is selected. As a general rule, the fuse current rating is selected as 25% larger than the load current or, in this case, $1\frac{1}{4}$ A (1 A + $\frac{1}{4}$ A). However, due to the high operating temperature, an additional 20% of the nominal load current will be added to the fuse current rating. Thus:

$$I_{\text{fuse}} = 1.25 I_L + 0.2 I_L = I_L(1.45)$$

TABLE 6–5 FUSING CHARACTERISTICS OF GLASS-CARTRIDGE FUSES AT 25°C

		Fuse Time to Open			
Type	*Time Range*	500% I_L	200% I_L	135% I_L	100% I_L
MDL	Delayed	3 s	25 s	≈1 h	∞
MDX	Delayed	2 s	12 s	≈1 h	∞
AGC	Medium	10 ms	5 s	≈1 h	∞
AGX	Fast	1 ms	500 ms	≈1 h	∞

Substituting $I_L = 1$ A:

$$I_{\text{fuse}} = 1.45(1) = 1.5 \text{ A}$$

∴ The selected fuse has the following rating:

Type MDX or MDL, 1.5 A, 125 V

(b) Because of the constant load current, a medium-action fuse of the AGC type is selected. Since the operating temperature is 25°C, no correction for temperature is necessary. Thus:

$$I_{\text{fuse}} = 1.25(I_L)$$

Substituting $I_L = 0.1$ A:

$$I_{\text{fuse}} = 1.25(0.1) = 125 \text{ mA} = \tfrac{1}{8} \text{ A}$$

∴ The selected fuse has the following rating:

Type AGC, $\tfrac{1}{8}$ A, 250 V

Observation A lower voltage rating may be used (e.g., 32 V); however, $\tfrac{1}{8}$-A type AGC fuses are only available with 250-V ratings.

Circuit Breakers

A circuit breaker, unlike a fuse, needs only to be *reset*, not replaced, when it is tripped by an overload. Figure 6–15 pictures an inexpensive manual reset circuit breaker for use in television receivers, vending machines, and audio amplifiers. This type of circuit breaker, like most, has a *trip-free* mechanism that allows the contacts to open under overload, even if the reset button is held closed. This circuit breaker is time delayed with a trip time of 10 s at 200% of the hold current rating. Like a fuse, the circuit breaker will carry the rated current indefinitely, and it will trip within a few seconds with a large overload. The time delay, however, is sufficient to allow for large initial currents in motors and other reactive loads without tripping the circuit breaker. Some circuit breakers incorporate both a toggle switch for power ON/OFF and a circuit breaker for overload protection all in one package.

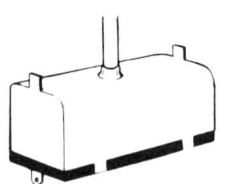

FIGURE 6–15
Circuit breaker used in commercial electronic equipment.

SELECTED TECHNICAL TERMS

Several of the technical terms used in this chapter are defined here to aid your understanding of them.

Average power: the power dissipated by the load. Also called real power or true power; in resistive loads it is equal to the product of current squared (I^2) and the resistance of the load. Unit of measure is the watt (W).

Derating: of the maximum performance rating of a device or electric system to ensure long-term operation. The reduction in performance rating introduces a *margin of safety* in the design.

Efficiency: the ratio of energy output to energy input; power output divided by power input.

PROBLEMS

Section 6–1

6–1 Determine the power rating of an electric iron if 12 A of current passes through the iron when it is connected to a 117-V ac source.

6–2 Determine the power rating of a fluorescent light if 855 mA of current passes through the light when it is connected to a 117-V ac source.

6–3 Determine the rate (in watts) at which an electric coffee maker converts electric energy to heat energy if 50 kJ of heat is produced in 1 min.

6–4 Determine the rate (in joules/second) at which a 120-V, 40-W lamp converts electric energy to heat and light energy.

6–5 Determine the power delivered to a 15-A air conditioner if it has a resistance of 15 Ω.

6–6 Determine the power delivered to a 120-V electric clock if it has a resistance of 620 Ω.

Section 6–2

6–7 Using the derating curve for the cermet-film resistor of Figure 6–4, approximate the power rating of the resistor operating at an ambient temperature of 110°C.

6–8 Using the derating curve for the 2-W carbon-composition resistor of Figure 6–5, approximate the power rating of the resistor operating at an ambient temperature of 90°C.

6–9 Specify the resistance, power rating, voltage rating, tolerance, and number code of a MIL-spec metal-film resistor operating at 80°C. The resistor drops 360 V to limit current in a circuit to 500 μA. Select the resistance from the E96 preferred resistor values and the wattage and voltage rating from Table 6–2. Use the curves of Figure 6–5 and a power derating factor of 2 in selecting the power rating.

6–10 Specify the resistance, power rating, voltage rating, tolerance, and color code of a carbon-film resistor operating at 75°C. The resistor drops 260 V to limit current in a circuit to 0.8 mA. Select the resistance from the E12 preferred resistor values and the wattage and voltage rating from

Table 6–2. Use the curves of Figure 6–5 and a power rating factor of 2 in selecting the power rating.

6–11 Using linear graph paper, construct the power derating curve for a 5-W power wire-wound resistor with the following specifications: power rating (nominal) at 125°C is 5 W, ambient operating temperature is −55°C to 275°C. From the derating curve, determine the power rating at 175°C.

6–12 Repeat Problem 6–11 for a 20-W power wire-wound resistor with the following specifications: power rating (nominal) at 150°C is 20 W, ambient operating temperature is −55°C to 325°C. From the derating curve, determine the power rating at 200°C.

Section 6–3

6–13 The output from a 230-V electric motor is 2 hp. (a) Determine the electric power into the motor from the line when 7.6 A flows into the motor. (b) Determine the motor's percent of efficiency. (c) Determine the cost of operation for 40 h at 9.8¢/kWh.

6–14 (a) Determine the percent of efficiency of a 117-V electric motor that develops 0.25 hp when 2.1 A passes through the motor. (b) Determine the cost of operating the motor for 8 h at 10.5¢/kWh.

6–15 A 20-A electric water heater is connected to a 220-V, 60-Hz ac line. The water inside the heater obtains 100 kJ of energy each minute the heater is in operation. (a) Determine the electric power into the heater from the line. (b) Determine the percent of efficiency of the water heater. (c) Determine the cost of operating the water heater for 30 days at 11.2¢/kWh if it operates 4 h/day.

6–16 A 220-V electric range is used to heat a kettle containing 1.5 liters of water to a boil in 9 min. The resistance of the heating element is 40 Ω. (a) Determine the percent of efficiency of the heating element if the water has obtained 475 kJ of heat energy. (b) Determine the cost of heating the water if electric energy costs 8.7¢/kWh.

6–17 Two energy-converting devices with efficiencies of 78% and 83% are cascaded. (a) Determine the power out of the system when 2.4 kW is put into the system. (b) Determine the overall percent of efficiency of the cascaded system.

6–18 A 10-hp electric motor, 230 V, 60 Hz ac, having an efficiency of 0.84 is coupled directly to a 440-V, 400-Hz alternator having an efficiency of 0.79. Determine (a) the power into the motor, (b) the overall efficiency of the system, and (c) the power out of the system.

Section 6-4

6-19 Determine the power dissipated at the contacts of a toggle switch with a contact resistance of 12 mΩ when 10 A of current is switched.

6-20 Determine the power dissipated at the contacts of a slide switch with a contact resistance of 30 mΩ when 3 A of current is switched.

6-21 Using the switch contact arrangements of Table 6-4, develop the schematic diagram of a circuit made up of a load (use a resistance symbol) connected to a switch connected either to a battery or an ac voltage source.

6-22 Modify the circuit of Figure 6-13 by placing a DPDT switch between S_1 and S_2. Wire this additional switch (S_3) so that the light may be controlled from any one of the three switches. Draw the schematic of the circuit.

6-23 Using the fusing characteristic curve of Figure 6-14, determine the time for a 2-A AGC fuse to open if an overload current of 6 A is in the circuit.

6-24 From the fusing characteristic curve of Figure 6-14, determine the time for an AGC fuse to open when it is carrying 150% of its rated current.

6-25 Specify the type and the nominal current rating of a glass-cartridge fuse for a circuit with 800 mA of current operating at an ambient temperature of 25°C.

7 The Series Circuit

7-1 DEFINITION OF A SERIES CIRCUIT
7-2 APPLICATION OF THE SERIES CIRCUIT RULES
7-3 CURRENT DIRECTION AND THE POLARIZATION OF CIRCUIT COMPONENTS
7-4 KIRCHHOFF'S VOLTAGE LAW
7-5 VOLTAGE SOURCES IN SERIES
7-6 FORMING AN EQUIVALENT SERIES CIRCUIT
7-7 NOTATION AND REFERENCES
7-8 VOLTAGE SOURCE CHARACTERISTICS

CHAPTER PREVIEW This chapter is the first of three chapters that, when completed, will allow you to apply certain rules, laws, and techniques to the solution of series circuits, parallel circuits, and series parallel networks. The behavior of a series circuit is described by the mathematical statement $E = IR$ (Ohm's law). This law, along with other laws, rules, and techniques, will be considered in this chapter.

PERFORMANCE OBJECTIVES Once you have read each section, worked each example with pencil, paper, and calculator, and answered each problem for every section, you will be able to:

☐ Determine the equivalent resistance of a series circuit.
☐ Apply Kirchhoff's voltage law to the solution of series circuits.
☐ Apply voltage notation to a series circuit.
☐ Determine the equivalent voltage of series aiding and opposing voltage sources.
☐ Differentiate between conventional current and electron flow in a series circuit.
☐ Polarize the source and the resistive components in a series circuit.
☐ Understand the characteristics of an ideal and a practical voltage source.

7–1 DEFINITION OF A SERIES CIRCUIT

A resistive series circuit is formed when each component in the circuit (voltage source and resistive loads) is connected in succession so that one end of a component is connected to an end of the next component, and so forth, until a single complete path is formed from the beginning to the end. A complete series circuit is shown in Figure 7–1.

Series Circuit Rules and Laws The following statements are the rules that form the basis for the analysis of series circuits. These concepts both define and describe the characteristics of the current, resistance, and voltage of the series circuit. Thus, in a series circuit:

☐ The circuit current, I, is the same throughout the entire series circuit at the same time. Stated mathematically:

$$I = I_1 = I_2 = I_3 = \cdots = I_n \quad \text{(A)} \qquad (7\text{–}1)$$

☐ The total resistance, R_T, of a series circuit is found by summing (adding) each of the resistances in the circuit. Stated mathematically:

$$R_T = R_1 + R_2 + R_3 + \cdots + R_n \quad (\Omega) \qquad (7\text{–}2)$$

Definition of a Series Circuit

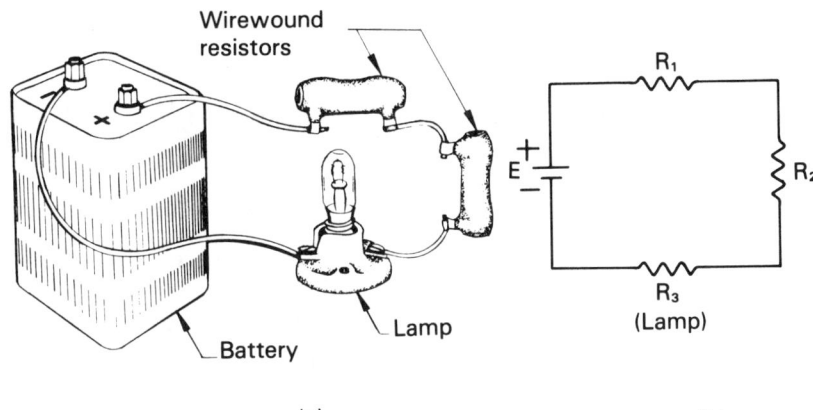

FIGURE 7-1 Series circuit made up of a battery (voltage source), two wirewound resistors, and a lamp (resistive loads).
(a) Pictorial diagram of connection.
(b) Schematic diagram with resistive loads shown and labeled with subscripted variables.

☐ The voltage rise in a series circuit must equal the sum of the individual voltage drops. Stated mathematically, where E represents a voltage rise (source) and V represents a voltage drop:

$$E = V_1 + V_2 + V_3 + \cdots + V_n \quad \text{(V)} \tag{7-3}$$

The following examples will demonstrate the use of the three fundamental rules for a series circuit.

EXAMPLE 7-1 Determine the current in R_1 and R_2 of Figure 7-2 when the ammeter indicates 2 A.

SOLUTION As noted in Equation 7-1, the current is the same throughout the series circuit; the current in R_1 and R_2 is 2 A.

$$\therefore \quad I = I_1 = I_2 = 2 \text{ A}$$

EXAMPLE 7-2 Determine the value of R_2 of Figure 7-2 when $R_1 = 10 \, \Omega$ and $R_T = 30 \, \Omega$.

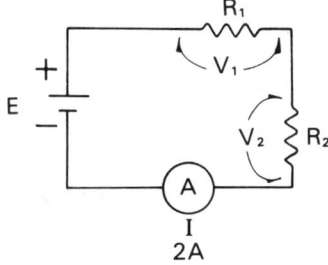

FIGURE 7-2
Circuit of Example 7-1.

SOLUTION Use Equation 7–2:

$$R_T = R_1 + R_2$$

Substitute and solve for R_2:

$$30 = 10 + R_2$$
$$R_2 = 30 - 10 = 20 \ \Omega$$

$$\therefore \quad R_2 = 20 \ \Omega$$

EXAMPLE 7–3 Determine the source voltage, E, of Figure 7–2 when $V_1 = 5$ V and $V_2 = 10$ V.

SOLUTION Use Equation 7–3:

$$E = V_1 + V_2$$

Observation The ammeter in this circuit is assumed to drop no voltage. Substitute:

$$E = 5 + 10 = 15 \ \text{V}$$

$\therefore \quad$ The source voltage is 15 V.

7–2 APPLICATION OF THE SERIES CIRCUIT RULES

Ohm's law will be used in conjunction with the three series circuit rules to analyze series circuits as demonstrated in the following set of examples.

EXAMPLE 7–4 Determine the current I_2 in resistance R_2 of the series circuit shown in Figure 7–3.

SOLUTION Since $I = I_1 = I_2 = I_3$, then $I = I_2$. From Ohm's law ($E = IR$), solve for I:

$$I = \frac{E}{R}$$

where $E = 45$ V and $R = R_T = R_1 + R_2 + R_3$. Solve for R:

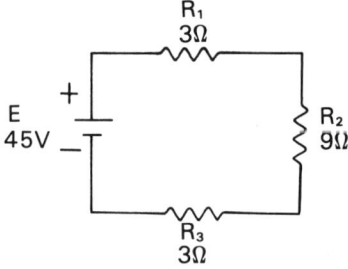

FIGURE 7–3
Circuit of Example 7–4.

Application of the Series Circuit Rules

$$R = R_T = 3 + 9 + 3 = 15 \, \Omega$$

Substitute into Ohm's law:

$$I = \frac{E}{R} = \frac{45}{15} = 3 \text{ A}$$

∴ The series circuit current is 3 A and $I_2 = 3$ A.

EXAMPLE 7-5 Using the circuit of Figure 7-3 and the values of the previous example, determine the voltage across R_1, R_2, and R_3.

SOLUTION Use Ohm's law to solve for V_1. Substitute $I = 3$ A and $R_1 = 3 \, \Omega$. Thus:

$$V_1 = IR_1 = (3)(3) = 9 \text{ V}$$

∴ $V_1 = 9$ V

Use Ohm's law to solve for V_2. Substitute $I = 3$ A and $R_2 = 9 \, \Omega$. Thus:

$$V_2 = IR_2 = (3)(9) = 27 \text{ V}$$

∴ $V_2 = 27$ V

Use Ohm's law to solve for V_3. Substitute $I = 3$ A and $R_3 = 3 \, \Omega$. Thus:

$$V_3 = IR_3 = (3)(3) = 9 \text{ V}$$

∴ $V_3 = 9$ V

Observation These calculations are easily checked since $E = V_1 + V_2 + V_3$. Thus:

$$45 \text{ V} = 9 \text{ V} + 27 \text{ V} + 9 \text{ V} \quad \text{and} \quad 45 \text{ V} = 45 \text{ V}$$

EXAMPLE 7-6 The lamp of Figure 7-4 is operated from the 9-V source. However, the lamp is a 3-V, 60-mA lamp, and it would quickly burn out if operated directly across the 9-V source. Determine the value of the series resistor, R_1, needed to drop

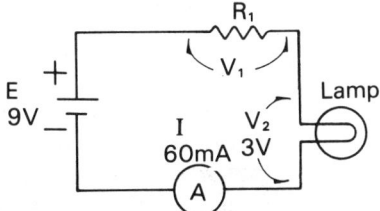

FIGURE 7-4
Circuit of Example 7-6.

the excess voltage and limit the current to 60 mA.

SOLUTION From Ohm's law ($V = IR$), $R_1 = V_1/I$. To solve for R_1, the values of V_1 and I are needed. Solve for V_1 using Equation 7-3 ($E = V_1 + V_2$). Thus:

$$V_1 = E - V_2$$

Substitute:

$$V_1 = 9 - 3 = 6 \text{ V}$$

Solve for R_1. I is given as 60 mA:

$$R_1 = \frac{V_1}{I} = \frac{6}{60 \times 10^{-3}} = 100 \text{ } \Omega$$

∴ A 100-Ω resistor is needed in series with the lamp to limit the circuit current to 60 mA.

The rules and laws for the analysis of the series circuit are summarized in Table 7-1.

7-3 CURRENT DIRECTION AND THE POLARIZATION OF CIRCUIT COMPONENTS

In the study of circuits, it is *conventional* to think of the circuit current (I) as flowing out of the positive terminal (+) of the voltage source and returning to the negative terminal (−). The rule for conventional current flow will be used for analyzing circuits in this text. Thus:

Conventional Direction for Current

> RULE 7-1. CONVENTIONAL CURRENT
>
> ☐ Conventional current in a resistance enters from the positive side and leaves from the negative side (+ to −).
> ☐ Conventional current in a source enters from the negative terminal and leaves from the positive terminal (− to +).

Polarizing the Circuit Components

Following the conventional current direction through the series circuit of Figure 7-5, you will notice that there is a **rise** in voltage across the source (i.e., − to +) and a **drop** in voltage across the resistance (i.e., + to −).

The polarity of a voltage source is usually established by the manufacturer of the source who clearly marks the source with positive (+) and negative (−) polarities. When this is the case, the direction of the circuit current is established by the manufac-

TABLE 7-1 SERIES CIRCUIT RULES AND LAWS

Current Direction and Polarization of Circuit Components

Equation	Comment
$I = I_1 = I_2 = I_3$	Current (I) is the same throughout the series circuit
$R_T = R_1 + R_2 + R_3$	Total resistance (R_T) of a series circuit is the sum of the individual resistances
$E = V_1 + V_2 + V_3$	Source voltage (E) is equal to the sum of the voltage drops (V)
$E = IR$	Ohm's law

turer's polarization of the source in conjunction with the rule of conventional current, as demonstrated in Example 7-7.

EXAMPLE 7-7 Polarize the resistances of Figure 7-6(a) using conventional current flow.

SOLUTION Applying the rule for conventional current results in the current arrow direction and polarities noted in Figure 7-6(b).

Observation Because the polarity of the voltage source was given in Figure 7-6(a), the conventional current direction was drawn so that the arrow enters the negative terminal and leaves the positive terminal (− to +). Once the current direction for the circuit is established, the polarization of the resistance follows Rule 7-1; conventional current enters a resistance from the positive side and leaves from the negative side (+ to −).

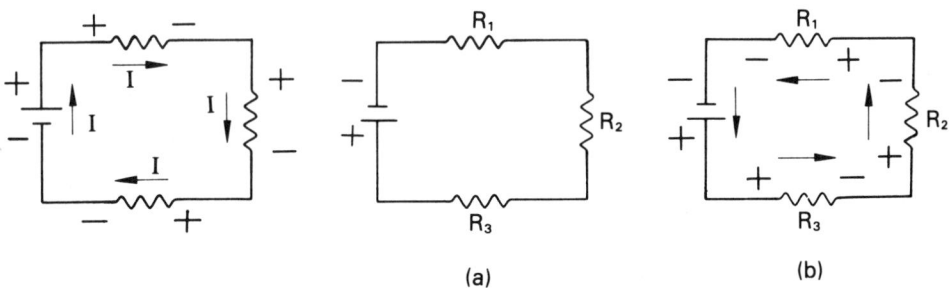

FIGURE 7-5 (left) When conventional current passes through a resistance, the resistance is polarized so the point of current entry is positive and the point of current exit is negative.
FIGURE 7-6 (right) Circuit of Example 7-7.

The Series Circuit

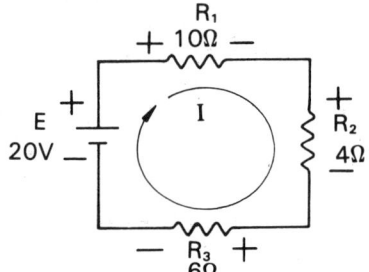

FIGURE 7-7
Circuit of Example 7-8.

7-4 KIRCHHOFF'S VOLTAGE LAW

As you now know, in a series circuit the voltage source equals the sum of the voltage drops. That is:

$$E = V_1 + V_2 + V_3 \tag{7-3}$$

Kirchhoff's Voltage Law (KVL)

Gustav Kirchhoff is credited with the discovery of this important circuit characteristic. Kirchhoff's voltage law (KVL) states that in any complete circuit the algebraic sum of the voltages (both rises and drops) must equal zero. Stated as an equation, where E represents a voltage rise (source) and V represents a voltage drop, Kirchhoff's voltage law becomes

$$E + V_1 + V_2 + V_3 = 0 \tag{7-4}$$

Applying Kirchhoff's Voltage Law to the Series Circuit

The following examples will use the rules for series circuits and conventional current as well as Ohm's and Kirchhoff's laws.

EXAMPLE 7-8 Apply Kirchhoff's voltage law to the circuit of Figure 7-7 to demonstrate that $E + V_1 + V_2 + V_3 = 0$.

SOLUTION Using Ohm's law, determine the circuit current I:

$$I = \frac{E}{R}$$

Substitute into Ohm's law, where

$$E = 20 \text{ V} \quad \text{and} \quad R = R_T = R_1 + R_2 + R_3$$
$$R = 10 + 4 + 6 = 20 \text{ } \Omega$$

and

$$I = \frac{E}{R} = \frac{20}{20} = 1 \text{ A}$$

Using Ohm's law, determine the voltage drop across each of the three resistances:

$$V_1 = IR_1 = (1)(10) = 10 \text{ V}$$

$V_2 = IR_2 = (1)(4) = 4$ V

$V_3 = IR_3 = (1)(6) = 6$ V

Apply Kirchhoff's voltage law by assigning a plus sign (+) to a voltage rise and a minus sign (−) to a voltage drop. Starting with R_1 of Figure 7-7 and following the conventional current around the circuit, we first encounter a voltage drop (+ to −) across R_1. V_1 is assigned a minus sign (−10 V). Continuing on, we next have a voltage drop (+ to −) across R_2. V_2 is assigned a minus sign (−4 V). Moving on, we find still another voltage drop (+ to −) across R_3. V_3 is assigned a minus sign (−6 V). Following the current through the voltage source, we experience a voltage rise (− to +). E is assigned a plus sign (+20 V). Thus, from Equation 7-4,

$$E + V_1 + V_2 + V_3 = 0$$

Substituting:

$$+20 \text{ V} + (-10 \text{ V}) + (-4 \text{ V}) + (-6 \text{ V}) = 0$$

$$0 = 0$$

∴ The algebraic sum of the voltages around the series circuit is zero.

Since $V = IR$, Equation 7-3 may be written in terms of I and R as Equation 7-5. Thus:

$$E = V_1 + V_2 + V_3 \quad \text{(V)} \quad (7-3)$$

and

$$E = IR_1 + IR_2 + IR_3 \quad \text{(V)} \quad (7-5)$$

EXAMPLE 7-9 Determine the current in a series circuit consisting of three carbon-film resistors of 100, 300, and 200 Ω connected to a voltage source of 30 V.

SOLUTION Use Equation 7-5:

$$E = IR_1 + IR_2 + IR_3$$

Substitute:

$$30 = 100I + 300I + 200I$$

Factor and solve for I:

$$30 = I(100 + 300 + 200)$$

$$30 = I(600)$$

$$I = \frac{30}{600} = 50 \text{ mA}$$

∴ The current in the series circuit is 50 mA.

EXAMPLE 7–10 Determine the value of one resistor (R_1) in a series circuit consisting of three resistances, R_1, $R_2 = 27 \ \Omega$, and $R_3 = 47 \ \Omega$ connected to a voltage source of 100 V. The current in the series circuit is 640 mA.

SOLUTION Use Equation 7–5 and substitute:

$$E = IR_1 + IR_2 + IR_3$$

$$100 = 0.64R_1 + 0.64(27) + 0.64(47)$$

$$100 = 0.64R_1 + 47.4$$

Solve for R_1:

$$R_1 = \frac{100 - 47.4}{0.64} = 82 \ \Omega$$

∴ The unknown resistor, R_1, is 82 Ω.

TABLE 7–2 VARIOUS FORMS OF KIRCHHOFF'S VOLTAGE LAW

Equation	Comment
$E = V_1 + V_2 + V_3$	Used to find the voltage drop across one resistor in a series circuit. Also useful in finding the source voltage (E) when all the voltage drops are known.
$E + V_1 + V_2 + V_3 = 0$	An alternative form of the previous equation where voltage rises (E) are assigned a positive (+) sign and voltage drops are assigned a negative (−) sign in relation to the conventional direction of current flow.
$E = IR_1 + IR_2 + IR_3$	Used to find circuit current (I) in the series circuit. Also used to find one resistance when I, E, and the other resistances are known.

Summary From the preceding examples, it has been seen that Kirchhoff's voltage law may be stated in several ways depending on the need or the application. Table 7–2 summarizes the various forms of KVL.

7–5 VOLTAGE SOURCES IN SERIES

Kirchhoff's voltage law may be applied to a complete circuit (this includes the series circuit) with more than one voltage source. Figure 7–8 pictures a series circuit with three voltage sources. In a series circuit with several sources, the sources may be connected *series aiding* or *series opposing*.

Series-Aiding and Series-Opposing Voltage Sources

Figure 7–9(a) pictures two voltage sources connected in *series aiding*. Notice in Figure 7–9(a) that the negative terminal (−) of one source is connected to the positive terminal (+) of the other source. When this is the case, the voltage sources are connected *series aiding*. The resultant voltage across terminals $(a - b)$ is equal to the summation of the two voltage sources. Thus:

$$E_{ab} = E_1 + E_2$$

EXAMPLE 7–11 Determine the voltage, E_{ab}, across terminals $(a - b)$ of Figure 7–9(a) when the voltage source $E_1 = 12$ V and voltage source $E_2 = 18$ V.

SOLUTION Since the voltage sources are connected series aiding,

$$E_{ab} = E_1 + E_2$$

Substitute:

$$E_{ab} = 12 + 18 = 30 \text{ V}$$

∴ E_{ab} is 30 V.

Figure 7–9(b) and (c) pictures two voltage sources connected in *series opposing*. Notice in Figure 7–9(b) and (c) that the two

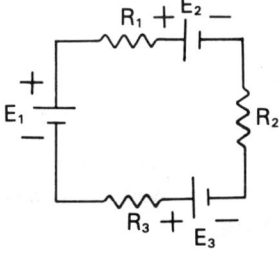

FIGURE 7–8
Series circuit with both series aiding and series opposing voltage sources.

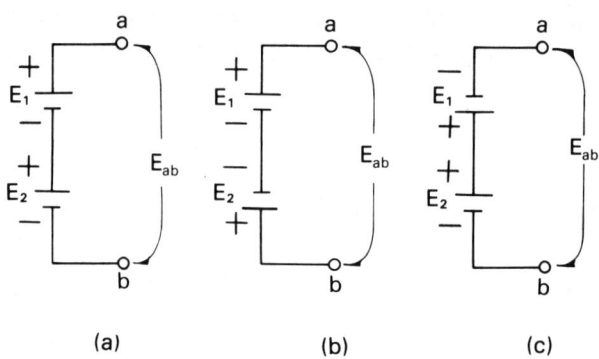

FIGURE 7-9
(a) Two voltage sources are connected series aiding when the negative terminal of one source is connected to the positive terminal of the other source.
(b) Two voltage sources are connected series opposing when the negative terminal of one source is connected to the negative terminal of the other source, or
(c) When the positive terminal of one source is connected to the positive terminal of the other source.

terminals connected together are the same polarity. The resultant voltage across terminals *a-b* is equal to the difference between the two sources. Thus:

$$E_{ab} = E_1 - E_2$$

When voltage sources are connected in series, the following general rule may be applied.

RULE 7-2. POLARITY OF THE RESULTANT VOLTAGE

When two voltage sources are connected *series aiding*, as in Figure 7-9(a), the polarity of the resultant voltage E_{ab} is determined by the polarity of either voltage source.
When two voltage sources are connected *series opposing*, as in Figure 7-9(b) and (c), the polarity of the *resultant* voltage E_{ab} is determined by the larger of the two voltage sources.

EXAMPLE 7-12 Determine the voltage, E_{ab}, across terminals *a-b* of Figure 7-9(b) when voltage source $E_1 = 6$ V and voltage source $E_2 = 11$ V.

SOLUTION Since the voltage sources are connected series opposing,

$$E_{ab} = E_1 - E_2$$

Voltage Sources in Series

Substitute:

$$E_{ab} = 6 - 11 = -5 \text{ V}$$

$\therefore E_{ab}$ is -5 V.

EXAMPLE 7-13 In the circuit of Figure 7-10, determine the polarity and the amount of E_{ab}.

SOLUTION Start by adding the series-aiding sources, E_1 and E_2. Thus:

$$E_1 + E_2 = 10 + 5 = 15 \text{ V}$$

From Figure 7-10(b):

$$E_{ab} = E_{1+2} - E_3$$

Substitute:

$$E_{ab} = 15 - 18 = -3 \text{ V}$$

$\therefore$ $E_{ab} = -3$ V and terminal a is negative in relation to terminal b which is positive, as shown in Figure 7-10(c).

Observation Since E_3 is larger than E_{1+2}, the polarity of terminals a-b is determined by the polarity of source E_3. Thus, terminal b is positive and terminal a is negative.

KVL for Multiple Voltage Sources in a Series Circuit

Kirchhoff's voltage law for a complete circuit with more than one voltage source is as follows:

$$E_1 + E_2 + E_3 + V_1 + V_2 + V_3 = 0 \tag{7-6}$$

The following examples demonstrate the application of Equation 7-6.

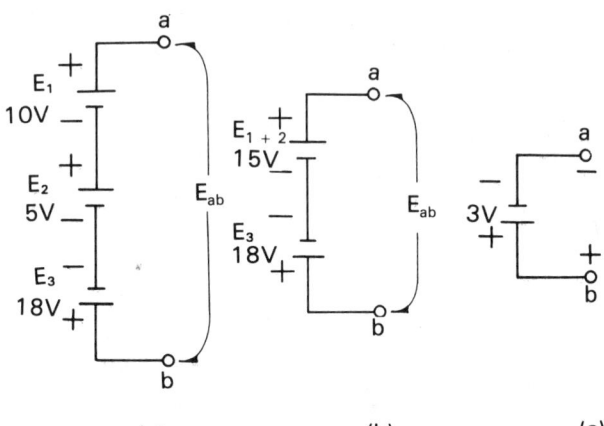

FIGURE 7-10
Circuit of Example 7-13.
(a) (b) (c)

EXAMPLE 7-14 Demonstrate that the algebraic sum of the voltage rises and the voltage drops equals zero in Figure 7-11.

SOLUTION Starting at R_1 and moving around the circuit in the direction of the current arrow, assign a plus sign (+) to a voltage rise (− to +) and a minus sign (−) to a voltage drop (+ to −). Thus:

$$-V_1 - E_2 - V_2 - V_3 + E_1 = 0$$

Observation Source E_2 is assigned a minus sign because the current passes from + to −, which is the definition of a voltage drop. Even though E_2 is a voltage source, it is treated as a voltage drop since it meets the definition of a voltage drop.
Substitute $V = IR$ in

$$-V_1 - V_2 - V_3 - E_2 + E_1 = 0$$

resulting in

$$-IR_1 - IR_2 - IR_3 - E_2 + E_1 = 0$$

Factor I:

$$-I(R_1 + R_2 + R_3) - E_2 + E_1 = 0$$

Substitute:

$$-1(10 + 8 + 2) - 10 + 30 = 0$$

$$-20 - 10 + 30 = 0 \text{ V}$$

∴ 0 V = 0 V and the algebraic sum of the voltage rises and the voltage drops equals zero.

EXAMPLE 7-15 Determine V_2, the voltage across R_2 of Figure 7-12(a), using Kirchhoff's voltage law.

SOLUTION Since E_1 and E_2 are series aiding, the direction of I is determined by the polarity of either source,

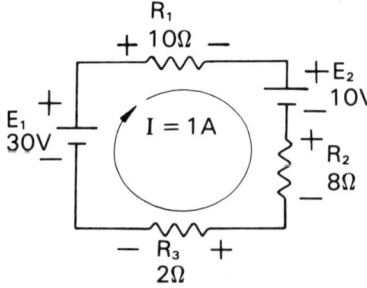

FIGURE 7-11
Circuit of Example 7-14. Direction of conventional current, I, is determined by E_1, the larger of the series opposing sources.

Voltage Sources in Series

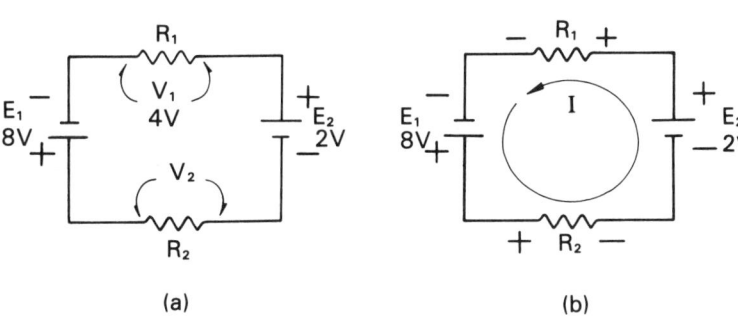

FIGURE 7-12 Circuits of Example 7-15.

as shown in Figure 7-12(b). Starting at E_1 in Figure 7-12(b) and following the current arrow, assign a plus (+) to a voltage rise (− to +) and a minus sign (−) to a voltage drop (+ to −). Thus:

$$+E_1 - V_2 + E_2 - V_1 = 0$$

Substitute:

$$8 - V_2 + 2 - 4 = 0$$

Solve for V_2:

$$8 + 2 - 4 = V_2$$

$$\therefore \quad V_2 = 6 \text{ V}$$

Conventional Current Direction In a series circuit with more than one voltage source, the circuit current direction is determined by applying the following rule.

RULE 7-3. CURRENT DIRECTION IN A SERIES CIRCUIT

If two voltage sources are connected *series aiding*, as shown in Figure 7-9(a), the conventional current direction is determined by the polarity of either voltage source.
If two voltage sources are connected *series opposing*, as shown in Figure 7-9(b), the conventional current direction is determined by the polarity of the larger source voltage.

The following examples will demonstrate how the direction of conventional current is selected in the series circuit with multiple voltage sources.

EXAMPLE 7-16 Determine the amount and the direction of the current in the series circuit of Figure 7-13(a).

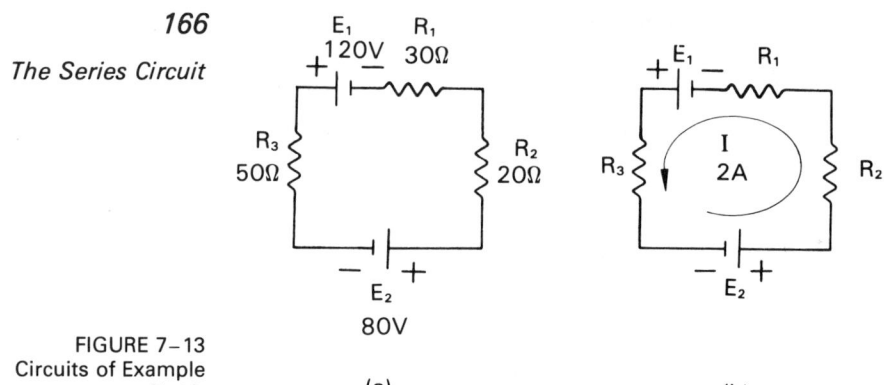

FIGURE 7–13
Circuits of Example 7–16.

SOLUTION Since E_1 and E_2 are connected series aiding, the direction of current is determined by the polarity of either source. Conventional current will move from − to + through the voltage sources as shown in Figure 7–13(b).

Determine the current in the series circuit from Ohm's law:

$$I = \frac{E}{R}$$

where $E = E_1 + E_2 = 120 + 80 = 200$ V and $R = R_T = R_1 + R_2 + R_3 = 30 + 20 + 50 = 100\Omega$:

$$I = \frac{200}{100} = 2.0 \text{ A}$$

∴ $I = 2$ A, as noted in Figure 7–13(b).

EXAMPLE 7–17 Determine the amount and direction of the current in the series circuit of Figure 7–14(a).

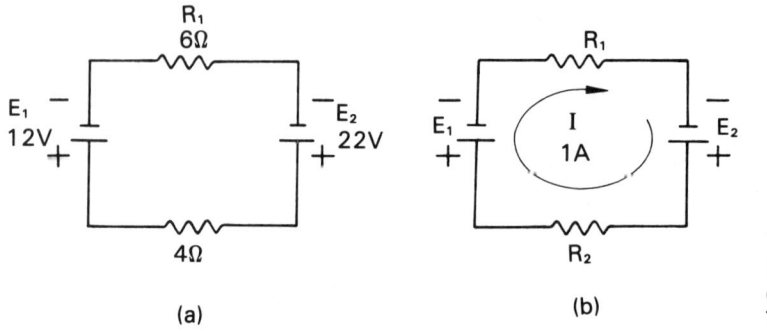

FIGURE 7–14
Circuits of Example 7–17.

Forming an Equivalent Series Circuit

SOLUTION Since E_1 and E_2 are connected series opposing and since E_2 is greater than E_1, the direction of conventional current is determined by the polarity of source E_2. Current will move from − to + through source E_2, as shown in Figure 7–14(b).

Determine the current in the series circuit from Ohm's law:

$$I = \frac{E}{R}$$

where $E = E_2 - E_1 = 22 - 12 = 10$ V and $R = R_T = R_1 + R_2 = 6 + 4 = 10$ Ω:

$$I = \frac{10}{10} = 1 \text{ A}$$

∴ $I = 1$ A, as noted in Figure 7–14(b).

7–6 FORMING AN EQUIVALENT SERIES CIRCUIT

A complete circuit, made up of several resistances connected in series with one or more voltage sources, may be simplified to an *equivalent series circuit* consisting of a single resistance and a single voltage source. The single resistance is the equivalent resistance of the series equivalent circuit. For purposes of identification, this resistance is labeled R_{eq}. The equivalent voltage source is labeled E_T and is formed by combining the voltage sources using the rules for series aiding and opposing voltage sources.

The equivalent series circuit has the same source voltage, current, total resistance, and power dissipation as the series circuit it was derived from.

EXAMPLE 7–18 Form the series equivalent circuit for the series circuit of Figure 7–15(a) and verify that the two circuits are equivalent by comparing the current in the circuit and the power dissipated by the loads.

SOLUTION Form R_{eq}:

$$R_{eq} = R_T = R_1 + R_2 = 50 + 100 = 150 \text{ Ω}$$

∴ $R_{eq} = 150$ Ω

Form E_T, the equivalent voltage:

$$E_T = E_1 + E_2 = 10 + 5 = 15 \text{ V}$$

∴ $E_T = 15$ V

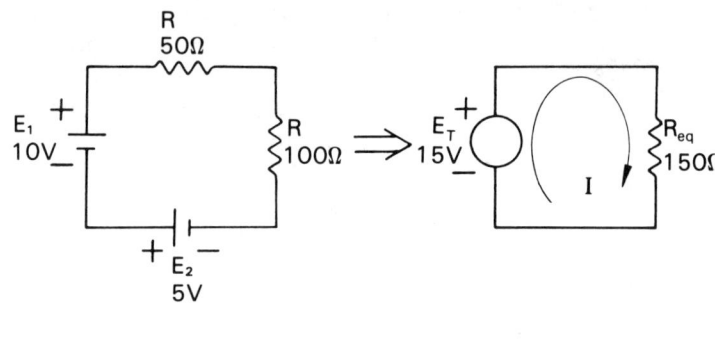

FIGURE 7-15
(a) Series circuit of Example 7-18.
(b) Series equivalent circuit.

Observation Figure 7-15(b) is the series equivalent circuit. Notice that the equivalent voltage source is indicated by a new symbol. From this point on in the text, this symbol will be used with equivalent circuits to indicate an equivalent voltage source. Compute I, the circuit current in each circuit. For Figure 7-15(a):

$$I = \frac{E_1 + E_2}{R_1 + R_2}$$

Substitute:

$$I = \frac{10 + 5}{50 + 100} = \frac{15}{150} = 0.1 \text{ A}$$

∴ $I = 100$ mA

For Figure 7-15(b):

$$I = \frac{E_T}{R_{eq}}$$

Substitute:

$$I = \frac{15}{150} = 0.1 \text{ A}$$

$I = 100$ mA

∴ The current in each circuit is the same.
Compute the power dissipated by R_1 and R_2 of Figure 7-15(a):

$$P = I^2(R_1 + R_2)$$

Substitute:

$$P = (0.1)^2(50 + 100) = (0.01)(150) = 1.5 \text{ W}$$

Forming an Equivalent Series Circuit

Compute the power dissipated by R_{eq} of Figure 7-15(b):

$$P = I^2 R_{eq}$$

Substitute:

$$P = (0.1)^2(150) = (0.01)(150) = 1.5\,\text{W}$$

∴ The power dissipated by the resistance in each of the circuits is the same.

From Example 7-18, you can see that the circuit current is easily determined from the series equivalent circuit. The concept of equivalent circuits is a very important concept in electronic circuit analysis. We will use this concept often in forthcoming chapters.

EXAMPLE 7-19 Form the series equivalent circuit for the series circuit of Figure 7-16(a) and then compute the circuit current.

SOLUTION Form R_{eq} of Figure 7-16(b):

$$R_{eq} = R_T = R_1 + R_2 + R_3$$

$$R_{eq} = 390 + 180 + 560 = 1130\,\Omega$$

∴ $R_{eq} = 1130\,\Omega$

Form E_T, the equivalent voltage, by first adding the series-aiding source and then subtracting the series-opposing source. Thus:

$$E_T = E_1 + E_3 - E_2 = 15 + 24 - 9$$

∴ $E_T = 30\,\text{V}$

Compute I, the circuit current:

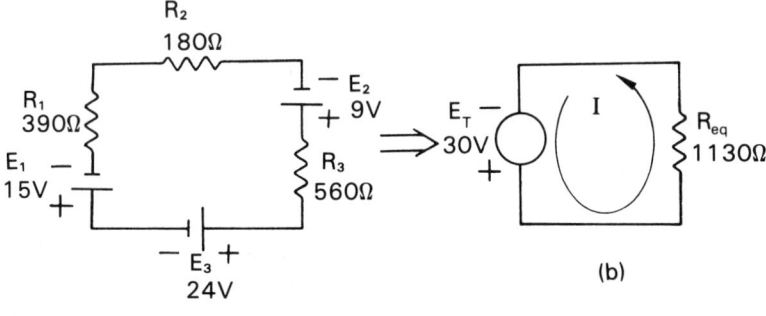

FIGURE 7-16
(a) Series circuit of Example 7-19.
(b) Series equivalent circuit.

$$I = \frac{E_T}{R_{eq}}$$

Substitute:

$$I = \frac{30}{1130} = 26.5 \text{ mA}$$

$$\therefore \quad I = 26.5 \text{ mA}$$

7-7 NOTATION AND REFERENCES

Voltage Notation

We are now ready to introduce a system of notation that will allow you to indicate both the magnitude (amount) and polarity (direction) of a voltage. The letters in Figure 7–17 are used to form a *symbolic* statement. The statement is written using a double-letter subscript, such as V_{ab}. The statement V_{ab} is read "the voltage from point a to point b." In this system of designating voltages, the second subscript letter is always the reference point.

The polarity of the designated voltage is assigned a positive sign (+) if the first point is positive in relation to the second point. Thus, V_{ab} of Figure 7–17 is +6 V since point a is positive in relation to point b, which is negative. However, V_{ba} is −6 V, because point b is negative in relation to point a, which is positive. Since the first point is negative in relation to the second point, the polarity of the designated voltage is assigned a negative sign (−).

EXAMPLE 7–20 Determine the following voltages in the circuit of Figure 7–17. (a) V_{ad}, (b) V_{da}, (c) V_{cd}, and (d) V_{cb}.

SOLUTION (a) The voltage V_{ad} is from point a to point d.

$$\therefore \quad V_{ad} = +12 \text{ V}$$

(b) The voltage V_{da} is from point d to point a.

$$\therefore \quad V_{da} = -12 \text{ V}$$

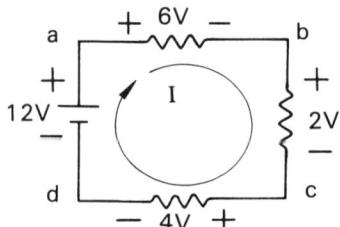

FIGURE 7–17
Letters are used to indicate voltage drops and voltage rises in the circuit.

(c) The voltage V_{cd} is from point c to point d.

$$\therefore \quad V_{cd} = +4 \text{ V}$$

(d) The voltage V_{cb} is from point c to point b.

$$\therefore \quad V_{cb} = -2 \text{ V}$$

Voltage Reference Symbols

When troubleshooting an electronic circuit, the voltages in the circuit are measured from various points in the circuit to a common reference point. This reference point may be the metal chassis or, as with printed wiring, a particular conductor may be used as the reference point. In many electronic devices, the reference point may be connected to an *earth ground* through the wiring of the line cord.

Several schematic symbols are used to indicate the reference point for voltage measurements. Figure 7–18 pictures three symbols used to indicate a *common connection point*. Unless otherwise noted on the schematic, the common connection point is the reference point. The first symbol of Figure 7–18(a) indicates a physical connection of the electric circuit to the metal chassis of the electronic device. The next symbol of Figure 7–18(b) indicates a physical connection of the electric circuit to the earth. This symbol indicates that the reference point is *grounded*. The last symbol of Figure 7–18(c) indicates only a common connection point, which is the reference point for voltage measurements. This symbol does not indicate a *ground* or *chassis* connection. Its only function is to indicate a common connection point in an electronic circuit, and it serves as a reference point for voltage measurements.

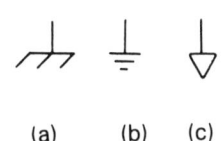

(a)   (b)   (c)

FIGURE 7–18
Common connection point symbols.
(a) Chassis connection.
(b) Ground connection.
(c) Common connection.

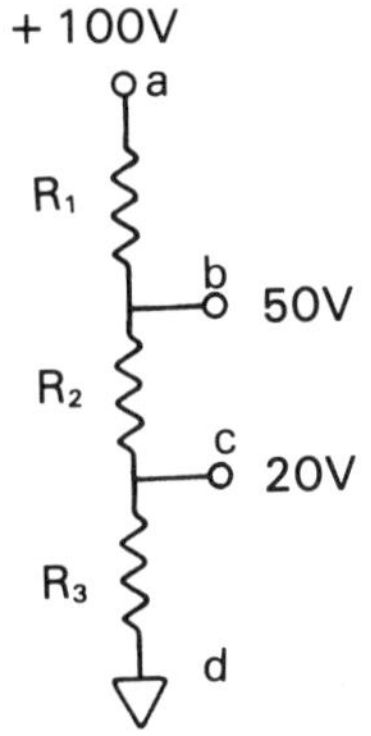

EXAMPLE 7–21 Determine the voltage across R_1, R_2, and R_3 of Figure 7–19.

SOLUTION Each of the voltages of Figure 7–19 is stated in relation to the reference point. Thus:

$$V_{ad} = +100 \text{ V}, \quad V_{bd} = +50 \text{ V}$$
$$V_{cd} = +20 \text{ V}$$

The voltage across R_1 is V_{ab}:

$$V_{ab} = V_1 = V_{ad} - V_{bd} = 100 - 50 = 50 \text{ V}$$

$\therefore \quad V_1$, the voltage across R_1, is 50 V.
The voltage across R_2 is V_{bc}:

$$V_{bc} = V_2 = V_{bd} - V_{cd} = 50 - 20 = 30 \text{ V}$$

FIGURE 7–19
Circuit of Example 7–21. Notice that point (a) is 100 V positive in relation to the common connection point (d).

∴ V_2, the voltage across R_2, is 30 V.
The voltage across R_3 is V_{cd}:

$$V_{cd} = V_3 = 20 \text{ V}$$

∴ V_3, the voltage across R_3, is 20 V.

Observation Check the calculations by adding the voltage drops. This sum should equal the source. Thus:

$$V_1 + V_2 + V_3 = 100$$

$$50 + 30 + 20 = 100 \text{ and } 100 \text{ V} = 100 \text{ V}$$

EXAMPLE 7–22 The circuit of Figure 7–20(a) has been designed to provide a range of voltages from point b to point d. Determine the range of V_{bd}.

SOLUTION First, redraw the circuit with the voltage source shown as in Figure 7–20(b) and then determine the two ends of the voltage range of V_{bd}, which are V_{ad} and V_{cd}.

$$V_{ad} = IR_{ad}$$

where

$$I = \frac{E}{R_T} = \frac{20}{100 + 500 + 400}$$

$$I = \frac{20}{1000} = 20 \text{ mA}$$

$$R_{ad} = 500 + 400 = 900 \text{ }\Omega$$

Substitute:

$$V_{ad} = IR_{ad} = (20 \times 10^{-3})(900) = 18 \text{ V}$$

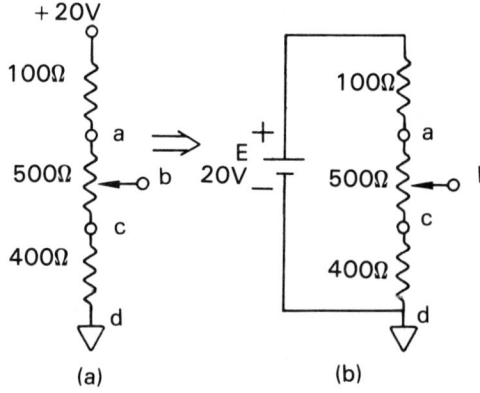

FIGURE 7–20
Circuits of Example 7–22.

173

Notation and References

Solve for V_{cd}:

$$V_{cd} = IR_{cd}$$

where $I = 20$ mA and $R_{cd} = 400\ \Omega$.
Substitute:

$$V_{cd} = IR_{cd} = (20 \times 10^{-3})(400) = 8\ \text{V}$$

∴ V_{bd} may be adjusted from a maximum of 18 V to a minimum of 8 V.

When voltages are stated in relation to a reference point, they may have a positive or a negative polarity depending on the placement of the reference point. This concept is demonstrated in Example 7–23.

EXAMPLE 7–23 Determine both the amount and the polarity of the voltages V_{ab}, V_{cb}, and V_{db}, that is, the voltages from points a, c, and d to the reference point b of Figure 7–21(a).

SOLUTION Draw a current arrow and polarize the resistances as shown in Figure 7–21(b).
Determine V_{ab}:
∴ $V_{ab} = +40$ V since point a is positive with respect to the reference point b.
Determine V_{cb}:
∴ $V_{cb} = -35$ V since point c is negative with respect to the reference point b.
Determine V_{db}:

$$V_{db} = V_{dc} + V_{cb} = -25\ \text{V} + (-35\ \text{V}) = -60\ \text{V}$$

∴ $V_{db} = -60$ V since point d is negative with respect to the reference point b.

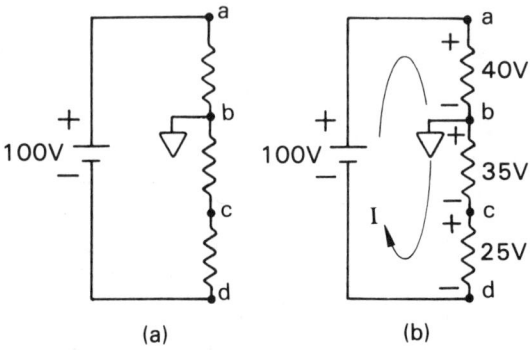

FIGURE 7–21
Circuits of Example 7–23.

Observation From this example, it may be seen that both positive and negative voltage drops are possible in the same circuit. This is the result of the reference point placement, since polarity depends upon the position of the reference point.

7-8 VOLTAGE SOURCE CHARACTERISTICS

Constant Voltage Source

A voltage source that maintains a constant terminal voltage for any amount of load current is an *ideal* constant voltage source. Although no actual device exists that will maintain a constant terminal voltage for all load currents, *practical* voltage sources do exist that will maintain a very constant voltage. These devices include commercial generators, batteries, and regulated power supplies. An ideal constant voltage source is symbolized as pictured in Figure 7-22(a).

Because resistance is always present in materials, the resistance of a practical voltage source must be considered before declaring the source ideal. For purposes of analysis, the practical voltage source may be thought of as an ideal voltage source in series with a resistance, as shown in Figure 7-22(b). The resistance in series with the ideal source represents the materials of the internal structure of the voltage source and is noted by the subscript *int*. R_{int} is the series **internal resistance** of the voltage source.

EXAMPLE 7-24 The following values of resistance are to be used for R_L in the circuit of Figure 7-23. Determine the voltage across the load for each load resistance and judge whether the voltage source, for all practical purposes, is providing a constant voltage across the load. Let R_L equal (a) 100 Ω (1000 × R_{int}), (b) 10 Ω (100 × R_{int}), (c) 2 Ω (20 × R_{int}), (d) 1 Ω (10 × R_{int}), and (e) 0.5 Ω (5 × R_{int}).

SOLUTION (a) Solve for V_L when R_L = 100 Ω (1000 × R_{int}):

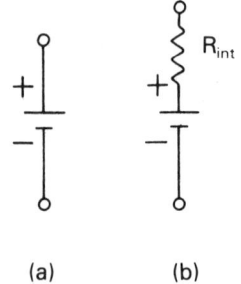

FIGURE 7-22
(a) Ideal dc constant voltage source.
(b) Practical dc voltage source with internal resistance of source included.

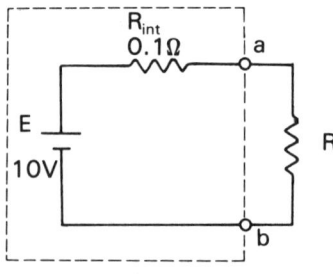

FIGURE 7-23
Practical voltage source, contained within dashed lines, consists of an ideal dc voltage source, E, in series with internal resistance, R_{int}. Practical voltage source E_{ab} may be considered an ideal voltage source when $V_L \approx E$.

Voltage Source Characteristics

$$V_L = IR_L$$

Substitute $\frac{E}{R_T}$ for I:

$$V_L = \frac{E}{R_T}(R_L) = \frac{10}{100.1}(100) = 9.99 \text{ V}$$

∴ The source is a constant voltage source.

(b) Solve for V_L when $R_L = 10\ \Omega$ ($100 \times R_{int}$):

$$V_L = IR_L$$

Substitute E/R_T for I:

$$V_L = \frac{E}{R_T}(R_L) = \frac{10}{10.1}(10) = 9.90 \text{ V}$$

∴ The source is a constant voltage source.

(c) Solve for V_L when $R_L = 2\ \Omega$ ($20 \times R_{int}$):

$$V_L = IR_L$$

Substitute E/R_T for I:

$$V_L = \frac{E}{R_T}(R_L) = \frac{10}{2.1}(2) = 9.52 \text{ V}$$

∴ The source is, for all practical purposes, a constant voltage source.

(d) Solve for V_L when $R_L = 1\ \Omega$ ($10 \times R_{int}$):

$$V_L = IR_L$$

Substitute E/R_T for I:

$$V_L = \frac{E}{R_T}(R_L) = \frac{10}{1.1}(1) = 9.09 \text{ V}$$

∴ The source is, for all practical purposes, a constant voltage source. However, it is becoming questionable at this load value ($R_L = 10R_{int}$) whether the source is providing the load with a constant voltage since 9.09 V versus 10.0 V represents a 9% difference.

(e) Solve for V_L when $R_L = 0.5\ \Omega$ ($5 \times R_{int}$):

$$V_L = IR_L$$

Substitute E/R_T for I:

$$V_L = \frac{E}{R_T}(R_L) = \frac{10}{0.6}(0.5) = 8.33 \text{ V}$$

∴ The source is not a constant voltage source.

TABLE 7-3 SUMMARY OF EXAMPLE 7-24

Ideal Voltage (V)	Available Load Voltage (V)	R_L (Ω)	Constant Voltage	% Difference (%)
10.00	9.99	100	Yes	−0.1
10.00	9.90	10	Yes	−1.0
10.00	9.52	2	Yes	−4.8
10.00	9.09	1	Yes/no	−9.1
10.00	8.33	0.5	No	−16.7

Observation From Table 7-3, which summarizes the calculations of this example, it may be seen that the *available load voltage* diminishes rapidly in value beyond $R_L = 2\ \Omega$ ($20 \times R_{int}$). In general, a practical voltage source will provide a constant voltage across a load if the load resistance is at least 20 times the internal series resistance of the voltage source.

Determining Series Internal Resistance The series internal resistance of a practical voltage source may be determined by measuring the voltage across the terminals of the source, first with the load removed, then with the load attached, as pictured in Figure 7-24. The two meter readings are then substituted into Equation 7-7 and the value for the series internal resistance is computed.

$$R_{int} = R_L \left(\frac{E_{oc}}{E_L} - 1\right) \quad (\Omega) \tag{7-7}$$

where R_{int} = internal source resistance

R_L = load resistance

E_{oc} = open-circuit voltage

E_L = voltage with the load connected

EXAMPLE 7-25 Determine the internal resistance (R_{int}) of Figure 7-24 if E_{oc} is 12 V and E_L is 10 V when R_L is 10 Ω.

SOLUTION Substitute in Equation 7-7:

$$R_{int} = R_L \left(\frac{E_{oc}}{E_L} - 1\right) = 10\left(\frac{12}{10} - 1\right) = 2\ \Omega$$

∴ The internal resistance of the voltage source is 2 Ω.

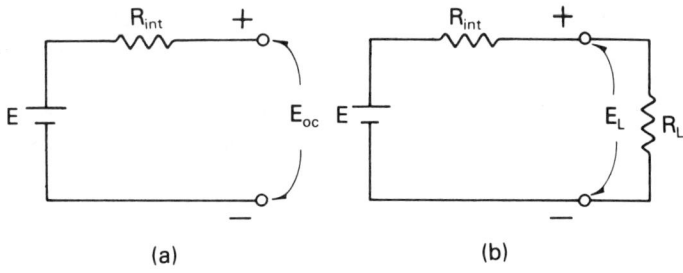

FIGURE 7-24 Internal resistance of voltage source may be determined by first measuring open circuit terminal voltage as shown in (a), then measuring terminal voltage with load connected as in (b). These measured voltages are then used with Equation 7-7 to compute internal resistance.

When analyzing circuits, it is usual to assume that the voltage source is ideal (zero internal resistance). Occasionally, though, the internal resistance must be considered in analyzing a circuit. When this is the case, we will include the voltage source internal resistance so you may understand the significance of the effects of the internal resistance on the performance of the circuit.

Open and Short Circuits Although not too common, circuit components may become open or short circuited. An open resistor or conductor is usually the result of excessive current causing the component or its connecting wiring to burn open, much like a fuse.

Usually the resistance of the wire used to connect the circuit components is extremely small compared to the component resistance, as noted in Figure 7-25(a). However, when the load is short circuited, as shown in Figure 7-25(b), the small resistance of the conductors is insufficient to limit current. The dramatic rise in the circuit current results in the overheating of both the conductor and the voltage source. The heat may be intense enough to cause the insulation of the conductor to melt and/or burn and the voltage source to be destroyed. The following example explores the result of short-circuiting the load of Figure 7-25.

EXAMPLE 7-26 Discuss the result of *shorting out* the load of Figure 7-25.

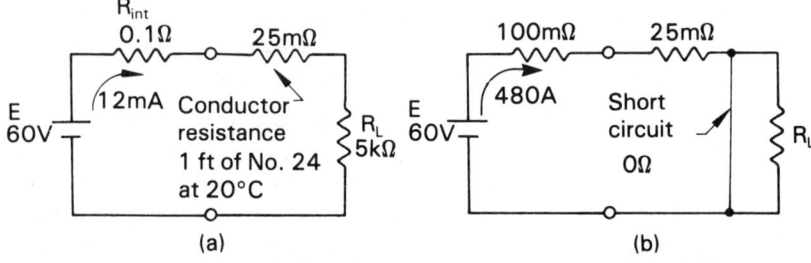

FIGURE 7-25 (a) Series circuit, operating normally; $I = 12$ mA. (b) The same circuit with the load short-circuited; I soars to 480 A!

SOLUTION If the 5-kΩ load resistance is shorted, the circuit current would be limited by only the internal resistance of the source and the resistance of the connecting wire. Without considering temperature effects on the resistance of the conductor:

$$I = \frac{E}{R_{int} + R_{cond}} = \frac{60}{100 \text{ m}\Omega + 25 \text{ m}\Omega}$$

$$I = \frac{60}{125 \text{ m}\Omega} = 480 \text{ A}$$

Determine the power dissipated by the internal resistance of the source:

$$P = I^2R = 480^2 (0.1) = 23 \text{ kW}$$

Determine the power dissipated by the conductor (wiring):

$$P = I^2R = 480^2 (0.025) = 5.8 \text{ kW}$$

Observation Since the connecting wire, which is #24 gauge, fuses (melts) at approximately 30 A, the source may not be destroyed by heat (and/or fire) as it would be required to dissipate 90 W before the wiring burned open. However, the voltage source would no doubt sustain damage owing to the large *overload* current.

Example 7–26 points out, once again, the importance of protecting the source from excessive current by providing a fuse or circuit breaker in series with the source.

EXAMPLE 7–27 Discuss the result of *opening* the load of Figure 7–25.

SOLUTION If the 5-kΩ load resistance is opened, the circuit current would drop to zero owing to the extremely high resistance of the open circuit. Additionally, the entire source voltage of 60 V would appear across the extremely high resistance of the open circuit.

Observation To locate an *open* in a series circuit, simply measure the voltage across each component in the circuit. The components having continuity will have a meter reading of zero volts. The open component will have a meter reading equal to the source voltage.

In summary, as you now know, an open circuit is characterized by an extremely high resistance resulting in zero current in the open circuit with the source voltage appearing across the open circuit. In contrast, a short circuit is characterized by nearly zero resistance, which results in an extremely high and sometimes destructive circuit current and nearly zero voltage drop across the short circuit.

SELECTED TECHNICAL TERMS

Several of the technical terms used in this chapter are defined here to aid your understanding of them.

Conventional current direction: an arbitrary choice for current direction in an electric circuit in which current is assumed to move from the positive terminal of the source, through the electric circuit, and back to the negative terminal of the source.
Equivalent circuit: a simplified circuit arrangement having the same characteristics as a more complicated circuit.
Ground: a connection, perhaps by accident, between a current-carrying conductor and the earth. A large conducting body, such as the earth, whose potential (voltage) is taken as zero.
KVL: Kirchhoff's voltage law; the algebraic sum of the voltages around a closed circuit is equal to zero.
Open circuit: a circuit that does not have a complete path for current flow. A closed circuit forms a continuous path for current; when the path is broken, the circuit is open.
Polarity: having a definite sign, such as positive or negative voltage.
Series: the connection of the components of an electric circuit so that a current flows in turn through each component along a single path.
Short circuit: a connection, perhaps by accident, of a low resistance between two points in a circuit providing a path through which all the current passes.

PROBLEMS

As part of the solution, it is suggested that you draw and label a schematic diagram for each problem.

Section 7–1

7–1 Four resistors are connected in series, and $R_1 = 10\ \Omega$, $R_2 = 22\ \Omega$, $R_3 = 15\ \Omega$, and $R_4 = 18\ \Omega$. Determine the total resistance.

7–2 Three resistances are connected in series, and $R_1 = 10\ \text{k}\Omega$, $R_2 = 7\ \text{k}\Omega$, and $R_3 = 3\ \text{k}\Omega$. Determine the total resistance.

7-3 Two resistors, R_1 and R_2, are connected in series. R_1 is color coded brown, black, brown, and silver. R_2 is color coded yellow, violet, brown, and gold. Determine the total resistance.

7-4 In Problem 7-1, the current in R_3 is measured and found to be 450 mA. Determine I_4 and I_1.

7-5 In Problem 7-2, the current in R_1 is 2 mA. Determine the current in R_2 and R_3.

7-6 Three resistances are connected in series: $R_1 = 50\ \Omega$, $R_2 = 25\ \Omega$, and $R_3 = 15\ \Omega$. If the current in R_3 is measured and found to be 0.5 A, then determine the current in R_1 and R_2.

7-7 Three resistances are connected in series: $R_T = 85\ \Omega$, $R_1 = 35\ \Omega$, $R_3 = 10\ \Omega$. Determine R_2.

7-8 Two resistors connected in series have a total resistance of 118 Ω. R_2 is color coded gray, red, and black. Determine the color code of R_1.

7-9 In the circuit of Problem 7-1, $V_1 = 20$ V, $V_2 = 44$ V, $V_3 = 30$ V, and $V_4 = 36$ V. Determine the source voltage, E.

7-10 In the circuit of Problem 7-2, $V_1 = 20$ V, $V_2 = 14$ V, and $V_3 = 6$ V. Determine the source voltage, E.

Section 7-2

7-11 Determine the current in the circuit of Problem 7-1 when $V_3 = 45$ V.

7-12 Determine the current in the circuit of Problem 7-2 when $V_1 = 50$ V.

7-13 In the circuit of Figure 7-26, $V_3 = 12$ V. Determine (a) I, (b) V_1, (c) V_2, (d) E, and (e) P_5.

7-14 Determine the source voltage (E) needed to cause 2.0 mA in the circuit of Figure 7-26.

7-15 If the power dissipated by R_4 in the circuit of Figure 7-26 is 96 mW, determine (a) I, and (b) E.

7-16 For the circuit of Figure 7-26, determine the total resistance, R_T, the current in the circuit, I, and voltage drops across each of the resistances when $E = 112.5$ V.

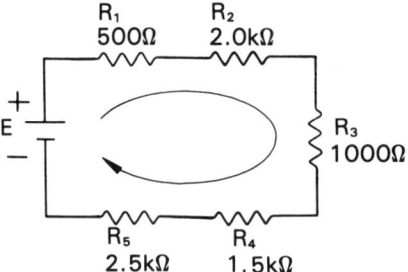

FIGURE 7-26
Circuit of Problems
7-13, 7-14, 7-15, and 7-16.

181

Problems

7-17 A heating element for an electric range is rated at 220 V. If the heating element, which has a resistance of 40 Ω, is located 500 ft from the 220-V source, determine (a) the voltage across the heating element, and (b) the voltage drop across the connecting wire. The conductors (wires) are No. 12 solid copper at 20°C.

7-18 Repeat Problem 7-17 using No. 8 connecting wire.

7-19 Three resistors are connected in series to a 36-V source. The first resistor is 20 Ω, the second resistor has a voltage of 15 V across it, the third resistor has a current of 55.6 mA passing in it. Determine (a) the color code of R_2, (b) the resistance of R_3, (c) the current in R_1, and (d) the power dissipated by R_2.

7-20 Four resistive electronic components are connected in series to a 60-V source. The voltage drop across R_1, R_2, and R_3 together is 45 V; the voltage drop across R_2, R_3, and R_4 together is 35 V; $R_2 = R_3$. If the total resistance is 500 Ω, determine the resistance of each of the four electronic components.

Section 7-3

7-21 Using the circuit of Figure 7-3, (a) polarize the resistances using conventional current flow, and (b) determine the circuit current.

7-22 The lamp of Figure 7-1(a) is specified as 6 V, 150 mA and the battery is 12 V. Assuming that the two wire-wound resistors are equal in resistive value, (a) determine the resistance of each resistor, and (b) polarize the resistances of the circuit of Figure 7-1(b) using conventional current flow.

Section 7-4

7-23 For Problem 7-21, apply Kirchhoff's voltage law to demonstrate that $E + V_1 + V_2 + V_3 = 0$. Remember to assign a plus sign (+) to a voltage rise and a minus sign (−) to a voltage drop.

7-24 For Problem 7-22, apply Kirchhoff's voltage law to dem-

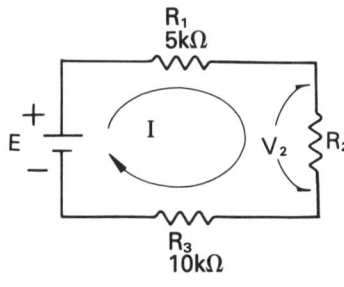

FIGURE 7-27
Circuit of Problems 7-27, 7-28, 7-29, and 7-30.

onstrate that $E + V_1 + V_2 + V_3 = 0$. Remember to assign a plus sign (+) to a voltage rise and a minus sign (−) to a voltage drop.

7-25 Three resistances, $R_1 = 1.5$ kΩ, $R_2 = 2.7$ kΩ, and R_3, are connected in series to an 80-V source. Determine R_3 when the circuit current is (a) 16 mA, (b) 4 mA, and (c) 12 mA.

7-26 Three resistors, R_1, $R_2 = 820$ Ω, and $R_3 = 560$ Ω, are connected in series to a 36-V source. Determine R_1 when the circuit current is (a) 20 mA, (b) 10 mA, and (c) 5 mA.

7-27 In the circuit of Figure 7-27, $V_2 = 8$ V and $E = 23$ V. Determine (a) V_1, (b) V_3, and (c) R_2.

7-28 In the circuit of Figure 7-27, $V_2 = 15$ V and $E = 72$ V. Determine (a) V_1, (b) V_3, and (c) R_2.

7-29 In the circuit of Figure 7-27, $P_2 = 100$ mW and $I = 4$ mA. Determine (a) V_1, (b) V_2, (c) V_3, and (d) E.

7-30 In the circuit of Figure 7-27, $P_2 = 50$ mW and $I = 2$ mA. Determine (a) V_1, (b) V_2, (c) V_3, and (d) E.

Section 7-5

7-31 In the circuit of Figure 7-8, $E_1 = 6$ V, $E_2 = 12$ V, $E_3 = 1.5$ V, and $R_1 = 10$ Ω, $R_2 = 30$ Ω, and $R_3 = 20$ Ω. An analog voltmeter is connected across R_2 to measure the voltage drop, V_2. Determine (a) how the meter leads are to be connected (positive-top, negative-bottom, or negative-top, positive-bottom), and (b) the meter reading.

7-32 In the circuit of Figure 7-8, $E_1 = 3$ V, $E_2 = 9$ V, and $E_3 = 6$ V. Determine (a) I, the circuit current, and (b) V_3.

7-33 Determine E_{ab} in the circuit of Figure 7-9(c) when $E_1 = 22$ V and $E_2 = 13$ V.

7-34 Determine E_{ab} in the circuit of Figure 7-9(a) when $E_1 = 12$ V and $E_2 = 9$ V.

7-35 Determine the amount and direction of the current in the series circuit of Figure 7-28 when $E_1 = 10$ V, $E_2 = 3$ V, and $E_3 = 15$ V.

Observation: Use cw to indicate clockwise direction and ccw to indicate counterclockwise direction.

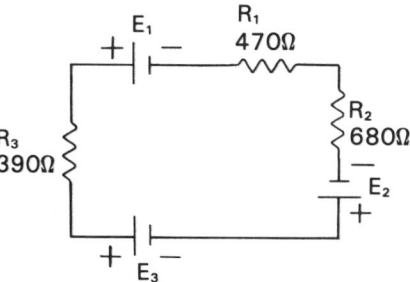

FIGURE 7-28
Circuit of Problems 7-35 and 7-36.

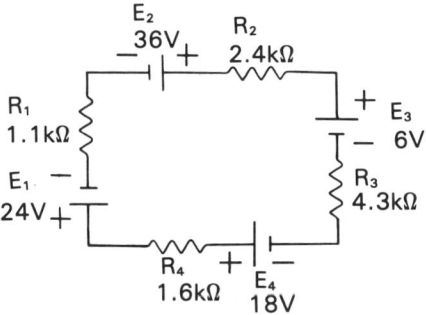

FIGURE 7-29
Circuit of Problem 7-39.

7-36 Determine the amount and direction of the current in the series circuit of Figure 7-28 when $E_1 = 12$ V, $E_2 = 6$ V, and $E_3 = 3$ V.

Section 7-6

7-37 Four resistors, $R_1 = 12$ kΩ, $R_2 = 30$ kΩ, $R_3 = 22$ kΩ, and $R_4 = 16$ kΩ, are connected in series to an 80-V source. Form the series equivalent circuit for this series circuit and verify that the two circuits are equivalent by comparing (a) the current in the circuit, and (b) the power dissipated by the loads.

7-38 Two resistors, $R_1 = 62$ kΩ and $R_2 = 47$ kΩ, are connected in series to a 200-V source. Form the series equivalent circuit for this series circuit and verify that the two circuits are equivalent by comparing (a) the current in the circuit, and (b) the power dissipated by the loads.

7-39 Form the series equivalent circuit for the series circuit of Figure 7-29 and then verify that the two circuits are equivalent by comparing the current in each circuit.

7-40 Form the series equivalent circuit for the series circuit of Figure 7-28 when $E_1 = 60$ V, $E_2 = 30$ V, and $E_3 = 10$ V. Verify that the two circuits are equivalent by comparing the current in each circuit.

Section 7-7

7-41 Determine both the magnitude (amount) and the polarity of the following voltages in the circuit of Figure 7-30 when $E_1 = 45$ V and $E_2 = 145$ V. (a) V_{ag}, (b) V_{cd}, (c) V_{fe}, and (d) V_{ga}.

7-42 Determine both the magnitude and the polarity of the following voltages in the circuit of Figure 7-30 when $E_1 = 80$ V and $E_2 = 20$ V. (a) V_{gf}, (b) V_{ed}, (c) V_{ab}, and (d) V_{de}.

7-43 Determine both the magnitude and the polarity of the following voltages in the circuit of Figure 7-30 when $E_1 = 45$ V and $E_2 = 145$ V. (a) V_{af}, (b) V_{ca}, (c) V_{ce}, and (d) V_{cf}.

7-44 Determine both the magnitude and the polarity of the fol-

184

The Series Circuit

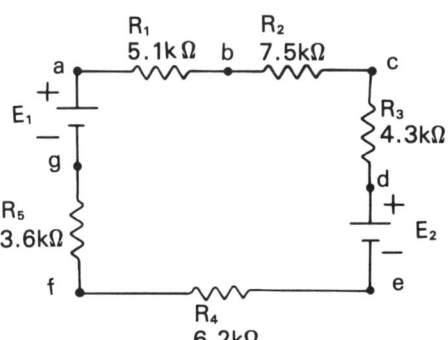

FIGURE 7–30
Circuit of Problems
7–41, 7–42, 7–43, and
7–44.

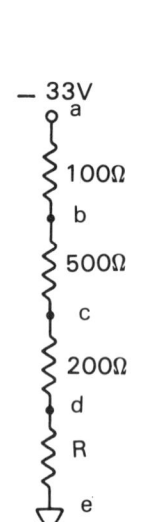

FIGURE 7–31
Circuit of Problems 7–45
and 7–46.

lowing voltages in the circuit of Figure 7–30 when $E_1 = 80$ V and $E_2 = 20$ V. (a) V_{ea}, (b) V_{ec}, (c) V_{ca}, and (d) V_{bg}.

7–45 For the circuit of Figure 7–31, determine voltages (a) V_{ae}, (b) V_{de}, and (c) V_{be} when $R = 300$ Ω.

7–46 For the circuit of Figure 7–31, determine voltages (a) V_{de}, (b) V_{ae}, and (c) V_{ce} when $R = 1000$ Ω.

7–47 For the circuit of Figure 7–32, determine voltages (a) V_{ae}, (b) V_{ce}, and (c) V_{de} when $R = 4$ Ω and $E = 3$ V.

7–48 For the circuit of Figure 7–32, determine voltages (a) V_{be}, (b) V_{ce}, and (c) V_{de} when $R = 16$ Ω and $E = 9$ V.

7–49 In the circuit of Figure 7–33, V_{ba} is measured and found to be $+15$ V and $R_1 + R_2 = 1000$ Ω. Determine the resistive values of R_1 and R_2.

7–50 In the circuit of Figure 7–33, V_{ba} is measured and found to be 0 V and $R_1 + R_2 = 1000$ Ω. Determine the resistive values of R_1 and R_2.

Section 7–8   7–51 A dry cell (battery) has an open-circuit terminal voltage of 1.5 V and an internal resistance of 400 mΩ. Determine (a) the terminal voltage with a 2-Ω load attached to the termi-

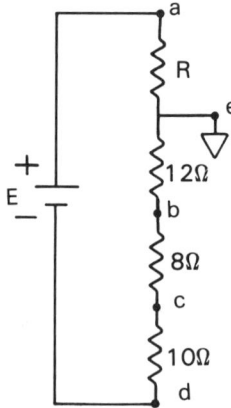

FIGURE 7–32
Circuit of Problems 7–47
and 7–48.

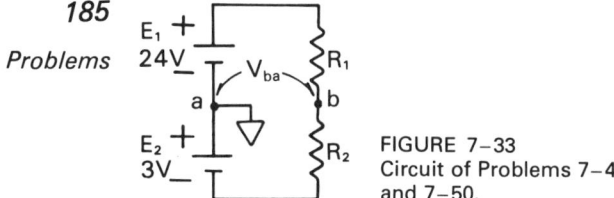

FIGURE 7-33
Circuit of Problems 7-49
and 7-50.

nals, (b) the power dissipated by the 2-Ω load, and (c) the efficiency of the system with the 2-Ω load connected.

7-52 A power supply has an open-circuit terminal voltage of 36 V and an internal resistance of 200 mΩ. An 8-Ω load is connected to the supply by wiring having a total resistance of 2 Ω. Determine (a) the voltage across the load, (b) the power lost to the internal resistance, (c) the power lost to the wiring, and (d) the efficiency of the system.

7-53 In the circuit of Figure 7-34, assume R_4 is open. Determine (a) the voltage from point d to the reference point, (b) the voltage across R_1, (c) V_3, (d) V_{be}, and (e) V_{ed}.

7-54 In the circuit of Figure 7-34, assume R_1 is open. Determine (a) V_{ce}, (b) V_{ae}, (c) V_{ab}, (d) V_5, and (e) V_{fa}.

7-55 In the circuit of Figure 7-34, determine the circuit current for the following conditions: (a) no shorts, no opens, (b) R_4 shorted, (c) point b shorted to the reference point, and (d) R_5 open.

7-56 In the circuit of Figure 7-34, determine V_{be} for the following conditions: (a) R_3 shorted, (b) R_1 open, (c) point d shorted to point b, and (d) R_2 open.

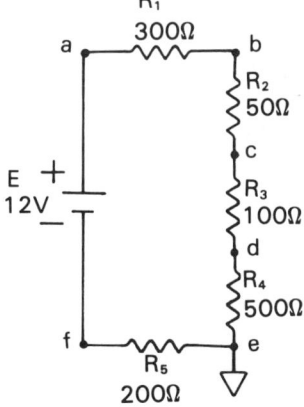

FIGURE 7-34
Circuit of Problems
7-53, 7-54, 7-55, and
7-56.

8 The Parallel Circuit

8-1 DEFINITION OF A PARALLEL CIRCUIT
8-2 APPLICATION OF THE PARALLEL CIRCUIT RULES
8-3 KIRCHHOFF'S CURRENT LAW
8-4 CURRENT SOURCES IN PARALLEL
8-5 FORMING AN EQUIVALENT CIRCUIT
8-6 CURRENT SOURCE CHARACTERISTICS

CHAPTER PREVIEW
Upon first encountering the parallel circuit—its laws, rules, and characteristics—you may find it similar in some ways to the series circuit, but in other ways you may find it different. Because of this *duality* between series and parallel circuits, it is very important that you do not learn the nature of a parallel circuit from a series circuit point of view. That is, the parallel circuit has unique characteristics that must be learned and understood in order to effectively analyze parallel-connected resistances. Although obvious, series rules are applied to the analysis of series circuits, and parallel rules are applied to the analysis of parallel circuits. In this chapter you will learn the rules, laws, and concepts of the parallel circuit.

PERFORMANCE OBJECTIVES
Once you have read each section, worked each example with pencil, paper, and calculator, and answered each problem for every section, you will be able to:

☐ Apply Kirchhoff's current law to the solution of parallel circuits.
☐ Use the concepts of conductance to solve parallel circuits.
☐ Form a series equivalent circuit of the parallel circuit.
☐ Understand the characteristics of an *ideal* and a *practical* current source.

8–1 DEFINITION OF A PARALLEL CIRCUIT

A resistive parallel circuit is formed when two or more resistive loads are connected directly to the source. Thus, a parallel connection is formed when two or more loads are connected between the same points (called *nodes*) in a circuit. When this is the case, the loads are said to be connected in parallel with each other. Figure 8–1 shows two ways of drawing parallel-connected loads.

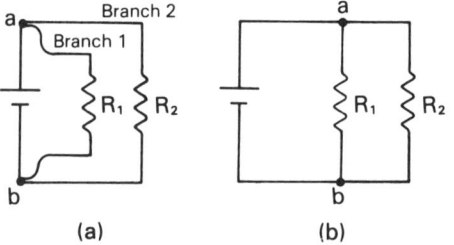

FIGURE 8–1
A two-*branch* parallel circuit is formed when two resistive loads, R_1 and R_2, are connected in parallel with each other and the voltage source. The connection points (a and b) are called nodes.

Parallel Circuit Rules and Laws

The following statements are the rules that form the basis for the analysis of parallel circuits. These concepts both define and describe the characteristics of the voltage, current, conductance, and resistance of the parallel circuit. Thus, in a parallel circuit:

☐ The voltage is the same across all components in the parallel circuit. Stated mathematically:

$$E = V_1 = V_2 = V_3 = \cdots = V_n \quad (V) \tag{8-1}$$

☐ The current entering the node (junction point) where the loads are connected in parallel must equal the current leaving the node. This concept is pictured in Figure 8-2. Stated mathematically:

$$I = I_1 + I_2 + I_3 + \cdots + I_n \quad (A) \tag{8-2}$$

☐ The total conductance, G_T, of a parallel circuit is found by summing (adding) each of the *branch* conductances in the circuit. Stated mathematically:

$$G_T = G_1 + G_2 + G_3 + \cdots + G_n \quad (S) \tag{8-3}$$

☐ Or stated in terms of the branch resistance, where $G = 1/R$:

$$G_T = \frac{1}{R_1} + \frac{1}{R_2} + \frac{1}{R_3} + \cdots + \frac{1}{R_n} \quad (S) \tag{8-4}$$

The following examples will demonstrate the use of the three fundamental rules for a parallel circuit.

EXAMPLE 8-1 Determine the voltage across R_1 and R_2 of Figure 8-3 when the source voltage is 10 V.

SOLUTION Since $E = V_1 = V_2$, the voltage is the same across all parts of the parallel circuit; the voltage across R_1 and R_2 is 10 V.

$$\therefore \quad E = V_1 = V_2 = 10 \text{ V}$$

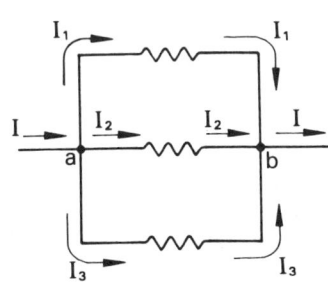

FIGURE 8-2
Current entering node a equals the sum of the currents leaving node a ($I = I_1 + I_2 + I_3$) and the sum of the currents entering node b equals the current leaving node b ($I_1 + I_2 + I_3 = I$).

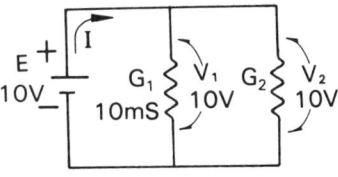

FIGURE 8-3
Circuit of Example 8-1 where $G_1 = 1/R_1$ and $G_2 = 1/R_2$.

EXAMPLE 8-2 Determine the value of G_2 of Figure 8-3 when $G_1 = 10$ mS and $G_T = 30$ mS.

SOLUTION Use Equation 8-3; substitute and solve for G_2:

$$G_T = G_1 + G_2$$
$$30 \text{ mS} = 10 \text{ mS} + G_2$$
$$G_2 = 30 \text{ mS} - 10 \text{ mS} = 20 \text{ mS}$$
$$\therefore \quad G_2 = 20 \text{ mS}$$

Observation The resistive value in ohms of R_2 is easily determined by taking the reciprocal of the conductance. Thus:

$$R_2 = \frac{1}{G_2} = \frac{1}{20} \text{ mS} = 50 \text{ }\Omega$$

EXAMPLE 8-3 Determine the circuit current (I) of Figure 8-3 when $I_1 = 100$ mA and $I_2 = 200$ mA.

SOLUTION Use Equation 8-2 and substitute:

$$I = I_1 + I_2$$
$$I = 100 \text{ mA} + 200 \text{ mA} = 300 \text{ mA}$$

$\therefore$ The circuit current is 300 mA.

EXAMPLE 8-4 Determine the total conductance (G_T) and the total resistance of the circuit of Figure 8-4.

SOLUTION Use Equation 8-4 and substitute:

$$G_T = \frac{1}{R_1} + \frac{1}{R_2} + \frac{1}{R_3}$$
$$G_T = \tfrac{1}{20} + \tfrac{1}{40} + \tfrac{1}{10} = 175 \text{ mS}$$

And

$$R_T = \frac{1}{G_T} = \frac{1}{175 \times 10^{-3}} = 5.71 \text{ }\Omega$$

$\therefore$ The total conductance of the three-branch parallel circuit is 175 mS and the total resistance is 5.71 Ω.

8-2 APPLICATION OF THE PARALLEL CIRCUIT RULES

Ohm's law will be used in conjunction with the three parallel circuit rules to analyze parallel circuits, as demonstrated in the following set of examples.

Application of the Parallel Circuit Rules

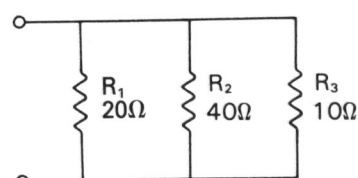

FIGURE 8-4
Circuit of Example 8-4.

EXAMPLE 8-5 Determine the current, I_3, in resistance R_3 of Figure 8-5.

SOLUTION From Ohm's law:

$$I_3 = \frac{V_3}{R_3}$$

However, $V_3 = V_1$ and

$$V_1 = I_1 R_1 = (2)(60) = 120 \text{ V}$$

So $I_3 = V_3/R_3$ when $V_3 = 120$ V and $R_3 = 1/G_3 = 1/50$ mS $= 20\Omega$.

$$I_3 = \frac{V_3}{R_3} = \frac{120}{20} = 6 \text{ A}$$

∴ $I_3 = 6$ A

Observation Ohm's law may be stated in terms of conductance as $E = I/G$, $I = EG$, and $G = I/E$. In this example, I_3 could have been stated as $I_3 = V_3 G_3$ instead of $I_3 = V_3/R_3$.

EXAMPLE 8-6 Determine the current in R_2 of Figure 8-5. Use the value of $I_1 = 2$ A, $I_3 = 6$ A, $E = 120$ V, and $G_T = 91.7$ mS.

SOLUTION $I_2 = I - (I_1 + I_3)$. Determine I:

$$I = \frac{E}{R_T}$$

where

$$R_T = \frac{1}{G_T}$$

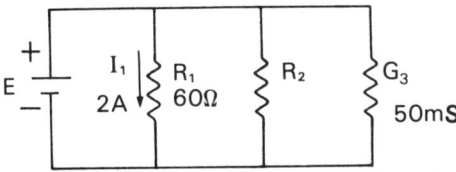

FIGURE 8-5
Circuit of Example 8-5.

Substituting $R_T = 1/G_T$:

$$I = \frac{E}{R_T} = \frac{E}{1/G_T} = EG_T$$

Solving for I:

$$I = EG_T = (120)(91.7 \times 10^{-3}) = 11 \text{ A}$$

Solve for I_2, where $I_2 = I - (I_1 + I_3)$:

$$I_2 = I - (I_1 + I_3) = 11 - (2 + 6) = 3 \text{ A}$$

$$\therefore \quad I_2 = 3 \text{ A}$$

EXAMPLE 8-7 The relay of Figure 8-6(a) is connected to a source that provides a constant current of 15 mA, as shown in Figure 8-6(b). The relay coil resistance is 500 Ω. For the relay to operate properly, the voltage drop across the coil resistance must range between 3.8 and 5.0 V. (a) Determine the current requirement for operation of the relay and (b) design a circuit to ensure the proper operation of the relay from the 15-mA current source.

SOLUTION (a) The range of current in the coil is found by Ohm's law:

$$I_{coil} = \frac{V_{coil}}{R_{coil}} = \frac{5}{500} = 10 \text{ mA}$$

And

$$I_{coil} = \frac{V_{coil}}{R_{coil}} = \frac{3.8}{500} = 7.6 \text{ mA}$$

∴ The current in the coil must range between 7.6 and 10 mA for proper operation of the relay.
(b) Since the current source is a constant 15 mA, any excess current must be *shunted* around the relay coil if the relay is to operate proper-

FIGURE 8-6
Circuit of Example 8-7.
(a) Schematic symbol for the relay which is an electrically operated switch. When current passes through the coil, a magnetic field is produced attracting the metal armature and opening or closing the contacts.
(b) The coil resistance is 500 Ω.
(c) A parallel resistance (R_s), called a *shunt* resistance, is needed to insure the operation of the relay.

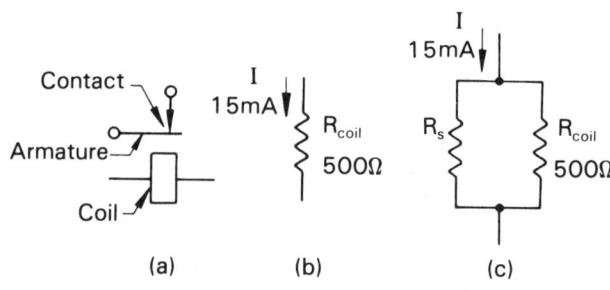

(a) (b) (c)

Kirchhoff's Current Law

ly. Figure 8–6(c) shows R_s in parallel with R_{coil}. The excess current is passed through R_s.

R_s is determined by Ohm's law:

$$R_s = \frac{V_{coil}}{I_s}$$

where

$$I_s = 15 \text{ mA} - I_{coil}$$

Substituting $I_{coil} = 10$ mA:

$$I_s = 15 \text{ mA} - 10 \text{ mA} = 5 \text{ mA}$$

And when $I_{coil} = 7.6$ mA:

$$I_s = 15 \text{ mA} - 7.6 \text{ mA} = 7.4 \text{ mA}$$

Compute R_s when $V_{coil} = 5$ V and $I_s = 5$ mA:

$$R_s = \frac{V_{coil}}{I_s} = \frac{5}{5 \times 10^{-3}} = 1 \text{ k}\Omega$$

And compute R_s when $V_{coil} = 3.8$ V and $I_s = 7.4$ mA:

$$R_s = \frac{V_{coil}}{I_s} = \frac{3.8}{7.4 \times 10^{-3}} = 514 \text{ }\Omega$$

∴ A 514- to 1000-Ω resistor must be connected in parallel with the relay coil to provide a shunt for the excess source current.

Observation A 750-Ω, $\frac{1}{2}$-W 5% carbon-composition resistor is selected for R_s. This value is selected because it is midway between 514 and 1000 Ω. The wattage rating is flexible, as $\frac{1}{8}$, $\frac{1}{4}$, or $\frac{1}{2}$ W is acceptable.

The rules and laws for the analysis of the parallel circuit are summarized in Table 8–1.

8–3 KIRCHHOFF'S CURRENT LAW

As you now know, in a parallel circuit the source current divides into the branch current, and $I = I_1 + I_2 + I_3$. Gustav Kirchhoff is credited with the discovery of this important circuit characteristic.

Kirchhoff's Current Law (KCL)

Kirchhoff's current law (KCL) states that at any node (junction) in a circuit the algebraic sum of the currents entering and leaving a node must equal zero. Stated as an equation, Kirchhoff's current law becomes

$$I + I_1 + I_2 + I_3 = 0 \qquad (8-5)$$

The Parallel Circuit

TABLE 8-1 PARALLEL CIRCUIT RULES AND LAWS

Equation	Comment
$E = V_1 = V_2 = V_3$	The voltage is the same across all parts of the parallel circuit
$I = I_1 + I_2 + I_3$	The source current is equal to the sum of the branch currents
$G = \dfrac{1}{R}$	Conductance is equal to the reciprocal of resistance
$R = \dfrac{1}{G}$	Resistance is equal to the reciprocal of conductance
$G_T = G_1 + G_2 + G_3$	The total conductance (G_T) of a parallel circuit is the sum of the individual branch conductances
$G_T = \dfrac{1}{R_1} + \dfrac{1}{R_2} + \dfrac{1}{R_3}$	The total conductance (G_T) of a parallel circuit is the sum of the reciprocals of the branch resistance
$E = IR$	Ohm's law
$I = EG$	Ohm's law

Applying Kirchhoff's Current Law to the Parallel Circuit

The following examples will use Kirchhoff's current law along with the rules for parallel circuits.

EXAMPLE 8-8 Apply Kirchhoff's current law to the circuit of Figure 8-7 to demonstrate that $I + I_1 + I_2 + I_3 = 0$ A at both node a and node b.

SOLUTION Assign a positive sign (+) to the currents entering the node and a negative sign (−) to the currents leaving the node.

At node a:

$$+I - I_1 - I_2 - I_3 = 0$$

Substitute:

$$12 - 5 - 3 - 4 = 0$$
$$0 = 0$$

And at node b:

$$-I + I_1 + I_2 + I_3 = 0$$
$$-12 + 5 + 3 + 4 = 0$$
$$0 = 0$$

∴ The algebraic sum of the current at any node is zero.

Kirchhoff's Current Law

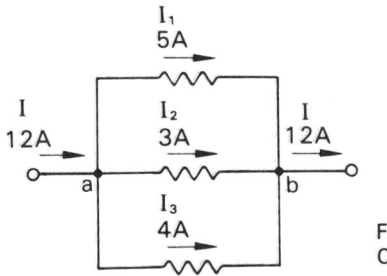

FIGURE 8-7
Circuit of Example 8-8.

Since $I = E/R$ and $I = EG$, Equation 8-2 may be written in terms of E and G as Equation 8-6 and in terms of E and R as Equation 8-7.

$$I = I_1 + I_2 + I_3 \quad \text{(A)} \tag{8-2}$$

and

$$I = EG_1 + EG_2 + EG_3 \quad \text{(A)} \tag{8-6}$$

or

$$I = \frac{E}{R_1} + \frac{E}{R_2} + \frac{E}{R_3} \quad \text{(A)} \tag{8-7}$$

EXAMPLE 8-9 Determine the source voltage of the circuit of Figure 8-8(a).

SOLUTION Use Equation 8-7:

$$I = \frac{E}{R_1} + \frac{E}{R_2} + \frac{E}{R_3}$$

Factor E:

$$I = E\left(\frac{1}{R_1} + \frac{1}{R_2} + \frac{1}{R_3}\right)$$

Solve for E:

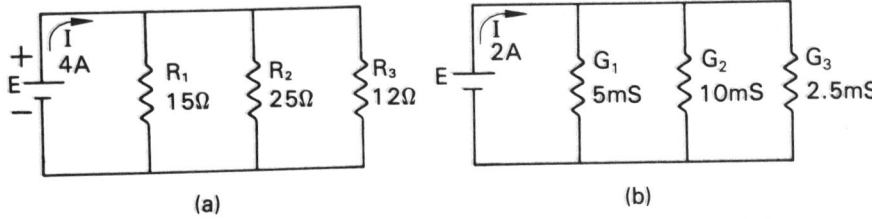

(a) (b)

FIGURE 8-8 (a) Circuit of Example 8-9. (b) Circuit of Example 8-10.

$$E = \frac{I}{1/R_1 + 1/R_2 + 1/R_3}$$

Substitute:

$$E = \frac{4}{1/15 + 1/25 + 1/12} = 21.1 \text{ V}$$

∴ The source voltage is 21.1 V.

EXAMPLE 8-10 Determine the source voltage of the circuit of Figure 8-8(b).

SOLUTION Use Equation 8-6:

$$I = EG_1 + EG_2 + EG_3$$

Factor E:

$$I = E(G_1 + G_2 + G_3)$$

Solve for E:

$$E = \frac{I}{G_1 + G_2 + G_3}$$

Substitute:

$$E = \frac{2}{5 \text{ mS} + 10 \text{ mS} + 2.5 \text{ mS}}$$

$$E = \frac{2}{17.5 \times 10^{-3}} = 114.3 \text{ V}$$

∴ The source voltage is 114 V.

From the preceding examples, it has been seen that Kirchhoff's current law may be stated in several ways depending on the need or the application. Table 8-2 summarizes the various forms of KCL.

8-4 CURRENT SOURCES IN PARALLEL

The current source is the *dual* of the voltage source. That is, the ideal voltage source provides a constant (fixed) voltage across the source terminals and a variable source current that depends on the value of the connected load. On the other hand, the ideal current source provides a constant (fixed) source current and a variable source terminal voltage that depends on the value of the connected load. Figure 8-9 shows both ideal and practical voltage and current sources.

Current sources may be connected in parallel either aiding or opposing. However, unequal current sources cannot be connected in series, as pictured in Figure 8-10, as this is a violation

Current Sources in Parallel

TABLE 8-2 VARIOUS FORMS OF KIRCHHOFF'S CURRENT LAW

Equation	Comment
$I = I_1 + I_2 + I_3$	Used to find the circuit current (I) when all the branch currents are known
$I + I_1 + I_2 + I_3 = 0$	An alternative form of the previous equation when currents entering a node are assigned a positive (+) sign and currents leaving a node are assigned a negative (−) sign
$I = EG_1 + EG_2 + EG_3$	Used to find the source current (I) of the parallel circuit
$I = \dfrac{E}{R_1} + \dfrac{E}{R_2} + \dfrac{E}{R_3}$	Same as the previous equation stated in terms of resistance instead of conductance

of Kirchhoff's current law (current into a node must equal the current out of the node; 2 A ≠ 5 A).

As a dual to this concept, unequal voltage sources cannot be connected in parallel without possible destruction to the sources. This is a violation of Kirchhoff's voltage law as applied to parallel circuits.

Parallel Aiding and Parallel Opposing Current Sources

Figure 8-11(a) shows two current sources connected *parallel aiding*. Notice that the current arrows are pointing in the same direction. The resultant current (I) is the summation of the two current sources. Thus:

$$I = I_1 + I_2 = 2 + 5 = 7 \text{ A}$$

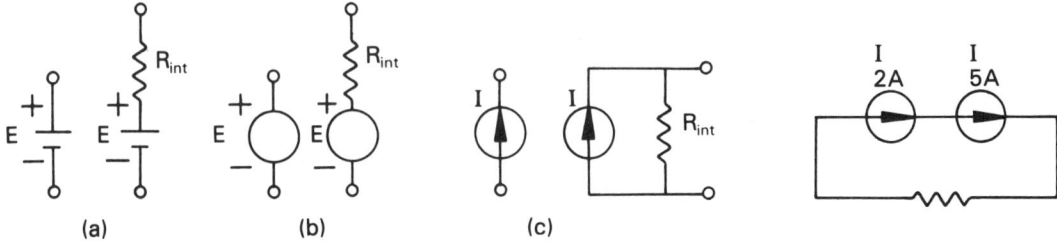

(a) (b) (c)

FIGURE 8-9 (left) Ideal (first symbol) and practical (second symbol) current and voltage sources. (a) Schematic symbol for a dc voltage source such as a battery. (b) Schematic symbol for an equivalent voltage source. (c) Schematic symbol for an equivalent current source where the arrow points in the direction of the conventional current.

FIGURE 8-10 (right) Unequal current sources cannot be connected in series because each source will provide a different constant current. The current must be the same throughout a series circuit; however, it appears that both 2 A and 5 A are in this circuit at the same time—a physical impossibility.

The Parallel Circuit

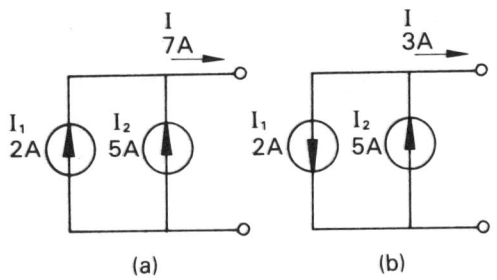

FIGURE 8-11
Aiding and opposing current sources.
(a) Parallel connected aiding current sources.
(b) Parallel connected opposing current sources.

Figure 8-11(b) pictures two current sources connected *parallel opposing*. Notice that the current arrows are pointing in the opposite direction. The resultant current (I) is the difference between the two current sources. Thus:

$$I = I_2 - I_1 = 5 - 2 = 3 \text{ A}$$

When current sources are connected in parallel, the following general rule may be applied.

RULE 8-1. DIRECTION OF THE RESULTANT CURRENT

☐ When two current sources are connected *parallel aiding*, the direction of the *resultant* current is determined by the direction of the current arrow of either source.

☐ When two current sources are connected *parallel opposing*, the direction of the *resultant* current is determined by the direction of the current arrow of the larger source.

EXAMPLE 8-11 Using the circuit of Figure 8-12, determine the current entering node b.

SOLUTION The current entering node b is equal to the resultant of I_1 and I_2. Thus:

$$I_b = I_1 - I_2 = 2 \text{ A}$$

∴ The current entering node b is 2 A.

8-5 FORMING AN EQUIVALENT CIRCUIT

A complete circuit, consisting of two or more resistances connected in parallel with one or more current sources or a voltage source, may be simplified to an *equivalent series* circuit consisting of a single resistance and a single current or voltage source. The equivalent resistance is labeled R_{eq}, the equivalent

Forming an Equivalent Circuit

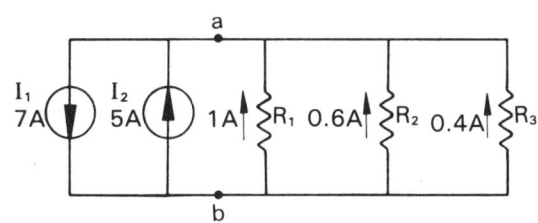

FIGURE 8–12
Circuit of Example 8–11.

voltage is labeled E_T, and the equivalent current is labeled I_T, as shown in Figure 8–13.

Computing the Equivalent Resistance of Parallel-Connected Resistances

The total resistance (R_T) of two or more resistances connected in parallel may be determined by the following equation.

$$R_T = \cfrac{1}{\cfrac{1}{R_1} + \cfrac{1}{R_2} + \cfrac{1}{R_3} + \cdots + \cfrac{1}{R_n}} \quad (\Omega) \qquad (8\text{–}8)$$

where R_T = total equivalent resistance of the parallel resistances ($R_T = R_{eq}$)
R_1, R_2, R_3, etc. = resistance of the parallel circuit branches

EXAMPLE 8–12 Form the series equivalent circuit for the parallel circuit of Figure 8–14(a) and determine the circuit current.

SOLUTION Form R_{eq}, where $R_T = R_{eq}$:

$$R_{eq} = R_T = \cfrac{1}{\cfrac{1}{R_1} + \cfrac{1}{R_2} + \cfrac{1}{R_3}}$$

Substitute:

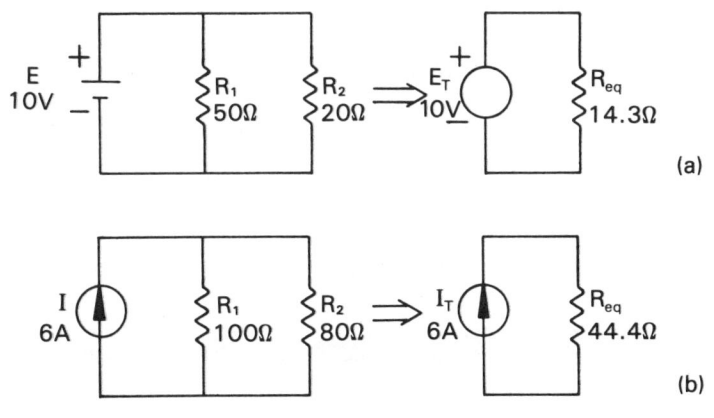

FIGURE 8–13
A parallel circuit may be formed into a *series equivalent circuit*.
(a) A series equivalent circuit formed from a parallel circuit with a voltage source.
(b) A series equivalent circuit formed from a parallel circuit with a current source.

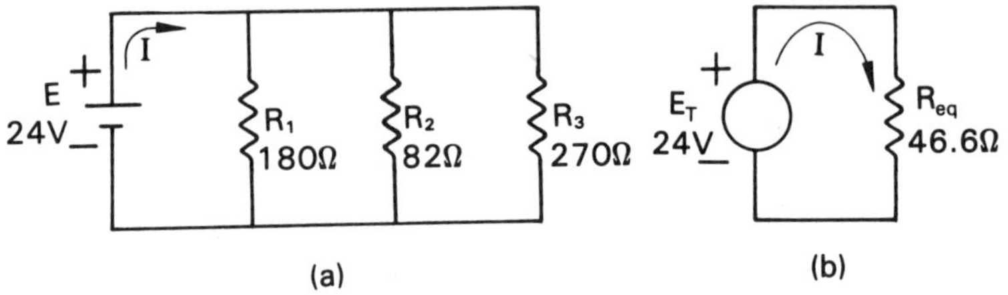

200

The Parallel Circuit

FIGURE 8–14
(a) Circuit of Example 8–12.
(b) The equivalent circuit.

$$R_{eq} = \frac{1}{\frac{1}{180} + \frac{1}{82} + \frac{1}{270}} = 46.6 \ \Omega$$

∴ $R_{eq} = 46.6 \ \Omega$

Form E_T, the equivalent voltage:

$$E_T = E = 24 \text{ V}$$

∴ $E_T = 24 \text{ V}$

Observation Figure 8–14(b) is the series equivalent circuit. Compute the circuit current:

$$I = \frac{E_T}{R_{eq}}$$

Substitute:

$$I = \frac{24}{46.6} = 515 \text{ mA}$$

∴ The circuit current I is 515 mA.

EXAMPLE 8–13 Form the series equivalent circuit for the parallel circuit of Figure 8–15(a).

SOLUTION Form R_{eq} when $R_T = R_{eq}$:

$$R_{eq} = R_T = \frac{1}{\frac{1}{R_1} + \frac{1}{R_2} + \frac{1}{R_3}}$$

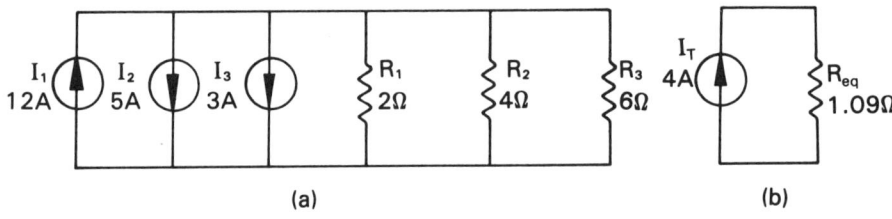

FIGURE 8–15 (a) Circuit of Example 8–13. (b) The equivalent circuit.

Forming an Equivalent Circuit

Substitute:

$$R_{eq} = \frac{1}{\frac{1}{2} + \frac{1}{4} + \frac{1}{6}} = 1.09 \text{ } \Omega$$

∴ The equivalent resistance R_{eq} is 1.09 Ω.
Form I_T, the equivalent current:

$$I_T = I_1 - I_2 - I_3 = 12 - 5 - 3 = 4 \text{ A}$$

∴ The equivalent current I_T is 4 A.

Observation Figure 8–15(b) is the series equivalent circuit.

The equivalent resistance of the parallel circuit may be computed from the total conductance by applying the following formula.

$$R_T = \frac{1}{G_T} \qquad (8-9)$$

where R_T = total equivalent resistance of the parallel resistance ($R_T = R_{eq}$)
G_T = summation of the branch conductances ($G_T = G_1 + G_2 + G_3$)

EXAMPLE 8–14 Form the equivalent series circuit of Figure 8–16 and determine the power dissipated by the equivalent resistance.

SOLUTION Form the equivalent resistance:

$$R_{eq} = R_T = \frac{1}{G_T}$$

Where $G_T = G_1 + G_2$:

$$G_T = 4 \text{ mS} + 2 \text{ mS} = 6 \text{ mS}$$

Substitute:

$$R_{eq} = \frac{1}{6 \times 10^{-3}} = 167 \text{ } \Omega$$

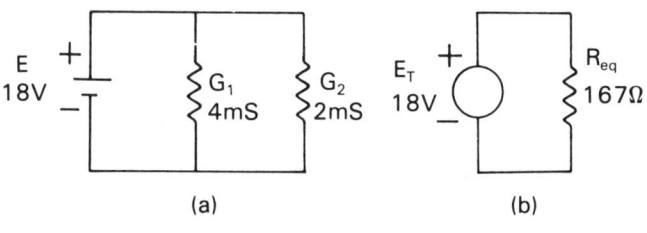

FIGURE 8–16
(a) Circuit of Example 8–14.
(b) The equivalent circuit.

$\therefore \quad R_{eq} = 167 \text{ }\Omega$

Form the equivalent source voltage:

$$E_T = E = 18 \text{ V}$$

$\therefore \quad E_T = 18 \text{ V}$

Solve for the power dissipation:

$$P = \frac{E_T^2}{R_{eq}}$$

Substitute:

$$P = \frac{18^2}{167} = 1.94 \text{ W}$$

$\therefore$ The power dissipated is 1.94 W.

8–6 CURRENT SOURCE CHARACTERISTICS

The ideal current source provides constant current to any load attached to the terminals of the source. The ideal current source, like its *dual*, the ideal voltage source, exists in theory only. The practical current source may be thought of as pictured in Figure 8–17(b), where the ideal current source is shunted by an internal resistance.

Certain electronic devices have constant-current characteriştics. Included among the devices are the constant current diode, the bipolar transistor, and the field-effect transistor (FET). The voltage versus current curve of a transistor, one of the constant-current solid-state devices, is pictured in Figure 8–18(a). This curve gives us an understanding of the magnitude of the internal resistance (R_{int}) associated with a practical current source. From Figure 8–18(b), it is further learned that a transistor may be represented as an ideal current source shunted by a resistance.

Internal Shunt Resistance of the Current Source

If the current source is to provide a constant current to an attached load, the *internal shunt resistance* of the current source must be very large when compared to the attached load resist-

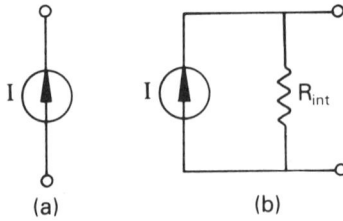

FIGURE 8–17
(a) Ideal current source.
(b) Practical current source.

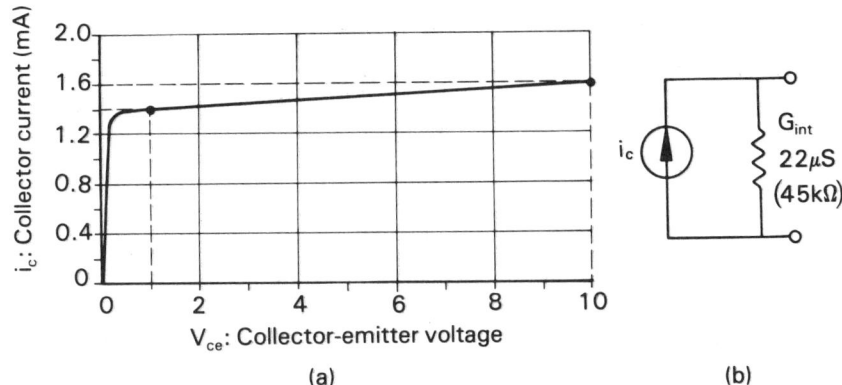

FIGURE 8–18 (a) The graphic estimation of R_{int} for a transistor. (b) The equivalent constant current source.

ance ($R_{int} \gg R_L$). Or, stated in terms of the load resistance, the load resistance must be very small when compared to the internal shunt resistance ($R_L \ll R_{int}$).

In general, a practical current source will provide a constant current in a load if the load resistance is no larger than 5% of the internal shunt resistance of the current source. Example 8–15 explores this concept.

EXAMPLE 8–15 Determine the range of load resistance values that result in constant current in the load by investigating the circuit of Figure 8–19 for load values ranging from 1% R_{int} (200 Ω) to 20% R_{int} (4 kΩ).

SOLUTION Select load values of 200 Ω (1% R_{int}), 500 Ω (2.5% R_{int}), 1 kΩ (5% R_{int}), 2 kΩ (10% R_{int}), and 4 kΩ (20% R_{int}).

Use the equation $I_L = IR_{int}/(R_L + R_{int})$ to determine I_L.

Observation This equation was derived in the following manner. From Ohm's law

$$I_L = \frac{V_L}{R_L}$$

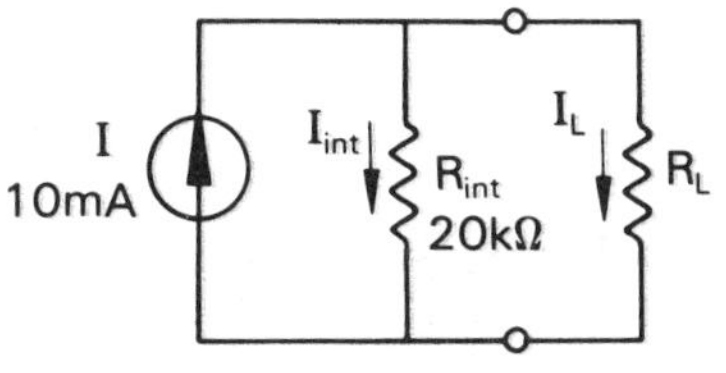

FIGURE 8–19
Circuit of Example 8–15.

However,

$$V_L = IR_T = \frac{I}{G_T} = \frac{I}{\frac{1}{R_{int}} + \frac{1}{R_L}}$$

Substituting

$$V_L = \frac{I}{\frac{1}{R_{int}} + \frac{1}{R_L}}$$

into $I_L = V_L/R_L$ results in:

$$I_L = \left(\frac{I}{\frac{1}{R_{int}} + \frac{1}{R_L}}\right)/R_L$$

Simplify by adding $\frac{1}{R_{int}} + \frac{1}{R_L}$ and then reciprocate:

$$I_L = \left(\frac{I}{\frac{R_L + R_{int}}{R_{int}R_L}}\right)/R_L = \left(\frac{IR_{int}R_L}{R_L + R_{int}}\right)/R_L$$

Factoring R_L results in the equation

$$I_L = \frac{IR_{int}}{R_L + R_{int}}$$

Determine I_L for $R_L = 200\ \Omega$ (1% R_{int}):

$$I_L = \frac{IR_{int}}{R_L + R_{int}}$$

$$I_L = \frac{(10 \times 10^{-3})(20 \times 10^3)}{200 + 20 \times 10^3} = 9.90\ mA$$

∴ The source is providing a constant current of ≈ 10 mA in the load.

Determine I_L for $R_L = 500\ \Omega$ (2.5% R_{int}):

$$I_L = \frac{IR_{int}}{R_L + R_{int}}$$

$$I_L = \frac{(10 \times 10^{-3})(20 \times 10^3)}{500 + 20 \times 10^3} = 9.76\ mA$$

∴ The source is providing a constant current of ≈ 10 mA in the load.

Determine I_L for $R_L = 1000\ \Omega$ (5% R_{int}):

Current Source Characteristics

$$I_L = \frac{IR_{int}}{R_L + R_{int}}$$

$$I_L = \frac{(10 \times 10^{-3})(20 \times 10^3)}{1000 + 20 \times 10^3} = 9.52 \text{ mA}$$

∴ The source is, for all practical purposes, providing a constant current of ≈ 10 mA in the load. Determine I_L for $R_L = 2000 \;\Omega$ (10% R_{int}):

$$I_L = \frac{IR_{int}}{R_L + R_{int}}$$

$$I_L = \frac{(10 \times 10^{-3})(20 \times 10^3)}{2000 + 20 \times 10^3} = 9.09 \text{ mA}$$

∴ It is becoming questionable at this load value ($R_L = 10\% \; R_{int}$) whether the source is providing a constant current to the load, since 9.10 mA versus 10.0 mA represents a 9% difference. Determine I_L for $R_L = 4000 \;\Omega$ (20% R_{int}):

$$I_L = \frac{IR_{int}}{R_L + R_{int}}$$

$$I_L = \frac{(10 \times 10^{-3})(20 \times 10^3)}{4000 + 20 \times 10^3} = 8.33 \text{ mA}$$

∴ The source is not providing a constant current in the load.

Observation From Table 8-3, which summarizes the calculations of this example, it may be seen that the current in the load has a marked reduction in value beyond $R_L = 1000 \;\Omega$ (5% R_{int}). Again, a practical current source will provide a constant current in a load if the load resistance is no larger than 5%

TABLE 8-3 SUMMARY OF EXAMPLE 8-15

Ideal Current (mA)	Available Load Current (mA)	R_L (ohms)	Constant Current	% Difference (%)
10	9.90	200	Yes	−1
10	9.76	500	Yes	−2.4
10	9.52	1000	Yes	−4.8
10	9.09	2000	Yes/no	−9.1
10	8.33	4000	No	−16.7

Determining the Internal Shunt Resistance

The internal shunt resistance of the practical current source may be determined by measuring both the *open-circuit voltage* across the open terminals of the source and the *short-circuit current* in a short circuit between the terminals, as shown in Figure 8–20. The two meter readings are then substituted into Ohm's law, and the internal shunt resistance is computed as demonstrated in the following example.

A word of caution is in order at this time. When working with electronic devices and circuits that act as current sources, it is **not**—we repeat—**not** a good practice to place a short circuit across the output. Unless the circuit has been specifically designed with short-circuit protection, the usual result is the destruction of the device or circuit.

EXAMPLE 8–16 Determine the internal resistance (R_{int}) of Figure 8–20 if E_{oc} is measured and found to be 50 V and I_{sc} is found to be 2 mA.

SOLUTION Substitute into Ohm's law:

$$R_{int} = \frac{E_{oc}}{I_{sc}} = \frac{50}{2 \times 10^{-3}} = 25 \text{ k}\Omega$$

∴ The internal shunt resistance, R_{int}, is 25 kΩ.

Observation Figure 8–20(c) pictures the value of R_{int} (25 kΩ) and the current source (2 mA).

When analyzing a circuit with a current source, the current source is usually assumed to be ideal (infinite internal resistance). On occasion, when the internal resistance is finite, it must be considered in the analysis of the circuit. When this is the case, the internal resistance will be included in the analysis of the circuit.

EXAMPLE 8–17 Determine the current in R_L and R_{int} of the circuit of Figure 8–21.

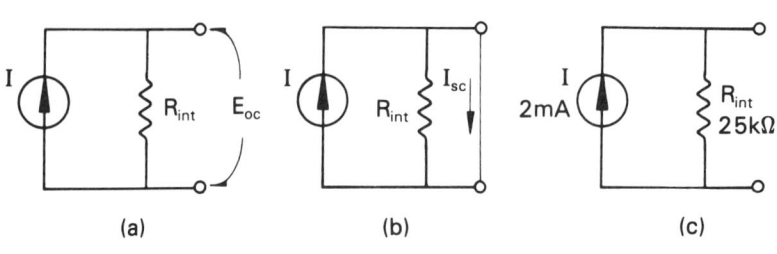

FIGURE 8–20 Determine R_{int} by: (a) measuring the open circuit terminal voltage, E_{oc}, (b) measuring the short circuit current, I_{sc}, and (c) computing R_{int} as done in Example 8–16.

Current Source Characteristics

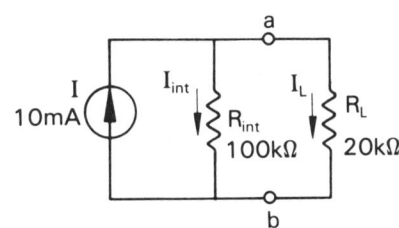

FIGURE 8-21
Circuit of Example 8-17.

SOLUTION Since R_{int} is only five times larger than R_L, R_{int} will be considered in the analysis of this circuit. Determine the voltage across R_L and R_{int} and then apply Ohm's law to determine the current in each resistance. Thus:

$$V_{ab} = IR_T$$

where

$$R_T = \frac{1}{\frac{1}{R_{\text{int}}} + \frac{1}{R_L}} = \frac{1}{\frac{1}{100 \text{ k}\Omega} + \frac{1}{20 \text{ k}\Omega}} = 16.7 \text{ k}\Omega$$

Substitute:

$$V_{ab} = IR_T = (10 \times 10^{-3})(16.7 \times 10^3)$$

$$V_{ab} = 166.7 \text{ V}$$

Determine I_{int}:

$$I_{\text{int}} = \frac{V_{ab}}{R_{\text{int}}} = \frac{166.7}{100 \times 10^3} = 1.67 \text{ mA}$$

Determine I_L:

$$I_L = \frac{V_{ab}}{R_L} = \frac{166.7}{20 \times 10^3} = 8.34 \text{ mA}$$

∴ The load current is 8.34 mA, while the current in the internal resistance is 1.67 mA.

Constructing a Constant-Current Source

In this section, we will explore the *design* of a constant-current source for a specific application. We will start by making a supposition.

Suppose the load attached to the current source of Figure 8-22(a) is actually a complex electronic circuit, which may be represented as a varying resistive load. Because of the nature of the electronic circuitry, it is important to the proper operation of the circuit that the current in the circuit be a constant 20 mA, even though its resistance varies from a low of 1.0 Ω to a high of 200 Ω.

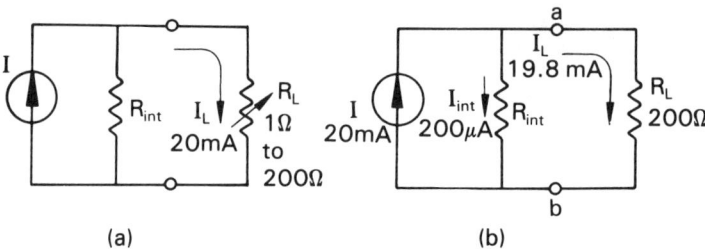

FIGURE 8–22 (a) An electronic circuit is attached to a current source. The electronic circuitry is represented by a variable resistance.
(b) The circuit used to determine the internal resistance, R_{int}.

In specifying the current source for this load, it is necessary that the stability of the load current be known since this will affect the selection of the value of R_{int}. The question is "how close to 20 mA does the load current need to be held?" For our purpose, let's assume the load current can vary ±200 µA, which is 1% of 20 mA.

The value for R_{int} is computed when the load is at its maximum resistance of 200 Ω, as pictured in Figure 8–22(b), and a maximum of 200 µA is in R_{int}. Thus:

$$R_{int} = \frac{V_{ab}}{I_{int}}$$

where

$$V_{ab} = I_L R_L = (19.8 \times 10^{-3})(200) = 3.96 \text{ V}$$

Substitute:

$$R_{int} = \frac{V_{ab}}{I_{int}} = \frac{3.96}{200 \times 10^{-6}} = 19.8 \text{ k}\Omega$$

Therefore, a current source of 20 mA with an internal resistance of 20 kΩ would meet the requirement of the electronic circuitry.

To carry this supposition further, let's construct a current source for this particular circuit. To do this, a voltage source will be placed in series with R_{int}, and a source voltage will be selected so that the current is 20 mA when the source terminals are short circuited, as shown in Figure 8–23(a). Thus:

$$E = I_{sc} R_{int} = (20 \times 10^{-3})(20 \times 10^3) = 400 \text{ V}$$

The complete circuit with the load connected is shown in Figure 8–23(b). All that remains to be done is to test the circuit to see that the load current is 20 mA ± 200 µA. Thus, for $R_L = 1$ Ω,

$$I_L = \frac{E}{R_{int} + R_L} = \frac{400}{20 \times 10^3 + 1} = 19.99 \text{ mA}$$

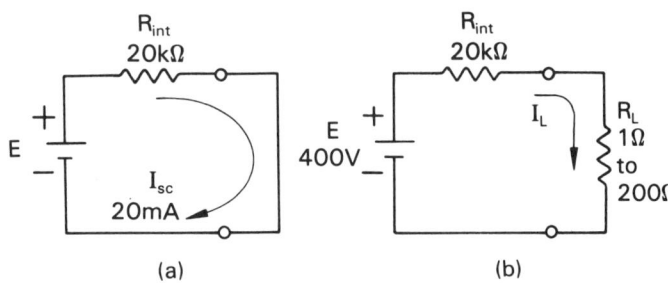

FIGURE 8–23 (a) The voltage source, E, is determined by Ohm's law where $E = I_{sc} R_{int}$. (b) A constant current results in the load, R_L, when a high value of resistance (R_{int}) is connected in series with the source voltage.

And for $R_L = 200\ \Omega$,

$$I_L = \frac{E}{(R_{int} + R_L)} = \frac{400}{20 \times 10^3 + 200} = 19.80\ \text{mA}$$

In conclusion, we see that the load current will vary between 19.80 and 19.99 mA, which meets the requirement of the electronic circuitry. Finally, as a word of caution, when working around current sources, be particularly careful of the high voltage across the open terminals. As you have experienced in Figure 8–23(b), the source voltage is usually very high and, as such, represents a shock hazard.

EXAMPLE 8–18 Develop a current source for a timing circuit that requires a constant current of 5 mA ± 50 μA. The timing circuit is represented as a resistive load of 1.5 kΩ.

SOLUTION Represent the practical current source as an ideal current source of 5 mA shunted by an unknown value of internal resistance, with 50 μA in the internal resistance. The current source is connected to a load resistance of 1.5 kΩ as pictured in Figure 8–24(a). Determine R_{int}:

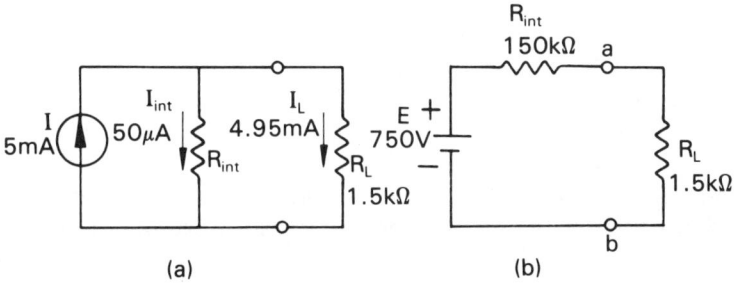

FIGURE 8–24
(a) Circuit of Example 8–18.
(b) The constant current circuit.

$$R_{\text{int}} = \frac{V_{ab}}{I_{\text{int}}}$$

where

$$V_{ab} = I_L R_L = (4.95 \times 10^{-3})(1.5 \times 10^3)$$

$$V_{ab} = 7.43 \text{ V}$$

Substitute:

$$R_{\text{int}} = \frac{V_{ab}}{I_{\text{int}}} = \frac{7.43}{50 \times 10^{-6}} = 0.15 \text{ M}\Omega$$

∴ The internal resistance is 150 kΩ.
Determine E in Figure 8-24(b) by short-circuiting points ab and letting $I_{\text{sc}} = 5$ mA:

$$E = I_{\text{sc}} R_{\text{int}} = (5 \times 10^{-3})(150 \times 10^3) = 750 \text{ V}$$

∴ The practical current source is constructed using a voltage source of 750 V in series with a resistance of 150 kΩ.

Check The circuit is tested to see that the 5 mA ± 50 µA requirement has been met. Thus:

$$I_L = \frac{E}{R_{\text{int}} + R_L} = \frac{750}{150\text{ k} + 1.5\text{ k}} = 4.95 \text{ mA}$$

We have seen in the examples of this section that a constant-current source can be constructed to meet the requirements of a particular application. However, it has also been seen that the voltage source can be very high in these circuits. It should be noted here that current sources are commercially manufactured that provide variable amounts of constant current at much lower terminal voltages. These pieces of electronic equipment are sold as power supplies having variable current limiting.

SELECTED TECHNICAL TERMS

Several of the technical terms used in this chapter are defined here to aid your understanding of them.

Branch: a conducting path between two nodes in a circuit.
KCL: Kirchhoff's current law; the algebraic sum of currents at a node in a circuit is zero.
Node: a point within a circuit where elements of the circuit are joined together.
Open-circuit voltage: the voltage across the output terminals of an electric circuit or device operating under normal conditions

Problems

Parallel circuits: constructed so that the current into the circuit divides among the branches.

Shunt: a synonym for parallel. In an ammeter, the low-value resistance connected in parallel across the meter movement. The shunt increases the range of the instrument.

PROBLEMS

As part of the solution, it is suggested that you draw and label a schematic diagram for each problem.

Section 8–1

8–1 Two resistances are connected in parallel. R_1 has a conductance of 50 mS and R_2 has a conductance of 25 mS. Determine (a) the total conductance, G_T, of the circuit, and (b) the equivalent resistance, R_T, of the circuit.

8–2 Determine the total conductance of a circuit consisting of a resistive device having a conductance of 2 mS connected in parallel with a resistive device having a conductance of 3.33 mS.

8–3 The total conductance of a two-branch parallel circuit is 833 μS and the conductance of branch 1 is 333 μS. Determine (a) the conductance of branch 2, and (b) the resistance of branch 2.

8–4 The total conductance of a three-branch parallel circuit is 50 μS. If $G_1 = 20$ μS and $G_2 = 10$ μS, determine (a) G_3 and (b) R_3.

8–5 The equivalent resistance, R_T, of a three-branch parallel circuit is 10 Ω. If $R_1 = 47$ Ω and $R_2 = 22$ Ω, determine R_3.

8–6 Determine the value of resistance that must be placed in parallel with 24 kΩ to yield an equivalent resistance, R_T, of 1.5 kΩ.

8–7 Two resistances, $R_1 = 820$ Ω and $R_2 = 680$ Ω, are connected in parallel with a 22-V source. Determine (a) V_1 and (b) V_2.

8–8 The voltage across branch 1 of a two-branch parallel circuit is 18 V. Determine the voltage across branch 2.

8–9 In the circuit of Figure 8–1, the current entering node a is 2 A. Determine the current entering node b.

8–10 In the circuit of Figure 8–2, the current leaving node b is 7 A. Determine the current entering node a.

8–11 In the circuit of Figure 8–1, the current in R_2 is 3 A. Determine the current in R_1 when the current leaving node b is 5 A.

8–12 In the circuit of Figure 8–2, current $I_1 = 300$ mA, $I_3 = 150$ mA, and $I = 800$ mA. Determine I_2.

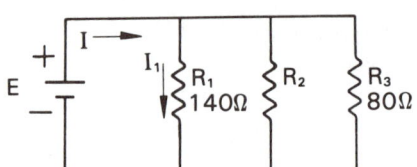

FIGURE 8–25
Circuit of Problems
8–13, 8–14, 8–15, and
8–16.

Section 8–2

8–13 In the circuit of Figure 8–25, $E = 100$ V and $R_2 = 20$ Ω. Determine (a) I_1, (b) R_T, (c) V_2, and (d) P_3.

8–14 For the values of Problem 8–13, show that $I = I_1 + I_2 + I_3$.

8–15 In the circuit of Figure 8–25, $I_1 = 72$ mA when $G_T = 39.6$ mS. Determine (a) the current in R_2, (b) the source current, I, and (c) the power dissipated by R_3.

8–16 In the circuit of Figure 8–25, $I_1 = 160$ mA. Determine (a) the voltage across R_3, (b) the resistive value of R_2, and (c) the circuit current, I, when $G_T = 29.6$ mS.

8–17 In a three-branch parallel circuit, the source current is 80 mA. Determine the current in each of the three branches when the resistance of the first branch is 2 kΩ, the second is 3 kΩ, and the third is 4 kΩ.

8–18 Repeat Problem 8–17 for a source current of 18 mA.

8–19 A 10-mA meter with a resistance of 25 Ω is used in a circuit having 50 mA of current ($I_T = 50$ mA). A shunt resistance, R_s, pictured in Figure 8–26, is placed in parallel with the meter to prevent the meter from being damaged by excessive current. Assuming 10 mA in the meter, determine the value of R_{shunt}.

8–20 Repeat Problem 8–19 for a circuit current, I_T, of 120 mA.

Section 8–3

8–21 Apply Kirchhoff's current law to the circuit of Figure 8–27 by assigning a positive sign (+) to the currents entering the node and a negative sign (−) to the current leaving the node. Show that $I + I_1 + I_2 + I_3 = 0$ for node a when $I = 880$ mA, $I_1 = 130$ mA, $I_2 = 410$ mA, and $I_3 = 340$ mA.

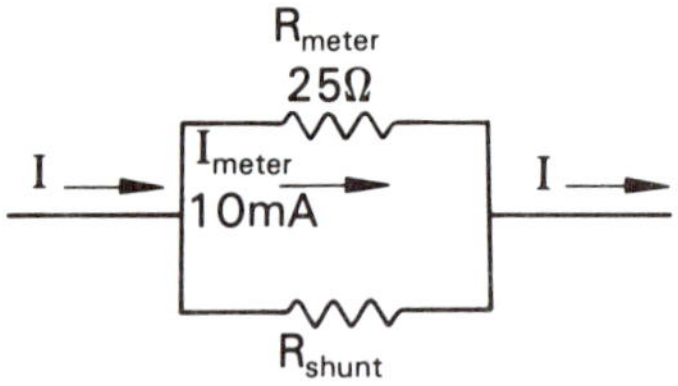

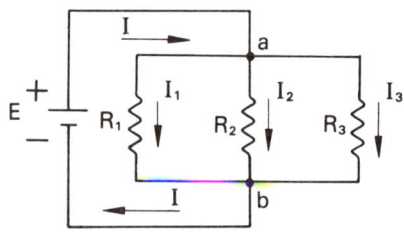

FIGURE 8–26
Circuit of Problems 8–19 and 8–20.

FIGURE 8–27 (right)   Circuit of Problems 8–21, 8–22, 8–23, and 8–24.

8-22 Repeat Problem 8-21 for node b.

8-23 Determine E, the source voltage in the circuit of Figure 8-27, when $I = 320$ mA and $R_1 = 1.5$ kΩ, $R_2 = 820$ Ω, and $R_3 = 2.2$ kΩ.

8-24 Determine E, the source voltage in the circuit of Figure 8-27, when $I = 28$ mA and $R_1 = 33$ kΩ, $R_2 = 56$ kΩ, and $R_3 = 27$ kΩ.

8-25 The sum of the currents in 100-, 500-, and 300-Ω resistors connected in parallel is 18 mA. Determine the current in each resistor.

8-26 The sum of the currents in a three-branch parallel circuit having conductances of 417, 370, and 625 μS is 740 μA. Determine the current in each branch.

8-27 The equivalent resistance, R_T, of three resistors connected in parallel is 810 Ω. Determine the current in each resistor and the total circuit current, I, when R_1 dissipates 200 mW, R_2 has a resistance of 3.9 kΩ, and R_3 has 20 V across it.

8-28 Two resistances are connected in parallel. The current in R_1 is 64 mA, the source voltage is 24 V, and the circuit current is 100 mA. Determine (a) R_2, (b) power dissipated by each resistance, and (c) G_T.

8-29 Determine the algebraic sum of the currents entering and leaving the node of Figure 8-28.

8-30 Determine I_3, the current in R_3 of Figure 8-28, when $I_1 = 3$ A, $I_2 = -2$ A, and $I_4 = -6$ A.

8-31 Determine I_4, the current in R_4 of Figure 8-28, when $I_1 = 2.5$ mA, $I_2 = 7$ mA and $I_3 = -4.5$ mA.

8-32 In the circuit of Figure 8-28, $I_1 = 5$ A, $I_2 = 3$ A, and $I_3 = -11$ A. Determine (a) I_4, (b) $I_3 + I_4$, and (c) show that $I_1 + I_2 + I_3 + I_4 = 0$.

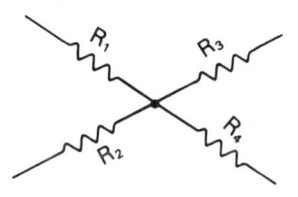

FIGURE 8-28
Circuit of Problems 8-29, 8-30, 8-31, and 8-32.

Section 8-4

8-33 Determine the resultant current of the parallel opposing current sources pictured in Figure 8-29 when $I = 36$ mA and $I' = 28$ mA.

8-34 Determine the resultant current of the parallel opposing current sources pictured in Figure 8-29 when $I = 236$ mA and $I' = 300$ mA.

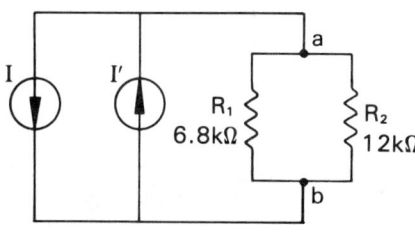

FIGURE 8-29
Circuit of Problems 8-33, 8-34, 8-35, and 8-36.

8-35 For the circuit conditions of Problem 8-33, Determine (a) V_{ab}, (b) I_1, (c) I_2, and (d) show that $I + I' + I_1 + I_2 = 0$ at node b.

8-36 For the circuit conditions of Problem 8-34, determine (a) V_{ba}, (b) I_1, (c) I_2, and (d) show that $I + I' + I_1 + I_2 = 0$ at node b.

Section 8-5

8-37 Form the series equivalent circuit for a parallel circuit consisting of two resistors, $R_1 = 8.2$ kΩ and $R_2 = 5.6$ kΩ, connected across a 28-V voltage source. Determine (a) R_{eq}, (b) I_T, and (c) E_T.

8-38 Form the series equivalent circuit for a parallel circuit consisting of three resistors, $R_1 = 1.6$ kΩ, $R_2 = 2.7$ kΩ, and $R_3 = 1.5$ kΩ, connected to a 10-mA current source. Determine (a) R_{eq}, (b) I_T, and (c) E_T.

8-39 Form the equivalent series circuit of the circuit of Figure 8-30 for $I = 120$ mA. Determine (a) I_T, (b) E_T, (c) R_{eq}, and (d) power dissipated by the equivalent resistance.

8-40 Form the equivalent series circuit of the circuit of Figure 8-30 for $I = 0.48$ A. Determine (a) I_T, (b) R_{eq}, (c) E_T, and (d) power dissipated by the equivalent resistance.

Section 8-6

8-41 Determine the range of load resistance values for the circuit of Figure 8-31(a) that will result in a constant current in the load when I is 30 mA and R_{int} is 15 kΩ.

8-42 Determine the range of load resistance values for the circuit of Figure 8-31(a) that will result in a constant current in the load when I is 6 mA and R_{int} is 50 kΩ.

8-43 Determine the internal resistance of the circuit of Figure 8-31(b) when the open-circuit terminal voltage is 58 V and the short-circuit current is 4 mA. Also determine (a) the power dissipated by the internal resistance when terminals a-b are shorted, and (b) the power dissipated by the internal resistance when terminals a-b are open.

8-44 Repeat Problem 8-43 for an open-circuit terminal voltage of 140 V and a short-circuit current of 7 mA.

8-45 Determine the current in R_L and R_{int} of the circuit of Figure 8-31(a) when I is 3 mA, the conductance of R_{int} is 25 μS, and R_L is 3.9 kΩ.

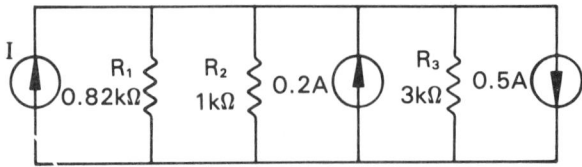

FIGURE 8-30
Circuit of Problems 8-39 and 8-40.

8-46 Determine the current in R_L and R_{int} of the circuit of Figure 8-31(a) when I is 8 mA, the conductance of R_{int} is 60 µS, and R_L is 6.8 kΩ.

8-47 Develop a current source from a voltage source and a resistance to provide 1 mA ± 20 µA of current through a 250-Ω meter movement. Specify the value of the voltage source and the series resistance.

8-48 Repeat Problem 8-47 for an 1800-Ω, 100 µA ± 3 µA meter movement.

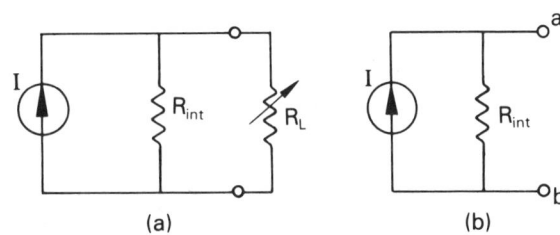

FIGURE 8-31
Circuit of Problems
8-41, 8-42, 8-43, 8-44,
8-45, and 8-46.

The Series-Parallel Network

9–1 FORMING AN EQUIVALENT CIRCUIT FOR THE SERIES-PARALLEL NETWORK
9–2 SOLVING SERIES-PARALLEL NETWORKS
9–3 POWER DISSIPATION IN THE SERIES-PARALLEL NETWORK
9–4 APPLICATIONS WITH THE SERIES-PARALLEL NETWORK

CHAPTER PREVIEW

Most electronic circuits are made up of a combination of series circuits and parallel circuits, rather than just a series or just a parallel circuit. When a series-parallel network* is analyzed, it is first broken down into sections consisting of series circuits and parallel circuits. The sections of the series-parallel network that are *series sections* are solved with the rules, laws, and techniques of series circuit analysis, while the sections of the series-parallel network that are *parallel sections* are solved with the rules, laws, and techniques of parallel circuit analysis. This chapter will provide you with an understanding of series-parallel networks by demonstrating the methods and techniques used for the solution of single-source series-parallel networks.

PERFORMANCE OBJECTIVES

Once you have read each section, worked each example with pencil, paper, and calculator, and answered each problem for every section, you will be able to:

☐ Solve single-source series-parallel networks.
☐ Form an equivalent circuit of a series-parallel network.
☐ Compute power dissipation in the network resistance.

9–1 FORMING AN EQUIVALENT CIRCUIT FOR THE SERIES-PARALLEL NETWORK

A series-parallel network may be formed into a series equivalent circuit consisting of an equivalent voltage (E_T) or current (I_T) source in series with the equivalent resistance (R_{eq}). Once the equivalent circuit is formed, the voltages across and the currents in the various sections of the series-parallel network may be computed. The following examples will introduce the methods and techniques used to form an equivalent series circuit from a series-parallel network.

EXAMPLE 9–1 Form the series-parallel network of Figure 9–1 into a series equivalent circuit.

SOLUTION Start the solution by representing the circuit as *blocks* connected in series and parallel, as shown in Figure 9–2.
Determine the value of the resistance contained within each block. Start with block A, which contains two series resistances, R_1 and R_5:

$$R_A = R_1 + R_5 = 10 + 2 = 12 \ \Omega$$

∴ The resistance within block A is 12 Ω.
Block B contains one resistance, R_2:

$$R_B = R_2 = 16 \ \Omega$$

---

* A network is made up of any number of series or parallel sections.

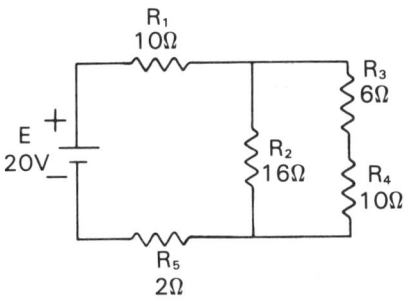
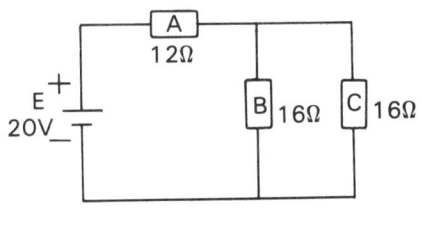

FIGURE 9–1 (left) Series-parallel network of Example 9–1.
FIGURE 9–2 (right) Network of Figure 9–1 is divided into three blocks.

∴ The resistance within block B is 16 Ω.
Block C contains two series resistances, R_3 and R_4:

$$R_C = R_3 + R_4 = 6 + 10 = 16 \text{ Ω}$$

∴ The resistance within block C is 16 Ω.

Observation The series-parallel network of Figure 9–1 may be represented as two sections. One section is a series section, while the other one is a parallel section. The series section is block A of Figure 9–2, and the parallel section is made up of blocks B and C. Form the two sections into an equivalent resistance:

$$R_{eq} = R_A + \cfrac{1}{\cfrac{1}{R_B} + \cfrac{1}{R_C}} = 12 + \cfrac{1}{\frac{1}{16} + \frac{1}{16}} = 20 \text{ Ω}$$

∴ The series equivalent resistance of the series-parallel network is 20 Ω.

Observation Figure 9–3(c) pictures the series equivalent circuit along with the original network and the block representation.

EXAMPLE 9–2 Form the series-parallel network of Figure 9–4 into a series equivalent circuit.

SOLUTION Organize the circuit into series and parallel blocks as shown in Figure 9–5. Determine the resistance contained within each block.
Block A contains two parallel resistances, R_1 and R_2:

$$R_A = R_1 \parallel R_2 = \cfrac{1}{\cfrac{1}{R_1} + \cfrac{1}{R_2}} = \cfrac{1}{\frac{1}{30} + \frac{1}{10}} = 7.5 \text{ Ω}$$

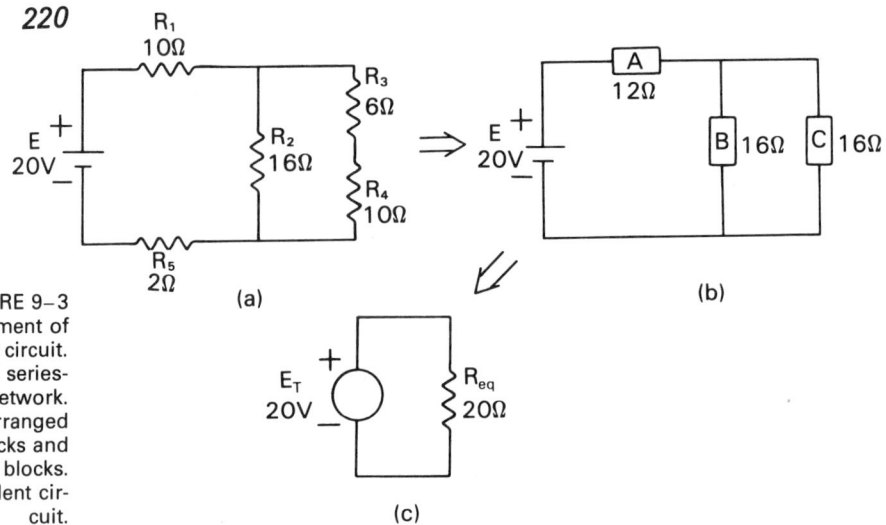

FIGURE 9-3
Steps in development of series equivalent circuit.
(a) Original series-parallel network.
(b) Network arranged into series blocks and parallel blocks.
(c) Series equivalent circuit.

Observation The notation $R_1 \| R_2$ is read; "R_1 in parallel with R_2."

∴ The resistance within block A is 7.5 Ω.
Block B contains two series resistances, R_3 and R_6:

$$R_B = R_3 + R_6 = 6 + 4 = 10 \text{ Ω}$$

∴ The resistance within block B is 10 Ω.
Block C contains two parallel resistances, R_4 and R_5:

$$R_C = R_4 \| R_5 = \frac{1}{\frac{1}{R_4} + \frac{1}{R_5}} = \frac{1}{\frac{1}{18} + \frac{1}{12}} = 7.2 \text{ Ω}$$

∴ The resistance within block C is 7.2 Ω.

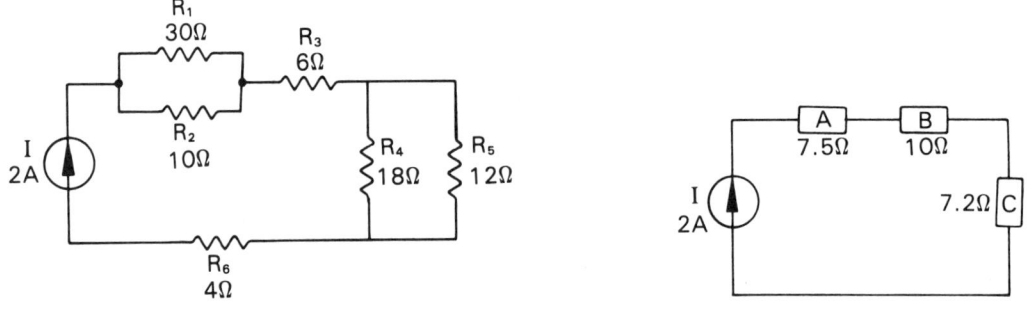

FIGURE 9-4 (left) Network of Example 9-2.
FIGURE 9-5 (right) Network of Figure 9-4 is divided into three blocks.

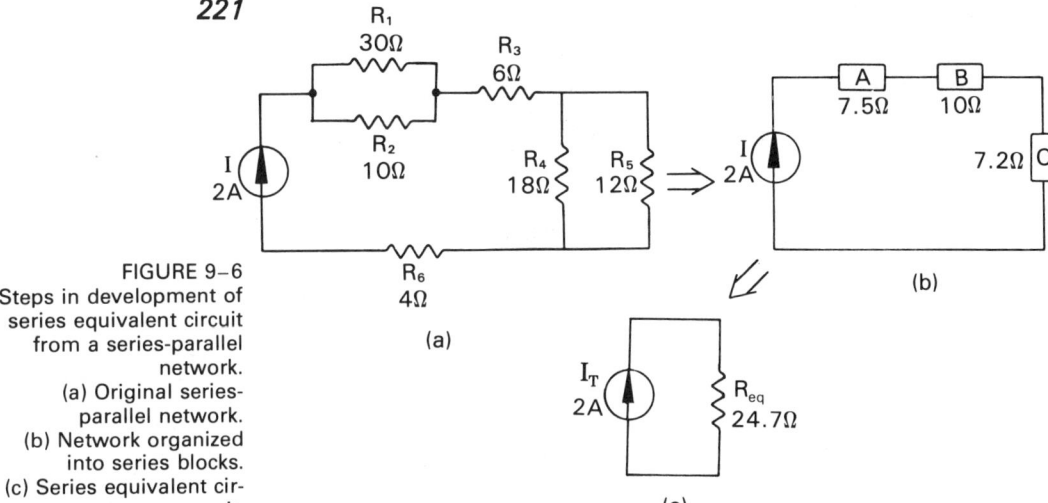

FIGURE 9-6
Steps in development of series equivalent circuit from a series-parallel network.
(a) Original series-parallel network.
(b) Network organized into series blocks.
(c) Series equivalent circuit.

Observation The series-parallel network of Figure 9-4 may be represented as three sections. Each section is in series with the other, as pictured in Figure 9-5. Blocks A, B, and C represent the three series sections.

Form the three series sections into an equivalent resistance:

$$R_{eq} = R_A + R_B + R_C$$

$$R_{eq} = 7.5 + 10 + 7.2 = 24.7 \ \Omega$$

∴ The series equivalent resistance of the series-parallel network is 24.7 Ω, as shown in Figure 9-6(c).

EXAMPLE 9-3 Form the series-parallel network of Figure 9-7 into a series equivalent circuit.

SOLUTION Organize the circuit into series and parallel

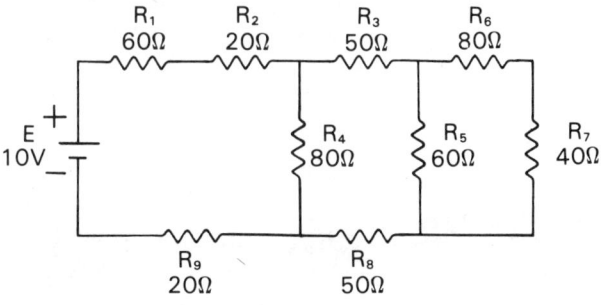

FIGURE 9-7
Network of Example 9-3.

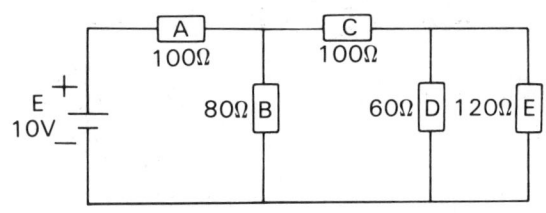

FIGURE 9-8
Network of Figure 9-7 is divided into five blocks.

blocks as shown in Figure 9-8. Determine the resistance contained within each block.
Block A contains three series resistances, R_1, R_2, and R_9:

$$R_A = R_1 + R_2 + R_9 = 60 + 20 + 20 = 100\ \Omega$$

∴ The resistance within block A is 100 Ω.
Block B contains one resistance, R_4:

$$R_B = R_4 = 80\ \Omega$$

∴ The resistance within block B is 80 Ω.
Block C contains two series resistances, R_3 and R_8:

$$R_C = R_3 + R_8 = 50 + 50 = 100\ \Omega$$

∴ The resistance within block C is 100 Ω.
Block D contains one resistance, R_5:

$$R_D = R_5 = 60\ \Omega$$

∴ The resistance within block D is 60 Ω.
Block E contains two series resistances, R_6 and R_7:

$$R_E = R_6 + R_7 = 80 + 40 = 120\ \Omega$$

∴ The resistance within block E is 120 Ω.

Observation The *blocked* series-parallel network (Figure 9-8) may be represented as two sections. One section, a series section, is block A. The other section, a parallel section, consists of two branches. Figure 9-9 pictures the two branches. Branch 1 is block B, while branch 2 is block C in series with block $D \parallel E$. Determine the resistance of section 2.
Start by computing the resistance of branch 2:

$$R_{\text{br}2} = R_C + R_D \parallel R_E$$

$$R_{\text{br}2} = 100 + \frac{1}{\frac{1}{60} + \frac{1}{120}} = 140\ \Omega$$

Forming an Equivalent Circuit for the Series-Parallel Network

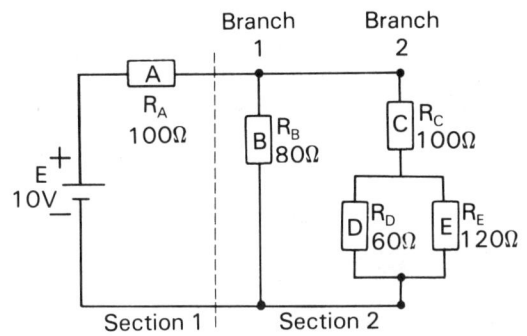

FIGURE 9-9
The *blocked* network of Figure 9-8 is sectioned into two sections.

∴ The resistance of branch 2 is 140 Ω.
Verify the resistance of branch 1:

$$R_{br1} = R_B = 80 \text{ Ω}$$

∴ The resistance of branch 1 is 80 Ω.
Form branch resistances 1 and 2 into a single series resistance:

$$R_{br1,2} = R_{br1} \| R_{br2} = \frac{1}{\frac{1}{80} + \frac{1}{140}} = 50.9 \text{ Ω}$$

∴ The resistance of section 2 is 50.9 Ω.

Observation Figure 9-10(a) pictures the resistance of section 1 in series with the resistance of section 2. Form the two sections into an equivalent resistance:

$$R_{eq} = R_A + R_{br1,2} = 100 + 50.9 = 151 \text{ Ω}$$

∴ The series equivalent resistance of the series-parallel network is 151 Ω, as shown in Figure 9-10(b).

When forming an equivalent circuit from a series-parallel network, it is sometimes helpful to draw in current arrows on the schematic to aid in determining which sections are connected in

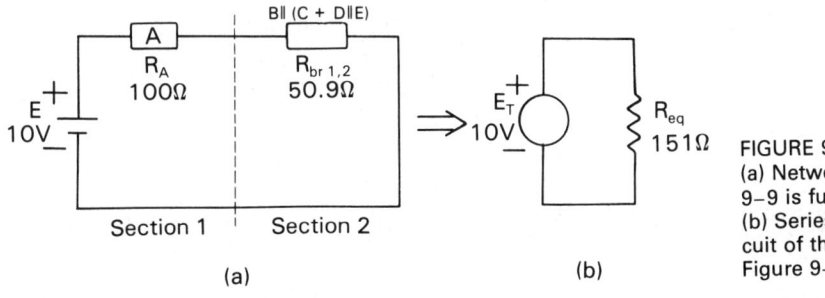

FIGURE 9-10
(a) Network of Figure 9-9 is further simplified.
(b) Series equivalent circuit of the network of Figure 9-7.

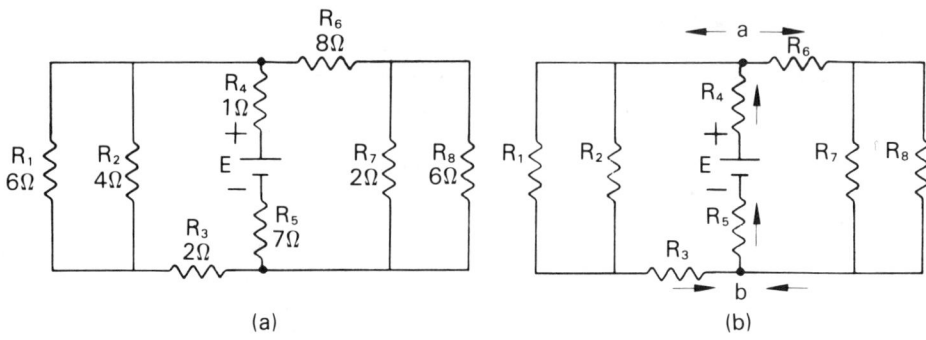

FIGURE 9–11 Network of Example 9–4.

series and which are connected in parallel. The circuit of Figure 9–11(a) is shown with current arrows inserted as Figure 9–11(b). When placing current arrows on a schematic, start at the source and move in the direction of the conventional current. While moving around the network, observe the nodes where the current separates into several branches. Because nodes are points where current separates, nodes then indicate the existence of a parallel section. Also, when the entire source current passes through a network resistance, that resistance is part of a series section. These concepts are demonstrated in Example 9–4.

EXAMPLE 9–4 Form an equivalent circuit of the series-parallel network of Figure 9–11(a).

SOLUTION Place current arrows on the schematic of the network, as shown in Figure 9–11(b), to aid in identifying the series and parallel sections of the network. Starting at the source, R_4 is part of a series section because the entire source current passes through it. The current separates at node a, signaling parallel sections. To the left of node a is one parallel section made up of $R_1 \parallel R_2$ in series with R_3. To the right of node a is another parallel section made up of R_6 in series with $R_7 \parallel R_8$. The branch currents recombine at node b, where the entire circuit current passes through R_5, indicating that R_5 is part of the series section. Figure 9–12 pictures the network of Figure 9–11 blocked and sectioned. Determine the value of the resistance contained within each block.

Block A contains two series resistances, R_4 and R_5:

$$R_A = R_4 + R_5 = 1 + 7 = 8 \ \Omega$$

Forming an Equivalent Circuit for the Series-Parallel Network

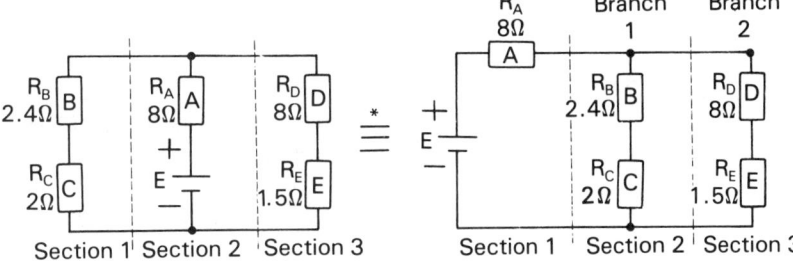

FIGURE 9-12 (a) Network of Figure 9-11 with series and parallel sections shown. (b) Sections have been arranged to add clarity to the analysis of the series-parallel network.

*Identically equal to.

∴ The resistance within block A is 8 Ω.
Block B contains two parallel resistances, R_1 and R_2:

$$R_B = R_1 \parallel R_2 = \frac{1}{\frac{1}{6} + \frac{1}{4}} = 2.4 \text{ Ω}$$

∴ The resistance within block B is 2.4 Ω.
Block C contains one resistance, R_3:

$$R_C = R_3 = 2 \text{ Ω}$$

∴ The resistance within block C is 2 Ω.
Block D contains one resistance, R_6:

$$R_D = R_6 = 8 \text{ Ω}$$

∴ The resistance within block D is 8 Ω.
Block E contains two parallel resistances, R_7 and R_8:

$$R_E = R_7 \parallel R_8 = \frac{1}{\frac{1}{2} + \frac{1}{6}} = 1.5 \text{ Ω}$$

∴ The resistance within block E is 1.5 Ω.
From Figure 9-12(b), compute the resistance of section 2 in parallel with section 3. First, determine the resistance of branch 1:

$$R_{\text{br1}} = R_B + R_C = 2.4 + 2 = 4.4 \text{ Ω}$$

∴ The resistance of branch 1 is 4.4 Ω.
Next, determine the resistance of branch 2:

$$R_{\text{br2}} = R_D + R_E = 8 + 1.5 = 9.5 \text{ Ω}$$

∴ The resistance of branch 2 is 9.5 Ω.
Finally, form branch resistances 1 and 2 into a single resistance:

226

The Series-Parallel Network

$$R_{br1,2} = R_{br1} \| R_{br2} = \frac{1}{\frac{1}{4.4} + \frac{1}{9.5}} = 3.0 \, \Omega$$

∴ The total resistance of the parallel sections (sections 1 and 3) is 3.0 Ω.

Form the series equivalent resistance of the network by combining section 2 with the total resistance of branches 1 and 2. Thus:

$$R_{eq} = R_A + R_{br1,2} = 8 + 3 = 11 \, \Omega$$

∴ The series equivalent resistance of the series-parallel network of Figure 9–11 is 11 Ω.

When forming an equivalent circuit for a series-parallel network, the degree of difficulty in analyzing the network has a lot to do with how you see the circuit configuration (arrangement) from the schematic. It is sometimes helpful to reconfigure the schematic so it provides you with better visibility of the makeup of the network. The circuit of Figure 9–13(a) has been redrawn as Figure 9–13(b). Notice how the schematic layout influences the degree of difficulty in analyzing the circuit.

To summarize the techniques and methods of forming an equivalent circuit for a series-parallel network, the following points are included:

☐ Sometimes it is necessary to redraw the schematic to add clarity to the circuit configuration.
☐ Look at a series-parallel network as sections of series and parallel circuits that have been joined to form the network.
☐ Block the network to aid in forming the equivalent resistance.
☐ Place current arrows onto the schematic of the network to aid in sectioning the network.

9–2 SOLVING SERIES-PARALLEL NETWORKS

The following examples will demonstrate techniques used to solve for various circuit parameters in a series-parallel network.

EXAMPLE 9–5 For the network pictured in Figure 9–14(a), determine the voltage across each resistance and the current in each resistance. Also, form the equivalent circuit of the network.

SOLUTION Redraw the network with the voltage source shown as in Figure 9–14(b). Form the equivalent circuit.

$$R_{eq} = R_1 + R_2 \| R_3 = 15 + \frac{1}{\frac{1}{24} + \frac{1}{16}} = 24.6 \, \Omega$$

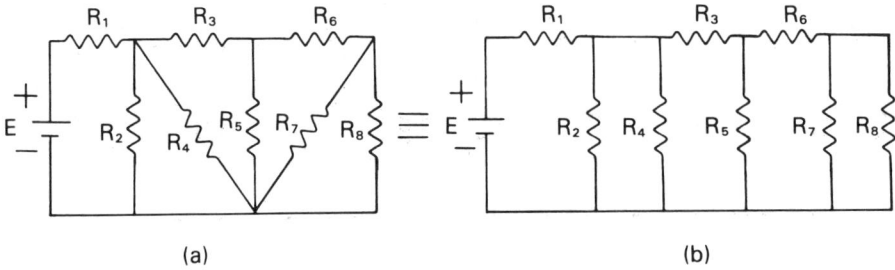

FIGURE 9-13 (a) Schematic of a series-parallel network. (b) Schematic redrawn to add clarity.

$$E_T = E = 12 \text{ V}$$

Compute I, the source current, from the equivalent circuit of Figure 9-14(c).

$$I = \frac{E_T}{R_{eq}} = \frac{12}{24.6} = 0.488 \text{ A}$$

Determine I_1 and V_1 from the network pictured in Figure 9-14(b). Since R_1 is in series, $I = I_1$. Thus:

$$I_1 = I = 0.488 \text{ A}$$

And

$$V_1 = I_1 R_1 = (0.488)(15) = 7.32 \text{ V}$$

Determine V_2 and V_3, the voltage across R_2 and R_3:

$$V_2 = V_3 = E - V_1 = 12 - 7.32 = 4.68 \text{ V}$$

And determine I_2 and I_3:

$$I_2 = \frac{V_2}{R_2} = \frac{4.68}{24} = 0.195 \text{ A}$$

$$I_3 = \frac{V_3}{R_3} = \frac{4.68}{16} = 0.293 \text{ A}$$

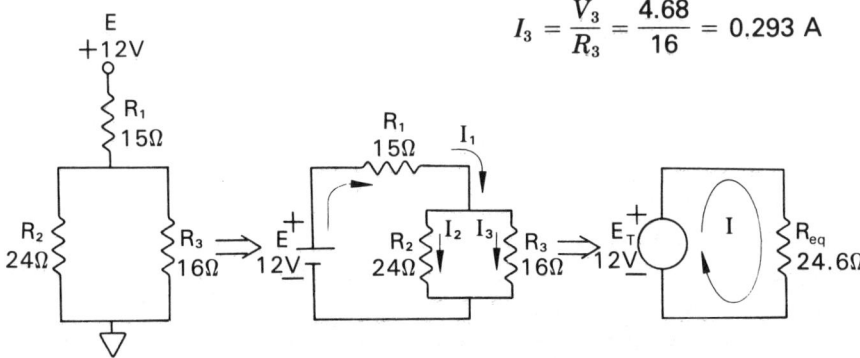

FIGURE 9-14 Network of Example 9-5.

$$\therefore \quad V_1 = 7.32 \text{ V}, \quad V_2 = 4.68 \text{ V}, \quad V_3 = 4.68 \text{ V}$$
$$I_1 = 0.488 \text{ A}, \quad I_2 = 0.195 \text{ A}, \quad I_3 = 0.293 \text{ A}$$

Observation These calculations are checked by applying Kirchhoff's laws. KCL: $I_1 = I_2 + I_3$ and 0.488 A = 0.195 A + 0.293 A. Also, KVL: $E = V_1 + V_2$ and 12 V = 7.32 V + 4.68 V.

EXAMPLE 9–6 Determine the source voltage E, the voltage across R_4 (V_4), and the current in R_3 (I_3) for the network of Figure 9–15(a).

SOLUTION The solution will begin by forming $R_3 \parallel R_4$ into an equivalent resistance so that V_4 may be computed as shown in Figure 9–15(b). Thus:

$$R_{3,4} = R_3 \parallel R_4 = \frac{1}{\frac{1}{100} + \frac{1}{60}} = 37.5 \text{ }\Omega$$

And

$$V_4 = IR_{3,4} = 2(37.5) = 75 \text{ V}$$

$\therefore$ The voltage across R_4 is 75 V.
The current I_3 is now determined. Since $R_4 \parallel R_3$, then $V_4 = V_3 = 75$ V. Thus:

$$I_3 = \frac{V_3}{R_3} = \frac{75}{100} = 0.75 \text{ A}$$

$\therefore$ The current in R_3 is 0.75 A.
The source voltage is computed from Figure 9–15(b) using KVL. Thus:

$$E = V_1 + V_2 + V_{3,4}$$

And

$$E = IR_1 + IR_2 + IR_{3,4}$$
$$E = I(R_1 + R_2 + R_{3,4})$$
$$E = 2(10 + 25 + 37.5) = 145 \text{ V}$$

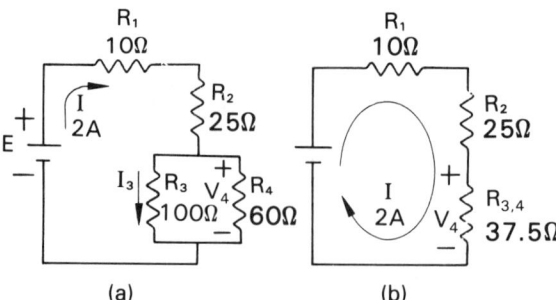

FIGURE 9–15
Network of Example 9–6.

Solving Series-Parallel Networks

∴ The source voltage is 145 V.

EXAMPLE 9-7 Determine the currents entering and leaving node *b* of the network of Figure 9-16(a).

SOLUTION Begin by blocking the circuit so that it may be formed into an equivalent circuit. Figure 9-16(b) pictures the circuit blocked and sectioned. Determine the resistance in each of the blocks:

$$R_A = R_1 = 200 \; \Omega$$

$$R_B = R_2 = 120 \; \Omega$$

$$R_C = R_3 = 160 \; \Omega$$

$$R_D = R_4 + R_5 \| R_6 = 140 + \frac{1}{\frac{1}{40} + \frac{1}{40}} = 160 \; \Omega$$

Determine the resistance in each of the sections:

$$R_{\text{sec}1} = R_A = 200 \; \Omega$$

$$R_{\text{sec}2} = R_B + R_C \| R_D$$

$$R_{\text{sec}2} = 120 + \frac{1}{\frac{1}{160} + \frac{1}{160}} = 200 \; \Omega$$

Form the resistance of sections 1 and 2 into the equivalent resistance of the network.

$$R_{\text{eq}} = R_T = R_{\text{sec}1} \| R_{\text{sec}2} = \frac{1}{\frac{1}{200} + \frac{1}{200}} = 100 \; \Omega$$

∴ The equivalent circuit, as shown in Figure 9-16(c), is made up of a current source of 5 mA in series with an equivalent resistance of 100 Ω.

Observation The voltage across R_{eq} in Figure 9-16(c) is equal to the voltage across sections 1 and 2 in Figure

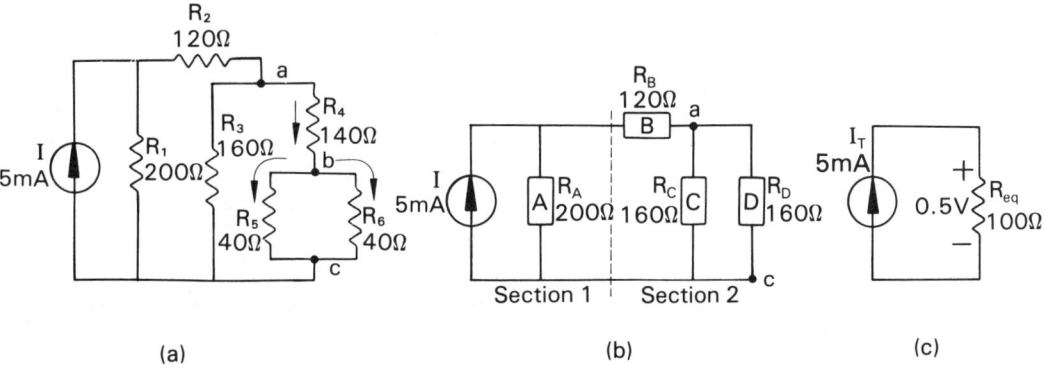

FIGURE 9-16 (a) Network of Example 9-7. (b) Block diagram. (c) Equivalent circuit.

The Series-Parallel Network

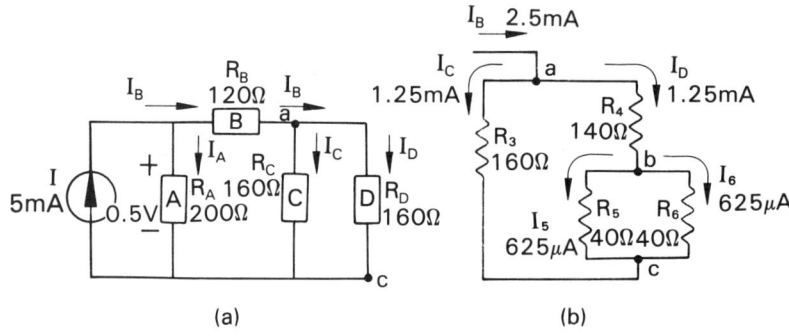

FIGURE 9–17 Current arrows have been added to the network of Figure 9–16 to aid in solving Example 9–7.

9–16(b), because sections 1 and 2 are connected in parallel. Thus:

$$V_{eq} = V_{sec1} = V_{sec2} = I(R_{eq})$$

$$V_{eq} = (5 \times 10^{-3})(100) = 0.5 \text{ V}$$

Determine the currents in the diagram of Figure 9–17(a), where $V_A = V_{sec1} = 0.5$ V. Thus:

$$I_A = \frac{V_A}{R_A} = \frac{0.5}{200} = 2.5 \text{ mA}$$

$$I_B = I - I_A = 5 \text{ mA} - 2.5 \text{ mA} = 2.5 \text{ mA}$$

$$I_C = \frac{V_{ac}}{R_c}$$

$$I_C = \frac{V_{sec2} - V_B}{160}$$

$$I_C = \frac{0.5 - I_B R_B}{160}$$

$$I_C = \frac{0.5 - (2.5 \times 10^{-3})(120)}{160}$$

$$I_C = 1.25 \text{ mA}$$

$$I_D = 1.25 \text{ mA}$$

Observation Since $R_C = R_D$ and $R_C \parallel R_D$, then $I_C = I_D = 1.25$ mA. The currents entering and leaving node b may now be computed. Figure 9–17(b) is a partial schematic of Figure 9–16(a) showing the current patterns into and out of node b. The current into node b is I_4, which equals I_D. The current out of node b is I_5 and I_6. Because $R_5 = R_6$, currents I_5 and I_6 are equal. Thus the current entering node b is

Solving Series-Parallel Networks

$I_4 = I_D = 1.25$ mA

and the current leaving node b is

$I_5 + I_6 = I_4 = 1.25$ mA

And

$$I_5 = \frac{V_{bc}}{R_5} = \frac{(I_4)(R_5 \parallel R_6)}{R_5}$$

$$I_5 = (1.25 \times 10^{-3}) \left(\frac{1}{\frac{1}{40} + \frac{1}{40}}\right) / 40 = 625 \ \mu A$$

$I_6 = I_4 - I_5$

$I_6 = (1.25 \times 10^{-3}) - (625 \times 10^{-6}) = 625 \ \mu A$

∴ The current entering node b is 1.25 mA and the two currents leaving node b are each 625 μA, as shown in Figure 9-17(b).

EXAMPLE 9-8 For the network in Figure 9-18(a), determine V_2, V_3, I_1, and I_6.

SOLUTION Redraw the network of Figure 9-18(a) as Figure 9-18(b) so that the circuit configuration can be understood. Block Figure 9-18(b) into three series resistances as shown in Figure 9-19, where

$R_A = R_1 = 3 \ \Omega$

$R_B = R_4 \parallel R_2 = \dfrac{1}{\frac{1}{9} + \frac{1}{8}} = 4.24 \ \Omega$

$R_C = R_3 \parallel R_5 \parallel R_6 = \dfrac{1}{\frac{1}{5} + \frac{1}{6} + \frac{1}{7}} = 1.96 \ \Omega$

Compute I from Figure 9-19:

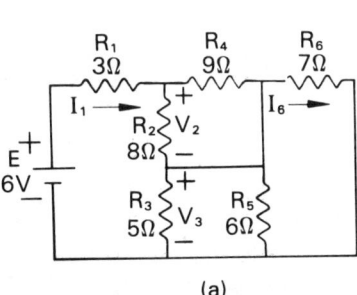

(a)

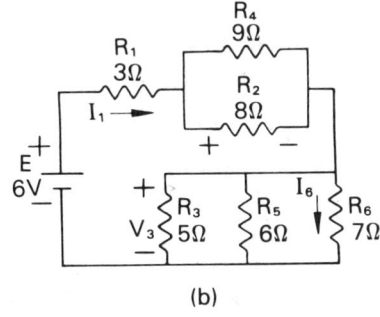

(b)

FIGURE 9-18
Network of Example 9-8.

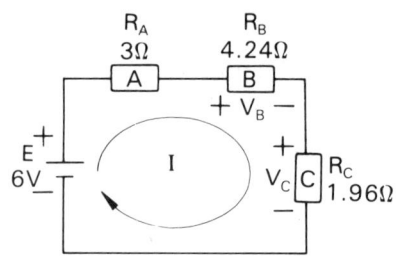

FIGURE 9-19
Network of Figure 9-18(b) is blocked so I, V_B, and V_C may be computed.

$$I = \frac{E}{R_T} = \frac{6}{3 + 4.24 + 1.96} = 0.652 \text{ A}$$

Determine V_2 and V_3. Since $R_B = R_4 \parallel R_2$, then $V_2 = V_B$, and since $R_C = R_3 \parallel R_5 \parallel R_6$, then $V_3 = V_C$. Thus:

$$V_2 = V_B = IR_B = 2.76 \text{ V}$$

And

$$V_3 = V_C = IR_C = 1.28 \text{ V}$$

∴ $V_2 = 2.76$ V and $V_3 = 1.28$ V

From Figure 9-18(b) it is learned that $I_1 = I$ of Figure 9-19:

$$I_1 = I = 0.652 \text{ A}$$

And since $V_3 = V_6$, then

$$I_6 = \frac{V_6}{R_6} = \frac{1.28}{7} = 0.183 \text{ A}$$

∴ $I_1 = 0.652$ A and $I_6 = 0.183$ A

9-3 POWER DISSIPATION IN THE SERIES-PARALLEL NETWORK

The power dissipated by each of the resistances in a network contributes to the total power dissipated by the entire network in the following manner.

$$P_T = P_1 + P_2 + P_3 + \cdots + P_n \tag{9-1}$$

where P_T = total power dissipated by the entire resistive network
P_1, P_2, etc. = power dissipated by the individual loads

From Equation 9-1, it can be inferred that power dissipation is independent of circuit configuration. That is, power dissipation is added whether the circuit elements (resistances and sources) are connected in series, parallel, or series-parallel. The following examples explore this concept.

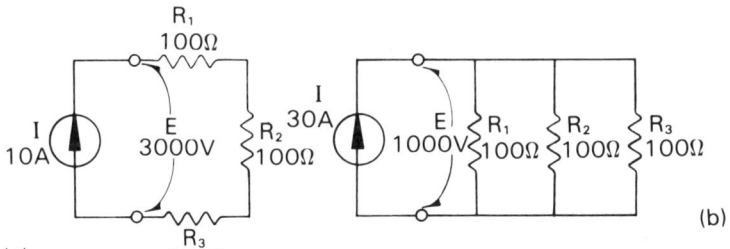

FIGURE 9-20
Circuit of Example 9-9.

EXAMPLE 9-9 For the circuits of Figure 9-20, determine the following:
(a) The current in each resistance.
(b) The power dissipated by each resistance.
(c) The power dissipated by all the loads in each circuit.
(d) The power delivered by the source.

SOLUTION (a) The current in each resistance in the circuit of Figure 9-20(a) is 10 A. The current in each resistance in the circuit of Figure 9-20(b) is also 10 A.

Observation Because each branch resistance is the same in the circuit of Figure 9-20(b), the source current divides equally in the branches.

(b) Because both the current and the resistance are the same in each of the resistances, the power dissipated by each resistance is also equal. Thus:

$$P = I^2R = (10)^2\, 100 = 10 \text{ kW}$$

∴ The power dissipated by each resistance in each circuit is 10 kW.

(c) The total power in the circuit of Figure 9-20(a) is

$$P_T = P_1 + P_2 + P_3$$

$$P_T = 10 \text{ kW} + 10 \text{ kW} + 10 \text{ kW} = 30 \text{ kW}$$

The total power in the circuit of Figure 9-20(b) is

$$P_T = P_1 + P_2 + P_3$$

$$P_T = 10 \text{ kW} + 10 \text{ kW} + 10 \text{ kW} = 30 \text{ kW}$$

∴ The total power dissipated in each circuit is 10 kW.

Observation The total power dissipated by a circuit (regardless of circuit configuration) is found by adding the individual powers. Power dissipation is independent of circuit configuration.

(d) The power delivered by the source of the circuit of Figure 9-20(a) is

$$P = IE = (10)(3000) = 30 \text{ kW}$$

The power delivered by the source of the circuit of Figure 9-20(b) is

$$P = IE = (30)(1000) = 30 \text{ kW}$$

∴ The power delivered by each circuit equals the power dissipated by the load in each circuit.

EXAMPLE 9-10 Determine the power delivered to the 300-Ω resistance in the network of Figure 9-21.

SOLUTION Block the network into a three-branch parallel circuit as diagrammed in Figure 9-21(b). Since P_4 (the power in the 300-Ω resistance R_4) is needed, only block C is of concern. Determine R_C:

$$R_C = R_4 \parallel R_5 + R_6 = \frac{1}{\frac{1}{300} + \frac{1}{400}} + 50 = 221 \text{ Ω}$$

Determine P_4:

$$P_4 = \frac{V_4^2}{R_4}$$

where $V_4 = E - V_6 = E - I_C R_6 = E - \left(\frac{E}{R_C}\right) R_6$

$$V_4 = 50 - \left(\frac{50}{221}\right) 50 = 38.7 \text{ V}$$

And

$$P_4 = \frac{V_4^2}{R_4} = \frac{38.7^2}{300} = 5 \text{ W}$$

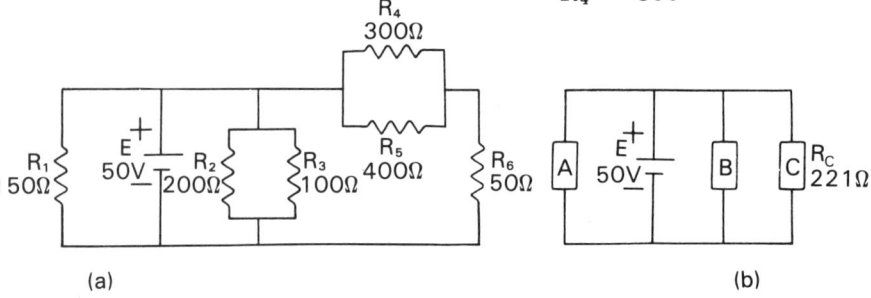

FIGURE 9-21 Network of Example 9-10.

Power Dissipation in the Series-Parallel Network

∴ The power delivered to the 300-Ω resistance R_4 is 5 W.

EXAMPLE 9-11 Determine both the power dissipated by R_4 and the resistance of R_4 in the network of Figure 9-22.

SOLUTION Determine the source power (P). Thus:

$$P = IE = (0.125)(10) = 1.25 \text{ W}$$

Determine the power of the parallel branches (P_p). Thus:

$$P_p = P - P_1 = P - I^2 R_1$$

$$P_p = 1.25 - (0.125^2)(40) = 0.625 \text{ W}$$

Determine the power of the branch with R_4. Thus:

$$P_{3+4} = P_p - P_2 = P_p - \frac{V_2^2}{R_2}$$

where

$$V_2 = E - V_1 = E - IR_1$$

$$V_2 = 10 - (0.125)(40) = 5 \text{ V}$$

and

$$P_{3+4} = P_p - \frac{V_2^2}{R_2} = 0.625 - \frac{(5^2)}{80} = 0.313 \text{ W}$$

Determine the power of P_4. Thus:

$$P_4 = P_{3+4} - P_3 = P_{3+4} - I_3^2 R_3$$

where

$$I_3 = I - I_2 = I - \frac{V_2}{R_2} = 0.125 - \frac{5}{80} = 62.5 \text{ mA}$$

and

$$P_4 = P_{3+4} - I_3^2 R_3$$

$$P_4 = 0.313 - (62.5 \times 10^{-3})^2 \, 30 = 0.196 \text{ W}$$

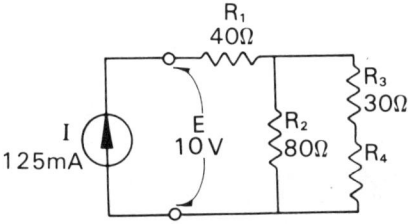

FIGURE 9-22
Network of Example 9-11.

∴ The power dissipated by R_4 is 196 mW. Determine the resistance of R_4. Thus:

$$P_4 = I_4^2 R_4$$

Solve for R_4:

$$R_4 = \frac{P_4}{I_4^2}$$

where $I_4 = I_3 = 62.5$ mA

$$R_4 = \frac{0.196}{(62.5 \times 10^{-3})^2} = 50 \ \Omega$$

∴ The resistance of R_4 is 50 Ω.

9–4 APPLICATIONS WITH THE SERIES-PARALLEL NETWORK

A very useful equation for determining the total resistance of two parallel-connected resistances is formulated as Equation 9–2.

$$R_T = \frac{R_1 R_2}{R_1 + R_2} \quad (\Omega) \tag{9-2}$$

where R_T = total equivalent resistance of the two parallel resistances
R_1 and R_2 = resistances of the two branches of the parallel circuit

Besides being used to determine R_T, Equation 9–2 is used to determine one branch resistance when R_T and the other branch resistance are known. Example 9–12 uses Equation 9–2 to determine the branch resistance.

EXAMPLE 9–12 Determine the value of resistor R_2 in the network of Figure 9–23. The network's total resistance (R_P) of 518 Ω was measured with a resistance bridge to an accuracy of ±1% of reading.

SOLUTION Let R_T represent the equivalent resistance of the parallel-connected resistors. Solve for the resistance of R_T. Thus:

$$R_T = R_P - R_3 = 518 - 180 = 338 \ \Omega$$

Solve for R_2:

$$R_T = \frac{R_1 R_2}{R_1 + R_2}$$

and

$$R_T R_1 + R_T R_2 = R_1 R_2$$

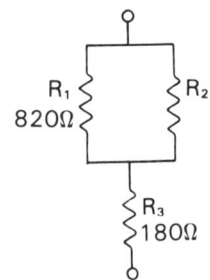

FIGURE 9–23
Network of Example 9–12.

Applications with the Series-Parallel Network

$$R_T R_1 = R_1 R_2 - R_T R_2$$
$$R_T R_1 = R_2(R_1 - R_T)$$
$$R_2 = \frac{R_T R_1}{R_1 - R_T}$$

Substitute circuit values:

$$R_2 = \frac{(338)(820)}{820 - 338} = 575 \ \Omega$$

∴ R_2 is 575 Ω.

Applications

EXAMPLE 9–13 Select a resistor value from the E24 preferred number series for R_3 of Figure 9–24 so that not more than 20% nor less than 16% of the source current will pass through R_3. Also, specify both the current in and the power dissipation of the selected value of R_3.

SOLUTION Since $R_2 \parallel R_3$, then $V_2 = V_3$. Substitute the Ohm's law equivalent $V = IR$ into $V_2 = V_3$. Thus:

$$V_2 = V_3$$

and

$$I_2 R_2 = I_3 R_3$$

However, when the current in R_3 is 20% I, then

$$I_2 = 0.8I \quad \text{and} \quad I_3 = 0.2I$$

Substitute:

$$I_2 R_2 = I_3 R_3$$
$$0.8 I R_2 = 0.2 I R_3$$

Solve for R_3:

$$R_3 = \frac{0.8 I R_2}{0.2 I} = 4 R_2$$

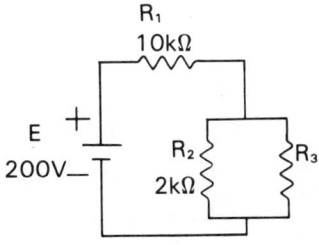

FIGURE 9–24
Network of Example 9–13.

Substitute the circuit value of R_2:

$$R_3 = 4R_2 = 4(2 \text{ k}\Omega) = 8 \text{ k}\Omega$$

∴ When R_3 is 8 kΩ, then 20% of the source current is in R_3.

Solve for R_3 when $I_3 = 0.16I$ and $I_2 = 0.84I$:

$$R_3 = \frac{0.84IR_2}{0.16I} = 5.25R_2$$

and

$$R_3 = 5.25R_2 = 5.25(2 \text{ k}\Omega) = 10.5 \text{ k}\Omega$$

∴ When R_3 is 10.5 kΩ, then 16% of the source current is in R_3.

Select a value from the E24 preferred number series. The choices are 8.2, 9.1, and 10 kΩ.

∴ A 9.1 kΩ ± 5% resistor is selected.

Observation A tolerance of ±5% is acceptable since 9.1 kΩ ±5% ranges from 8645 to 9555 Ω, which is within the specified range of 8000 to 10 500 Ω. Determine the nominal current in the 9.1-kΩ resistor. Thus:

$$I_3 = \frac{V_3}{R_3}$$

where

$$V_3 = E - V_1 = E - IR_1$$

and

$$I = \frac{E}{R_T} = \frac{E}{R_1 + \dfrac{1}{\dfrac{1}{R_2} + \dfrac{1}{R_3}}}$$

$$I = \frac{200}{10 \text{ k}\Omega + \dfrac{1}{\dfrac{1}{2 \text{ k}\Omega} + \dfrac{1}{9.1 \text{ k}\Omega}}}$$

$$I = \frac{200}{11.64 \text{ k}\Omega} = 17.2 \text{ mA}$$

Substitute:

$$V_3 = E - IR_1$$

$$V_3 = 200 - (17.2 \times 10^{-3})(10 \times 10^3) = 28.0 \text{ V}$$

Applications with the Series-Parallel Network

and

$$I_3 = \frac{V_3}{R_3} = \frac{28.0}{9.1 \times 10^3} = 3.1 \text{ mA}$$

∴ The current in the selected resistance is 3.1 mA. Compute the power dissipated by R_3.

$$P_3 = I_3^2 R_3 = (3.1 \times 10^{-3})^2 (9.1 \times 10^3)$$
$$P_3 = 87.5 \text{ mW}$$

∴ The power dissipated by R_3 is 87.5 mW.

Observation The final selected resistor is a 9.1 kΩ ±5%, ¼-W metal-film, carbon-film, or carbon-composition resistor.

Example 9–14 explores the effects of an open circuit, a short circuit, and normal operation of a series-parallel network.

EXAMPLE 9–14 For the network of Figure 9–25, determine the resistance between, the voltage across, and the current in terminals a-b for each of the three switch positions.

Position 1:

SOLUTION $R_{ab} = 0$ Ω. The resistance of a short circuit is virtually 0 Ω since the only resistance between terminals a-b is the switch contact resistance of a few milliohms and the resistance of the conductor (>10 milliohms).

$V_{ab} = 0$ V. Even though there is current in the short circuit, there is virtually no voltage (only a few millivolts) developed across the low resistance of the short circuit.

$I_{ab} = 0.30$ A. The current in the short circuit is the current in R_3. I_{ab} is determined in the following manner.

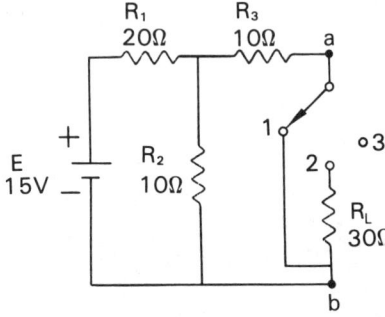

FIGURE 9–25
Network of Example 9–14.

$$I_{ab} = I_3 = \frac{V_3}{R_3}$$

where

$$V_3 = E - V_1 = E - IR_1$$

and

$$I = \frac{E}{R_T} = \frac{15}{20 + \dfrac{1}{\frac{1}{10} + \frac{1}{10}}} = 0.60 \text{ A}$$

Substitute:

$$V_3 = E - IR_1 = 15 - (0.6)20 = 3 \text{ V}$$

Substitute:

$$I_{ab} = \frac{V_3}{R_3} = \frac{3}{10} = 0.3 \text{ A}$$

Position 2:
$R_{ab} = 30 \; \Omega$. The contact and conductor resistance is insignificant when compared to the 30 ohms of resistance.

$V_{ab} = 3.21$ V. The voltage across terminals a-b is the voltage across R_L, which is determined as follows:

$$V_{ab} = V_L = I_L R_L$$

where

$$I_L = \frac{V_2}{R_3 + R_L}$$

and

$$V_2 = E - V_1 = E - IR_1$$

Solve for I:

$$I = \frac{E}{R_T} = \frac{E}{R_1 + \dfrac{1}{\dfrac{1}{R_2} + \dfrac{1}{R_3 + R_L}}}$$

$$I = \frac{15}{20 + \dfrac{1}{\frac{1}{10} + \frac{1}{40}}} = 0.536 \text{ A}$$

Substitute:

Applications with the Series-Parallel Network

$V_2 = E - IR_1 = 15 - (0.536)(20) = 4.28 \text{ V}$

Substitute:

$$I_L = \frac{V_2}{R_3 + R_L} = \frac{4.28}{10 + 30} = 0.107 \text{ A}$$

Substitute:

$V_{ab} = V_L = I_L R_L = (0.107)(30) = 3.21 \text{ V}$

$I_{ab} = 0.107$ A. The current in the load is the current in terminals a-b; therefore, $I_L = I_{a-b}$. I_L was determined during the calculation for V_{ab}.

Position 3:
$R_{ab} = \infty \ \Omega$. The resistance of the open circuit is extremely large ($>10^{15}$) and depends upon the insulating material used in the construction of the switch, as well as the temperature and moisture content in the air surrounding the switch contacts.

$V_{ab} = 5$ V. The voltage across terminals a-b is equal to the voltage across R_2. Because no current passes through R_3, V_3 (the voltage across R_3) is zero, and terminal a has the same voltage as the top end of R_2. V_{ab} was determined from the following:

$$V_{ab} = V_2 = IR_2 = \left(\frac{E}{R_T}\right) R_2$$

where

$R_T = R_1 + R_2 = 20 + 10 = 30 \ \Omega$

Substitute:

$$V_{ab} = V_2 = \left(\frac{E}{R_T}\right) R_2 = \left(\frac{15}{30}\right) 10 = 5.0 \text{ V}$$

$I_{ab} = 0$ A. Even though the potential across the switch contacts is 5 V, there is virtually zero current (<picoamperes) through the high resistance of the insulating material.

Observation The circuit conditions determined in this example are summarized in Table 9–1.

Sometimes a seemingly simple network may require the application of several circuit analysis techniques before the de-

The Series-Parallel Network

TABLE 9-1 SUMMARY FOR TERMINALS a–b OF FIGURE 9-25

Switch Position	R_{ab} (Ω)	V_{ab} (V)	I_{ab} (A)
1	0	0	0.30
2	30	3.2	0.107
3	∞	5	0

sired circuit parameter is determined. Example 9-15 is such a case.

EXAMPLE 9-15 The network of Figure 9-26 is powered from a 120-V ac source. The circuit components are housed in a sealed chassis, and only one of the two conductors connecting R_2 is accessible. It is known that R_1 is 10 Ω and R_3 is 30 Ω. Without opening (cutting) the accessible conductor or breaking the seal on the chassis, determine the value of R_2.

SOLUTION Measure the current in R_2 by attaching a *clamp-on* ammeter to the accessible conductor. The conductor need not be opened to use the clamp-on ammeter. Instead, this instrument fits around the current-carrying conductor. The ammeter reads 4 A. Therefore, $I_2 = 4$ A. Determine R_2:

$$R_2 = \frac{V_2}{I_2}$$

where

$$V_2 = V_3 = I_3 R_3$$

Determine I_3 using KVL. Thus:

$$E = V_1 + V_3$$

and

$$E = I_1 R_1 + I_3 R_3$$

However,

$$I_1 = I_3 + I_2 = I_3 + 4$$

Substitute $I_1 = I_3 + 4$. Thus:

$$E = I_1 R_1 + I_3 R_3$$

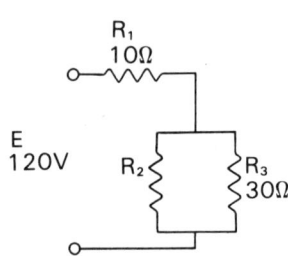

FIGURE 9-26
Network of Example 9-15.

$$E = (I_3 + 4)R_1 + I_3R_3$$

Substitute the circuit values and solve for I_3:

$$120 = (I_3 + 4)10 + I_330$$
$$120 = 10I_3 + 40 + 30I_3$$
$$40I_3 = 80$$
$$I_3 = 2 \text{ A}$$

Determine V_2:

$$V_2 = V_3 = I_3R_3 = 2(30) = 60 \text{ V}$$

Solve for R_2:

$$R_2 = \frac{V_2}{I_2} = \frac{60}{4} = 15 \text{ }\Omega$$

$\therefore$ R_2 has a resistance of 15 Ω.

Most of the time we are not too concerned with the resistance of the conductors (wiring) in a circuit. However, when large amounts of energy are moved in the conductors, their resistance may be significant. Example 9–16 explores this possibility.

EXAMPLE 9–16 A large electrically operated security gate is installed along with an electric light. The motor for the gate requires at least 110 V and 10 A of current for its operation. The light requires at least 110 V and 2 A of current for its operation. The source voltage at the power panel is 117 V. The 2-in. electrical conduit carrying number 14 gauge electrical wiring between the power panel and the gate is 630 ft long. Determine the voltage across the motor and light.

SOLUTION Draw a schematic of the circuit using resistive symbols to represent the wiring, motor, and lamp, as shown in Figure 9–27. From the wire table, determine the resistance of the number 14 copper conductor.

$$R_{\text{con}} = \frac{2.525 \text{ }\Omega}{1000 \text{ ft}}$$

Compute the resistance for the two 630-ft conductors. Thus:

$$R_{\text{con}} = \frac{2.525 \text{ }\Omega}{1000 \text{ ft}} \times 2(630) \text{ ft} = 3.18 \text{ }\Omega$$

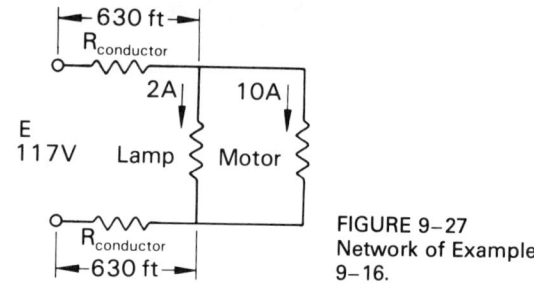

FIGURE 9-27
Network of Example 9-16.

Determine the voltage drop across the conductors. Thus:

$$V_{con} = IR_{con}$$

where

$$I = I_{lamp} + I_{motor} = 2 + 10 = 12 \text{ A}$$

Substitute:

$$V_{con} = IR_{con} = 12(3.18) = 38.2 \text{ V}$$

Determine the voltage across the motor and lamp. Thus:

$$V_{motor} = V_{lamp} = E - V_{con} = 117 - 38.2$$
$$V_{motor} = 78.8 \text{ V}$$

∴ The conductor resistance of 3.18 Ω is too large for this application.

Observation A large-diameter, lower-resistance wire is needed. The wiring cannot have a resistance greater than (117 − 110 V)/12 A or 0.583 Ω for the 1260 ft of wiring.

Determine a suitable wire gauge number for the wiring. Start by determining the resistance per 1000 ft. Thus:

$$\frac{0.583 \text{ }\Omega}{1260 \text{ ft}} = \frac{R \text{ }\Omega}{1000 \text{ ft}}$$

Solve for R, the resistance for 1000 ft:

$$R = \frac{(0.583)(1000)}{1260} = 0.463 \text{ }\Omega$$

∴ A number 6 gauge wire with 0.395 Ω per 1000 ft is selected for this application.

SELECTED TECHNICAL TERMS

Two technical terms used in this chapter are defined here to aid your understanding of them.

Network: a number of electric circuits joined together to form a system of interrelated circuits.
Series-parallel: an arrangement of the circuit elements so that some are in series and some are in parallel with each other.

PROBLEMS

As part of the solution, it is suggested that you draw and label a schematic diagram for each problem.

Section 9–1

9-1 Form the series-parallel network of Figure 9–28(a) into a series equivalent circuit. Let $E = 10$ V, $R_1 = 50$ Ω, $R_2 = 100$ Ω, and $R_3 = 100$ Ω.

9-2 Form the series-parallel network of Figure 9–28(a) into a series equivalent circuit. Let $E = 48$ V, $R_1 = 27$ kΩ, $R_2 = 43$ kΩ, and $R_3 = 16$ kΩ.

9-3 Form the series-parallel network of Figure 9–28(a) into a series equivalent circuit and determine the current in the circuit. Let $E = 18$ V, $R_1 = 5.6$ kΩ, $R_2 = 47$ kΩ, and $R_3 = 4.7$ kΩ.

9-4 Form the series-parallel network of Figure 9–28(a) into a series equivalent circuit and determine (a) the current in the circuit, and (b) the power dissipated by the equivalent resistance. Let $E = 60$ V, $R_1 = 0.75$ MΩ, $R_2 = 2.7$ MΩ, and $R_3 = 3.3$ MΩ.

9-5 Form the network of Figure 9–28(b) into an equivalent circuit and determine the voltage V_{ab} when $I = 12$ mA, $R_1 = 240$ Ω, $R_2 = 1.6$ kΩ, and $R_3 = 1.5$ kΩ.

9-6 Form the network of Figure 9–28(b) into an equivalent circuit and determine the voltage V_{ba} when $I = 32$ mA, $R_1 = 1.3$ kΩ, $R_2 = 2.2$ kΩ, and $R_3 = 3.0$ kΩ.

9-7 Form the network of Figure 9–29 into an equivalent circuit and determine the current in the circuit when $E = 22$ V, $R_4 = 1.5$ kΩ, and $R_5 = 910$ Ω.

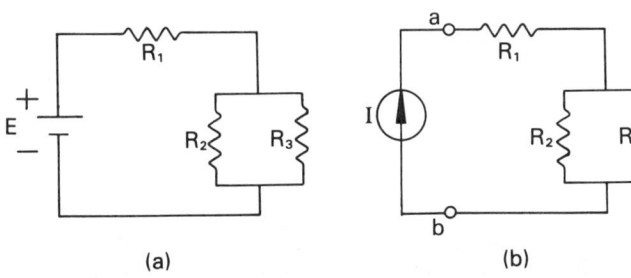

FIGURE 9–28
(a) Network of Problems 9–1, 9–2, 9–3, and 9–4.
(b) Network of Problems 9–5 and 9–6.

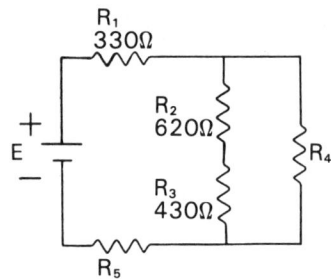

FIGURE 9-29
Network of Problems 9-7 and 9-8.

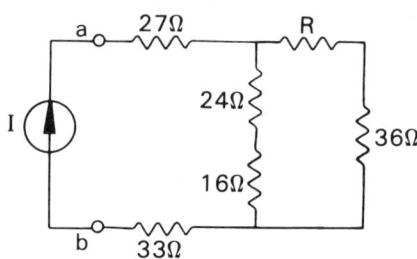

FIGURE 9-30
Network of Problems 9-9, 9-10, 9-11, and 9-12.

9-8 Form the network of Figure 9-29 into an equivalent circuit and determine (a) the current in the circuit, and (b) the power dissipated by the equivalent resistance when $E = 80$ V, $R_4 = 120$ Ω, and $R_5 = 68$ Ω.

9-9 For $I = 200$ mA and $R = 4$ Ω, form the network of Figure 9-30 into an equivalent circuit and determine V_{ba}.

9-10 For $I = 72$ mA and $R = 18$ Ω, form the network of Figure 9-30 into an equivalent circuit and determine V_{ab}.

9-11 Determine the source current, I, of Figure 9-30 when $V_{ab} = 10$ V and $R = 24$ Ω.

9-12 Determine the source current, I, of Figure 9-30 when $V_{ba} = -30$ V and $R = 8.2$ Ω.

Section 9-2

9-13 For the network of Figure 9-31, determine the current into node c when $E = 9$ V and $R_1 = R_2 = 2$ kΩ.

9-14 For the network of Figure 9-31, determine (a) the current into node a, (b) the currents into node b, and (c) the current into node c when $E = 36$ V, $R_1 = 12$ kΩ, and $R_2 = 8.2$ kΩ.

9-15 For the network of Figure 9-31, determine (a) I_1 and (b) V_{bc} when $E = 15$ V, $R_1 = 10$ kΩ, and $R_2 = 6.2$ kΩ.

9-16 For the network of Figure 9-31, determine (a) V_{ab} and (b) V_{bc} when $E = 18$ V, $R_1 = 3$ kΩ, and $R_2 = 4.7$ kΩ.

9-17 For the network of Figure 9-32, determine (a) the source voltage, E, and (b) the circuit current, I, when $R_1 = 3.3$ kΩ, $R_5 = 2.7$ kΩ, and $V_6 = 8$ V.

9-18 For the network of Figure 9-32, determine (a) the source voltage, E, and (b) the circuit current, I, when $R_1 = 1.5$ kΩ, $R_5 = 820$ Ω, and $V_{bc} = 3$ V.

9-19 For the network of Figure 9-32, determine R_1 when $V_{bd} = 12$ V, $R_5 = 1.8$ kΩ, and $E = 18$ V.

9-20 For the network of Figure 9-32, determine R_5 when $V_{ac} = 7$ V, $R_1 = 1$ kΩ, and $I_5 = 2.27$ mA.

9-21 Determine the source current, I, of the network pictured in Figure 9-33 when $V_{ad} = 94$ V.

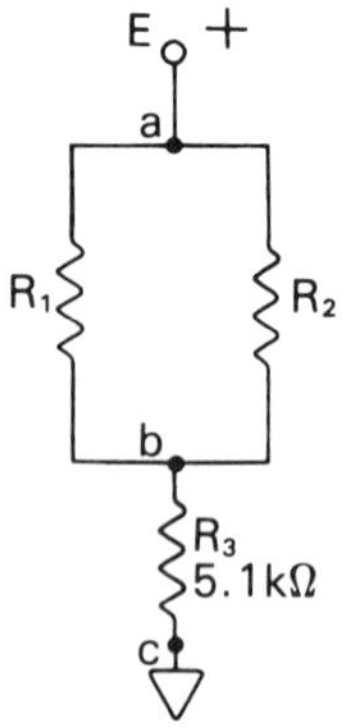

FIGURE 9-31
Network of Problems 9-13, 9-14, 9-15, and 9-16.

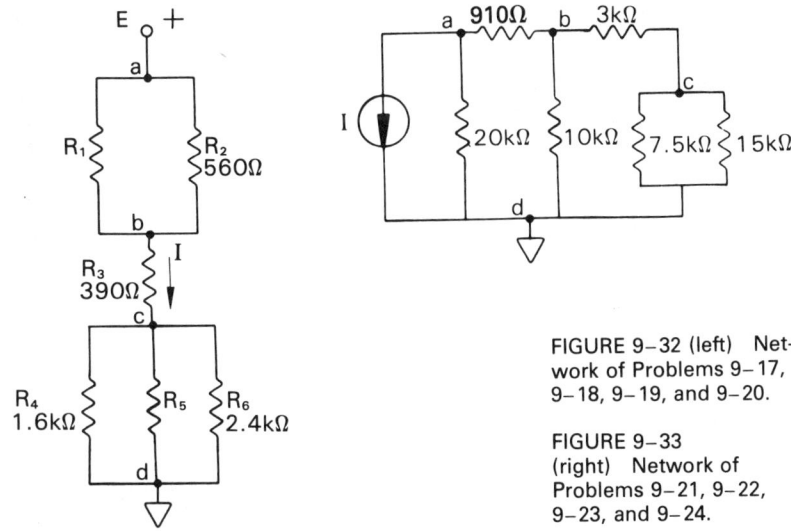

FIGURE 9-32 (left) Network of Problems 9-17, 9-18, 9-19, and 9-20.

FIGURE 9-33 (right) Network of Problems 9-21, 9-22, 9-23, and 9-24.

9-22 Determine the current(s) entering and leaving node c of the network pictured in Figure 9-33 when I is 54 mA.

9-23 For the network of Figure 9-33, determine V_{cd} when $I = 28$ mA.

9-24 For the network of Figure 9-33, determine V_{ba} when $I = 16$ mA.

9-25 For the network of Figure 9-34, determine the current leaving node a when $E = 27$ V and $R = 10$ Ω.

9-26 For the network of Figure 9-34, determine the current entering node c when $E = 15$ V and R is 36 Ω.

9-27 Determine V_{bc} in the network of Figure 9-34 when $E = 9$ V and $R = 82$ Ω.

9-28 Determine the voltage across the 10-Ω resistor in the network of Figure 9-34 when $E = 60$ V and $R = 22$ Ω.

Section 9-3

9-29 Determine the power delivered to the 510-Ω resistor in the network of Figure 9-35.

9-30 Determine the power delivered to the 220-Ω resistor in the network of Figure 9-35.

9-31 Determine the power delivered to the 510-Ω resistor when nodes a and c are short circuited in the network of Figure 9-35.

9-32 Determine the power delivered to the 220-Ω resistor in the network of Figure 9-35 when nodes d and e are short circuited.

9-33 For the network of Figure 9-36, determine (a) V_{bc}, (b) R,

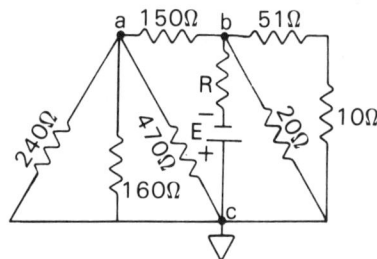

FIGURE 9-34 Network of Problems 9-25, 9-26, 9-27, and 9-28.

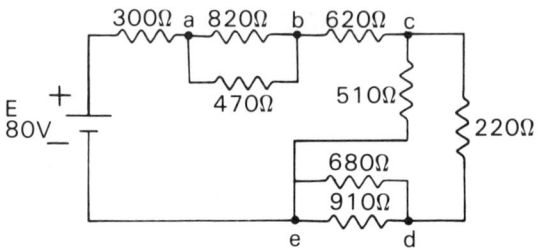

FIGURE 9-35 Network of Problems 9-29, 9-30, 9-31, and 9-32.

and (c) P_R when $V_{ac} = 15$ V.

9-34 For the network of Figure 9-36, determine (a) V_{bc}, (b) R, and (c) P_R when $V_{ac} = 13.5$ V.

Section 9-4

9-35 A 4-kW motor is connected in parallel with a 500-W light. The motor and light are located 500 ft from a 240-V source. Determine the wire size needed to maintain the voltage drop in the line to no more than 5% of the source. Assume the conductors to be copper operating at 20°C.

9-36 Repeat Problem 9-35, but limit the voltage drop in the line to no more than 10% of the source.

9-37 An electric iron is connected in parallel with a 250-W lamp. The electric conductors are 12-gauge solid copper at 20°C. Determine the distance between the load and source if the voltage falls from 117 V at the source to 105 V at the load and the current in the iron is 12 A.

9-38 A voltmeter placed across the 80-Ω resistor in Figure 9-37 indicates 5 V. Determine what, if anything, is shorted or opened in the network.

9-39 A voltmeter placed across the 120-Ω resistor in Figure 9-37 indicates 7.5 V. Determine what, if anything, is shorted or opened in the network.

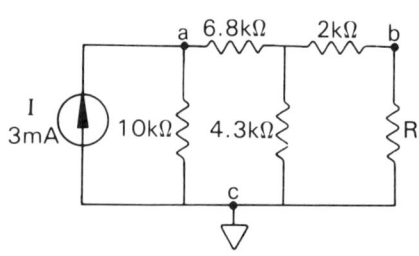

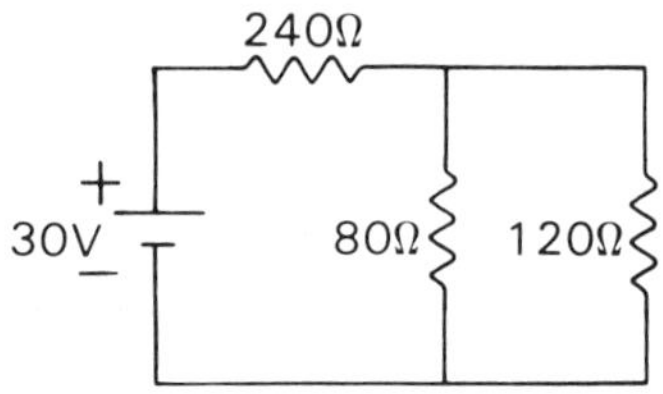

FIGURE 9-36 (left) Network of Problems 9-33 and 9-34.
FIGURE 9-37 (right) Network of Problems 9-38, 9-39, and 9-40.

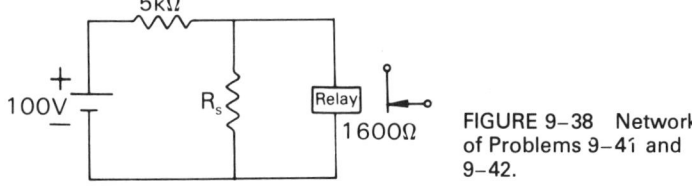

FIGURE 9-38 Network of Problems 9-41 and 9-42.

9-40 A voltmeter placed across the 240-Ω resistor in Figure 9-37 indicates 30 V. Determine what, if anything, is shorted or opened in the network.

9-41 Select a resistor from the E24 preferred number series for R_s of Figure 9-38 so that 7.5 mA ± 10% passes through the relay coil. Also, select the power rating of the resistor.

9-42 Select a resistor from the E24 preferred number series for R_s of Figure 9-38 so that 5 mA ± 10% passes through the relay coil. Also, select the power rating of the resistor.

10 Voltage and Current Dividers

10-1 VOLTAGE DIVIDERS
10-2 VOLTAGE DIVIDER APPLICATIONS
10-3 CURRENT DIVIDERS
10-4 CURRENT DIVIDER APPLICATIONS
10-5 LOADED AND UNLOADED CONDITIONS
10-6 APPROXIMATIONS

CHAPTER PREVIEW The process of designing an electronic circuit may simply be described as the control of the voltages and currents that are applied to specific electronic devices. In studying about *active* electronic devices, such as transistors and integrated circuits, it is learned that these devices are designed to operate with specific voltages and currents. These sophisticated solid-state devices, as well as the simple electric light bulb, will not work properly if too small or too large a voltage is applied. In fact, applying too high a voltage may cause the device to *burn up*. Since every electronic device has both a voltage and a current limitation as well as an *optimum* (best) operating range, the design of an electronic circuit must involve the proper selection of voltage and current for each electronic device in the circuit.

Because size and cost are major factors in a circuit design, circuits must be operated from a limited number of power sources. In most consumer products (transistor radios, electronic games, toys, etc.), only one or two different batteries (voltage levels) are used to power the product. However, in complex electronic equipment, many power sources may be used. Typical dc power-supply voltage levels include ±5, ±12, ±15, ±18, and ±28 V.

Since power supplies are one of the most expensive parts of an electronic system, only a few main voltage sources are ever used. However, electronic circuitry with many solid-state devices requires many different voltage levels for proper operation. This, then, is why the knowledge of voltage dividers is most important. With an understanding of voltage division, any number of voltage levels may be provided from one power supply, thus ensuring proper circuit operation at a minimal cost. This chapter explores both voltage and current dividers.

PERFORMANCE OBJECTIVES Once you have read each section, worked each example with pencil, paper, and calculator, and answered each problem for every section, you will be able to:

☐ Analyze and design voltage dividers.
☐ Analyze and design current dividers.
☐ Understand the effects of *loading* and *unloading* a network.
☐ Make approximations of circuit parameters without formal calculations.

10–1 VOLTAGE DIVIDERS In previous chapters, you learned how voltage and current are related to resistance through Ohm's law. With the knowledge of Ohm's law and an understanding of equivalent resistance of a

Voltage Dividers

series circuit, you have all the tools necessary to analyze and design voltage dividers.

Figure 10–1 shows two circuits that are very similar in that each has a 10-V source. However, the circuit of Figure 10–1(b) provides a second voltage of +5 V. Both circuits have 10 mA of current flowing through them. Computing the voltage dropped across each resistance results in the following voltages.

From Figure 10–1(a):

$V_1 = (10\text{ mA})(1\text{ k}\Omega)$

$V_1 = (10 \times 10^{-3})(1 \times 10^3)$

$V_1 = 10\text{ V}$

From Figure 10–1(b):

$V_2 = (10\text{ mA})(500\text{ }\Omega)$

$V_2 = 10 \times 10^{-3} \times 500$

$V_2 = 5\text{ V}$

$V_3 = (10\text{ mA})(500\text{ }\Omega)$

$V_3 = 10 \times 10^{-3} \times 500$

$V_3 = 5\text{ V}$

$V_{2+3} = (10\text{ mA})(500\text{ }\Omega + 500\text{ }\Omega)$

$V_{2+3} = 10 \times 10^{-3} \times 1000$

$V_{2+3} = 10\text{ V}$

From the preceding calculation for the circuit of Figure 10–1(a), it is now apparent that the circuit provides only +10 V. Additionally, the calculations for the circuit of Figure 10–1(b) indicate that two voltages (+10 V and 5 V) are available.

Voltage-divider analysis and design is like building a flight of stairs up a hillside. In building a flight of stairs, you have several design options to consider:

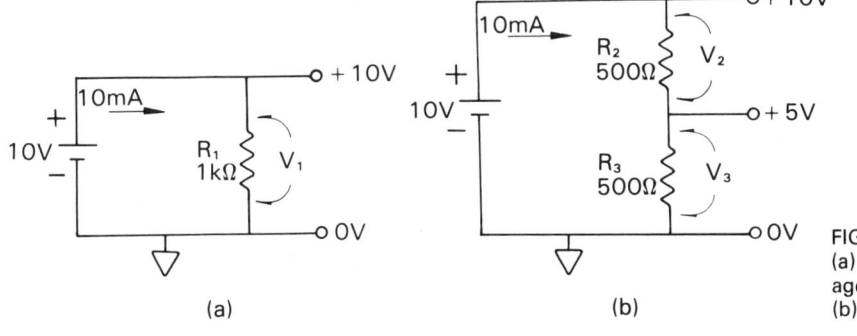

FIGURE 10–1
(a) Single output voltage.
(b) Two output voltages.

Voltage and Current Dividers

☐ If the height of each step is constant, the number of steps required is determined by dividing the height of the flight of stairs by the height of one step.
☐ If the number of steps of constant height is known, the height of each step is known.
☐ If the height of each step is different (for variety), the number of steps will vary.

Similarly, in designing a voltage divider, you have several design options to consider:

☐ If the amount of each voltage step is constant, the total voltage will equal the number of required steps times the value of each step.
☐ If the number of voltages is of equal value, the value of each step is the total voltage divided by the number of steps.
☐ If the amount of the individual voltage steps is different, the total voltage will equal the sum of the individual voltage steps.

In designing a voltage divider, you need a clear understanding of the number of voltages, the level of each voltage, and the total voltage to be applied to the divider.

Voltage-Divider Analysis

The analysis of voltage dividers will start with the case shown in Figure 10–2. First the current flow (I) is computed; then the voltage drop across each resistor is computed by applying Ohm's law. This is demonstrated in Example 10–1.

EXAMPLE 10–1 Determine the voltage drop across R_1 and R_2 of Figure 10–2.

SOLUTION Compute the circuit current, I:

$$I = \frac{E}{R_T}$$

$$E = 24 \text{ V}$$

$$R_T = R_1 + R_2 = 10 \text{ k}\Omega + 30 \text{ k}\Omega = 40 \text{ k}\Omega$$

Substitute 24 V for E and 40 kΩ for R_T:

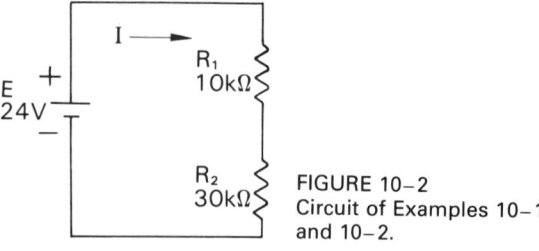

FIGURE 10–2
Circuit of Examples 10–1 and 10–2.

Voltage Dividers

$$I = \frac{24}{40 \times 10^3} = 0.60 \times 10^{-3} = 0.60 \text{ mA}$$

Determine V_1 and V_2:

$$V_1 = IR_1 = (0.60 \times 10^{-3})(10 \times 10^3) = 6.0 \text{ V}$$

$$V_2 = IR_2 = (0.60 \times 10^{-3})(30 \times 10^3) = 18 \text{ V}$$

$$\therefore \quad V_1 = 6 \text{ V} \quad \text{and} \quad V_2 = 18 \text{ V}$$

Notice in Example 10–1 that the voltage drop across each resistor is proportional to the resistance. That is, R_1 is one third of R_2, and V_1 is one third of V_2. Because the voltage drop in a series circuit is proportional to the values of the resistance, another method of calculating voltage drops may be developed.

From Kirchhoff's law, we know that in a series circuit $E = V_1 + V_2$, and from Ohm's law, $E = IR_T = I\Sigma R$. We also know that $I = I_1 = I_2$ since current is constant throughout the series circuit. Thus:

$$I = \frac{E}{R_T}, \quad I_1 = \frac{V_1}{R_1}, \quad I_2 = \frac{V_2}{R_2}$$

Since $I = I_1$ and $I = I_2$,

$$\frac{E}{R_T} = \frac{V_1}{R_1}$$

Transposing results in

$$\frac{E}{V_1} = \frac{R_T}{R_1} \tag{10-1}$$

and

$$\frac{E}{V_2} = \frac{R_T}{R_2} \tag{10-2}$$

Also $I_1 = I_2$; then

$$\frac{V_1}{R_1} = \frac{V_2}{R_2}$$

Transposing results in

$$\frac{V_1}{V_2} = \frac{R_1}{R_2} \tag{10-3}$$

These three equations tell us that **the ratio of any two voltages in a series circuit is equal to the ratio of their respective resistances.** This is the *voltage-divider principle*. Each of these equations will be used in Example 10–2 to rework Example 10–1.

EXAMPLE 10-2 Compute the voltage drop across R_1 and R_2 of Figure 10-2. Use Equations 10-1, 10-2, and 10-3.

SOLUTION From Figure 10-2, $E = 24$ V, $R_1 = 10$ kΩ, $R_2 = 30$ kΩ, and $R_T = 40$ kΩ. Solving for V_1 from Equation 10-1,

$$\frac{E}{V_1} = \frac{R_T}{R_1}$$

$$V_1 = \frac{ER_1}{R_T}$$

Substitute:

$$V_1 = \frac{(24)(10 \times 10^3)}{40 \times 10^3}$$

∴ $V_1 = 6.0$ V

From Equation 10-2,

$$\frac{E}{V_2} = \frac{R_T}{R_2}$$

$$V_2 = \frac{ER_2}{R_T}$$

Substitute:

$$V_2 = \frac{(24)(30 \times 10^3)}{40 \times 10^3}$$

∴ $V_2 = 18$ V

From Equation 10-3,

$$\frac{V_1}{V_2} = \frac{R_1}{R_2}$$

$$V_2 = \frac{V_1 R_2}{R_1}$$

Substitute:

$$V_2 = \frac{(6.0)(30 \times 10^3)}{10 \times 10^3}$$

∴ $V_2 = 18$ V

Example 10-2 demonstrates that the voltage-divider principle may be applied to any two voltage drops in a series circuit. This principle may be stated in a general formula of voltage division as

$$\frac{E}{V_n} = \frac{R_T}{R_n} \qquad (10\text{-}4)$$

where E = source voltage
V_n = voltage drop across R_n
R_T = summation of all the series resistances ($R_T = \Sigma R$)
R_n = selected resistance of the series circuit

A very useful form of the voltage-divider formula for a known source voltage is stated as Equation 10-5.

VOLTAGE-DIVIDER EQUATION

$$V_n = \frac{ER_n}{\Sigma R} \quad (\text{V}) \qquad (10\text{-}5)$$

where V_n = voltage dropped across R_n in volts
E = source voltage in volts
R_n = selected resistance in the voltage divider in ohms
ΣR = summation of resistance ($R_1 + R_2 + \cdots + R_n$) in the series voltage divider in ohms

---

Example 10-3 will use Equation 10-5 to determine the voltage division in Figure 10-3.

EXAMPLE 10-3 Determine the voltage across R_1 (V_{ab}), R_2 (V_{bc}), and R_3 (V_{cd}) in Figure 10-3(a).

SOLUTION Substitute into Equation 10-5; let $\Sigma R = 100$ kΩ and solve for V_{ab}.

$$V_n = \frac{ER_n}{\Sigma R}$$

$$V_{ab} = \frac{(20)(20 \times 10^3)}{100 \times 10^3}$$

$$V_{ab} = 4.0 \text{ V}$$

Solve for V_{bc}:

$$V_{bc} = \frac{20(10 \times 10^3)}{100 \times 10^3}$$

$$V_2 = 2.0 \text{ V}$$

Solve for V_{cd}:

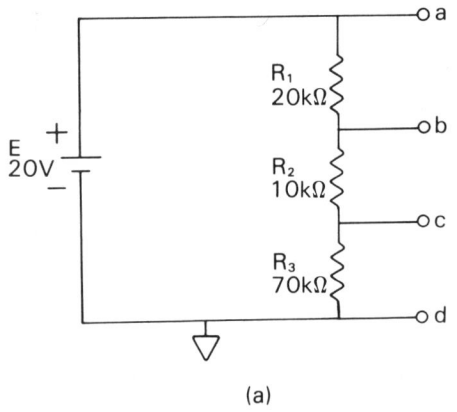
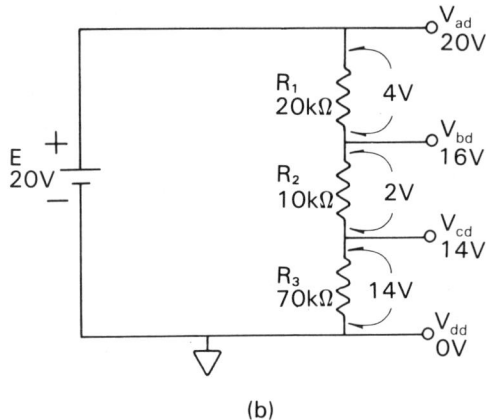

(a)   (b)

FIGURE 10-3 (a) Circuit of Example 10-3. (b) Output voltages of Example 10-3.

$$V_{cd} = \frac{20(70 \times 10^3)}{100 \times 10^3}$$

$$V_{cd} = 14 \text{ V}$$

∴ $V_1 = 4.0$ V, $V_2 = 2.0$ V, $V_3 = 14.0$ V

Check Apply Kirchhoff's voltage law:

$$E = V_{ab} + V_{bc} + V_{cd}$$

$$20 = 4.0 + 2.0 + 14$$

$$20 \text{ V} = 20 \text{ V}$$

In Example 10-3 the voltage drop across each resistance in the divider was computed. The voltage from each point in the divider *with respect to the reference point* is shown as Figure 10-3(b).

Notice that

$V_{ad} = 20$ V $(V_{ab} + V_{bc} + V_{cd} = 4$ V $+ 2$ V $+ 14$ V $= 20$ V$)$

$V_{bd} = 16$ V $(V_{bc} + V_{cd} = 2$ V $+ 14$ V $= 16$ V$)$

$V_{cd} = 14$ V $(V_{cd} = 14$ V$)$

Given a source voltage, any desired voltage may be obtained by computing the appropriate resistance for the voltage divider. This concept is demonstrated in Example 10-4.

EXAMPLE 10-4 Given the circuit of Figure 10-4(a) with a source of 20 V and an R_T of 10 kΩ, determine the values of R_1 and R_2 in Figure 10-4(b) so that $R_1 + R_2 = 10$ kΩ, $V_1 = 15$ V, and $V_2 = 5$ V.

Voltage Dividers

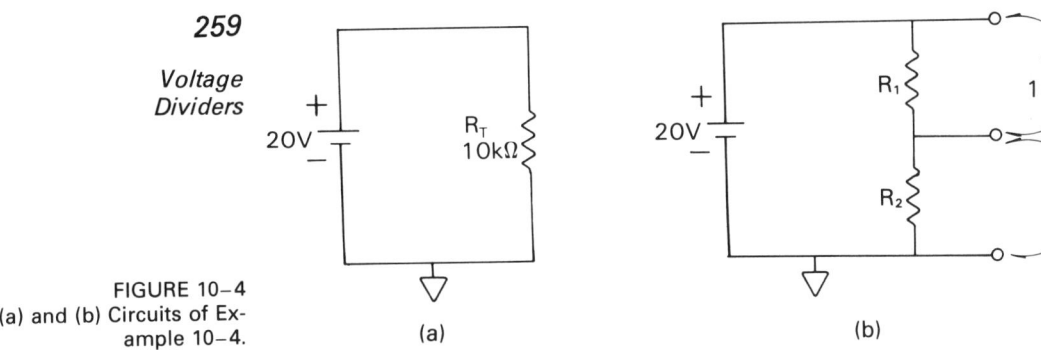

FIGURE 10-4 (a) and (b) Circuits of Example 10-4.

SOLUTION Use Equation 10-4:

$$\frac{E}{V_n} = \frac{R_T}{R_n}$$

From Figure 10-4(b), let $E = E$, $V_n = V_1$, $R_1 = R_n$, and $R_T = 10$ kΩ. Thus:

$$\frac{E}{V_1} = \frac{R_T}{R_1}$$

Solve for R_1 and substitute:

$$R_1 = \frac{V_1 R_T}{E} = \frac{(15)(10 \times 10^3)}{20} = 7.5 \times 10^3$$

$$\therefore \quad R_1 = 7.5 \text{ k}\Omega$$

Solve for R_2 and substitute given

$$R_2 = \frac{V_2 R_T}{E} = \frac{(5)(10 \times 10^3)}{20} = 2.5 \times 10^3$$

$$\therefore \quad R_2 = 2.5 \text{ k}\Omega$$

Check $R_T = R_1 + R_2$

10 kΩ = 7.5 kΩ + 2.5 kΩ

10 kΩ = 10 kΩ

Since $R_T = R_1 + R_2$ in a series circuit, it is then possible to determine the resistances in a voltage divider given the source voltage, the sum of the resistances in the divider, and the desired voltage level. This concept is demonstrated in Example 10-5.

EXAMPLE 10-5 Given the circuit of Figure 10-5, determine the value of R_1 and R_2 when $R_T = 150$ kΩ.

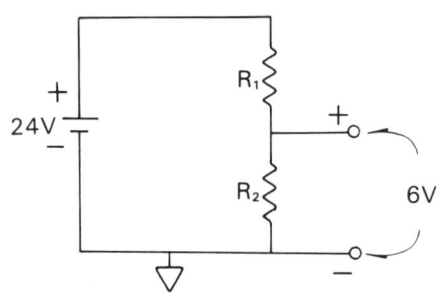

FIGURE 10-5
Circuit of Example 10-5.

SOLUTION Solve for R_2 in the following equation (Equation 10-2) and substitute:

$$\frac{E}{V_2} = \frac{R_T}{R_2}$$

$$R_2 = \frac{V_2 R_T}{E} = \frac{(6)(150 \times 10^3)}{24} = 37.5 \times 10^3$$

∴ $R_2 = 37.5$ kΩ

Determine R_1 for $R_T = R_1 + R_2$

$$R_1 = R_T - R_2$$

Substitute:

$$R_1 = 150 \text{ k}\Omega - 37.5 \text{ k}\Omega$$

∴ $R_1 = 112.5$ kΩ

It is not necessary to memorize each of the formulas stated in this section. Instead, remember the general principle: *In a voltage divider, the voltages dropped are directly proportional to the value of the resistances.*

$$\frac{V_n}{R_n} = \frac{V_m}{R_m}$$

10-2 VOLTAGE-DIVIDER APPLICATIONS

Because most schematics do not specify the voltages at every point in the circuit, you must have the ability to calculate the voltage at the unspecified points. This ability will greatly assist in analyzing and *troubleshooting* a circuit. By knowing how to calculate the division of voltage in a series circuit, you can determine if the circuit is malfunctioning by comparing the voltage measurements to the voltage calculations.

EXAMPLE 10-6 Determine if the circuit of Figure 10-6 is operating properly when measured voltages are

Voltage Divider Applications

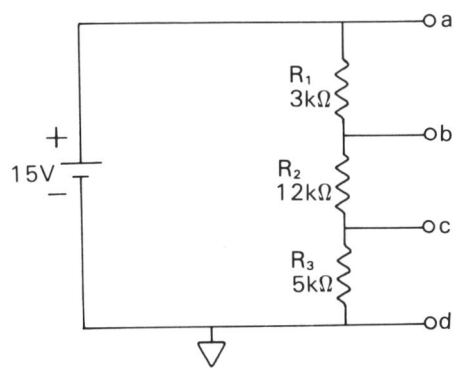

FIGURE 10-6
Circuit of Example 10-6.

$$V_{ad} = 15\,V, \quad V_{bd} = 12.35\,V, \quad V_{cd} = 4.41\,V$$

SOLUTION Determine V_{ad}:

$V_{ad} = 15\,V$ (the source voltage)

Calculate V_{bd} from:

$$\frac{V_{bd}}{R_{bd}} = \frac{E}{R_T}$$

$$V_{bd} = \frac{E R_{bd}}{R_T} = \frac{(15)(17 \times 10^3)}{20 \times 10^3} = 12.75\,V$$

∴ V_{bd} measured $\neq V_{bd}$ calculated (12.35 V $\neq$ 12.75 V).

Calculate V_{cd} from:

$$V_{cd} = \frac{E R_{cd}}{R_T} = \frac{(15)(5 \times 10^3)}{20 \times 10^3} = 3.75\,V$$

∴ V_{cd} measured $\neq V_{cd}$ calculated (4.41 V $\neq$ 3.75 V). Thus, the circuit is not operating properly.

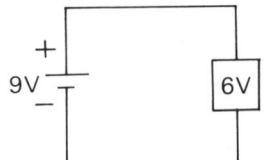

FIGURE 10-7
What needs to be added to this circuit for proper operation?

As previously mentioned, resistors are used to control the voltage in different parts of a series circuit. Let's assume that an electronic device requires a source of 6 V at a current of 1 mA. If the device is connected to the 9 V source, as shown in Figure 10-7, it would malfunction. What can be done to operate this device properly?

By inserting a resistor in series with the device, as pictured in Figure 10-8, the correct operating characteristics are established. From Kirchhoff's law, we know that there is 3 V across R and 1 mA of current is flowing through R. Therefore, $R = E/I = 3\,V/1\,mA = 3\,k\Omega$. So, if a 3-k$\Omega$ resistor is placed in series

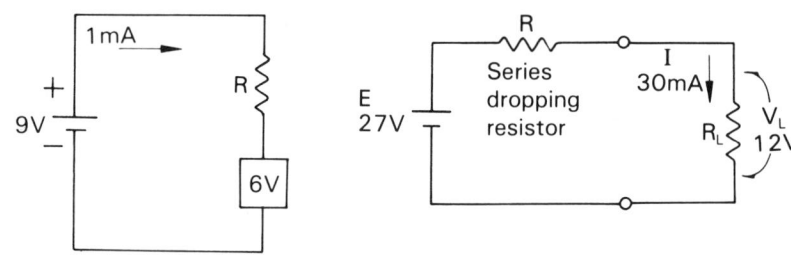

FIGURE 10-8 (left) By adding a series dropping resistor to the circuit of Figure 10-7, the proper operating point is achieved.
FIGURE 10-9 (right) Circuit of Example 10-7.

with the device, proper operation is ensured. A resistor used in this manner is called a *series dropping resistor*.

EXAMPLE 10-7 An electronic preamplifier requires an operating voltage of 12 V at a current of 30 mA. Determine the value of the series dropping resistor required to operate this device from a 27-V source.

SOLUTION Represent the circuit as a resistance (R_L) of

$$R_L = \frac{12 \text{ V}}{30 \text{ mA}} = 400 \text{ }\Omega$$

From Figure 10-9,

$$R = \frac{V_R}{I} = \frac{E - V_L}{I} = \frac{27 - 12}{30 \times 10^{-3}} = 500$$

$\therefore \quad R = 500 \text{ }\Omega$

Complete the design by computing the power rating of R:

$$P_R = I^2 R = (30 \times 10^{-3})^2 (500) = 400 \times 10^{-3}$$

$\therefore \quad P_R = 450$ mW

Observation A 510 $\Omega \pm 5\%$ carbon-composition resistor rated at 1 W would be a satisfactory selection. A large wattage rating is selected to provide a lower operating temperature, thus ensuring a stable resistance value and reliable operation.

EXAMPLE 10-8 The potentiometer of Figure 10-10(a) is used to *trim* (adjust) the circuit for proper operation. In the circuit of Figure 10-10(a), this is achieved when V_{ab} is adjusted to 2.2 V. After this adjustment, it is desired that the potentiometer be replaced with two fixed metal-film resistors, as shown in Figure 10-10(b). Determine the values of R_1 and R_2.

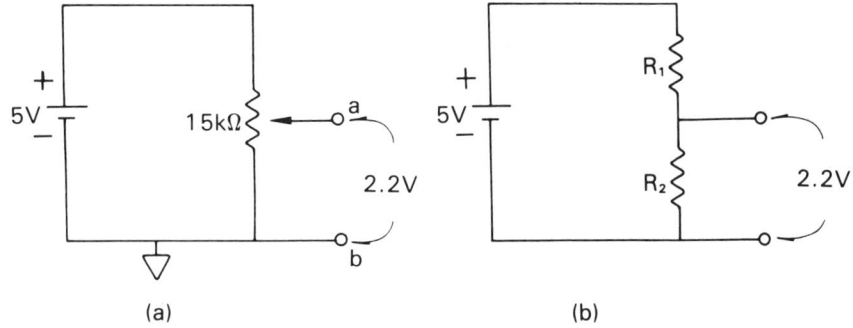

FIGURE 10-10 (a) A potentiometer is used to determine desired output voltage. (b) Fixed resistors are used to replace the potentiometer.

SOLUTION The voltage across $R_2 = 2.2$ V, $E = 5$ V, and $R_T = 15$ kΩ. Determine R_2 from

$$R_2 = \frac{V_2 R_T}{E}$$

$$R_2 = \frac{(2.2)(15 \times 10^3)}{5} = 6.6 \times 10^3$$

$\therefore \quad R_2 = 6.6$ kΩ

Determine R_1 from

$$R_1 = R_T - R_2$$

$$R_1 = 15 \text{ k}\Omega - 6.6 \text{ k}\Omega$$

$\therefore \quad R_1 = 8.4$ kΩ

Compute the power rating of R_1 and R_2:

$$P_1 = \frac{V_1^2}{R_1} = \frac{(5 - 2.2)^2}{8.4 \times 10^3} = 933 \ \mu\text{W}$$

$$P_1 = 933 \ \mu\text{W}$$

$$P_2 = \frac{V_2^2}{R_2} = \frac{2.2^2}{6.6 \times 10^3} = 733 \ \mu\text{W}$$

$\therefore \quad R_1$ is selected as an 8.45 kΩ ± 1% $\frac{1}{8}$-W metal-film resistor, and R_2 is selected as a 6.65 kΩ ± 1% $\frac{1}{8}$-W metal-film resistor.

10-3 CURRENT DIVIDERS

In this section you will be concerned with learning how to control the amount of current in two or more parallel branches. Because many electronic devices and circuits are *current operative*, that is, they depend on current for their operation, the knowledge of how to control circuit current is very important.

Voltage and Current Dividers

Determining how much current is in a parallel circuit can be accomplished through calculations using Ohm's law and Kirchhoff's current law. Figure 10–11 shows a simple current source and two resistors in parallel. From Kirchhoff's current law, we know that $I_T = I_1 + I_2$. We also know that $E = V_1 = V_2$, since the voltage is the same across two resistors in parallel.

Thus, in a parallel circuit

$$E = I_T R_{eq} = \frac{I_T}{G_T}, \quad V_1 = I_1 R_1 = \frac{I_1}{G_1}, \quad V_2 = I_2 R_2 = \frac{I_2}{G_2}$$

where

$$R_{eq} = \frac{1}{G_T}, \quad R_1 = \frac{1}{G_1}, \quad R_2 = \frac{1}{G_2}$$

Since $E = V_1$ and $E = V_2$, then by substitution

$$E = V_1$$

$$\frac{I_T}{G_T} = \frac{I_1}{G_1}$$

Transposing results in

$$\frac{I_T}{I_1} = \frac{G_T}{G_1} \tag{10-6}$$

and

$$E = V_2$$

$$\frac{I_T}{I_2} = \frac{G_T}{G_2} \tag{10-7}$$

Also, $V_1 = V_2$; thus

$$\frac{I_1}{G_1} = \frac{I_2}{G_2}$$

Transposing results in

$$\frac{I_1}{I_2} = \frac{G_1}{G_2} \tag{10-8}$$

These three equations tell us that **the ratio of any two currents in a parallel circuit is equal to the ratio of their respective conductances.** This is the *current-divider* principle.

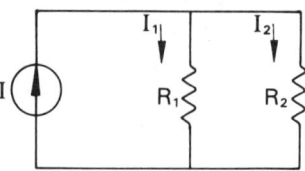

FIGURE 10–11
Simple current divider circuit of Example 10–9.

EXAMPLE 10-9 Determine the current through R_2 of Figure 10-11 when $I_T = 100$ mA, $G_1 = 5$ mS, and $G_2 = 2$ mS.

SOLUTION Substitute into Equation 10-7 where $G_T = G_1 + G_2 = 7$ mS and solve for I_2. Thus:

$$\frac{I_T}{I_2} = \frac{G_T}{G_2}$$

$$\frac{100 \times 10^{-3}}{I_2} = \frac{7 \times 10^{-3}}{2 \times 10^{-3}}$$

$$I_2 = \frac{(100 \times 10^{-3})(2 \times 10^{-3})}{7 \times 10^{-3}} = 28.6 \times 10^{-3}$$

$$\therefore \quad I_2 = 28.6 \text{ mA}$$

Since the voltage is constant across a parallel circuit ($E = V_n$), any two currents are directly proportional to the conductance they pass through. This concept is stated as a general formula of current division as

$$\frac{I_T}{I_n} = \frac{G_T}{G_n} \qquad (10-9)$$

A very useful form of the current-divider formula for a known source current is stated as Equation 10-10.

CURRENT-DIVIDER EQUATION

$$I_n = \frac{IG_n}{\Sigma G} \qquad (10-10)$$

where I_n = current in G_n
I = source current in amperes
G_n = conductance of one branch in siemens
ΣG = summation of conductances ($G_1 + G_2 + \cdots + G_n$) in the parallel current divider in siemens

---

EXAMPLE 10-10 Use Equation 10-10 to determine currents I_1 and I_2 of Figure 10-12.

SOLUTION Determine ΣG:

$$\Sigma G = G_1 + G_2 = \frac{1}{R_1} + \frac{1}{R_2}$$

$$\Sigma G = \frac{1}{8 \times 10^3} + \frac{1}{2 \times 10^3} = 125 \ \mu S + 500 \ \mu S$$

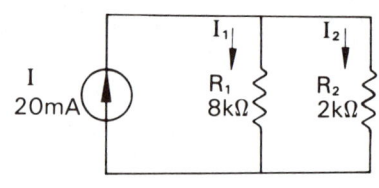

FIGURE 10-12
Circuit of Example 10-10.

$$\Sigma G = 625 \ \mu S$$

Substitute and solve for I_1 in Equation 10-10:

$$I_1 = \frac{IG_1}{\Sigma G} = \frac{(20 \times 10^{-3})(125 \times 10^{-6})}{625 \times 10^{-6}}$$

∴ $I_1 = 4$ mA

Substitute and solve for I_2:

$$I_2 = \frac{IG_2}{\Sigma G} = \frac{(20 \times 10^{-3})(500 \times 10^{-6})}{625 \times 10^{-6}}$$

∴ $I_2 = 16$ mA

Check $I = I_1 + I_2$

20 mA = 4 mA + 16 mA

20 mA = 20 mA

For a circuit with any number of parallel branches, as shown in Figure 10-13,

$$\frac{I_T}{G_T} = \frac{I_1}{G_1} = \frac{I_2}{G_2} = \frac{I_3}{G_3} = \cdots = \frac{I_n}{G_n}$$

If the values of all the conductances along with one of the currents is known, then any of the other currents may be calculated. Also, if the current in each branch and one of the conductances is known, then the values of all the remaining conductances may be calculated.

EXAMPLE 10-11 Figure 10-14 shows a parallel circuit with three branches. Using this circuit, calculate the current in each branch.

SOLUTION Calculate G_T:

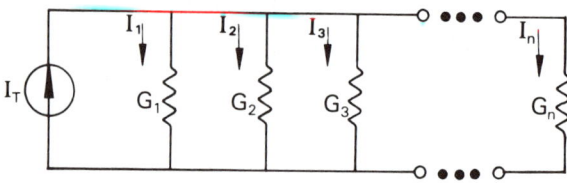

FIGURE 10-13
Current source with "n" branches.

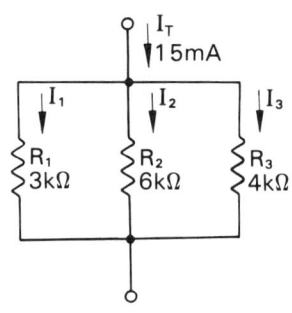

FIGURE 10-14
Three-branch circuit of Example 10-11.

$$G_T = G_1 + G_2 + G_3$$

$$G_T = \frac{1}{3 \times 10^3} + \frac{1}{6 \times 10^3} + \frac{1}{4 \times 10^3}$$

$$G_T = 0.750 \times 10^{-3} = 750 \ \mu S$$

Determine I_1 from $I_1/G_1 = I_T/G_T$:

$$I_1 = \frac{I_T G_1}{G_T} = \frac{(15 \times 10^{-3})(333 \times 10^{-6})}{750 \times 10^{-6}}$$

$$\therefore \quad I_1 = 6.66 \text{ mA}$$

Determine I_2 from $I_2/G_2 = I_T/G_T$:

$$I_2 = \frac{I_T G_2}{G_T} = \frac{(15 \times 10^{-3})(167 \times 10^{-6})}{750 \times 10^{-6}}$$

$$\therefore \quad I_2 = 3.34 \text{ mA}$$

Determine I_3 from $I_3/G_3 = I_T/G_T$:

$$I_3 = \frac{I_T G_3}{G_T} = \frac{(15 \times 10^{-3})(250 \times 10^{-6})}{750 \times 10^{-6}}$$

$$\therefore \quad I_3 = 5.0 \text{ mA}$$

Check
$$I_T = I_1 + I_2 + I_3$$

15 mA = 6.66 mA + 3.34 mA + 5.0 mA

15 mA = 15 mA

It is important that you do *not* try to memorize all the divider formulas. Instead, remember the fundamental principle of current division: *in a current divider, the currents are directly proportional to the conductances.* Thus:

$$\frac{I_n}{G_n} = \frac{I_m}{G_m}$$

Voltage and Current Dividers

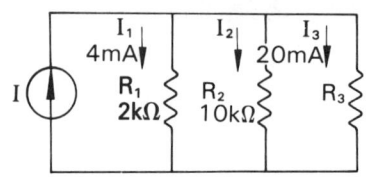

FIGURE 10-15
Circuit of Example 10-12.

Since current in a parallel branch is directly proportional to conductance, it must also be inversely proportional to resistance. Thus:

$$\frac{I_1}{I_2} = \frac{R_2}{R_1} \qquad (10\text{-}11)$$

Example 10-12 uses Equation 10-11 along with previously learned concepts to determine the values of resistance and current of Figure 10-15.

EXAMPLE 10-12 Determine the value of I, I_2, and R_3 of Figure 10-15.

SOLUTION Use Equation 10-11 and solve for I_2. Thus:

$$\frac{I_1}{I_2} = \frac{R_2}{R_1}$$

$$I_2 = \frac{I_1 R_1}{R_2} = \frac{(4 \times 10^{-3})(2 \times 10^3)}{10 \times 10^3}$$

$$\therefore \quad I_2 = 0.80 \text{ mA}$$

Solve for I:

$$I = I_1 + I_2 + I_3$$

$$I = 4 \text{ mA} + 0.80 \text{ mA} + 20 \text{ mA}$$

$$I = 24.8 \text{ mA}$$

Since $V_3 = V_1$, then $I_3 R_3 = I_1 R_1$ and

$$\frac{I_3}{I_1} = \frac{R_1}{R_3}$$

Solve for R_3 and substitute:

$$R_3 = \frac{I_1 R_1}{I_3} = \frac{(4 \times 10^{-3})(2 \times 10^3)}{20 \times 10^{-3}}$$

$$\therefore \quad R_3 = 400 \text{ }\Omega$$

As a final reminder to yourself, remember that voltage and resistance in a series voltage-divider circuit have the same relationship as current and conductance in a parallel current-divider circuit. Thus $V_1/V_2 = R_1/R_2$ is true for a series circuit and $I_1/I_2 = G_1/G_2$ is true for a parallel circuit.

10-4 CURRENT-DIVIDER APPLICATIONS

Many circuits require that the current flow through each device in the circuit be controlled for optimum operation. In designing or analyzing circuits with *current sources*, a familiarity with the current-divider principle is essential.

EXAMPLE 10-13 A high-fidelity amplifier is *current limited* in its output to a maximum of 3 A of current. The amplifier will *output* its full 3 A of current when driving a loudspeaker of 8 Ω or less. If a 16-Ω loudspeaker is connected in parallel with an 8-Ω loudspeaker, (a) determine the total current flow in both speakers, (b) compute the maximum current that will flow in the 16-Ω loudspeaker, and (c) compute the power dissipated in each speaker.

SOLUTION Figure 10-16 shows the schematic representation of the problem. (a) Because the amplifier is current limited to 3 A and the speaker load is less than 8 Ω, the current flow ($I_T = I_1 + I_2$) is 3 A of current.

$\therefore \quad I_T = 3$ A

(b) Determine the current in the 16-Ω speaker. Use

$$I_2 = \frac{I_T G_2}{\Sigma G}$$

where

$\Sigma G = G_1 + G_2 = \frac{1}{8} + \frac{1}{16} = 0.125 + 0.0625$

$\Sigma G = 188$ mS

Substitute:

$$I_2 = \frac{I_T G_2}{\Sigma G} = \frac{(3)(62.5 \times 10^{-3})}{188 \times 10^{-3}}$$

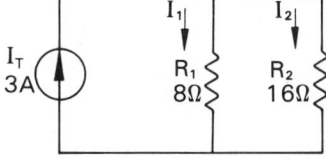

FIGURE 10-16
Schematic representation of an amplifier driving two speakers in parallel.

(c) Compute the power in the 16-Ω speaker:

$$P_2 = I_2^2 R_2 = (1)^2(16) = 16 \text{ W}$$

Compute the power in the 8-Ω speaker:

$$I_1 = I_T - I_2 = 3 - 1 = 2 \text{ A}$$

$$P_1 = I_1^2 R_1 = (2)^2(8) = 32 \text{ W}$$

∴ 32 W is dissipated by the 8-Ω speaker and 16 W is dissipated by the 16-Ω speaker.

Observation The power dissipated with both the 8-Ω and the 16-Ω speakers connected to the amplifier is 48 W. However, if only the 8-Ω speaker is connected, then 72 W is dissipated $[P = I^2R = (3)^2(8) = 72 \text{ W}]$. Since the amplifier is current limited to 3 A, the power dissipation depends upon the equivalent speaker resistance attached to the amplifier. Thus:

$$P = I^2 R_{eq} \quad \text{and} \quad P = 9R_{eq}$$

Because the equivalent resistance of the parallel combination of speaker loads decreased while the current remained constant at 3 A, the total power delivered to the speakers also decreased.

EXAMPLE 10–14 A meter with a coil resistance (R_m) of 500 Ω indicates *full scale* when 1 mA of current flows in it. What must be done so that it can measure 10 mA at full scale?

SOLUTION To accomplish this without ruining the meter, place a shunt (R_s) across the meter as shown in Figure 10–17. By properly selecting R_s, 9 mA of current is shunted (diverted) through the shunt, while 1 mA flows through the meter. Thus:

$$I_T = I_m + I_s$$

and

$$I_s = I_T - I_m = 10 \text{ mA} - 1 \text{ mA} = 9 \text{ mA}$$

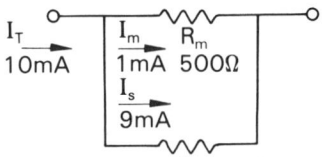

FIGURE 10–17 Schematic representation of an ammeter. By adding a shunt (R_s), the range of the meter circuit is increased from 1 mA to 10 mA.

Current Divider Applications

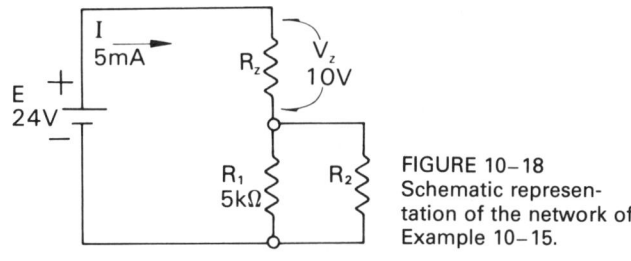

FIGURE 10-18 Schematic representation of the network of Example 10-15.

From the current-divider principle

$$\frac{I_s}{G_s} = \frac{I_m}{G_m} \quad \text{and} \quad I_s R_s = I_m R_m$$

Solving for R_s,

$$R_s = \frac{I_m R_m}{I_s} = \frac{(1 \times 10^{-3})(500)}{9 \times 10^{-3}} = 55.56 \, \Omega$$

∴ A resistor with a value of 55.56 Ω must be placed in shunt with the meter.

Observation A resistance of 55.56 Ω is a standard value in the E192 preferred number series for standard resistance values.

EXAMPLE 10-15 The device labeled R_z in Figure 10-18 drops a constant voltage of 10 V across it when 5 mA passes through it. A resistance, R_2, must be added in parallel with R_1 to achieve 5 mA of current, the desired operating condition. Determine the value of R_2.

SOLUTION Under the desired operating condition, $V_z = 10$ V and $I_z = I = 5$ mA. Using Kirchhoff's voltage law, determine V_1:

$$V_1 = V_2 = E - V_z$$
$$V_1 = V_2 = 24 \text{ V} - 10 \text{ V} = 14 \text{ V}$$

Determine I_1 using Ohm's law:

$$I_1 = \frac{V_1}{R_1} = \frac{14}{5 \times 10^3} = 2.8 \times 10^{-3} = 2.8 \text{ mA}$$

Compute I_2 from

$$I_2 = I - I_1 = 5 \text{ mA} - 2.8 \text{ mA} = 2.2 \text{ mA}$$

Calculate R_2 from:

$$\frac{I_1}{I_2} = \frac{R_2}{R_1}$$

$$R_2 = \frac{I_1 R_1}{I_2} = \frac{(2.8 \times 10^{-3})(5 \times 10^3)}{2.2 \times 10^{-3}}$$

∴ $R_2 = 6.36 \text{ k}\Omega$

Observation A 6.2 kΩ ± 5% resistor would be selected for this application.

10–5 LOADED AND UNLOADED CONDITIONS

Voltage and current sources are often treated as ideal when, in fact, they are not. A situation that is commonly encountered is when a load is attached to a voltage source and the voltage is found to be less than previously calculated or measured.

A voltage source is considered to be *unloaded* when it is in an open-circuit condition ($I = 0$ A), as shown in Figure 10–19. However, a more common situation is when the voltage source is *loaded*, as shown in Figure 10–20. In this case, the output or load voltage depends upon the ratio between the *source resistance*, R_s, and the load resistance, R_L. The load voltage may be determined by the voltage-divider formula.

$$V_L = \frac{E R_L}{R_s + R_L} \qquad (10\text{–}12)$$

where V_L = voltage across the load resistance in volts
E = source voltage in volts
R_L = load resistance in ohms
R_s = source resistance in ohms

Obviously, as R_L changes, so will V_L. As R_L gets very large, approaching an open circuit (∞ Ω), V_L approaches E. On the other hand, if R_L gets very small, approaching 0 Ω or a short-circuit condition, V_L approaches 0 V. This, of course, assumes that the source can support a current equal to $I = E/R_s$ without destroying itself.

EXAMPLE 10–16 Calculate V_L for the various values of R_L in the circuit shown in Figure 10–21 when $R_s = 50$ Ω.

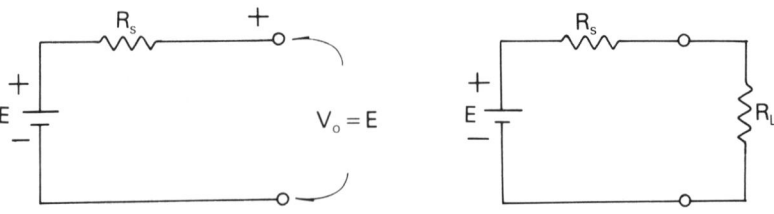

FIGURE 10–19 (left) Unloaded voltage source.
FIGURE 10–20 (right) Loaded voltage source.

Loaded and Unloaded Conditions

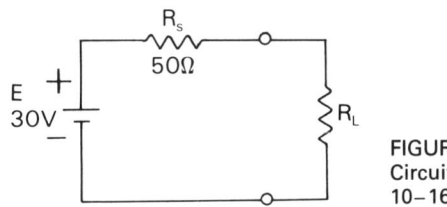

FIGURE 10-21
Circuit of Example 10-16.

SOLUTION When $R_L = 0 \ \Omega$ (short circuit),

$$V_L = \frac{ER_L}{R_s + R_L} = \frac{(30)(0)}{50 + 0} = 0 \text{ V}$$

∴ $V_L = 0$ V when $R_L = 0 \ \Omega$.
When $R_L = 10 \ \Omega$,

$$V_L = \frac{(30)(10)}{50 + 10} = 5 \text{ V}$$

∴ $V_L = 5$ V when $R_L = 10 \ \Omega$.
When $R_L = 50 \ \Omega$,

$$V_L = \frac{(30)(50)}{50 + 50} = 15 \text{ V}$$

∴ $V_L = 15$ V when $R_L = 50 \ \Omega$.
When $R_L = 1$ kΩ,

$$V_L = \frac{(30)(1000)}{50 + 1000} = 28.57 \text{ V}$$

∴ $V_L = 28.57$ V when $R_L = 1$ kΩ.
When $R_L = 10$ kΩ,

$$V_L = \frac{(30)(10 \times 10^3)}{50 + 10 \times 10^3} = 29.85 \text{ V}$$

∴ $V_L = 29.85$ V when $R_L = 10$ kΩ.
When $R_L = \infty$ (open circuit),

∴ $V_L = E = 30$ V

Example 10-16 shows that the load or output voltage approaches the source voltage as the load resistance increases. Only when the load resistance (R_L) is very large relative to the source resistance $(R_L \gg R_s)$ does the load voltage approach the value of the source voltage. One must not forget that in most circuits a finite (measurable) source resistance exists, and therefore the loaded and unloaded voltages will be different.

At some time you may have attached a load resistance to a voltage divider as shown in Figure 10-22 and observed the volt-

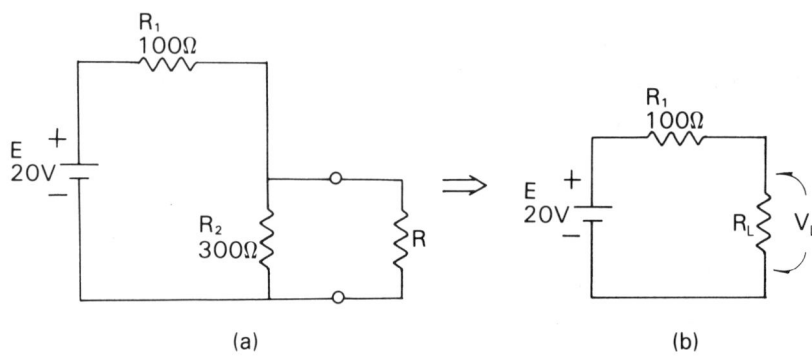

FIGURE 10–22 (a) and (b) Loading a circuit changes the voltages in the circuit.

age fall. You then may have wondered why the voltage did not equal the original open-circuit voltage. Example 10–17 explores this situation.

EXAMPLE 10–17 Using the circuit of Figure 10–22, where $R_1 = 100\ \Omega$ and $E = 20$ V, determine V_L for the following values of R. (a) $R = \infty\ \Omega$ (open circuit), (b) $R = 1000\ \Omega$, (c) $R = 100\ \Omega$, (d) $R = 0\ \Omega$ (short circuit).

SOLUTION (a) Determine R_L when $R = \infty\ \Omega$ (open circuit):

$$R_L = R \parallel R_2$$

$$R_L = R_2 = 300\ \Omega$$

Substitute and solve for V_L when $R_1 = R_s = 100\ \Omega$:

$$V_L = \frac{ER_L}{R_s + R_L} = \frac{(20)(300)}{100 + 300} = 15\text{ V}$$

∴ $V_L = 15$ V when $R = \infty\ \Omega$ (open circuit).

(b) Determine R_L when $R = 1000\ \Omega$:

$$R_L = R \parallel R_2$$

$$R_L = \frac{1}{\frac{1}{R} + \frac{1}{R_2}} = \frac{1}{\frac{1}{1000} + \frac{1}{300}} = 231\ \Omega$$

Substitute and solve for V_L when $R_1 = R_s = 100\ \Omega$:

$$V_L = \frac{ER_L}{R_s + R_L} = \frac{(20)(231)}{100 + 231} = 14\text{ V}$$

∴ $V_L = 14$ V when $R = 1000\ \Omega$.

(c) Determine R_L when $R = 100\ \Omega$:

Loaded and Unloaded Conditions

$$R_L = R \| R_2$$

$$R_L = \frac{1}{\frac{1}{100} + \frac{1}{300}} = 75 \, \Omega$$

Substitute and solve for V_L when $R_1 = R_s = 100 \, \Omega$:

$$V_L = \frac{ER_L}{R_s + R_1} = \frac{(20)(75)}{100 + 75} = 8.57 \text{ V}$$

∴ $V_L = 8.57$ V when $R = 100 \, \Omega$.

(d) Determine R_L when $R = 0 \, \Omega$ (short circuit):

$$R_L = R \| R_2$$

$$R_L = R = 0 \, \Omega$$

Substitute and solve for V_L when $R_1 = R_s = 100 \, \Omega$:

$$V_L = \frac{ER_L}{R_s + R_L} = \frac{(20)(0)}{100 + 0} = 0 \text{ V}$$

∴ $V_L = 0$ V when $R = 0 \, \Omega$ (short circuit).

When a voltage divider is *loaded*, the resistance across the terminals forms an equivalent resistance with the divider resistance, causing the output voltage to decrease. For this reason, a voltmeter must have a high internal resistance to accurately measure the voltage in a circuit. If it does not, then it would not measure the true voltage because it would load the circuit. This concept is explored in Example 10–18.

EXAMPLE 10–18 Determine the effect that a voltmeter, with an internal resistance of 20 kΩ, will have on the circuit of Figure 10–23 when it is used to measure the output voltage, V_o.

SOLUTION Theoretically, V_o should equal

$$V_o = \frac{ER_2}{R_1 + R_2} = \frac{(20)(5 \text{ k}\Omega)}{15 \text{ k} + 5 \text{ k}\Omega}$$

$$V_o = \frac{(20)(5 \times 10^3)}{20 \times 10^3} = 5 \text{ V}$$

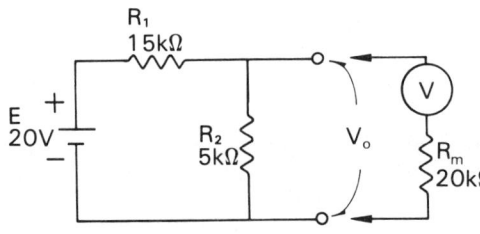

FIGURE 10–23
Effect of connecting a voltmeter to a circuit.

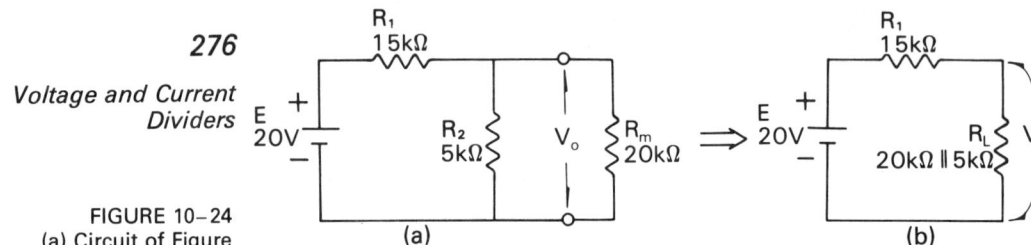

FIGURE 10-24
(a) Circuit of Figure 10-23 with voltmeter attached.
(b) Resistances R_2 and R_m are formed into an equivalent load, R_L.

However, when the voltmeter is connected to the circuit to measure V_o, as pictured in Figure 10-24(a), the equivalent circuit looks like the one shown in Figure 10-24(b), and R_L is

$$R_L = R_m \parallel R_2$$

$$R_L = \frac{1}{\frac{1}{5 \times 10^3} + \frac{1}{20 \times 10^3}} = 4 \text{ k}\Omega$$

The measured output, V_o', is then

$$V_o' = \frac{ER_L}{R_1 + R_L} = \frac{(20)(4 \text{ k}\Omega)}{15 \text{ k}\Omega + 4 \text{ k}\Omega}$$

$$V_o' = \frac{(20)(4 \times 10^3)}{19 \times 10^3}$$

$$\therefore \quad V_o' = 4.21 \text{ V}$$

Observation The measurement is in error by -15.8% due to the loading of the voltage divider by the voltmeter.

In designing a loaded voltage divider for two or more voltage levels, particular attention is paid to the selection of the stabilization or *bleeder* resistor. Figure 10-25 pictures a typical loaded divider showing the location of the bleeder resistor. As a *rule of thumb*, the current through the bleeder resistor is *picked* between 10% and 20% of the total load current. Example 10-19 demonstrates this concept.

EXAMPLE 10-19 Determine the value of the divider resistors R_1, R_2, and R_B in the circuit of Figure 10-25.

SOLUTION Using the information of Figure 10-25:
Step 1: Label the schematic with current arrows indicating the direction of the current flow through the divider and loads as shown in Figure 10-26.
Step 2: Compute I_B by letting $I_B = 0.2 \, I_L$ (20% of I_L), where

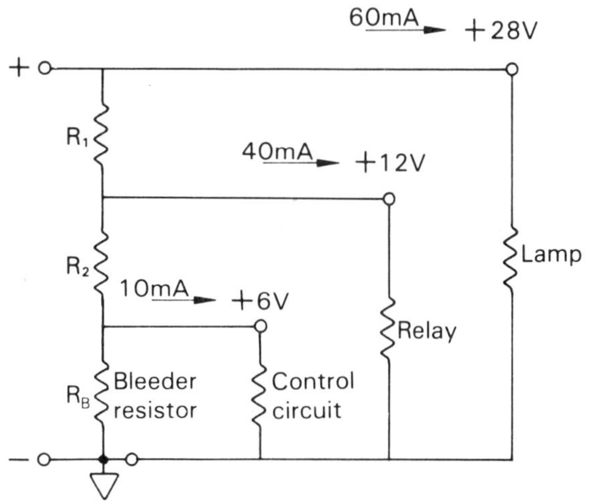

Loaded and Unloaded Conditions

FIGURE 10-25
Loaded voltage divider network with bleeder resistor R_B labeled.

$$I_L = 10 \text{ mA} + 40 \text{ mA} + 60 \text{ mA} = 110 \text{ mA}$$

∴ $I_B = 0.2\, I_L = 0.2\, (110 \text{ mA}) = 22 \text{ mA}$

Step 3: Determine I_2 and I_1:

$$I_2 = I_B + 10 \text{ mA} = 22 \text{ mA} + 10 \text{ mA}$$

∴ $I_2 = 32 \text{ mA}$

$$I_1 = I_2 + 40 \text{ mA} = 32 \text{ mA} + 40 \text{ mA}$$

∴ $I_1 = 72 \text{ mA}$

Check Does I_T entering equal I_T leaving?

$$I_T \text{ (entering)} = I_T \text{ (leaving)}$$

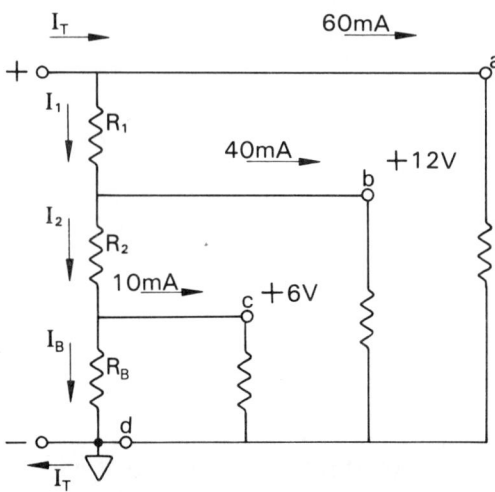

FIGURE 10-26
Proper labeling of circuit parameters prevents confusion when solving complex circuits.

$$60 + I_1 = I_B + 10 + 40 + 60$$
$$60 + 72 = 22 + 10 + 40 + 60$$
$$132 \text{ mA} = 132 \text{ mA}$$

Step 4: Compute R_1, R_2, and R_B:

$$R_1 = \frac{V_1}{I_1}$$

$$V_1 = V_{ab} = 28 \text{ V} - 12 \text{ V} = 16 \text{ V}$$

$$I_1 = 72 \text{ mA}$$

$$R_1 = \frac{16}{72 \times 10^{-3}} = 222 \text{ }\Omega$$

$$P_1 = V_1 I_1 = (16)(72 \times 10^{-3}) = 1.2 \text{ W}$$

∴ R_1 is selected as a 220 Ω ± 5% 3-W wirewound resistor.

$$R_2 = \frac{V_2}{I_2}$$

$$V_2 = V_{bc} = 12 \text{ V} - 6 \text{ V} = 6 \text{ V}$$

$$I_2 = 32 \text{ mA}$$

$$R_2 = \frac{6}{32 \times 10^{-3}} = 188 \text{ }\Omega$$

$$P_2 = V_2 I_2 = (6)(32 \times 10^{-3}) = 192 \text{ mW}$$

∴ R_2 is selected as a 180 Ω ± 5% ½-W carbon-composition or carbon-film resistor.

$$R_B = \frac{V_B}{I_B}$$

$$V_B = V_{cd} = 6 \text{ V} - 0 \text{ V} = 6 \text{ V}$$

$$I_B = 22 \text{ mA}$$

$$R_B = \frac{6}{22 \times 10^{-3}} = 273 \text{ }\Omega$$

$$P_B = V_B I_B = (6)(22 \times 10^{-3}) = 132 \text{ mW}$$

∴ R_B is selected as a 270 Ω ± 5% ½-W carbon-composition or carbon-film resistor.

10-6 APPROXIMATIONS

A focal point throughout this book, and in most textbooks, is on the accuracy of calculations. We usually calculate answers to three or more significant figures. In reality, this precision is not

Approximations

necessary because of a number of factors. First, most resistors are accurate to only 5% or 10% and, as you know, even when precision 1% resistors are used, only certain standard values are available. In practice, resistors must be selected from these standard values. Like resistors, capacitors and inductors have a tolerance of 10% to 20%. Recognizing these limitations allows the engineer and technician to make approximations in circuit design and analysis. Obviously, there is no point in designing a circuit with high precision (0.01%) components if they are not readily available or are too expensive to manufacture.

Approximation Rule A rule commonly used by engineers and technicians is the *ten-to-one* rule. Simply stated, when resistors are connected in **series,** if one resistor is ten or more times larger than the other resistor, then the **smaller** value may be ignored. In similar terms, when resistors are connected in **parallel,** if one resistor is ten or more times larger than the other resistor, then the **larger** value may be ignored. These ideas, along with concepts from earlier chapters, will be demonstrated in the following examples. Table 10–1 summarizes these concepts.

EXAMPLE 10–20 Given the circuit of Figure 10–27, determine the voltage drop across R_2.

SOLUTION Since R_2 is more than ten times greater than R_1 ($R_2 \gg R_1$),

$$R_T \approx R_2 \quad \text{and} \quad V_2 \approx 10 \text{ V}$$

Check
$$V_2 = \frac{ER_2}{R_1 + R_2} = \frac{(10)(20 \times 10^3)}{21 \times 10^3} = 9.52 \text{ V}$$

$$\% \text{ error} = \frac{10 - 9.52}{9.52} \times 100 = +5\%$$

Observation A 5% error is quite acceptable since 5% resistors will be used in this circuit.

TABLE 10–1 SUMMARY OF THE TEN-TO-ONE RULE

Series circuit	$R_1 \gg R_2$	then	$R_T = R_1$
	$R_2 \gg R_1$	then	$R_T = R_2$
Parallel circuit	$R_1 \gg R_2$	then	$R_{eq} = R_2$
	$R_2 \gg R_1$	then	$R_{eq} = R_1$

Note: $\gg$ is read "much greater than" and indicates at least a ten-to-one relationship. The term *"an order of magnitude"* is used to indicate a ten-to-one relationship.

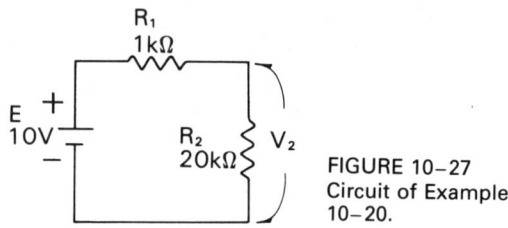

FIGURE 10-27
Circuit of Example 10-20.

EXAMPLE 10-21 Determine the approximate equivalent resistance of the three resistors pictured in Figure 10-28.

SOLUTION Since the 5-kΩ and 6-kΩ resistors are approximately equal, their equivalent will be about one-half the 5-kΩ resistor, or 2.5 kΩ. This equivalent resistance is more than ten times the 100-Ω resistor ($R_{eq} \gg R_{100}$).

∴ The equivalent of the three resistors in parallel is 100 Ω.

∴ The exact value is

$$R_{eq} = \frac{1}{\frac{1}{100} + \frac{1}{5000} + \frac{1}{6000}} = 96.5 \text{ }\Omega$$

$$\% \text{ error} = \frac{100 - 96.5}{96.5} \times 100 = +3.6\%$$

Observation Again, a quick mental approximation is quite accurate.

EXAMPLE 10-22 From the network shown in Figure 10-29, (a) estimate the approximate output voltage (V_o), and (b) calculate the exact output voltage (V_o).

SOLUTION (a) Using the ten-to-one rule, the following is true:

R_1 is large in relation to R_2

∴ R_2 may be ignored.

R_3 is small in relation to R_4

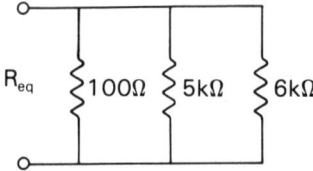

FIGURE 10-28
Determine approximate equivalent resistance, R_{eq}, without detailed calculations.

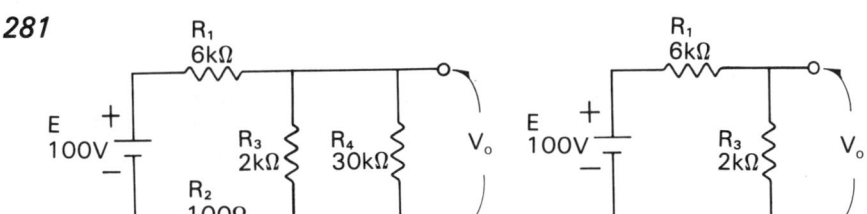

FIGURE 10-29 (left)   Estimate approximate voltage of V_o in Example 10-22.
FIGURE 10-30 (right) The approximate equivalent circuit of Example 10-22.

∴ R_4 may be ignored.
The resulting circuit is pictured as Figure 10-30. Applying voltage-divider concepts results in

$$V_o = \frac{(100)(2 \text{ k}\Omega)}{8 \text{ k}\Omega} = \tfrac{1}{4}(100) = 25 \text{ V}$$

SOLUTION (b) The exact equivalent circuit is formed by adding R_1 and R_2 and calculating the parallel equivalent of R_3 and R_4. Thus:

$$R_1 + R_2 = 6 \text{ k}\Omega + 0.1 \text{ k}\Omega = 6.1 \text{ k}\Omega$$

and

$$R_3 \parallel R_4 = R_{eq} = \frac{1}{\frac{1}{2 \text{ k}\Omega} + \frac{1}{30 \text{ k}\Omega}} = 1.88 \text{ k}\Omega$$

The more exact equivalent circuit is pictured as Figure 10-31. Solving for V_o results in

$$V_o = \frac{(100)(1.88 \text{ k}\Omega)}{6.1 \text{ k}\Omega + 1.88 \text{ k}\Omega} = \frac{(100)(1.88 \times 10^3)}{7.98 \times 10^3}$$

∴ $V_o = 23.6 \text{ V}$

Check % error $= \dfrac{25 - 23.6}{23.6} \times 100 = +5.9\%$

Observation Again, the error is well within tolerance using only solutions obtained with approximations

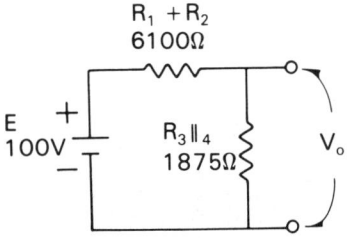

FIGURE 10-31
The exact equivalent circuit of Example 10-22.

and few calculations.

EXAMPLE 10–23 Determine the approximate voltage dropped across R_2 in Figure 10–32.

SOLUTION Examining the three resistors, it is evident that both R_2 and R_3 are significantly larger than R_1; hence, we ignore R_1. Since R_2 and R_3 do not differ by a factor of 10, they must be taken into account. The approximate voltage across R_2 is

$$V_2 = \frac{ER_2}{R_2 + R_3} = \frac{(30)(5 \times 10^3)}{15 \times 10^3}$$

$$\therefore V_2 = 10 \text{ V}$$

Check The exact value is

$$V_2 = \frac{ER_2}{R_1 + R_2 + R_3}$$

$$V_2 = \frac{(30)(5 \times 10^3)}{60 + 5 \times 10^3 + 10 \times 10^3}$$

$$V_2 = 9.96 \text{ V}$$

Observation % error $= \dfrac{10 - 9.96}{9.96} \times 100 = +0.4\%$

Summary With the use of electronic calculators, there is a tendency for one to plug in values without thinking about the significance of the calculations. Before starting a detailed analysis of a circuit, the technician or engineer should consider the overall purpose of what he or she is trying to do and also consider what equipment or components are available, as well as the accuracy to which the calculations need to be made. Some questions to consider before calculating are as follows:

☐ What are the tolerances of the components in the circuit?
☐ What is the operating range of the active devices (transistors, operational amplifiers) in the circuit?

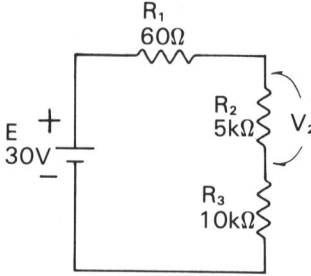

FIGURE 10–32
Circuit of Example 10–23.

SELECTED TECHNICAL TERMS

Several of the technical terms used in this chapter are defined here to aid your understanding of them.

Bleeder resistor: connected across a power supply, it serves to discharge or *bleed off* the charge on the filter capacitors once the supply is turned off.

Current limited: the maximum current that a source may provide in a load. Some commercial power supplies have circuitry that automatically limits the current to a constant amount.

Full scale: the maximum value of the measured quantity that an instrument is calibrated to indicate.

Loaded: in reference to a source, the source is loaded when a load is connected across its terminals and the source is supplying power to the load.

Order of magnitude: ten times greater than, as in 100 Ω is an *order of magnitude* greater than 10 Ω. One-tenth as great, as in 100 Ω is an *order of magnitude* less than 1000 Ω.

Source resistance: the resistance presented to the input terminals of a circuit by any source of electric energy. The source resistance of an ideal voltage source is zero ohms, whereas that of an ideal current source is an infinite resistance.

Unloaded: in reference to a source, the source is unloaded when no load is connected across the terminals and no power is supplied to the load.

PROBLEMS

As part of the solution, it is suggested that you draw and label a schematic diagram for each problem.

Sections 10–1 and 10–2

10–1 Two resistors, $R_1 = 400\ \Omega$ and $R_2 = 100\ \Omega$, are connected in series with a voltage source of 20 V. Apply the voltage-divider equation to determine the voltage drops, V_1 and V_2, across each resistor.

10–2 Three resistors, $R_1 = 250\ \Omega$, $R_2 = 100\ \Omega$, and $R_3 = 400\ \Omega$, are connected in series with a voltage source of 24 V. Determine the voltage drops across each resistor by applying the voltage-divider equation.

10–3 For a certain circuit to operate properly, the sum of two resistors, R_1 and R_2, must equal 50 kΩ. If the resistors

284

Voltage and Current Dividers

FIGURE 10–33
Circuit of Problem 10–5.

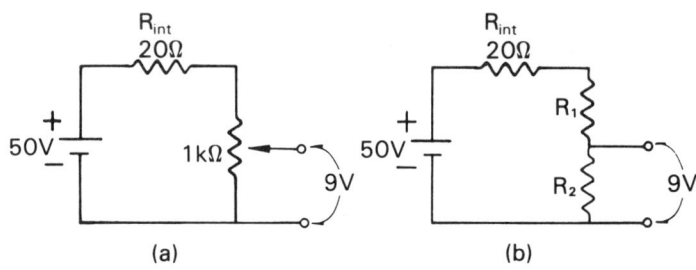

(a)   (b)

are connected in series with a 12-V dc source, determine the values of R_1 and R_2 if $V_1 = 9$ V and $V_2 = 3$ V.

10–4 Two resistors, $R_1 = 20$ kΩ and R_2, are connected in series with a 30-V source. Determine the value of R_2 when $V_2 = 12$ V.

10–5 A voltage source of 50 V with an internal resistance of 20 Ω is connected in series with a 1-kΩ potentiometer, as shown in Figure 10–33(a). The potentiometer is adjusted to obtain 9-V output. (a) Determine the value of the fixed resistances, R_1 and R_2 of Figure 10–33(b), to exactly replace the 1-kΩ potentiometer. (b) Select values of resistors for R_1 and R_2 from the E96 preferred series of ±1% industrial resistor values. (c) Determine the power dissipation of the selected resistors.

10–6 Replace the 1-kΩ potentiometer of Figure 10–33(a) with a 100-Ω potentiometer and repeat Problem 10–5.

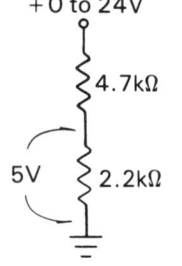

FIGURE 10–34
Circuit of Problem 10–7.

10–7 A variable 0- to 24-V source is connected to a 4.7-kΩ resistor in series with a 2.2-kΩ resistor to ground, as pictured in Figure 10–34. (a) Determine the voltage source setting when the voltage drop across the 2.2-kΩ resistor, measured from the 2.2-kΩ resistor to ground, is 5 V. (b) Determine the power dissipated by each of the resistors.

10–8 Replace the 2.2-kΩ resistor with a 10-kΩ resistor and repeat Problem 10–7.

10–9 Select the value of a series dropping resistor needed to ensure proper operation of a dc relay connected in series with a 24-V source. The relay is *actuated* when 50 mA passes through the 120-Ω relay coil. Make your selection from a carbon-film E24 preferred series of ±5% MIL-spec resistor values. (a) State the color code found on the body of the device. (b) Assuming wattage ratings of ½, 1,

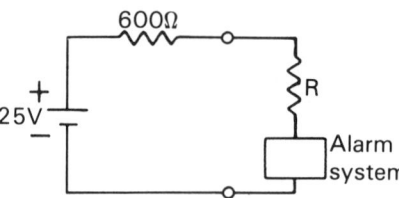

FIGURE 10–35
Schematic for Problem 10–10.

and 2 W, select an appropriate wattage rating and state why you selected that rating.

10-10 An electronic alarm system requires that 15 V be connected across its power terminals and that the voltage source supply 10 mA of current. If the voltage source is 25 V with an internal resistance of 600 Ω, then (a) determine the additional series dropping resistor needed for proper operation of the system, and (b) specify the resistor from the E12 series of preferred values. Figure 10-35 is the schematic of the circuit.

Sections 10-3 and 10-4

10-11 A current source of 125 mA is connected in parallel with three resistors, $R_1 = 2$ kΩ, $R_2 = 5$ kΩ, and $R_3 = 3$ kΩ. Apply the current-divider equation to determine the current in each resistor.

10-12 An ammeter has 100 mV across its terminals when a full-scale current of 50 mA is passing through it. Determine the value of shunt resistance needed to convert this ammeter to a 250-mA full-scale meter.

10-13 Two parallel-connected resistors, R_1 and R_2, have an equivalent resistance of 150 Ω. Determine the values of R_1 and R_2 when the current in R_1 is three times greater than the current in R_2.

10-14 A certain 300-Ω dc relay, having a *pull-in* current (operating current) of 20 mA ± 2 mA, is connected in series with a 33-mA constant current source. (a) Determine the value of resistance needed to be placed in parallel across the 300-Ω relay coil to shunt the excess current, thereby ensuring proper operation of the relay. (b) Select an appropriate carbon-film resistor from the E12 series of preferred values. (c) Select an appropriate wattage rating for the resistor from $\frac{1}{8}$, $\frac{1}{4}$, or $\frac{1}{2}$ W. (d) State your reason for your choice of wattage rating. (e) Using the nominal value of the selected resistor, determine if the current in the relay is within 20 mA ± 2 mA.

10-15 A 5-mA, full-scale analog meter movement has an internal resistance of 10 Ω, as indicated in Figure 10-36. Determine the values of shunt resistance needed to convert this meter to a multirange meter with ranges of (a) 100 mA, (b) 500 mA, and (c) 1 A.

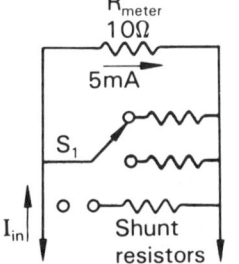

FIGURE 10-36
Circuit of Problem 10-15 where I_{in} may be 5 mA, 100 mA, 500 mA, or 1 A. The range switch (S_1) is switched to accommodate the different current ranges being measured by the meter.

Section 10-5

10-16 A power supply with a 10-Ω source resistance measures 100 V across its terminals in an open-circuit condition. Determine the terminal voltage with the following load resistances connected to the output terminals of the supply: (a) 10 kΩ, (b) 500 Ω, and (c) 50 Ω.

Voltage and Current Dividers

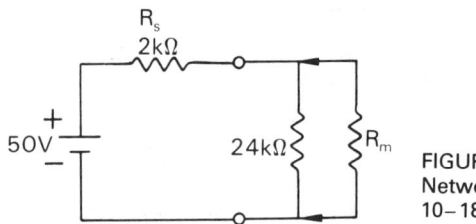

FIGURE 10–37
Network of Problems
10–18 and 10–19.

10–17 A voltage source measures 24 V in an open-circuit condition and 23.9 V when a 150-Ω load resistor is connected to its output terminals. Determine the value of the source resistance.

10–18 A voltmeter with a 50-kΩ internal resistance (R_m) is used to measure the voltage drop across a 24-kΩ resistor, as noted in Figure 10–37. The voltage source is 50 V with a 2-kΩ source resistance. (a) Determine the voltmeter reading. (b) Determine the percent of error between the true value and the measured value of voltage, where % error = [(measured value − true value)/(true value)] × 100%.

10–19 Repeat Problem 10–18 using a meter with an internal resistance (R_m) of 200 kΩ.

10–20 Determine the current in the 4-kΩ load resistor connected to the 10-mA current generator of Figure 10–38 when the source resistance (R_s) is 7.5 kΩ.

10–21 Determine the value of the source resistance, R_s, in Figure 10–38 that will cause the current in the 4-kΩ load to be 97% of the 10-mA source current.

10–22 Determine the voltage drop across a 500-Ω load resistance connected to a 25-mA current source having a 2-kΩ source resistance.

10–23 Determine the voltage across the unloaded terminals of the current source of Problem 10–22.

10–24 (a) Determine the value of the voltage-divider resistors, R_1 and R_B, in the voltage-divider circuit of Figure 10–39, assuming the current in the bleeder resistor (R_B) is 10% of the total load current, V_{ac} = 30 V, and V_{bc} = 12 V. (b)

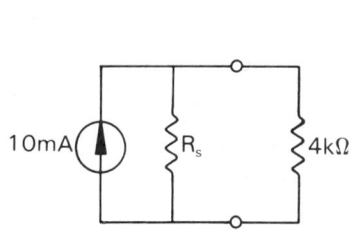
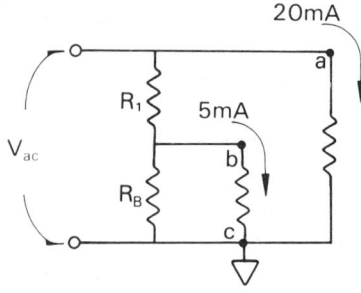

FIGURE 10–38
(left) Circuit of Problems 10–20 and 10–21.
FIGURE 10–39
(right) Network of Problems 10–24 and 10–25.

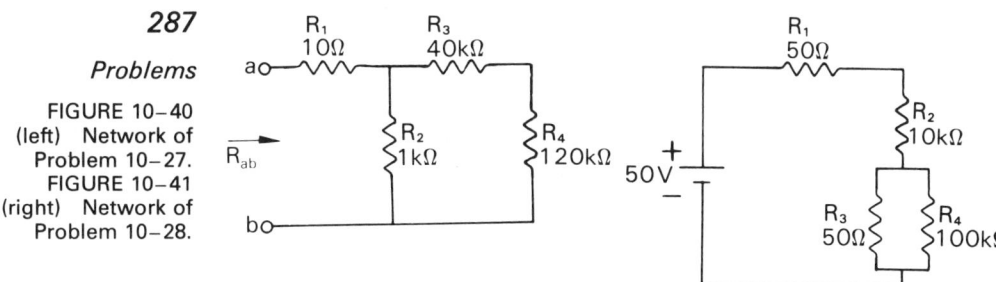

FIGURE 10-40 (left) Network of Problem 10-27.
FIGURE 10-41 (right) Network of Problem 10-28.

Select resistor values from the E24 series of preferred values. (c) Specify the wattage rating of each resistor; assume $\frac{1}{4}$, $\frac{1}{2}$, 1, and 2 W availability and derate by a factor of 2.

10-25 Repeat Problem 10-24 for $V_{ac} = 24$ V and $V_{bc} = 10$ V. Assume the current in the bleeder resistor R_B is 15% of the total load current.

Section 10-6

10-26 Approximate the equivalent resistance of the following resistors connected in parallel: (a) $R_1 = 100$ kΩ, (b) $R_2 = 75$ kΩ, and (c) $R_3 = 1$ kΩ.

10-27 (a) Determine the approximate equivalent resistance, R_{ab}, of the network shown in Figure 10-40. (b) Determine the percent of error between the approximate answer and the exact answer.

10-28 (a) Approximate the voltage across R_2 in Figure 10-41. (b) Approximate the equivalent resistance of the network.

10-29 Approximate the current in R_3 of Figure 10-42.

10-30 Approximate the equivalent resistance as seen by the current source of Figure 10-42.

10-31 Determine the percent of error between the approximate answer in Problem 10-29 and the exact answer.

10-32 Determine the output voltage, V_o, of the network in Figure 10-43.

FIGURE 10-42 (below) Network of Problem 10-29.
FIGURE 10-43 (right) Network of Problem 10-32.

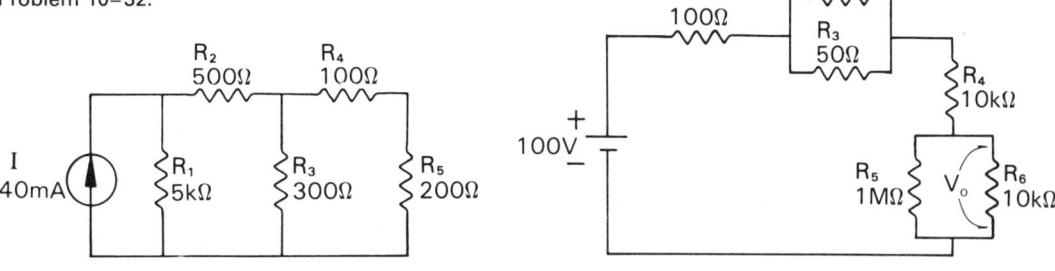

11 Equivalent Circuits and the Network Theorems

11-1 SUPERPOSITION THEOREM
11-2 THEVENIN'S THEOREM
11-3 NORTON'S THEOREM
11-4 CONVERSION OF THEVENIN'S AND NORTON'S EQUIVALENT CIRCUITS
11-5 MAXIMUM POWER TRANSFER THEOREM
11-6 MILLMAN'S THEOREM
11-7 APPLYING THE NETWORK THEOREMS

CHAPTER PREVIEW

The analysis of electrical circuits could be very time consuming if it were not for a number of simplifications that have been developed by various engineers and mathematicians. By using the concepts developed by these people, complex circuits are simplified into equivalent circuits by applying the *network theorems*. The following network theorems are covered and explained in this chapter:

☐ **Superposition theorem,** which shows how circuits with more than one source can be simplified by treating the circuit as several separate circuits.
☐ **Thevenin's theorem,** which allows complex circuits to be represented as an equivalent circuit composed of one voltage source and one resistor.
☐ **Norton's theorem,** which allows complex circuits to be represented as an equivalent circuit composed of one current source and one resistor.
☐ **Millman's theorem,** which evolved from Norton's theorem, provides for calculation of the voltage across two points in a network.
☐ **Maximum power transfer theorem,** which sets up the conditions for the transfer of maximum power from one circuit to another.

Because each of these theorems is an important tool to be used in analyzing circuits, a full understanding of each theorem will be needed for application now and in future courses relating to active devices.

PERFORMANCE OBJECTIVES

Once you have read each section, worked each example with pencil, paper, and calculator, answered each problem of every section, you will be able to:

☐ Select the most appropriate network theorem to solve a particular network.
☐ Apply each network theorem to solve a given network.
☐ Use a combination of the theorems to simplify the solution of a given network.
☐ Determine the value of the load for maximum power transfer.
☐ Convert between a Thevenin's equivalent circuit and a Norton's equivalent circuit.
☐ Form an equivalent circuit of a network.

11–1 SUPERPOSITION THEOREM

The superposition theorem is very useful in solving networks where more than one voltage or current source is present. The superposition theorem states that the voltage or current that is

Definition present in a component (element) is equal to the algebraic sum of the voltages or currents that exist independently in that component as a result of each of the sources. For example, if source E_1 of Figure 11–1(a) causes a 5-V drop across resistor R_4, and source E_2 of Figure 11–1(b) causes a 2-V drop with opposite polarities then when both sources $E_1 + E_2$ are applied at the same time, the voltage across R_4 will be their algebraic sum, 5 V − 2 V = 3 V, as noted in Figure 11–1(c).

THEOREM 11–1. SUPERPOSITION THEOREM

In a linear bilateral network, the voltage across any circuit element or the current in any circuit element is equal to the algebraic sum of the voltages or currents that exist in that component when each source is considered independently of the others.

Additional Concepts Up until this time, we have been concentrating on analyzing circuits with a single source. We learn from the superposition theorem that a circuit with many sources can be solved by algebraically adding the voltage across or current in each component. When calculating the results for each source, the other

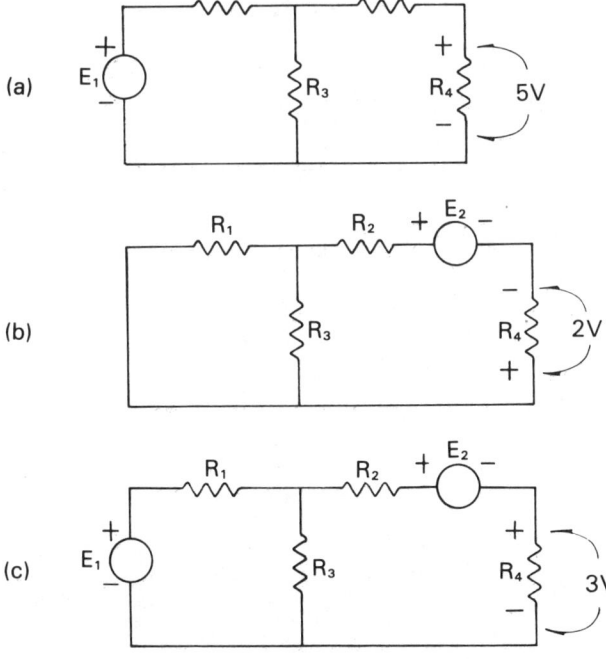

FIGURE 11–1
(a) Source E_1 causes a 5-V drop across R_4.
(b) Source E_2 causes a 2-V drop of opposite polarity across R_4.
(c) Voltage drop across R_4 due to both E_1 and E_2 is the algebraic sum of (a) and (b).

292

Equivalent Circuits and the Network Theorems

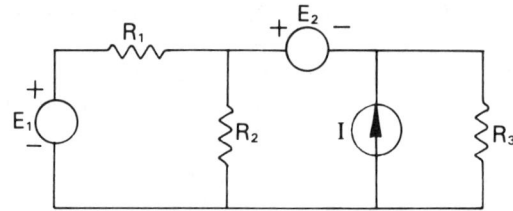

FIGURE 11–2
Network of Example 11–1.

sources must be set to zero and replaced by their internal resistance. In other words, **voltage sources are replaced by a short circuit (0 Ω) and current sources by an open circuit (∞ Ω)**. This concept is amplified in Example 11–1.

EXAMPLE 11–1 Given the circuit of Figure 11–2, draw the individual circuits used to make the calculations for each source.

SOLUTION The circuit of voltage source E_1 is shown as Figure 11–3(a). The circuit of voltage source E_2 is shown as Figure 11–3(b). The circuit of current source I_1 is shown as Figure 11–3(c).

Observation Notice in Figures 11–3(a, b, c) that the voltage sources are replaced with short circuits and the current source is replaced with an open circuit.

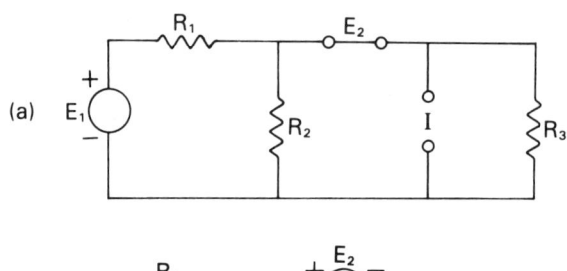

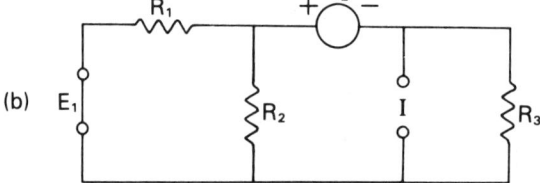

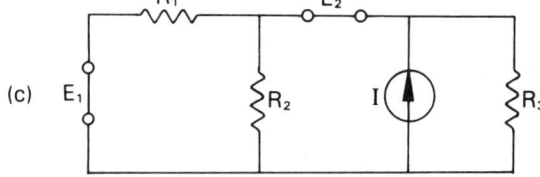

FIGURE 11–3
(a) Network with E_1 ON and other sources OFF.
(b) Network with E_2 ON and other sources OFF.
(c) Network with I ON and other sources OFF.

Superposition Theorem

Once the technique used to draw the individual circuits for calculating the effects of each source is learned, it becomes a simple task to analyze each circuit and determine the results caused by all the combined sources. A key concept in using the superposition theorem is the notation of the polarity of each voltage and current in the circuit. This notation is very important since the end result is the **algebraic** sum of the individual results.

As noted in the statement of the theorem, certain circuit conditions must be satisfied for superposition to be applicable. The components in the circuit must be *linear* and *bilateral*. A linear component is one whose current is always proportional to the voltage dropped across it. Also, a linear component is one whose voltage drop across the component is always proportional to the current through it. Circuit components such as resistors, capacitors, and inductors, when thought of as being ideal, are linear devices.

A bilateral component is one that will have the same current for opposite polarities of voltages. The current will, of course, have opposite flow direction. A diode is not a bilateral device because current flows when voltage is applied in one direction, but not in the opposite direction. Active devices such as amplifiers are not bilateral because a signal is amplified in one direction, but not in the other. Circuit components such as resistors, capacitors, and inductors are bilateral devices.

Applying Superposition This section will deal with several examples showing the application of the superposition theorem to circuits with voltage sources and current sources.

EXAMPLE 11-2 Using the superposition theorem, determine the voltage drop across each resistor in the network of Figure 11-4.

SOLUTION First draw the equivalent circuits for each source operating independently. Then calculate the voltage drops caused by each source. Finally, determine the resultant voltages by algebraically adding the individual voltages across each resistor.

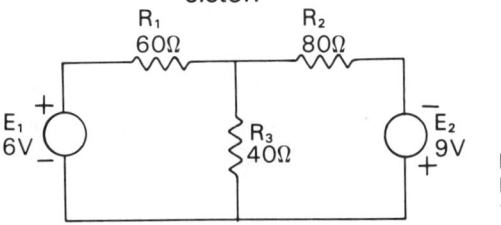

FIGURE 11-4
Network of Example 11-2.

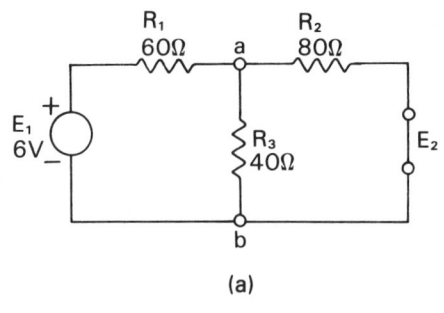

(a)

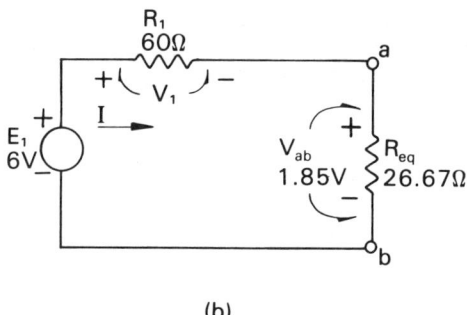

(b)

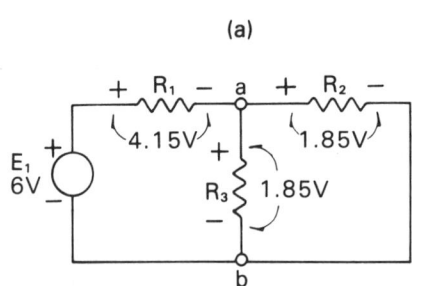

(c)

FIGURE 11-5
(a) Network with E_1 ON and E_2 OFF.
(b) Solving for the contribution to V_{ab} caused by E_1.
(c) Voltage across each resistor caused by E_1.

Source E_1 Figure 11–5(a):
Replace the 9-V source E_2 with a short circuit as shown in Figure 11–5(a).
Form $R_2 \parallel R_3$ into an equivalent resistance, R_{eq}, as shown in Figure 11–5(b), where

$$R_{eq} = \frac{R_2 \times R_3}{R_2 + R_3} = \frac{(80)(40)}{120} = 26.67 \;\Omega$$

Establish current direction and polarize the circuit as indicated in Figure 11–5(b).
Determine V_1 of Figure 11–5(b) by applying the voltage-divider rules. Thus:

$$V_1 = \frac{E_1 R_1}{R_T} \quad \text{and} \quad R_T = R_1 + R_{eq}$$

$$R_T = 60 + 26.67 = 86.67 \;\Omega$$

Substitute into the voltage-divider formula:

$$V_1 = \frac{(6)(60)}{86.67}$$

$$\therefore \quad V_1 = 4.15 \text{ V}$$

Determine V_{ab} of Figure 11–5(b) by Kirchhoff's voltage law. Thus:

$$E_1 = V_1 + V_{ab}$$

Superposition Theorem

and

$$V_{ab} = E_1 - V_1$$

Substitute:

$$V_{ab} = 6 - 4.15$$

$$\therefore \quad V_{ab} = 1.85 \text{ V}$$

Observation Figure 11–5(a) is labeled with the values computed and the polarities established using source E_1. This is shown as Figure 11–5(c).

Source E_2 Figure 11–6(a):
Replace the 6-V source E_1 with a short circuit as shown in Figure 11–6(a).
Form $R_1 \| R_3$ into an equivalent resistance, R_{eq}, as shown in Figure 11–6(b), where

$$R_{eq} = \frac{R_1 R_3}{R_1 + R_3} = \frac{(60)(40)}{100} = 24.0 \text{ }\Omega$$

Establish current direction and polarize the circuit as indicated in Figure 11–6(b).
Determine V_2 of Figure 11–6(b) by applying the voltage-divider rules. Thus:

$$V_2 = \frac{E_2 R_2}{R_T} \quad \text{and} \quad R_T = R_{eq} + R_2$$

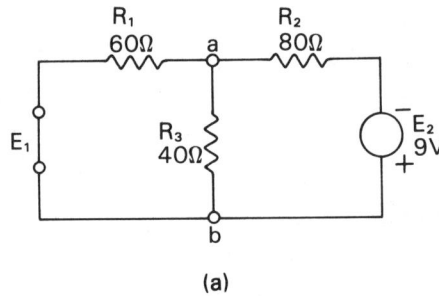

(a)

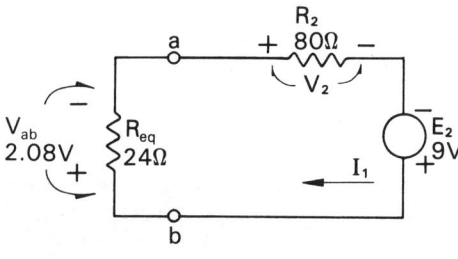

(b)

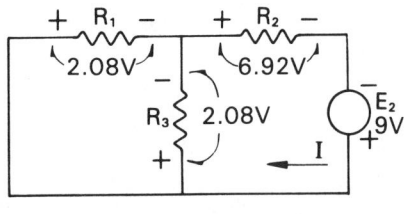

(c)

FIGURE 11–6
(a) Network with E_2 ON and E_1 OFF.
(b) Solving for the contribution to V_{ab} caused by E_2.
(c) Voltage across each resistor caused by source E_2.

Equivalent Circuits and the Network Theorems

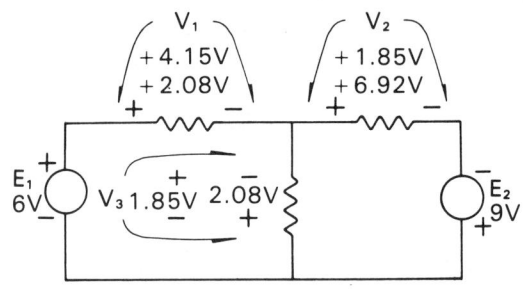

(a)

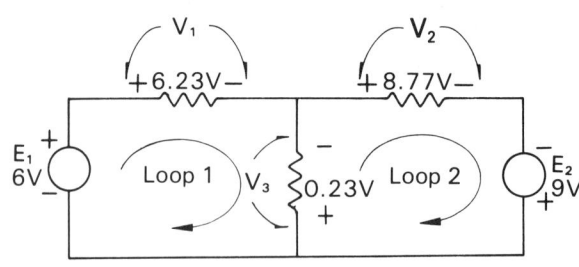

(b)

FIGURE 11-7 (a) Network showing voltage contributions to each resistor with both E_1 and E_2 ON. (b) Resultant voltage drops across each resistor after applying superposition theorem to Example 11-2.

$$R_T = 24.0 + 80.0 = 104\,\Omega$$

Substitute into the voltage-divider formula:

$$V_2 = \frac{(9)(80)}{104}$$

∴ $V_2 = 6.92$ V

Determine V_{ab} of Figure 11-6(b) by Kirchhoff's voltage law. Thus:

$$E_2 = V_2 + V_{ab}$$
$$V_{ab} = E_2 - V_2$$

Substitute:

$$V_{ab} = 9 - 6.92$$

∴ $V_{ab} = 2.08$ V

Observation Figure 11-6(a) is labeled with the values computed and the polarities established using source E_2. This is shown in Figure 11-6(c).

Composite of source E_1 and E_2:
Place both the value and the polarity of each of the computed voltage drops from Figures 11-5(c) and 11-6(c) onto the network diagram as shown in Figure 11-7(a). Using the voltages

of Figure 11-7(a), algebraically add each pair of voltages to determine the actual voltage drops. Thus:

$$V_1 = 4.15 + 2.08 = 6.23 \text{ V}$$

$$V_2 = 1.85 + 6.92 = 8.77 \text{ V}$$

$$V_3 = 1.85 - 2.08 = -0.230 \text{ V}$$

∴ The composite schematic diagram of the circuit is shown in Figure 11-7(b) with the combined voltages and polarities indicated.

Observation When voltages of opposite polarities are combined, the larger voltage determines the polarity.

Check Check the results of Figure 11-7(b) by applying Kirchhoff's voltage law.

Moving around loop 1 results in

$$E_1 = V_1 + V_3$$

$$6 = 6.23 - 0.230$$

∴ 6 V = 6 V

Moving around loop 2 results in

$$E_2 = V_3 + V_2$$

$$9 = 0.230 + 8.77$$

9 V = 9 V

EXAMPLE 11-3 Using the superposition theorem, determine the current in each resistor in the network of Figure 11-8.

SOLUTION First, draw the equivalent circuits for each source operating independently. Then, calculate the currents due to each source. Finally, determine the resultant currents by algebraically adding the individual currents in each resistor.

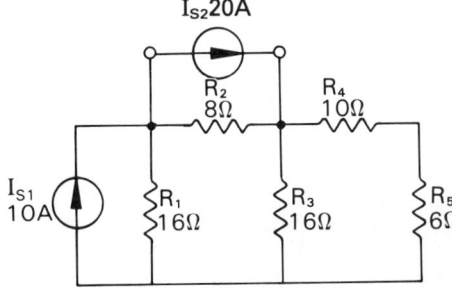

FIGURE 11-8
Network of Example 11-3.

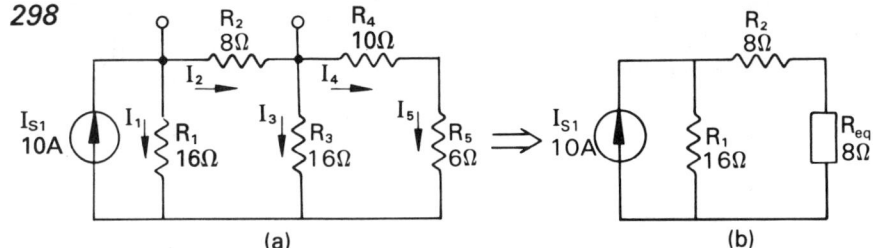

FIGURE 11-9 Network of Example 11-3 and its equivalent with I_{s1} ON and I_{s2} OFF.

Source I_{s1}, Figure 11-9(a):
With I_{s2} open circuit, as shown in Figure 11-9(a), calculate the current in each resistor due to source I_{s1} by determining the equivalent resistance of $R_3 \parallel (R_4 + R_5)$. By inspection

$$R_4 + R_5 = 10 + 6 = 16 \ \Omega$$

Since $R_3 = 16 \ \Omega$,

$$R_3 \parallel (R_4 + R_5) = 16 \ \Omega \parallel 16 \ \Omega = 8 \ \Omega$$

∴ The equivalent resistance of $R_3 \parallel (R_4 + R_5)$ is 8 Ω.

Redraw Figure 11-9(a) as Figure 11-9(b), where

$$R_{eq} = R_3 \parallel (R_4 + R_5) = 16 \parallel 16 \Rightarrow 8 \ \Omega$$

Observation Since $R_2 + R_{eq} = 16 \ \Omega$, as seen in Figure 11-9(b), then 10 A from source I_{s1} will divide equally between the two branches: 5 A will flow in R_1 and 5 A will also flow in R_2.

NOTE: While viewing Figure 11-10, it is noted that the 5-A current in R_2 divides equally between R_3 and R_{4-5}, resulting in 2.5 A of current in each branch.

∴ The currents due to source I_{s1} are noted in Figure 11-10.

Check Applying Kirchhoff's current law,

$$I_{s1} = I_1 + I_2 = I_1 + I_3 + I_5$$

10 A = 5 A + 5 A = 5 A + 2.5 A + 2.5 A

10 A = 10 A = 10 A

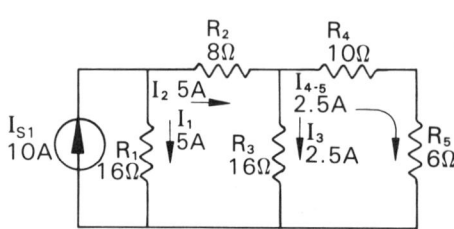

FIGURE 11-10
Current in each resistor due to I_{s1}.

Superposition Theorem

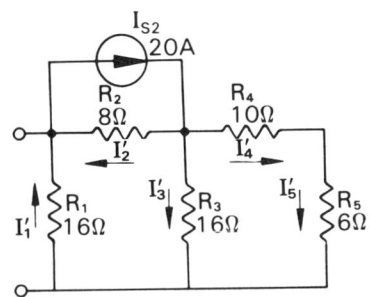

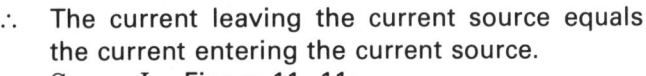

FIGURE 11-11
Network with I_{s2} ON and I_{s1} OFF.

∴ The current leaving the current source equals the current entering the current source.

Source I_{s2}, Figure 11-11:

The current due to source I_{s2} is now calculated by setting I_{s1} to zero (I_{s1} is open circuit), as shown in Figure 11-11. The noted current directions have been determined from the polarity of the source. Simplify the circuit by determining the equivalent resistance R_{eq1} of $R_4 + R_5$ in parallel with R_3. Thus:

$$R_{eq1} = R_3 \parallel (R_4 + R_5) = 16 \parallel 16 \Rightarrow 8\ \Omega$$

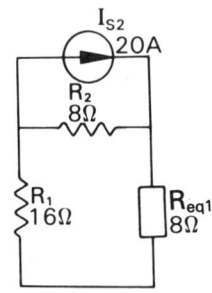

FIGURE 11-12
Equivalent circuit of Figure 11-11.

∴ The resultant is shown in Figure 11-12 as R_{eq1}. Combine R_{eq1} with R_1 to form R_{eq2}. Thus:

$$R_{eq2} = R_{eq1} + R_1 = 8 + 16 = 24\ \Omega$$

∴ The resultant is shown in Figure 11-13 as R_{eq2}. Apply the current-divider equation to the circuit of Figure 11-13 and solve for I_2' and I_{eq2}. Thus:

$$I_2' = \frac{(I_{s2})(G_2)}{\Sigma G}$$

where $I_{s2} = 20$ A, $G_2 = 1/R_2$ and $\Sigma G = (1/R_2) + (1/R_{eq2})$

Solving,

$$\Sigma G = \frac{1}{R_2} + \frac{1}{R_{eq2}} = \frac{1}{8} + \frac{1}{24} = 0.167\ \text{S}$$

$$I_2' = \frac{(I_{s2})(G_2)}{\Sigma G} = \frac{(20)(0.125)}{0.167} = 15\ \text{A}$$

and

$$I_{eq2} = \frac{I_{s2} G_{Req2}}{\Sigma G} = \frac{(20)(41.7 \times 10^{-3})}{0.167} = 5.0\ \text{A}$$

∴ $I_2' = 15$ A and $I_{eq2} = 5.0$ A

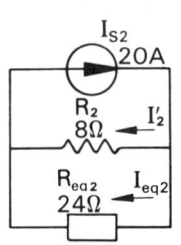

FIGURE 11-13
Current flow in the equivalent circuit of Figure 11-12.

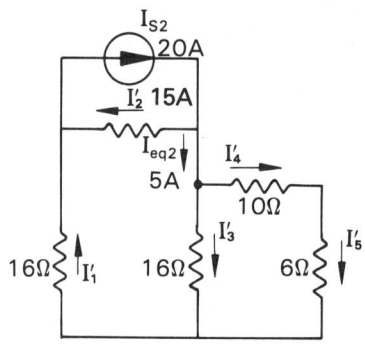

FIGURE 11-14
Current flow in network
of Figure 11-11.

Replacing R_{eq2} with the actual circuitry as shown in Figure 11-14, determine the current I_1', I_3', I_4', and I_5':

By inspection,

$I_4' = I_5'$, because series currents are equal

$I_3' = I_4'$, because $R_3 = R_4 + R_5 = 16 \, \Omega$

and

$I_{eq2} = I_1'$ because the current entering a node must equal the current leaving

$\therefore \ I_3' = I_4' = I_5' = I_{eq2}/2 = 2.5$ A, and $I_1' = 5$ A.

Therefore, $I_3' = 2.5$ A, $I_4' = 2.5$ A, $I_5' = 2.5$ A, and $I_1' = 5$ A, as noted in Figure 11-15.

Check Applying Kirchhoff's current law,

$I_{s2} = I_2' + I_3' + I_4' = I_1' + I_2'$

20 A = 15 A + 2.5 A + 2.5 A = 5 A + 15 A

20 A = 20 A = 20 A

$\therefore$ The current leaving the current source equals the current entering the current source.

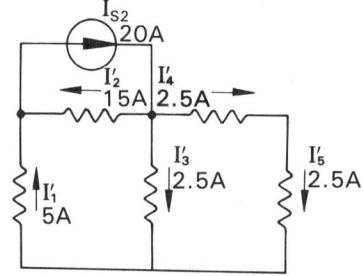

FIGURE 11-15
Current in each resistor due to I_{s2}.

Thevenin's Theorem

Composite of source I_{s1} and I_{s2}:
Place both the value and the current arrow of each of the computed currents from Figures 11-10 and 11-15 onto the network diagram as shown in Figure 11-16. Using the currents of Figure 11-16, algebraically add each pair of currents to determine the actual currents. Thus:

$$I_1 = 5 - 5 = 0 \text{ A}$$

$$I_2 = 15 - 5 = 10 \text{ A}$$

$$I_3 = 2.5 + 2.5 = 5 \text{ A}$$

$$I_4 = 2.5 + 2.5 = 5 \text{ A}$$

$$I_5 = 2.5 + 2.5 = 5 \text{ A}$$

∴ The composite schematic diagram of the circuit is shown in Figure 11-17 with the combined currents and polarities noted.

Observation
☐ Currents in the same direction add and assume the direction of either current.
☐ Currents in the opposite direction subtract and assume the direction of the larger current.

11-2 THEVENIN'S THEOREM

Definition

Thevenin's theorem, named after a French engineer, simplifies circuit calculations by treating any complex circuit as a combination of a voltage source and an equivalent series resistance. Gaining an understanding of *Thevenin's equivalent circuit* is very important since many concepts in electronics use this theorem. As an example, the specification for a piece of test equipment may indicate the *input* or *output* resistance of the equipment. When this is the case, the specification is actually stating the Thevenin's equivalent resistance of the input or output circuit.

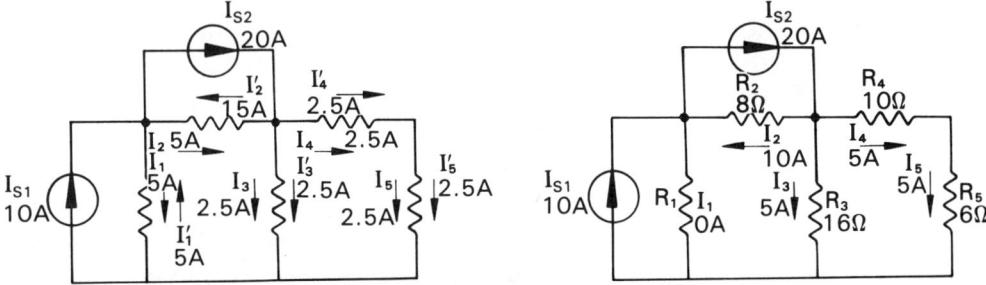

FIGURE 11-16 (left) Currents in each resistor due to both I_{s1} and I_{s2}.
FIGURE 11-17 (right) Solution to Example 11-3 by applying superposition theorem.

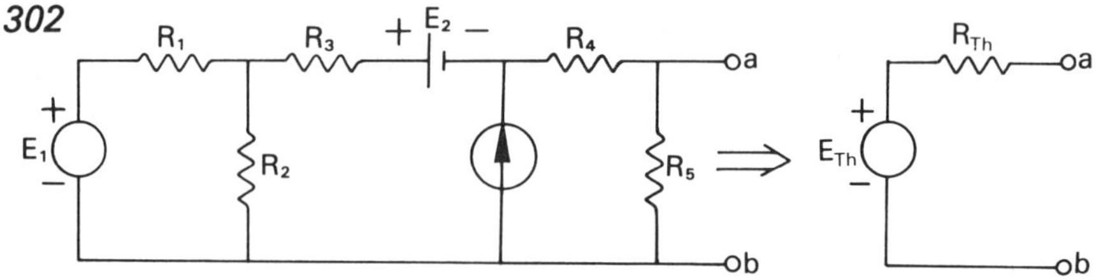

FIGURE 11–18 Simplification of a network through use of Thevenin's theorem.

THEOREM 11–2. THEVENIN'S THEOREM

Any linear two-terminal bilateral network can be represented by an electrically equivalent circuit consisting of one voltage source and one series impedance. (For the present, consider impedance to mean resistance.)

Additional Concepts

As the theorem states, any linear network of resistors and sources can be represented by a single voltage source (E_{Th}) and a single resistor (R_{Th}) when viewed from a given set of terminals. These ideas are depicted in Figure 11–18.

For your understanding of Thevenin's theorem, you must realize that the two circuits pictured in Figure 11–18 are equivalent only when they are viewed from terminals *a-b* looking toward the source. Obviously, their internal structures are entirely different; however, the voltage V_{ab} is the same in both circuits. Also, the current in a conductor connected between terminals *a-b* will be identical in each circuit.

The equivalent source is called *Thevenin's equivalent voltage source*, while the equivalent resistance is called *Thevenin's equivalent resistance*.

Applying Thevenin's Theorem

Given two terminals in a network, the Thevenin's equivalent circuit is formed in the following order:

☐ First, label the given terminals.
☐ Determine the Thevenin's equivalent voltage source (E_{Th}) by calculating the open-circuit voltage between the labeled terminals.
☐ Determine the Thevenin's equivalent resistance (R_{Th}) by setting all voltage and current sources to zero by replacing the voltage sources with short circuits and the current sources with open circuits. Then calculate the resistance between the labeled terminals.

303
Thevenin's Theorem

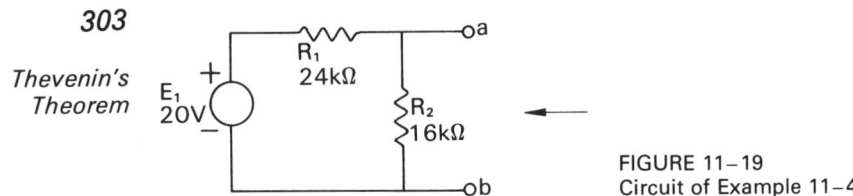

FIGURE 11–19
Circuit of Example 11–4.

☐ Finally, form the Thevenin's equivalent circuit.

Example 11–4 demonstrates the preceding outlined procedure for forming Thevenin's equivalent circuit.

EXAMPLE 11–4 Form the Thevenin's equivalent circuit for the network of Figure 11–19.

SOLUTION View the circuit from terminals a-b by looking toward the source. Thevenin's equivalent voltage (E_{Th}) is V_{ab} of Figure 11–19. Compute E_{Th}:

$$E_{Th} = V_{ab} = \frac{E_1 R_2}{R_T}$$

where

$$R_T = R_1 + R_2 = 24 \text{ k}\Omega + 16 \text{ k}\Omega = 40 \text{ k}\Omega$$

Substitute:

$$E_{Th} = \frac{20(16 \times 10^3)}{40 \times 10^3} = 8 \text{ V}$$

∴ $E_{Th} = 8$ V, as shown in Figure 11–20(a). Determine Thevenin's equivalent resistance by setting the voltage source to zero and calculating the resistance between terminals a and b as shown in Figure 11–20(b). Compute R_{Th}:

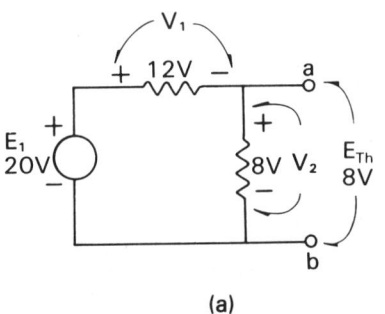
(a)

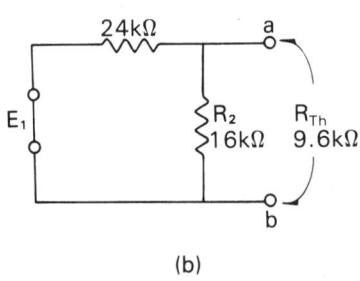
(b)

FIGURE 11–20
(a) Determining Thevenin's equivalent voltage.
(b) Determining Thevenin's equivalent resistance.

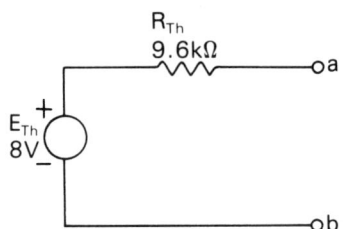

FIGURE 11–21
Thevenin's equivalent circuit of Example 11–4.

$$R_{Th} = R_{ab} = R_1 \parallel R_2 = \frac{1}{\frac{1}{R_1} + \frac{1}{R_2}}$$

where $R_1 = 24$ kΩ and $R_2 = 16$ kΩ. Thus:

$$R_{Th} = \frac{1}{\frac{1}{24 \times 10^3} + \frac{1}{16 \times 10^3}} = 9.60 \text{ k}\Omega$$

$\therefore \quad R_{Th} = 9.60$ kΩ

Observation The Thevenin's equivalent circuit is formed as shown in Figure 11–21. Figure 11–22 shows the circuit of Figure 11–19 and its Thevenin's equivalent.

EXAMPLE 11–5 Form the Thevenin's equivalent circuit for the network in Figure 11–23 at the specified terminals a-b.

SOLUTION Determine the Thevenin's equivalent resistance (R_{Th}) between terminals a and b with source I set to zero.
NOTE: An ideal current source, when set to zero, is equivalent to an open circuit, as shown in Figure 11–24. Thus, from Figure 11–25,

$$R_{cd} = R_3 \parallel (R_1 + R_2) = \frac{1}{\frac{1}{2000} + \frac{1}{1500}} = 857 \text{ }\Omega$$

and

$$R_{Th} = R_{ab} = R_4 + R_{cd}$$

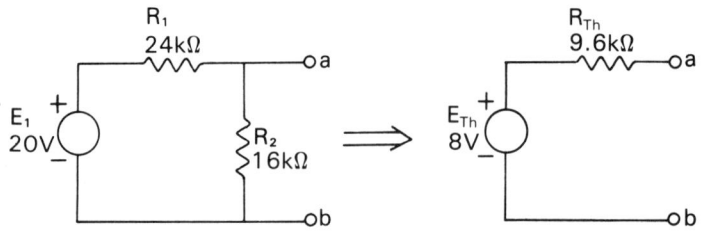

FIGURE 11–22
Solution to Example 11–4.

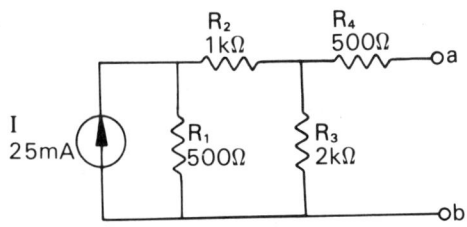
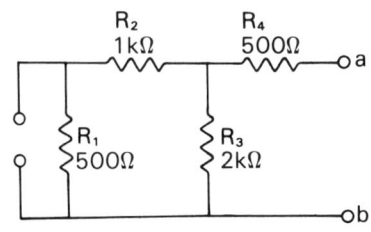

FIGURE 11-23 (left) Network of Example 11-5.
FIGURE 11-24 (right) Solving for Thevenin's equivalent resistance by setting $I = 0$.

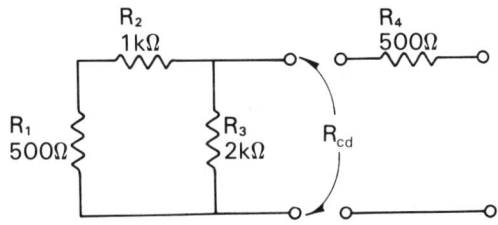
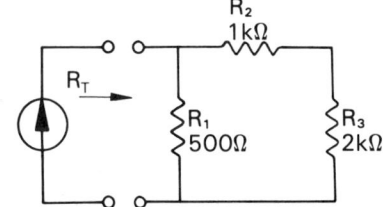

FIGURE 11-25 (left) Solving for R_{cd}.
FIGURE 11-26 (right) Circuit to solve for V_3.

Substitute and solve:

$$R_{Th} = 500 + 857 = 1357 \ \Omega$$

$$\therefore \quad R_{Th} = 1357 \ \Omega$$

The Thevenin's equivalent voltage source (E_{Th}) equals the open circuit voltage V_{ab}, which also equals the voltage V_3 developed across R_3 by the current from source I, as noted in Figure 11-26.
NOTE: Because no current is in R_4, there is no voltage across R_4; therefore, $E_{Th} = V_{ab} = V_3$. Thus, from Figure 11-27:

$$E_{Th} = V_{ab} = V_3 = (I_{2-3})(R_3)$$

The current flowing through R_2 and R_3 (I_{2-3}) of Figure 11-27 is calculated with the current-divider rule as

$$I_{2-3} = \frac{IG_{2-3}}{\Sigma G} = \frac{(25 \ \text{mA})(0.33 \ \text{mS})}{2.33 \ \text{mS}} = 3.54 \ \text{mA}$$

and

$$E_{Th} = V_{ab} = V_3 = (I_{2-3})(R_3)$$
$$E_{Th} = (3.54 \ \text{mA})(2 \ \text{k}\Omega) = 7.08 \ \text{V}$$

$$\therefore \quad E_{Th} = 7.08 \ \text{V}$$

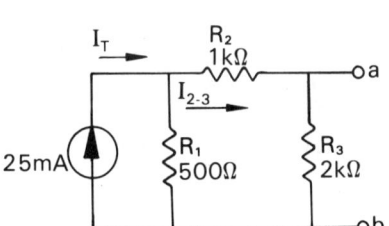

FIGURE 11-27
Using current division to determine $V_3 = V_{ab}$.

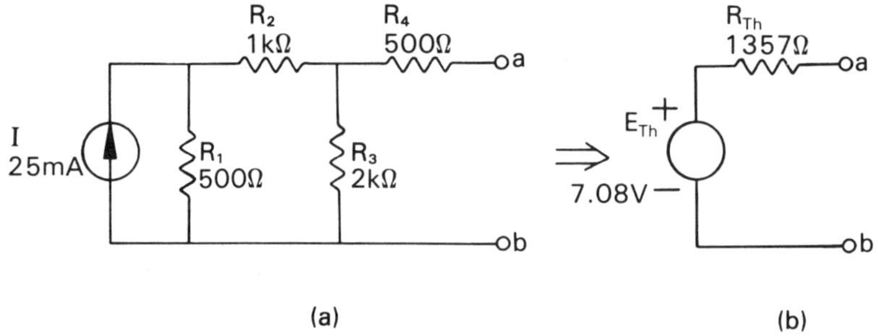

FIGURE 11–28
(a) Network of Example 11–5.
(b) Thevenin's equivalent circuit of Example 11–5.

Observation The Thevenin's equivalent circuit is formed as shown in Figure 11–28(b).

The power of Thevenin's theorem is apparent in the case where the load resistance is changed to different values of resistance. In this case, the load may be as simple as a single resistance or as complex as a combination of resistances. Once the load is removed from the network and the remaining network is formed into a Thevenin's equivalent circuit, then the load may be reattached and voltage, current, or power may be computed.

It is the ability to remove a portion of the network in which voltage or current is to be found that makes Thevenin's theorem unique. Once the *load* is removed from the network, it has no role in forming the remaining network into a Thevenin's equivalent circuit.

In general, the procedure for applying Thevenin's theorem to a network is as follows.

GUIDELINE 11–1. FORMING THEVENIN'S EQUIVALENT CIRCUIT

1. Remove the load from the network and label the open-circuit terminals of the remaining network.
2. Determine the voltage across the open-circuit network terminals. This is E_{Th} of the Thevenin's equivalent circuit.
3. Determine the resistance across the open-circuit network terminals with all sources set to zero. This is R_{Th} of the Thevenin's equivalent circuit.
4. Form the Thevenin's equivalent circuit.
5. Attach the load to the Thevenin's equivalent circuit.
6. Compute the desired load parameter (voltage, current, or power).

Thevenin's Theorem

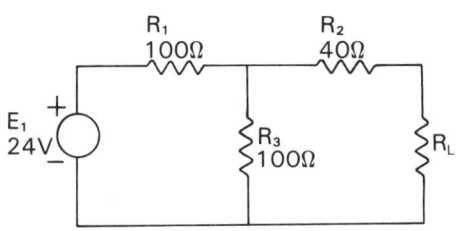

FIGURE 11-29
Network of Example 11-6.

Example 11-6 will demonstrate the application of Thevenin's theorem to determine voltage across the load and power dissipation in the load.

EXAMPLE 11-6 Using Figure 11-29, determine (a) the voltage drop across the load R_L, and (b) the power dissipated in R_L when $R_L = 20\ \Omega$ and $100\ \Omega$.

SOLUTION Form the Thevenin's equivalent circuit of the network pictured in Figure 11-29 by removing the load R_L from the network as shown in Figure 11-30(a). Compute E_{Th}:

$$E_{Th} = V_{ab} = V_3 = \frac{24(100)}{200} = 12\ V$$

∴ $E_{Th} = 12\ V$

From Figure 11-30(b) compute R_{Th}:

$$R_{Th} = R_2 + R_1 \parallel R_3$$

$$R_{Th} = 40 + \frac{1}{\frac{1}{100} + \frac{1}{100}} = 40 + 50 = 90\ \Omega$$

∴ $R_{Th} = 90\ \Omega$

Form the Thevenin's equivalent circuit as shown in Figure 11-31(a).

(a) Attach R_L to the Thevenin's equivalent circuit and determine V_L when $R_L = 20\ \Omega$ as shown in Figure 11-31(b). Thus:

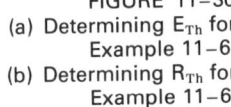

FIGURE 11-30
(a) Determining E_{Th} for Example 11-6.
(b) Determining R_{Th} for Example 11-6.

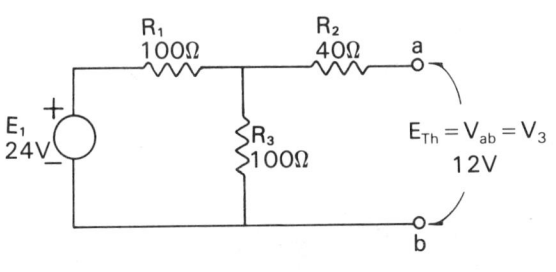

(a)

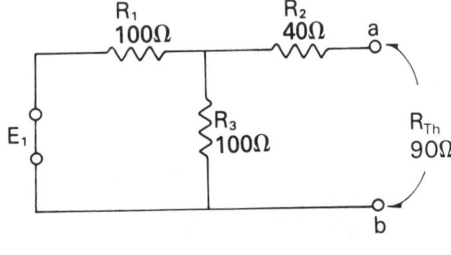

(b)

Equivalent Circuits and the Network Theorems

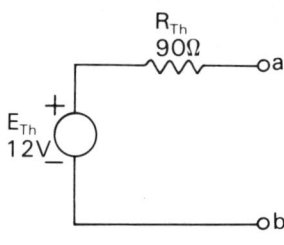

(a)

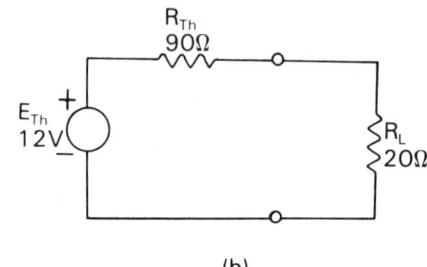

(b)

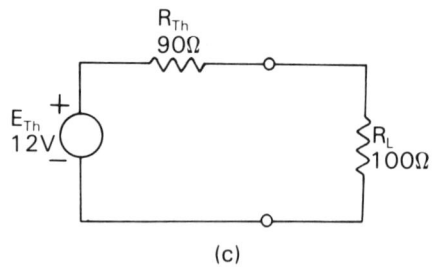

(c)

FIGURE 11–31
(a) Thevenin's equivalent circuit of Example 11–6.
(b) Equivalent circuit for $R_L = 20\ \Omega$.
(c) Equivalent circuit for $R_L = 100\ \Omega$.

$$V_L = \frac{(E_{Th})(R_L)}{\Sigma R}$$

where

$$\Sigma R = R_{Th} + R_L = 90 + 20 = 110\ \Omega$$

Substitute:

$$V_L = \frac{(12)(20)}{110} = 2.18\ \text{V}$$

∴ $V_L = 2.18$ V when $R_L = 20\ \Omega$.

(b) Determine the power dissipated in the load when $R_L = 20\ \Omega$:

$$P = \frac{V_L^2}{R_L} = \frac{2.18^2}{20} = 238\ \text{mW}$$

∴ The power dissipation in the load is 238 mW when $R_L = 20\ \Omega$.

(a) Determine V_L when R_L is increased to 100 Ω as pictured in Figure 11–31(c). Thus:

$$V_L = \frac{(E_{Th})(R_L)}{\Sigma R} = \frac{(12)(100)}{190} = 6.32\ \text{V}$$

∴ $V_L = 6.32$ V when R_L is increased to 100 Ω.

(b) Determine the power dissipated in the load when $R_L = 100\ \Omega$:

$$P = \frac{V_L^2}{R_L} = \frac{6.32^2}{100} = 399\ \text{mW}$$

Thevenin's Theorem

∴ The power dissipation in the load is 399 mW when $R_L = 100\ \Omega$.

Observation It is obviously easier to calculate V_L and P_L for various values of R_L by first forming the Thevenin's equivalent circuit rather than repeatedly recalculating each case with the original complex circuit.

Summary Any circuit, no matter how complex it may be, can be *"Thevenin-ized"* with respect to any two terminals as a voltage source (E_{Th}) and a series resistance (R_{Th}), provided the network is linear and bilateral.

The ability to view a network as a Thevenin's equivalent circuit is very important since many applications involve anticipating the result of connecting a load to the output terminals of an electronic circuit. The electronic circuit may be an amplifier, a power supply, or a function generator, or any other electronic circuit whose purpose it is to deliver power (voltage and current) to a load. Example 11–7 explores this application.

EXAMPLE 11–7 A power supply, represented as the square in Figure 11–32, is used to supply power to a variety of loads. Determine if the supply is capable of providing at least 3 A of current in a 2-Ω load if the supply is limited to a short-circuit current of 10 A.

SOLUTION Determine the Thevenin's equivalent voltage (E_{Th}) by placing a voltmeter across the output terminals of the power supply as shown in Figure 11–32(b). Assume the voltmeter reads 10.0 V.

∴ E_{Th} of the Thevenin's equivalent circuit is 10.0 V. Determine the Thevenin's equivalent resistance (R_{Th}) from the manufacturer's specification of short-circuit current. Thus:

$$R_{Th} = \frac{E_{Th}}{I_{sc}}$$

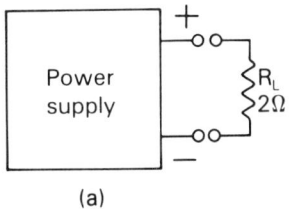

(a)

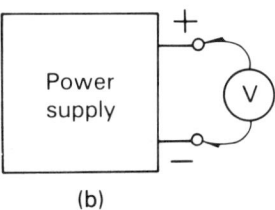

(b)

FIGURE 11–32
Circuit of Example 11–7.

Equivalent Circuits and the Network Theorems

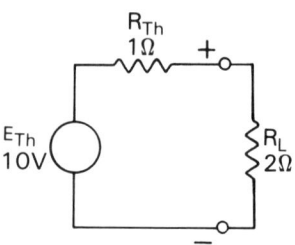

FIGURE 11–33
Thevenin's equivalent circuit of Example 11–7.

where

$$I_{sc} = 10 \text{ A}$$

Substitute:

$$R_{Th} = \frac{10.0}{10} = 1 \text{ }\Omega$$

∴ R_{Th} of the Thevenin's equivalent circuit is 1 Ω. Form the Thevenin's equivalent circuit as shown in Figure 11–33. Attach the 2-Ω load and compute the current in the load. Thus:

$$I_L = \frac{E_{Th}}{R_{Th} + R_L} = \frac{10}{3} = 3.33 \text{ A}$$

∴ The power supply can provide 3 A in a 2-Ω load.

Observation The power supply would be unable to provide 3 A in a 3-Ω load since $I_L = E_{Th}/(R_{Th} + R_L) = 2.5$ A.

Point of Reference In Chapters 7 to 9, considerable emphasis was placed on determining the series equivalent resistance R_{eq} of networks attached to voltage and current sources. In this case, the point of reference was from the two terminals of the source looking into the load, as indicated in Figure 11–34(a).

In forming the Thevenin's equivalent circuit, the point of reference is, as you now know, from the two terminals of the load looking into the source, as indicated in Figure 11–34(b).

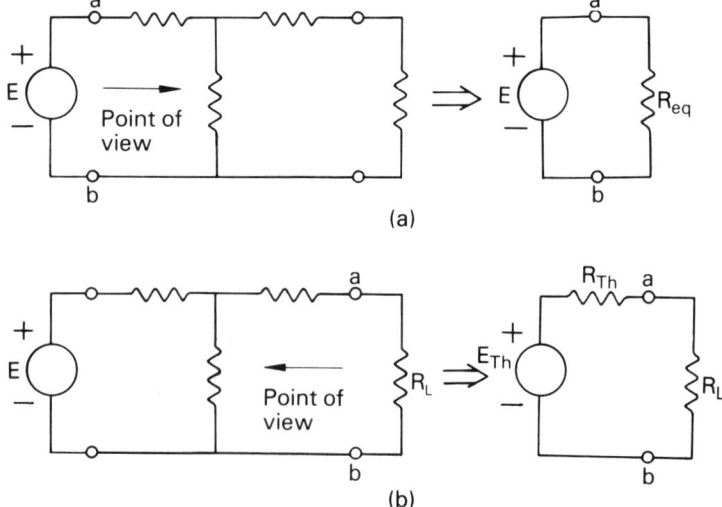

FIGURE 11–34
(a) Equivalent resistance as seen by the source.
(b) Equivalent resistance as seen by the load.

11-3 NORTON'S THEOREM

Definition

In the previous section, we showed how a network could be represented by a Thevenin's equivalent circuit consisting of a single voltage source and a single series resistance. The same network can also be represented by a Norton's equivalent circuit consisting of a single current source and a single parallel resistance. The current source is called the *Norton's equivalent current source*, and the parallel resistance is called the *Norton's equivalent resistance*.

THEOREM 11-3. NORTON'S THEOREM

Any linear two-terminal bilateral network can be represented by an electrically equivalent circuit consisting of one current source and one parallel impedance. (For the present, consider impedance to mean resistance.)

Norton's theorem is the dual of Thevenin's theorem. In Norton's equivalent circuit, the current source (I_N) has replaced the Thevenin's voltage source, and the equivalent resistance (R_N) is in parallel with the current source instead of in series with the voltage source, as indicated in Figure 11-35.

Applying Norton's Theorem

The Norton's equivalent resistance, R_N, is computed by setting all voltage and current sources to zero by replacing the voltage sources with short circuits and the current sources with open circuits. The equivalent resistance (R_N) between the two selected terminals is then calculated, as was done with the Thevenin's equivalent resistance.

In Thevenin's theorem, the equivalent voltage source was equal to the open-circuit voltage present at the two selected terminals; in Norton's theorem, the equivalent current source is equal to the short-circuit current between the two selected terminals. Example 11-8 uses the same network previously used with Thevenin's theorem; this network will now be used to demonstrate Norton's equivalent circuit.

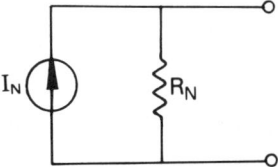

FIGURE 11-35
Norton's equivalent circuit.

Equivalent Circuits and the Network Theorems

EXAMPLE 11-8 Form the Norton's equivalent circuit for the network of Figure 11-36.

SOLUTION Determine Norton's equivalent current (I_N) by shorting terminals a and b as pictured in Figure 11-37(a). Thus:

$$I_N = \frac{E}{R_1}$$

Substitute:

$$I_N = \frac{20}{24 \times 10^3} = 0.833 \times 10^{-3}$$

$$\therefore \quad I_N = 0.833 \text{ mA}$$

Determine Norton's equivalent resistance (R_N) by setting the voltage source to zero and calculating the resistance between terminals a and b, as shown in Figure 11-37(b). Thus:

$$R_N = R_{ab} = R_1 \parallel R_2$$

Substitute:

$$R_N = \frac{1}{\dfrac{1}{24 \times 10^3} + \dfrac{1}{16 \times 10^3}} = 9.6 \text{ k}\Omega$$

$$\therefore \quad R_N = 9.6 \text{ k}\Omega$$

Observation Notice that the Norton's equivalent resistance is computed in the same manner as the Thevenin's equivalent resistance in the preceding section. $R_N \equiv R_{Th}$.

$\therefore$ The Norton equivalent circuit is formed by a current source of 0.833 mA and a parallel resistance of 9.6 kΩ, as shown in Figure 11-38.

EXAMPLE 11-9 Apply Norton's theorem to the network of Figure 11-39 to determine the current (I_2) in R_2.

FIGURE 11-36
Network of Example 11-8.

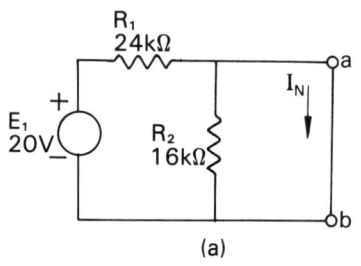
(a)

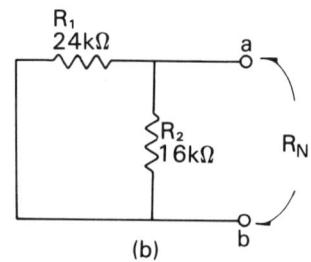
(b)

FIGURE 11-37
(a) Determining Norton's equivalent current.
(b) Determining Norton's equivalent resistance.

Norton's Theorem

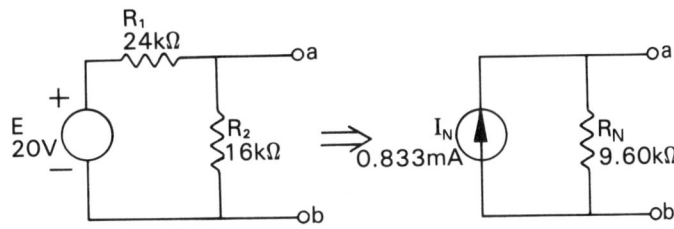

FIGURE 11-38
Norton's equivalent circuit of Example 11-8.

SOLUTION Remove R_2, R_3, and I from the network at points a-b of Figure 11-39. Apply Norton's theorem to E in series with R_1 to form the Norton's equivalent circuit. Compute the Norton's equivalent current (I_N) by shorting points a-b. The current in the short circuit is I_N.

$$I_N = \frac{E}{R_1} = \frac{10}{5} = 2 \text{ A}$$

∴ $I_N = 2$ A

Determine R_N across the open-circuit network by setting the voltage source to zero and calculating the Norton's equivalent resistance between points a-b. Thus:

$$R_N = R_1 = 5 \text{ }\Omega$$

∴ $R_N = 5$ Ω

Observation The Norton's equivalent circuit is formed as shown in Figure 11-40.
Attach R_1, R_3, and I to the terminals a-b as shown in Figure 11-41.
Redraw Figure 11-41 so the two current sources are together, as shown in Figure 11-42. Combine the current sources:

$$I_T = I - I_N = 5 - 2 = 3 \text{ A}$$

∴ $I_T = 3$ A in the direction (downward) of the larger current source I, as shown in Figure

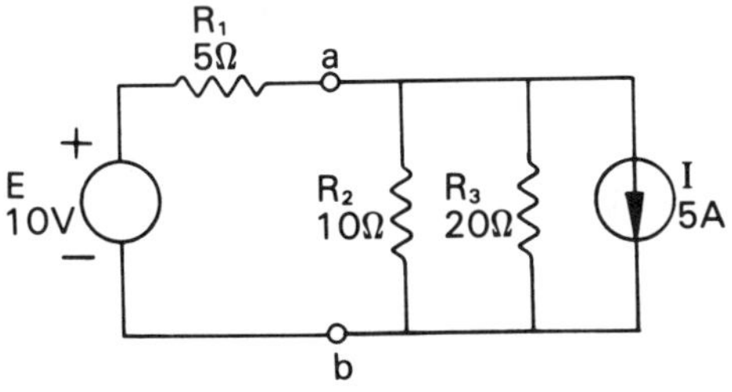

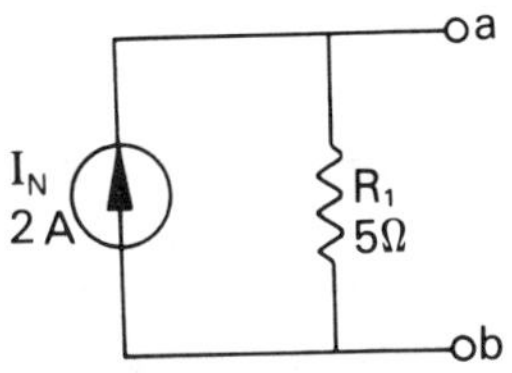

FIGURE 11-39 (left)
Network of Example 11-9.

FIGURE 11-40 (right)
Norton's equivalent circuit of E and R_1 of Figure 11-39.

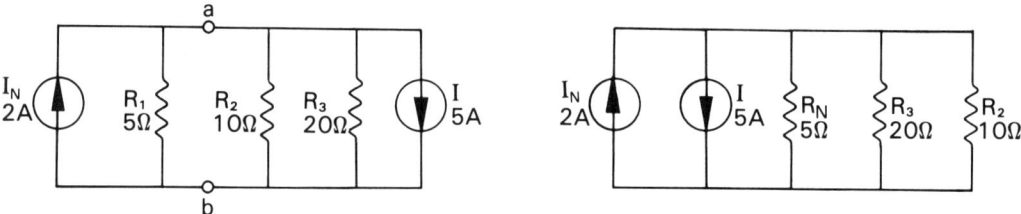

FIGURE 11–41 (left) Replacing E and R_1 of Figure 11–39 with the Norton's equivalent circuit of Figure 11–40.
FIGURE 11–42 (right) Rearrangement of Figure 11–41.

11–43. Apply the current-divider rule to calculate I_2 of Figure 11–43:

$$I_2 = \frac{I_T G_2}{\Sigma G} = \frac{(3)(\frac{1}{10})}{\frac{1}{5} + \frac{1}{20} + \frac{1}{10}} = 857 \text{ mA}$$

$$\therefore \quad I_2 = 857 \text{ mA}$$

Comparison of Thevenin's and Norton's Equivalent Circuits

EXAMPLE 11–10 Compare the Thevenin's equivalent circuit of Figure 11–44(a) to the Norton's equivalent circuit of Figure 11–44(b). Determine if the two circuits are equivalent when various loads are attached to terminals a-b by comparing the voltage, current, and power in each for (a) open-circuit voltage, (b) short-circuit current and (c) power dissipation in a 1000-Ω load.

SOLUTION (a) Compare open-circuit voltage, V_{ab}, across terminals a-b. Thus:

Thevenin's Equivalent	Norton's Equivalent
$V_{ab} = E_{Th} = 8$ V	$V_{ab} = I_N R_N$
	$V_{ab} = (833 \times 10^{-6})(9.6 \times 10^3) = 8$ V

$\therefore$ The open-circuit voltage V_{ab} is identical.

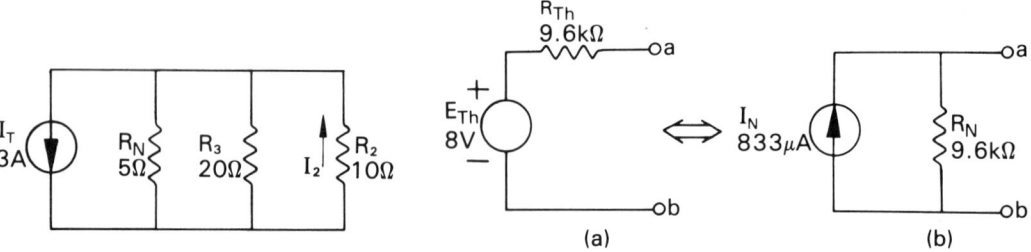

FIGURE 11–43 (left) Combining the current sources of Figure 11–42.
FIGURE 11–44 (right) (a) Thevenin's equivalent circuit of Example 11–10. (b) Norton's equivalent circuit of Example 11–10.

(b) Compare short-circuit current, I_{sc}, between terminals a-b. Thus with terminals a-b shorted:

Thevenin's Equivalent	Norton's Equivalent
$I_{sc} = \dfrac{E_{Th}}{R_{Th}}$	$I_{sc} = I_N = 833 \ \mu A$
$I_{sc} = \dfrac{8}{9.6 \times 10^3} = 833 \ \mu A$	

∴ The short-circuit current, I_{sc}, is identical.

(c) Compare power dissipated, P_L, in a 1000-Ω load connected between terminals a-b. Thus with $R_L = 1000 \ \Omega$:

Thevenin's Equivalent	Norton's Equivalent
$P_L = I_L^2 R_L$	$P_L = I_L^2 R_L$
$I_L = \dfrac{E_{Th}}{R_T}$	$I_L = \dfrac{I_N G_L}{\Sigma G}$
$I_L = \dfrac{8}{9600 + 1000}$	$I_L = \dfrac{(833 \times 10^{-6})(\frac{1}{1000})}{\frac{1}{9.6 \times 10^3} + \frac{1}{1000}} = 755 \ \mu A$
$I_L = 755 \ \mu A$	
$P_L = (755 \times 10^{-6})^2(1000)$	$P_L = (755 \times 10^{-6})^2(1000)$
$P_L = 570 \ \mu W$	$P_L = 570 \ \mu W$

∴ The power dissipation, P_L, in a 1000-Ω load is identical.

Observation The same answer for the two circuits will be obtained for any value of R_L between 0 Ω and $\infty \ \Omega$. Thus, the two are truly equivalent.

11–4 CONVERSION OF THEVENIN'S AND NORTON'S EQUIVALENT CIRCUITS

Once Thevenin's and Norton's equivalent circuits are formed, it is simple to convert from one equivalent circuit to the other. The relationship between the two theorems is shown in Figure 11–45. When applying these theorems, the selection of which equivalent to use depends on the circuit parameter to be determined. Often a student will become more familiar or adept at forming one equivalent circuit and will then convert to the other equivalent when it is needed.

EXAMPLE 11–11 Given the Thevenin's equivalent circuit of Figure 11–46(a), convert to a Norton's equivalent circuit.

FIGURE 11–45
Converting between Thevenin's and Norton's equivalent circuits.

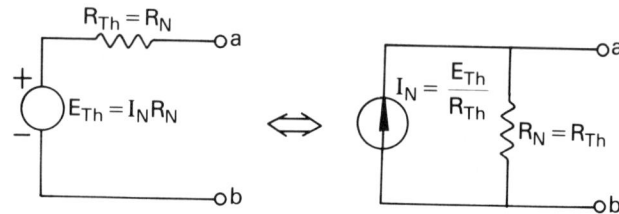

SOLUTION Determine R_N:

$$R_N = R_{Th} = 4.0 \text{ k}\Omega$$

$\therefore \quad R_N = 4.0 \text{ k}\Omega$

Determine I_N by shorting terminals a-b as shown in Figure 11–46(b) and computing the current through the short circuit. Thus:

$$I_N = \frac{E_{Th}}{R_{Th}}$$

Substitute:

$$I_N = \frac{20}{4 \times 10^3} = 5.0 \text{ mA}$$

$\therefore \quad I_N = 5.0 \text{ mA}$

Form the Norton's equivalent circuit as pictured in Figure 11–47(b).

Observation Notice that the current arrow in the current source points in the same direction as the positive terminal of the voltage source.

EXAMPLE 11–12 Convert the Norton's equivalent circuit shown in Figure 11–48 to a Thevenin's equivalent circuit.

SOLUTION Determine R_{Th}:

$$R_{Th} = R_N = 10 \text{ k}\Omega$$

$\therefore \quad R_{Th} = 10 \text{ k}\Omega$

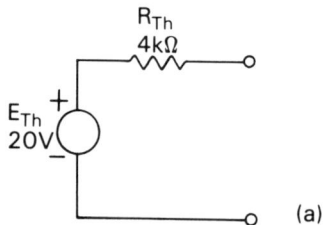

(a)

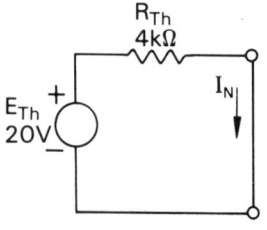

(b)

FIGURE 11–46
Determining Norton's equivalent current from Thevenin's equivalent circuit of Example 11–11.

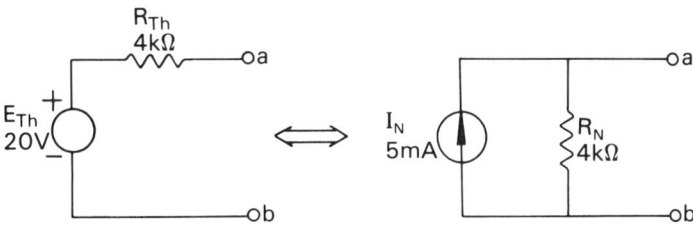

FIGURE 11-47
Equivalent Thevenin's and Norton's circuits of Example 11-11.

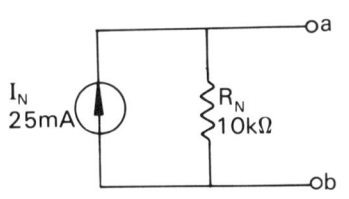

FIGURE 11-48
Circuit of Example 11-12.

Determine E_{Th} where

$$E_{Th} = I_N R_N$$

Substitute:

$$E_{Th} = (25 \times 10^{-3})(10 \times 10^3) = 250 \text{ V}$$

∴ $E_{Th} = 250$ V

Observation Form the equivalent Thevenin's circuit as pictured in Figure 11-49(b).

Notice that the Norton's current of Figure 11-49(a) results when terminals a-b of Figure 11-49(b) are short circuited. That is, $I_N = 250/(10 \times 10^3) = 25$ mA.

Summary Calculating or deriving Thevenin's and Norton's equivalent circuits from a network is not difficult if you separate the problem into two parts. The first part is the same for either equivalent circuit. Calculate the resistance ($R_N = R_{Th}$) looking into the selected two terminals with all sources set to zero (short-circuit voltage sources, open-circuit current sources). The equivalent resistance determined in this step is either the Thevenin or Norton equivalent resistance ($R_{Th} = R_N$).

The second part involves calculating either the open-circuit voltage or the short-circuit current at the two terminals. If the Thevenin equivalent voltage source is needed, calculate the open-circuit voltage at the selected two terminals. In a multiple-source circuit, this will involve using the superposition theorem to calculate the total voltage at the output, which is the

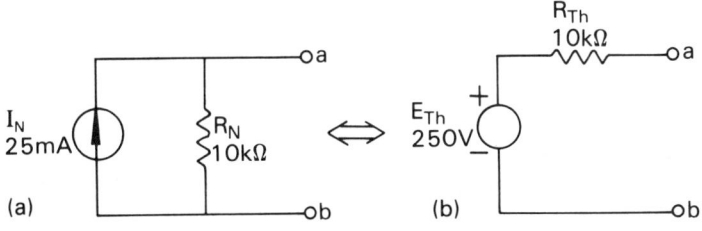

FIGURE 11-49
Conversion from Norton's to Thevenin's equivalent circuits in Example 11-12.

Equivalent Circuits and the Network Theorems

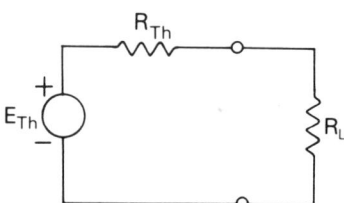

FIGURE 11-50
Maximum power transfer occurs when $R_{Th} = R_L$.

sum of the voltages generated by each source. If Norton's equivalent current source is desired, calculate the total current that would flow in a short circuit across the selected two terminals. Again, in the case of multiple sources, use the superposition theorem to determine the resultant current.

11-5 MAXIMUM POWER TRANSFER THEOREM

Definition

The maximum power transfer principle states that, given a linear dc bilateral network represented by a Thevenin's equivalent circuit, the maximum power will be transferred to a resistive load when the load resistance is equal to the Thevenin's equivalent resistance. Figure 11-50 pictures a Thevenin's equivalent circuit with a resistive load, R_L, attached.

THEOREM 11-4. MAXIMUM POWER TRANSFER THEOREM

Maximum power will be transferred from the source through a linear bilateral resistive network to the load when the resistance of the load is equal to the Thevenin's equivalent resistance of the network.

$$R_L = R_{Th} \quad \text{or} \quad R_L = R_N \quad \text{since} \quad R_N = R_{Th}$$

Applying the Maximum Power Transfer Theorem

As stated in Theorem 11-4, maximum power is developed in a resistive load when the load resistance equals the Thevenin's equivalent resistance of the network to which the load is attached. The Thevenin's equivalent resistance is also called the *source resistance*, R_s, since it is in series with the Thevenin's equivalent source voltage. Looking at Figure 11-51, you see a Thevenin's equivalent circuit with a load resistance (R_L) attached. Maximum power will be developed in the load when the load resistance (R_L) is adjusted to 50 Ω, the resistance of the source resistance (R_s). Example 11-13 explores the maximum power transfer theorem.

Maximum Power Transfer Theorem

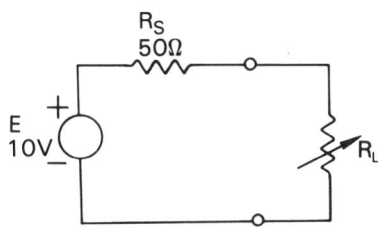

FIGURE 11–51
Schematic of an experimental set-up to demonstrate maximum power transfer.

EXAMPLE 11–13 Construct a graph of load power (ordinate) as a function of load resistance (abscissa) by constructing a table of values of load resistance (R_L), load current (I), load power (P_L), source power (P_s) and transfer efficiency (η_t).

SOLUTION From Figure 11–51, select values of R_L equal to 0 Ω, 100 Ω, and ∞ Ω. Compute I, P_L, P_s and η_t for each case, where

$$I = \frac{E}{R_T} = \frac{10}{50 + R_L} \quad \text{and} \quad P_L = I^2 R_L$$

Substitute $R_L = 0$ Ω (short circuit):

$$I = \frac{10}{50} = 0.20 \text{ A}$$

$$P_L = (0.20)^2(0) = 0 \text{ W}$$

Substitute $R_L = 100$ Ω:

$$I = \tfrac{10}{150} = 0.067 \text{ A}$$

$$P_L = (0.067)^2(100) = 0.449 \text{ W}$$

Substitute $R_L = \infty$ Ω (open circuit):

$$I = \frac{10}{\infty} \Rightarrow 0 \text{ A}$$

$$P = (0)^2(\infty) \Rightarrow 0 \text{ W}$$

Observation Since R_L is open circuit (∞ Ω), no current flows and there cannot be any power dissipated. In the extreme cases ($R_L = 0$ Ω and $R_L = \infty$ Ω), no power is transferred from the source to the load. Table 11–1 summarizes the calculation for load current (I) and load power (P_L). Compute source power (P_s) and transfer efficiency (η_t), where

$$P_s = I^2(50 + R_L) \quad \text{and} \quad \eta_t = \frac{P_L}{P_s}$$

Substitute $R_L = 0$ Ω:

TABLE 11-1 TABULAR DATA FOR THE POWER TRANSFER CURVE OF FIGURE 11-52 FOR THE CIRCUIT OF FIGURE 11-51

Load Resistance R_L (Ω)	Load Current $I = 10/(50 + R_L)$ (A)	Load Power $P_L = I^2 R_L$ (W)	Source Power $P_s = I^2(50 + R_L)$ (W)	Transfer Efficiency $\eta_t = P_L/P_s \times 100$ (%)
0*	0.200*	0.000*	2.00	00.0
10	0.167	0.279	1.67	16.7
20	0.143	0.409	1.43	28.6
30	0.125	0.469	1.25	37.5
40	0.111	0.493	1.11	44.4
50	**0.100**	**0.500**	**1.00**	**50.0**
60	0.091	0.497	0.91	54.6
70	0.083	0.482	0.83	58.1
80	0.077	0.474	0.77	61.6
90	0.071	0.454	0.71	63.9
100	0.067	0.449	0.67	67.0
1000	0.010	0.100	0.11	90.9
∞†	0.000†	0.000†	0.00	—

* Short-circuit value.
† Open-circuit value.

$$P_s = (0.20)^2(50) = 2.0 \text{ W}$$

$$\eta_t = \frac{0}{2.0} = 0\%$$

Substitute $R_L = 100$ Ω:

$$P_s = (0.067)^2(150) = 0.67 \text{ W}$$

$$\eta_t = \frac{0.449}{0.67} \times 100 = 67\%$$

Substitute $R_L = \infty$ Ω:

$$P_s = (0)^2 (\infty) \Rightarrow 0 \text{ W}$$

$$\eta_t = \frac{0}{0} \Rightarrow \text{no power transfer}$$

Observation The graph of load power as a function of load resistance (Figure 11-52) shows that maximum power does indeed occur when the load resistance *matches* (equals) the source resistance of 50 Ω.

Notice from the values of Table 11-1 that the maximum power transfer of 0.5 W does not result in the best transfer efficiency. Maximum transfer of power is the main concern in many

Maximum Power Transfer Theorem

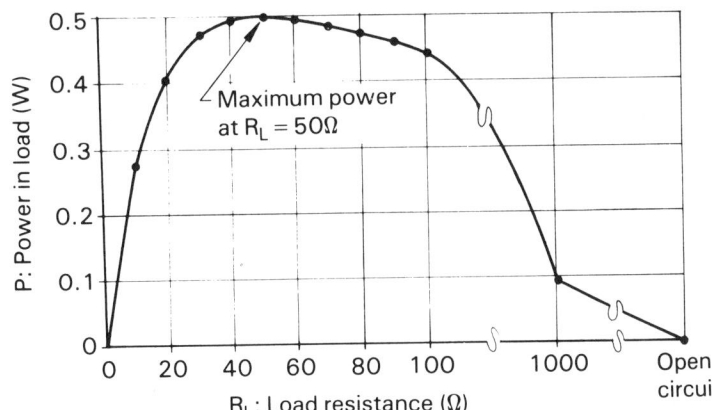

FIGURE 11-52
Plot of the data from Table 11-2.

types of electronic circuits, especially those related to communications systems. The major technical problem in circuits and systems operating at radio frequencies (RF) centers around *matching* the output resistance to the input resistance to obtain the maximum transfer of power from one circuit into another. That is, maximum power is transferred from one electronic circuit to the next when the load resistance is equal to the source resistance. The efficiency of transferring power from one circuit to the next at radio frequencies is 50%, as noted in Figure 11-53.

A practical example of matching source to load occurs in virtually every home. On the back of every television receiver are one or two antenna terminals labeled 75 ohms or 300 ohms. This label indicates the type of antenna and cable to use to achieve maximum power transfer. Without proper matching, the RF signal level would be reduced, causing less power to reach the load, which in the case of the television set results in a lower-quality picture.

When selecting a load for maximum transfer of power, it is not always necessary to form a Thevenin's equivalent circuit.

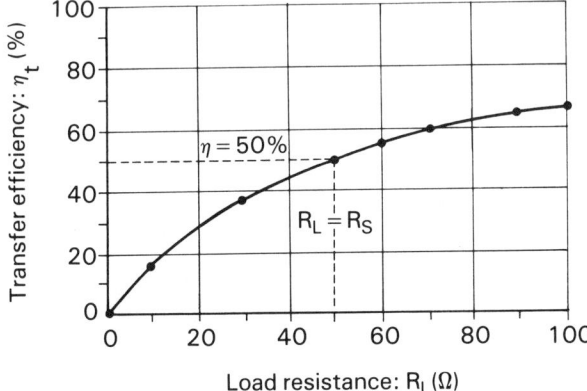

FIGURE 11-53
Power transfer efficiency as a function of load resistance.

Equivalent Circuits and the Network Theorems

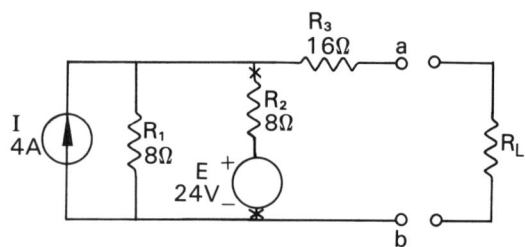

FIGURE 11–54
Network of Example 11–14.

Since the Norton's equivalent resistance equals the Thevenin's equivalent resistance ($R_N = R_{Th}$), maximum power in the load will also occur when the load resistance equals the Norton's equivalent resistance ($R_L = R_N$). This concept is demonstrated in Example 11–14.

EXAMPLE 11–14 For the network of Figure 11–54, determine the value of load resistance, R_L, that must be connected to terminals a-b to ensure maximum power transfer from the network into the load.

SOLUTION Remove R_2 and E from the network and form a Norton's equivalent circuit as shown in Figure 11–55, where

$$R_N = R_2 = 8 \; \Omega$$

and

$$I_N = \frac{E}{R_2} = \frac{24}{8} = 3 \text{ A}$$

∴ $R_N = 8 \; \Omega$ and $I_N = 3$ A

Connect the Norton's equivalent circuit into the network in place of the voltage source, as shown in Figure 11–56. Combine the current sources and the parallel resistance. Thus:

$$I_T = I + I_N = 4 + 3 = 7 \text{ A}$$

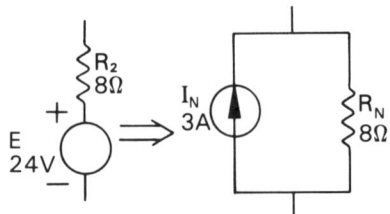
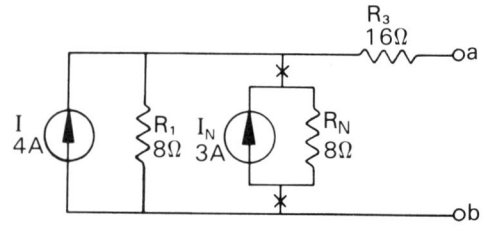

FIGURE 11–55 (left)   Convert voltage source E and resistance R_2 to a Norton's equivalent circuit.
FIGURE 11–56 (right)   Replace E and R_2 of Figure 11–54 with Norton's equivalent circuit of Figure 11–55.

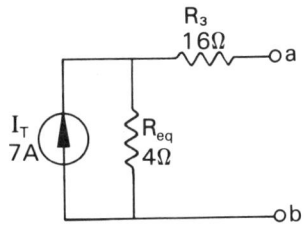

FIGURE 11-57
Combining current sources I and R_N of Figure 11-56.

and
$$R_{eq} = R_1 \parallel R_N = 4 \, \Omega$$

∴ $I_T = 7$ A and $R_{eq} = 4 \, \Omega$, as shown in Figure 11-57.

Compute the Norton's equivalent resistance looking into terminals $a\text{-}b$. Thus:

$$R_{ab} = R_3 + R_{eq} = 16 \, \Omega + 4 \, \Omega = 20 \, \Omega$$

∴ $R_L = R_{ab} = R_N = 20 \, \Omega$ for maximum power transfer.

11-6 MILLMAN'S THEOREM

Definition

Millman's theorem is best described as a procedure used to analyze a network made up of two or more parallel branches containing sources and resistances. A Thevenin's equivalent circuit may be formed by applying the specified procedure of Millman's theorem. The required steps are both summarized as Guideline 11-2 and pictured in Figure 11-58.

GUIDELINE 11-2. APPLYING MILLMAN'S
THEOREM TO
FORM THEVENIN'S
EQUIVALENT CIRCUIT

1. Remove the load R_L from the network as noted in Figure 11-58(a).
2. Convert all the voltage sources to current sources as shown in Figure 11-58(b). Each branch is treated as a Thevenin's equivalent circuit, which is then converted to a Norton's equivalent circuit.
3. Algebraically add the parallel current sources to form a single current source as pictured in Figure 11-58(c).
4. Form the branch resistances into an equivalent resistance as noted in Figure 11-58(c).
5. Transform the Norton's equivalent circuit of Figure 11-58(c) into the Thevenin's equivalent circuit of Figure 11-58(d).
6. Reconnect the load, R_L.

Millman's theorem, in essence, states that the voltage across the two selected terminals of a network, such as that pictured in Figure 11-58(a), is equal to the summation of each branch voltage source divided by the resistance of that branch, all divided

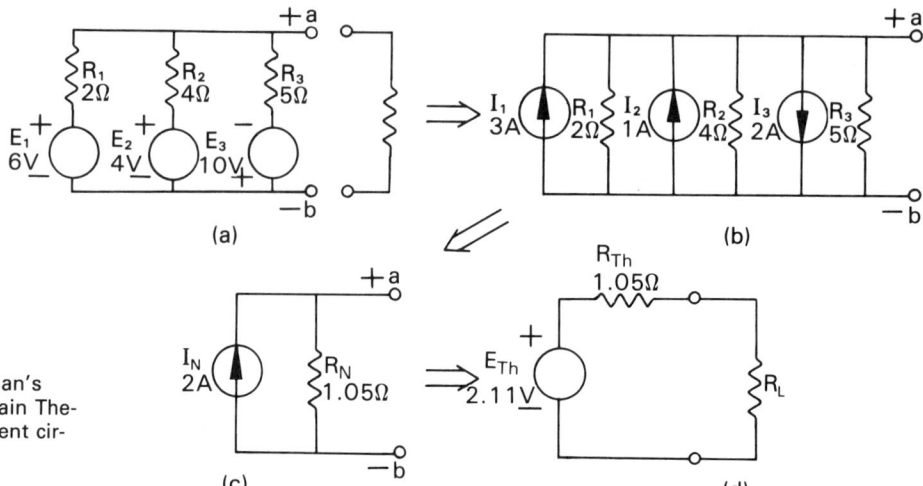

FIGURE 11-58
Applying Millman's theorem to obtain Thevenin's equivalent circuit.

by the total conductance of the branches. Millman's theorem is stated mathematically as Equation 11-1.

$$V_M = \frac{\dfrac{E_1}{R_1} + \dfrac{E_2}{R_2} + \dfrac{E_3}{R_3} + \cdots + \dfrac{E_n}{R_n}}{\dfrac{1}{R_1} + \dfrac{1}{R_2} + \dfrac{1}{R_3} + \cdots + \dfrac{1}{R_n}} \quad (V) \quad (11-1)$$

Or when the network is expressed in conductance rather than resistance,

$$V_M = \frac{E_1 G_1 + E_2 G_2 + E_3 G_3 + \cdots + E_n G_n}{G_1 + G_2 + G_3 + \cdots + G_n} \quad (V) \quad (11-2)$$

where V_M = voltage across the two selected terminals
$V_M = E_{Th}$
E_1, E_2, E_3, etc. = voltage source of each branch
R_1, R_2, R_3, etc. = branch resistance
G_1, G_2, G_3, etc. = branch conductance

Applying Millman's Theorem When applying Equation 11-1 or 11-2, the sign of each voltage source depends on the polarity of the two selected terminals. Also, in applying Equation 11-1 or 11-2, there may not be a voltage source in a particular branch. When this is the case, substitute 0 V into the equation. The following examples demonstrate the usefulness of the equation of Millman's theorem.

EXAMPLE 11-15 Apply Equation 11-1 to the circuit of Figure 11-58(a) to form the Thevenin's equivalent circuit of Figure 11-58(d).

324   **SOLUTION** Substitute into Equation 11-1. Thus:

$$V_M = \frac{\frac{6}{2} + \frac{4}{4} - \frac{10}{5}}{\frac{1}{2} + \frac{1}{4} + \frac{1}{5}} = 2.11 \text{ V}$$

$$\therefore \quad E_{Th} = V_M = 2.11 \text{ V}$$

Determine R_{Th} from the total conductance:

$$R_{Th} = \frac{1}{\frac{1}{R_1} + \frac{1}{R_2} + \frac{1}{R_3}} = \frac{1}{\frac{1}{2} + \frac{1}{4} + \frac{1}{5}} = 1.05 \text{ }\Omega$$

$$\therefore \quad R_{Th} = 1.05 \text{ }\Omega$$

Observation Thevenin's equivalent circuit of Figure 11–58(d) is easily formed by applying Millman's theorem as stated in Equation 11–1.

EXAMPLE 11–16 Determine the voltage across terminals a-b of Figure 11–59 using Millman's theorem (Equation 11–1).

SOLUTION Substitute into Equation 11–1. Thus:

$$V_{ab} = V_M = \frac{\frac{10}{100} + \frac{0}{40} - \frac{6}{30}}{\frac{1}{100} + \frac{1}{40} + \frac{1}{30}} = -1.46 \text{ V}$$

$$\therefore \quad V_{ab} = -1.46 \text{ V}$$

Observation Since V_{ab} is negative, the polarity is opposite to that pictured in Figure 11–59.

EXAMPLE 11–17 Determine the power dissipated in resistance R_3 of Figure 11–60.

SOLUTION Compute the voltage across R_3 by applying Equation 11–1 to Figure 11–61, the redrawn circuit of Figure 11–60.

Observation Notice that the load terminals have been arbitrarily assigned a polarity.
Substitute:

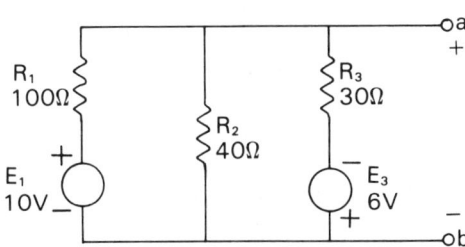

FIGURE 11–59
Network of Example 11–16.

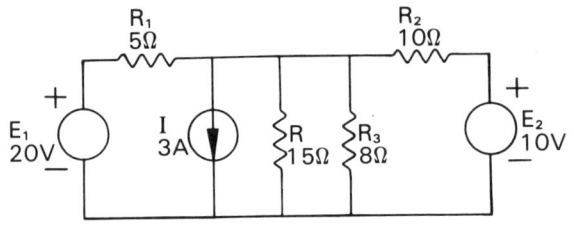

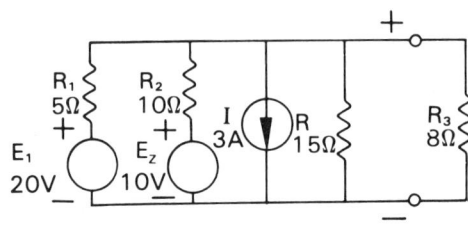

FIGURE 11-60 (above) Network of Example 11-17.

FIGURE 11-61 (right) Rearrangement of network in Figure 11-60.

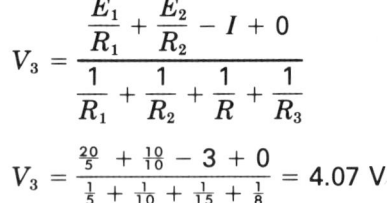

FIGURE 11-62 Thevenin's equivalent circuit is often used to represent a piece of test equipment.

$$V_3 = \frac{\frac{E_1}{R_1} + \frac{E_2}{R_2} - I + 0}{\frac{1}{R_1} + \frac{1}{R_2} + \frac{1}{R} + \frac{1}{R_3}}$$

$$V_3 = \frac{\frac{20}{5} + \frac{10}{10} - 3 + 0}{\frac{1}{5} + \frac{1}{10} + \frac{1}{15} + \frac{1}{8}} = 4.07 \text{ V}$$

Observation The selected polarity of the load terminals was correct since the voltage V_3 is positive. Also notice that the current source was substituted directly into the numerator of the equation since $I = E/R$.

Compute the power dissipation in R_3:

$$P = \frac{V_3^2}{R_3} = \frac{(4.07)^2}{8} = 2.07 \text{ W}$$

∴ The power dissipated in R_3 is 2.07 W.

11-7 APPLYING THE NETWORK THEOREMS

Specifications of electronic equipment such as power supplies, amplifiers, and signal generators will state that they have a certain output impedance (for the present, consider impedance to mean resistance). By knowing the specified output resistance as well as the open-circuit voltage, the Thevenin's equivalent circuit of the electronic equipment may be formed as shown in Figure 11-62.

Even though the actual electronic device may have a very complicated circuit design, it can be represented as a simple equivalent circuit. This then makes it very easy to analyze the circuit when a load is attached to the output.

EXAMPLE 11-18 A speech amplifier has an 8-Ω output with an open-circuit voltage of 16 V. Determine the

FIGURE 11-63
Thevenin's equivalent circuit of an amplifier and speaker.

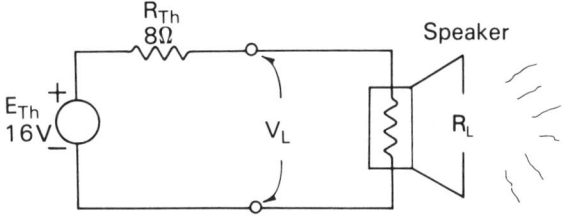

power delivered to (a) a 4-Ω speaker, (b) an 8-Ω speaker, and (c) a 16-Ω speaker when each speaker is attached to the output terminals of the amplifier.

SOLUTION Apply Thevenin's theorem by drawing the equivalent as shown in Figure 11-63.

(a) Determine output power (P_o) for $R_L = 4\ \Omega$.

$$P_O = \frac{V_L^2}{R_L}$$

Apply the voltage-divider rule:

$$V_L = \frac{E_{Th}R_L}{R_{Th} + R_L} = \frac{(16)(4)}{12} = 5.33\ \text{V}$$

Substitute and solve for P_o:

$$P_o = \frac{5.33^2}{4} = 7.1\ \text{W}$$

∴ The power in the 4-Ω speaker is 7.1 W.

(b) Determine output power (P_o) for $R_L = 8\ \Omega$.

$$P_O = \frac{V_L^2}{R_L}$$

Apply the voltage-divider rule:

$$V_L = \frac{E_{Th}R_L}{R_{Th} + R_L} = \frac{(16)(8)}{16} = 8\ \text{V}$$

Substitute and solve for P_o:

$$P_o = \frac{8^2}{8} = 8\ \text{W}$$

∴ The power in the 8-Ω speaker is 8 W.

(c) Determine output power (P_o) for $R_L = 16\ \Omega$.

$$P_O = V_L^2/R_L$$

Apply the voltage-divider rule:

$$V_L = \frac{E_{Th}R_L}{R_{Th} + R_L} = \frac{(16)(16)}{24} = 10.67 \text{ V}$$

Substitute and solve for P_o:

$$P_o = \frac{10.67^2}{16} = 7.1 \text{ W}$$

∴ The power in the 16-Ω speaker is 7.1 W.

All the theorems may be needed to solve a single complex circuit, as demonstrated in Example 11–19.

EXAMPLE 11–19 Determine the maximum power possible in resistance R_4 of Figure 11–64 when R_4 is attached to the circuit.

SOLUTION For maximum power transfer, it is known that the resistive value of R_4 must equal the equivalent resistance of the network as viewed looking toward the source. The boxed network of Figure 11–64 can be represented as a Thevenin's equivalent circuit by applying the superposition theorem.

The Thevenin's equivalent resistance is determined by setting the voltage and current sources equal to zero by opening the current source and shorting the voltage source as shown in Figure 11–65. Thus:

$$R_{Th} = R_3 + \frac{1}{\frac{1}{R_1} + \frac{1}{R_2}} = 50 + \frac{1}{\frac{1}{80} + \frac{1}{40}}$$

∴ $R_{Th} = 76.67 \text{ Ω}$

The Thevenin's equivalent voltage, V_{ab}, is determined by first applying the superposition theorem to the voltage source as shown in Fig-

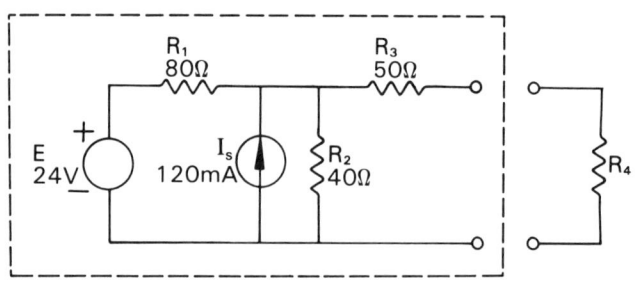

FIGURE 11–64
Network of Example 11–19.

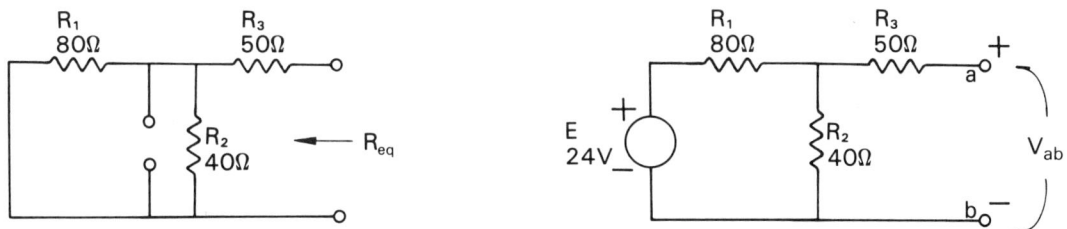

FIGURE 11-65 (left)   Determining Thevenin's resistance of the network of Figure 11-64.
FIGURE 11-66 (right) Determining voltage contribution to V_{ab} due to voltage source E.

ure 11-66, then to the current source as shown in Figure 11-67, and then algebraically adding the two voltages. From the voltage source,

$$V_{ab} = V_{R2} = \frac{ER_2}{R_1 + R_2} = \frac{(24)(40)}{80 + 40} = 8\text{ V}$$

And from the current source (Figure 11-67),

$$V_{ab} = I(R_1 \parallel R_2)$$

$$V_{ab} = (120 \times 10^{-3})\left(\frac{1}{\frac{1}{80} + \frac{1}{40}}\right) = 3.2\text{ V}$$

Observation Since V_{ab} has the same terminal polarities for each source, the Thevenin's equivalent voltage is the sum of the two source voltages.

$$\therefore \quad E_{Th} = 8\text{ V} + 3.2\text{ V} = 11.2\text{ V}$$

Observation The equivalent for the circuit within the dashed lines of Figure 11-64 is shown as Figure 11-68. Applying the maximum power transfer theorem results in the resistance R_4 being equal to $R_{Th} = 76.67\text{ }\Omega$, as noted in Figure 11-69. Determine the power in R_4:

$$P_4 = \frac{V_4^2}{R_4}$$

Determine V_4:

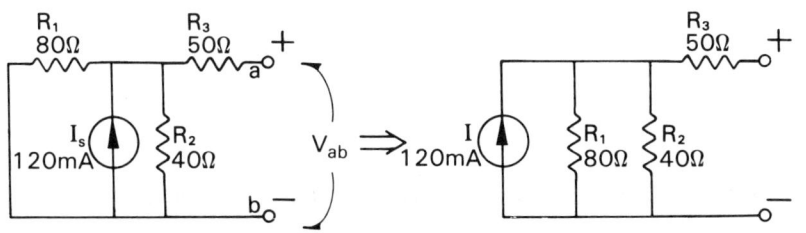

FIGURE 11-67 Determining voltage contribution V_{ab} due to source I_s.

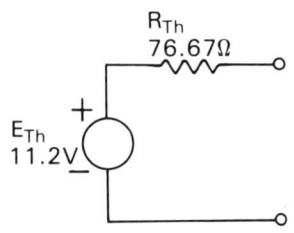

FIGURE 11–68
Resultant Thevenin's equivalent circuit after applying superposition theorem.

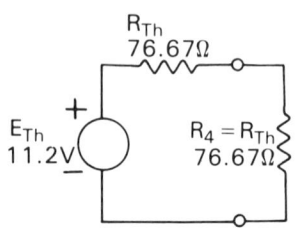

FIGURE 11–69
Applying maximum power transfer theorem.

$$V_4 = \frac{E_{Th}}{2} = \frac{11.2}{2} = 5.6 \text{ V}$$

Substitute:

$$P_4 = \frac{5.6^2}{76.67} = 0.409 \text{ W}$$

∴ The maximum power in R_4 is 409 mW.

Another approach to the solution of Example 11–17 is taken in Example 11–20.

EXAMPLE 11–20 Determine the maximum power possible in resistance R_4 of Figure 11–70(a) when R_4 is attached to the circuit.

SOLUTION Convert the 24-V source and resistor R_1 (80 Ω) into a Norton's equivalent circuit as pictured in Figure 11–70(b). Thus:

$$I_N = \frac{E}{R_1} = \frac{24}{80} = 0.3 \text{ A}$$

∴ $I_N = 0.3$ A

and

$$R_N = R_1 = 80 \text{ Ω}$$

∴ $R_N = 80$ Ω

Combine the two current sources:

$$I_T = 0.3 + 0.12 = 0.42 \text{ A}$$

Form R_1 and R_2 into an equivalent resistance:

$$R_{eq} = R_1 \parallel R_2 = \frac{1}{\frac{1}{80} + \frac{1}{40}} = 26.67 \text{ Ω}$$

Construct a new circuit with I_T and R_{eq} as shown in Figure 11–71.

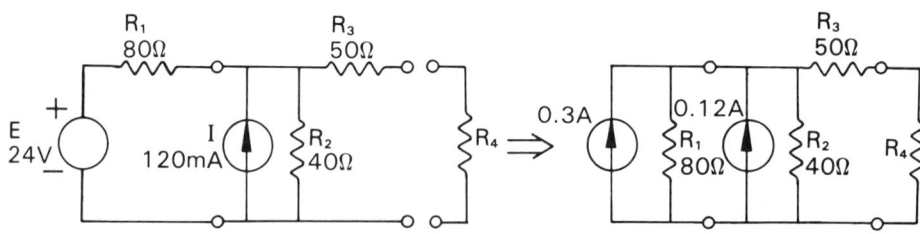

FIGURE 11–70 Replace voltage source E and R_1 with its Norton's equivalent.

Applying the Network Theorems

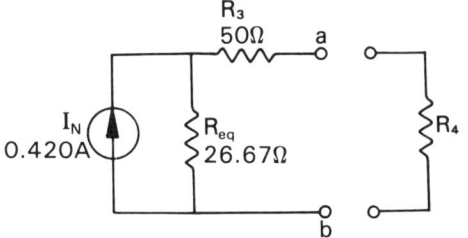

FIGURE 11-71 Resultant after combining current sources of Figure 11-70 and substituting R_{eq} in place of the parallel combination of R_1 and R_2.

Set the current source of Figure 11-71 to zero (open circuit) as shown in Figure 11-72(a) and form the Norton's equivalent circuit of Figure 11-72(b). Thus:

$$R_N = R_3 + R_{eq} = 50 + 26.67 = 76.67 \ \Omega$$

∴ $R_N = 76.67 \ \Omega$

The Norton current is equal to the current in a short circuit between terminals a-b of Figure 11-71. Apply the current-divider rule:

$$I_N = \frac{I_T G_3}{G_3 + G_{eq}} = \frac{(0.420)\left(\frac{1}{50}\right)}{\frac{1}{50} + \frac{1}{26.67}} = 0.146 \ \text{A}$$

∴ $I_N = 0.146 \ \text{A}$

To achieve maximum power, R_4 must equal $R_N = 76.67 \ \Omega$. Determine the power in R_4 of Figure 11-73:

$$P_4 = I_4^2 R_4$$

Determine I_4:

$$I_4 = \frac{I_N}{2} = \frac{0.146}{2} = 0.073 \ \text{A}$$

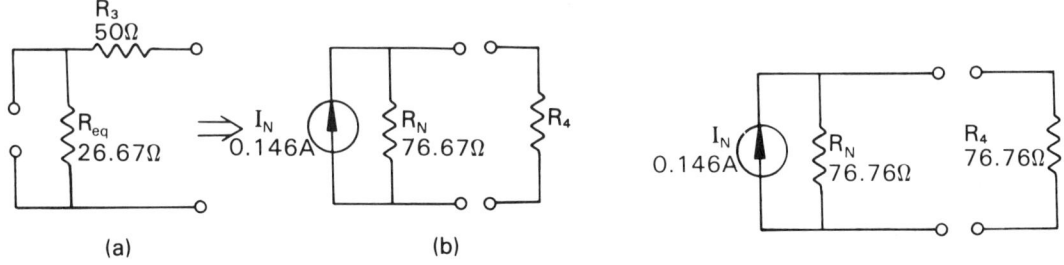

FIGURE 11-72 (left) Norton's equivalent circuit of Figure 11-71.
FIGURE 11-73 (right) Maximum power transfer occurs when $R_4 = R_N$.

332

Equivalent Circuits and the Network Theorems

Substitute:

$$P_4 = (0.073)^2(76.67) = 0.409 \text{ W}$$

∴ The maximum power in R_4 is 409 mW.

There are usually several approaches to the solution of a given problem. Which approach you may select depends to a large degree upon how you perceive the problem and your knowledge of the various network theorems. After learning all the various theorems, you may find yourself predominantly using one or two theorems.

Millman's theorem can be used to demonstrate the operation of an *analog adder* circuit such as pictured in Figure 11–74(a). In the equation for Millman's theorem, Equation 11–1, the Millman's voltage (V_M) is a direct function of the sum of the individual source voltages. If all the source voltages and resistances are equal, Equation 11–1 becomes Equation 11–3. Thus:

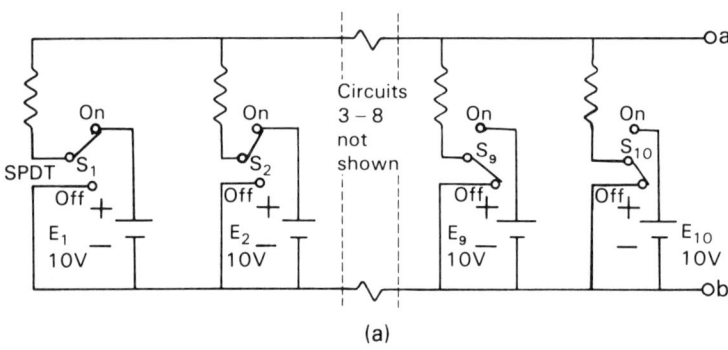

FIGURE 11–74
Analog adder circuit.
(a) Schematic of the circuit.
(b) Switch *truth table*.

				Switch condition						Output
1	2	3	4	5	6	7	8	9	10	voltage (V_{ab})
OFF	OFF	OFF	OFF	OFF	OFF	OFF	OFF	OFF	OFF	0 V
ON	OFF	OFF	OFF	OFF	OFF	OFF	OFF	OFF	OFF	1 V
ON	ON	OFF	OFF	OFF	OFF	OFF	OFF	OFF	OFF	2 V
ON	ON	ON	OFF	OFF	OFF	OFF	OFF	OFF	OFF	3 V
ON	ON	ON	ON	OFF	OFF	OFF	OFF	OFF	OFF	4 V
ON	ON	ON	ON	ON	OFF	OFF	OFF	OFF	OFF	5 V
ON	ON	ON	ON	ON	ON	OFF	OFF	OFF	OFF	6 V
ON	ON	ON	ON	ON	ON	ON	OFF	OFF	OFF	7 V
ON	ON	ON	ON	ON	ON	ON	ON	OFF	OFF	8 V
ON	ON	ON	ON	ON	ON	ON	ON	ON	OFF	9 V
ON	ON	ON	ON	ON	ON	ON	ON	ON	ON	10 V

(b)

$$V_M = \frac{\frac{E}{R} + \frac{E}{R} + \frac{E}{R} + \cdots + \frac{E_n}{R_n}}{\frac{1}{R} + \frac{1}{R} + \frac{1}{R} + \cdots + \frac{1}{R_n}}$$

Simplify by factoring:

$$V_M = \frac{n\left(\frac{E}{R}\right)}{m\left(\frac{1}{R}\right)} = \left(\frac{nE}{R}\right)\left(\frac{R}{m}\right)$$

$$V_M = \frac{nE}{m} \quad \text{(V)} \tag{11-3}$$

where V_M = voltage across the output terminals
E = voltage of each source
n = number of ON voltage sources
m = number of identical parallel branches in the adder

Suppose there are 10 identical parallel circuits of the type shown in Figure 11-74(a) and only 4 are turned ON while the remaining 6 are set to zero (OFF).

The voltage across the output terminals would then be 4 V as computed using Equation 11-3. Thus:

$$V_{ab} = V_M = \frac{nE}{m}$$

Substituting,

$$V_{ab} = \frac{4(10)}{10} = 4 \text{ V}$$

Therefore, the output voltage is 4 V, and it is a direct function of the number of ON sources, as indicated by Figure 11-74(b).

Another concept of an analog adder is pictured in Figure 11-75. Notice that each branch has a different voltage source.

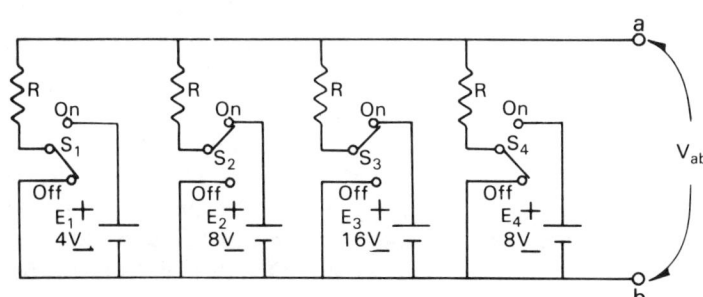

FIGURE 11-75
Adder circuit of Example 11-21.

The equation used to describe the operation of this circuit is a variation of Equation 11–1, Millman's theorem. Thus:

$$V_M = V_{ab} = \frac{\dfrac{E_1}{R} + \dfrac{E_2}{R} + \dfrac{E_3}{R} + \dfrac{E_4}{R}}{\dfrac{1}{R} + \dfrac{1}{R} + \dfrac{1}{R} + \dfrac{1}{R}}$$

Factoring,

$$V_M = \frac{\dfrac{1}{R}(E_1 + E_2 + E_3 + E_4)}{4/R}$$

Simplifying,

$$V_M = \left(\frac{1}{R}\right)\left(\frac{R}{4}\right)(E_1 + E_2 + E_3 + E_4)$$

Thus:

$$V_M = \frac{E_1 + E_2 + E_3 + E_4}{4} \tag{11-4}$$

where $E_1 = 4$ V, $E_2 = 8$ V, $E_3 = 16$ V, and $E_4 = 8$ V.

By switching S_1 through S_4 ON and OFF, any voltage from 0 to 9 V is obtainable (in 1-V steps) across terminals a-b of Figure 11–75.

EXAMPLE 11–21 Using Equation 11–4, determine V_{ab} of Figure 11–75 for the following switch settings: S_1 OFF, S_2 ON, S_3 ON, and S_4 OFF.

SOLUTION Substitute, $E_1 = 0$, $E_2 = 8$ V, $E_3 = 16$ V, and $E_4 = 0$ V into Equation 11–4:

$$V_{ab} = V_M = \frac{0 + 8 + 16 + 0}{4} = 6 \text{ V}$$

∴ $V_{ab} = 6$ V when S_1 is off, S_2 is on, S_3 is on, and S_4 is off.

The network theorems are very useful in establishing desired conditions in the load. Example 11–22 further demonstrates the application of the network theorems.

EXAMPLE 11–22 Design the circuit of Figure 11–76 so that a minimum of 5 V is across resistance R_4 when either S_1 or S_2 is ON. However, the voltage across resistance R_4 is not to exceed 8 V when both S_1 and S_2 are ON. Since the controlling

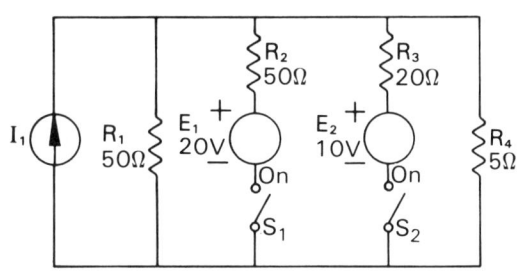

FIGURE 11-76
Network of Example 11-22.

parameter of the voltage across R_4 is the current source (I_1), determine the value of the current source to meet the design criteria.

SOLUTION First, determine the voltage across R_4 when S_1 and S_2 are ON. This will allow the maximum value of I_1 to be determined without exceeding 8 V across R_4. Next, determine the minimum value of I needed to provide at least 5 V across R_4 when either S_1 or S_2 is ON. Finally, with an understanding of both the maximum and minimum values of I, the operating value of I is selected.

Determine I_1 maximum, the current causing V_4 to be 8 V, by switching both S_1 and S_2 ON and setting I_1 to zero (open circuit). Change the two Thevenin's equivalent circuits to Norton's equivalent circuits. Thus:

$$I_2 = \frac{E_1}{R_2} = \frac{20}{50} = 0.40 \text{ A}$$

and

$$I_3 = \frac{E_2}{R_3} = \frac{10}{20} = 0.50 \text{ A}$$

The Norton's equivalent circuit of Figure 11-76 is shown as Figure 11-77(a).

Form I_2 and I_3 and the four resistances into an equivalent circuit. Thus:

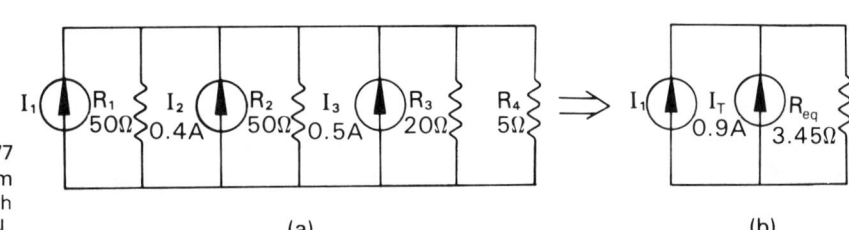

FIGURE 11-77
Determining maximum value of I_1 when both S_1 and S_2 are ON.

(a)   (b)

$I_T = I_2 + I_3 = 0.4 + 0.5 = 0.9$ A

and

$$R_{eq} = \frac{1}{\frac{1}{R_1} + \frac{1}{R_2} + \frac{1}{R_3} + \frac{1}{R_4}}$$

$$R_{eq} = \frac{1}{\frac{1}{50} + \frac{1}{50} + \frac{1}{20} + \frac{1}{5}} = 3.45 \ \Omega$$

From Figure 11–77(b), when $V_4 = 8$ V then

$(I_1 + I_T)(R_{eq}) = 8$ V

And solving for I_1 and substituting:

$$I_1 = \frac{8}{R_{eq}} - I_T = \frac{8}{3.45} - 0.9 = 1.42 \text{ A}$$

∴ The maximum value of I_1 is 1.42 A.

Next determine the minimum value of I_1 so that at least 5 V is across R_4 of Figure 11–76 when either of the two switches is ON. Start by switching S_1 ON and S_2 OFF. Thus:

$I_2 = \frac{20}{50} = 0.4$ A

and

$I_3 = 0$ A

Figure 11–78(a) pictures this condition. Now form the equivalent circuit of Figure 11–78(b) and compute I_1 for $V_4 = V_{eq} = 5$ V.

$I_T = I_2 + I_3 = 0.4 + 0 = 0.4$ A

and

$$R_{eq} = \frac{1}{\frac{1}{R_1} + \frac{1}{R_2} + \frac{1}{R_4}} = \frac{1}{\frac{1}{50} + \frac{1}{50} + \frac{1}{5}} = 4.17 \ \Omega$$

From Figure 11–78(b), when $V_4 = V_{eq} = 5$ V then

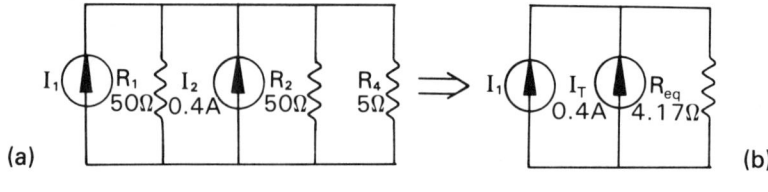

FIGURE 11–78
Determining minimum value of I_1 when S_1 is ON and S_2 is OFF.

Applying the Network Theorems

$$(I_1 + I_T)(R_{eq}) = 5 \text{ V}$$

And solving for I_1 and substituting:

$$I_1 = \frac{5}{R_{eq}} - I_T = \frac{5}{4.17} - 0.4 = 0.8 \text{ A}$$

∴ I_1 equals a minimum of 0.8 A when S_1 is OFF and S_2 is ON and $V_4 = 5$ V.

Now switch S_1 OFF and S_2 ON in Figure 11-76 and determine I_1 for $V_4 = 5$ V. Thus:

$$I_2 = 0 \text{ A}$$

and

$$I_3 = \tfrac{10}{20} = 0.5 \text{ A}$$

Figure 11-79(a) pictures this condition. Form the equivalent circuit of Figure 11-79(b) and compute I_1 for $V_4 = V_{eq} = 5$ V.

$$I_T = I_2 + I_3 = 0 + 0.5 = 0.5 \text{ A}$$

and

$$R_{eq} = \frac{1}{\frac{1}{R_1} + \frac{1}{R_3} + \frac{1}{R_4}} = \frac{1}{\frac{1}{50} + \frac{1}{20} + \frac{1}{5}} = 3.70 \text{ }\Omega$$

From Figure 11-79(b), when $V_4 = V_{eq} = 5$ V then

$$(I_1 + I_T)(R_{eq}) = 5 \text{ V}$$

And solving for I_1 and substituting:

$$I_1 = \frac{5}{R_{eq}} - I_T = \frac{5}{3.70} - 0.5 = 0.85 \text{ A}$$

∴ I_1 must be a minimum of 0.85 A when S_1 is ON and S_2 is OFF and $V_4 = 5$ V.

Observation From the previous calculations, the maximum value for I_1 is 1.42 A and the minimum value of I_1 must be greater than both 0.80 and 0.85 A. The

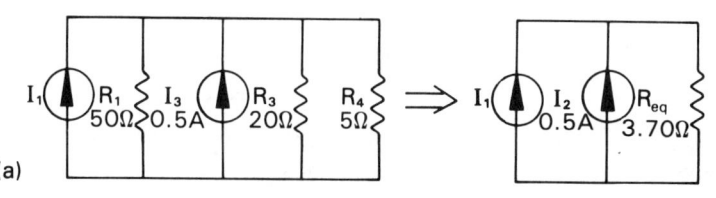

FIGURE 11-79
Determining minimum value of I_1 when S_1 is OFF and S_2 is ON.

value of current I_1 is therefore selected midway between these two extremes. Thus:

$$I_1 = \frac{I_{max} - I_{min}}{2} + I_{min}$$

$$I_1 = \frac{1.42 - 0.85}{2} + 0.85 = 1.14 \text{ A}$$

$$\therefore \quad I_1 = 1.14 \text{ A}$$

PROBLEMS As part of the solution, it is suggested that you draw and label a schematic diagram for each problem.

Section 11–1 Apply the superposition theorem to the following:
- 11–1 Determine the voltage drop across R_2 in the network of Figure 11–80.
- 11–2 Determine the current through R_1 in the network of Figure 11–80.
- 11–3 Determine the voltage across R_2 when the voltage source, E, in the network of Figure 11–80 is changed to a 15-mA current source, I_s, pointing toward R_1.
- 11–4 Determine the current in R_1 of the network pictured in Figure 11–80 using the conditions of Problem 11–3.
- 11–5 Determine the voltage drop across R_5 in the network of Figure 11–81.
- 11–6 Determine the current in R_4 of the network of Figure 11–81.
- 11–7 With the remaining sources appropriately terminated, determine the voltage at point A of Figure 11–82 due only to the +12-V source only.
- 11–8 With the remaining sources appropriately terminated, determine the voltage at point A of Figure 11–82 due only to the −5-V source.

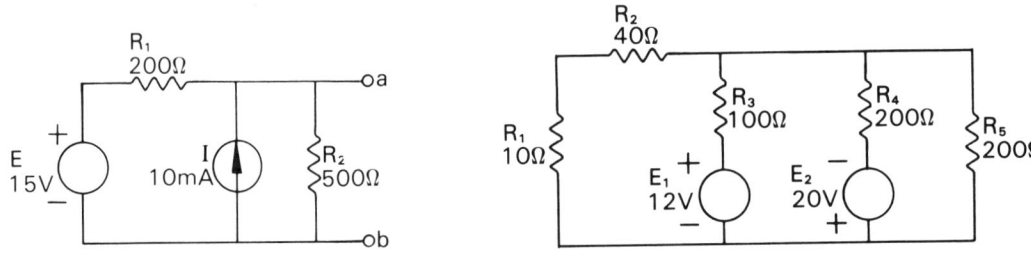

FIGURE 11–80 (left) Network of Problems 11–1, 11–2, 11–14, 11–27, and 11–31.
FIGURE 11–81 (right) Network of Problems 11–5, 11–6, 11–15, 11–28, and 11–32.

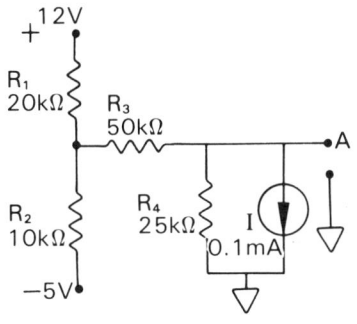
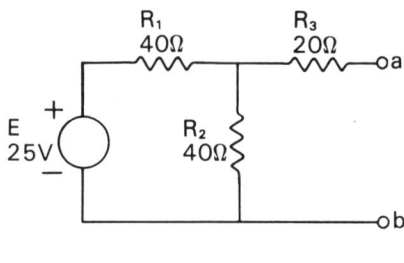

FIGURE 11–82 (left) Network of Problems 11–7, 11–8, 11–9, 11–16, and 11–29.
FIGURE 11–83 (right) Network of Problem 11–10.

11–9 Determine the voltage from point A to common due to all three sources in the network of Figure 11–82.

Section 11–2

11–10 Form the Thevenin's equivalent circuit of the network of Figure 11–83, looking into terminals a-b.

11–11 Form the Thevenin's equivalent circuit of the network of Figure 11–84, looking into terminals a-b.

11–12 Form the Thevenin's equivalent circuit of the network of Figure 11–85, looking into terminals a-b.

11–13 Determine the current in resistance R_L in the network of Figure 11–86 when (a) $R_L = 100\ \Omega$, (b) $R_L = 200\ \Omega$, and (c) $R_L = 500\ \Omega$.

11–14 Determine the Thevenin's equivalent circuit of the network of Figure 11–80 looking into terminals a-b.

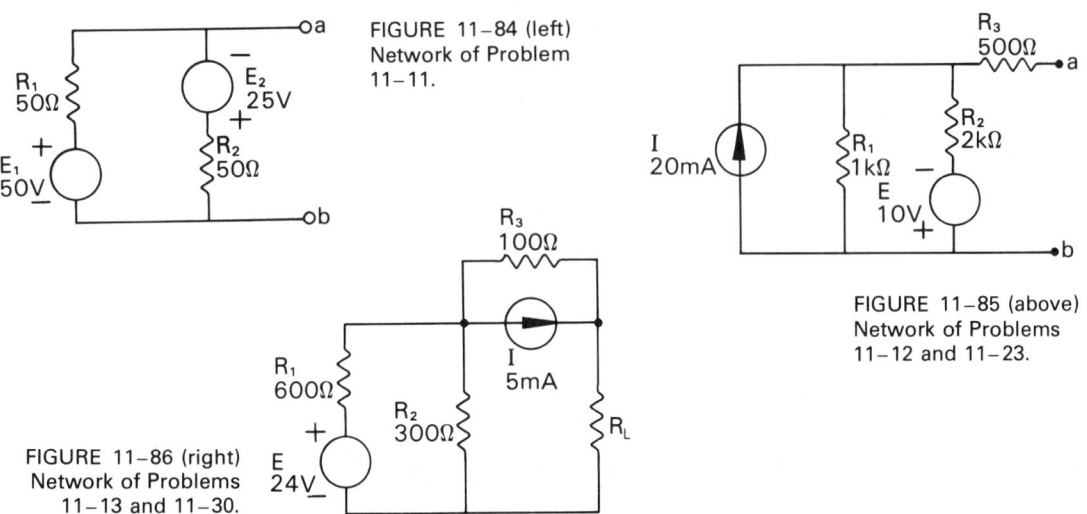

FIGURE 11–84 (left) Network of Problem 11–11.

FIGURE 11–85 (above) Network of Problems 11–12 and 11–23.

FIGURE 11–86 (right) Network of Problems 11–13 and 11–30.

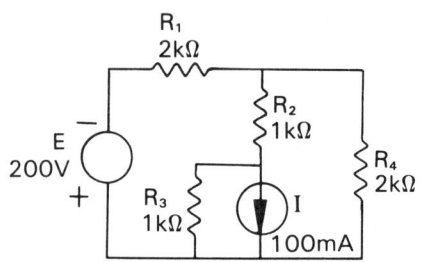

FIGURE 11-87
Network of Problems
11-17 and 11-26.

11-15 Apply Thevenin's theorem to the network of Figure 11-81 to determine the voltage drop across R_1.

11-16 Apply Thevenin's theorem to the network of Figure 11-82 to determine the voltage across a 5-kΩ resistor connected from point A to common.

Section 11-3

11-17 Apply Norton's theorem to the network of Figure 11-87 and determine the current in R_4.

11-18 Determine the Norton's equivalent circuit of the network of Figure 11-88, looking into terminals a-b.

11-19 Apply Norton's theorem to the network of Figure 11-89 to determine the current in R_L.

11-20 Form the Norton's equivalent circuit of the network of Figure 11-23, looking into terminals a-b.

Section 11-4

11-21 Convert the Thevenin's equivalent circuit formed in Problem 11-11 into a Norton's equivalent circuit.

11-22 Convert the Thevenin's equivalent circuit formed in Problem 11-10 into a Norton's equivalent circuit.

11-23 Convert the voltage source of $E = 10$ V and the series resistance $R_2 = 2$ kΩ of the network of Figure 11-85 into a Norton's equivalent circuit and determine the current in R_3 when terminals a-b are shorted.

11-24 Convert the Norton's equivalent circuit formed in Problem 11-18 into a Thevenin's equivalent circuit.

11-25 Convert the Norton's equivalent circuit formed in Problem 11-20 into a Thevenin's equivalent circuit.

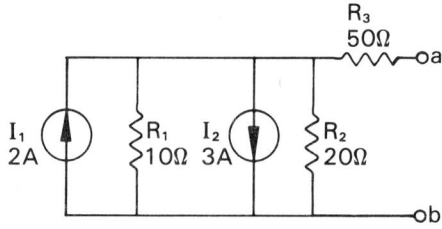

FIGURE 11-88
Network of Problems
11-18 and 11-33.

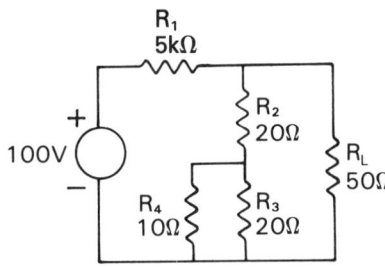

FIGURE 11-89
Network of Problem
11-19.

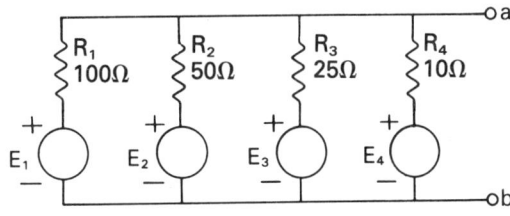

FIGURE 11-90
Network of Problems
11-34, 11-35, 11-36,
and 11-37.

11-26 In the network of Figure 11-87, let the current source equal 50 mA. Change the current source $I = 50$ mA and the parallel resistance $R_3 = 1$ kΩ into a Thevenin's equivalent circuit and determine the voltage across R_4.

Section 11-5

11-27 Determine the value of the load resistor (R_L) to be connected between terminals a-b of Figure 11-80 for maximum power transfer in R_L.

11-28 Determine the value of R_5 in the network of Figure 11-81 necessary for maximum power transfer in R_5.

11-29 For the network of Figure 11-82, determine the value of the load resistor (R_L) to be connected between point A and common for maximum power transfer.

11-30 For the network of Figure 11-86, determine the value of R_L for maximum power transfer.

Section 11-6

11-31 Using Millman's theorem, determine the voltage V_{ab} in the network of Figure 11-80.

11-32 Using Millman's theorem, determine the voltage across R_5 in the network of Figure 11-81.

11-33 Apply Millman's theorem to the network of Figure 11-88 to determine the voltage V_{ab}.

11-34 Determine V_{ab} of the network of Figure 11-90 when each voltage source is 10 V.

11-35 For the network of Figure 11-90, determine V_{ab} when $E_1 = E_2 = 20$ V and $E_3 = E_4 = 10$ V.

11-36 Determine V_{ab} of the network of Figure 11-90 when $E_1 = +10$ V, $E_2 = +20$ V, $E_3 = -50$ V and $E_4 = +10$ V.

11-37 Determine V_{ab} of the network of Figure 11-90 when all the voltage sources of Problem 11-36 are doubled.

12 Loop and Node Analysis

12-1 LOOP ANALYSIS
12-2 ESTABLISHING THE NUMBER OF LOOP EQUATIONS
12-3 WRITING AND SOLVING LOOP EQUATIONS
12-4 NODE ANALYSIS
12-5 ESTABLISHING THE NUMBER OF NODE EQUATIONS
12-6 WRITING AND SOLVING NODE EQUATIONS
12-7 SELECTING NODE OR LOOP ANALYSIS

CHAPTER PREVIEW Solving for circuit parameters in complex networks is simplified by establishing a systematic method for setting up and solving several circuit equations. Two of the most frequently used methods are loop analysis and node analysis. Loop analysis applies Kirchhoff's voltage law, whereas node analysis applies Kirchhoff's current law.

For circuits where there are more than two or three loop or node equations, the required calculations become rather tedious. However, with the now common use of computers, this systematic method of solving complex networks becomes routine.

PERFORMANCE OBJECTIVES Once you have read each section, worked each example with pencil, paper, and calculator, and worked each problem for every section, you will be able to:

☐ Set up and write loop and node equations.
☐ Solve simultaneous equations using determinants.
☐ Comprehend much of the technical literature relating to circuit analysis.
☐ Continue your study of more complex networks.

12-1 LOOP ANALYSIS

Loop analysis is an application of Kirchhoff's voltage law which states that the algebraic sum of all the voltages in a closed loop is equal to zero. A loop is defined as a closed circuit path from some point in the circuit back to the same point.

Several loops have been drawn in Figure 12-1:

☐ Loop A consists of the voltages resulting from the current passing through E_1, R_1, and R_2.
☐ Loop B consists of the voltages resulting from the current passing through R_2, R_3, and R_4.
☐ Loop C consists of the voltages resulting from the current passing through E_1, R_1, R_3, and R_4.

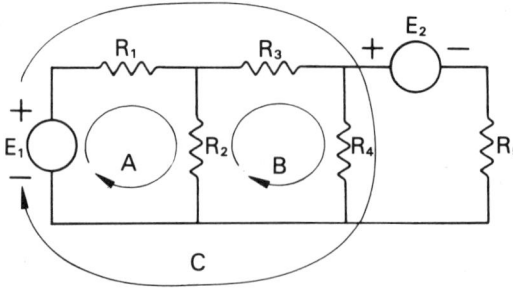

FIGURE 12-1
Three possible choices for loops.

Establishing the Number of Loop Equations

345

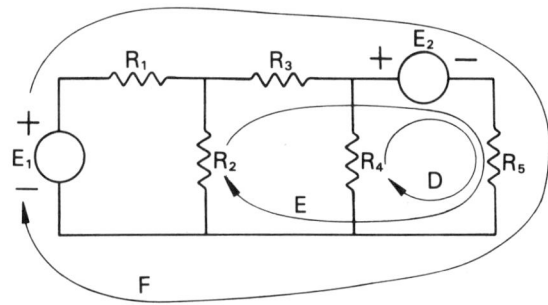

FIGURE 12-2
Three other loops.

There are six loops possible. Can you identify the other loops? Remember that a loop is any closed path. The other three loops are shown in Figure 12-2:

☐ Loop D consists of the voltage drops across resistors R_4 and R_5 and the voltage source E_2.
☐ Loop E consists of the voltage drops across R_2, R_3, R_5, and E_2.
☐ Loop F consists of the voltage drops across E_1, R_1, R_3, E_2, and R_5.

From the previous demonstration, you have learned that there are a total of six different loops for the network of Figure 12-1. The next step in applying loop analysis to this network is to determine how many of the six loops are required to analyze the network. Once this is known, the loop equations are written, and the solution for the desired unknown quantities can be determined.

12-2 ESTABLISHING THE NUMBER OF LOOP EQUATIONS

In networks that are relatively simple in structure, it is easy to determine how many loop equations are required for the solution of the circuit parameters. However, when the networks are complex, Equation 12-1 will be helpful in determining the number of required equations.

$$N_\ell = N_b - N_j + N_s \tag{12-1}$$

where N_ℓ = number of required loop equations
N_b = number of branches in the circuit, where a *branch* is defined as any active or passive element
N_j = number of junctions, where a junction is any point (node) where two or more elements are connected together
N_s = number of separate parts to the circuit, as shown in Figure 12-6

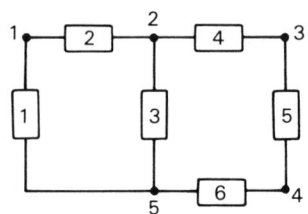

FIGURE 12–3
Network of Example 12–1.

EXAMPLE 12–1 How many loop equations are required for the solution of the network of Figure 12–3?

SOLUTION Figure 12–3 shows a circuit with 6 branches, 5 junctions, and 1 part. Thus:

$$N_\ell = N_b - N_j + N_s = 6 - 5 + 1 = 2$$

∴ Two loop equations are needed to solve this network.

EXAMPLE 12–2 How many loop equations are required for the solution of the network of Figure 12–4?

SOLUTION Figure 12–4 has 7 branches, 5 junctions, and 1 part. Thus:

$$N_\ell = N_b - N_j + N_s = 7 - 5 + 1 = 3$$

∴ Three loop equations are needed to solve this network.

EXAMPLE 12–3 How many loop equations are required for the solution of the network of Figure 12–5?

SOLUTION Figure 12–5 has 9 branches, 6 junctions, and 1 part. Thus:

$$N_\ell = N_b - N_j + N_s = 9 - 6 + 1 = 4$$

∴ Four loop equations are needed to solve this network.

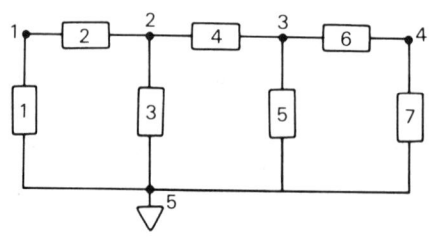 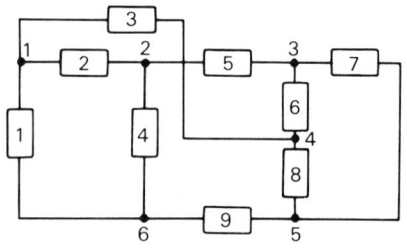

FIGURE 12–4 (left)   Network of Example 12–2.
FIGURE 12–5 (right) Network of Example 12–3.

Establishing the Number of Loop Equations

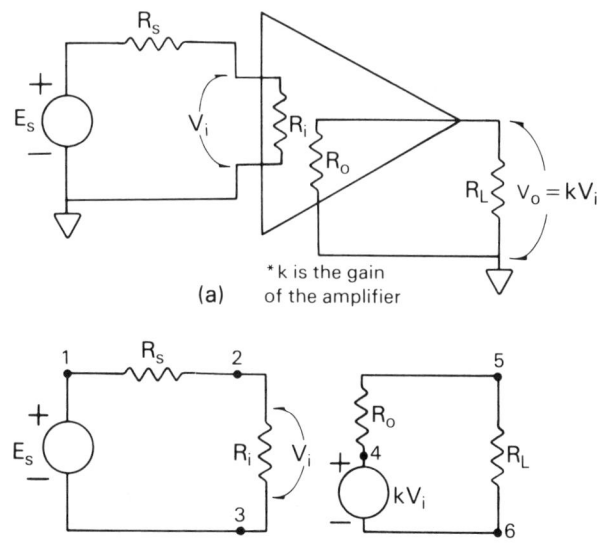

FIGURE 12-6
Amplifier circuit equivalent with two separate parts.

An example of a network with more than one separate part is the equivalent circuit of an amplifier as shown in Figure 12-6. This equivalent circuit shows that the output voltage of the amplifier is a function of the input voltage. If we assume that $k = 10$, then the output voltage source (kv_i) is 10 times the voltage developed across R_i.

To determine the number of loop equations required, apply Equation 12-1. Thus:

$$N_\ell = N_b - N_j + N_s$$

From Figure 12-6(b),

$$N_j = 6, \qquad N_b = 6, \qquad N_s = 2$$

Substitute:

$$N_\ell = N_b - N_j + N_s = 6 - 6 + 2 = 2 \text{ loops}$$

The two required loops for the solution of the circuit of Figure 12-6 are pictured in Figure 12-7. From Figure 12-7, the following loop equations may be written.

For loop 1,

$$E_s - v_s - v_i = 0$$

For loop 2,

$$kv_i - v_o - v_L = 0$$

The next section will deal with writing and solving loop equations.

348

Loop and Node Analysis

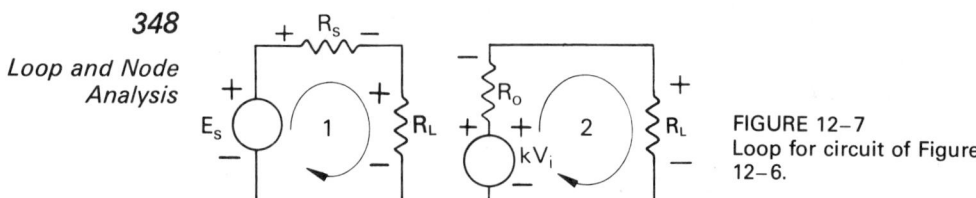

FIGURE 12-7
Loop for circuit of Figure 12-6.

12-3 WRITING AND SOLVING LOOP EQUATIONS

Even though there may be several loops for a given network, only the minimum number of loops need be selected. The selection of loops must be made so that every circuit element (every source and every resistance) is included in at least one of the traced loops.

In the network of Figure 12-8, six loops are shown. However, by applying Equation 12-1, it is determined that only three loops are required for the solution of the network. Loops A, B, and D are selected for the solution of the network because, together, these three loops include every circuit element at least once in one of the traced loops. Among the other acceptable combinations of loops would be A, B, E, and A, B, F, etc. However, selecting loops A, B, C would not be satisfactory because not all the circuit elements are included by the three traced loops (E_2 and R_5 are missing).

Assuming a Loop Current Direction

The choice of the current direction within a loop is arbitrary. But once the loop direction is chosen, certain guidelines must be followed for establishing the voltage polarities of the elements within that loop. Hence:

☐ The direction of the assumed loop current is **always** considered to be **positive**. Thus, for conventional current, the positive side of the resistance is the side where the current enters and the negative side is where the current leaves. In other words, there is a voltage drop when current passes through a resistor in the direction of the assumed loop current.

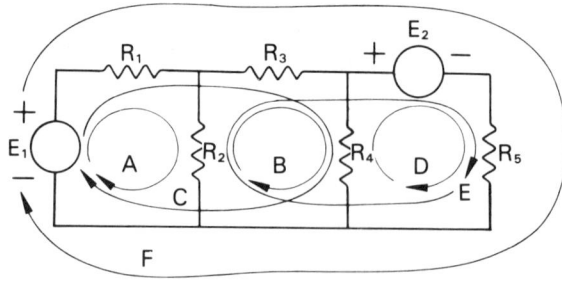

FIGURE 12-8
All possible loops are shown in this figure.

Writing and Solving Loop Equations

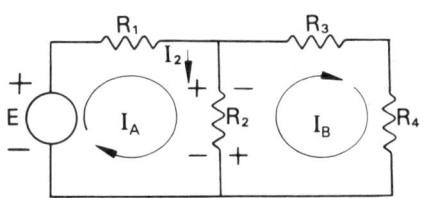

FIGURE 12-9
Current through R_2 is the difference between I_A and I_B.

☐ The polarity of a voltage source remains unchanged regardless of the direction of the chosen loop current.
☐ When the assumed loop current passes through a voltage source from $-$ to $+$, there is a voltage rise, which is assigned a positive sign.
☐ When the assumed loop current passes through a voltage source from $+$ to $-$, there is a voltage drop, which is assigned a negative sign.

Specifying Currents

Although loop currents have been discussed, the actual current in a resistor is the algebraic sum of the loop currents passing through a given resistor. For example, in Figure 12-9, the current I_2 through R_2 is the difference between currents I_A and I_B. In Figure 12-10, because of the choice of current directions, the current I_2 is the sum of currents I_A plus I_B.

Going back to Figure 12-9, when writing the Kirchhoff's voltage law equation for loop A, we consider the voltage drop across R_2 as $(I_A - I_B)R_2$. On the other hand, when writing the equation for loop B, the voltage drop across R_2 is $(I_B - I_A)R_2$.

For Figure 12-10, the voltage drop across R_2 for either loop A or loop B is $(I_A + I_B)R_2$.

Writing Loop Equations

Once the loops have been established, the loop equation may be written. Writing a loop equation is essentially writing out Kirchhoff's voltage law. Consider the two loops shown in Figure 12-11. From Kirchhoff's voltage law, you recall the statement that the algebraic sum of the voltage sources and voltage drops

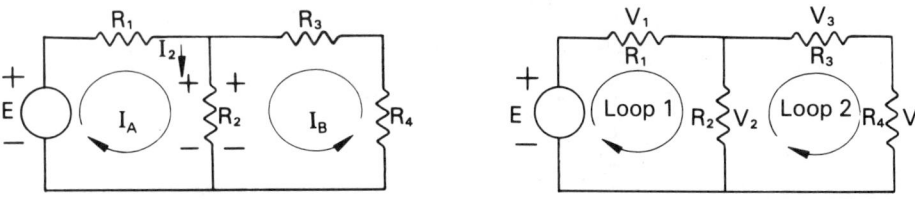

FIGURE 12-10 (left) Current through R_2 is the sum of I_A and I_B.
FIGURE 12-11 (right) Establishing desired loops to solve the circuit.

in a closed loop must equal zero. From KVL, the following is obtained for loop 1:

$$E - V_1 - V_2 = 0$$

And for loop 2,

$$-V_2 - V_3 - V_4 = 0$$

If I_1 is the assumed loop current in loop 1 of Figure 12–11, then I_1 is positive, and from Ohm's law

$$V_1 = I_1 R_1 \quad \text{and} \quad V_2 = (I_1 - I_2)R_2$$

Since R_2 has "*two*" currents in it, both I_1 and I_2 are included with R_2. As noted in Figure 12–12, I_1 and I_2 are opposing currents. Because I_1 is the assumed loop current and the assumed loop current is **always positive,** then in this case I_1 is positive, and since I_2 opposes I_1, then I_2 is negative.

If I_2 is the assumed loop current in loop 2 of Figure 12–11, then I_2 is positive, and from Ohm's law

$$V_3 = I_2 R_3, \quad V_4 = I_2 R_4, \quad V_2 = (I_2 - I_1)R_2$$

Since R_2 has "*two*" currents in it, both I_2 and I_1 are included with R_2. As noted in Figure 12–12, I_2 and I_1 are opposing currents. In this case, I_2 is positive (it is the assumed current), and since I_1 opposes I_2, then I_1 is negative.

Substitute the IR equivalents into the loop equation of loop 1 and loop 2. *For loop 1,*

$$E - V_1 - V_2 = 0$$

and

$$E - I_1 R_1 - (I_1 - I_2)R_2 = 0$$

Solve for E:

$$E = I_1 R_1 + (I_1 - I_2)R_2$$

Distribute R_2:

$$E = I_1 R_1 + I_1 R_2 - I_2 R_2$$

Factor I_1:

$$E = I_1(R_1 + R_2) - I_2 R_2$$

For loop 2,

$$-V_2 - V_3 - V_4 = 0$$

and

$$-(I_2 - I_1)R_2 - I_2 R_3 - I_2 R_4 = 0$$

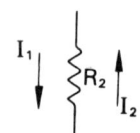

FIGURE 12–12
Loop currents oppose each other in this example.

Writing and Solving Loop Equations

Distribute R_2:

$$0 = -I_2R_2 + I_1R_2 - I_2R_3 - I_2R_4$$

Factor I_2 and change the sign to simplify:

$$0 = I_2(R_2 + R_3 + R_4) - I_1R_2$$

Once the loop equations for a particular network have been stated in terms of the circuit current, the system of equations is solved using the techniques for the solution of *simultaneous equations*. The following examples demonstrate the application of loop equations for the solution of resistive networks.

EXAMPLE 12–4 Solve for the loop current I_1 and I_2 in the network of Figure 12–13.

SOLUTION Write a loop equation for loop 1 using KVL:

$$E - V_1 - V_2 = 0$$

and

$$E - I_1R_1 - (I_1 - I_2)R_2 = 0$$

Solve for E:

$$E = I_1R_1 + (I_1 - I_2)R_2$$

Substitute:

$$10 = I_1 5 + (I_1 - I_2)10$$

Expand and combine like terms:

$$10 = 5I_1 + 10I_1 - 10I_2$$

$$10 = 15I_1 - 10I_2$$

For loop 2, using KVL,

$$-V_3 - V_4 - V_2 = 0$$

and

$$-I_2R_3 - I_2R_4 - (I_2 - I_1)R_2 = 0$$

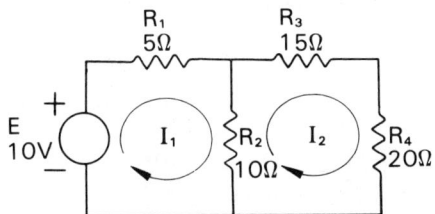

FIGURE 12–13
Network of Example 12–4.

Change signs and interchange members:

$$0 = I_2R_3 + I_2R_4 + (I_2 - I_1)R_2$$

Substitute:

$$0 = I_2 15 + I_2 20 + (I_2 - I_1)10$$

Expand and combine like terms:

$$0 = 15I_2 + 20I_2 + 10I_2 - 10I_1$$

$$0 = 45I_2 - 10I_1$$

∴ The system of loop equations is

$$\begin{cases} 10 = 15I_1 - 10I_2 & (1) \\ 0 = -10I_1 + 45I_2 & (2) \end{cases}$$

Solve the system of equations using the substitution method. Thus:
From Equation 2, solve for I_1:

$$0 = -10I_1 + 45I_2$$

$$10I_1 = 45I_2$$

$$I_1 = 4.5I_2 \quad (3)$$

Substitute $I_1 = 4.5I_2$ into Equation 1 and solve for I_2.

$$10 = 15I_1 - 10I_2$$

$$10 = 15(4.5I_2) - 10I_2$$

$$10 = 67.5I_2 - 10I_2$$

$$57.5I_2 = 10$$

∴ $$I_2 = \frac{10}{57.5} = 0.174 \text{ A}$$

Substitute $I_2 = 0.174$ A into Equation 3 and solve for I_1:

∴ $I_1 = 4.5I_2 = (4.5)(0.174) = 0.783$ A

Observation The methods of determinants can be used, instead of the substitution method, for solving the system of loop equations. Thus:

$$\begin{cases} 10 = 15I_1 - 10I_2 \\ 0 = -10I_1 + 45I_2 \end{cases}$$

Evaluate the determinant:

$$\Delta = \begin{vmatrix} 15 & -10 \\ -10 & 45 \end{vmatrix}$$

$$\Delta = (15)(45) - (-10)(-10) = 575$$

Solve for I_1:

$$I_1 = \frac{\begin{vmatrix} 10 & -10 \\ 0 & 45 \end{vmatrix}}{\Delta}$$

$$I_1 = \frac{(10)(45) - (0)(-10)}{575} = 0.783 \text{ A}$$

Solve for I_2:

$$I_2 = \frac{\begin{vmatrix} 15 & 10 \\ -10 & 0 \end{vmatrix}}{\Delta}$$

$$I_2 = \frac{(15)(0) - (-10)(10)}{575} = 0.174 \text{ A}$$

EXAMPLE 12-5 Determine the voltage drops across each resistance in the network of Figure 12-14.

SOLUTION Begin by selecting the directions for the current loops. The conventional current direction has been selected for each loop, since it is apparent from the voltage source polarity that the current will flow in this direction. Notice that the currents in R_3 are aiding. Write the loop equation for the first loop using KVL:

Loop 1:

$$E_1 - V_1 - V_3 = 0$$

and

$$E_1 - I_1 R_1 - (I_1 + I_2) R_3 = 0$$

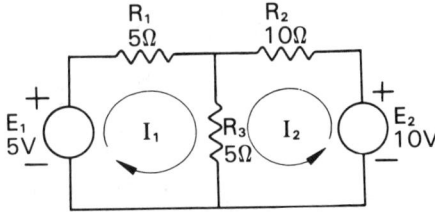

FIGURE 12-14
Network of Example 12-5.

Loop and Node Analysis

Solve for E_1:
$$E_1 = I_1R_1 + (I_1 + I_2)R_3$$

Substitute circuit values:
$$5 = 5I_1 + 5(I_1 + I_2)$$

Expand and combine like terms:
$$5 = 5I_1 + 5I_1 + 5I_2$$
$$5 = 10I_1 + 5I_2$$

Loop 2:
$$E_2 - V_2 - V_3 = 0$$

and
$$E_2 - I_2R_2 - (I_2 + I_1)R_3 = 0$$

Solve for E_2:
$$E_2 = I_2R_2 + (I_2 + I_1)R_3$$

Substitute circuit values:
$$10 = 10I_2 + 5(I_2 + I_1)$$

Expand and combine like terms:
$$10 = 10I_2 + 5I_2 + 5I_1$$
$$10 = 5I_1 + 15I_2$$

∴ The system of equations is
$$\begin{cases} 5 = 10I_1 + 5I_2 \\ 10 = 5I_1 + 15I_2 \end{cases}$$

Use determinants to solve the system of equations. Thus:
$$\Delta = \begin{vmatrix} 10 & 5 \\ 5 & 15 \end{vmatrix}$$
$$\Delta = (10)(15) - (5)(5) = 125$$

Solve for I_1:
$$I_1 = \frac{\begin{vmatrix} 5 & 5 \\ 10 & 15 \end{vmatrix}}{\Delta}$$
$$I_1 = \frac{(5)(15) - (10)(5)}{125} = 0.200 \text{ A}$$

Solve for I_2:

Writing and Solving Loop Equations

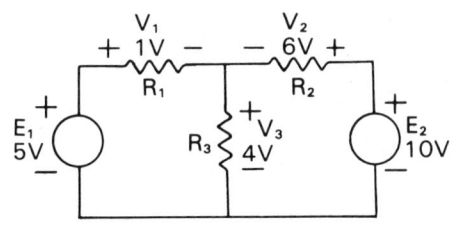

FIGURE 12–15
Polarities of voltage drops are shown.

$$I_2 = \frac{\begin{vmatrix} 10 & 5 \\ 5 & 10 \end{vmatrix}}{\Delta}$$

$$I_2 = \frac{(10)(10) - (5)(5)}{125} = 0.600 \text{ A}$$

Solve for the voltages across the resistances with Ohm's law:

$$V_1 = I_1 R_1 = (0.2)(5) = 1 \text{ V}$$
$$V_2 = I_2 R_2 = (0.6)(10) = 6 \text{ V}$$
$$V_3 = (I_1 + I_2) R_3 = (0.2 + 0.6)(5) = 4 \text{ V}$$

∴ $V_1 = 1$ V, $V_2 = 6$ V, and $V_3 = 4$ V.

Observation The polarities of the voltage drops are noted in Figure 12–15. A quick check with KVL around any loop shows that the algebraic sum of the voltages is zero and the solution is correct. A loop of particular interest is made up of the circuit elements E_1, R_1, R_2, and E_2. The KVL equation for this loop is

$$+E_1 - V_1 + V_2 - E_2 = 0$$
$$5 - 1 + 6 - 10 = 0$$
$$0 = 0$$

EXAMPLE 12–6 Determine the voltage across R_5 in the network of Figure 12–16.

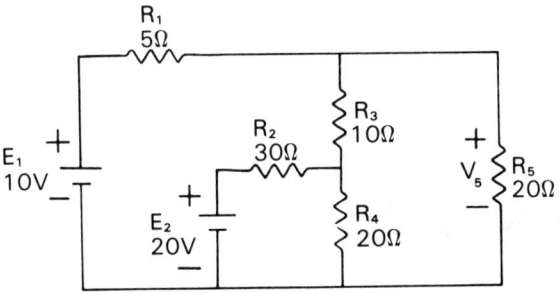

FIGURE 12–16
Network of Example 12–6.

SOLUTION Apply Equation 12–1 to determine the number of required loops.

Observation As you gain experience, you will be able to determine the number of required loops by inspection. Thus:

$$N_\ell = N_b - N_j + N_s$$

In this case $N_b = 7$, $N_j = 5$, and $N_s = 1$:

$$N_\ell = 7 - 5 + 1 = 3$$

∴ Three loops are needed. Figure 12–17 shows one possible choice of loops.

Observation The loop currents are arbitrarily selected in a clockwise direction (cw). Once current direction is established, the rules for the polarity of the voltage drop and voltage rise are in force, as are the rules for conventional current:

☐ The current through a resistance polarizes the resistance and the voltage across the resistance so that the point of entry is positive and the point of exit is negative. Because of this polarization, a negative sign (−) is assigned to the voltage across a resistance in the KVL equation.

☐ The current through a voltage source does not change the polarity of the terminals of the source, but it does cause the voltage in the KVL equation to be assigned a plus (+) sign when the current moves from the (−) terminal to the (+) terminal in the source.

☐ The current through a voltage source causes the voltage in the KVL equation to be assigned a minus (−) sign when the current moves from the (+) terminal to the (−) terminal in the source.

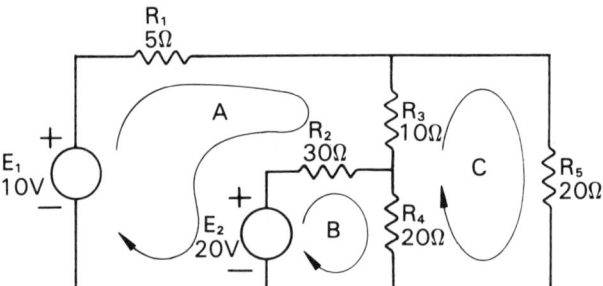

FIGURE 12–17
Selected loops for circuit of Example 12–6.

Writing and Solving Loop Equations

LOOP A: Following the current arrow of loop A (Figure 12–17) around the loop, the following KVL equation results.

$$+E_1 - V_1 - V_3 - V_2 - E_2 = 0$$

Substituting the Ohm's law equivalent ($V = IR$), the equation may be stated in terms of the loop currents. Thus:

$$E_1 - I_A R_1 - (I_A - I_C)R_3 - (I_A - I_B)R_2 - E_2 = 0$$

Distribute:

$$E_1 - I_A R_1 - I_A R_3 + I_C R_3 - I_A R_2 + I_B R_2 - E_2 = 0$$

Factor I_A:

$$E_1 - I_A(R_1 + R_2 + R_3) + I_B R_2 + I_C R_3 - E_2 = 0$$

Substitute circuit values:

$$10 - I_A(5 + 30 + 10) + I_B 30 + I_C 10 - 20 = 0$$

Simplify and arrange in the form

$$k = aI_A + bI_B + cI_C$$

$\therefore \quad 10 = -45I_A + 30I_B + 10I_C$ is the loop equation for loop A.

LOOP B: Following the current arrow of loop B (Figure 12–17) around the loop, the following KVL equation results.

$$+E_2 - V_2 - V_4 = 0$$

Substituting the Ohm's law equivalent ($V = IR$), the equation may be stated in terms of the loop currents. Thus:

$$E_2 - (I_B - I_A)R_2 - (I_B - I_C)R_4 = 0$$

Distribute:

$$E_2 - I_B R_2 + I_A R_2 - I_B R_4 + I_C R_4 = 0$$

Factor I_B:

$$E_2 - I_B(R_2 + R_4) + I_A R_2 + I_C R_4 = 0$$

Substitute circuit values:

$$20 - I_B(30 + 20) + I_A 30 + I_C 20 = 0$$

Simplify and arrange in the form

$$k = aI_A + bI_B + cI_C$$

∴ $-20 = 30I_A - 50I_B + 20I_C$ is the loop equation for loop B.

LOOP C: Following the current arrow of loop C (Figure 12–17) around the loop, the following KVL equation results.

$$-V_3 - V_5 - V_4 = 0$$

Substituting the Ohm's law equivalent ($V = IR$), the equation may be stated in terms of the loop currents. Thus:

$$-(I_C - I_A)R_3 - I_C R_5 - (I_C - I_B)R_4 = 0$$

Distribute:

$$-I_C R_3 + I_A R_3 - I_C R_5 - I_C R_4 + I_B R_4 = 0$$

Factor I_C:

$$-I_C(R_3 + R_4 + R_5) + I_A R_3 + I_B R_4 = 0$$

Substitute circuit values:

$$-I_C(10 + 20 + 20) + I_A 10 + I_B 20 = 0$$

Simplify and arrange in the form

$$k = aI_A + bI_B + cI_C$$

∴ $0 = 10I_A + 20I_B - 50I_C$ is the loop equation for loop C.

The complete system of equations for the network is:

$$\begin{cases} 10 = -45I_A + 30I_B + 10I_C \\ -20 = 30I_A - 50I_B + 20I_C \\ 0 = 10I_A + 20I_B - 50I_C \end{cases}$$

Apply the techniques of determinants to the system of equations:

$$\Delta = \begin{vmatrix} -45 & 30 & 10 \\ 30 & -50 & 20 \\ 10 & 20 & -50 \end{vmatrix}$$

$$\Delta = -112\,500 + 6000 + 6000 + 5000 + 18\,000 + 45\,000 = -32\,500$$

Solve for I_A:

Writing and Solving Loop Equations

$$I_A = \frac{\begin{vmatrix} 10 & 30 & 10 \\ -20 & -50 & 20 \\ 0 & 20 & -50 \end{vmatrix}}{\Delta}$$

$$I_A = \frac{25\,000 + 0 - 4000 + 0 - 4000 - 30\,000}{-32\,500}$$

$$I_A = \frac{-13\,000}{-32\,500} = 0.40 \text{ A}$$

Solve for I_B:

$$I_B = \frac{\begin{vmatrix} -45 & 10 & 10 \\ 30 & -20 & 20 \\ 10 & 0 & -50 \end{vmatrix}}{\Delta}$$

$$I_B = \frac{-45\,000 + 2000 + 0 + 2000 + 0 + 15\,000}{-32\,500}$$

$$I_B = \frac{-26\,000}{-32\,500} = 0.80 \text{ A}$$

Solve for I_C:

$$I_C = \frac{\begin{vmatrix} -45 & 30 & 10 \\ 30 & -50 & -20 \\ 10 & 20 & 0 \end{vmatrix}}{\Delta}$$

$$I_C = \frac{0 - 6000 + 6000 + 5000 - 18\,000 + 0}{-32\,500}$$

$$I_C = \frac{-13\,000}{-32\,500} = 0.40 \text{ A}$$

∴ $I_A = 0.40$ A, $I_B = 0.80$ A, and $I_C = 0.40$ A.

Check To check the results, the values for I_A, I_B, and I_C are substituted into each of the equations. Thus:

$$10 = -45I_A + 30I_B + 10I_C$$
$$10 = -18 + 24 + 4$$
$$10 \text{ V} = 10 \text{ V}$$

Observation The two remaining equations also check. We invite you to satisfy yourself that this is indeed true. The voltage across R_5 (V_5) is calculated by Ohm's law. Thus:

Loop and Node Analysis

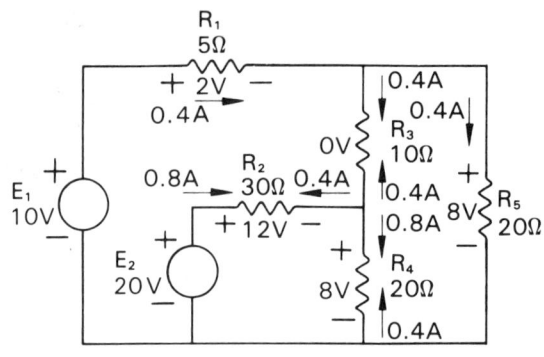

FIGURE 12-18 Individual loop currents in each resistor of Example 12-6 are shown.

$$V_5 = I_C R_5 = (0.4)(20) = 8 \text{ V}$$

$$\therefore \quad V_5 = 8 \text{ V}$$

Observation The voltages across the remaining resistors are calculated by first determining the resultant current in each resistor. Figure 12-18 shows the currents in each resistance as well as the polarized voltage across each resistance. Table 12-1 summarizes the currents and voltages in the circuit.

Summary To solve a network with Kirchhoff's voltage law and loop analysis, use the following procedure.

☐ Determine the number of loops required for solution.
☐ Configure the loops so that all circuit elements are included in at least one of the traced loops.
☐ Write the Kirchhoff's voltage law (KVL) equation for each loop and set the algebraic sum equal to zero.
☐ Assign a polarity (+ or −) to each voltage term in the KVL equation. The polarity assignment is made with regard to the

TABLE 12-1 SUMMARY OF THE CURRENTS AND VOLTAGES IN THE NETWORK OF FIGURE 12-18

Resistance	I_A (A)	I_B (A)	I_C (A)	Resultant (A)	× R (Ω)	= Voltage (V)
R_1	0.4	—	—	0.4	5	2
R_2	0.4	0.8	—	0.4	30	12
R_3	0.4	—	0.4	0	10	0
R_4	—	0.8	0.4	0.4	20	8
R_5	—	—	0.4	0.4	20	8

direction of the assumed loop current in the following manner:
- ▽ When the head of the assumed loop current arrow passes through a voltage source from the negative terminal (−) to the positive terminal (+), a (+) sign is assigned to that term in the KVL equation.
- ▽ When the head of the assumed loop current arrow passes through a voltage source from the positive terminal (+) to the negative terminal (−), a (−) sign is assigned to that term in the KVL equation.
- ▽ When the head of the assumed loop current arrow passes through a resistance, the resistance and the accompanying voltage across it are polarized from positive (+) to negative (−), and a (−) sign is assigned to that term in the KVL equation.

☐ The assumed current in the loop being worked with is always positive. And:
- ▽ When this current passes through a resistance along with another current that is an aiding current, each current is considered to be positive (+).
- ▽ When this current passes through a resistance along with another current that is an opposing current, the opposing current is considered to be negative (−).

12-4 NODE ANALYSIS

Kirchhoff's current law (KCL) is basically a statement of the law of conservation of charge; that is, "at any junction or node of an electrical circuit, the sum of the currents entering the node must equal the sum of the currents leaving the node." If this statement were not true, there would be a buildup of charge at the node.

Node analysis is based on Kirchhoff's current law. Figure 12-19 shows a series-parallel network with nodes labeled a, b, c, and d. The currents entering and leaving each node are also labeled and have the following (KCL) equations.

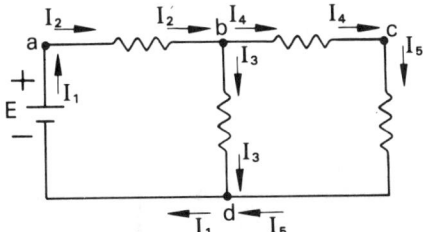

FIGURE 12-19
Node voltages and currents are determined through node analysis.

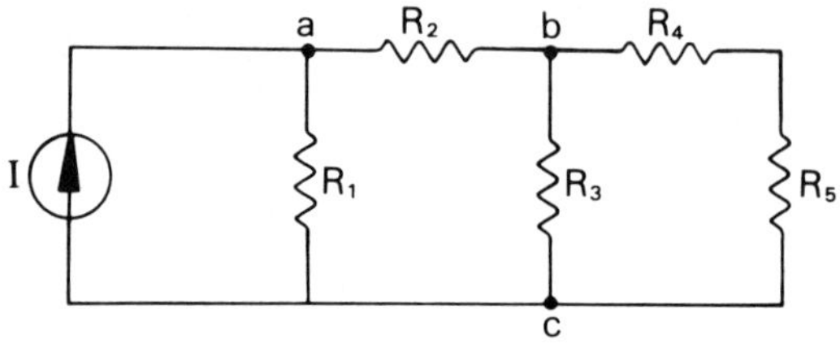

362

Loop and Node Analysis

FIGURE 12-20
Network of Example 12-7.

Node a:	$I_1 = I_2$	or	$I_1 - I_2 = 0$
Node b:	$I_2 = I_3 + I_4$	or	$I_2 - I_3 - I_4 = 0$
Node c:	$I_4 = I_5$	or	$I_4 - I_5 = 0$
Node d:	$I_3 + I_5 = I_1$	or	$I_3 + I_5 - I_1 = 0$

It is important that you note that currents entering a node are positive (+) and that currents leaving a node are negative (−).

12-5 ESTABLISHING THE NUMBER OF NODE EQUATIONS

In general, the number of node equations, N_n, needed for the solution of a network is determined by Equation 12-2:

$$N_n = N_j - N_s \tag{12-2}$$

where N_n = number of node equations
N_j = number of junctions where three or more elements are connected
N_s = number of separate parts

EXAMPLE 12-7 Determine the number of node equations needed to solve the network of Figure 12-20.

SOLUTION There are three junctions, a, b, and c, and the circuit consists of one part. Therefore, the number of node equations is

$$N_n = N_j - N_s = 3 - 1 = 2$$

∴ Two node equations are needed to solve the network.

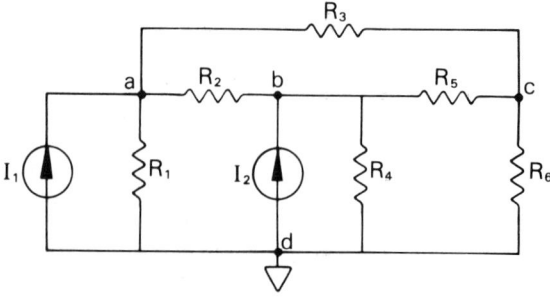

FIGURE 12-21
Network of Example 12-8.

EXAMPLE 12-8 Determine the number of node equations needed to solve the network of Figure 12-21.

SOLUTION This circuit has four junctions, a, b, c, and d, and consists of one part. Hence:

$$N_n = N_j - N_s = 4 - 1 = 3$$

12-6 WRITING AND SOLVING NODE EQUATIONS

Once it is decided to use node equations to solve a particular network, the next steps are to determine the number of node equations needed and to select the *reference node*.

In theory, any node in the network may be chosen as the reference node; however, the common or ground point is the most practical choice for the reference node. Once the reference node has been selected, the remaining nodes, which are the independent nodes, are used in the solution of the *node voltages* (the voltage from the independent node to the reference node). Once the node voltages have been determined, the actual currents in the circuit elements may be determined, resulting in the complete solution of the network.

As with loop analysis, the loop current direction can be arbitrarily set in any direction desired as long as the rules for polarizing the circuit elements are followed. Additionally, the terms in the KCL equation must also be assigned positive and negative signs based on the rules for the assumed current direction. Once the node voltages are determined, the signs of the voltage solutions will indicate whether the assumed current was moving in the same direction as the actual circuit current. A negative sign with a solved voltage means that the actual direction of the current is opposite to the direction of the assumed current.

EXAMPLE 12-9 Solve for the currents indicated in the network of Figure 12-22.

SOLUTION An inspection of the network shows that there are 2 major nodes.

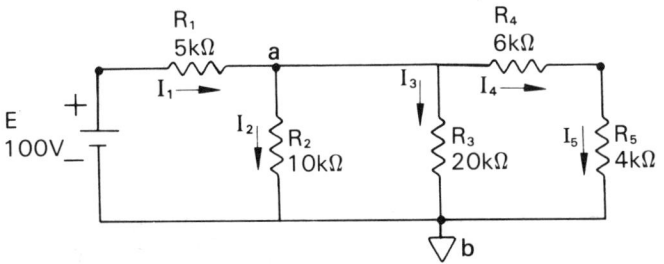

FIGURE 12-22
Network of Example 12-9.

Loop and Node Analysis

Observation A *major node* is a node with three or more branches connected at one point. Nodes a and b are major nodes.

Select the common point as the reference node, leaving node a as the only major node for which a KCL equation is written. Thus:

$$I_1 = I_2 + I_3 + I_4$$

And substituting the Ohm's law equivalent,

$$\frac{V_1}{R_1} = \frac{V_2}{R_2} + \frac{V_3}{R_3} + \frac{V_{4,5}}{R_4 + R_5}$$

Express the voltage in terms of the node voltages (voltage from the independent nodes to the reference node):

$$\frac{V_{ad} - V_{bd}}{R_1} = \frac{V_{bd}}{R_2} + \frac{V_{bd}}{R_3} + \frac{V_{bd}}{R_4 + R_5}$$

Factor V_{bd} in the right member:

$$\frac{V_{ad} - V_{bd}}{R_1} = V_{bd}\left(\frac{1}{R_2} + \frac{1}{R_3} + \frac{1}{R_4 + R_5}\right)$$

Substitute the circuit values where $V_{ad} = E = 100$ V:

$$\frac{100 - V_{bd}}{5 \text{ k}\Omega} = V_{bd}\left(\frac{1}{10 \text{ k}\Omega} + \frac{1}{20 \text{ k}\Omega} + \frac{1}{6 \text{ k}\Omega + 4 \text{ k}\Omega}\right)$$

and

$$\frac{100}{5 \text{ k}\Omega} - \frac{V_{bd}}{5 \text{ k}\Omega} = V_{bd}(0.25 \times 10^{-3})$$

$$20 \times 10^{-3} = V_{bd}(0.25 \times 10^{-3}) + V_{bd}(0.2 \times 10^{-3})$$

Factor V_{bd}:

$$20 \times 10^{-3} = V_{bd}(0.25 \times 10^{-3} + 0.2 \times 10^{-3})$$

Solve for V_{bd}:

$$V_{bd} = \frac{20 \times 10^{-3}}{0.45 \times 10^{-3}} = 44.4 \text{ V}$$

∴ V_{bd} is 44.4 V and $V_{bd} = V_2 = V_3 = V_{4,5} = 44.4$ V. Solve for $I_1, I_2, I_3, I_4,$ and I_5:

$$I_1 = \frac{V_{ad} - V_{bd}}{R_1} = \frac{100 - 44.4}{5 \text{ k}\Omega} = 11.1 \text{ mA}$$

$$I_2 = \frac{V_{bd}}{R_2} = \frac{44.4}{10 \text{ k}\Omega} = 4.44 \text{ mA}$$

$$I_3 = \frac{V_{bd}}{R_3} = \frac{44.4}{20 \text{ k}\Omega} = 2.22 \text{ mA}$$

$$I_4 = I_5 = \frac{V_{bd}}{R_4 + R_5} = \frac{44.4}{10 \text{ k}\Omega} = 4.44 \text{ mA}$$

$\therefore \quad I_1 = 11.1$ mA, $I_2 = 4.44$ mA, $I_3 = 2.22$ mA, $I_4 = 4.44$ mA, and $I_5 = 4.44$ mA.

Observation $\quad V_{cd} = I_5 R_5 = (4.44 + 10^{-3})(4 \times 10^3) = 17.8$ V

Checking the KCL equation results in

$$I_1 = I_2 + I_3 + I_4$$

11.1 mA = 4.44 + 2.22 + 4.44

11.1 mA = 11.1 mA

12–7 SELECTING NODE OR LOOP ANALYSIS

Whether you use node or loop analysis depends mainly on which technique would be easier to use in a particular circuit. If one method requires fewer equations than the other to solve, it would be the preferred method.

EXAMPLE 12–10 Determine which method would be preferable in the solution of the network of Figure 12–23.

SOLUTION Upon examination, it is evident that three loop equations would be needed to solve this problem. However, using nodal analysis would require only one KCL equation, since there is one major node. Thus:

$$I_1 = I_2 + I_3 + I_4$$

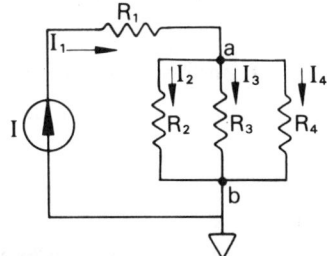

FIGURE 12–23
Network of Example 12–10.

And, stated in terms of the node voltage, where

$$V_{ab} = V_2 = V_3 = V_4$$

$$I_2 = \frac{V_{ab}}{R_2}, \quad I_3 = \frac{V_{ab}}{R_3}, \quad I_4 = \frac{V_{ab}}{R_4}$$

Substituting into $I_1 = I_2 + I_3 + I_4$,

$$I_1 = \frac{V_{ab}}{R_2} + \frac{V_{ab}}{R_3} + \frac{V_{ab}}{R_4}$$

Factor:

$$I_1 = V_{ab}\left(\frac{1}{R_2} + \frac{1}{R_3} + \frac{1}{R_4}\right)$$

And $I = I_1$:

$$I_1 = I = V_{ab}\left(\frac{1}{R_2} + \frac{1}{R_3} + \frac{1}{R_4}\right)$$

∴ Given values of I, R_2, R_3, and R_4, then V_{ab} may be determined.

Observation Once V_{ab} is known, then I_2, I_3, and I_4 may be computed using Ohm's law.

The following are reasons for considering the nodal method of circuit analysis:

☐ The sources are mainly current sources.
☐ The desired solutions are voltages.
☐ The number of node equations is less than the number of loop equations.

The following are reasons for considering the use of the loop method of circuit analysis:

☐ The sources are mainly voltage sources.
☐ The desired solutions are currents.
☐ The number of loop equations is less than the number of node equations.

EXAMPLE 12–11 Solve for V_1, V_2, V_3, V_4, and V_5 of the network of Figure 12–24.

SOLUTION Write the KCL equation for major node a:

$$I_1 = I_2 + I_3$$

And, stated in terms of the node voltage, where

$$I_2 = \frac{V_{ab}}{R_2} \quad \text{and} \quad I_3 = \frac{V_{ab}}{R_3 + R_4 + R_5}$$

Selecting Node or Loop Analysis

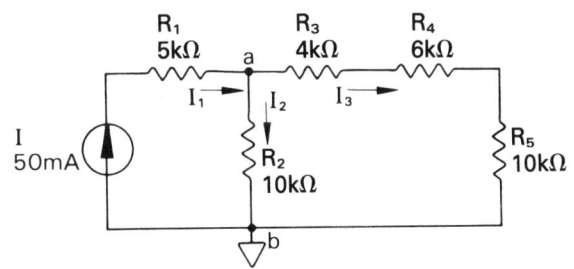

FIGURE 12-24
Network of Example 12-11.

$$I_1 = \frac{V_{ab}}{R_2} + \frac{V_{ab}}{R_3 + R_4 + R_5}$$

Substitute and solve for V_{ab}, where $I_1 = I = 50$ mA:

$$50 \times 10^{-3} = V_{ab} \left(\frac{1}{10 \text{ k}\Omega} + \frac{1}{4 \text{ k}\Omega + 6 \text{ k}\Omega + 10 \text{ k}\Omega} \right)$$

$$V_{ab}(150 \times 10^{-6}) = 50 \times 10^{-3}$$

$$V_{ab} = \frac{50 \times 10^{-3}}{150 \times 10^{-6}} = 333.3 \text{ V}$$

Determine I_2 and I_3:

$$I_2 = \frac{V_{ab}}{R_2} = \frac{333.3}{10 \text{ k}\Omega} = 33.33 \text{ mA}$$

$$I_3 = I_1 - I_2 = 50 \text{ mA} - 33.33 \text{ mA} = 16.67 \text{ mA}$$

Knowing I_3, compute V_3, V_4, and V_5 using Ohm's law. Thus:

$$V_3 = I_3 R_3 = (16.67 \text{ mA})(4 \text{ k}\Omega) = 66.7 \text{ V}$$

$$V_4 = I_3 R_4 = (16.67 \text{ mA})(6 \text{ k}\Omega) = 100 \text{ V}$$

$$V_5 = I_3 R_5 = (16.67 \text{ mA})(10 \text{ k}\Omega) = 166.7 \text{ V}$$

Determine V_2:

$$V_2 = V_{ab} = 333.3 \text{ V}$$

Determine V_1:

$$V_1 = I_1 R_1 = IR_1 = (50 \text{ mA})(5 \text{ k}\Omega) = 250 \text{ V}$$

∴ $V_1 = 250$ V, $V_2 = 333.3$ V, $V_3 = 66.7$ V, $V_4 = 100$ V, and $V_5 = 166.7$ V.

EXAMPLE 12-12 Determine V_{ac} and V_{bc} in the network of Figure 12-25.

SOLUTION *Node a:* Write the KCL equation for node a:

368
Loop and Node Analysis

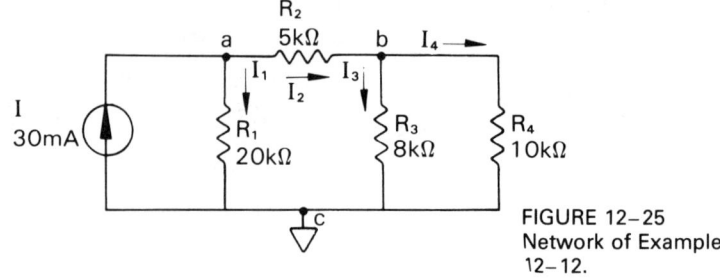

FIGURE 12-25
Network of Example 12-12.

$$I = I_1 + I_2$$

And, stated in terms of the node voltages,

$$I = \frac{V_{ac}}{R_1} + \frac{V_{ac} - V_{bc}}{R_2}$$

Substitute and solve:

$$30 \times 10^{-3} = \frac{V_{ac}}{20 \text{ k}\Omega} + \frac{V_{ac} - V_{bc}}{5 \text{ k}\Omega}$$

$$30 \times 10^{-3} = \frac{V_{ac}}{20 \text{ k}\Omega} + \frac{V_{ac}}{5 \text{ k}\Omega} - \frac{V_{bc}}{5 \text{ k}\Omega}$$

Factor:

$$30 \times 10^{-3} = V_{ac}\left(\frac{1}{20 \text{ k}\Omega} + \frac{1}{5 \text{ k}\Omega}\right) - \frac{V_{bc}}{5 \text{ k}\Omega}$$

$$30 \times 10^{-3} = V_{ac}(0.25 \times 10^{-3}) - V_{bc}(0.20 \times 10^{-3})$$

$$\therefore\ 30 \times 10^{-3} = (0.25 \times 10^{-3})V_{ac} - (0.20 \times 10^{-3})V_{bc}$$

for node a.

Node b: Write the KCL equation for node b:

$$I_2 = I_3 + I_4$$

And, stated in terms of the node voltages,

$$\frac{V_{ac} - V_{bc}}{R_2} = \frac{V_{bc}}{R_3} + \frac{V_{bc}}{R_4}$$

Substitute and solve:

$$\frac{V_{ac} - V_{bc}}{5 \text{ k}\Omega} = \frac{V_{bc}}{8 \text{ k}\Omega} + \frac{V_{bc}}{10 \text{ k}\Omega}$$

$$\frac{V_{ac}}{5 \text{ k}\Omega} - \frac{V_{bc}}{5 \text{ k}\Omega} = V_{bc}\left(\frac{1}{8 \text{ k}\Omega} + \frac{1}{10 \text{ k}\Omega}\right)$$

$$V_{ac}(0.2 \times 10^{-3}) - V_{bc}(0.2 \times 10^{-3}) - V_{bc}(0.225 \times 10^{-3}) = 0$$

Interchange the members and combine like terms:

Selecting Node or Loop Analysis

$$0 = (0.2 \times 10^{-3})V_{ac} - (0.425 \times 10^{-3})V_{bc}$$

∴ $0 = (0.2 \times 10^{-3})V_{ac} - (0.425 \times 10^{-3})V_{bc}$ for node b.

The complete system of equations for the network is:

$$\begin{cases} 30 \times 10^{-3} = (0.25 \times 10^{-3})V_{ac} - (0.20 \times 10^{-3})V_{bc} \\ 0 = (0.2 \times 10^{-3})V_{ac} - (0.425 \times 10^{-3})V_{bc} \end{cases}$$

Apply the techniques of determinants to the system of equations:

$$\Delta = \begin{vmatrix} 0.25 \times 10^{-3} & -0.20 \times 10^{-3} \\ 0.20 \times 10^{-3} & -0.425 \times 10^{-3} \end{vmatrix}$$

$$\Delta = -0.1063 \times 10^{-6} + 0.04 \times 10^{-6}$$
$$\Delta = -66.3 \times 10^{-9}$$

Solve for V_{ac}:

$$V_{ac} = \frac{\begin{vmatrix} 30 \times 10^{-3} & -0.20 \times 10^{-3} \\ 0 & -0.425 \times 10^{-3} \end{vmatrix}}{\Delta}$$

$$V_{ac} = \frac{-12.75 \times 10^{-6} + 0}{-66.3 \times 10^{-9}} = 192.3 \text{ V}$$

Solve for V_{bc}:

$$V_{bc} = \frac{\begin{vmatrix} 0.25 \times 10^{-3} & 30 \times 10^{-3} \\ 0.20 \times 10^{-3} & 0 \end{vmatrix}}{\Delta}$$

$$V_{bc} = \frac{0 - 6.0 \times 10^{-6}}{-66.3 \times 10^{-9}} = 90.5 \text{ V}$$

∴ $V_{ac} = 192.3$ V and $V_{bc} = 90.5$ V.

Observation Check the solution by substituting into

$$I = \frac{V_{ac}}{R_1} + \frac{V_{ac} - V_{bc}}{R_2}$$

$$30 \text{ mA} = \frac{192.3}{20 \text{ k}\Omega} + \frac{192.3 - 90.5}{5 \text{ k}\Omega}$$

$$30 \text{ mA} = 9.62 \text{ mA} + 20.4 \text{ mA}$$

$$30 \text{ mA} = 30 \text{ mA}$$

EXAMPLE 12–13 For the network of Figure 12–26, determine I_1, I_2, I_3, I_4, and I_5 using both the node method and the loop method.

370
Loop and Node Analysis

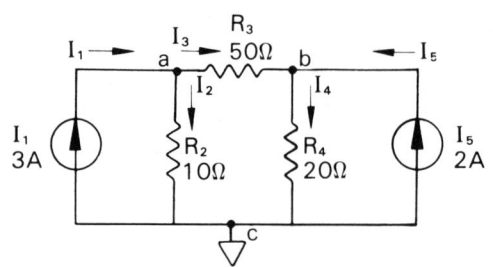

FIGURE 12-26
Network of Example 12-13.

Observation At first glance, the obvious choice appears to be nodal analysis since there are two current sources and three loops.

SOLUTION *Node Analysis*
Node a: Write the KCL equation for node a:

$$I_1 = I_2 + I_3$$

And, stated in terms of the node voltage,

$$I_1 = \frac{V_{ac}}{R_2} + \frac{V_{ac} - V_{bc}}{R_3}$$

Substitute and solve:

$$3 = \frac{V_{ac}}{10} + \frac{V_{ac} - V_{bc}}{50}$$

$$3 = \frac{V_{ac}}{10} + \frac{V_{ac}}{50} - \frac{V_{bc}}{50}$$

Factor:

$$3 = V_{ac}\left(\frac{1}{10} + \frac{1}{50}\right) - \frac{V_{bc}}{50}$$

$$3 = 0.12 V_{ac} - 0.02 V_{bc}$$

∴ $3 = 0.12 V_{ac} - 0.02 V_{bc}$ for node a.

Node b: Write the KCL equation for node b:

$$I_3 + I_5 = I_4$$

And, stated in terms of the node voltage,

$$\frac{V_{ac} - V_{bc}}{R_3} + I_5 = \frac{V_{bc}}{R_4}$$

Substitute and solve:

$$\frac{V_{ac} - V_{bc}}{50} + 2 = \frac{V_{bc}}{20}$$

Selecting Node or Loop Analysis

$$2 = \frac{V_{bc}}{20} - \frac{V_{ac}}{50} + \frac{V_{bc}}{50}$$

$$2 = V_{bc}\left(\frac{1}{20} + \frac{1}{50}\right) - \frac{V_{ac}}{50}$$

$$2 = 0.07 V_{bc} - 0.02 V_{ac}$$

∴ $2 = -0.02 V_{ac} + 0.07 V_{bc}$ for node b.

The complete system of equations for the network is

$$\begin{cases} 3 = 0.12 V_{ac} - 0.02 V_{bc} \\ 2 = -0.02 V_{ac} + 0.07 V_{bc} \end{cases}$$

Apply the techniques of determinants to the system of equations:

$$\Delta = \begin{vmatrix} 0.12 & -0.02 \\ -0.02 & 0.07 \end{vmatrix}$$

$$\Delta = 8.4 \times 10^{-3} - 0.4 \times 10^{-3} = 8 \times 10^{-3}$$

Solve for V_{ac} and V_{bc}:

$$V_{ac} = \frac{\begin{vmatrix} 3 & -0.02 \\ 2 & 0.07 \end{vmatrix}}{\Delta}$$

$$V_{ac} = \frac{0.21 + 0.04}{8 \times 10^{-3}} = 31.25 \text{ V}$$

$$V_{bc} = \frac{\begin{vmatrix} 0.12 & 3 \\ -0.02 & 2 \end{vmatrix}}{\Delta}$$

$$V_{bc} = \frac{0.24 + 0.06}{8 \times 10^{-3}} = 37.5 \text{ V}$$

∴ $V_{ac} = 31.25$ V and $V_{bc} = 37.5$ V, as noted on the network pictured in Figure 12-27.

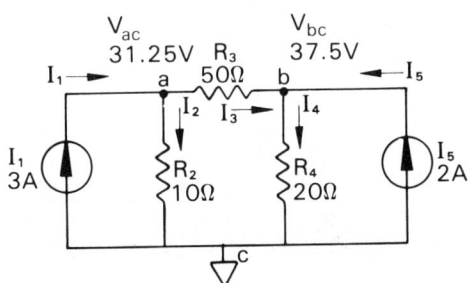

FIGURE 12-27
Node voltages are determined in Example 12-13.

Loop and Node Analysis

Determine the currents I_1, I_2, I_3, I_4, and I_5. Thus:

$I_1 = 3$ A   (current source)

$$I_2 = \frac{V_{ac}}{R_2} = \frac{31.25}{10} = 3.125 \text{ A}$$

$$I_3 = \frac{V_{ac} - V_{bc}}{R_3} = \frac{31.25 - 37.5}{50} = -0.125 \text{ A}$$

Observation Because of the minus sign (-0.125 A), the actual current direction in R_3 is opposite to the assumed current direction pictured in Figure 12-27.

$$I_4 = \frac{V_{bc}}{R_4} = \frac{37.5}{20} = 1.875 \text{ A}$$

$I_5 = 2$ A   (current source)

$\therefore$ $I_1 = 3$ A, $I_2 = 3.125$ A, $I_3 = -0.125$ A, $I_4 = 1.875$ A, and $I_5 = 2$ A.

Observation Check the solution by substituting into

$$I_1 = I_2 + I_3$$

$$3 = 3.125 - 0.125$$

$$3 = 3$$

And

$$I_3 + I_5 = I_4$$

$$-0.125 + 2 = 1.875$$

$$1.875 = 1.875$$

SOLUTION *Loop Analysis:* From the circuit configuration of Figure 12-28, it appears that a system of three KVL equations will be needed for the three specified loops. However, loop currents I_A and I_C are already known since, by definition, the current in the loop is equal to the current source. Thus:

$$I_A = I_1 = 3 \text{ A} \quad \text{and} \quad I_C = I_5 = 2 \text{ A}$$

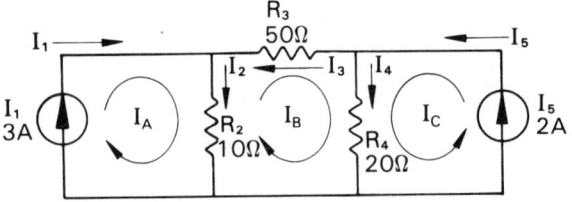

FIGURE 12-28
Example 12-13 is solved using loop analysis.

The loop equation for I_B is

$$-V_2 - V_3 - V_4 = 0$$

And

$$-10(I_B - I_A) - 50I_B - 20(I_b + I_c) = 0$$
$$-10I_B + 10I_A - 50I_B - 20I_B - 20I_C = 0$$
$$-80I_B + 10I_A - 20I_C = 0$$

Substitute $I_A = 3$ A and $I_C = 2$ A:

$$-80I_B + 30 - 40 = 0$$
$$-80I_B = 10$$
$$I_B = \frac{-10}{80} = -0.125 \text{ A}$$

Determine I_1, I_2, I_3, I_4, and I_5:

$I_1 = 3$ A   (current source)

$I_2 = I_A - I_B = 3 + 0.125 = 3.125$ A

$I_3 = -I_B = 0.125$ A

$I_4 = I_C + I_B = 2 - 0.125 = 1.875$ A

$I_5 = 2$ A   (current source)

$\therefore$ $I_1 = 3$ A, $I_2 = 3.125$ A, $I_3 = 0.125$ A, $I_4 = 1.875$ A, and $I_5 = 2$ A.

Observation Since the loop method proved easier than the node method, this particular network is counter to the rules that would indicate the use of the node method. As you gain more experience, you will become more adept at selecting the easier method.

PROBLEMS As part of the solution, it is suggested that you draw and label a schematic diagram for each problem.

Sections 12–1, 12–2, and 12–3

12–1 Write the loop equation for the network of Figure 12–29.

12–2 Show all the loops possible for the network of Figure 12–30.

12–3 Write equations for the loops determined in Problem 12–2.

12–4 List the equations required to solve for V_3 in the network of Figure 12–30.

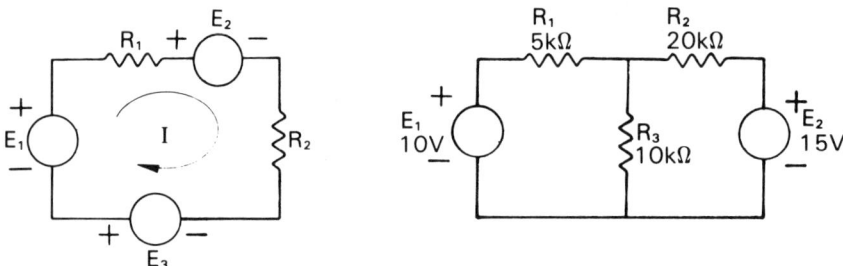

FIGURE 12-29 (left)   Network of Problem 12-1.
FIGURE 12-30 (right) Network of Problems 12-2, 12-3, and 12-4.

12-5 Determine the number of loop equations required to solve the network of Figure 12-31.
12-6 Write the set of loop equations necessary to solve the network of Figure 12-31.
12-7 Determine the voltage drops across R_1, R_2, R_3, and R_4 in the network of Figure 12-32.
12-8 Determine the current I in Figure 12-33.

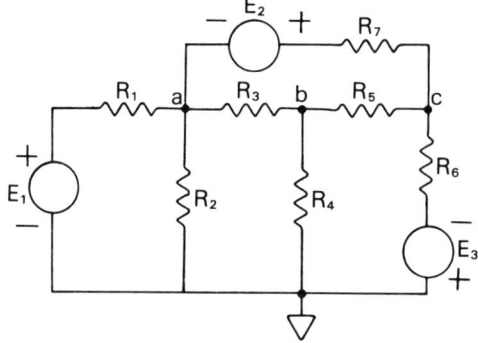

FIGURE 12-31
Network of Problems
12-5, 12-6, 12-9, and
12-10.

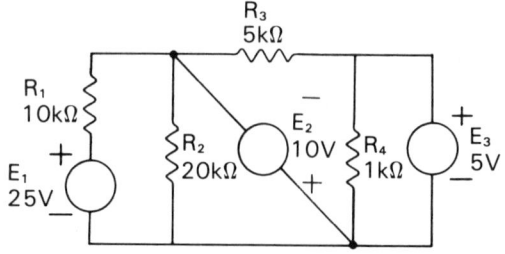

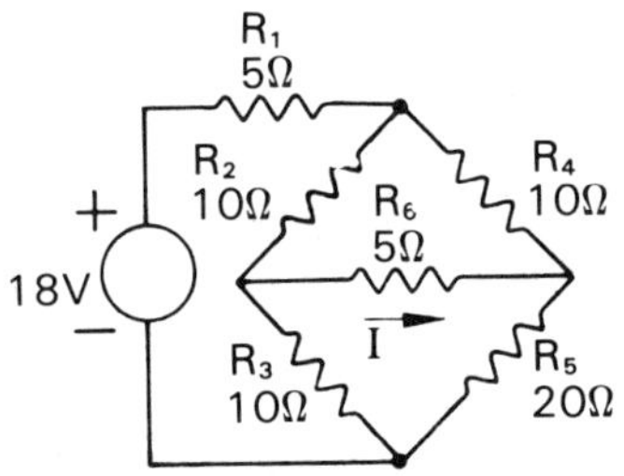

FIGURE 12-32 (left)   Network of Problem 12-7.
FIGURE 12-33 (right) Network of Problem 12-8.

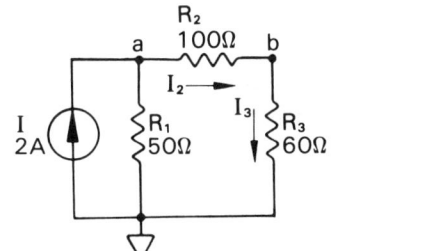
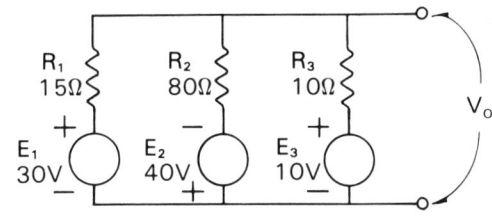

FIGURE 12–34 (left)   Network of Problem 12–11.
FIGURE 12–35 (right)   Network of Problem 12–12.

Sections 12–4, 12–5, and 12–6

12–9 Determine the number of node equations required to solve the network of Figure 12–31.

12–10 Write the node equations required to solve the network of Figure 12–31.

12–11 Write the node equations for nodes a and b of Figure 12–34.

12–12 Using nodal analysis, determine V_o in the network of Figure 12–35.

12–13 Determine the number of node equations required to solve the network of Figure 12–36.

12–14 Determine the voltage drops across $R_1, R_2, R_3,$ and R_4 in the network of Figure 12–36.

12–15 Solve for the voltage across R_5 in the network of Figure 12–20 when $I = 3$ A, $R_1 = 5\ \Omega$, $R_2 = 5\ \Omega$, $R_3 = 20\ \Omega$, $R_4 = 5\ \Omega$, and $R_5 = 10\ \Omega$.

12–16 Use nodal analysis to determine V_o in Figure 12–37.

Section 12–7

12–17 Determine the number of equations needed to solve for V_o in the network of Figure 12–38 by (a) loop analysis and (b) node analysis.

12–18 Solve for V_o in the network of Figure 12–38 using the method with the least number of required equations.

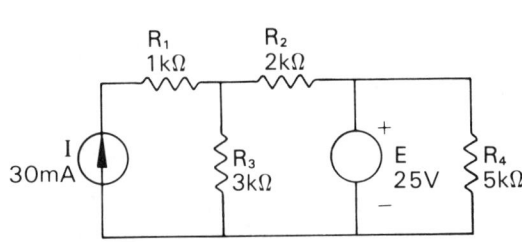
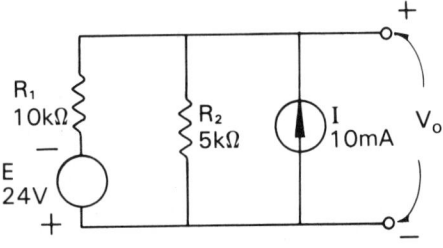

FIGURE 12–36 (left)   Network of Problems 12–13 and 12–14.
FIGURE 12–37 (right)   Network of Problem 12–16.

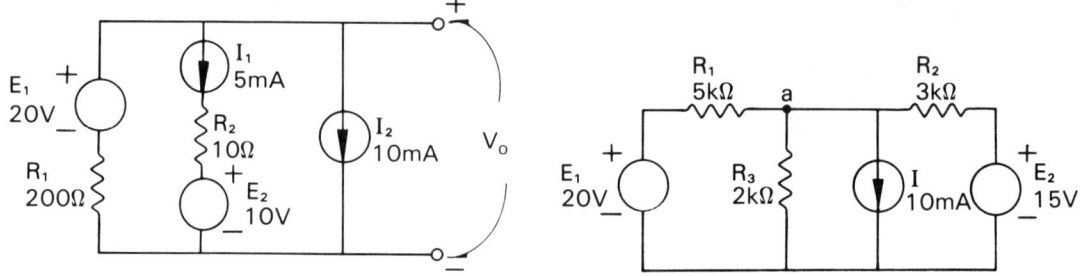

FIGURE 12-38 (above, left)   Network of Problems 12-17 and 12-18.
FIGURE 12-39 (above, right) Network of Problem 12-19.

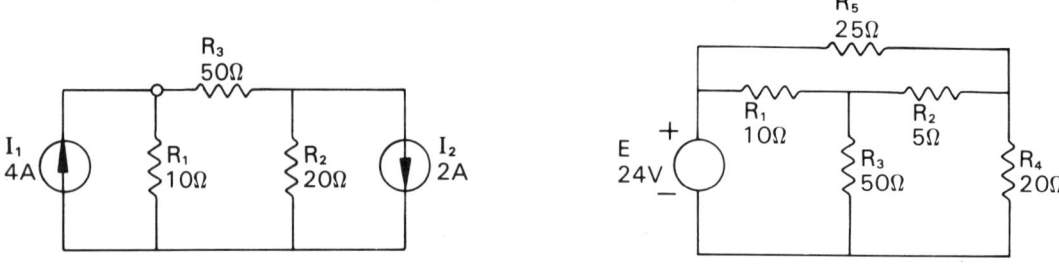

FIGURE 12-40 (left)   Network of Problems 12-20 and 12-21.
FIGURE 12-41 (right) Network of Problems 12-22 and 12-23.

12-19 Determine the voltage at node a in the network of Figure 12-39 using either loop or node analysis.

12-20 Use either node or loop analysis to determine the voltages across each resistor in the network of Figure 12-40.

12-21 Reverse the current source I_2 in the network of Figure 12-40 and solve for the voltage across R_3.

12-22 Using loop analysis, determine the voltage drop across R_4 in the network of Figure 12-41.

12-23 Using node analysis, determine the voltage drop across R_4 in the network of Figure 12-41.

12-24 Determine the voltage across R_3 (V_3) in the network of Figure 12-42.

12-25 Determine both the magnitude (amount) and the direction of the current in resistance R_1 in the network of Figure 12-42.

12-26 Determine the voltage across R_3 (V_3) in the network of Figure 12-43.

12-27 Determine the voltage across R_4 (V_4) in the network of Figure 12-43.

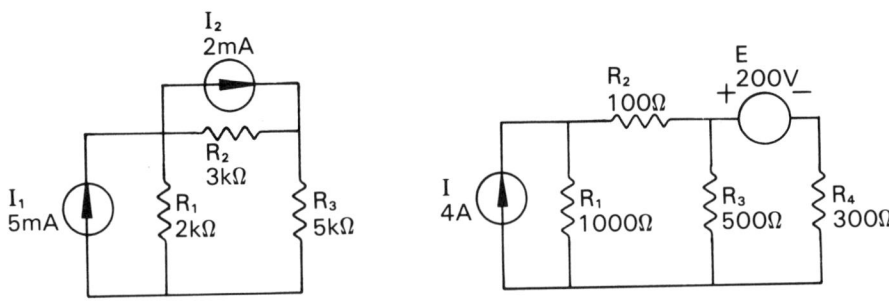

FIGURE 12-42 (left)   Network of Problems 12-24 and 12-25.
FIGURE 12-43 (right)   Network of Problems 12-26 through 12-29.

12-28 Determine the current in R_4 (I_4) of Figure 12-43.
12-29 Determine the current in R_2 (I_2) of Figure 12-43.

13 Capacitance

13-1 ELECTROSTATICS
13-2 CAPACITANCE
13-3 CAPACITOR
13-4 DIELECTRIC
13-5 CONNECTING CAPACITORS IN SERIES AND PARALLEL
13-6 CAPACITOR APPLICATIONS, CHARACTERISTICS, AND SPECIFICATIONS

CHAPTER PREVIEW This chapter begins with another look at electrostatics; Coulomb's law, permittivity, electric field strength, and flux density are introduced. You will learn about capacitance and dielectrics, as well as about a very important electrical device, the capacitor. The chapter concludes with a brief look at sources of stray capacitance.

PERFORMANCE OBJECTIVES Once you have read each section, worked each example with pencil, paper, and calculator, and answered each problem for every section, you will be able to:

☐ Use Coulomb's law to determine the mutual force between two point charges.
☐ Differentiate between various capacitor types.
☐ Use tables of dielectric constants and dielectric strengths.
☐ Compute the capacitance of a capacitor.
☐ Determine the equivalent capacitance of capacitors in series and parallel.
☐ Understand such terms as permittivity and electric flux density.

13–1 ELECTROSTATICS Because the properties of a capacitor are related to the storage of a charge and a potential difference, we must once again turn our attention to electric charge.

In Chapter 2, we noted that the electric field between electric charges was a field force. Let's suppose that a small positively charged particle is introduced into the electric field between the oppositely charge objects, as shown in Figure 13–1. From previous understanding, it would be expected that the particle would be repelled by the positively charged object and attracted by the negatively charged object. Thus, the particle would move along the line of force to the negatively charged objective.

Coulomb's Law The electric force, of attraction and repulsion, between pairs of charged particles was studied by the eighteenth-century scien-

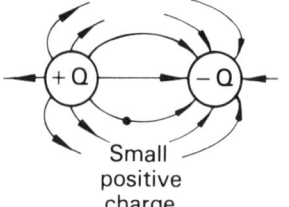

FIGURE 13–1
A small positive test charge is introduced into the electric field between the opposing charges to determine the polarity of the charges. The positive test charge is attracted to the negative charged object and repelled by the positive charged objective.

Electrostatics

tist, Charles Coulomb. He determined that the electric force between two charged particles is directly proportional to the product of the magnitude of their electric charge and inversely proportional to the square of the distance between them. This observation, now known as *Coulomb's law*, is stated mathematically as

$$F \propto \frac{Q_1 Q_2}{d^2}$$

It has since been learned that the magnitude of the electric forces also depends on the medium (materials between the charges). The greatest force is experienced when the particles are in a vacuum. Thus, Coulomb's law stated as an equation is

$$F = \frac{kQ_1 Q_2}{d^2} \quad (N) \tag{13-1}$$

where F = force in newtons (N)
Q_1 and Q_2 = charges in coulombs (C)
d = distance between the pair of charges in meters (m)
k = proportionality constant in Nm²/C². For a vacuum (approximately air) between the charged particles $k = 9 \times 10^9$ Nm²/C².

EXAMPLE 13-1 Determine the force between two small particles in a vacuum 12 cm apart. The particles have charges of -4.6×10^{-6} C and 1.8×10^{-7} C.

SOLUTION Substitute into Coulomb's law:

$$F = \frac{kQ_1 Q_2}{d^2}$$

$$F = \frac{(9 \times 10^9)(-4.6 \times 10^{-6})(1.8 \times 10^{-7})}{(0.12)^2}$$

$$\therefore \quad F = -0.52 \text{ N}$$

Observation A negative value of force indicates that the charged particles have opposite polarity and that they attract one another. A positive value of force indicates that the charged particles have like polarity and that they oppose one another.

Permittivity The proportionality constant, k, in Coulomb's law (Equation 13-1) is often expressed as $k = 1/(4\pi\epsilon)$, which provides for different mediums between the charged particles. The Greek letter epsilon (ϵ) is the permittivity of the surrounding medium. The permittivity indicates the degree to which the medium can resist

Capacitance

the movement of electric charge within it, thus allowing more lines of force between the charges. The importance of this physical quantity will become apparent in the next section. The permittivity of a vacuum (ϵ_0) is $\epsilon_0 = 8.85 \times 10^{-12}$ C²/Nm². Thus, Coulomb's law for the force between a pair of charged particles in a vacuum (or air) may be stated as

$$F = \frac{Q_1 Q_2}{4\pi\epsilon_0 d^2} \quad (N) \tag{13-2}$$

Potential Difference

In previous work with circuits, the word voltage (potential difference) has been used to describe the difference in potential across a resistive load. Because the concept of potential difference plays an important role in understanding capacitance, we have included the following discussion.

Potential difference, as you know, is measured in a unit called a volt. The volt is a derived unit equal to 1 joule per coulomb (1 volt = 1 joule/1 coulomb). That is, it takes 1 joule of work (energy) to move 1 coulomb of charge between two points with a potential difference of 1 volt.

To gain a conceptual understanding of this definition, let's take a small, positively charged particle ($+Q$) and place it in the electric field of a negatively charged object ($-Q$), as pictured in Figure 13–2. From Coulomb's law, we know that there is an attractive force between the pair of unlike charges. To overcome the attractive force and move the charged particle from point A to point B, it will take work from an outside source. Moving charge $+Q$ to point B increases its potential energy. Thus, electric potential difference (voltage) is defined as the work per unit charge between two points in an electric field done by an external source in moving the charge against the Coulomb force of the electric field:

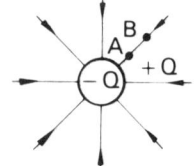

FIGURE 13–2
Work is done in moving charge $+Q$ from point A to point B.

$$V = \frac{W}{Q} \quad (J/C) \tag{13-3}$$

The potential difference (emf) generated by a battery or other source is the result of moving charge (electrons) onto the negative terminal from the positive terminal. The action creates an excess of electrons on the negative terminal and a deficiency on the positive terminal. In the battery, the potential difference across the terminals of the battery is created by the work done by the chemical action of the battery. Thus, the potential difference (voltage) of a source results from the work done in transferring a unit charge from one terminal to the other.

We have seen in our work with circuits that the potential difference of the source is used to move charge carrier (current) through the load.

EXAMPLE 13-2 Due to the mechanical work of a generator, 40 C of charge (electrons) is moved from one terminal to the other within the generator. Determine the potential difference between the terminals if 480 J of work is done in transferring the charge.

SOLUTION Use Equation 13-3.

$$V = \frac{W}{Q} = \frac{480 \text{ J}}{40 \text{ C}} = 12 \text{ J/C}$$

∴ The potential difference is 12 V across the terminals of the generator.

Electric Field Strength

The electric field is represented by lines of force. That is, lines of force give a visual representation of an electric field. The arrowheads indicate direction, and the field strength is indicated by the density of the lines of force.

Because capacitance is concerned with parallel plates rather than point charges, we will shift our attention to the electric field that exists between pairs of metal plates (conductors) as shown in Figure 13-3(a). From Coulomb's law, we know that the charges on the plates will exert an attractive force on the plates. This force, of course, depends on the size of the charges on the plates, which, in turn, determines the *strength* of the electric field between the plates.

The electric field strength, $\mathscr{E}$, represented by a script E, is numerically equal to the potential difference between the plates divided by the distance separating the plates. Thus

$$\mathscr{E} = \frac{V}{d} \quad \text{(V/m)} \tag{13-4}$$

FIGURE 13-3
(a) Electric field represented by lines of force between two metal plates.
(b) A few lines of force extend outside the plate area, producing an effect called *fringing*.

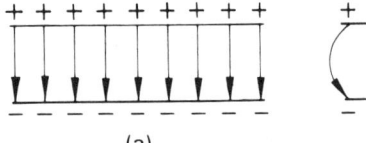

EXAMPLE 13–3 Determine the electric field strength between two metal plates with a potential difference of 100 V for (a) a plate separation of 4 cm and (b) a plate separation of 1 cm.

SOLUTION Use Equation 13–4:

$$(a) \; \mathscr{E} = \frac{V}{d} = \frac{100}{4 \times 10^{-2}} = 2.5 \text{ kV/m}$$

∴ The field strength is 2500 V/m for a plate separation of 4 cm.

$$(b) \; \mathscr{E} = \frac{V}{d} = \frac{100}{1 \times 10^{-2}} = 10 \text{ kV/m}$$

∴ The field strength is 10 kV/m for a plate separation of 1 cm.

Observation The electric field strength is increased by decreasing the plate separation.

Because the electric field strength is a measure of the force exerted on the charged plates, field strength may be expressed in units of newtons per coulomb (N/C) in addition to volts per meter (V/m). Thus

$$\mathscr{E} = \frac{F}{Q} = \frac{V}{d} \quad \text{(N/C)} \tag{13–5}$$

EXAMPLE 13–4 Express the field strength of the previous example in N/C (newtons per coulomb).

SOLUTION (a) 2500 V/m = 2.5 kN/C
(b) 10 kV/m = 10 kN/C

EXAMPLE 13–5 Determine the force on a free electron ($Q = 1.6 \times 10^{-19}$ C) between two parallel plates separated by 2 cm when 200 V is applied across the plates.

SOLUTION Use Equation 13–5 and solve for F:

$$\frac{F}{Q} = \frac{V}{d} \quad \text{and} \quad F = \frac{QV}{d}$$

Substitute:

$$F = \frac{(1.6 \times 10^{-19})(200)}{2 \times 10^{-2}} = 1.6 \times 10^{-15} \text{ N}$$

∴ The force on the electron is 1.6×10^{-15} N.

Electric Flux Density

The density (degree of concentration) of the lines of force, called the *electric flux density*, is the number of lines of force per unit area. Thus

$$D = \frac{\Psi}{A} \quad (C/m^2) \tag{13-6}$$

where D = flux density in coulombs per square meter (C/m²)
Ψ = electric flux in coulombs (C); because the total number of lines (Ψ) varies directly with the amount of electric charge (Q), the two quantities may be equated thus: $\Psi \equiv Q$.
A = area in square meters (m²)

The number of lines of force and, consequently, the flux density are directly proportional to the field strength. Thus, $D \propto \mathscr{E}$ and

$$D = \epsilon \mathscr{E} \tag{13-7}$$

where the permittivity, ϵ, of the medium is the proportionality constant. For a vacuum between the plates or in air, the permittivity is $\epsilon_0 = 8.85 \times 10^{-12}$ C²/Nm². Thus

$$D = \epsilon_0 \mathscr{E} = \frac{\Psi}{A} \quad (C/m^2) \tag{13-8}$$

13-2 CAPACITANCE

The two metal plates of Figure 13-4(a) are separated by a distance, d. *Both* plates have no net charge, which results in no potential difference between the plates. Assume that a small positive quantity of charge ($+Q$) is removed from plate p_1 and placed on plate p_2. Work is done in moving the charge by some outside agent, and the potential difference between p_2 and p_1 has been raised. Additionally, there is a coulomb force between the oppositely charged plates, and plate p_1 is negative in relation to plate p_2, as noted in Figure 13-4(b). Again, another small positive

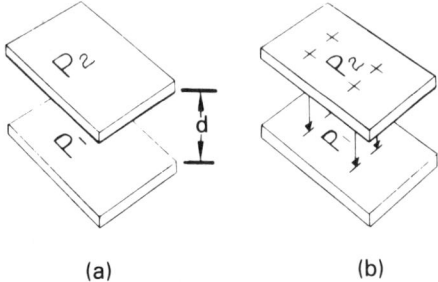

FIGURE 13-4
(a) Two uncharged metal plates separated by a distance (d).
(b) Work is done to move a positive charge from P₁ to P₂, thereby increasing the potential difference between the plates.

charge is taken from plate p_1 and placed on plate p_2. The potential difference between p_2 and p_1 has increased an amount proportional to the charge moved, and the Coulomb force is larger and the work done is greater. It may be noted that the potential difference between the charged plates is directly proportional to the charge moved. Thus

$$Q \propto V$$

The proportionality constant between the net charge (Q) on the plates and the corresponding potential difference (V) is called *capacitance* (C). Capacitance is usually measured in farads rather than coulombs per volts. That is,

---
1 farad = 1 coulomb/1 volt

Thus

$$Q = CV \quad (C) \tag{13-9}$$

and

$$C = \frac{Q}{V} \quad (F) \tag{13-10}$$

where C = capacitance in farads (F)
Q = net charge on the plates in coulombs (C)
V = potential difference in volts (V)

EXAMPLE 13-6 Determine the capacitance of two metal plates when 10 mC of charge is deposited on the plates with a potential of 100 V across the plates.

SOLUTION Use Equation 13-10 and substitute:

$$C = \frac{Q}{V} = \frac{10 \times 10^{-3}}{100} = 1 \times 10^{-4} \text{ F}$$

∴ The capacitance is 100 μF.

Observation For practical application, capacitance is usually expressed in microfarads (μF) or picofarads (pF).

13-3 CAPACITOR Capacitance is used to indicate the change in potential (voltage) between conductors when a net charge is deposited on the conductors. A device having the property of capacitance is called a *capacitor*.

Energy Stored by Capacitors

A capacitor stores electric potential energy for short periods of time. The energy stored by a capacitor is equal to one-half the product of the potential difference and the quantity of charge. Thus

$$W = \tfrac{1}{2}VQ \quad (J) \tag{13-11}$$

where W = potential energy in joules (J)
V = potential difference in volts (V)
Q = quantity of charge in coulombs (C)

By substituting $Q = CV$ from Equation 13-8 into Equation 13-11, the following equation results.

$$W = \tfrac{1}{2}VQ = \tfrac{1}{2}V(CV)$$
$$W = \tfrac{1}{2}CV^2 \quad (J) \tag{13-12}$$

EXAMPLE 13-7 Determine the energy stored by a 50-μF capacitor with a voltage of 300 V.

SOLUTION Use Equation 13-12 and substitute:

$$W = \tfrac{1}{2}CV^2 = \frac{(50 \times 10^{-6})(300^2)}{2} = 2.25 \text{ J}$$

∴ The stored or potential energy is 2.25 J.

Observation The capacitor does not dissipate energy since the stored energy is returned to the circuit to do work. The work done is equal to the energy stored by the capacitor.

Parallel Plate Capacitor

The simplest form of capacitor is made up of two closely spaced parallel metal plates separated by air or an insulator. The plates are charged by connecting an external voltage source, such as a battery or dc power supply, to do work by charging the metal plates with equal quantities of opposite polarity charges.

Since capacitors have a relatively large plate area (A) and small distance (d) between the plates, the lines of force (flux) of the electric field are parallel to each other and perpendicular to the plates. Except for insignificantly small regions at the ends of the parallel plates, as noted in Figure 13-3(b), where the electric flux *fringes* (moves outside the plates), the electric field is present only in the region between the metal plates. The field strength, $\mathcal{E}$, between the plates, has a magnitude described by Equation 13-8 as

$$\mathcal{E} = \frac{Q}{\epsilon_0 A} \tag{13-13}$$

where $Q \equiv \Psi$ and ϵ_0 is the permittivity of a vacuum (approximately air), $\epsilon_0 = 8.85 \times 10^{-12}$, and A is the plate interface area (m²).

An expression of the capacitance of a capacitor relating the plate area (A), the spacing between the plates (d), and the permittivity of air ($\epsilon_0 = 8.85 \times 10^{-12}$) may be derived from Equations 13–13, 13–10, and 13–5. Thus

$$\mathscr{E} = \frac{Q}{\epsilon_0 A}, \qquad \mathscr{E} = \frac{V}{d}, \qquad C = \frac{Q}{V}$$

Solve for C in terms of ϵ_0, A, and d. Thus

$$\mathscr{E} = \frac{Q}{\epsilon_0 A} = \frac{V}{d}$$

and

$$Q = \frac{V(\epsilon_0 A)}{d}$$

Substitute for Q in

$$C = \frac{Q}{V} = \frac{(V\epsilon_0 A)/d}{V} = \frac{V\epsilon_0 A}{dV}$$

Factor V:

$$C = \frac{\epsilon_0 A}{d} \quad \text{(F)} \qquad\qquad (13\text{–}14)$$

where C = capacitance in farads (F)
ϵ_0 = permittivity of a vacuum (or air) of 8.85×10^{-12} C²/Nm²
A = plate interface area in square meters (m²)
d = distance between plates in meters (m)

From Equation 13–14, it is learned that capacitance depends directly on plate area. The larger the plate area, the greater the capacitance. Capacitance also depends directly on the permittivity of the medium between the plates. An increase in the permittivity will allow a greater net charge on the plates and a larger capacitance. Finally, it is learned that capacitance is inversely related to the distance between the plates. The narrower the spacing between the plates, the larger the capacitance.

EXAMPLE 13–8 Determine the capacitance of a capacitor made up of parallel metal plates with an interface area of 2 cm², separated by 0.5 mm of air (1 cm² = 1 × 10⁻⁴ m²).

SOLUTION Use Equation 13-14 and substitute:

$$C = \frac{\epsilon_0 A}{d}$$

$$C = \frac{(8.85 \times 10^{-12})(2 \times 10^{-4})}{0.5 \times 10^{-3}}$$

$$C = 3.54 \times 10^{-12} \text{ F}$$

∴ The capacitance of the capacitor is 3.54 pF.

EXAMPLE 13-9 A capacitor is constructed of two parallel metal plates with an interface area of 6 cm² separated by a distance of 0.1 mm. Determine (a) the capacitance (C) of the capacitor with air between the plates, (b) the charge (Q) on each plate when a source of 24 V is applied to the plates, as shown in Figure 13-5, (c) the magnitude of the electric field strength ($\mathscr{E}$) between the plates, (d) the flux density (D) between the plates, and (e) the energy (W) stored by the capacitor.

SOLUTION (a) $C = \dfrac{\epsilon_0 A}{d} = \dfrac{(8.85 \times 10^{-12})(6 \times 10^{-4})}{0.1 \times 10^{-3}}$

∴ $C = 53.1$ pF

(b) $Q = CV = (53.1 \times 10^{-12})(24)$

∴ $Q = 1.274$ nC

(c) $\mathscr{E} = \dfrac{V}{d}$

$\mathscr{E} = 24/(0.1 \times 10^{-3})$

∴ $\mathscr{E} = 240$ kV/m

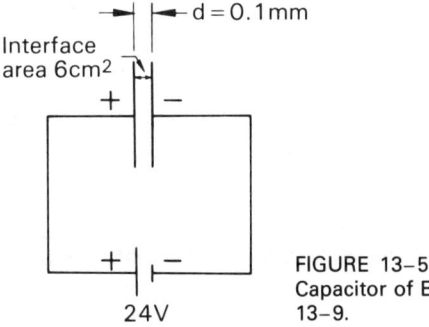

FIGURE 13-5
Capacitor of Example 13-9.

or

$$\mathscr{E} = \frac{Q}{\epsilon_0 A}$$

$$\mathscr{E} = \frac{1.274 \times 10^{-9}}{8.85 \times 10^{-12} \times 6 \times 10^{-4}}$$

∴ $\mathscr{E} = 240$ kV/m

(d) $D = \epsilon_0 \mathscr{E}$

$$D = (8.85 \times 10^{-12})(240 \times 10^3)$$

∴ $D = 2.12$ μC/m²

or

$$D = \frac{\Psi}{A}$$

$$D = \frac{1.274 \times 10^{-9}}{6 \times 10^{-4}}$$

∴ $D = 2.12$ μC/m²

(e) $W = \frac{1}{2}CV^2$

$W = \frac{1}{2}(53.1 \times 10^{-12})(24^2)$

∴ $W = 15.3$ nJ

or

$W = \frac{1}{2}VQ$

$W = \frac{1}{2}(24)(1.274 \times 10^{-9})$

∴ $W = 15.3$ nJ

13–4 DIELECTRIC

Practical capacitors are not operated within a vacuum, but are instead operated with a *dielectric* between their plates. A dielectric is an insulator in which an electric field may be established with little or no leakage current. With the placement of a dielectric in the space between the plates, the permittivity (ϵ) of the dielectric increases.

Increasing Capacitance In the following discussion, we will investigate two parallel plate capacitors of the same physical dimensions. That is, area (A) equals 4×10^{-4} m², plate separation (d) is 1×10^{-3} m, and po-

391

Dielectric

tential difference between the plates is 100 V. The only difference in the two capacitors is the permittivity of the dielectric (ϵ) between the metal plates. One case will have air ($\epsilon = 8.855 \times 10^{-12}$); the other will have mica ($\epsilon = 4.425 \times 10^{-11}$). The field strength is the same for each dielectric since $\mathcal{E}$ depends on voltage (potential difference) and plate spacing, $\mathcal{E} = V/d$ (Equation 13-4). So each capacitor will have a field strength of $\mathcal{E} = 100/1 \times 10^{-3} = 100$ kV/m. However, the flux density (D) will vary, because it depends on the permittivity of the dielectric, ϵ (a variable), and the field strength $\mathcal{E}$ (a constant); $D = \epsilon\mathcal{E}$ (Equation 13-7). Thus, for an air dielectric,

$$D = \epsilon\mathcal{E} = (8.85 \times 10^{-12})(100 \times 10^3)$$
$$D = 885 \text{ nQ/m}^2$$

For a mica dielectric,

$$D = \epsilon\mathcal{E} = (4.425 \times 10^{-11})(100 \times 10^3)$$
$$D = 4425 \text{ nQ/m}^2$$

Comparing the two calculations for electric flux density, we see that the capacitor with the mica dielectric has a flux density 5 times greater than the capacitor with an air dielectric. Furthermore, the capacitance of the capacitor with a mica dielectric will also be increased by the same factor of 5 over that of the capacitance of the capacitor with an air dielectric. Thus for the air dielectric,

$$D = \frac{\Psi}{A} \quad \text{and} \quad \Psi \equiv Q = DA$$

$$Q = 885.5 \times 10^{-9} \frac{C}{m^2} \times 4 \times 10^{-4} m^2 = 3.54 \times 10^{-10} \text{ C}$$

$$C = \frac{Q}{V}$$

$$C = \frac{3.54 \times 10^{-10}}{100} = 3.54 \times 10^{-12}$$

$$\therefore \quad C = 3.54 \text{ pF}$$

For the mica dielectric

$$D = \frac{\Psi}{A} \quad \text{and} \quad \Psi \equiv Q = DA$$

$$Q = 4425 \times 10^{-9} \frac{C}{m^2} \times 4 \times 10^{-4} m^2 = 1.77 \times 10^{-9} \text{ C}$$

$$C = \frac{Q}{V}$$

$$C = \frac{1.77 \times 10^{-9}}{100} = 17.7 \times 10^{-12}$$

$$\therefore\ C = 17.7\ \text{pF}$$

Thus, the capacitance of the capacitor with the mica dielectric is greater than the capacitance of the capacitor with the air dielectric by a factor of 5 (17.4 pF/3.54 pF = 5).

To summarize, when different dielectrics are placed between a pair of parallel metal plates with a constant potential difference across the plates, the amount of charge will vary depending on the permittivity of the dielectric between the plates. The larger the value of permittivity, the greater will be the charge on the plates, the greater the electric flux density, and, consequently, the greater the capacitance of the capacitor.

Dielectric Constant The increase by a factor of 5 in the capacitance of the capacitor with a mica dielectric over the capacitance of the capacitor with an air dielectric is, of course, due to the insertion of a dielectric other than air (or a vacuum) between the plates. This multiplying factor, which is always greater than 1, is called the *dielectric constant*. The value of the dielectric constant, indicated by the Greek letter kappa (κ), depends only on the kind of insulating material used between the metal plates of the capacitor. Values of the dielectric constants for selected materials are given in Table 13–1.

Computing Capacitance We know from our previous work with capacitors that capacitance is a function of interface plate area (A), plate spacing (d), and permittivity (ϵ_0). Equation 13–14 of the previous section stated

$$C = \frac{\epsilon_0 A}{d}\ \text{(F)} \qquad (13\text{–}14)$$

where ϵ_0 is the permittivity of free space (vacuum or air). With a dielectric having a dielectric constant greater than 1, the equation for capacitance becomes

$$C = \frac{\kappa \epsilon_0 A}{d}\ \text{(F)} \qquad (13\text{–}15)$$

where $\kappa \epsilon_0 = \epsilon$ is the permittivity of the dielectric. The dielectric constant is defined as the ratio of ϵ to ϵ_0. Thus

$$\kappa = \frac{\epsilon}{\epsilon_0}\ \text{(no units)} \qquad (13\text{–}16)$$

Dielectric

TABLE 13-1 TYPICAL VALUES OF DIELECTRIC CONSTANTS OF SELECTED MATERIALS

Material	Dielectric Constant (κ)
Air	1.0006
Bakelite (plastic)	7
Barium-strontium titanate	≈7500
Glass	5 to 10
Mica	5 to 9
Oil	3
Paper (waxed)	2.5
Porcelain	6 to 10
Teflon	2
Vacuum	1.0000

Since $\epsilon_0 = 8.85 \times 10^{-12}$, a practical equation for computing capacitance is

$$C = \frac{8.85 \kappa A}{10^{12} d} \quad \text{(F)}$$

$$C = \frac{8.85 \kappa A}{d} \quad \text{(pF)}$$

(13-17)

EXAMPLE 13-10 Determine the capacitance of a parallel plate capacitor with a plate area of 1.6 cm² and a plate separation of 0.3 mm when a dielectric of (a) air, (b) waxed paper, and (c) barium-strontium titanate is used between the plates.

SOLUTION (a) Use Equation 13-17 and a dielectric constant of 1.0006 for air from Table 13-1.

$$C = \frac{8.85 \kappa A}{d} \quad \text{(pF)}$$

$$C = \frac{8.85 \times 1.0006 \times 1.6 \times 10^{-4}}{0.3 \times 10^{-3}}$$

∴ $C = 4.72$ pF for an air dielectric.

(b) Use Equation 13-17 and a dielectric constant of 2.5 for waxed paper from Table 13-1.

$$C = \frac{8.85 \kappa A}{d} \quad \text{(pF)}$$

$$C = \frac{8.85 \times 2.5 \times 1.6 \times 10^{-4}}{0.3 \times 10^{-3}}$$

(c) Use Equation 13–17 and a dielectric constant of 7500 for the Ba-Sr titanate from Table 13–1.

$$C = \frac{8.85 \kappa A}{d} \quad \text{(pF)}$$

$$C = \frac{8.85 \times 7500 \times 1.6 \times 10^{-4}}{0.3 \times 10^{-3}}$$

∴ $C = 35\,400$ pF $= 35.4$ nF for a barium-strontium titanate dielectric.

Dielectric Strength

The atoms within the dielectric between the charged metal plates are acted upon by the electric field force of the charge. The electrons within the molecules of the dielectric are attracted by the positive charge of the positive plate, while the protons of the molecules are attracted by the negative charge of the negative plate. As a result, both the electrons and the protons are displaced (moved) off center from one another creating a *dipole*. That is, the negative charge of the electrons and the positive charge of the protons are separated by a small distance, as pictured in Figure 13–6(b). The action of displacing the center of charges in the presence of an electric field is called *polarization* of the molecule.

The amount of the polarization of the atom is a function of the field strength, specifically the potential difference between the plates, since the plate spacing is constant and $\mathscr{E} = V/d$. As the voltage is increased across the plates, at some point a few of the covalent bonds are broken, causing a leakage current in the dielectric and thereby lowering the resistivity of the material. With an added increase of the potential difference between the plates, a value will be reached where the dielectric will rupture,

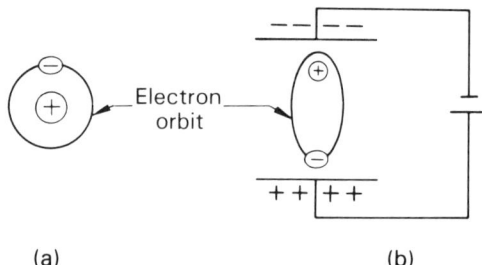

FIGURE 13–6 (a) The center of the negative charge of the orbiting electrons corresponds with the center of the positive charge in the nucleus. (b) Under the influence of an electric field, the negative and positive centers of charge are displaced (moved) from one another.

395

Dielectric

resulting in a short circuit between the plates. At this point, the capacitor has no capacitance, since the dielectric has lost its insulating properties. The value of electric field strength at which the dielectric loses its insulating properties is called the *dielectric strength* of the dielectric material. Dielectric strength is usually expressed as volts per millimeter (V/mm). Table 13–2 lists the dielectric strengths of selected materials used as dielectrics in the manufacture of capacitors. Additional dielectric strengths are listed in Table 4–2, page 60.

EXAMPLE 13–11 Determine the maximum voltage that a 0.1-mm-thick mica dielectric can withstand. Also, determine a safe *working* voltage for the mica dielectric.

SOLUTION Use Table 13–2.

$$\frac{200\ 000\ V}{\cancel{mm}} \times 0.1\ \cancel{mm} = 20\ kV$$

∴ The maximum voltage is 20 kV.
For a working voltage, derate the maximum by at least a factor of 10.

$$\frac{20\ kV}{10} = 2\ kV$$

∴ The working voltage would be less than 2 kV.

Observation Because the insulating materials used as dielectrics are subject to thermal runaway, the working voltage of the dielectric must be conservative enough to provide for an elevated ambient temperature equal to or greater than 85°C.

TABLE 13–2 TYPICAL VALUES OF DIELECTRIC STRENGTHS OF SELECTED MATERIALS AT 25°C

Material	*Dielectric Strength (V/mm)*
Air	3000
Barium-strontium titanate	4000
Glass	80 000
Mica	200 000
Paper (waxed)	50 000
Teflon	40 000

Capacitance

FIGURE 13-7
(a) Capacitors connected in parallel.
(b) Capacitors connected in series.

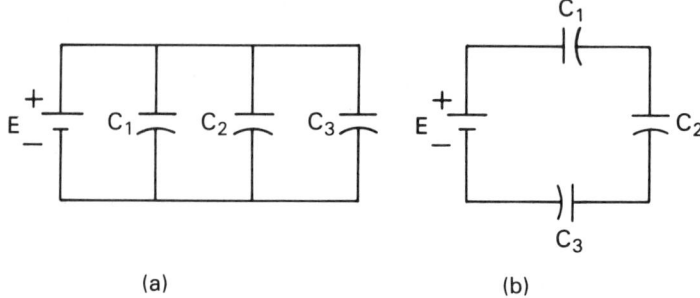

(a)　　　　　　　(b)

13-5 CONNECTING CAPACITORS IN SERIES AND PARALLEL

Capacitors connected in parallel, as shown in Figure 13-7(a), result in an increase in the circuit capacitance due to an increase in the area on which charge is deposited by the voltage source. The total charge (Q_T) deposited on the capacitor plates is equal to the sum of the charge of each capacitor. Thus

Parallel

$$Q_T = Q_1 + Q_2 + Q_3 \quad (C) \tag{13-18}$$

From Equation 13-9,

$$Q = CV$$

Substituting into Equation 13-18,

$$C_T E = C_1 V_1 + C_2 V_2 + C_3 V_3$$

However, $E = V_1 = V_2 = V_3$ in parallel circuit. Thus

$$C_T E = C_1 E + C_2 E + C_3 E$$

Dividing both sides by E and factoring results in

$$C_T = C_1 + C_2 + C_3 \quad (F) \tag{13-19}$$

EXAMPLE 13-12 Determine the equivalent capacitance of the three capacitors, connected in parallel, of Figure 13-7(a) when $C_1 = 10 \ \mu F$, $C_2 = 50 \ \mu F$, and $C_3 = 25 \ \mu F$.

SOLUTION Use Equation 13-19.

$$C_T = C_1 + C_2 + C_3 = 10 \ \mu F + 50 \ \mu F + 25 \ \mu F$$

$$\therefore \quad C_T = 85 \ \mu F$$

Series Capacitors connected in series, as shown in Figure 13-7(b), result in a decrease in the circuit capacitance due to what appears to be a widening of the spacing between the plates. Because of the series connection, the current in the capacitors during the initial charging of the plates is the same in each capacitor at any given time. Thus, the charge deposited on each capacitor is the same and

Connecting Capacitors in Series and Parallel

$$Q_T = Q_1 = Q_2 = Q_3 \quad (C) \tag{13-20}$$

From Equation 13-9,

$$V = \frac{Q}{C}$$

And in a series circuit

$$E = V_1 + V_2 + V_3$$

Substituting results in

$$\frac{Q_T}{C_T} = \frac{Q_1}{C_1} + \frac{Q_2}{C_2} + \frac{Q_3}{C_3}$$

However, $Q_T = Q_1 = Q_2 = Q_3$ in a series circuit. Thus

$$\frac{Q_T}{C_T} = \frac{Q_T}{C_1} + \frac{Q_T}{C_2} + \frac{Q_T}{C_3}$$

Dividing both sides by Q_T and factoring results in

$$\frac{1}{C_T} = \frac{1}{C_1} + \frac{1}{C_2} + \frac{1}{C_3}$$

Reciprocating both sides results in

$$C_T = \frac{1}{\frac{1}{C_1} + \frac{1}{C_2} + \frac{1}{C_3}} \quad (F) \tag{13-21}$$

NOTE: The equation form for capacitors in series is the same as resistors in parallel.

EXAMPLE 13-13 Determine the equivalent capacitance of the three capacitors, connected in series, of Figure 13-7(b) when $C_1 = 50$ pF, $C_2 = 25$ pF, and $C_3 = 10$ pF.

SOLUTION Use Equation 13-21.

$$C_T = \frac{1}{\frac{1}{C_1} + \frac{1}{C_2} + \frac{1}{C_3}} = \frac{1}{\frac{1}{50 \text{ pF}} + \frac{1}{25 \text{ pF}} + \frac{1}{10 \text{ pF}}}$$

$$\therefore \quad C_T = 6.25 \text{ pF}$$

EXAMPLE 13-14 For the circuit of Figure 13-8, (a) compute the equivalent capacitance (C_T), (b) determine the charge on each capacitor and determine the total charge, (c) determine the potential difference (voltage) across each capacitor, and (d)

398

Capacitance

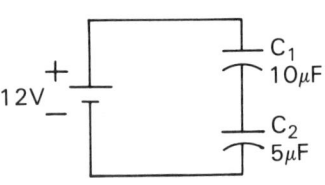

FIGURE 13-8
Circuit of Example 13-14.

determine the energy stored by the capacitors of the circuit.

SOLUTION (a) Use Equation 13-21:

$$C_T = \frac{1}{\frac{1}{10\ \mu F} + \frac{1}{5\ \mu F}}$$

∴ $C_T = 3.33\ \mu F$

(b) Since $Q = CV$ and $Q_T = Q_1 = Q_2$ in a series circuit, then

$$Q_T = C_T E = 3.33\ \mu F\ (12\ V)$$
$$Q_T = (3.33 \times 10^{-6})(12)$$

∴ $Q_T = Q_1 = Q_2 = 40\ \mu C$

(c) $V_1 = \dfrac{Q_1}{C_1} = \dfrac{40\ \mu C}{10\ \mu F} = 4\ V$

$V_2 = \dfrac{Q_2}{C_2} = \dfrac{40\ \mu C}{5\ \mu F} = 8\ V$

$V_1 = 4\ V,\quad V_2 = 8\ V$

∴ $E = V_1 + V_2 = 12\ V$

(d) $W = \tfrac{1}{2}CV^2$

$$W_T = \tfrac{1}{2}C_T E^2 = \frac{(3.33 \times 10^{-6})(12)^2}{2}$$

∴ $W_T = 240\ \mu J$

13-6 CAPACITOR APPLICATIONS, CHARACTERISTICS, AND SPECIFICATIONS

Over the years, the use of capacitors has become varied and specialized, resulting in no one type of capacitor being able to serve all the needs of the electronics industry. Instead, many types of capacitors are available, each designed to accommodate a specific set of applications.

Applications

The many specialized applications of capacitors include starting motors, blocking dc current, passing ac current, filtering unwanted signals, tuning circuits to a specific frequency, shifting phase, timing, suppressing ignition and other noise, coupling active circuits together, and bypassing signals.

Characteristics

Capacitors are commonly classified by their dielectric material since the characteristic of capacitors is mainly due to the properties of the dielectric. The following glossary of capacitors is summarized in the family tree of Figure 13-9.

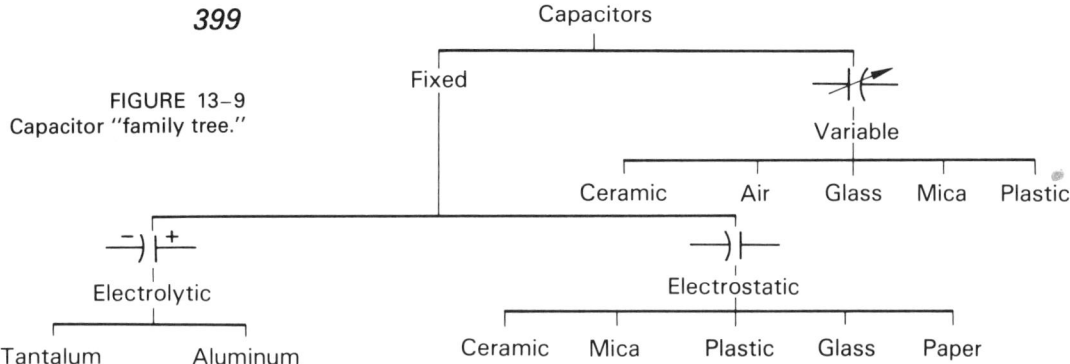

FIGURE 13-9 Capacitor "family tree."

The fixed *electrostatic* capacitors are made up of two metal conductors separated by a dielectric. They are characterized by very high resistance to leakage current in their dielectric. The fixed electrostatic capacitors include ceramic (barium-strontium titanate and others), mica, glass, paper, and the plastic film capacitors.

Ceramic capacitors: are extremely small for a given capacitance due to their high κ; used in consumer entertainment products for bypass and coupling; extremely good high-frequency properties (to 200 MHz); low in cost; available in disc, as pictured in Figure 13-10(a), and monolithic (multilayers of metal and ceramic fired into a single block) packages.

Mica capacitors: are able to withstand very high voltages due to high dielectric strength; extremely good high-frequency properties (to 200 MHz); used in radio and telecommunication applications; good temperature stability.

Paper capacitors: oil-filled paper foil capacitors have excellent high-voltage (greater than 600 V), high-capacitance characteristics; used below 50 kHz due to substantial high-frequency losses; used in uncritical power supply applications because of wide variation in capacitance due to temperature, aging, and frequency; manufactured by rolling alternate layers of aluminum foil and paper in ribbon form and sealing in an oil-filled container.

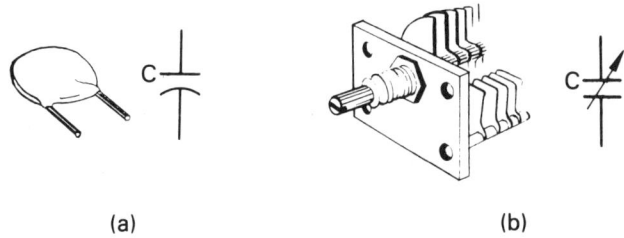

FIGURE 13-10
(a) Fixed ceramic disc capacitor.
(b) Variable air capacitor.

Plastic film capacitors: include such plastic dielectric materials as polyester (Mylar), polypropylene, polystyrene, polysulphone, polycarbonate, and polytetrafluoroethylene (TEF Teflon); similar to paper capacitors in construction, the only major difference being the metallization of the plate directly onto the dielectric; Teflon capacitors may be operated to 250°C; polystyrene capacitors have very low leakage, close tolerance, and good high-frequency properties; as a group, the film capacitors provide a wide choice in cost versus performance.

The fixed *electrolytic* capacitors are made up of two metal plates that have a definite polarity separated by a thin metal oxide dielectric, which is formed on the anode or positive plate. An electrolyte (a current-carrying liquid or solid) makes up the cathode or negative plate. The polarity marking on the package of electrolytic capacitors must be observed when connecting a dc source to the device. A misconnection will result in the capacitor acting as a short circuit. Electrolytic capacitors are characterized by a small leakage current in the dielectric. The oxide layers found on the anode have dielectric constants (κ) between 8 and 25. The electrolytic capacitors include aluminum and tantalum.

Aluminum electrolytic capacitors: When compared to capacitors of similar capacity, aluminum electrolytics have higher working voltage, higher ripple current rating, and lower cost per capacitance-volt (μF-V); are very susceptible to deterioration by certain degreasing compounds including Freon, trichlorethylene, and carbon tetrachloride; these and other halogenated hydrocarbon solvents should not be used to clean circuit boards when aluminum electrolytic capacitors are attached.

Tantalum electrolytic capacitors: are generally superior to aluminum electrolytic in tolerance specification, leakage current, length of life, temperature range, frequency range, size to capacitance, and resistance to mechanical shock and vibration; are available as foil types, solid dry types, and wet slug types.

The *variable* capacitor is designed so that the plate area is variable, as in the case of the rotary air capacitor pictured in Figure 13-10(b), or the plate spacing is adjustable, as in the trimmer capacitor. In the manufacture of variable capacitors, materials such as air, mica, or ceramic are used as dielectrics.

Specifications When specifying capacitors, the most needed items include the following:

☐ Dielectric type: mica, tantalum, ceramic, and so on.
☐ Capacitance: commercially expressed in either microfarads (μF) or picofarads (pF).

- ☐ Working voltage: maximum voltage at which a capacitor may be operated continuously at a specified temperature.
- ☐ Capacitance tolerance: a percentage of the nominal value; may also be given as MRV (minimum rated value).
- ☐ Physical size: length, diameter, lead spacing, and so on.

Stray Capacitance

Since any pair of conductors separated by air or other dielectric form a capacitor, small amounts of capacitance (1 to 10 pF) are present between conductors within a wire cable, as well as between circuit wiring and metal chassis or between adjacent or opposite conductors on a printed circuit board. Capacitance resulting from these and other sources is called *stray capacitance* and is usually unwanted as it has not been planned for in the design of the circuit. Stray capacitance is sometimes used in tuned circuits as part of the tuning capacitance.

PROBLEMS

Section 13–1

13–1 Determine the mutual force between two electric charges 4 cm apart if the charges are 8.01×10^{-14} C and 1.92×10^{-13} C.

13–2 Determine the mutual force between two electric charges 7 cm apart if the charges are -80 pC and 120 pC.

13–3 Determine the force of repulsion between a charge of -20 nC and one electron (-1.60×10^{-19} C) 0.5 cm apart.

13–4 Determine the force of attraction between a charge of $+130$ μC and a single electron (-1.60×10^{-19} C) 0.18 m apart.

13–5 Determine the electric field strength between two metal conductors spaced 0.8 mm apart having a potential difference of 100 V between them.

13–6 Determine the electric field strength between the plates of a mica capacitor when the capacitor is connected across a 500-V source if the dielectric is 0.2 mm thick.

13–7 Determine the force on a free electron between two parallel plates separated by 0.5 cm when the plates are connected to a 48-V source.

13–8 Determine the force on a free electron between two parallel plates separated by 1.6 cm when a potential difference of 1000 V is measured across the plates.

13–9 Determine the electric flux density of an electrolytic capacitor with a plate area of 0.1 m², having 1 mC of charge on its plates.

Capacitance

13-10 Determine the electric flux density of a mica capacitor with a plate area of 0.25 cm², having 0.48 μC of charge on its plates (1 cm² = 1 × 10⁻⁴ m²).

Section 13-2

13-11 Determine the charge on a 0.5-μF capacitor when the voltage across the plates is 300 V.

13-12 Determine the charge that may be stored on a 22-μF, 30-V tantalum electrolytic capacitor.

13-13 The voltage across an unbranded capacitor is 80 V and the charge on the capacitor is 160 μC. Determine the capacitance of the capacitor.

13-14 Determine the capacitance of an unmarked capacitor if 10 μC of charge deposited on the plates raises the potential difference between the plates to 200 V.

Section 13-3

13-15 Determine the energy stored by a 6800-μF aluminum electrolytic filter capacitor when 60 V is measured across its terminals.

13-16 Determine the energy stored by a 0.15-μF Mylar capacitor when 180 V is placed across its leads.

13-17 A 1-μF capacitor is charged to 100 V by a dc power supply. Determine the energy removed from the capacitor when it is discharged to 40 V.

13-18 The 820-μF capacitor in an electronic strobe is charged to 250 V. Determine the energy stored by the capacitor.

13-19 Determine the capacitance of a capacitor made up of parallel metal plates with an interface area of 52 cm² separated by 3 mm of air.

13-20 A capacitor charged to 35 V is made up of two parallel metal plates with an interface area of 10 cm² separated by a distance of 2.5 mm.
  (a) Determine the capacitance when air is between the plates.
  (b) Determine the charge on each plate.
  (c) Determine the electric field strength.
  (d) Determine the electric flux density.

13-21 A capacitor is constructed of two parallel metal plates with an interface area of 5.5 cm². The plates are separated by a 0.15-mm air dielectric. Determine the voltage across the plates when the electric flux density is 4 μC/m².

13-22 For the capacitor of Problem 13-21, determine (a) the magnitude of the electric field strength and (b) the energy stored by the capacitor.

Section 13–4 13–23 Using the values in Tables 13–1 and 13–2, determine the capacitance and specify a working voltage rating (using a derating factor of 10) for a paper capacitor having a plate area of 0.06 m² and a dielectric thickness of 0.05 mm.

13–24 From the values in Tables 13–1 and 13–2, determine the capacitance and specify a working voltage rating (using a derating factor of 10) for a Teflon capacitor having a plate area of 200 cm² and a dielectric thickness of 0.025 mm.

Section 13–5 13–25 Two aluminum electrolytic capacitors, $C_1 = 50\ \mu F$ and $C_2 = 15\ \mu F$, are connected in parallel across a 50-V source.
 (a) Determine the equivalent capacitance of the capacitors.
 (b) Determine the total charge on the capacitors.
 (c) Determine the total energy stored by the capacitors.

13–26 A 0.01-μF ceramic capacitor is connected in series with a 0.05-μF mica capacitor. The two capacitors are connected in series across a 200-V source.
 (a) Determine the voltage across each capacitor.
 (b) Determine the equivalent capacitance of the capacitors.

13–27 Determine the voltage across each capacitor in the circuit of Figure 13–11.

13–28 Determine the voltage across each capacitor in the circuit of Figure 13–12.

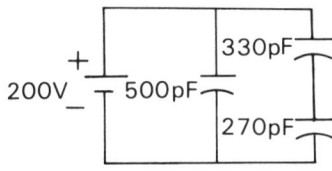

FIGURE 13–11 Circuit of Problem 13–27.

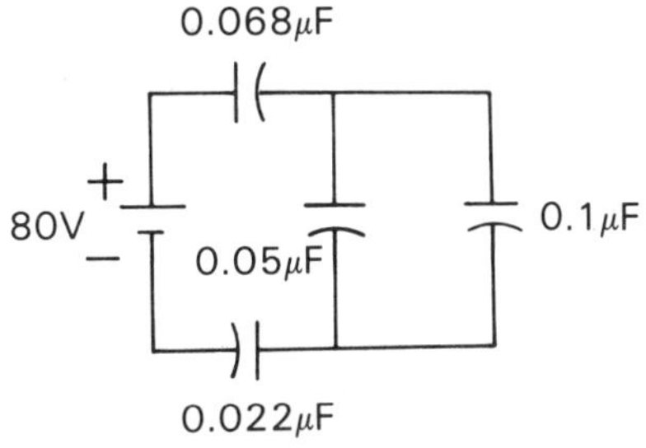

FIGURE 13–12 (right) Circuit of Problem 13–28.

14 RC Transient Circuits

14-1 CHARGE AND DISCHARGE OF CAPACITORS
14-2 RC TIME CONSTANT
14-3 INITIAL AND FINAL CONDITIONS OF RC CIRCUITS
14-4 INSTANTANEOUS VALUES OF RC CIRCUITS
14-5 APPLICATIONS

CHAPTER PREVIEW This chapter will give you an understanding of what happens to the parameters in an RC circuit when step changes in voltage or current sources are applied to the circuit. A knowledge of various relationships among the circuit parameters is extremely important, particularly in computer circuits, since voltages change in discrete or digital steps. In addition, you will learn how to calculate the exact voltage in a circuit at any given time after the application of a step voltage source.

PERFORMANCE OBJECTIVES Once you have read each section, worked each example with pencil, paper, and calculator, and answered each problem of every section, you will be able to:

☐ Understand qualitatively the charging or discharging of an RC circuit.
☐ Determine the initial and final circuit conditions of an RC circuit after a step source has been applied.
☐ Understand the concept of the time constant.
☐ Calculate the instantaneous voltage or current at any time during the transient period.

14–1 CHARGE AND DISCHARGE OF CAPACITORS A capacitor by itself in an electric circuit has a relatively limited application. It can block current and it can store energy by being charged to some voltage. However, when it is used in conjunction with a resistor, many useful applications exist.

Since a capacitor charges at some rate, which is a function of the charging current, then by controlling the current, it can be determined how long it will take to charge a capacitor to a given voltage. You can compare the charging of a capacitor with the filling of a bathtub. For a given size of tub, you can fill it up to a prescribed level in various amounts of time depending on the flow of water into the tub.

If you turn the faucet on to a slow dribble, it may take many minutes, or even hours, to fill the tub. On the other hand, with both the hot and cold faucets turned full on, it will only take a few minutes to fill the tub. A direct analogy can be made by comparing the size of the tub with capacitance, the level of water with voltage, and the rate of water flow with current. In this chapter, we will investigate the physical and mathematical relationships between these parameters.

In charging a capacitor, as the charge accumulates in the capacitor, the voltage across the plates is ever increasing and opposite in polarity to the applied source. Consider the case of a

RC Time Constant

10-V source charging a capacitor that initially has zero voltage across it. At the first instant of time, there is a 10-V difference across the capacitor, resulting in a certain amount of current. As time passes, charge accumulates on the plates of the capacitor and voltage increases across the capacitor. The voltage difference between the voltage source and the voltage across the capacitor becomes less. Less voltage difference will, of course, result in less current and, consequently, less charge is delivered to the capacitor. An analogy to the charging of a capacitor is the inflating of a tire with a hand pump. Initially, it is very easy to operate the pump because there is no air in the tire. However, as the tire fills up, there is more and more pressure in the tire, which works against the acceptance of additional air. Eventually, assuming the tire doesn't blow out, a point will be reached where a state of equilibrium exists and you stop pumping. The equation that describes the interrelationship between a forcing source and a device being "filled up" will be discussed in the following sections.

14-2 RC TIME CONSTANT

Time Constant Defined

The rate at which an RC circuit charges or discharges is determined by the time constant of the circuit. By definition, the time constant, symbolized by the Greek letter tau (τ), is the time in seconds for a dc voltage or current to increase to 63.2% of its final value, or to decrease to 36.8% of its initial value. The increase or decrease is in response to a change in the conditions of the circuit. Stated mathematically,

$$\tau = RC \quad \text{(s)} \tag{14-1}$$

It is important to note that it is the combination of R and C that produces the time constant. In the following examples, each combination of R and C has the same time constant, and each charges or discharges at the same rate, even though the individual values of R and C are significantly different.

☐ $R = 1 \text{ k}\Omega$, $C = 1 \text{ }\mu\text{F}$, $\tau = (1 \times 10^3)(1 \times 10^{-6}) = 1$ ms
☐ $R = 200 \text{ }\Omega$, $C = 5 \text{ }\mu\text{F}$, $\tau = (200)(5 \times 10^{-6}) = 1$ ms
☐ $R = 40 \text{ k}\Omega$, $C = 25 \text{ nF}$, $\tau = (40 \times 10^3)(25 \times 10^{-9}) = 1$ ms

Once again, the bathtub analogy can be used. The time it takes for a bathtub to be filled depends on the rate of flow of water (current), which depends on the faucet setting (resistance) and how large a bathtub (capacitance). It takes just as long for a large-capacity tub (large capacitance) to be filled by a wide open faucet (low resistance) as for a small-capacity tub (small capacitance) with a faucet turned to a low level (high resistance).

RC Transient Circuits

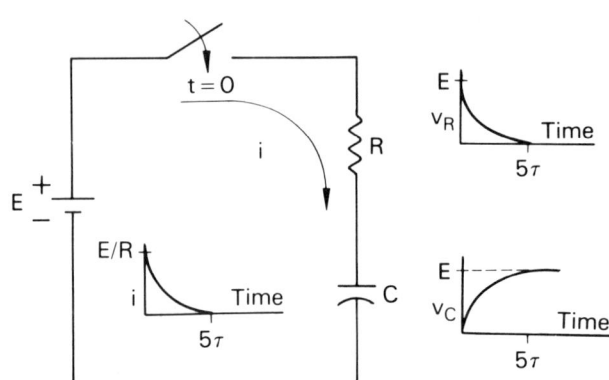

FIGURE 14–1
RC circuit with a switch being closed at time t = 0 seconds. The wave forms indicate the response of the circuit parameters to the closing of the switch during the *transient* period.

RC Transient Behavior

When a capacitor is connected in series with a resistance, a switch, and a dc source of voltage (as shown in Figure 14–1), the circuit current is at its maximum the instant the switch is closed and then decreases to zero as the voltage across the capacitor rises to equal E, the source voltage. The decrease in circuit current and the rise in capacitor voltage is not a linear function, but is an exponential function, as noted by the *transient behavior* shown in the small wave forms of voltage and current in Figure 14–1. Notice that lowercase lettering is used to indicate current (i) and voltage (v). This is done because these parameters are a function of time during this transient period of time.

Determining Transient Time

As you now know, upon closing the switch in Figure 14–1, a period of transient circuit behavior is started. The time between the switch closing and stable circuit conditions is called the *transient time*, and it depends on the size of R and C as described by the statement of the time constant $\tau = RC$ (Equation 14–1). As the capacitor takes on charge, the transient voltage across the capacitor plates can be described in terms of the circuit's time constant.

After one time constant, the capacitor voltage has reached 63.2% of the charging voltage, E. After the second time constant, the capacitor voltage has reached 86.5% of the charging voltage, E. After the third time constant, the capacitor voltage has reached 95.0% of the charging voltage, E. After the fourth time constant, the capacitor voltage has reached 98.2% of the charging voltage, E. After the fifth time constant, the capacitor voltage has reached 99.3% of the charging voltage, E. By now you have realized that 100% of the charging source, E, is never going to be reached. This is indeed the case; however, for practical purposes, it has been agreed that the time represented by five time constants is sufficient to reach the final value. Thus, the period of time of the transient behavior may be stated as

$$t_{\text{trans}} = 5\tau \quad \text{(s)} \tag{14-2}$$

where t_{trans} = period of transient circuit behavior in seconds (s)
5τ = five time constants where $\tau = RC$

EXAMPLE 14-1 For the circuit of Figure 14-1, $R = 100$ kΩ and $C = 100$ μF. Determine the amount of time required for the circuit to come to its steady-state values.

SOLUTION Substitute into Equation 14-2:

$$t_{\text{trans}} = 5\tau = 5(100 \times 10^3)(100 \times 10^{-6}) = 50 \text{ s}$$

∴ The period of transient behavior is 50 seconds when $R = 100$ kΩ and $C = 100$ μF.

In summary, the time constant $\tau = RC$ determines how quickly or slowly a particular circuit will reach its final or steady-state value. A short time constant means that it will take a relatively short time for the transition from one condition to another. On the other hand, a long time constant means that it will take a relatively long time to change from one condition to another. In the latter sections of this chapter, we will apply exponential equations to the transient behavior of the RC circuit.

14-3 INITIAL AND FINAL CONDITIONS OF RC CIRCUITS

Initial Conditions

To understand the behavior of an RC circuit under transient conditions, we must first know the voltages and currents of each component immediately before and after a switch is closed or the operating condition of the source is changed. By knowing the values of voltage and current prior to the transient and after the transient, it then becomes a fairly straightforward task to write the equations for the parameters during the transient period.

When working with capacitive circuits, it must be remembered that the voltage across a capacitor cannot change instantaneously. Recall that the amount of charge that is moved is a function of the current and the duration of time over which the current flows. For voltage to change instantaneously, the amount of charge must also change instantaneously. By definition, $i = \Delta Q/\Delta t$. If $\Delta t = 0$ and $\Delta Q = k$, then $i = k/0$, which is undefined according to mathematicians, but for our purpose it can be thought of as an infinite current. Obviously, this is impossible; hence the conclusion that the voltage across a capacitor

410

RC Transient Circuits

cannot change instantaneously. By knowing this fact, it becomes relatively easy to determine the initial voltages in an RC circuit. In the remainder of the chapter, we will be using the symbol $t = 0^-$ as the instant of time just before $t = 0$, and $t = 0^+$ as the instant of time just after $t = 0$.

EXAMPLE 14-2 Determine the circuit current and the voltages across R and C of Figure 14-2 before and after the switch is closed. Assume there is no initial charge on C before the switch is closed.

SOLUTION Before the switch is closed, no voltage exists across R and C since $i = 0$ A. At the instant after the switch closes, $t = 0^+$, we know that v_C is still zero since voltage cannot change instantaneously. However, Kirchhoff's voltage law must be satisfied, so v_R must equal E at $t = 0^+$.

∴ At $t = 0^-$, $v_C = 0$ and $v_R = 0$; at $t = 0^+$, $v_C = 0$ V and $v_R = E = 20$ V.

The current at $t = 0^+$ is then

$$i = \frac{v_R}{R} = \frac{20}{5 \text{ k}\Omega} = 4 \text{ mA}$$

∴ At $t = 0^-$, $i = 0$ A; at $t = 0^+$, $i = 4$ mA.

EXAMPLE 14-3 Determine v_1, v_2, and v_C of Figure 14-3 at $t = 0^+$, assuming no initial charge on C before the switch is closed. Also determine i_C at $t = 0^+$.

SOLUTION At $t = 0^+$, $v_C = 0$ since there was no voltage there before the switch was closed. But then $v_2 = 0$ V since it is in parallel with C and $v_C = 0$. By Kirchhoff's law, v_1 must equal E.

∴ At $t = 0^+$, $v_1 = 10$ V, $v_2 = 0$, and $v_C = 0$ V.

The currents at $t = 0^+$ are

$$i_1 = \frac{10}{100} = 100 \text{ mA}$$

$$i_2 = \frac{0}{300} = 0 \text{ A}$$

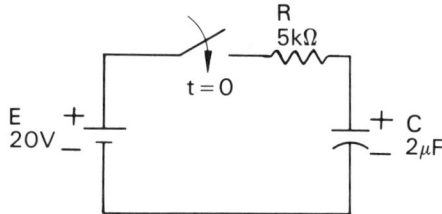

FIGURE 14-2
Circuit of Example 14-2.

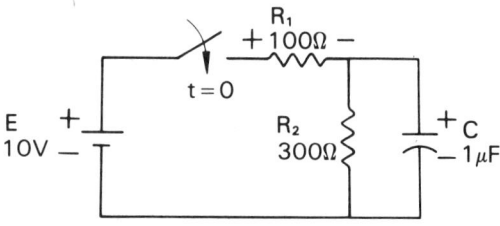

FIGURE 14-3
Circuit of Example 14-3.

∴ $i_C = 100$ mA since $i_C = i_1 - i_2$.

EXAMPLE 14-4 Assume that C in Figure 14-2 has an initial charge and that $v_C = +5$ V before the switch is closed. Determine the voltages and currents at $t = 0^+$.

SOLUTION Because $v_C(0^-) = +5$ V, then $v_C(0^+) = +5$ V. Applying KVL, determine v_R:

$$E = v_R + v_C$$

$$v_R(0^+) = E - v_C(0^+) = 20 \text{ V} - 5 \text{ V} = 15 \text{ V}$$

∴ At $t = 0^+$, $v_R = 15$ V and $v_C = 5$ V.
The current at $t = 0^+$ is then

$$i_R(0^+) = \frac{v_R}{R} = \frac{15}{5 \text{ k}\Omega} = 3 \text{ mA}$$

∴ At $t = 0^+$, $i_R = 3$ mA and $i_C = 3$ mA since current is the same in a series circuit.

EXAMPLE 14-5 Assume that C in Figure 14-3 has an initial charge and that $v_C(0^-) = 3$ V. Determine the voltages and currents at $t = 0^+$.

SOLUTION Since C and R_2 are connected in parallel, $v_C = v_2$. Thus

$$v_2(0^+) = v_C(0^+) = 3 \text{ V}$$

Applying KVL, determine v_1:

$$v_1(0^+) = E - v_C(0^+) = 10 \text{ V} - 3 \text{ V} = 7 \text{ V}$$

∴ At $t = 0^+$, $v_C = v_2 = 3$ V and $v_1 = 7$ V.
The currents at $t = 0^+$ are

$$i_2(0^+) = \frac{3}{300} = 10 \text{ mA}$$

$$i_1(0^+) = \frac{7}{100} = 70 \text{ mA}$$

From Kirchhoff's current law,

$$i_1(0^+) = i_2(0^+) + i_C(0^+)$$
$$i_C(0^+) = i_1(0^+) - i_2(0^+)$$
$$i_C(0^+) = 70 \text{ mA} - 10 \text{ mA} = 60 \text{ mA}$$

∴ At $t = 0^+$, $i_1 = 70$ mA, $i_2 = 10$ mA, and $i_C = 60$ mA.

Final Conditions In an RC circuit, the final conditions exist after all voltages and current have stopped changing. Theoretically, after an infinite amount of time, everything has reached a state of equilibrium. Practically, *steady-state* or *final conditions* are reached in a finite amount of time ($\approx 5\tau$). The easiest way to determine final conditions in a simple RC circuit is to assume that an open circuit exists wherever there is a capacitor.

EXAMPLE 14-6 Determine the final voltages and currents for the circuit of Figure 14-2.

SOLUTION With the switch closed, the capacitor will eventually charge to a voltage equal to the source. At that time, $i = 0$ and $v_R = 0$ and $v_C = E = 20$ V.

∴ At $t = \infty$, $v_C = 20$ V, $v_R = 0$ V, and $i = 0$ A.

EXAMPLE 14-7 Determine the final voltages and currents for the circuit of Figure 14-3.

SOLUTION Assuming that C is an open circuit, the final conditions will be as shown in Figure 14-4. Apply the voltage-divider rule to determine v_1 and v_2.

$$v_1(\infty) = \frac{(10)(100)}{100 + 300} = 2.5 \text{ V}$$

$$v_2(\infty) = \frac{(10)(300)}{100 + 300} = 7.5 \text{ V}$$

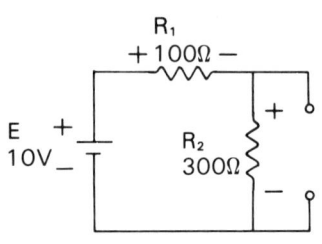

FIGURE 14-4
Circuit of Example 14-7.

Since R_2 is connected in parallel with C, then

$$v_C(\infty) = v_2(\infty) = 7.5 \text{ V}, \quad i_C = 0$$

∴ At $t = \infty$, $v_1 = 2.5$ V, $v_2 = 7.5$ V, and $v_C = 7.5$ V. The currents at $t = \infty$ are

$$i_1(\infty) = \frac{2.5}{100} = 25 \text{ mA}$$

$$i_2(\infty) = \frac{7.5}{300} = 25 \text{ mA}$$

Since C is open circuit at $t = \infty$,

$$i_C(\infty) = 0 \text{ A}$$

∴ At $t = \infty$, $i_1 = 25$ mA, $i_2 = 25$ mA, and $i_C = 0$ A.

Voltage Division Between Two Series Capacitors We mentioned in the previous examples that C is assumed to be an open circuit in order to calculate the final voltage conditions. However, this assumption can be made only when there is one capacitor or an equivalent of one capacitor in the circuit.

EXAMPLE 14–8 Determine the final voltage across C_1 and C_2 in the circuit of Figure 14–5.

SOLUTION For a final steady-state condition to exist, the amount of charge on C_1 must equal the amount of charge on C_2; otherwise, a potential difference would exist and current would continue to flow in the circuit. From our work with capacitance, we know that $Q_1 = C_1 V_1$ and $Q_2 = C_2 V_2$ and, for a final steady-state condition, the entire source voltage of 10 V divides between C_1 and C_2. Since $Q_1 = Q_2$ at $t = \infty$,

$$C_1 V_1 = C_2 V_2$$

and

$$\frac{C_1}{C_2} = \frac{V_2}{V_1}$$

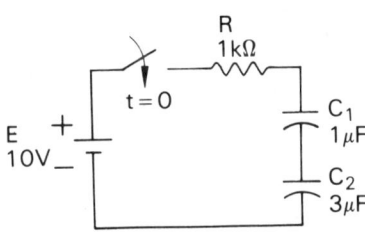

FIGURE 14–5
Circuit of Example 14–8.

Substituting for C_1 and C_2, solve for V_1:

$$\frac{1 \ \mu F}{3 \ \mu F} = \frac{V_2}{V_1}$$

$$V_1 = 3 V_2$$

Apply KVL:

$$V_1 + V_2 = 10 \text{ V}$$

Substituting $V_1 = 3V_2$ into $V_1 + V_2 = 10$ results in

$$3V_2 + V_2 = 10 \text{ V}$$

$$4V_2 = 10 \text{ V}$$

$$V_2 = 2.5 \text{ V}$$

$$V_1 = 10 \text{ V} - 2.5 \text{ V} = 7.5 \text{ V}$$

∴ At $t = \infty$, $V_1 = 7.5$ V and $V_2 = 2.5$ V.

Observation The voltage divides inversely to the value of the capacitance. Thus, $C_1 < C_2$; however, $V_1 > V_2$.

A formula for determining the voltage division between two capacitors results from the fact that the charges on series-connected capacitors are equal ($Q_1 = Q_2$) when the circuit is in equilibrium. Thus

$$V_x = \frac{EC_y}{C_x + C_y} \quad \text{(V)} \tag{14-3}$$

where V_x = voltage across capacitor C_x in volts (V)
E = charging source voltage in volts (V)
C_x = capacitance of C_x in farads (F)
C_y = capacitance of C_y in farads (F)

NOTE: This formula is limited to two capacitors in series.

Charge and Discharge RC Wave Forms

The following examples will serve to provide you with an insight into the division of voltage in the elementary RC circuit during the transient period of time.

EXAMPLE 14-9 Assume that the circuit shown in Figure 14-6 initially has no voltage applied to it and the switch is in position 1. At $t = 0$, the switch is moved to position 2. Determine the voltage wave form across resistor R and capacitor C.

SOLUTION From the previous discussions, the initial voltage across C is zero and remains at zero at the instant the switch is moved. But then the capacitor starts to charge, and after approximately five time constants, $v_C = E$. The voltage across R, v_R, must satisfy Kirchhoff's voltage law; hence its voltage wave shape must be such that when added to v_C the total equals E. This is shown in Figure 14-7.

EXAMPLE 14-10 Determine what happens when the switch remains in position 2, in Figure 14-6, for a long time ($t > 5\tau$) and then is moved to position 1. Draw the wave forms for v_C and v_R.

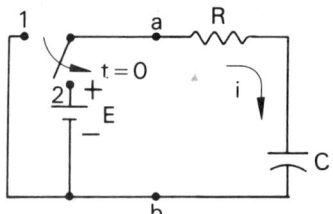

FIGURE 14-6
Circuit of Example 14-9.

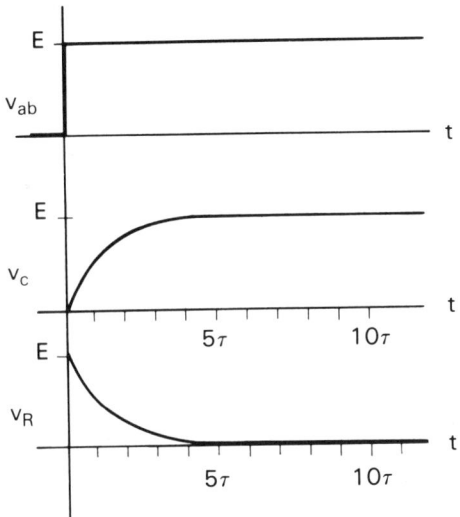

FIGURE 14-7
Voltage wave forms of Example 14-9.

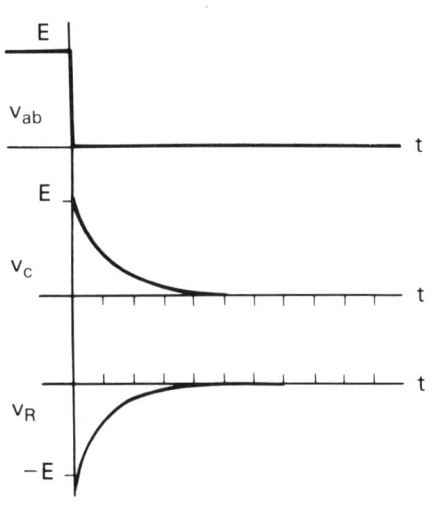

FIGURE 14-8
Voltage wave forms of Example 14-10.

SOLUTION We know that v_C starts out at E and then discharges or *decays* to zero. However, from KVL, $v_R + v_C = 0$ at $t = 0^+$. For this to be a true statement, v_R must start out at $-E$, as pictured in Figure 14-8.

Observation This makes sense because the circuit at $t = 0^+$ looks like Figure 14-9. The current direction is opposite to the current direction in the previous case when the capacitor was charging. Now that it is discharging, the current is flowing out of C.

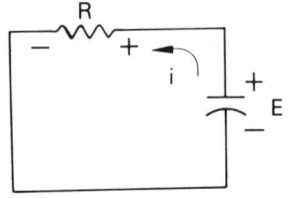

FIGURE 14-9
Voltage polarization and current direction of Example 14-10.

Once the initial and final conditions of an RC circuit are known, then determining the wave shape and writing the equations describing the transient conditions can be carried out.

14-4 INSTANTANEOUS VALUES OF RC CIRCUITS

Determining the exact voltage in an RC circuit at any time, t, during the transient period is quickly done with the aid of the $\boxed{e^x}$ key on the calculator. On some calculators, this same function is carried out by first depressing the $\boxed{\text{INV}}$ key, followed immediately by the $\boxed{\text{ln}}$ key.

Charging the RC Circuit

The transient period is defined as that period of time between the initial and final conditions in the RC circuit. The general equation for the voltage across the capacitor as it charges during the transient period is

$$v_C = E[1 - e^{-t/(RC)}] \quad (V) \tag{14-4}$$

where v_C = voltage across the capacitor at any time (t) during the transient period
E = source of the charging voltage applied to C
RC = time constant (τ) with R in ohms (Ω) and C in farads (F)
e = natural number (2.718)

Observation. In Equation 14-4, when $t = 0$ s,

$$e^{-t/(RC)} = e^{-0} = 1$$

and the equation becomes

$$v_C = E(1 - e^{-0}) = E(1 - 1) = 0$$

Therefore, at $t = 0$, $v_C = 0$ V.
In Equation 14-4, when $t = \infty$ or steady-state voltage ($\tau \geq 5\,RC$),

$$e^{-t/(RC)} = e^{-\infty} = \frac{1}{e^{\infty}} = 0$$

and the equation becomes

$$v_C = E(1 - e^{-\infty}) = E(1 - 0) = E$$

Therefore, at $t = \infty$, $v_C = E$.

EXAMPLE 14-11 Determine the voltage across a 1000-pF capacitor 1 μs after connecting the capacitor to a 5-V source in series with a 2-kΩ resistor.

SOLUTION Use Equation 14-4 and substitute:

$$v_C = E(1 - e^{-t/(RC)})$$
$$v_C = 5[1 - e^{-(1 \times 10^{-6})/(2 \times 10^3 \times 1 \times 10^{-9})}]$$
$$v_C = 5(1 - e^{-0.5}) = 5(1 - 0.607) = 1.97 \text{ V}$$
$$\therefore \quad v_C = 1.97 \text{ V at } t = 1\ \mu\text{s}.$$

Equation 14-4 may also be used to determine the time to charge a capacitor to a particular voltage, as demonstrated in the following example.

EXAMPLE 14-12 Determine the time to charge a 0.05-μF capacitor to 7 V when the capacitor is connected in series with a 500-Ω resistor to a 12-V battery.

SOLUTION Use Equation 14-4 and substitute:

Instantaneous Values of RC Circuits

$$v_C = E[1 - e^{-t/(RC)}]$$

$$7 = 12[1 - e^{-t/(500 \times 5 \times 10^{-8})}]$$

Divide both members by 12:

$$0.583 = 1 - e^{-t/(2.5 \times 10^{-5})}$$

Subtract 1 from both members and change sign:

$$-0.417 = -e^{-t/(2.5 \times 10^{-5})}$$

$$0.417 = e^{-t/(2.5 \times 10^{-5})}$$

Take the natural log $\boxed{\ln}$ of each member:

$$\ln(0.417) = \frac{-t}{2.5 \times 10^{-5}} \ln(e)$$

$$-0.8747 = \frac{-t}{2.5 \times 10^{-5}}$$

Change signs and solve for t:

$$t = (0.8747)(2.5 \times 10^{-5}) = 2.187 \times 10^{-5}$$

∴ The time for charging the capacitor from 0 V to 7 V is 21.87 μs.

Discharging the RC Circuit

The general equation for discharging a capacitor from an initial voltage, E_C, to the steady-state or final voltage, v_C, is

$$v_C = E_C e^{-t/(RC)} \quad \text{(V)} \tag{14-5}$$

where v_C = voltage across the capacitor at any time (t) during the transient period
 E_C = voltage across the capacitor when the discharge is started
 RC = time constant (τ) with R in ohms and C in farads
 e = natural number (2.718)

EXAMPLE 14-13 A 25-μF capacitor is charged to 100 V. The capacitor is then discharged through a 100-kΩ resistor connected across its terminals. Determine the voltage across the capacitor 6 s after the resistor is connected.

SOLUTION Use Equation 14-5 and substitute:

$$v_C = E_C e^{-t/(RC)}$$

$$v_C = 100 \ e^{-6/(100 \times 10^3 \times 25 \times 10^{-6})}$$

$$v_C = 100 \ e^{-2.4} = 9.07 \text{ V}$$

∴ At $t = 6$ s, $v_C = 9.07$ V.

Equation 14–5 can also be used to design an RC timing circuit, as demonstrated in the following example.

EXAMPLE 14–14 Determine the value of resistance needed to discharge a 20-μF capacitor, which has been charged to 10 V, down to 5 V in 2 s.

SOLUTION Use Equation 14–5 and substitute:

$$v_C = Ee^{-t/(RC)}$$
$$5 = 10e^{-2/(R20 \times 10^{-6})}$$

Solve for R:

$$0.5 = e^{-2/(R20 \times 10^{-6})}$$

$$\ln(0.5) = \frac{-2}{R20 \times 10^{-6}} \ln(e)$$

$$-0.693 = \frac{-2}{R20 \times 10^{-6}}$$

$$R = \frac{-2}{-0.693 \times 20 \times 10^{-6}} = 1.443 \times 10^5$$

$$\therefore \quad R = 144.3 \text{ k}\Omega$$

Observation If the timing in this circuit is critical, a 143-kΩ resistor $\pm 1\%$ is used; otherwise, a 150-kΩ resistor $\pm 5\%$ is satisfactory.

Current in the RC Circuit From an earlier discussion, we now know that initially the current in the RC series circuit is high and then diminishes to zero as the capacitor takes on charge. Initially, at $t = 0^+$, the voltage across R (v_R) is equal to E (the source voltage), and the voltage across C (v_C) equals zero. However, as time progresses to $t = \infty$, the current (i) diminishes, as does the voltage across R (v_R). Since Kirchhoff's voltage law must be satisfied in any circuit, including an RC series circuit, during the period of transient behavior, then

$$E = v_C + v_R$$

And solving for v_C,

$$v_C = E - v_R$$

However, from Equation 14–4, v_C is also equal to

$$v_C = E[1 - e^{-t/(RC)}] = E - Ee^{-t/(RC)}$$

Comparing the KVL statement $v_C = E - v_R$ to the preceding statement, $v_C = E - Ee^{-t/(RC)}$, we see that

Instantaneous Values of RC Circuits

$$v_R = Ee^{-t/(RC)} \quad (V) \tag{14-6}$$

From Ohm's law,

$$v_R = iR$$

Substituting iR for v_R in Equation 14-6 results in

$$iR = Ee^{-t/(RC)}$$

And solving for i,

$$i = \frac{E}{R} e^{-t/(RC)} \quad (A) \tag{14-7}$$

EXAMPLE 14-15 Close the switch and determine the current in the circuit of Figure 14-1 at (a) $t = 0^+$ s and (b) $t = 8$ s when $E = 80$ V, $R = 50$ kΩ, and $C = 200$ μF.

SOLUTION (a) Use Equation 14-7 and substitute:

$$i = \frac{E}{R} e^{-t/(RC)}$$

$$i = \frac{80}{50 \times 10^3} e^{-0/RC} = \frac{80}{50 \times 10^3} (1)$$

∴ $i = 1.6$ mA at $t = 0^+$.

(b) Use Equation 14-7 and substitute:

$$i = \frac{E}{R} e^{-t/(RC)}$$

$$i = \frac{80}{50 \times 10^3} e^{-8/(50 \times 10^3 \times 200 \times 10^{-6})}$$

$$i = (1.6 \times 10^{-3})e^{-0.8} = 719 \text{ μA}$$

∴ $i = 719$ μA at $t = 8$ s.

Table 14-1 summarizes the equations of the transient period during both charge and discharge of the capacitor.

Applying Thevenin's Theorem to the RC Circuit

From the previous discussion for the development of Equations 14-6 and 14-7, we learned that the RC circuit equations satisfy all the circuit rules and laws, and we can now solve for the circuit parameters of a simple series RC circuit connected to a voltage source. This may appear to you to be a very limited ability since most "real" electronic circuits have more than one resistor, one capacitor, and one source. However, recall that The-

TABLE 14–1 CHARGE AND DISCHARGE EQUATIONS FOR SERIES EQUIVALENT RC CIRCUITS

Instantaneous:	Charge	Discharge
Circuit current	$i = \dfrac{E}{R} e^{-t/(RC)}$	$i = \dfrac{E_C}{R} e^{-t/(RC)}$
Capacitance voltage	$v_C = E[1 - e^{-t/(RC)}]$	$v_C = E_C e^{-t/(RC)}$
Resistance voltage	$v_R = E e^{-t/(RC)}$	$v_R = E_C e^{-t/(RC)}$

venin's and Norton's theorems allow us to form single equivalent circuits from complex networks.

EXAMPLE 14–16 Determine the output voltage (V_{ab}) of the circuit pictured in Figure 14–10(a) at $t = 10$ ms after the switch is closed.

SOLUTION By *Thevenizing* the source and resistive part of the circuit, as shown in Figure 14–10(b), a series equivalent circuit of one resistor and one capacitor will result, as shown in Figure 14–10(c). Thus

$$R_{Th} = R_1 \| R_2 = \dfrac{1}{\dfrac{1}{500} + \dfrac{1}{1000}} = 333 \ \Omega$$

and

$$E_{Th} = \dfrac{ER_2}{R_1 + R_2} = \dfrac{(10)(1000)}{1500} = 6.67 \text{ V}$$

Solving for V_{ab} at $t = 10$ ms,

$$V_{ab} = v_C = E_{Th}[1 - e^{-t/(R_{Th}C)}]$$
$$V_{ab} = 6.67[1 - e^{-(10 \times 10^{-3})/(333 \times 50 \times 10^{-6})}]$$
$$V_{ab} = 6.67(1 - 0.548) = 3.01 \text{ V}$$

∴ The voltage across the capacitor at $t = 10$ ms is $V_{ab} = 3.01$ V.

14–5 APPLICATIONS

Familiarity with the behavior of RC circuits is important for your understanding of the operation of a number of electronic circuits. RC circuits may be designed to perform simple, noncritical timing functions, as in the turn signal in an automobile, or to perform complex, precise timing functions, as in computers. Many integrated circuit timers leave an R or a C for the user to select so the circuit may be tailored to a specific application.

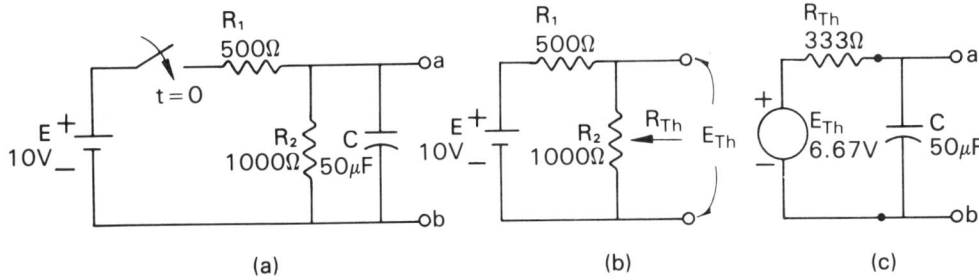

FIGURE 14–10 (a) Circuit of Example 14–16. (b) Forming Thevenin's equivalent circuit. (c) The Thevenized RC circuit of Example 14–16.

Power supply circuits are designed to have a short charging time and a long discharging time.

EXAMPLE 14–17 The circuit of Figure 14–11 represents a dc power supply. Determine both the charging and discharging time constants for the circuit.

SOLUTION With the switch closed, the capacitor is charged by the dc source, E. During this time, the Thevenin's equivalent resistance across points a-b consists of the 10-Ω resistor in parallel with the 1-kΩ resistor. Thus

$$R_{Th} = R_1 \parallel R_2 = \frac{1}{\frac{1}{10} + \frac{1}{1000}} = 9.9 \ \Omega$$

And the charge time constant, τ_C, is

$$\tau_C = R_{Th}C = (9.9)(10 \times 10^{-3}) = 99 \times 10^{-3} \text{ s}$$

∴ The charge time constant is 99 ms.

With the switch open, the capacitor discharges through resistance R_2 because R_1 is in an open circuit. The discharge time constant, τ_D, is

$$\tau_D = R_2C = (1000)(10 \times 10^{-3}) = 10 \text{ s}$$

∴ The discharge time constant is 10 s.

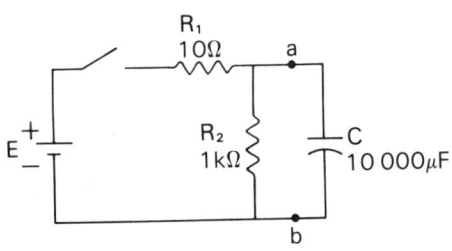

FIGURE 14–11
Circuit of Example 14–17.

Observation The capacitor will charge at a rate approximately 100 times faster than it will discharge.

Effects of Capacitor Leakage Current

When using electrolytic capacitors in timing circuits, the effect of the inherent leakage current must be considered. Figure 14–12 pictures an electrolytic capacitor shunted by a resistance representing the current leakage path. The value of this resistance depends on the voltage across the capacitor and the leakage current in the capacitor. The value of R_{leak} is usually high (>100 kΩ).

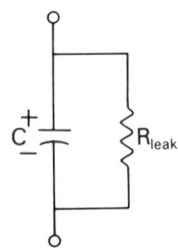

FIGURE 14–12
An electrolytic capacitor is pictured with a parallel resistance to indicate the path of the leakage current.

EXAMPLE 14–18 Determine the time constant for the circuit pictured in Figure 14–13(a), where C is considered to be perfect, and then for Figure 14–13(b), where R_{leak} represents the capacitor leakage in the practical circuit. The capacitor leakage current was measured at 32 V (63.2% of E or the v_C at $\tau = 1$) and found to be 10 μA.

SOLUTION For Figure 14–13(a), $\tau = RC$:

$$\tau = (1 \times 10^6)(10 \times 10^{-6}) = 10 \text{ s}$$

∴ The theoretical time constant is 10 s.
For Figure 14–13(b) $\tau = R_{Th}C$.

Observation Assume R_{leak} to be a constant value during the transient period equal to 32 V/10 μA (3.2 MΩ). Determine the Thevenin's equivalent resistance for the circuit of Figure 14–13(b):

$$R_{Th} = R_{leak} \parallel R$$

$$R_{Th} = \frac{1}{\dfrac{1}{3.2 \times 10^6} + \dfrac{1}{1 \times 10^6}} = 762 \text{ kΩ}$$

$$\tau = R_{Th}C = (762 \times 10^3)(10 \times 10^{-6}) = 7.6$$

∴ The practical time constant is 7.6 s.

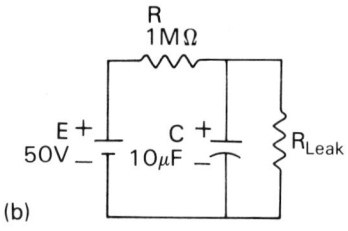

FIGURE 14–13
Circuit of Example 14–18.

423

Applications

Observation It is best to use low-leakage tantalum or electrolytic capacitors when an RC timing application requires large values of capacitance.

Effects of Stray Capacitance

In many types of computer circuits, timing signals in the form of pulses, square waves, and the like, must have very fast rise times and specific wave forms in order to trigger the circuits. The fast rise time of the triggering pulses prevents extraneous and slowly varying random signals from giving the computer false information. However, a long time constant RC circuit will tend to round off the square waves and pulses, as shown in Figure 14–14, thereby increasing the rise time. To prevent waveform distortion, it would be best to avoid using capacitors in the circuits. Unfortunately, all wires and components have inherent resistance and stray capacitance, which contribute to increasing the time constants. Although stray capacitance may be small, in an absolute sense, it is very large from a relative sense as computers now operate at very high frequencies and in very short periods of time.

EXAMPLE 14–19 Determine the time constant of a 10-ft length of cable used to interconnect two computer circuits. The cable has a resistance of 40 mΩ/ft and a capacitance of 10 pF/ft.

SOLUTION Determine R and C of the 10-ft cable:

$$R = \frac{40 \text{ m}\Omega}{\text{ft}} \times 10 \text{ ft} = 400 \text{ m}\Omega$$

$$C = \frac{10 \text{ pF}}{\text{ft}} \times 10 \text{ ft} = 100 \text{ pF}$$

Compute the time constant:

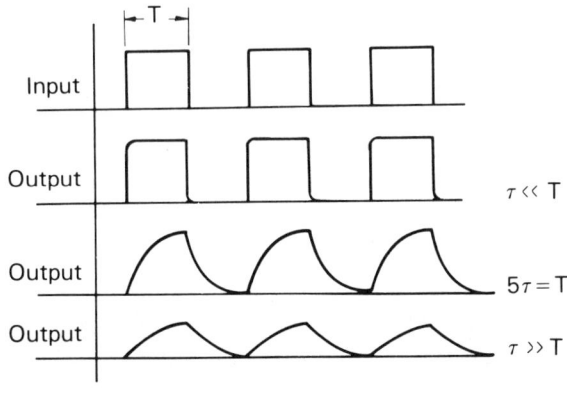

FIGURE 14–14
The effect of different values of circuit time constants (τ) on a dc rectangular pulse input with pulse width of T seconds.

PROBLEMS Determine the time constants of the following series circuits:

Section 14-2
14-1 $R = 47$ kΩ, $C = 200$ pF.
14-2 $R = 500$ Ω, $C = 0.1$ μF.
14-3 $R = 500$ kΩ, $C = 100$ pF.
14-4 $R = 10$ Ω, $C = 20\ 000$ μF.
14-5 A time constant of 3 s is desired; determine the value of R when C is 500 μF.
14-6 A wire has a resistance of 10 mΩ/ft and a capacitance of 50 pF/ft. If the wire is 50 ft long, determine the time constant of the wire.
14-7 With the switch closed, determine the time constant of the circuit shown in Figure 14-15 if $R_1 = 500$ Ω, $R_2 = 2$ kΩ, and $C = 200$ μF.

Section 14-3
14-8 Determine the voltage across R and C of Figure 14-16 at $t = 0^+$ if the switch moves from position 1 to position 2 at $t = 0$. Assume the dc source, E, is 20 V.
14-9 After 10 time constants in position 2 of Problem 14-8, the switch is changed to position 1. Determine the voltages across R and C the instant after the change. Indicate the direction of current flow immediately after the switch is changed from 2 to 1.
14-10 With the switch of Figure 14-16 in position 2, consider the source to be a pulse generator that produces a periodic dc rectangular pulse with equal on and equal off times (see Table 6-2). The pulse has a positive amplitude of 20 V peak and a period for one cycle of 10 τ of the RC components in the circuit of Figure 14-16. Draw the

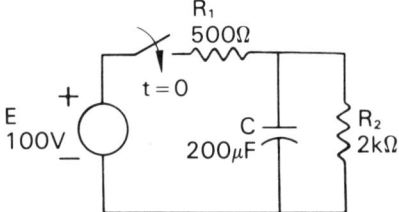
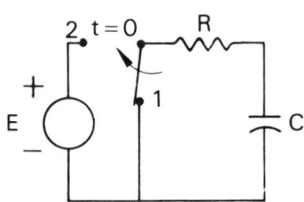

FIGURE 14-15 (left) Circuit of Problems 14-7, 14-11, 14-17, and 14-18.

FIGURE 14-16 (right) Circuit of Problems 14-8, 14-9, and 14-10.

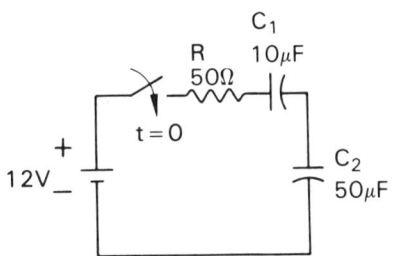

FIGURE 14-17
Circuit of Problem
14-12.

wave form of v_C for three cycles of the source voltage. Place time on the horizontal axis and v_C on the vertical axis.

14-11 With the switch closed at $t = 0$, determine the initial ($t = 0^+$) and final voltages ($t = \infty$) across R_1, R_2, and C in Figure 14-15.

14-12 Determine the voltages across each component at $t = 0^+$ and $t = \infty$ for the circuit of Figure 14-17.

Section 14-4

14-13 A 10-V battery is connected to a series combination of a 5-kΩ resistor and a 2-μF capacitor at time $t = 0$. Draw the wave shape showing the initial voltage and the voltages at times $t = 10$ ms, 20 ms, and 50 ms across the resistor and the capacitor. Place time on the horizontal axis and voltage on the vertical axis.

14-14 A 500-μF capacitor that has been charged to 25 V is connected across a 40-kΩ resistor. Draw and label the wave shape with the voltage at $t = 0$, 1τ, 3τ, and 5τ.

14-15 Determine the resistor value needed to be connected in series with a 10-μF capacitor to achieve 8 V across the capacitor 50 ms after the circuit is connected to a 10-V source.

14-16 Draw and label the voltage wave forms of v_1 and v_2 for the time from $t = 0$ to $t = 5\tau$ of Problem 14-11.

14-17 Draw and label the current wave forms of i_1, i_2, and i_C for the circuit in Figure 14-15 from $t = 0$ to $t = 6\tau$.

14-18 With the switch closed and the variables in the circuit of Figure 14-15 having reached their final values, the source is then shorted. Draw and label the voltage and current wave forms in the circuit from the time the source is shorted until all voltages and currents reach their new steady-state conditions.

14-19 Determine the voltage drop across C of Figure 14-2 at the following times after the switch is closed: (a) 1 ms, (b) 5 ms, (c) 10 ms, (d) 25 ms, and (e) 1 s.

14-20 Determine the current flowing through C of Figure 14-2 at the same time intervals of Problem 14-19.

14-21 The component Q in Figure 14-18 appears to be an open circuit until the voltage across it reaches 7 V. It then "fires" and appears as a short circuit and discharges the capacitor to zero volts almost instantaneously. It then acts like an open circuit again and the action repeats itself. Determine the value for R so that the circuit fires every 5 s.

14-22 Draw and label the wave shape of time (horizontal axis) versus the voltage (vertical axis) across the capacitor for three "firings" of the circuit of Problem 14-21.

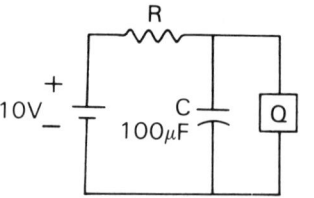

FIGURE 14-18
Circuit of Problems
14-21 and 14-22.

15 Magnetism

15-1 MAGNETS
15-2 MAGNETIC FIELD CHARACTERISTICS
15-3 SELECTED MAGNETIC QUANTITIES
15-4 MAGNETIC PROPERTIES OF MATERIALS

CHAPTER PREVIEW In handling permanent magnets, you may have observed that some magnets are stronger than others or that a magnet will attract objects made of iron and steel while not attracting glass or plastic. Perhaps you have noticed that a weak magnet could be formed when a needle or other small steel object was stroked with a permanent magnet. In this chapter, we will explore some of the principles behind these observations. We will also be naming and giving quantity to the properties of magnetism.

PERFORMANCE OBJECTIVES Once you have read each section, worked each example with pencil, paper, and calculator, and answered each problem for every section, you will be able to:

☐ Apply the appropriate SI units to magnetic quantities.
☐ Understand the characteristics of the magnetic field.
☐ Make calculations with magnetic quantities.
☐ Relate how magnetic materials are magnetized.
☐ Understand the meaning of a magnetic hysteresis loop.

15–1 MAGNETS

Natural Magnets Historians have evidence that ancient civilizations had knowledge of *natural* magnets. Natural magnets were suspended by a thread and used for navigation. This weak natural magnet, called magnetite (an oxide of iron), would align itself in a north-south direction. This device was the forerunner of the modern *compass*. Even today, natural magnets are called lodestones, meaning leading stone or compass.

We know that the end of the lodestone pointing in a northerly direction is the north-seeking end of the magnet, which is called the *north pole* (N). The other end is the south-seeking end of the magnet, which is called the *south pole* (S).

Artificial Magnets Artificial magnets (called permanent magnets) are manufactured from alloys of iron and are much stronger than natural magnets. Like natural magnets, permanent magnets attract iron and other *magnetic* materials such as nickel and cobalt. Besides being able to attract magnetic materials, the poles of a permanent magnet can also attract and repel other magnetic poles. This ability of magnetic poles to both attract and repel is the basis for operation of electric motors, loudspeakers, indicating meters, and magnetic clutches. From observing the interaction of permanent magnets, the following is determined:

> Like magnetic poles repel.
> Unlike magnetic poles attract.

15-2 MAGNETIC FIELD CHARACTERISTICS

The magnetic field, like the electric field, is a field force. A magnetic field around the poles of a magnet has a sphere of influence in which it acts upon magnetic materials with a force of attraction or repulsion. The magnetic force diminishes by the square of the distance from the pole of a magnet.

Magnetic Lines of Force

The invisible magnetic field may be represented by *magnetic lines of force* enabling us to form a visual image of the magnetic field. As noted in Figure 15-1, the lines of force move in a definite direction, and they form a closed loop by moving out of the north pole, returning to the south pole, and then passing through the magnet. As pictured, the magnetic lines of force surrounding the magnet tend to fan out when they pass through nonmagnetic material such as air, paper, and copper. However, the magnetic lines of force are concentrated in magnetic materials, as indicated in Figure 15-2. The lines of force have a repulsive action between them when they are moving in the same direction. It is this repulsive force that will not permit magnetic lines of force to cross over one another.

Magnetic lines of force will take the most direct path and the path of least opposition between the north and south poles of a magnet. In Figure 15-2, we noted the lines of force were pictured passing through the steel watch case, rather than in the air surrounding the case. Since magnetic lines of force always take the path of least opposition, it is reasonable to assume then that magnetic materials, like iron, present less opposition to lines of force than do nonmagnetic materials, like air. The opposition to the flow of magnetic lines of force is called *reluctance*. For all practical purposes, the reluctance of all nonmagnetic materials

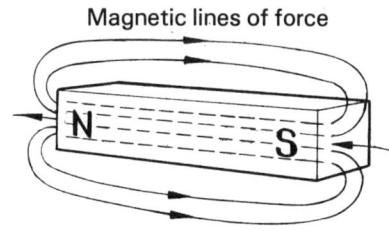

FIGURE 15-1 Magnetic lines of force form a complete loop by traveling through the material of the magnet, exiting into the air (or other medium) at the north pole, and reentering the magnet at the south pole.

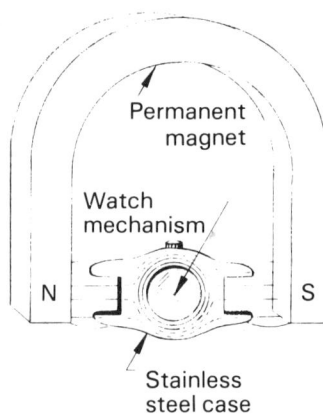

FIGURE 15-2
The stainless steel (an iron alloy) watch case conducts the magnetic field through it, protecting the watch mechanism from the effects of the magnetic lines of force.

is the same; thus, the movement of lines of force is the same through air, glass, plastic, and so on. Because iron and steel have a low reluctance, more lines of force will pass through this magnetic material than through a similar-sized nonmagnetic material.

Magnetic Induction When soft iron is placed in the magnetic field of a permanent magnet, it will assume a magnetic polarity, as noted in Figure 15-3. The soft iron is temporarily magnetized by *magnetic inductions*. The soft iron will have little or no magnetism when it is removed from the magnetic field. Soft iron has low *retentivity*. However, steel, as well as an alloy of iron, aluminum, nickel, and cobalt (called alnico), retain their magnetism because of their high retentivity. Materials with high retentivity are used for permanent magnets.

Summary
- ☐ A magnet always has two poles.
- ☐ Unlike poles attract; like poles repel.
- ☐ Each magnetic line of force forms a separate closed loop.
- ☐ Magnetic lines of force move from the north pole to the south pole.
- ☐ Magnetic lines of force moving in the same direction repel each other.
- ☐ Magnetic lines of force do not cross over one another.

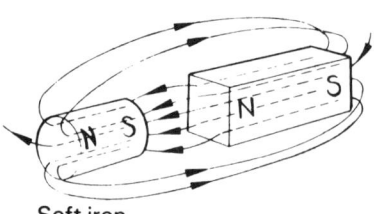

FIGURE 15-3
Soft iron has become a temporary magnet by a process called magnetic induction.

Selected Magnetic Quantities

- ☐ Magnetic lines of force always travel the shortest distance possible.
- ☐ Magnetic materials have a lower reluctance than nonmagnetic materials.

15-3 SELECTED MAGNETIC QUANTITIES

Magnetic Flux (Φ)

The lines of force of a magnetic field taken as a whole are called the *magnetic flux*. Magnetic flux is symbolized with the Greek letter phi, Φ, and the SI unit for magnetic flux is the weber (Wb).

Flux Density (B)

The force of the magnetic field is directly proportional to the concentration of the flux. That is, magnetic fields are stronger when they occupy small areas. The number of magnetic lines of flux per unit area is an indication of the flux density. The flux density of a magnetic field is symbolized with the letter B, and the SI unit for magnetic flux density is the tesla (T). Since flux density is flux per unit area, the following mathematical statement may be made.

$$B = \frac{\Phi}{A} \quad (T) \tag{15-1}$$

where B = magnetic flux density in teslas (T)
Φ = magnetic flux in webers (Wb)
A = area normal to the magnetic flux in square meters (m²)

From Equation 15-1 we learn the following:

---
1 tesla = 1 weber/square meter

EXAMPLE 15-1 Determine the flux density of the watch case pictured in Figure 15-2 if the flux in the case is 20 μWb. Assume the average cross section to be 1 cm² (1 cm² = 1 × 10⁻⁴ m²).

SOLUTION Use Equation 15-1.

$$B = \frac{\Phi}{A}$$

$$B = \frac{20 \times 10^{-6}}{1 \times 10^{-4}} = 0.2 \text{ Wb/m}^2$$

$$B = 0.2 \text{ Wb/m}^2 = 0.2 \text{ T}$$

∴ The flux density is 0.2 tesla.

432

Magnetism

As a point of reference for your understanding of flux density, the density of the magnetic field at the poles of a small toy magnet is approximately 50 μWb/m² (50 μT), while the lodestone used by ancient navigators had a flux density at the poles on the order of 10 mWb/m² (10 mT). Some permanent magnets made of special alloys have flux density of 1 Wb/m² (1 T) or more.

Magnetomotive Force ($\mathscr{F}$)

In Figure 15-3, we see the magnetic field represented by magnetic lines of force that form closed loops around and through both the magnetic materials and the nonmagnetic materials (air). The path traced by these loops is a *magnetic circuit*.

To establish magnetic flux in a magnetic circuit, a *magnetomotive force* (mmf) must be applied. The magnetomotive force results when current travels through a coil of wire. The magnetomotive force of a magnetic circuit is symbolized with the script letter $\mathscr{F}$, and the SI unit for magnetomotive force is the ampere (A).

The magnetomotive force increases by 1 ampere for every complete turn the current traverses. Therefore, magnetomotive force is multivalued since it increases for each turn traveled by the current. This concept is unlike its counterpart, electromotive force, in the electric circuit, which is a fixed voltage. Thus, a one-turn coil with 10 A has a mmf of 10 A while a four-turn coil with 10 A has an mmf of 40 A. Stated mathematically,

$$\mathscr{F} = NI \quad (A) \tag{15-2}$$

where $\mathscr{F}$ = magnetomotive force (mmf) in amperes (A)
N = number of turns traversed by the current in the coil (no units)
I = current in the coil in amperes (A)

EXAMPLE 15-2 Determine the magnetomotive force of the coil pictured in Figure 15-4.

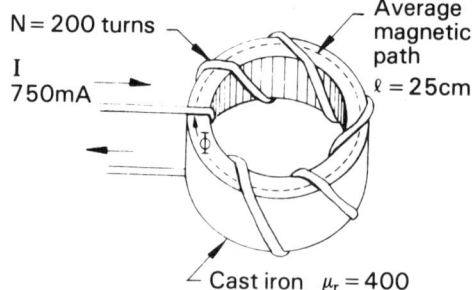

FIGURE 15-4
Coil of Example 15-2 and cast iron ring of Example 15-5.

Selected Magnetic Quantities

SOLUTION Use Equation 15-2.

$$\mathscr{F} = NI$$

$$\mathscr{F} = (200)(0.75) = 150 \text{ A}$$

∴ The magnetomotive force is 150 A.

Magnetic Field Strength (H)   Magnetic field strength takes into consideration the length of the path taken by the magnetic flux in forming a closed loop. In Figure 15-5, there are two magnets made in the shape of a ring. Each ring has the same cross-sectional area, and each is initially magnetized with the same magnetomotive force. The only difference between them is the average length of the magnetic path, as noted in Figure 15-5. Because the flux in the 5-cm-diameter magnet travels a shorter distance when compared to the distance in the 10-cm-diameter magnet, the magnetic field strength (H) is greater in the shorter path. This concept is verified in the mathematical statement for magnetic field strength:

$$H = \frac{NI}{\ell} \quad \text{(A/m)} \tag{15-3}$$

where H = magnetic field strength of the magnetic path in amperes per meter of path length (A/m)
NI = magnetomotive force in amperes (A)
ℓ = average magnetic path length in meters (m)

EXAMPLE 15-3   Determine the magnetic field strength for each of the ring magnets if each was originally magnetized by winding 25 turns around the ring and pasing 4 A through the turns.

SOLUTION (a) For Figure 15-5(a), $\ell_1 = 0.157$ m:

$$H = \frac{NI}{\ell}$$

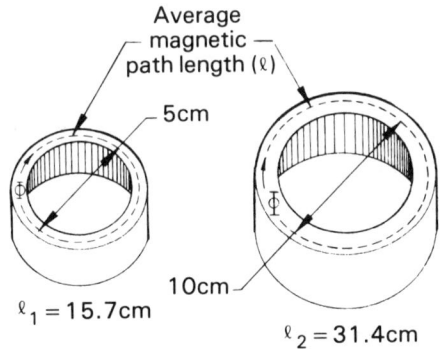

(a)   (b)

FIGURE 15-5
Two ring magnets have equal cross-sectional area and are magnetized with equal magnetomotive force. However, because the path lengths are different, the magnetic field strength, H, will differ as will the magnetic flux density, B (B ∝ H).

$$H = \frac{25(4)}{0.157} = 637 \text{ A/m}$$

∴ The field strength for the 5-cm-diameter magnet is 637 A/m.

(b) For Figure 15-5(b), $\ell_2 = 0.314$ m:

$$H = \frac{NI}{\ell}$$

$$H = \frac{25(4)}{0.314} = 318 \text{ A/m}$$

∴ The field strength for the 10-cm-diameter magnet is 318 A/m.

Observation All parameters being equal except length, the shorter the magnet path, the greater the magnetic field strength.

Permeability (μ) We have just learned that the magnetic field strength (H) varies inversely with length for a constant magnetomotive force. However, when field strength varies, the flux density (B) also varies.

If we look at the two magnetic paths of Figure 15-5 as magnetic circuits, then we can compare the application of an mmf (magnetomotive force) in a magnetic circuit to the application of an emf (voltage) to an electric circuit. From our understanding of the electric circuit, we know that, when equal voltages are applied to similar wires, more current is in a shorter conductor than is in a longer conductor. Thus, when an equal magnetomotive force (mmf) is applied to similar magnetic paths, the shorter path will have the greater flux density (B) and the longer magnetic path will have a lesser flux density. From the preceding discussion, it is learned that the magnetic flux density (B) is directly proportional to the magnetic field strength. Thus

$$B \propto H$$

and

$$B = \mu H \quad (\text{T}) \tag{15-4}$$

Solving for μ,

$$\mu = \frac{B}{H} \quad \left(\frac{\text{Wb}}{\text{A} \cdot \text{m}} = \frac{\text{T} \cdot \text{m}}{\text{A}}\right) \tag{15-5}$$

where μ = permeability of the material in the magnetic path in Wb/(A · m)
B = magnetic flux density in T
H = magnetic field strength in A/m

Selected Magnetic Quantities

The permeability of a substance is an indication of its ability to become magnetized when acted on by a magnetomotive force.

For a given field strength, the magnetic flux density, B (Equation 15–4), will change by a factor, μ, which is related to the permeability of free space (air), μ_0, by a unitless number called the relative permeability, μ_r. The relative permeability for nonmagnetic materials (glass, air, paper, etc.) is 1 ($\mu_r = 1$), and for magnetic materials (iron, steel, nickel, etc.) the relative permeability is usually greater than 100 ($\mu_r > 100$). Thus, the permeability of a substance (μ) is related to the permeability of free space, μ_0, in the following manner:

$$\mu = \mu_r \mu_0 \tag{15-6}$$

where μ = permeability of the material in the magnetic path in Wb/(A · m)

μ_r = relative permeability, which depends on the magnetization curve of the particular substance (no units)

μ_0 = permeability of free space where $\mu_0 = 4\pi \times 10^{-7}$ Wb/(A · m)

EXAMPLE 15–4 Determine the magnetic flux density in each of the average magnetic paths of the ring magnets of Figure 15–5. The relative permeability (μ_r) of the material is 390. Use the magnetic field strengths previously calculated in Example 15–3.

SOLUTION (a) For $\ell_1 = 0.157$ m and $H = 637$ A/m of Figure 15–5(a), solve for μ. Thus

$$\mu = \mu_r \mu_0$$
$$\mu = (390)(4\pi \times 10^{-7}) = 4.9 \times 10^{-4}$$

Solve for the flux density, B:

$$B = \mu H$$
$$B = (4.9 \times 10^{-4})(637) = 0.312 \text{ T}$$

∴ The flux density is 0.312 T for the 5-cm-diameter ring.

(b) For $\ell_2 = 0.314$ m and $H = 318$ A/m of Figure 15–5(b), solve for B. Thus

$$B = \mu H$$
$$B = (4.9 \times 10^{-4})(318) = 0.156 \text{ T}$$

Observation The flux density is 0.156 T for the 10-cm-diameter ring.

The flux density in a magnetic circuit (path) depends directly on the mmf and the type of material in the path (air, iron, etc.) and inversely on the length of the path (shorter path length, greater flux densities). To summarize:

$$B = \mu \frac{\mathscr{F}}{\ell} \quad (T) \tag{15-7}$$

EXAMPLE 15-5 Determine the magnetic flux density (B) in the cast-iron ring being magnetized, as pictured in Figure 15-4.

SOLUTION Determine the magnetomotive force:

$$\mathscr{F} = NI = 200 \times 0.75 = 150 \text{ A}$$

Determine the permeability (μ) of the cast iron:

$$\mu = \mu_r \mu_0 = (400)(4\pi \times 10^{-7}) = 5 \times 10^{-4}$$

Determine the magnetic flux density:

$$B = \frac{\mu \mathscr{F}}{\ell}$$

$$B = \frac{(5 \times 10^{-4})(150)}{0.25} = 0.3 \text{ T}$$

∴ The magnetic flux density is 0.3 T.

15-4 MAGNETIC PROPERTIES OF MATERIALS

Atomic Theory of Magnetism

To understand why certain materials can be magnetized, we once again turn to the inner structure of the atom. In Chapter 2 we introduced the idea of potential energy levels within the atom and that predictable numbers of electrons resided in these levels. We also noted in Figure 2-5 that electrons not only orbit around the nucleus, they also spin around their own axis. It is theorized that the magnetic property of magnetic materials is largely due to the *spin* of the electrons. Apparently, each electron acts like a magnet as it orbits around the nucleus. Each electron within the atom spins either clockwise, called *spin up* (north pole up), or counterclockwise, called *spin down* (north pole down). In the atoms of nonmagnetic materials, the electrons occur in pairs having one spin-up and one spin-down electron. Because of this pairing, the magnetic effect is canceled and the substance is nonmagnetic and has a relative permeability of 1.

Magnetic Properties of Materials

In magnetic materials, the relative permeability is much greater than 1, typically greater than 100. These materials are called *ferromagnetic* materials, and they include iron (Fe), cobalt (Co), nickel (Ni), and alloys of these materials. In the atoms of ferromagnetic materials, more electrons spin in one direction than in the other. The result of unpaired electrons is a large magnetization within ferromagnetic solids when acted on by an external magnetic field.

Consider the electron spin of the 26 electrons in an atom of iron. The K shell has 2 electrons with one spin up and one spin down, the L shell has 8 electrons with four spin up and four spin down, the M shell has 14 electrons with nine spin up and five spin down, and the N shell has the 2 valence electrons with one spin up and one spin down. The iron atom has four unpaired electrons in the M shell and, therefore, has ferromagnetic properties. From this discussion, you might conclude that all ferromagnetic materials are permanent magnets; this, of course, is not the case. The reason is that the atoms of ferromagnetic materials interact with one another, causing neighboring atoms to line up parallel to each other in a manner such that their magnetic fields aid each other. This group of atoms is called a *domain*. The ferromagnetic material contains many, many domains, each being a very small magnet. In unmagnetized ferromagnetic material, these domains have the direction of their magnetic field oriented in a random pattern. Consequently, there is no external magnetic field.

Magnetization of Ferromagnetic Materials

The curve of Figure 15–6 is the magnetization curve for a ferromagnetic material. The characteristic shape of the magnetization curve is explained by considering the behavior of the domains of the ferromagnetic material when a magnetic field is applied. An unmagnetized ferromagnetic solid, Figure 15–7(a), has a net magnetic density (B) of zero teslas, as indicated by point 0 on the magnetization curve. With the applicaton of an external magnetic field to the ferromagnetic material, the flux den-

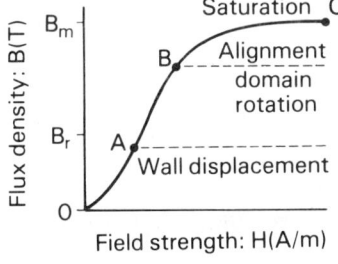

FIGURE 15–6
Typical magnetization curve for a ferromagnetic material. All magnetic materials have nonlinear magnetization curves.

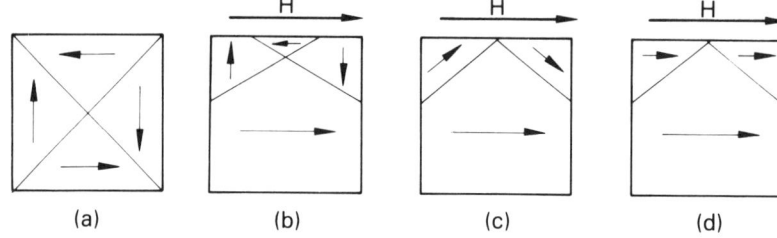

FIGURE 15-7
Domain growth. (a) Unmagnetized ferromagnetic material. Note: arrows indicate direction of domain magnetization and magnetic flux density, B. (b) Domains parallel to the applied field strength, H, increase in size as the domain walls are displaced. (c) Domains rotate as H is increased. (d) Domains are aligned parallel to the direction of H and the material is saturated.

sity (B) within the solid material increases as the applied field strength (H) increases. For small values of H (point A of the magnetization curve), the domain walls are displaced as noted in Figure 15-7(b). If the applied magnetic strength, H, is removed at this point, the walls move back, and the flux density would once again return to zero. H is increased from point A to point B on the magnetization curve; a value of H is reached where the direction of the domain magnetization rotates, as pictured in Figure 15-7(c). At this point, the change becomes irreversible. Some flux density (B_r) will now be present when the applied magnetic field strength, H, is removed. At point C of the magnetization curve, all the domains have rotated and are aligned parallel to H, as indicated in Figure 15-7(d). The material is fully magnetized and it is said to be *saturated*. Increasing H does not increase B beyond the maximum flux density B_m.

Hysteresis Removing the external field strength, H, results in many of the domains remaining aligned with one another. Because of this alignment, some *residual magnetism*, or *remanence* (B_r), remains in the material.

When the applied field strength, H, is removed, the demagnetization curve is not curve C-0 of Figure 15-6, but is instead the curve of Figure 15-8.

Starting at point C and reducing the applied field H to zero amperes per meter results in the movement of the flux density from B_m to B_r at point D. To reduce B_r to zero, the magnetic field strength, H, of the applied field must be reversed to a value equal to $-H_c$ (point E), which is the *coercive force*. Increasing the value of the reverse magnetic field strength, $-H$, results in the ferromagnetic material becoming fully magnetized (point F) in the opposite direction.

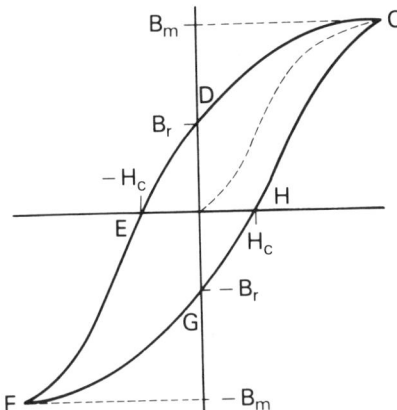

FIGURE 15-8
Hysteresis loop.

Continuing on around the curve from F to G, the value of $-H$ is reduced to zero at G, where the material has a residual magnetism, $-B_r$. At a value equal to H_c, the material is once again demagnetized (point H). From points H to C, the applied field strength, H, increases until at point C the material is once again saturated with a flux density equal to B_m.

From Figure 15-8, it is obvious that the curve of the magnetization does not follow the curve of the demagnetization of the ferromagnetic material; the magnetic flux density (B) lags the applied magnetic field strength (H). This lag is called *hysteresis*, and the loop is called a *hysteresis loop*.

Ferromagnetic materials having large hysteresis loops (Figure 15-9) are *magnetically hard* materials. These materials are used to make permanent magnets (alnico), memory cores (manganese-zinc ferrite), and ceramic permanent magnets (barium ferrites). They are characterized by a large coercive force ($-H_c$) and a remanence ratio (B_r/B_m) nearly equal to unity.

Ferromagnetic materials having small hysteresis loops (Figure 15-9) are *magnetically soft* materials. These materials are used as cores in transformers (mumetal) and as adjustable cores in rf communication coils (ferrite). They are characterized by very small coercive force ($-H_c$).

PROBLEMS

Section 15-1

15-1 Relate the similarities and differences between natural and manufactured permanent magnets.
15-2 List several magnetic materials.
15-3 List several nonmagnetic materials.
15-4 Name some devices that use permanent magnets for their operation.

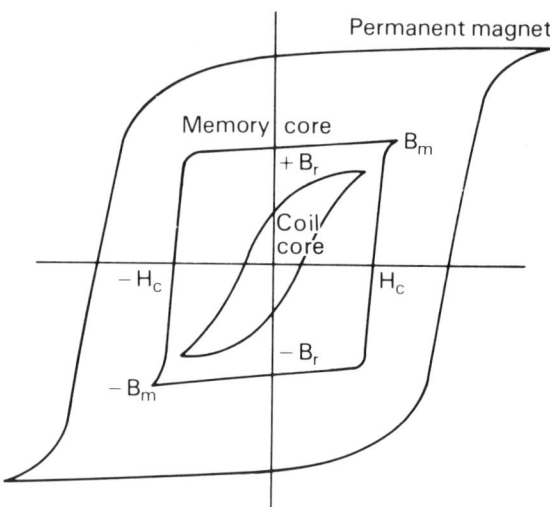

FIGURE 15-9
Hysteresis loops of various ferrite ferromagnetic materials.

Section 15-2 15-5 Indicate which statements are true. Rewrite the statements that are false so that they will be true.
(a) Magnetic lines of force have a definite direction from the south pole to the north pole.
(b) The field force of a magnet is independent of distance.
(c) Unlike poles repel each other.
(d) Magnetic lines of force can be seen with an electron microscope.
(e) Nonmagnetic materials have a greater reluctance than magnetic materials.
(f) Magnetic lines of force start at the north pole and end at the south pole of a permanent magnet.

Section 15-3 15-6 Match one item from the list at the right to one item at the left. Some items in the right list may be used more than once.

(1) Flux
(2) H
(3) Wb/m²
(4) NI/ℓ
(5) Wb/(A · m)
(6) NI
(7) B
(8) B/H
(9) A
(10) μ

(a) Flux density
(b) mmf
(c) T · m/Wb
(d) $\mathscr{F}$
(e) Φ
(f) $\mu_r \mu_0$
(g) T
(h) Field strength
(i) Wb
(j) A/m

Problems

15-7 Determine the magnetic flux in an iron bar having a rectangular cross section of 1.5 cm by 3 cm with a flux density of 200 mT.

15-8 Determine the magnetic flux in a round permanent magnet having a diameter of 1.2 cm ($A = \pi d^2/4$) with a flux density of 0.8 T.

15-9 Determine the magnetic flux density at the poles of a horseshoe magnet if the magnet produces a flux of 12 mWb and the cross-sectional area is 3 cm².

15-10 Determine the magnetic flux density in the soft iron, 2-cm-diameter cylinder of Figure 15-3 if the flux in it is 600 μWb.

15-11 Determine the magnetomotive force of the coil pictured in Figure 15-4 when the current is increasd to 1.2 A.

15-12 Determine the magnetomotive force of the coil pictured in Figure 15-4 when the number of turns is increased to 300.

15-13 Determine the magnetic field strength in a magnetic cicuit 20 cm long when the mmf is 60 A.

15-14 Determine the magnetic field strength in a magnetic circuit 16 cm long when the mmf is 5 A.

15-15 Determine the current in an 80-turn coil if the field strength of a 27-cm magnetic cicuit is 150 A/m.

15-16 Determine the number of turns needed to produce a magnetic field strength of 400 A/m in a 36-cm magnetic circuit if the current in the coil around the core is 600 mA.

15-17 Determine the relative permeability of air if its permeability (μ) is 1.26 μWb/(A · m).

15-18 Determine the permeability of an alloy of iron (steel) with a relative permeability of 28 000.

15-19 Determine the permeability of the soft iron cylinder of Figure 15-3 if the field strength is 200 A/m and the flux density is 1 T.

15-20 Determine the relative permeability of the soft iron cylinder of Problem 15-19.

15-21 Determine the magnetic flux density in a 12-cm long solenoid wound on a cardboard core when 80 turns of wire are wound around the cardboard core and a current of 0.5 A is in the coil.

15-22 Determine the magnetic flux density in an 8-cm-long iron core when 40 turns of wire are wound around the iron core ($\mu_r = 6000$) and a current of 100 mA is in the coil.

Section 15-4

15-23 Two ferromagnetic materials are being considered for permanent magnets. The first, a ferrite material, has a saturation flux density (B_m) of 0.5 T and a remanent flux density (B_r) of 0.45 T. The second, an iron alloy material, has a saturation flux density (B_m) of 0.8 T and a remanent flux density (B_r) of 0.4 T. (a) Determine the remanence ratio for each, and (b) select one of the materials to be the permanent magnet.

NOTE: Refer to the magnetization curves of Figure 16–7 to solve the following problems.

15–24 Approximate the magnetic flux density in transformer steel for an applied field strength of 200 A/m.

15–25 Determine μ and μ_r for cast iron when $H = 1200$ A/m.

15–26 Construct a curve of μ_r (vertical axis) versus H (horizontal axis) for mild steel. Comment on the statement, "permeability is nonlinear for ferromagnetic materials."

16 Electromagnetism and the Magnetic Circuit

16-1 ELECTROMAGNETISM
16-2 SOLENOID
16-3 SERIES MAGNETIC CIRCUITS
16-4 AIR GAP IN THE MAGNETIC CIRCUIT
16-5 SYMMETRICAL PARALLEL MAGNETIC CIRCUITS
16-6 ADDITIONAL MAGNETIC UNITS

CHAPTER PREVIEW In this chapter, concepts learned in the previous chapter are used to solve magnetic circuits. Magnetic circuits, which are mainly constructed of iron and other ferromagnetic materials, are analogous to electric circuits. This similarity will be apparent in the techniques used to solve magnetic circuits.

This chapter will round out your knowledge of magnetism by providing you with information on electromagnetism, the solenoid, and the solution of magnetic circuits.

PERFORMANCE OBJECTIVES Once you have read each section, worked each example with pencil, paper, and calculator, and answered each problem for every section, you will be able to:

☐ Understand the role that moving charge (electric current) plays in electromagnetism.
☐ Apply the right-hand rule.
☐ Relate how the interaction of magnetic fields produces motion in such devices as electric motors.
☐ Make computations with the solenoid.
☐ Solve series and parallel magnetic circuits.
☐ Understand the need for an air gap in a magnetic circuit.
☐ Understand magnetic units other than the SI units.

16–1 ELECTROMAGNETISM

A magnetic field results when an electric charge is moved. Because an electric current is the movement of electric charge, a magnetic field is set up around all conductors in which current is present. This discovery was made in 1819 by a Danish physicist, Hans Oersted, when he placed a compass next to a current-carrying conductor. The compass was deflected from its usual north-south alignment. The direction of the magnetic field around a current-carrying conductor is defined by the right-hand rule.

Right-hand Rule The magnetic flux around a current-carrying conductor has an assigned direction, whicn is determined by the *right-hand rule*. To apply the rule, hold the conductor in the right hand, as shown in Figure 16–1, with the thumb pointing in the direction of the conventional current; the curled fingers indicate the direction of the magnetic flux. Like the electric field, the magnetic field occupies the entire space around the conductor.

Force on Current-carrying Conductors Figure 16–2 pictures a single current-carrying conductor positioned normal (at right angles) to the page with the current indicated by the dot or tip of the current arrow coming out of the

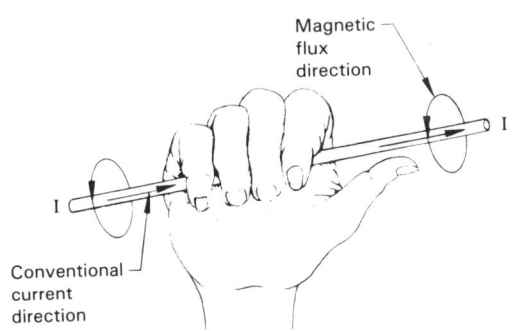

FIGURE 16-1
Right-hand rule being used to determine the direction of the magnetic field around a current-carrying conductor.

page. The direction of flux around the conductor, by the right-hand rule, is ccw as indicated.

The external magnetic field, from the poles of the permanent magnet, combines with the flux from the current-carrying conductor. Above the conductor, the field directions are opposing, which results in a reduction of the flux density in this region. Below the conductor, the field directions are aiding, which results in an increase of the flux density in this region. The magnetic force, as pictured, is directed away from the region of increased flux density and toward the region of lessened flux density. The conductor would, in this case, be thrust upward in the direction indicated by the bold arrow.

For purposes of discussion, it is convenient to imagine that the magnetic flux encircling the current-carrying conductor is an inseparable part of the conductor. Thus, two parallel conductors will be attracted or repelled depending on the direction of the currents in the conductors. Parallel conductors with currents in the opposite direction will be pushed apart by the repelling force of like magnetic fields, whereas parallel conductors with currents in the same direction will be pulled together by the attracting force of unlike magnetic fields.

The force acting on a wire in a magnetic field is given by Ampere's law, which relates electrical and mechanical phenomena. The law states that the force (F), in newtons, on a current-carrying conductor is equal to the product of current (I) in am-

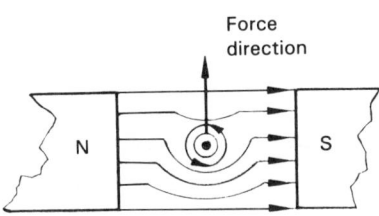

FIGURE 16-2
The flux of the current-carrying conductor increases the flux density below the conductor and decreases the flux density above the conductor. Note: the current in the conductor is moving out of the page.

peres, flux density (B) in teslas, and conductor length (ℓ) in meters. Thus

$$F = IB\ell \quad (\text{N}) \tag{16-1}$$

It is this law that allows the definition of the ampere as a base unit of the SI system. Recall from Chapter 3 that the ampere was defined as the amount of constant current required to produce 2×10^{-7} newtons of force per meter of length in two parallel conductors in a vacuum separated from each other by 1 meter.

Torque on a Current Loop

The rotation of a motor or the deflection of a moving coil (D'Arsonval) meter is a result of the torque or twisting motion that results when a loop of wire carrying current is placed into a magnetic field. Figure 16-3(a) pictures a top view of a loop of wire in a magnetic field. Notice that the parallel conductor segments carry opposing current. Investigation of the forces on the loop [Figure 16-3(b)] reveals that equal and opposite forces are set up, causing the loop to align perpendicular to the magnetic flux. Thus, the loop will be rotated from a horizontal position between the fixed magnetic poles to a vertical position. Again, we see that the magnetic force (F) is directed toward the region where the magnetic flux density is at its smallest value. It is this principle that explains the operation of the electric motor.

The torque (τ) in newton meters, resulting from a coil with N number of turns, positioned horizontally in the flux of a constant magnetic field, is equal to the product of the current in the coil (I) in amperes, the flux density (B) in teslas of the constant magnetic field, and the area (A) in square meters of the current-carrying loop. Thus

$$\tau = NIBA \quad (\text{N·m}) \tag{16-2}$$

EXAMPLE 16-1 Determine the torque on a 100-turn coil positioned horizontally between the magnetic poles of a permanent magnet with a flux density of 2 T. The current in the coil is 500 mA, and the loop is 8 cm long and 4 cm wide (area = $0.08 \times 0.04 = 3.2 \times 10^{-3}$ m²).

FIGURE 16-3
(a) Current-carrying loop in a constant magnetic field, as viewed from the top.
(b) The same loop viewed head on. Notice the conductor with a dot indicating the tip of the current arrow and the conductor with a cross indicating the tail of the current arrow.

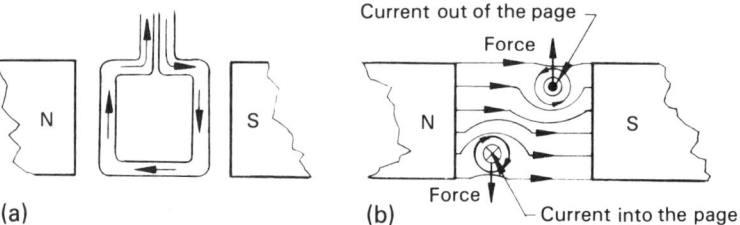

SOLUTION Use Equation 16-2:

$$\tau = NIBA$$

$$\tau = 100(0.5)(2)(3.2 \times 10^{-3}) = 0.32 \text{ N·m}$$

∴ The magnetic torque is 0.32 N·m.

16-2 SOLENOID

Electromagnet

When a conductor carrying current is formed into a series of coils, the magnetic field of each loop links with its neighbor to form an electromagnet, which, like the bar magnet, has an internal flux coming out of the north pole and returning to the south pole, as pictured in Figure 16-4. The direction of the magnetic flux, as well as the direction of the north pole, is determined by the right-hand rule for an electromagnet. The rule is applied as follows. Position the coil in the palm of the right hand so that the fingers curl in the same direction as the direction of the current in the turns of wire. The thumb will then point in the direction of the flux and the north pole of the electromagnet.

Flux Density in the Solenoid

The coil pictured in Figure 16-4 is classed as a *solenoid* because its length is much greater than its diameter (length ≫ diameter). The flux density in the mid-section of the solenoid is uniform. However, toward the ends of the solenoid, the flux density is only about one-half as great as in the center due to the loss of magnetic lines of force along the sides of the solenoid. This loss is called *flux leakage*.

The flux density (B) in the center of the solenoid is equal to the product of the number of turns (N) in the coil, the current (I) in the coil, the permeability (μ) of the core material (material in the center of the coil), all divided by the length (ℓ), in meters, of the core material within the solenoid. Thus

$$B = \frac{NI\mu}{\ell} \quad \text{(T)} \tag{16-3}$$

EXAMPLE 16-2 Determine the flux density at the center and at the ends of an air-core solenoid of 280 turns

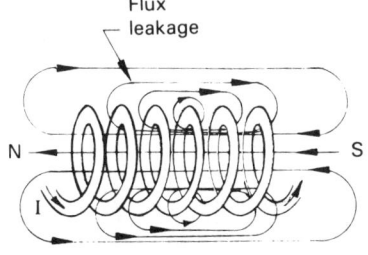

FIGURE 16-4
Current-carrying conductor of the coil or solenoid forms an electromagnet.

when the current in the coil is 300 mA. The solenoid (11 cm in length) is constructed on a nonmagnetic 1.2-cm-diameter coil form.

SOLUTION Determine the μ of the air core:

$$\mu = \mu_r\mu_0 = (1)(4\pi \times 10^{-7}) = 4\pi \times 10^{-7}$$

Substitute into Equation 16-3:

$$B = \frac{NI\mu}{\ell} = \frac{(280)(0.3)(4\pi \times 10^{-7})}{0.11} = 960 \ \mu T$$

∴ The flux density at the ends of the solenoid is approximately one-half that at the center or 960 $\mu T/2 \approx 480 \ \mu T$.

By placing an iron core into the solenoid, the flux density is increased. Furthermore, because the permeability of the iron core channels the flux, the flux leakage is, for practical purposes, eliminated, and the flux density is uniform at both the center and ends of the solenoid.

Toroid A solenoid wound on a ring core is a *toroid*. The toroid is constructed by winding wire around a doughnut-shaped core, producing a closed magnetic path, as pictured in Figure 16-5. The very uniform magnetic flux is almost totally contained within the core, with very little flux outside the toroid.

EXAMPLE 16-3 The toroid of Figure 16-5 has 420 turns wound on a ferrite core with a relative permeability of 1500. The core has an average radius of 2.5 cm and a rectangular cross section of 0.8 cm by 1.5 cm. Determine (a) the magnetomotive force ($\mathscr{F}$) for a coil current of 0.3 A, (b) the field strength (H), (c) the flux density (B), and (d) the flux (Φ) in the core.

SOLUTION Use Equations from Chapter 15:
(a) Solve for the mmf:

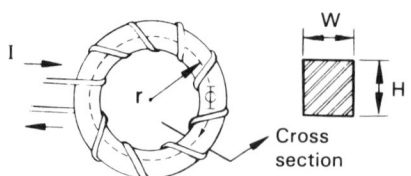

FIGURE 16-5
Toroid for Example 16-3.

$$\mathcal{F} = NI \quad (15-2)$$
$$\mathcal{F} = (420)(0.3) = 126 \text{ A}$$

∴ The magnetomotive force is 126 A.

(b) Solve for the field strength:

$$H = \frac{NI}{\ell} = \frac{\mathcal{F}}{\ell} \quad (15-3)$$

Compute the length of the magnetic circuit, which is the mean circumference of the ring core in meters.

$$l = C = \pi d = \pi 2r = \pi 2(2.5 \times 10^{-2})$$
$$\ell = 0.157 \text{ m}$$

$$H = \frac{\mathcal{F}}{\ell} = \frac{126}{0.157} = 803 \text{ A/m}$$

(c) Solve for the flux density:

$$B = \frac{NI\mu}{\ell} = \mu H \quad (15-5)$$

Compute the permeability of the core material:

$$\mu = \mu_r \mu_0 = (1500)(4\pi \times 10^{-7})$$
$$\mu = 1.89 \times 10^{-3}$$
$$B = \mu H = (1.89 \times 10^{-3})(803) = 1.52 \text{ T}$$

∴ The flux density throughout the toroid is 1.52 T.

(d) Solve for the flux:

$$\Phi = BA \quad (15-1)$$

Compute the cross-sectional area of the core in square meters when $A = $ width $\times$ height:

$$A = wh = (0.8 \times 10^{-2})(1.5 \times 10^{-2})$$
$$A = 120 \times 10^{-6} \text{ m}^2$$
$$\Phi = BA = (1.52)(120 \times 10^{-6}) = 182 \text{ }\mu\text{Wb}$$

∴ The flux in the core of the toroid is 182 μWb.

Solenoid Applications An electromagnetic device employing a coil and a movable iron core, as pictured in Figure 16-6, is called a solenoid (not to be confused with the coil of the same name). The principle of operation uses the electromagnetic properties of the current-carrying coil to translate electric energy into linear mechanical motion. The movable core, called the *plunger*, is spring loaded

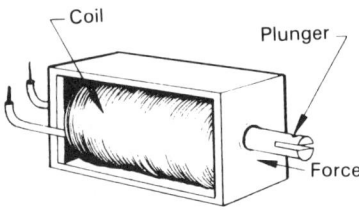

FIGURE 16-6
Electromagnetic device called a solenoid.

and offset from the center of the coil. The plunger moves inward and centers itself in the coil when current is passed through the coil.

Commercial solenoids are available with "strokes" (movement) of a few millimeters to 10 cm (a fraction of an inch to 4 inches). The force of the pull of the plunger is from 0.2 N to 500 N ($\approx$1 oz to 100 lb). Electromechanical actuators, like the solenoid, are widely used as contactors (a magnetically operated type of switch used with large currents to automatically make and break a frequently used circuit). Solenoids are also used to actuate valves, brakes, and machine tools.

16-3 SERIES MAGNETIC CIRCUITS

An understanding of the magnetic circuit and its characteristics is of importance because the underlying principle of the design and operation of many electromagnetic devices may be understood once a knowledge of the magnetic circuit is gained. Devices employing a magnetic circuit include inductors, relays, transformers, electric motors, electromagnets, meter movements, and loudspeakers.

Linear Magnetic Circuit

The linear magnetic circuit is solved in a manner much like the linear electric circuit. The *magnetic Ohm's law* is the counterpart to the familiar Ohm's law formula. Thus

$$\mathscr{F} = \Phi \mathscr{R} \quad \text{(A)} \tag{16-4}$$

where $\mathscr{F}$ = magnetomotive force (NI) in amperes
Φ = flux in webers
$\mathscr{R}$ = reluctance in amperes per weber

In the magnetic circuit, the magnetomotive force sets up magnetic lines of force, which are collectively called magnetic flux. The number of lines of force depends on the *reluctance* of the material in the magnetic circuit. The reluctance of the material ($\mathscr{R}$) to the establishment of magnetic lines depends upon the permeability of the material (μ) in the magnetic path, the cross-sectional area of the core material (A) in square meters, and the length (ℓ) of the magnetic circuit in meters. Thus

Series Magnetic Circuits

$$\mathcal{R} = \frac{\ell}{\mu A} \quad \text{(A/Wb)} \tag{16-5}$$

There is no standard derived unit for reluctance, thus the use of the compound unit of amperes per weber (A/Wb).

Equation 16-4, $\mathcal{F} = \Phi\mathcal{R}$, may be applied to an air-core solenoid because the permeability of air is constant and not dependent on the field strength. Since the permeability is constant, the reluctance will also be constant. However, when a ferromagnetic material (such as iron) is used as the magnetic path, then the permeability depends on the magnetization curve (see Figures 15-6 and 16-7), which relates the magnetic field strength to the flux density. Because of the nonlinear characteristic of this curve, the permeability (which is the function of field strength) will be nonlinear, resulting in the reluctance also being nonlinear. Thus, the equation $\mathcal{F} = \Phi\mathcal{R}$ may only be applied to linear (air-core) magnetic circuits. Nonlinear circuits are treated by applying the BH magnetization curve of the ferromagnetic material.

EXAMPLE 16-4 Determine the current needed to set up a flux of 1.09 μWb at the center of a 280 turn air-core solenoid. The solenoid, which is wound on a nonmagnetic coil form, is 15 cm long and 1.2 cm in diameter.

Observation Because the reluctance of the magnetic path within the solenoid is much greater than the re-

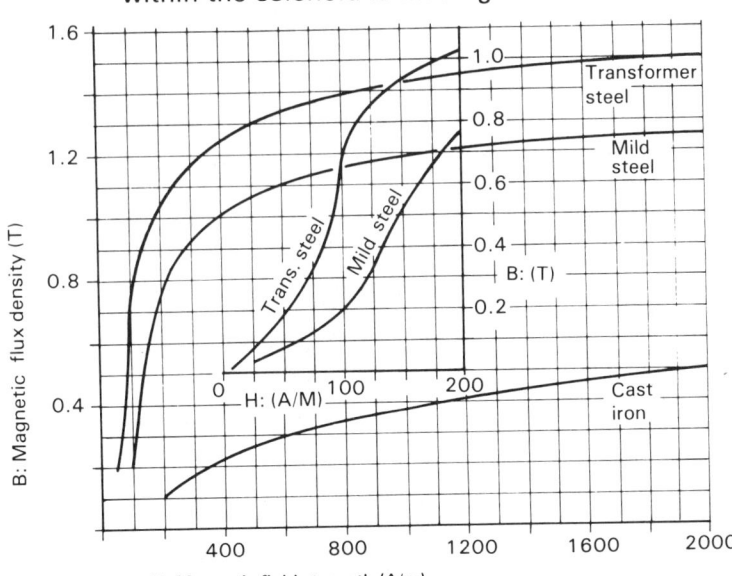

FIGURE 16-7
BH magnetization curves for selected ferromagnetic materials.

luctance of the magnetic path outside the solenoid, we may assume that the reluctance for the entire magnetic circuit is the reluctance within the length of the solenoid. The length of the magnetic path then becomes the length of the solenoid.

SOLUTION Determine the reluctance of the magnetic path. Thus

$$\mathcal{R} = \frac{\ell}{\mu A}$$

Substitute $\ell = 0.15$ m, $\mu = \mu_r \mu_0 = (1)(4\pi \times 10^{-7})$, and $A = \pi d^2/4 = 1.13 \times 10^{-4}$ m:

$$\mathcal{R} = \frac{0.15}{4\pi \times 10^{-7} \times 1.13 \times 10^{-4}}$$

$$\mathcal{R} = 1.056 \times 10^9 \text{ A/Wb}$$

Substitute into Equation 16–4 and solve for the mmf:

$$\mathcal{F} = \Phi\mathcal{R} = (1.09 \times 10^{-6})(1.056 \times 10^9)$$
$$\mathcal{F} = 1.15 \times 10^3 \text{ A}$$

Since $\mathcal{F} = NI$,

$$I = \frac{\mathcal{F}}{N} = \frac{1.15 \times 10^3}{280} = 4.11 \text{ A}$$

∴ The current in the solenoid is 4.11 A for a flux of 1.09 μWb at its center.

Nonlinear Magnetic Circuit Because of the nonlinear nature of the magnetization of ferromagnetic materials, a technique using the BH magnetization curves (Figure 16–7) for the magnetic path of the magnetic circuit is used in conjunction with a rule similar to Kirchhoff's voltage law for a closed loop. Thus

$$\mathcal{F} = H_1\ell_1 + H_2\ell_2 + H_3\ell_3 \quad (16\text{–}6)$$

Stating Equation 16–6 in terms of the current in the magnetizing coil results in

$$NI = H_1\ell_1 + H_2\ell_2 + H_3\ell_3 \quad (16\text{–}7)$$

A sequel to Kirchhoff's current law for the constant magnetic flux in a series magnetic circuit is

$$\Phi = \Phi_1 = \Phi_2 = \Phi_3 \quad (16\text{–}8)$$

Series Magnetic Circuits

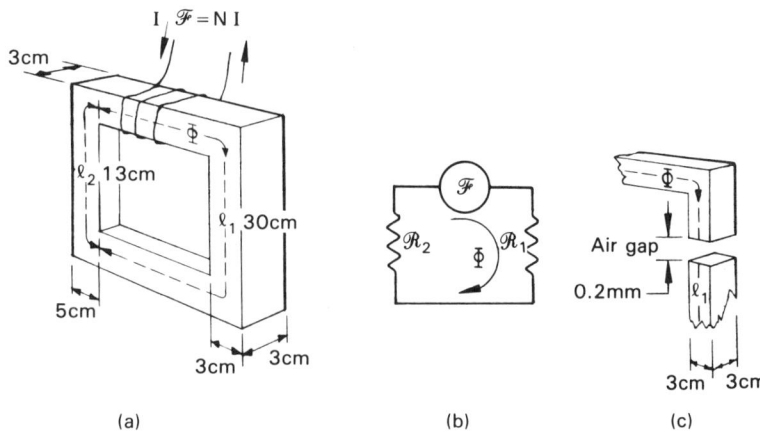

FIGURE 16-8 (a) Magnetic circuit for Example 16-5. (b) Schematic of the series magnetic circuit using electric circuit symbols to represent magnetomotive force ($\mathscr{F}$), flux (Φ), and reluctance ($\mathscr{R}$). (c) Air gap in ℓ_1 of (a) for Example 16-7.

EXAMPLE 16-5 For the series magnetic circuit of Figure 16-8, determine the current in the 200-turn coil so that a flux of 900 μWb is present in the mild steel of the magnetic circuit.

Observation The flux densities are different for paths ℓ_1 and ℓ_2 due to the difference in cross-sectional area. This, in turn, will result in two different values of field strength (H_1 and H_2) from the curve of Figure 16-7 for mild steel. Remember, μ is not constant in ferromagnetic materials.

SOLUTION Determine the flux density in each path:
Where $A_1 = 3$ cm $\times$ 3 cm,

$$B_1 = \frac{\Phi}{A_1} = \frac{900 \times 10^{-6}}{9 \times 10^{-4}} = 1.0 \text{ T}$$

Where $A_2 = 5$ cm $\times$ 3 cm,

$$B_2 = \frac{\Phi}{A_2} = \frac{900 \times 10^{-6}}{1.5 \times 10^{-3}} = 0.6 \text{ T}$$

Use the mild steel curve of Figure 16-7; enter the vertical axis at the computed values of B_1 and B_2, move horizontally to the curve and project vertically down, reading the values of H_1 and H_2 from the horizontal axis. Thus

$$H_1 = 400 \text{ A/m} \quad \text{and} \quad H_2 = 165 \text{ A/m}$$

Substitute into $NI = H_1\ell_1 + H_2\ell_2$:

$$NI = H_1\ell_1 + H_2\ell_2$$

$$200I = 400 \times 0.3 + 165 \times 0.13$$
$$200I = 141$$
$$I = \frac{141}{200} = 0.707 \text{ A}$$
$$\therefore \quad I = 707 \text{ mA}$$

EXAMPLE 16–6 Determine the current in the 800-turn coil needed to establish a flux of 0.32 mWb in the magnetic circuit of Figure 16–9. The cross section of 8 cm² is uniform throughout the magnetic circuit. Let ℓ_{ab} = 13 cm, ℓ_{bc} = 6 cm, ℓ_{cd} = 11 cm, and ℓ_{da} = 6 cm.

SOLUTION Determine the flux density for each material in the magnetic circuit:

$$B_{\text{cast}} = B_{\text{steel}} = \frac{\Phi}{A} = \frac{0.32 \times 10^{-3}}{8 \times 10^{-4}} = 0.4 \text{ T}$$

Observation The flux density is uniform throughout the magnetic circuit of Figure 16–9 because of the uniform cross-sectional area.

Determine H_{cast} and H_{steel} from Figure 16–7:

$$H_{\text{cast}} = 1150 \text{ A/m} \quad \text{and} \quad H_{\text{steel}} = 85 \text{ A/m}$$

$$NI = H_1 \ell_1 + H_2 \ell_2$$

and

$$NI = H_{\text{cast}} \ell_{ab} + H_{\text{steel}} \ell_{bcda}$$
$$800I = 1150(0.13) + 85(0.06 + 0.11 + 0.06)$$
$$800I = 149.5 + 19.55 = 169$$
$$I = \frac{169}{800} = 0.211 \text{ A}$$
$$\therefore \quad I = 211 \text{ mA}$$

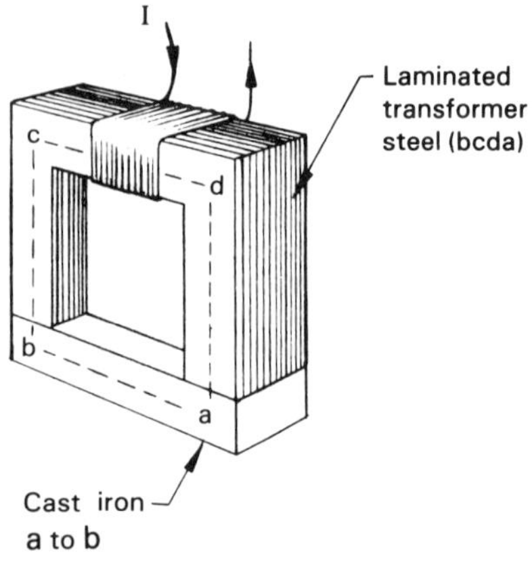

FIGURE 16–9 Magnetic circuit of Example 16–6.

Observation Notice the wide variation between the field strength in the two materials. Of the 169 A of magnetomotive force ($NI = 0.211$ A $\times$ 800 T) applied to the magnetic circuit, approximately 150 A (88%) goes into setting up the flux density in the cast iron, while only approximately 20 A (12%) goes into setting up the flux density in the transformer steel.

16–4 AIR GAP IN THE MAGNETIC CIRCUIT

When an air gap is introduced into the magnetic circuit, the magnetomotive force must be greatly increased to maintain a given flux density. Thus, more turns or an increased current, or both, are needed in the coil.

Effect of an Air Gap

The effect of even a very small air gap (0.1 mm) is to alter the slope of the BH curve. The slope of the curve will be much less, indicating an apparent lowering of the permeability of the material. Transformer steel, with relative permeability in the thousands, will have relative permeabilities in the hundreds with the addition of an air gap.

Since $B_m = \mu H = \mu NI/\ell$, a decrease in μ will allow for an increase in I for a given saturation flux density (B_m) when N and ℓ are constant. Thus, the air gap is added to increase the reluctance of the magnetic circuit, thereby preventing the core from being saturated by excessively large currents in the coil. Besides the air gap designed into a magnetic circuit, an air gap is needed for mechanical clearance between moving parts in electric motors, meters, and loudspeakers. Thus, out of necessity, most magnetic circuits will include an air gap.

The air gap is treated as an additional series reluctance in the magnetic circuit calculations and, although fringing (a spreading out of the flux at the air gap) is present, it may be ignored if the air gap is small compared to the cross section of the core. If this is not the case, a compensating factor of 10% or more is added to the area in the calculation of the flux density of the air gap. For our purposes, it will be assumed that the flux is uniform throughout the magnetic circuit, including the air gap, and that the area of the air gap is taken as the area of the cross section of the magnetic material.

EXAMPLE 16–7 The magnetic circuit of Figure 16–8 has an air gap of 0.2 mm (0.008 in.) introduced in the cross section of the ℓ_1 path, as shown in Figure 16–8(c). Determine the current in the 200-turn coil so that a flux of 900 μWb is present in the mild steel magnetic circuit (including the air gap).

SOLUTION Apply Equation 16–7, providing for the two different cross sections and the air gap. Thus

$$NI = H_1\ell_1 + H_2\ell_2 + H_{\text{air}}\ell_{\text{air}}$$

Determine the flux density for each path. Where $A_1 = 3 \text{ cm} \times 3 \text{ cm}$,

$$B_1 = \frac{\Phi}{A_1} = \frac{900 \times 10^{-6}}{9 \times 10^{-4}} = 1.0 \text{ T}$$

$$B_{air} = \frac{\Phi}{A_1} = \frac{900 \times 10^{-6}}{9 \times 10^{-4}} = 1.0 \text{ T}$$

Where $A_2 = 5 \text{ cm} \times 3 \text{ cm}$,

$$B_2 = \frac{\Phi}{A_2} = \frac{900 \times 10^{-6}}{1.5 \times 10^{-3}} = 0.6 \text{ T}$$

Determine H_1 and H_2 from the mild steel curve of Figure 16–7:

$$H_1 = 400 \text{ A/m} \quad \text{and} \quad H_2 = 165 \text{ A/m}$$

Determine the field strength of the air gap (H_{air}) from $B = \mu H$:

$$H_{air} = \frac{B}{\mu_0} = \frac{1.0}{4\pi \times 10^{-7}} = 7.96 \times 10^5 \text{ A/m}$$

Substitute:

$$NI = H_1 \ell_1 + H_2 \ell_2 + H_{air} \ell_{air}$$

$$200I = 400 \times 0.3 + 165 \times 0.13 + 7.96 \times 10^5 \times 0.2 \times 10^{-3}$$

$$200I = 120 + 21.5 + 159$$

$$I = \frac{301}{200} = 1.5 \text{ A}$$

$$\therefore \quad I = 1.5 \text{ A}$$

Observation Notice how the introduction of this small air gap into the magnetic circuit increased the mmf. Without the 0.2-mm (8-mil) air gap, the mmf is $200 \times 0.707 = 141$ A (from Example 16–5); with the air gap the mmf is 300 A for the same 900-μWb flux.

Force Resulting from an Air Gap As you have previously learned, magnetic lines of force form paths as short as possible. With the addition of an air gap in a magnetic circuit, the magnetic lines of force will exert a force on the magnetic surfaces on either side of the air gap in an attempt to shorten the path by closing the gap. This force may be determined by the following formula, which relates the flux density (B) of the magnetic circuit in teslas, the area (A) of the air gap in square meters, and the permeability of air (μ_0). Thus

$$F = \frac{B^2 A}{2\mu_0} \text{ (N)} \tag{16-9}$$

457

Symmetrical Parallel Magnetic Circuits

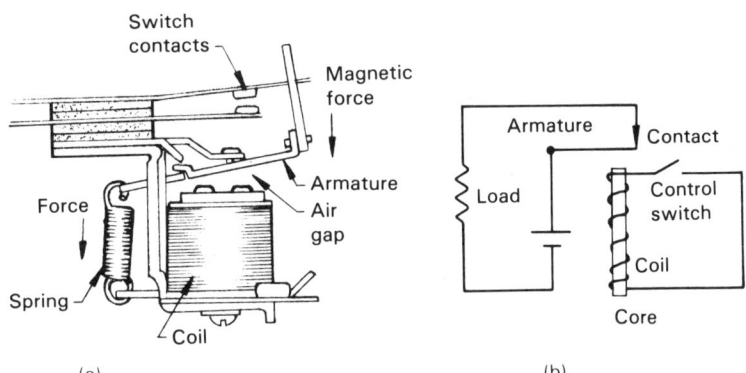

FIGURE 16-10 (a) The armature type of relay consists of an iron core coil, a pivoting armature, a return spring, and a set of contacts. (b) Schematic of a relay circuit.

The closing force, resulting from an air gap in a magnetic circuit, is the basis of operation of the *relay*. The relay, an electromagnetic switch, has one or more sets of electric contacts (switch contacts) attached to a pivoted armature, as pictured in Figure 16-10. A solenoid coil with a soft iron core acts as the source of magnetomotive force for the flux in the magnetic circuit. When the coil is energized by the application of a current, the armature moves against a spring to close the air gap. It is this motion that operates the switch contacts. Once the current is removed from the coil, the magnetic surfaces are separated by the relaxing of the spring.

16-5 SYMMETRICAL PARALLEL MAGNETIC CIRCUITS

In this section we will deal with the symmetrical, two-path parallel magnetic circuit in which the flux produced by the magnetomotive force is divided equally between two branches ($\Phi_T = \Phi_1 + \Phi_2$), as pictured in Figure 16-11. Although this is admittedly a special case, its occurrence is the usual case in parallel

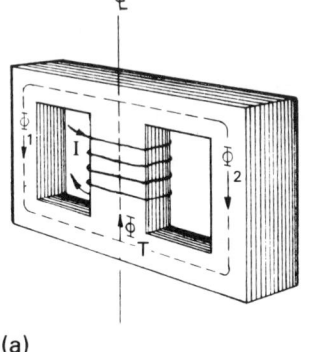
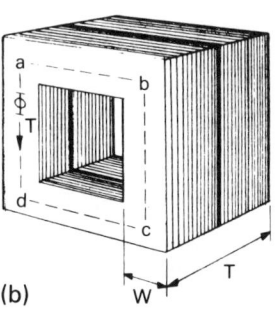

FIGURE 16-11 (a) In this parallel magnetic circuit, the flux divides so that $\Phi_T = \Phi_1 + \Phi_2$. (b) When analyzing symmetric, two path magnetic circuits, the circuit is cut along the center line ($\mathcal{C}$) and stacked one behind the other. (a)   (b)

magnetic designs. The following procedure will serve to solve this case of parallel magnetic circuits.

In Figure 16–11(b), imagine that the parallel magnetic circuit is cut along the center line of the center section of the magnetic circuit. The two identical magnetic circuits are then stacked one behind the other. The magnetic circuit, as pictured in Figure 16–11(b), is then treated as a series magnetic circuit with a flux equal to Φ_T, which is the sum of the flux in the two identical circuits. The use of this technique is demonstrated in the following example.

EXAMPLE 16–8 The magnetic circuit of Figure 16–11 is used with a transformer and has been designed with two magnetic paths to reduce the magnetic loss due to flux leakage. As pictured in Figure 16–11(a), a 1200-turn coil is wrapped around the center section of the laminated transformer steel core. For a magnetic circuit of Figure 16–11(b), the average magnetic path, ℓ_{abcd}, is 18 cm and the cross section (width by thickness) is 2 cm by 8 cm, which is uniform throughout the path. Determine the current in the coil so that a magnetic flux of 1.04 mWb is produced in both the left and right leg of the magnetic circuit.

SOLUTION Use Figure 16–11(b) as the composite magnetic circuit. Determine Φ_T:

$$\Phi_T = \Phi_1 + \Phi_2$$
$$\Phi_T = 1.04 \text{ mWb} + 1.04 \text{ mWb} = 2.08 \text{ mWb}$$

Determine the flux density in the composite magnetic circuit of Figure 16–11(b), where $A = 2 \text{ cm} \times 8 \text{ cm}$:

$$B = \frac{\Phi_T}{A} = \frac{2.08 \times 10^{-3}}{16 \times 10^{-4}} = 1.3 \text{ T}$$

Determine the field strength from the BH curve for transformer steel (Figure 16–7).

$$H = 500 \text{ A/m}$$

Substitute:

$$NI = H\ell_{abcd}$$
$$1200I = 500(0.18)$$

16-6 ADDITIONAL MAGNETIC UNITS

$$I = \frac{90}{1200} = 0.075 \text{ A}$$

$$\therefore \quad I = 75 \text{ mA}$$

In relating your understanding of magnetism to the "real world," you will encounter the agreed-upon SI units along with the now obsolete units of the CGS and the English systems. To be both functional and literate, you will need to translate these units into your realm of understanding. We have provided Table 16-1 for that purpose.

EXAMPLE 16-9 Convert the following magnetic units to SI units: (a) 4000 maxwells, (b) 8 oersteds, and (c) 12 000 gauss.

SOLUTION Use Table 16-1:

(a) $4000 \text{ Mx} \times \frac{1}{10^8} \frac{\text{Wb}}{\text{Mx}} = 40 \text{ }\mu\text{Wb}$

$\therefore$ 40 microwebers = 4000 maxwells

(b) $8 \text{ Oe} \times \frac{1}{0.0126} \frac{\text{A/m}}{\text{Oe}} = 635 \text{ A/m}$

$\therefore$ 635 amperes/meter = 8 oersteds

(c) $12\ 000 \text{ G} \times \frac{1}{10^4} \frac{\text{T}}{\text{G}} = 1.2 \text{ T}$

$\therefore$ 1.2 teslas = 12 000 gauss

PROBLEMS

Section 16-1

16-1 Figure 16-12(a) represents the cross section of a current-carrying conductor. The cross indicates the tail of the current arrow. Determine the direction of the magnetic flux around the wire. Use the letters cw (clockwise) or ccw (counterclockwise) to indicate the directions.

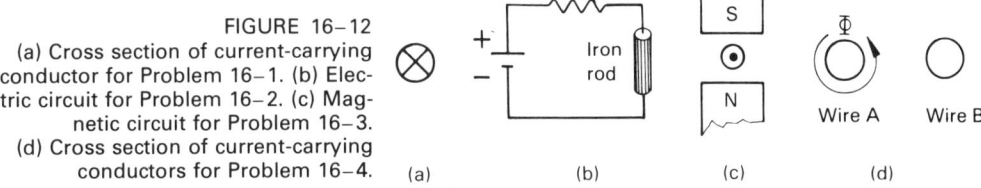

FIGURE 16-12
(a) Cross section of current-carrying conductor for Problem 16-1. (b) Electric circuit for Problem 16-2. (c) Magnetic circuit for Problem 16-3. (d) Cross section of current-carrying conductors for Problem 16-4.

TABLE 16–1 COMMON QUANTITIES, SYMBOLS, AND UNITS OF THE SI, CGS, AND ENGLISH SYSTEMS OF MAGNETISM

Quantity	Symbol	SI Unit	CGS Unit	English Unit	Unit Relation
Flux	Φ	weber (Wb)	maxwell (Mx)	lines	1 Wb = 10^8 Mx = 10^8 lines
Flux density	B	tesla (T)	gauss (G)	lines/in.^{2}	1 T = 10^4 G = 6.45 × 10^4 lines/in.^{2}
Field strength	H	A/m or A·t/m*	oersted (Oe)	A/in.	1 A/m = 0.0126 Oe = 0.0254 A/in.
Magnetomotive force	$\mathscr{F}$	A or A·t*	gilbert (Gb)	A or A·t	1 A·t = 1.26 Gb
Permeability of air	μ_0	Wb/A·m	G/Oe	lines/A·in.	$4\pi \times 10^{-7}$ Wb/A·m = 1 G/Oe = 3.2 lines/A·in.

* The ampere-turns is a descriptive unit; the ampere is the SI unit.

Problems

16-2 Determine the direction of the flux around the iron rod of Figure 16-12(b). Use the letters cw or ccw to indicate the direction.

16-3 For the magnetic system pictured in Figure 16-12(c), determine the direction of the force on the current-carrying conductor. The dot in the center of the cross section of the conductor is the tip of the current arrow.

16-4 The direction of the magnetic flux around wire A of Figure 16-12(d) is indicated by the curved arrow. If wire B is attracted to wire A, then (a) using either a dot or cross, indicate the direction of current in both wire A and B; (b) indicate the direction of flux around wire B, using the letters cw or ccw.

16-5 Determine the force on a 12-cm-long wire that has a current of 8 A in it. The current-carrying conductor is located in a magnetic flux with a density of 0.280 T.

16-6 Determine the torque on an 80-turn loop (coil) positioned horizontally between the poles of a permanent magnet with a flux density of 0.8 T. The current in the 12-cm by 8-cm loop is 2 A.

16-7 Determine the current in the coil of Problem 16-6 when the torque is measured as 0.5 N·m.

Section 16-2

16-8 Approximate the flux density at the end of a 1-cm-diameter air-core solenoid constructed by winding 320 turns of wire over a length of 14 cm when the current in the coil is 1.4 A.

16-9 An air-core toroid is constructed by winding 800 turns of wire onto a round doughnut-shaped nonconductive form with an average diameter of 12 cm. The cross section of the toroid is also round, having a radius of 1.4 cm. Determine (a) the magnetomotive force for a coil current of 0.75 A, (b) the field strength, (c) the flux density, and (d) the flux in the core.

Section 16-3

16-10 Using the same physical dimensions as the toroid of Problem 16-9, determine the flux in the core when the current in the coil is 542 mA and the core material is cast iron (BH curve, Figure 16-7).

16-11 Determine the reluctance of the magnetic circuit of the toroid of Problem 16-9.

16-12 Determine the current in the toroid of Problem 16-10 for a flux of 180 μWb if the cast-iron core has a 1-mm air gap in its cross section.

16-13 Determine the current in the 120-turn coil of the magnetic circuit of Figure 16-8 needed to set up a flux of 1.31 mWb in the transformer steel of the path.

16-14 Determine the current in the 310-turn coil of the magnetic circuit of Figure 16-9 needed to produce a flux of 500 µWb. The magnetic path of the laminated steel (ℓ_{bcda}) is 18 cm, while the path for the cast iron (ℓ_{ab}) is 8 cm. The cross section of 10 cm² is uniform throughout the magnetic circuit.

16-15 Repeat Problem 16-14 with the cast-iron path (ℓ_{ab}) replaced with mild steel.

Section 16-4

16-16 A 200-turn toroid is constructed on a mild steel core having a 1-mm air gap. The core (shaped as pictured in Figure 16-5) has a radius of 4.5 cm and a cross section of 1 cm × 2 cm (W × H). Determine the current in the toroid so that a flux of 220 µWb is in the magnetic circuit of the toroid.

16-17 The magnetic circuit of Problem 16-14 has a 0.8-mm air gap added to it. Determine the current in the 310-turn coil.

16-18 Determine the force exerted on the air gap in Example 16-7 by the magnetic field in its attempt to close the gap.

16-19 The magnetic path of a relay is 13 cm long with an average cross-sectional area of 0.275 cm². The armature frame and core are made from mild steel; the 6-V, 3200-turn coil is wound from 800 feet of no. 30 gauge solid round enameled wire. Determine (a) the flux density in the magnetic circuit of the relay when the relay is actuated, and (b) the force exerted on the armature to keep it closed.

Section 16-5

16-20 For the magnetic circuit of Figure 16-11, determine the current in the 120-turn coil to produce 200 µWb of flux in the left and right legs of the cast-iron magnetic circuit. The cross-sectional area of the magnetic path, $abcda$ of Figure 16-11(b), is a constant 12 cm², and the length, ℓ_{abcda}, is 40 cm.

16-21 For the magnetic circuit of Figure 16-11, determine the current, I, in the 500-turn coil so that a magnetic flux of 350 µWb is produced in the center section of the mild steel magnetic circuit pictured in Figure 16-11(a). The cross-sectional area of the magnetic path $badc$ of Figure 16-11(b) is 6 cm², and the length, ℓ_{badc}, is 14 cm. The cross-sectional area of the magnetic path bc is 3 cm², and the length, ℓ_{bc}, is 4 cm.

Section 16-6 16-22 Convert the following magnetic units to SI units: (a) 6.2×10^4 lines, (b) 20 oersteds, and (c) 600 gilberts.

16-23 Convert the following magnetic units to SI units: (a) 5280 maxwells, (b) 50 A/in., and (c) 2300 gauss.

17 Inductance

17-1 ELECTROMAGNETIC INDUCTION
17-2 FLUX LINKAGE
17-3 FARADAY'S LAW
17-4 LENZ'S LAW
17-5 SELF-INDUCTION
17-6 INDUCTANCE OF A COIL
17-7 ENERGY STORED BY INDUCTORS
17-8 INDUCTORS IN SERIES AND PARALLEL

CHAPTER PREVIEW With the completion of this chapter, you will have learned about the last of three properties of an electric circuit. You know about resistance and capacitance, and now you are to study about inductance. You have learned that resistance depends on the ability of electrons to move through a material. In your study of capacitance, you learned that the capacitance depends on electric fields; and in this chapter, you will learn that inductance depends on the magnetic field created by the movement of electrons in a conductor.

The chapter begins with electromagnetic induction; Faraday's law, Lenz's law, mutual induction, self-induction, and flux linkage are introduced. You will then learn about inductance and electrical components called *inductors*. The chapter concludes with a brief discussion of how the equivalent inductance of series and parallel connected inductors is determined.

PERFORMANCE OBJECTIVES Once you have read each section, worked each example with pencil, paper, and calculator, and answered each problem for every section, you will be able to:

☐ Use Faraday's law to determine the induced voltage in a conductor or coil.
☐ Apply Lenz's law to determine the polarity of the induced voltage.
☐ Compute the inductance of an inductor from physical parameters.
☐ Determine the equivalent inductance of inductors in series and parallel.

17-1 ELECTROMAGNETIC INDUCTION

We have learned that the application of a voltage to a conductor results in a current in the conductor. The current, which is made up of electrons in motion, produces a magnetic field around the conductor. The reverse of this process is also true, as was independently discovered in 1831 by Joseph Henry and Michael Faraday. Each demonstrated that a voltage is *induced* in a conductor that is moved through a magnetic field. Further experimentation has proved that this phenomenon also occurs in a stationary conductor when a varying magnetic field surrounds the conductor.

Electromagnetic induction will result when there is net motion between the magnetic field and the conductor in a manner such that magnetic lines of force are cut. Figure 17-1 demonstrates both cases. In Figure 17-1(a), the conductor is moved through a

Flux Linkage

467

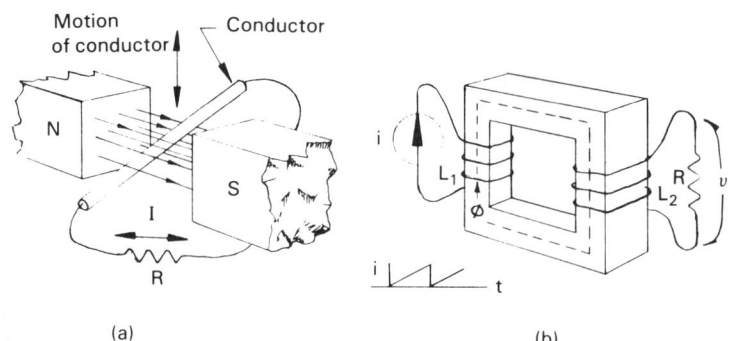

FIGURE 17-1
(a) Electromagnetic induction is the name given to the generation of a voltage when a conductor cuts across magnetic lines of flux.
(b) Mutual induction is the name given to the phenomenon of a varying current in L_1 causing a voltage in L_2.

magnetic flux so that the conductor is cutting across the line of force. This action induces a voltage in the conductor. In Figure 17-1(b), a varying flux is set up in the core by the varying current passing in the left-hand coil. The varying flux, ϕ, circulating in the magnetic path *links* the turns of the right-hand coil, inducing a voltage in this coil. This cause-and-effect relationship between two coils sharing the same magnetic path is called *mutual induction*.

17-2 FLUX LINKAGE

As you have just learned, electromagnetic induction occurs when there is a change in the *flux linkage*. To aid in your understanding of this concept, consider Figure 17-2(a). This figure illustrates one turn in a conductor with one line of force linking the turn ($1 \times 1 = 1$ flux linkage). An increase in the number of turns in close proximity with one another, as pictured in Figure 17-2(b), increases the flux linkage ($2 \times 2 = 4$ flux linkages). Thus, in a coil surrounded by lines of force, the number of flux linkages is the product of the number of turns and the number of magnetic lines of force linking the turns.

(a)

(b)

FIGURE 17-2
Flux linkage (a) in a single turn, (b) in a coil.

**17-3
FARADAY'S
LAW**

In performing his experiment of electromagnetic induction, Faraday determined that the average induced voltage is equal to the product of the number of turns (N) linked by the magnetic flux and the time rate of change of the magnetic flux ($\Delta\phi/\Delta t$). Thus

$$e = N \frac{\Delta\phi}{\Delta t} \quad \text{(V)} \tag{17-1}$$

This equation is known as *Faraday's law*.

EXAMPLE 17-1 A 10-turn coil, wound one turn on top of the other, has an average radius of 6 cm. Determine the average voltage induced in the coil when the flux density in the magnetic path moves from 20 to 500 mT in 4 ms.

SOLUTION Determine the area and the initial and final magnetic flux. Use $B = \Phi/A$:

$$A = \pi r^2 = \pi(6 \times 10^{-2})^2 = 1.13 \times 10^{-2} \text{ m}^2$$

$$\Phi_1 = B_1 A = (20 \times 10^{-3})(1.13 \times 10^{-2})$$
$$\Phi_1 = 2.26 \times 10^{-4} \text{ Wb}$$

$$\Phi_2 = B_2 A = (500 \times 10^{-3})(1.13 \times 10^{-2})$$
$$\Phi_2 = 56.5 \times 10^{-4} \text{ Wb}$$

Substitute into Faraday's law:

$$e = \frac{N(\Delta\phi)}{\Delta t}$$

$$e = \frac{10(56.5 \times 10^{-4} - 2.26 \times 10^{-4})}{4 \times 10^{-3}} = 13.6 \text{ V}$$

∴ The instantaneous value of the induced voltage, e, is 13.6 V.

**17-4
LENZ'S LAW**

To satisfy the law of conservation of energy, the motion of the conductor in Figure 17-1(a) must be acted on by an opposing force. If this were not the case, the current moving through the load, R, producing heat would result from no work being done in moving the conductor through the lines of flux. Likewise, the flux linking L_2 must be opposed; otherwise, the current in L_2 and R would result from no work being done to move the flux through L_2. Heinrich Lenz reasoned that this must indeed be the case, and he formulated a statement known today as *Lenz's law:* the current resulting from the induced voltage caused by a change

469

Lenz's Law

in flux linkage will set up a magnetic field to **oppose** any change in the original flux causing the induced voltage.

From Lenz's law, we learn that the opposition is in the form of an opposing magnetic field. By using Lenz's law along with the right-hand rule, it is possible to determine the direction of the current in the conductor resulting from the induced voltage.

Thus, Faraday's law provides for the amount (magnitude) of the induced voltage and Lenz's law provides for the polarity (direction) of the current resulting from the induced voltage.

EXAMPLE 17-2 Determine the direction of the current induced in the conductor of Figure 17-1(a) when the conductor is moving down.

SOLUTION Since the conductor is being moved down, the opposition must be directed upward. From the discussion in Chapter 16 of Figure 16-2, the flux around the conductor must be counterclockwise (ccw), as pictured in Figure 17-3(a), to cause an increase in the flux density below the conductor and a decrease in the flux density above, thereby producing an upward directed force. Applying the right-hand rule results in a current coming toward you out of the paper.

∴ When the conductor is moved downward, the current is out of the paper.

EXAMPLE 17-3 Determine the direction of the flux opposing the original flux in Figure 17-1(b) and the direction of the induced current in L_2.

SOLUTION Since the original flux moves in a cw direction, as shown in Figure 17-1(b), the opposing flux (ϕ_2) must move in a ccw direction, as illustrated in Figure 17-3(b). Applying the right-hand rule to

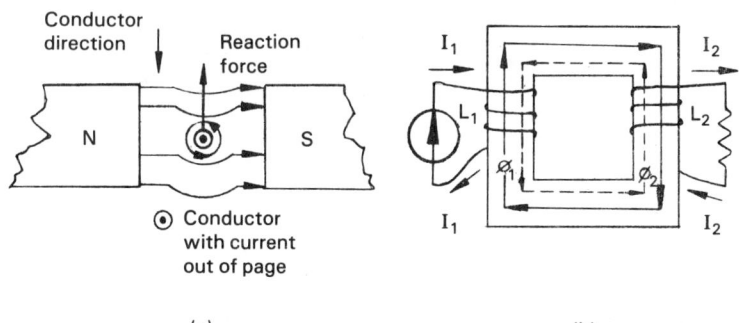

FIGURE 17-3
(a) Downward motion of the conductor is opposed by upward force due to the induced current.
(b) Induced current in L_2 sets up a flux, ϕ_2, that opposes the original flux, ϕ_1.

(a)   (b)

L_2 with the thumb pointing in the direction of the opposing flux, ϕ_2, results in an induced current moving down through the resistance as pictured.

∴ The induced current and flux have directions as shown in Figure 17–3(b).

Now that the concepts of flux linkage, Faraday's law, Lenz's law, and mutual inductance are understood, the following example may be worked.

EXAMPLE 17–4 For the magnetic circuit of Figure 17–3(b), determine the average induced voltage in the 500 turns of L_2 when the current I_1 rises from 0.2 to 0.4 A in 1 ms in the 100 turns of L_1. The magnetic path is 20 cm and the cross section of the mild steel core is 5 cm². Use the BH curve of Figure 16–7.

SOLUTION Apply Faraday's law, but first determine the flux that corresponds to each current. Let $I_1 = 0.2$ A and $I_1' = 0.4$ A.

$$H_1 = \frac{N_1 I_1}{\ell} = \frac{(100)(0.2)}{0.2} = 100 \text{ A/m}$$

$B_1 = 0.2$ T   (Figure 16–7)

$\phi_1 = B_1 A = (0.2)(5 \times 10^{-4}) = 1 \times 10^{-4}$ Wb

and

$$H_1' = \frac{N_1 I_1'}{\ell} = \frac{(100)(0.4)}{0.2} = 200 \text{ A/m}$$

$B_1' = 0.75$ T   (Figure 16–7)

$\phi_1' = B_1' A = (0.75)(5 \times 10^{-4}) = 3.75 \times 10^{-4}$ Wb

The time rate of change of flux is

$$\frac{\Delta \phi}{\Delta t} = \frac{(\phi_1' - \phi_1)}{1 \times 10^{-3}} = \frac{(3.75 \times 10^{-4}) - (1 \times 10^{-4})}{1 \times 10^{-3}}$$

$$\frac{\Delta \phi}{\Delta t} = \frac{2.75 \times 10^{-4}}{1 \times 10^{-3}}$$

Solve for the induced voltage:

$$e = N_2 \frac{\Delta \phi}{\Delta t} = 500(2.75 \times 10^{-4}/1 \times 10^{-3})$$

$e = 137.5$ V

Self-Induction

∴ The induced voltage across L_2 is 137.5 V.

Observation The induced voltage will only appear across L_2 when the current is changing in L_1. When the current in L_1 is steady, there is no change in the flux linkage and no induced voltage across L_2.

17-5 SELF-INDUCTION

Thus far, we have learned about electromagnetic induction, which results in a voltage being induced in a conductor when lines of force are cut. We also learned about mutual induction, which results in a voltage being induced in one coil as a result of varying current in another coil. And now we are about to learn about *self-induction*, the term given to the situation when a voltage is induced within a coil or conductor by a varying current within the coil or conductor. Self-induction is present in a coil when the varying flux of one turn links with its neighboring turns, inducing a voltage in them which opposes the change in the current in the coil producing the changing flux. Each induced voltage in each turn is connected in series with every other induced voltage and, in combination, they collectively assume a polarity that opposes the change in current in the coil (Lenz's law). This induced voltage is often called the *counter-electromotive force* (cemf), and it may be determined by Faraday's law [$e = N(\Delta\phi/\Delta t)$].

The induced voltage across the terminals of a coil is proportional to the time rate of change of current, $e \propto (\Delta i/\Delta t)$, and $e = L(\Delta i/\Delta t)$. The proportionality constant, L, is the self-inductance of the circuit. Thus

$$e = L \frac{\Delta i}{\Delta t} \quad \text{(V)} \tag{17-2}$$

where L is the self-inductance (usually called *inductance*) in the SI unit, the henry (H).

EXAMPLE 17-5 A counter-emf of 4 V is across a solenoid when the current in the coil increases from 100 to 900 mA in 200 μs. Determine the inductance of the solenoid in henrys.

SOLUTION From Equation 17-2, solve for L:

$$L = \frac{e}{\Delta i/\Delta t}$$

Substitute, where $\Delta i = (900 \text{ mA} - 100 \text{ mA}) = 800$ mA and $\Delta t = 200$ μs. Thus

$$L = \frac{4}{(800 \times 10^{-3})/(200 \times 10^{-6})} = 1 \times 10^{-3}$$

∴ The inductance of the solenoid is 1 mH.

Inductor Types An electric circuit component that has self-inductance is called an *inductor*. Figure 17-4 pictures two fixed inductors. In addition to the fixed air-core inductor and the fixed iron-core inductor, inductors are also available with adjustable ferrite cores. This type of inductor is called a *variable* inductor.

17-6 INDUCTANCE OF A COIL

The inductance of a solenoid or a toroid may be determined from the physical parameters of the inductor if the permeability of the magnetic circuit is relatively constant. This is the case with nonmagnetic materials, such as air, as well as with ferromagnetic materials in the linear portion of the magnetization curve (BH curve). The derivation of the equation for inductance, L, begins with a definition of flux from Equation 15-1:

$$\Phi = BA \quad (15-1)$$

However, B is equal to μH (Equation 15-4). Substituting into Equation 15-1 results in

$$\Phi = \mu H A$$

And $H = NI/\ell$ (Equation 15-3); thus

$$\Phi = \frac{\mu NI}{\ell} A$$

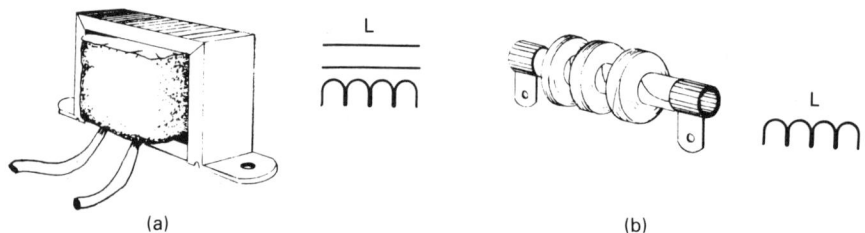

(a) (b)

FIGURE 17-4 Fixed inductors:
(a) Iron core for audio frequency (AF) applications.
(b) Air-core for radio frequency (RF) applications.

473

Inductance of a Coil

Since a variation in current in the inductor in terms of time results in a variation in flux in terms of time,

$$\frac{\Delta \phi}{\Delta t} = \frac{\mu N (\Delta i / \Delta t)}{\ell} A \quad \text{(Wb)} \qquad (17\text{–}3)$$

By equating Equations 17–1 and 17–2 to each other, inductance may be defined in terms of the time rate of change in flux resulting from the time rate of change in current. Thus

$$e = N \frac{\Delta \phi}{\Delta t} \quad (17\text{–}1) \quad \text{and} \quad e = L \frac{\Delta i}{\Delta t} \quad (17\text{–}2)$$

Then

$$L \frac{\Delta i}{\Delta t} = N \frac{\Delta \phi}{\Delta t}$$

Solve for L:

$$L = \frac{N(\Delta \phi / \Delta t)}{(\Delta i / \Delta t)} = N \frac{\Delta \phi}{\Delta i} \quad \text{(H)} \qquad (17\text{–}4)$$

However, from Equation 17–3,

$$\Delta \phi = \frac{\mu N \, \Delta i A}{\ell}$$

Substituting into Equation 17–4 for $\Delta \phi$,

$$L = \frac{N(\mu N \, \Delta i A)/\Delta i}{\ell}$$

$$L = \frac{\mu N^2 A}{\ell} \quad \text{(H)} \qquad (17\text{–}5)$$

From this equation for inductance, we see that the inductance of an inductor depends directly on the cross-sectional area, A, of the magnetic path in square meters, the permeability, μ, of the magnetic path in henrys per meter (H/m), and the number of turns in the solenoid or toroid. Inductance also depends inversely on the length, ℓ, of the magnetic path in meters. (*Note:* For air-core solenoids, the length of the coil is the assumed length of the magnetic path.)

EXAMPLE 17–6 Determine the inductance of an iron-core inductor with a μ of 3.8 mWb/A·m. The core has a cross-sectional area of 8 cm² and an average magnetic path length of 20 cm. The core is wrapped with 200 turns of wire.

SOLUTION Substitute into Equation 17-5:

$$L = \frac{\mu N^2 A}{\ell} = \frac{(3.8 \times 10^{-3})(200^2)(8 \times 10^{-4})}{0.2}$$

$$L = 0.61 \text{ H}$$

∴ The inductance of the inductor is 0.61 H.

17-7 ENERGY STORED BY INDUCTORS

An inductor stores energy in the magnetic field generated by the current in the inductor. When the current in the inductor increases, the work done in overcoming the induced voltage increases, as does the energy stored in the magnetic field. However, when the source goes to zero, the field around the conductors collapses, producing a self-induced voltage that opposes the change. Energy that was stored in the magnetic field is returned as work by opposing the diminishing circuit current. The energy stored by an inductor in equilibrium is equal to one-half the product of the inductance of the inductor, L, and the current, I, squared. Thus

$$W = \tfrac{1}{2} L I^2 \quad \text{(J)} \tag{17-6}$$

EXAMPLE 17-7 Determine the energy stored in the magnetic field of a 610-mA current in a 0.35-H inductor.

SOLUTION Use Equation 17-6:

$$W = \tfrac{1}{2} L I^2 = \tfrac{1}{2}(0.35)(0.610^2) = 65.1 \times 10^{-3} \text{ J}$$

∴ The energy stored by the inductor is 65.1 mJ.

Core and Coil Losses

In the practical inductor, some of the work done is lost in heat and is not returned to the circuit. The source of this lost energy is in the core and the coil of the inductor.

The coil has resistance inherent in the copper wire, which dissipates energy. Additional energy loss occurs when some flux fails to link the turns. This flux *leakage* uses energy to establish it, but does no work in return. Core losses include hysteresis (losses caused by the retentivity of the core material) and *eddy currents* (currents induced in the magnetic core). The eddy currents oppose the flux in the core. Core losses produce heat and waste energy. These losses are lessened by using a laminated core (lessens eddy currents) made up of thin steel coated with an insulating material and by adding a small amount of silicon (lessens hysteresis) to the alloy of steel.

17-8 INDUCTORS IN SERIES AND PARALLEL

Since inductance opposes the change in the current in an electric circuit, it may be treated in a manner similar to resistance. Thus, for inductors connected in series with *no* coupling between the coils, the total inductance is the sum of each inductor's value. Therefore,

$$L_T = L_1 + L_2 + L_3 + \cdots + L_n \quad \text{(H)} \tag{17-7}$$

EXAMPLE 17-8 Determine the equivalent inductance of three inductors connected in series when $L_1 = 15$ mH, $L_2 = 47$ mH, and $L_3 = 27$ mH.

SOLUTION Use Equation 17-7:

$$L_T = L_1 + L_2 + L_3 = 15 \text{ mH} + 47 \text{ mH} + 27 \text{ mH}$$

$$\therefore \quad L_T = 89 \text{ mH}$$

Inductors connected in parallel without any coupling between them are treated in a manner similar to resistors connected in parallel. Thus, inductors connected in parallel have a total inductance equal to the reciprocal of the sum of the reciprocals. Therefore,

$$L_T = \frac{1}{1/L_1 + 1/L_2 + 1/L_3 + \cdots + 1/L_n} \quad \text{(H)} \tag{17-8}$$

EXAMPLE 17-9 Determine the equivalent inductance of three inductors connected in parallel when $L_1 = 220$ mH, $L_2 = 100$ mH, and $L_3 = 82$ mH.

SOLUTION Use Equation 17-8:

$$L_T = \frac{1}{1/L_1 + 1/L_2 + 1/L_3}$$

$$L_T = \frac{1}{1/0.22 + 1/0.1 + 1/0.082}$$

$$\therefore \quad L_T = 37.4 \text{ mH}$$

PROBLEMS
Section 17-3

17-1 Determine the change in flux in a 400-turn toroid when an average induced voltage of 40 V is measured across the terminals and the time for the change in flux is 0.3 s.

17-2 Determine the number of turns in a solenoid when a time rate of change of 50 mWb/s results in an induced voltage of 2.8 V.

Inductance

17-3 Determine the average voltage induced in a 200-turn, 18-cm-long, 14-mm-diameter solenoid when the flux in the magnetic path moves from 8 mWb to 48 mWb in 600 µs.

17-4 Determine the average voltage induced in a 700 turn air-core toroid 20 cm in diameter, having a cross-sectional area of 12.8 cm², when the flux density varies from 120 mT to 770 mT in 5 ms.

Section 17-4

17-5 Assuming the wire in Figure 17-5(a) is moved between the pair of magnetic poles, determine the direction of motion (either up or down) by applying Lenz's law. The induced current, as indicated by the dot, is moving out of the page.

17-6 The conductor of Figure 17-5(b) is moved to the left between the pair of magnetic poles, as indicated by the arrow. Determine the direction of the induced current using Lenz's law. Indicate the current direction by marking "dot" (current out of the page) or "cross" (current into the page).

17-7 The conductor of Figure 17-5(c) is moved to the right between the pair of magnetic poles, as indicated by the arrow. Using Lenz's law, determine the north and south magnetic poles when the current is moving into the page, as indicated by the cross.

17-8 The conductor of Figure 17-5(d) is moved up between the pair of magnetic poles, as indicated by the arrow. Using Lenz's law, (a) determine the direction of the reaction force, (b) state the direction of the magnetic field around the conductor (cw or ccw), and (c) determine the direction of the induced current (dot or cross).

Section 17-5

17-9 Determine the average voltage induced across the terminals of a 10-H inductor if the time rate of change of current is 15 A/s.

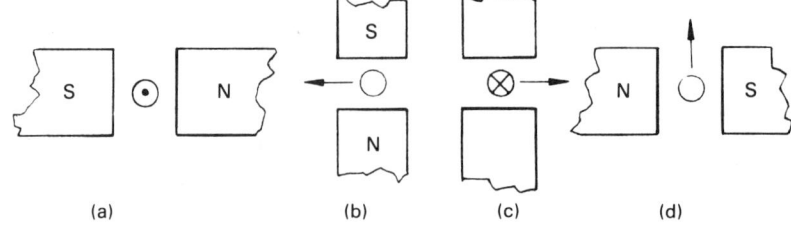

FIGURE 17-5
(a) Condition for Problem 17-5.
(b) Condition for Problem 17-6.
(c) Condition for Problem 17-7.
(d) Condition for Problem 17-8.

Problems

17-10 Determine the average induced voltage across a 1-mH choke (inductor) when the time rate of change of current is 5 mA/μs.

17-11 A counter-emf of 20 V is induced across a solenoid when the current in the windings decreases from 1.2 A to 400 mA in 200 ms. Determine the inductance of the solenoid.

17-12 Determine the inductance of a coil that has an induced voltage of 6 V across its terminals when the current increases from 110 mA to 730 mA in 500 μs.

Section 17-6

17-13 Determine the inductance of a 120-turn air-core toroid with a circular cross section of 15 cm² and a mean circumference of 30 cm.

17-14 If the toroid of Problem 17-13 is wound on a cast-iron core with a relative permeability of 400, determine the inductance of the toroid.

17-15 An 800-turn, 100-mH air-core inductor is to be used in an application requiring an 84-mH inductor. Assuming no change in length, determine how many turns must be removed from the inductor in order to achieve the desired inductance.

17-16 A 900-turn tapped inductor is wound so that the inductance from one end to the tap is 800 μH and from that same end to the other end is 1000 μH. Determine the number of turns in the 800 μH part of the inductor.

Sections 17-7 and 17-8

17-17 For the inductors in Figure 17-6(a), $L_1 = 15$ mH, $L_2 = 39$ mH, and $L_3 = 47$ mH. (a) Determine the total inductance of the inductors. (b) Determine the energy stored in each inductor when a dc current source of 50 mA is attached to terminals a-b.

17-18 For the inductors in Figure 17-6(b), $L_1 = 270$ mH, $L_2 = 680$ mH, and $L_3 = 100$ mH. (a) Determine the total inductance of the inductors. (b) Determine the energy stored

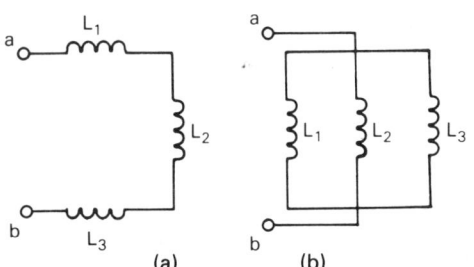

FIGURE 17-6
(a) Series connected inductor for Problem 17-17.
(b) Parallel connected inductor for Problems 17-18, 17-19, and 17-20.

in each inductor when a dc current source of 2.2A is attached to terminals a-b.

17-19 For the inductors in Figure 17-6(b), $L_1 = 2$ H, $L_2 = 10$ H, and $L_3 = 1$ H. (a) Determine the total inductance of the inductors. (b) Determine the total energy stored in the inductors when a dc current source of 750 mA is attached to terminals a-b.

17-20 For the inductors in Figure 17-6(b), $L_1 = 330$ μH, $L_2 = 82$ μH, and $L_3 = 120$ μH. (a) Determine the total inductance of the inductors. (b) Determine the total energy stored in the inductors when a dc current source of 10 mA is attached to terminals a-b.

18 RL Transient Circuits

18-1 CHARGE AND DISCHARGE OF INDUCTORS
18-2 L/R TIME CONSTANT
18-3 INITIAL AND FINAL CONDITIONS OF RL CIRCUITS
18-4 INSTANTANEOUS VALUES OF RL CIRCUITS
18-5 APPLICATIONS

CHAPTER PREVIEW This chapter will give you an understanding of what happens to the parameters in an RL circuit when step changes of voltage or current are applied to the circuit. Like the RC circuit of Chapter 14, this chapter will focus on the transient period after the application of a step source to the RL circuit by making calculations of instantaneous circuit values.

PERFORMANCE OBJECTIVES Once you have read each section, worked each example with pencil, paper, and calculator, and answered each problem of every section, you will be able to:

☐ Understand qualitatively the charging and discharging of an RL circuit.
☐ Determine the initial and final circuit conditions of an RL circuit after the application of a step source.
☐ Calculate the instantaneous voltage or current at any time during the transient period.

18–1 CHARGE AND DISCHARGE OF INDUCTORS From the previous chapter, you have learned that a relationship exists between a time-varying current flowing through an inductor and the emf developed across its terminals. The polarity of this emf is such that it will oppose any change in current. The physical response of an inductor opposing any change in current can be compared with the response of a capacitor opposing any change in voltage.

Charge In the circuit of Figure 18–1, when the switch closes at $t = 0$, the current attempts to increase because of the applied source voltage. However, the increasing current generates an emf that opposes the source voltage, thereby decreasing the net applied voltage. Because the net applied voltage is reduced, the current in the inductor will increase at a slower rate. In reality, the current will initially increase at one rate and then, as time passes,

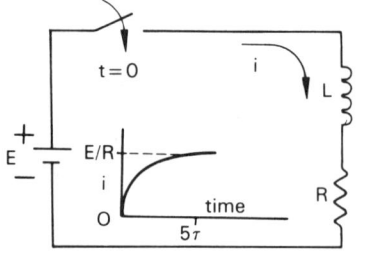
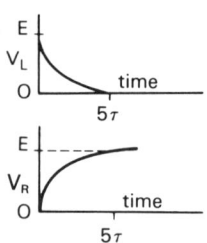

FIGURE 18–1
Simple RL circuit with switch closing at t = 0 seconds. Wave form shows response of circuit parameters to closing of switch.

L/R Time Constant

the rate diminishes until equilibrium is reached.

Looking at the wave forms of Figure 18-1 when the switch is closed ($t = 0$), there is initially zero current flowing in the circuit since the current in an inductor cannot change a finite amount in zero time. Because there is zero current, the voltage across the resistor, v_R, must be zero by Ohm's law. Then the voltage across the inductor, v_L, must equal the source voltage, E, by Kirchhoff's law. As the current increases, the voltage across the resistor must increase proportionately, again by Ohm's law. Then the voltage across the inductor must be decreased by the amount that the voltage across the resistor increased, since $E = v_L + v_R$. This action of increasing current and redistributing voltages continues until the current has increased to its maximum value (i_{mx}), where $i_{mx} = I = E/R$.

At this time, V_L must be zero and the current will have reached a constant steady-state value. We see from the curves of Figure 18-1 that the rate at which current increases after $t = 0$ follows an exponential curve.

Discharge

In the case where initially a constant current is flowing and then the voltage source is set to zero, the reverse will occur. Without an applied source, the current will decrease because the stored energy in the inductor ($W = \frac{1}{2}LI^2$) will dissipate through the resistor. As energy is dissipated, the current will decrease. This, in turn, lowers the voltage across the resistor, and since Kirchhoff's voltage law must be satisfied, the voltage across the inductor is also lowered. As the current decreases, an emf is generated that opposes the negative change in current. Thus, the current will decrease following a diminishing exponential curve.

18-2 L/R TIME CONSTANT

Time Constant Defined

In resistive-inductive circuits, there is a time constant just as in resistive-capacitive circuits. As mentioned in the previous section, a relationship exists between the rate at which the current changes in an inductor and the voltage across the inductor. This rate is determined by the time constant of the circuit. For RL circuits, it is the time it takes for the current to increase to 63.2% of its final value, or to decrease to 36.8% of its initial value.

In the circuit of Figure 18-1, the maximum current possible, assuming the inductor is ideal, is $I = E/R$. At $t = 0$,

$$v_L = E = L\frac{\Delta i}{\Delta t} \quad \text{or} \quad \frac{\Delta i}{\Delta t}(0^+) = \frac{E}{L}$$

By defining the time constant, τ, as the time to reach the final value of current when the rate of change is assumed constant at its initial value, the following expression for the time constant results.

$$I = \frac{\Delta i}{\Delta t}(0^+)(\tau)$$

Substitute $I = E/R$ and $(\Delta i/\Delta t)(0^+) = E/L$:

$$\frac{E}{R} = \frac{E}{L}(\tau)$$

Solve for τ:

$$\tau = \frac{L}{R} \quad \text{(s)} \tag{18-1}$$

Relation of R and L to τ Upon close observation of the expression $\tau = L/R$, it may be noted that as the inductance increases, so does the time constant. This is logical since the larger the inductance, the greater the opposition to a change in the current flow, yielding a slower rate of change. An increase in the value of R will decrease the time constant. Physically, increasing R decreases the final value of current in the circuit. Thus, the difference between the initial current and final current is less; consequently, the time necessary for the current to reach its final value will be less (remember that the initial rate of change of current is independent of the value of R).

Computing L/R Time Constants

EXAMPLE 18-1 Determine the time constant of a circuit composed of an inductor of 25 mH and a resistor of 4.3 kΩ.

SOLUTION Substitute into Equation 18-1:

$$\tau = \frac{L}{R} = \frac{25 \times 10^{-3}}{4.3 \times 10^3} = 5.81 \times 10^{-6} \text{ s}$$

∴ The time constant is 5.81 μs.

EXAMPLE 18-2 Determine the time constant of the circuit shown in Figure 18-2(a).

Observation To determine the time constant of a circuit with more than one L and/or more than one R, the equivalent circuit must first be obtained.

SOLUTION Thevenize the current source and R_1 across points ab. The equivalent circuit is shown in Figure 18-2(b). Combine R_{Th} and R_2 of Figure 18-2(b):

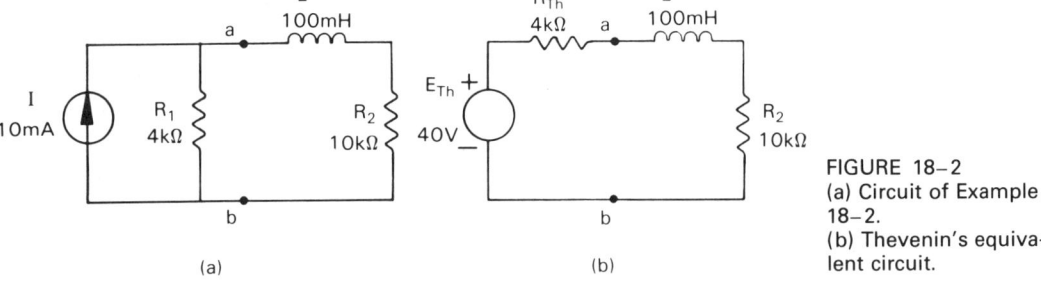

FIGURE 18-2
(a) Circuit of Example 18-2.
(b) Thevenin's equivalent circuit.

$$R_{eq} = R_{Th} + R_2 = 4\text{ k}\Omega + 10\text{ k}\Omega = 14\text{ k}\Omega$$

Compute the time constant:

$$\tau = \frac{L}{R_{eq}} = \frac{100 \times 10^{-3}}{14 \times 10^3} = 7.14 \times 10^{-6} \text{ s}$$

∴ The time constant is 7.14 μs.

EXAMPLE 18-3 Determine the time constant of the circuit shown in Figure 18-3(a).

SOLUTION Simplifying this circuit will require several steps. First, form the Thevenin's equivalent circuit of the source E and resistors R_1 and R_2. Then combine it with L_1, L_2, and R_3.

$$E_{Th} = E_s \frac{R_2}{R_1 + R_2} = 20 \text{ V} \frac{8\text{ k}\Omega}{2\text{ k}\Omega + 8\text{ k}\Omega}$$

$$E_{Th} = 20 \text{ V } (0.8) = 16 \text{ V}$$

$$R_{Th} = R_1 \parallel R_2 = \frac{R_1 R_2}{R_1 + R_2} = \frac{(2\text{ k}\Omega)(8\text{ k}\Omega)}{2\text{ k}\Omega + 8\text{ k}\Omega}$$

$$R_{Th} = 1.6\text{ k}\Omega$$

Observation The circuit looks like Figure 18-3(b). Since all the components are now in series, as shown in Fig-

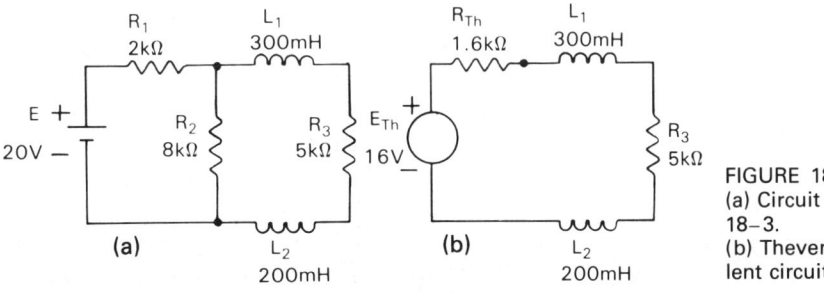

FIGURE 18-3
(a) Circuit of Example 18-3.
(b) Thevenin's equivalent circuit.

ure 18–3(b), it is an easy matter to combine them into equivalent values. Thus

$$R_{eq} = R_{Th} + R_3 = 1.6 \text{ k}\Omega + 5 \text{ k}\Omega = 6.6 \text{ k}\Omega$$

$$L_{eq} = L_1 + L_2$$

$$L_{eq} = 300 \text{ mH} + 200 \text{ mH} = 500 \text{ mH}$$

Compute the time constant:

$$\tau = \frac{L_{eq}}{R_{eq}} = \frac{500 \times 10^{-3}}{6.6 \times 10^3} = 75.8 \times 10^{-6} \text{ s}$$

∴ The time constant is 75.8 µs.

18–3 INITIAL AND FINAL CONDITIONS OF RL CIRCUITS

The ideal inductor is considered to have zero resistance. Therefore, under steady-state conditions, think of the inductor as a short circuit. Contrast this with the capacitor, which, under steady-state conditions, is considered to be an open circuit. When a sudden change occurs in an RL circuit, the current tends to remain the same before and after the step change in the source. This is due to the emf developed across the inductor, which opposes a sudden change in current. The following assumptions are made for the purposes of circuit analysis.

☐ The current remains unchanged in an inductor from time $t = 0^-$ to time $t = 0^+$.
☐ The voltage across an inductor may change instantaneously from one value to another and is calculated from Kirchhoff's law.
☐ The inductor appears as a short circuit under steady-state conditions.
☐ The symbol $t = 0^-$ is the instant of time just before $t = 0$, and $t = 0^+$ is the instant of time just after $t = 0$.

EXAMPLE 18–4 Determine the initial and final voltages and currents of the RL circuit in Figure 18–4(a).

SOLUTION Before the switch is closed, there is no current flowing in the circuit; at $t = 0$, the switch closes. Since the current cannot change instantaneously, the current, i, is still zero at $t = 0^+$. Apply Kirchhoff's law:

$$E = v_L + v_R$$

Since $v_R = iR$ and $i = 0$ A, then at $t = 0^+$:

$$v_R = 0 \text{ V} \quad \text{and} \quad v_L = E = 20 \text{ V}$$

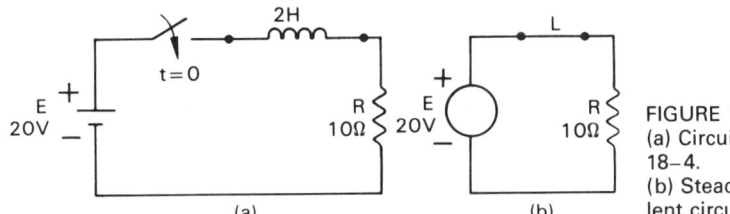

FIGURE 18-4
(a) Circuit of Example 18-4.
(b) Steady state equivalent circuit.

∴ At $t = 0^+$, $v_L = E = 20$ V, $v_R = 0$ V, and $i = 0$ A. After more than five time constants, $t = \infty$, the circuit reaches steady-state conditions, and the inductor can be considered to be a short circuit. Thus, from Figure 18-4(b),

$$E = v_L + v_R$$

Since $v_L = 0$ V at $t = \infty$,

$$E = v_R = 20 \text{ V} \quad \text{and} \quad i = I = \frac{E}{R} = \frac{20}{10} = 2 \text{ A}$$

∴ At $t = \infty$, $v_L = 0$ V, $v_R = E = 20$ V, and $i = 2$ A.

EXAMPLE 18-5 The RL circuit of Figure 18-5 is in a steady-state condition ($t = \infty$). At $t = 0$, the switch is moved to position 2. Determine the voltages and currents at $t = 0^+$ and $t = \infty$ for the switch in position 2.

SOLUTION At $t = 0^-$, the current is equal to

$$i = \frac{E}{R} = \frac{10}{10} = 1 \text{ A}$$

and

$$v_R = iR = 1(10) = 10 \text{ V}$$

Observation The voltage across L (v_L) at $t = 0^-$ is zero since the current was not changing at that time. Thus $v_L = 0$ V.

∴ At $t = 0^-$, $i = 1$ A, $v_R = 10$ V, and $v_L = 0$ V.

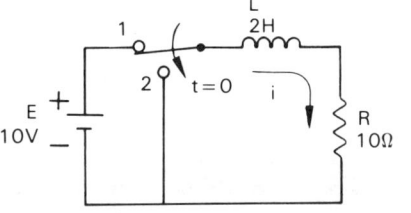

FIGURE 18-5
Circuit of Examples 18-5, 18-6, and 18-7.

486

RL Transient Circuits

At $t = 0^+$, the current in L and the circuit remain at $i = 1$ A. However, to satisfy Kirchhoff's law, $v_L = v_R = 10$ V.

Observation The voltage drop across L at $t = 0^+$ is equal but opposite in polarity to v_R since Kirchhoff's law still must be satisfied. Note: v_L is the source.

∴ At $t = 0^+$, $i = 1$ A, $v_R = -10$ V, and $v_L = +10$ V.

Observation At $t = \infty$, all the energy originally stored in the inductor has dissipated in R.

∴ At $t = \infty$, $i = 0$ A, $v_R = 0$ V, and $v_L = 0$ V.

18–4 INSTANTANEOUS VALUES OF RL CIRCUITS

You have already learned how to calculate instantaneous values in an RC circuit, and the procedure for performing the calculations in an RL circuit is basically the same, since the format of the equations is identical. The only major difference is that in the RL circuit the current rather than the voltage will charge or discharge.

Fall Equation for Current

For a circuit where an initial steady-state current, I, exists and then decays (falls) toward zero, the relationship between instantaneous current, i, at any time, t, and the L/R time constant, $\tau = L/R$, is expressed as

$$i = Ie^{-t/\tau} \quad \text{(A)} \qquad (18-2)$$

where $\tau = L/R$ and $I = E/R$.

EXAMPLE 18–6 Determine the current in the circuit of Figure 18–5 0.5 s after the switch is moved from position 1 to position 2.

SOLUTION Determine the steady-state current, I, at $t = 0^-$:

$$I = \frac{E}{R} = \frac{10}{10} = 1 \text{ A}$$

Determine the L/R time constant, τ:

$$\tau = \frac{L}{R} = \frac{2}{10} = 0.2 \text{ s}$$

Substitute into Equation 18–2 at $t = 0.5$ s:

$$i = Ie^{-t/\tau} = 1e^{-0.5/0.2} = 0.082 \text{ A}$$

∴ At $t = 0.5$ s, $i = 82$ mA.

EXAMPLE 18–7 For the circuit of Figure 18–5, let $L = 15$ mH, $R = 1$ kΩ, and $E = 12$ V. Using these given val-

Instantaneous Values of RL Circuits

ues, determine the current in the circuit 20 μs after the switch is moved from position 1 to position 2.

SOLUTION Determine the steady-state current, I, at $t = 0^-$:

$$I = \frac{E}{R} = \frac{12}{1 \times 10^3} = 12 \text{ mA}$$

Determine the L/R time constant, τ:

$$\tau = \frac{L}{R} = \frac{15 \times 10^{-3}}{1 \times 10^3} = 15 \text{ μs}$$

Substitute into Equation 18–2 at $t = 20$ μs:

$$i = Ie^{-t/\tau} = (12 \times 10^{-3})e^{\frac{-(20 \times 10^{-6})}{15 \times 10^{-6}}}$$

$$i = 3.16 \times 10^{-3} \text{ A}$$

∴ At $t = 20$ μs, $i = 3.16$ mA.

Rise Equation for Current

For a circuit where either a voltage or current source is applied to an RL circuit (as shown in Figure 18–6), the instantaneous circuit current, i, will rise toward the steady-state current, I. The relationship between the instantaneous current, i, at any time, t, and the L/R time constant, $\tau = L/R$, is expressed as

$$i = I(1 - e^{-t/\tau}) \quad \text{(A)} \tag{18-3}$$

where $I = E/R$ and $\tau = L/R$.

EXAMPLE 18–8 For the circuit of Figure 18–6(a), the current source of $I_N = 50$ mA, $R_N = 2$ kΩ is applied to an inductor of 100 mH. Determine the current in the inductor 80 μs after the switch is closed.

Observation Initially, zero current is in the inductor at $t = 0^-$. After the switch is closed, at $t = 0^+$, the current in the inductor is still zero; however, it has started its exponential rise toward a maximum value of 50 mA ($i = 50$ mA), which is the source current, I_N.

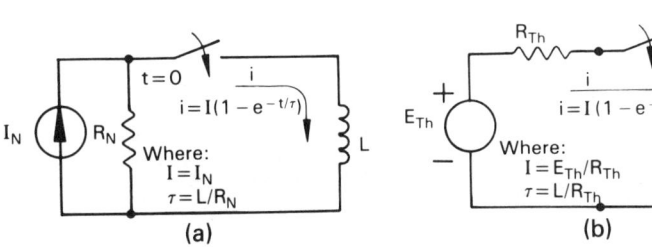

FIGURE 18–6
(a) Inductor connected to equivalent current source with charge equation indicated.
(b) Inductor connected to equivalent voltage source with charge equation indicated.

SOLUTION Use Equation 18–3. First determine I and τ; then substitute:

$$I = I_N = 50 \text{ mA}$$

and

$$\tau = \frac{L}{R_N} = \frac{100 \times 10^{-3}}{2 \times 10^3} = 50 \text{ }\mu\text{s}$$

Substitute into Equation 18–3:

$$i = I(1 - e^{-t/\tau}) = 50 \times 10^{-3}[1 - e^{\frac{-(80 \times 10^{-6})}{(50 \times 10^{-6})}}]$$

$$i = 50 \times 10^{-3}(1 - 0.202) = 39.9 \times 10^{-3}$$

∴ At $t = 80$ μs, $i = 39.9$ mA

EXAMPLE 18–9 In the circuit of Figure 18–7, determine at what instant of time the voltage across R_2 is equal to 5 V.

Observation Upon investigation of the circuit, it is determined that the maximum voltage across R_2 at $t = \infty$ is 10 V. Thus, by the voltage-divider equation,

$$V_2 = \frac{ER_2}{R_1 + R_2} = \frac{(12)10 \text{ k}\Omega}{12 \text{ k}\Omega} = 10 \text{ V}$$

∴ The voltage across R_2 will increase from 0 to 10 V.

SOLUTION Since no specific equation for voltage across a resistance has thus far been given, we will select an appropriate one. To aid in this selection, we will first sketch the anticipated wave form using our understanding of the behavior of the circuit.

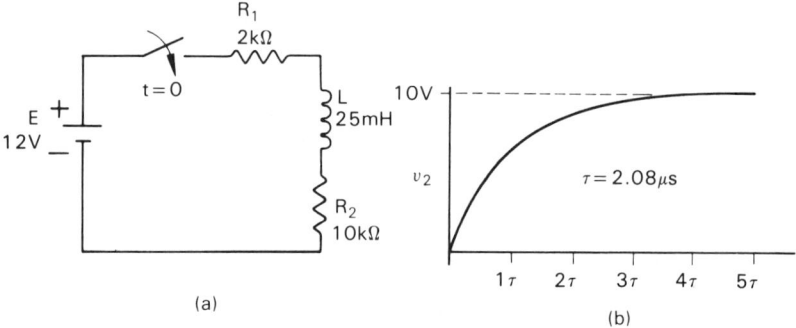

FIGURE 18–7
(a) Circuit of Example 18–9.
(b) Voltage wave form, v_2, across R_2 from $t = 0$ to $t = 5\tau$.

Instantaneous Values of RL Circuits

In this case, the voltage increases exponentially, as pictured in Figure 18–7(b), from 0 to 10 V in five time constants. Select Equation 18–3, which describes the charging current for its form, and replace the reference to current with voltage symbols. Thus

$$i = I(1 - e^{-t/\tau}) \Rightarrow v = V(1 - e^{-t/\tau})$$

And for our needs

$$v_2 = V_2(1 - e^{-t/\tau})$$

where $v_2 = 5$ V, $V_2 = 10$ V, and $\tau = 25$ mH/12 kΩ = 2.08 μs. Substitute and solve for time, t:

$$5 = 10(1 - e^{-t/2.08 \times 10^{-6}})$$

Divide both members by 10:

$$0.5 = 1 - e^{-t/2.08 \times 10^{-6}}$$

Subtract one from each member and change signs:

$$0.5 = e^{-t/2.08 \times 10^{-6}}$$

Take the natural logarithm of each member:

$$\ln(0.5) = \frac{-t}{2.08 \times 10^{-6}} \ln(e)$$

Solve for t and change signs:

$$t = -2.08 \times 10^{-6} \ln(0.5) = 1.44 \times 10^{-6} \text{ s}$$

$\therefore \quad t = 1.44 \;\mu$s

Observation The solution of 1.44 μs for the voltage to rise to 5 V across R_2 is reasonable since we know that 6.3 V corresponds to a time of 2.08 μs. That is, the voltage is 63% of its final value in one time constant. Thus, 2.08 μs (1 τ) corresponds to 6.3 V (63% of 10 V).

Summary The most systematic way of solving RC and RL transient circuits is to first determine the initial and final values of voltage and current. Next, sketch the anticipated wave form. Finally, develop an appropriate equation to solve for the desired unknown.

The general equation forms for rise (charge) and fall (discharge) of current or voltage in either an RC (where $\tau = RC$) or

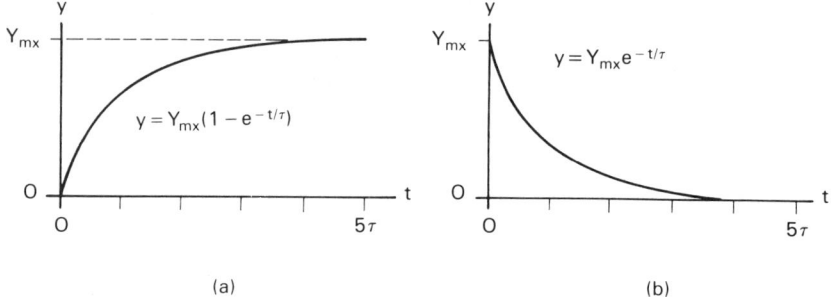

FIGURE 18-8 (a) Wave form and instantaneous equation for a charging variable.
(b) Wave form and instantaneous equation for a discharging variable.

an RL (where $\tau = L/R$) circuit are given in Figure 18-8 and repeated here. Thus, for rise (charge),

$$y = Y_{mx}(1 - e^{-t/\tau}) \qquad \text{Figure 18-8(a)}$$

And for fall (discharge),

$$y = Y_{mx}e^{-t/\tau} \qquad \text{Figure 18-8(b)}$$

18-5 APPLICATIONS

RL circuits are not as common as RC circuits because of the size and cost of inductors. Because of these factors, RL circuits are rarely used in timing circuits. Instead, RC circuits are easily synthesized to operate in these and similar applications at considerable cost and space saving. RL and RC circuits are both used in filter applications and resonant circuits.

High-Voltage Transients

One common use for RL circuits is to generate high-voltage spikes. This is possible because the current tends to remain constant over small time periods in an RL circuit. This concept is demonstrated in the following example.

EXAMPLE 18-10 The switch in Figure 18-9(a) is in position 1 for a time long enough for the circuit to have reached steady state. The switch is then moved to position 2, as shown in Figure 18-9(b). Determine the voltage across L at $t = 0^-$ and $t = 0^+$.

SOLUTION At $t = 0^-$, the current i is

$$i = \frac{E}{R} = \frac{10}{10} = 1 \text{ A}$$

Applications

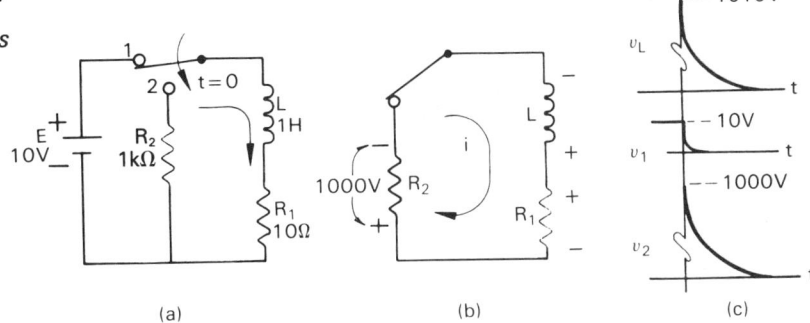

FIGURE 18-9
(a) Circuit of Example 18-10.
(b) Voltage polarities at t = 0⁺. *Note:* v_L is now the source.
(c) Wave forms of voltages at t = >0 with switch as in (b).

∴ The voltage across L is zero, and $v_R = 10$ V.
At $t = 0^+$, since the current does not change instantaneously, the voltage across R_1 remains

$$v_1(0^+) = i(0^+)\, R_1 = 1\text{ A}(10\ \Omega) = 10\text{ V}$$

The voltage across R_2 is

$$v_2(0^+) = i(0^+)\, R_2 = 1\text{ A}(1000\ \Omega) = 1000\text{ V}$$

By Kirchhoff's law,

$$v_L(0^+) = v_1(0^+) + v_2(0^+)$$

$$v_L(0^+) = 10\text{ V} + 1000\text{ V} = 1010\text{ V}$$

∴ $v_L = 1.01$ kV when the source, E, is 10 V.

Observation The important point to note is that a very large instantaneous voltage was obtained with only a small voltage source. Of course, the voltage will decay rapidly, but the fact remains that for an instant of time a large voltage is present. The wave forms of v_L, v_1, and v_2 are depicted in Figure 18-9(c).

One of the most common applications for this technique of creating high voltage is in the automobile ignition system. A simplified schematic of an ignition system is shown in Figure 18-10. When the engine is running, the breaker points in the distributor are constantly opening and closing. While the points are closed, the primary coil is charging to a high-current condi-

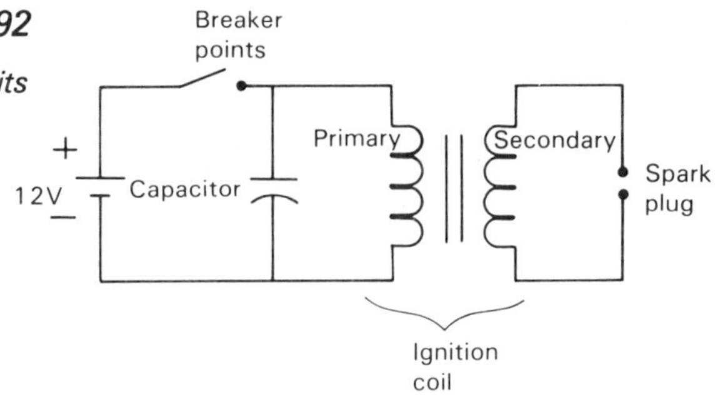

FIGURE 18-10 Simplified schematic of automobile ignition system.

tion. When the points are open, a large resistance is created (open circuit), similar to the previous example, which causes a voltage in the range of a few hundred volts to appear across the primary winding. In addition, the transformer action of the ignition coil boosts the voltage at the output of the secondary winding to more than 20 000 V. This voltage generates the spark required to ignite the fuel. The capacitor is needed to prevent the points from burning owing to the large spark that would appear there in the absence of the capacitor. Thus, the capacitor stores energy that would otherwise form a spark. This action both lengthens the point life and improves the magnitude of the output voltage.

When working with electrical or electronic inductive circuits using devices such as motors and transformers, care must be taken not to instantaneously interrupt the flow of curent in the device in an unsafe manner. A typical example of this would be to unplug a vacuum cleaner from the wall socket without first turning the appliance off. A spark will undoubtedly be created at the wall socket, which could be either hazardous to you or detrimental to the device.

PROBLEMS

Sections 18-1 and 18-2

18-1 Determine the time constant of the following:
(a) $L = 3$ H, $R = 50$ Ω
(b) $L = 50$ mH, $R = 2.7$ kΩ
(c) $L = 300$ mH, $R = 8$ Ω
(d) $L = 250$ μH, $R = 700$ Ω

18-2 Determine the value of R for a time constant of 100 ms when L is 35 mH.

18-3 Determine the value of L for a time constant of 50 μs when R is 10 kΩ.

18-4 Determine the time constant of the circuit shown in Figure 18-11.

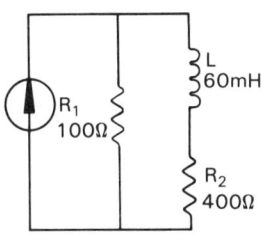

FIGURE 18–11
Circuit of Problems 18–4 and 18–5.

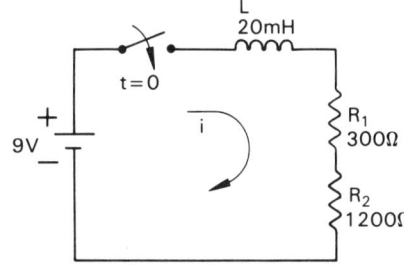

FIGURE 18–12 (right)
Circuit of Problems 18–6 and 18–16.

18–5 For Figure 18–11, the value of R_1 is changed to 1 kΩ and L to 100 mH (R_2 is unchanged). Determine the time constant of the new circuit.

Section 18–3

18–6 Using Figure 18–12, determine the initial and final voltages and currents of the following circuits:
(a) $i(0^+)$ and $i(\infty)$
(b) $v_L(0^+)$ and $v_L(\infty)$
(c) $v_1(0^+)$ and $v_1(\infty)$
(d) $v_2(0^+)$ and $v_2(\infty)$

18–7 If the switch in Figure 18–13 is initially in position 1 for a long period of time ($t > 5\tau$) and is then switched to position 2 at $t = 0$, determine the following at $t = 0^+$ and $t = \infty$.
(a) $i_1(0^+)$, $i_1(\infty)$
(b) $i_2(0^+)$, $i_2(\infty)$
(c) $v_{R_1}(0^+)$, $v_{R_1}(\infty)$
(d) $v_{R_2}(0^+)$, $v_{R_2}(\infty)$
(e) $v_{R_3}(0^+)$, $v_{R_3}(\infty)$

18–8 Draw a sketch of the wave forms of the following parameters of Figure 18–13 with appropriate voltage and current amplitudes. Don't worry about the actual time constants at this time.
(a) i_{L_1} (b) v_{R_1} (c) v_{L_2}

18–9 Determine the time constant of the circuit in Figure 18–14 when the switch is in position 1.

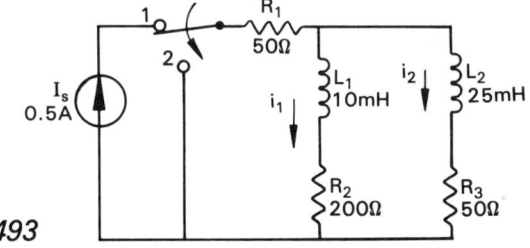

FIGURE 18–13
Circuit of Problems 18–7 and 18–8.

RL Transient Circuits

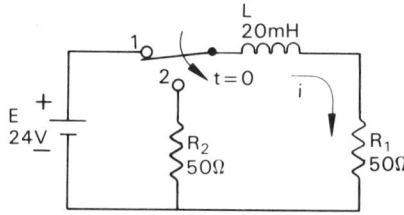

FIGURE 18–14
Circuit of Problems 18–9 through 18–14, 18–17, and 18–18.

18–10 Determine the time constant of the circuit of Figure 18–14 with the switch in position 2.

18–11 For the circuit of Figure 18–14, move the switch to position 2 at $t = 0$ and determine the current i at $t = 0^-$ and $t = 0^+$.

18–12 For the circuit of Figure 18–14, move the switch to position 2 at $t = 0$ and determine the following voltages.
(a) $v_L(0^-)$ (d) $v_1(0^+)$
(b) $v_L(0^+)$ (e) $v_2(0^-)$
(c) $v_1(0^-)$ (f) $v_2(0^+)$

18–13 Sketch the wave form of the following parameters in the circuit of Figure 18–14 when the switch is moved from 1 to 2. Label the initial and final values and the time constants:
(a) i (b) v_L (c) v_1 (d) v_2

18–14 Determine the wave forms of the parameters of Problem 18–13 when, after a time duration greater than five time constants, the switch is moved back to position 1.

18–15 Determine the following currents and voltages at $t = 0^-$ and $t = 0^+$ of Figure 18–15: (a) v_{L_1}, (b) v_{L_2}, (c) v_{R_1}, (d) v_{R_2}, (e) i_1, (f) i_2.

Sections 18–4 and 18–5

18–16 For the circuit of Figure 18–12, determine the value of v_1 20 μs after the switch is closed.

18–17 In the circuit of Figure 18–14, determine the current, i, 100 μs after the switch is moved to position 2. Assume that the switch was in position 1 for a time long enough for the circuit to reach steady state.

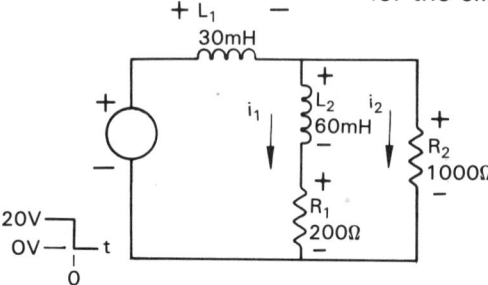

FIGURE 18–15
Circuit of Problem 18–15.

495

Problems

18-18 For the circuit of Figure 18-14, assume that the switch is in position 2 and is then moved back to position 1 10 s later. Determine the value of v_1 100 μs after the change occurs.

18-19 A solenoid requires 500 mA in its coil before it can actuate. The solenoid has an inductance of 250 mH and a resistance of 6.0 Ω. Determine how long it will take for the solenoid to actuate once a 9-V battery is applied.

18-20 The solenoid of Problem 18-19 remains activated until the current drops below 350 mA. How long after the battery is removed will the solenoid deactivate? Assume that in removing the battery the leads are shorted together.

18-21 A 5-V rectangular dc pulse with equal on and off times of 100 μs is applied to a loudspeaker with a 0.25-mH inductance and a dc resistance of 7 Ω. Draw a detailed current wave form for two cycles of the dc pulse showing current and time values.

18-22 For the circuit of Figure 18-16, determine the voltage v_L at time $t = 22$ ms if the duration of the pulse, T, is 20 ms.

18-23 Draw and label the current wave forms of i_1 and i_2 in Problem 18-22 from $t = 0$ ms to $t = 40$ ms.

18-24 Determine the voltage v_L at time $t = 6$ ms if $T = 3$ ms in Figure 18-16.

18-25 Draw and label the current wave forms of i_1 and i_2 in Problem 18-24 from $t = 0$ ms to $t = 6$ ms.

18-26 The SPST switch of Figure 18-4(a) has a practical resistance of 1 MΩ in the open position. Determine the voltage developed across the switch when it is opened after it had been closed for $>5\tau$.

18-27 To prevent large voltage spikes when an inductive circuit is turned off, a discharge resistor is often placed in parallel with the inductor. Determine the value of discharge resistor to be used if the insulation of the inductor in Figure 18-4(a) can safely handle 1000 V.

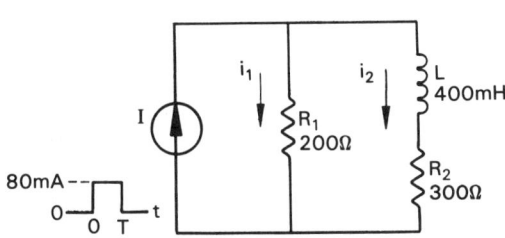

FIGURE 18-16
Circuit of Problems 18-22 through 18-25.

III AC Circuits, Concepts, and Devices

This section of the book builds onto your understanding of dc circuits and concepts by adding another dimension to circuit concepts—time. Concepts such as reactance, impedance, and admittance will enhance your knowledge of circuits. Besides ac circuits and networks, you will learn about power, decibels, resonance, transformers, and three-phase systems. Once you master ac circuits, you will then be able to comprehend advanced topics in electronics. These topics include filters, frequency domain, and active devices (amplifiers).

19 Alternating Current

19-1 ELEMENTARY GENERATOR AND AC TERMINOLOGY
19-2 SINE WAVE
19-3 PHASE RELATION OF R, L, AND C ELEMENTS
19-4 EFFECTIVE AND AVERAGE VALUE OF CURRENT AND VOLTAGE
19-5 AVERAGE POWER

CHAPTER PREVIEW

An ac generator produces a time-varying voltage. When a load is placed across an ac source, the resulting current is also time-varying. This chapter introduces you to alternating current by exploring the sine-wave properties of alternating current. You will learn the names and units used to describe the sinusoidal nature of alternating current. You will also study how the varying amplitudes of the voltage and the current are related to the sine wave. The chapter concludes with a discussion of how average power is determined from sinusoidal voltage and current.

PERFORMANCE OBJECTIVES

Once you have read each section, worked each example with pencil, paper, and calculator, and answered each problem of every section, you will be able to:

☐ Recall the names and units associated with the sine wave.
☐ Write the instantaneous equation of ac voltage and current for a resistive, inductive, and capacitive load.
☐ Compute the effective values of periodic wave forms.
☐ Determine the power dissipated by reactive components.
☐ Compute the inductive and capacitive reactance.
☐ **Determine the effective and average value of sinusoidal wave forms, as well as various dc wave forms.**
☐ Compute the average power dissipated by a resistive load.

19-1 ELEMENTARY GENERATOR AND AC TERMINOLOGY

When a conductor is rotated through a uniform magnetic flux, as pictured in Figure 19-1(a), a voltage is induced in it. If the conductor is rotated at a constant rate, the induced voltage is an alternating voltage having a sinusoidal wave form, as pictured in Figure 19-1(b).

A generator is a machine in which many electric conductors are rotated through a magnetic field to produce ac voltage.

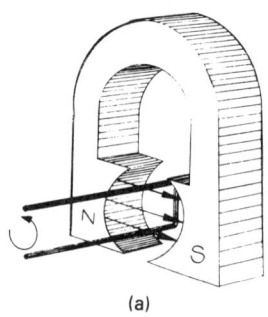
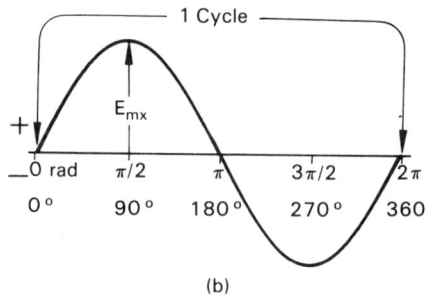

FIGURE 19-1
(a) As the conductor is rotated through the magnetic lines of force, a voltage is induced within the conductor.
(b) The voltage has a sinusoidal wave shape when the conductor is rotated at a constant rate.

501 Within the generator, the mechanical energy, in the form of the rotating conductor, is converted to electric energy in the form of an ac voltage.

Frequency In the elementary generator of Figure 19–1(a), the number of times in a second that the conductor is rotated through a complete revolution to produce a voltage is called the *frequency* of the voltage. As an example, suppose the conductor of Figure 19–1(a) is rotated 600 times at a constant rate through the magnetic lines of force in 10 s; the frequency of the resulting sine-wave voltage is 60 cycles per second. The SI unit for frequency is the hertz (Hz). Thus, a frequency of 60 cycles per second is expressed as 60 Hz.

Period The *period* of the induced voltage wave form is the time it takes for one complete cycle to be made by the conductor through the magnetic flux. Thus, the period of the voltage induced when the conductor is rotated 60 times a second (60 Hz) through the magnetic flux is $\frac{1}{60}$ or 16.67 ms. The relationship between frequency (f) in hertz and period (T) in seconds is expressed in Equation 19–1.

$$T = \frac{1}{f} \text{ (s)} \tag{19–1}$$

EXAMPLE 19–1 Determine the frequency and the period of an induced voltage if a conductor is rotated through a magnetic flux at a constant rate of 200 revolutions in 5 s.

SOLUTION Express 200 revolutions in 5 s in hertz:

$$\frac{200 \text{ rev}}{5 \text{ s}} = 40 \text{ cycles per second} = 40 \text{ Hz}$$

Determine the period of the voltage wave form. Use Equation 19–1:

$$T = \frac{1}{f} = \frac{1}{40} = 25 \text{ ms}$$

∴ The frequency is 40 Hz and the period is 25 ms.

Angular Velocity Each time the conductor of Figure 19–1(a) rotates through one cycle, the conductor travels a circular distance of 2π radians (2π = 6.28 rad = 360°). Since the conductor rotates at a constant rate, the rate of travel, called the *angular velocity*, is also a constant. The angular velocity is expressed in radians per second (rad/s). Thus, a wave generated by a conductor traveling through 1000 rad in 10 s would have an angular velocity of 1000 rad/10 s or 100 rad/s. The following equation relates angular velocity (ω) in radians per second to angular displacement (α) in radians and time (t) in seconds.

$$\omega = \frac{\alpha}{t} \quad \text{(rad/s)} \tag{19-2}$$

EXAMPLE 19-2 Express an angular displacement of 60 rad in 5 s as an angular velocity in radians per second.

SOLUTION Use Equation 19-2:

$$\omega = \frac{\alpha}{t} = \frac{60 \text{ rad}}{5 \text{ s}} = 12 \text{ rad/s}$$

∴ The angular velocity is 12 rad/s.

Frequency (f) is related to angular velocity by Equation 19-3 where 2π (6.28 rad) is the number of radians traversed by each cycle of the wave forms and f is the number of cycles per second. The Greek letter omega (ω) is used exclusively for angular velocity. Thus

$$\omega = 2\pi f \quad \text{(rad/s)} \tag{19-3}$$

EXAMPLE 19-3 Determine the angular velocity of a 60-Hz sinusoidal voltage wave.

SOLUTION Use Equation 19-3:

$$\omega = 2\pi f = 2\pi 60 = 377 \text{ rad/s}$$

∴ The angular velocity of a 60-Hz wave is 377 rad/s.

19-2 SINE WAVE

If the conductors within the generator producing an ac voltage are turned at a constant rate, the periodic wave form resulting from this action will also have a constant frequency. Furthermore, since the induced voltage in the rotating conductor depends directly on the rate at which the conductor moves through the magnetic lines of force, a constant angular velocity will yield a wave form with the same maximum amplitude for each cycle of the periodic sinusoidal voltage wave form. Thus, the wave form pictured in Figure 19-1(b) repeats indefinitely and has a constant frequency, period, and maximum amplitude (E_{mx}).

Time and Displacement

The motion of the rotating conductor in the magnetic flux and the maximum induced voltage (E_{mx}) may be represented by a rotating vector that has one end fixed, as shown in Figure 19-2(a). The length of the vector is E_{mx} and the rate of its rotation is equal to the angular velocity (ω) of the conductor. The direction of rotation is counterclockwise (ccw). From Equation 19-2, the displacement (α) in radians is a direct function of time since the angular velocity (ω) is a constant. Thus

$$\alpha = \omega t \quad \text{(rad)} \tag{19-4}$$

Figure 19-2(a) represents the rotating conductor of the generators and Figure 19-2(b) represents the induced voltage in the conductor. From Figure 19-2, an equation that relates the *instantaneous* voltage (e) to the amplitude of the wave (E_{mx}) and the angular displacement (α) may be derived. Applying the concepts of trigonometry to the triangle of Figure 19-2(a), we write a sine function relating the instantaneous voltage (e) to the angular displacement (α) and the amplitude of the wave E_{mx}. Thus

$$\sin(\alpha) = \frac{e}{E_{mx}}$$

Solve for e:

$$e = E_{mx} \sin(\alpha) \quad \text{(V)} \tag{19-5}$$

Since the angular displacement (α) is related to time by Equation 19-4, $\alpha = \omega t$, then

$$e = E_{mx} \sin(\omega t) \quad \text{(V)} \tag{19-6}$$

Equations 19-5 and 19-6 provide a means of determining the amount of induced voltage in the conductor at any position of the conductor in the magnetic field. Equation 19-5 relates instantaneous voltage to the angular displacement from the reference axis, while Equation 19-6 relates the instantaneous voltage to time. Looking once again at Figure 19-2, you may now

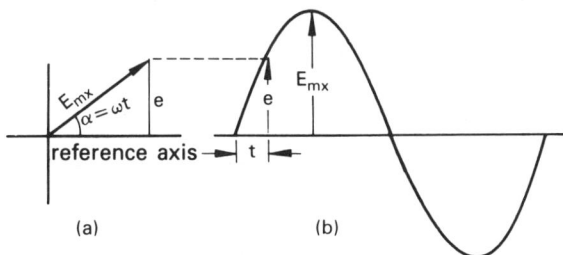

FIGURE 19-2
The vector (a) displays the conditions of the periodic wave (b) at time t.

EXAMPLE 19–4 Determine the instantaneous voltage (e) of a 400-Hz ac source (generator) that has a maximum amplitude (E_{mx}) of 68 V for the following times: (a) $t_1 = 0.4$ ms, (b) $t_2 = 1.25$ ms, and (c) $t_3 = 1.8$ ms.

SOLUTION Use Equation 19–4. Let $\omega = 2\pi f = 2\pi 400$.

(a) $e = E_{mx} \sin(\omega t)$

$e = 68 \sin(2\pi 400 \times 0.4 \times 10^{-3})$

$e = 68 \sin(1.01 \text{ rad}) = 57.4$ V

∴ $e = 57.4$ V at $t = 0.4$ ms

(b) $e = 68 \sin(2\pi 400 \times 1.25 \times 10^{-3})$

$e = 68 \sin(\pi \text{ rad}) = 0$ V

∴ $e = 0$ V at $t = 1.25$ ms

(c) $e = 68 \sin(2\pi 400 \times 1.8 \times 10^{-3})$

$e = 68 \sin(4.524 \text{ rad}) = -66.8$ V

∴ $e = -66.8$ V at $t = 1.8$ ms

Observation Format the calculator to operate on angles expressed in radians by depressing the RAD or DRG key before taking the sine.

EXAMPLE 19–5 Determine the angular displacement (α) from the reference axis in radians and degrees for each of the given times of Example 19–4. *Note:* 1 radian ≈ 57.3°.

SOLUTION Use Equation 19–4. Let $\omega = 2\pi f = 2\pi 400 = 2513$ rad/s.

(a) $t_1 = 0.4$ ms

$\alpha = \omega t = 2513(0.4 \times 10^{-3})$

∴ $\alpha = 1.01$ rad $= 57.9°$

(b) $t_2 = 1.25$ ms

$\alpha = \omega t = 2513(1.25 \times 10^{-3})$

∴ $\alpha = 3.14$ rad $= 180°$

(c) $t_3 = 1.8$ ms

$$\alpha = \omega t = 2513(1.8 \times 10^{-3})$$

$$\therefore \quad \alpha = 4.52 \text{ rad} = 259°$$

EXAMPLE 19-6 Express the induced voltage in a conductor as a function of time (t) when the conductor is rotated through the magnetic lines of force at a frequency of 60 Hz. The amplitude of the voltage (E_{mx}) is 165 V.

SOLUTION Use Equation 19-3. Determine ω for a frequency of 60 Hz:

$$\omega = 2\pi f = 2\pi 60 = 377 \text{ rad/s}$$

Substitute into Equation 19-6:

$$e = E_{mx} \sin(\omega t)$$

$$\therefore \quad e = 165 \sin(377t) \text{ V}$$

19-3 PHASE RELATION OF R, L, AND C ELEMENTS

Resistance

The ac generator is represented in Figure 19-3(a) by the sine wave within the source symbol. When a resistive load, R, is connected across the terminals of the ac source, an ac current passes through the resistance causing an ac voltage to be dropped across the resistance, as pictured in Figure 19-3(a). When the curve picturing the current in the resistance is placed over the curve depicting the voltage across the resistance, as shown in Figure 19-3(b), it will be noted that the two waves start and finish together. Furthermore, both waves reach their maximum positive amplitude and their maximum negative amplitude together. When the current and voltage pass through corresponding points in their cycle at the same time, the two waves are said to be *in phase*.

The displacement angle between the voltage and current wave is called the *phase angle*. For a resistance, the voltage and current are in phase and the phase angle is zero ($\theta = 0°$). The Greek letter theta (θ) is used to designate the phase angle in this text.

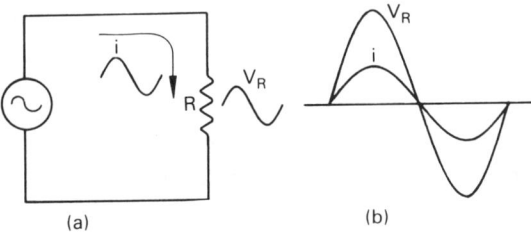

FIGURE 19-3
When an alternating current (i) passes through a resistance (a), the resulting voltage (v_R) is in phase with the current (b), and the phase angle is 0°.

The equations describing the wave forms of current and voltage of the resistor in Figure 19–3(b) are

$$v_R = E_{mx} \sin(\omega t) \quad \text{(V)} \tag{19-7a}$$

$$i = I_{mx} \sin(\omega t) \quad \text{(A)} \tag{19-7b}$$

EXAMPLE 19–7 Write the instantaneous equations for voltage and current in the resistance of the circuit for Figure 19–3(a) when the 60-Hz source has a maximum amplitude of 22 V and $R = 10\ \Omega$.

SOLUTION Since the angular velocity is the same for both the current and voltage,

$$\omega = 2\pi f = 2\pi 60 = 377\ \text{rad/s}$$

Substitute:

$$v_R = E_{mx} \sin(\omega t)$$

$$\therefore \quad v_R = 22 \sin(377t)\ \text{V}$$

By Ohm's law, $I_{mx} = E_{mx}/R$:

$$I_{mx} = \frac{22}{10} = 2.2\ \text{A}$$

Substitute:

$$i = I_{mx} \sin(\omega t)$$

$$\therefore \quad i = 2.2 \sin(377t)\ \text{A}$$

To summarize, when an ac voltage source is applied to a resistive load, the current and voltage are in phase, the phase angle is zero ($\theta = 0°$), and Ohm's law ($E_{mx} = I_{mx}R$) may be applied.

Inductance As you know, the inductance of an inductor is measured in units of henrys (H). When an alternating current from an ac sinusoidal source passes through an inductor, the inductor "reacts" to the current by producing a counter-emf that opposes the current in the inductor. This opposition is termed the *inductive reactance* of the inductor. The inductive reactance, X_L (in ohms), of an inductor, L (in henrys), depends on the frequency of the ac current and the size of the inductor, as described in the following equation.

$$X_L = \omega L = 2\pi f L \quad (\Omega) \tag{19-8}$$

From Equation 19–8, $X_L = 0\ \Omega$ when $f = 0$; that is, direct current ($f = 0$ Hz) will have no opposition from the inductor. However, when the frequency of the ac current is high, the opposition to the current will be very large. Because of the frequency dependence of the inductive reactance, an inductor may be used to *filter out* unwanted frequencies, while allowing wanted frequencies to pass unopposed. This concept is demonstrated in the next example.

EXAMPLE 19-8 Determine the inductive reactance of a 50-mH RF choke coil (inductor) to an ac current with the following frequencies: (a) 0 Hz (dc) and (b) 10 MHz (rf).

SOLUTION Use Equation 19-8:

(a) $X_L = 2\pi f L = 2\pi(0)(50 \times 10^{-3}) = 0\ \Omega$

∴ $X_L = 0\ \Omega$ at 0 Hz

(b) $X_L = 2\pi f L = 2\pi(10 \times 10^6)(50 \times 10^{-3})$

$X_L = 3.14\ M\Omega$

∴ $X_L = 3.14\ M\Omega$ at 10 MHz

The inductor derives its opposition to alternating current from its counter-emf, and this same property causes the voltage across the inductor to *lead* the current passing through the inductor. From your work in Chapter 18, you learned that a sudden change in voltage across an inductor did not produce a corresponding change in current in the inductor. Instead, the change in current lagged behind the change in voltage. It is, of course, this behavior that gives the inductor its unique property of opposing a change in current. Thus, when the curve representing the current in the inductor of Figure 19-4(a) is placed on the same axis as the curve picturing voltage across the inductor, as shown in Figure 19-4(b), it is seen that the voltage wave *leads* (has a more positive amplitude at the origin than the current wave) the current wave by 90° and the phase angle, θ, is +90°. The equations describing the wave forms of current and voltage of the inductor in Figure 19-4(b) are

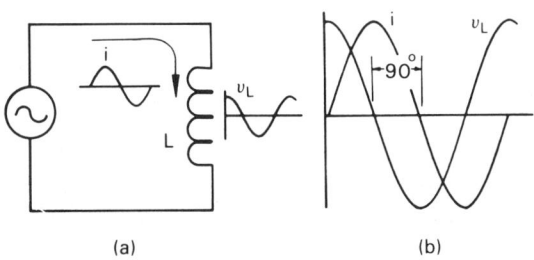

(a)   (b)

FIGURE 19-4
When an ac sinusoidal current (i) is in an inductor (a), the voltage (v_L) leads the current (b) by 90°.

$$v_L = E_{mx} \sin\left(\omega t + \frac{\pi}{2}\right) \text{ (V)}$$

$$i = I_{mx} \sin(\omega t) \text{ (A)}$$

Since ωt is in radians (the SI unit for plane angles), the phase angle of 90° must also be expressed in radians as $\pi/2$. Thus, 2π rad = 360°, π rad = 180°, and $\pi/2$ rad = 90°.

EXAMPLE 19–9 Write the instantaneous equations for voltage and current in the inductor of the circuit of Figure 19–4(a) when the 400-Hz source has a maximum amplitude of 165 V and L is 0.25 H.

SOLUTION Determine the angular velocity:

$$\omega = 2\pi f = 2\pi 400 = 2513 \text{ rad/s}$$

Substitute:

$$v_L = E_{mx} \sin\left(\omega t + \frac{\pi}{2}\right)$$

$$\therefore \quad v_L = 165 \sin\left(2513t + \frac{\pi}{2}\right) \text{ V}$$

By Ohm's law, $I_{mx} = E_{mx}/X_L$ and $X_L = \omega L$:

$$I_{mx} = \frac{E_{mx}}{\omega L} = \frac{165}{2513 \times 0.25} = 0.263 \text{ A}$$

Substitute:

$$i = I_{mx} \sin(\omega t)$$

$$\therefore \quad i = 0.263 \sin(2513t) \text{ A}$$

To summarize, the application of an ac sinusoidal voltage to an inductor, as in Figure 19–4(a), causes an ac current with the same angular velocity to pass through the inductor, which reacts to the passage of current ($X_L = \omega L$) by opposing the current. This process causes a phase shift between the current and voltage of the inductor, resulting in a phase angle (θ) of 90°. Thus, in a perfect inductor (one without resistance), voltage leads current by 90°. Because the inductor is a linear bilateral device, Ohm's law, in the form $E_{mx} = I_{mx}X_L$, may be used to compute the circuit conditions.

Capacitance As you know, the capacitance of a capacitor is measured in units of farads (F). When an alternating current from an ac sinusoidal source passes through a capacitor, the capacitor reacts by not instantaneously changing voltage across the plates of the

Phase Relation of R, L, and C Elements

capacitor; thus, an opposition to the current is set up. This opposition is called the *capacitive reactance*, X_C, of the capacitor. The capacitive reactance, X_C (in ohms), of a capacitor, C (in farads), depends on the frequency of the ac current and the size of the capacitor, as described in the following equation.

$$X_C = \frac{1}{\omega C} = \frac{1}{2\pi f C} \quad (\Omega) \tag{19-9}$$

From Equation 19–9, $X_C = \infty\ \Omega$ when $f = 0$; that is, direct current ($f = 0$ Hz) will not pass in a capacitor. However, when the frequency of the ac current is high, the opposition to the current will be very small. Because of the frequency dependence of the capacitive reactance, a capacitor may be used to block direct current, yet allow alternating current to pass unopposed. This idea is explored in the following example.

EXAMPLE 19–10 Determine the capacitive reactance of a 10-μF capacitor to an ac current with the following frequencies: (a) 0 Hz (dc) and (b) 10 kHz (AF).

SOLUTION Use Equation 19–9:

(a) $X_C = \dfrac{1}{2\pi f C} = \dfrac{1}{2\pi 0 \times 10 \times 10^{-6}} = \infty\ \Omega$

∴ X_C is an open circuit at $f = 0$ Hz.

(b) $X_C = \dfrac{1}{2\pi f C} = \dfrac{1}{2\pi 10 \times 10^3 \times 10 \times 10^{-6}}$

$X_C = 1.59\ \Omega$

∴ $X_C = 1.59\ \Omega$ at 10 kHz.

The capacitor derives its opposition to alternating current from its inability to take on or give up charge instantaneously. It is this property that causes the voltage across the capacitor to *lag* the current passing through the capacitor. From your work in Chapter 14, you learned that a sudden change in current in a capacitor did not produce a corresponding change in voltage across the capacitor. Instead, the change in voltage lagged behind the change in current. It is, of course, this behavior that gives the capacitor its unique property of opposing a change in voltage. Thus, when the curve representing the current in the capacitor of Figure 19–5(a) is placed on the same axis as the curve depicting the voltage across the capacitor, as noted in Figure 19–5(b), it is seen that the voltage *lags* (has a less positive amplitude at the origin than the current wave) the current wave by 90°, and the phase angle, θ, is $-90°$. The equations describing

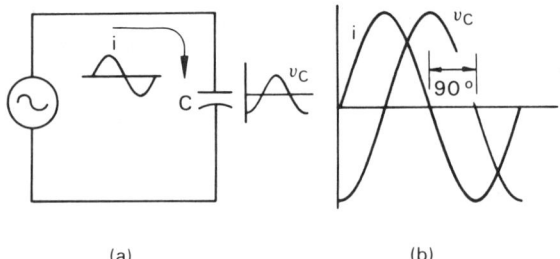

FIGURE 19-5
When an ac sinusoidal current (i) is in a capacitor (a), the voltage (v_C) lags the current (b) by 90°.

the wave forms of current and voltage of the capacitor in Figure 19-5(b) are

$$v_C = E_{mx} \sin\left(\omega t - \frac{\pi}{2}\right) \quad (V)$$

$$i = I_{mx} \sin(\omega t) \quad (A)$$

Note that $-90°$ is represented in radians as $-\pi/2$.

EXAMPLE 19-11 Write the instantaneous equation for voltage and current in the capacitor of the circuit of Figure 19-5(a) when the 2-kHz source has a maximum amplitude of 12 V and $C = 0.22\ \mu F$.

SOLUTION Determine the angular velocity:

$$\omega = 2\pi f = 2\pi 2 \times 10^3 = 12.57\ \text{krad/s}$$

Substitute:

$$v_C = E_{mx} \sin\left(\omega t - \frac{\pi}{2}\right)$$

$$\therefore\quad v_C = 12 \sin\left(12.57 \times 10^3 t - \frac{\pi}{2}\right)\ V$$

By Ohm's law, $I_{mx} = E_{mx}/X_C$ and $X_C = 1/(\omega C)$:

$$I_{mx} = E_{mx}(\omega C)$$
$$I_{mx} = (12)(12.57 \times 10^3)(0.22 \times 10^{-6})$$
$$I_{mx} = 33.2\ \text{mA}$$

Substitute:

$$i = I_{mx} \sin(\omega t)$$

$$\therefore\quad i = 33.2 \sin(12.57 \times 10^3 t)\ \text{mA}$$

To summarize, the application of an ac sinusoidal voltage to a capacitor, as in Figure 19-5(a), causes the capacitor to react ($X_C = 1/\omega C$). This process causes the voltage to lag behind the current by 90°, which results in a phase angle (θ) of $-90°$. Be-

Phase Relation of R, L, and C Elements

cause the capacitor is a linear bilateral device, Ohm's law, in the form $E = IX_C$, is used to compute the circuit condition of a perfect capacitive circuit (one without resistance).

Summary When writing a system of instantaneous equations, one equation is written as the reference equation, while the other equation contains the phase angle. Either equation may be selected as the reference equation; however, since current is constant throughout a series circuit, the current is usually selected as the reference in a series circuit. The following equations are the generalized equations for the instantaneous voltage and current.

$$e = E_{mx} \sin(\omega t \pm \theta) \quad (V) \tag{19-10}$$

$$i = I_{mx} \sin(\omega t \pm \theta) \quad (A) \tag{19-11}$$

Special care must be taken when evaluating the sine of $\omega t \pm \theta$. Both ωt and θ must be in the same units before adding and evaluating. This is particularly troublesome because θ may be given in radians or degrees while ωt is always in radians.

EXAMPLE 19-12 From the diagram of the instantaneous voltage and current of Figure 19-6, write the instantaneous equations for voltage and current at an angular velocity of 377 rad/s. The amplitudes of the waves are $E_{mx} = 110$ V, $I_{mx} = 3$ A, and the phase angle, θ, is 36°.

SOLUTION Select voltage as the reference; thus, the voltage equation will have zero phase angle:

$$e = E_{mx} \sin(\omega t \pm \theta)$$

$$\therefore \quad e = 110 \sin(377t + 0) \text{ V}$$

The current equation will indicate the lagging phase angle; convert $-36°$ to radians:

$$\theta = \frac{-36° \pi}{180°} = -0.628 \text{ radians}$$

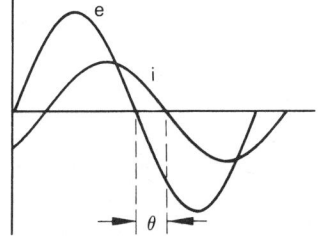

FIGURE 19-6
The wave forms of voltage and current for Example 19-12.

Alternating Current

Substitute:

$$i = I_{mx} \sin(\omega t \pm \theta)$$

$$\therefore \quad i = 3 \sin(377t - 0.628) \text{ A}$$

In Figure 19–6, we see that current is lagging voltage by some phase angle, θ. This observation may also be stated as voltage leading current by the same phase angle, θ. The phase relationship between current and voltage in a series equivalent circuit is an indication of the components in the circuit. For example, with the current wave as the reference, we know that $\theta = 0°$ indicates a resistive circuit, $\theta = 90°$ indicates an inductive circuit, and $\theta = -90°$ indicates a capacitive circuit. That is, voltage is coincident with current in a resistive circuit ($\theta = 0°$), voltage leads current in an inductive circuit ($0° < \theta \leq 90°$), and voltage lags current in a capacitive circuit ($-90° \leq \theta < 0°$).

Because voltage leads current in the circuit represented by the wave forms of Figure 19–6, the circuit must contain an inductor. Table 19–1 summarizes these concepts.

EXAMPLE 19–13 Determine the instantaneous current when the voltage has completed 1.5 rad of its cycle. The voltage, $E_{mx} = 50$ V, lags the current, $I_{mx} = 0.9$ A, by 18°.

SOLUTION Because the instantaneous current *tracks* (moves) with the voltage, the current has also completed 1.5 rad of its cycle. Write the equation for current with voltage as the reference.

TABLE 19–1 SUMMARY OF THE EFFECTS OF ALTERNATING CURRENT ON RESISTANCE, INDUCTANCE, AND CAPACITANCE

Property	Phase (θ)	Name	Opposition Unit	Dependency on Frequency	Ohm's Law
Resistance	E and I are in phase: $\theta = 0°$	Resistance	Ohm	None	$E = IR$
Inductance	E leads I by 90°: $\theta = 90°$	Inductive reactance	Ohm	$X_L = 2\pi f L$	$E = IX_L$
Capacitance	E lags I by 90°: $\theta = -90°$	Capacitive reactance	Ohm	$X_C = \dfrac{1}{2\pi f C}$	$E = IX_C$

$$i = I_{\text{mx}} \sin(\omega t + \theta)$$

Observation The phase angle, θ, is indicated as a positive angle because voltage is assumed to be the reference and voltage lags current. Therefore, when voltage is at zero, current is ahead of voltage (leads) by 18°. Substitute $\alpha = \omega t = 1.5$ rad, $\theta = 18° = 18°\pi/180° = 0.314$ rad, and $I_{\text{mx}} = 0.9$ A:

$$i = I_{\text{mx}} \sin(\omega t + \theta)$$
$$i = 0.9 \sin(1.5 + 0.314)$$
$$\therefore \quad i = 0.874 \text{ A}$$

Observation Remember to format your calculator in radians by depressing the angular mode key $\boxed{\text{RAD}}$ or $\boxed{\text{DRG}}$ before taking the sine.

EXAMPLE 19–14 The voltage across a capacitor lags the current in the capacitor by 90°. Determine the instantaneous voltage across the capacitor at $t = 11$ ms if the angular velocity of the sine wave is 400 rad/s and $E_{\text{mx}} = 60$ V.

SOLUTION Use Equation 19–10. Let $v_C = e$, $\theta = -90° = -90°\pi/180° = -1.57$ rad, and $\omega = 400$ rad/s.

$$e = E_{\text{mx}} \sin(\omega t \pm \theta)$$
$$v_C = 60 \sin(400 \times 11 \times 10^{-3} - 1.57)$$
$$\therefore \quad v_C = 18.4 \text{ V at } t = 11 \text{ ms}$$

19–4 EFFECTIVE AND AVERAGE VALUE OF CURRENT AND VOLTAGE

In order to determine the power produced in a resistive load, the effective value of current and/or voltage must be known. The effective value of a periodic alternating current or voltage may be precisely determined with calculus. However, since we assume no knowledge of calculus, an alternative method for approximating the effective value of alternating current or voltage is used.

An approximate value of effective current may be determined from the graph of the wave form of the current or voltage by dividing the area bounded by the curve (wave form) into many small segments or *increments*. The ordinate (vertical axis value) of each increment of area is *squared* and added to the squared values of all the other increments within the wave form. This value is then divided by the number of increments to determine the *mean* value (average value). Finally, the square *root* is taken to yield the effective value of current.

From the previous description, you may rightly conclude that the effective value of any periodic wave form is the result of taking the square **root** of the **mean** of the **squared** ordinate values of current or voltage of both the positive and negative portions of the wave. Because of this process, the effective value is also called the *root mean square* or simply the rms value.

The meaning of the words squared, mean, and root are explored in the following narrative example. **Squared** refers to the *sum* of the terms that have been squared. Suppose we have four values of current, 3, 2, 4, and 5 A. The sum of the squared values would be 9 + 4 + 16 + 25 or 54 A. The **mean** of the squared values would be the average value, which is found by dividing the sum of the squared terms by the number of terms, or $\frac{54}{4}$ = 13.5 A. The **root** of the mean of the squared terms is found by taking the second root of the mean value, or $\sqrt{13.5}$ = 3.67 A.

Unless otherwise specified, all reference to voltage or current will imply effective or rms value of current or voltage. The capital letters I for current and V or E for voltage imply that the effective (rms) value is meant.

Approximating the Effective (RMS) Quantities from Wave Forms

In the following examples, we will approximate the effective current of several wave forms commonly used in electronic curcuits. Although effective current is being determined in the following examples, the effective voltage may also be approximated by the demonstrated techniques. Equation 19–12 is a statement of the procedure used to compute the root mean square (rms) or effective value of current.

$$I = \sqrt{\frac{i_1^2(t_1 - t_0) + i_2^2(t_2 - t_1) + \cdots + i_n^2(t_n - t_{n-1})}{\Sigma T}} \quad (A) \quad (19\text{--}12)$$

where I = approximate effective current (rms) in amperes
i_n = *average instantaneous amplitude* (height) of the current wave at time increment $(t_n - t_{n-1})$
$t_n - t_{n-1}$ = one time increment of the wave = T
ΣT = total number of time increments

When the time increment $t_n - t_{n-1}$ is 1, Equation 19–12 may be stated as Equation 19–13.

$$I = \sqrt{\frac{i_1^2 + i_2^2 + \cdots + i_n^2}{\Sigma T}} \quad (A) \quad (19\text{--}13)$$

EXAMPLE 19–15 Approximate the effective current of one cycle of the periodic sinusoidal current wave form of Figure 19–7.

Effective and Average Value of Current and Voltage

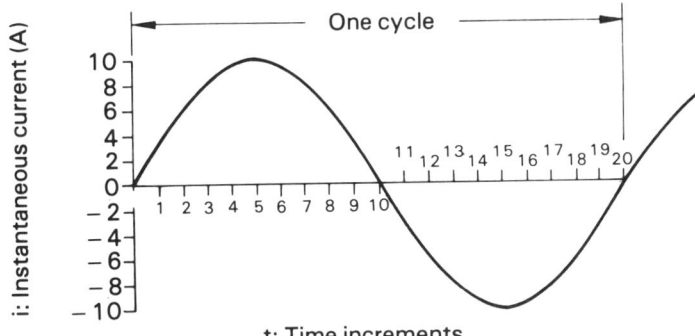

FIGURE 19-7 Periodic ac sinusoidal wave form with one cycle divided into 20 time increments for Example 19-15.

SOLUTION Divide the horizontal axis into several increments of time. After labeling the increments, the time axis (horizontal axis) appears as pictured in Figure 19-7. Select values of instantaneous current (vertical axis) within each increment of time at the midpoint of the time increment. Construct a table of values as shown in Table 19-2.

Substitute into Equation 19-13:

$$I = \sqrt{\frac{4 + 16 + 49 + 81 + 90 + \cdots}{20}}$$

Observation Because of space limitations, only the first five terms from Table 19-2 are shown. The other 15 terms would, of course, be added to the indicated terms. The first term is computed in the following manner:

$$i_1 = 2 \quad \text{and} \quad i_1^2 = 4; \quad t_1 - t_0 = 1 - 0 = 1$$

TABLE 19-2 VALUES OF INSTANTANEOUS CURRENT (i) FOR THE WAVE FORM OF FIGURE 19-7

Increment	i	i^2	$i^2(t_n - t_{n-1})$	Increment	i	i^2	$i^2(t_n - t_{n-1})$
1	2	4	4(1)	11	-2	4	4(1)
2	4	16	16(1)	12	-4	16	16(1)
3	7	49	49(1)	13	-7	49	49(1)
4	9	81	81(1)	14	-9	81	81(1)
5	9.5	90	90(1)	15	-9.5	90	90(1)
6	9.5	90	90(1)	16	-9.5	90	90(1)
7	9	81	81(1)	17	-9	81	81(1)
8	7	49	49(1)	18	-7	49	49(1)
9	4	16	16(1)	19	-4	16	16(1)
10	2	4	4(1)	20	-2	4	4(1)

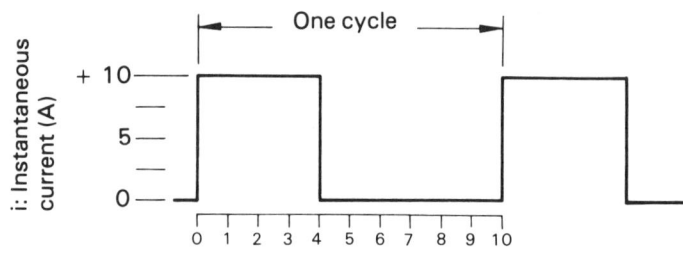

FIGURE 19-8 Periodic dc rectangular pulse wave form with one cycle divided into 10 time increments for Example 19-16.

Substitution into $i_1^2(t_1 - t_0)$ results in $4(1) = 4$. The process is repeated for each term within the radical. Thus:

$$I = \sqrt{\frac{960}{20}} = \sqrt{48} = 6.93$$

∴ The effective value of current for the wave forms of Figure 19-7 is 6.93 A.

Observation Applying the methods of calculus to the wave form of Figure 19-7 results in $I = 7.07$ A. The approximate value of 6.93 A is −2% in error.

EXAMPLE 19-16 Approximate the effective current of one cycle of the periodic rectangular pulse wave form shown in Figure 19-8.

SOLUTION Substitute into Equation 19-13:

$$I = \sqrt{\frac{10^2 + 10^2 + 10^2 + 10^2 + 0 + \cdots + 0}{10}}$$

$$I = \sqrt{\frac{400}{10}} = \sqrt{40} = 6.32$$

∴ The effective current is 6.32 A.

Observation The effective current of the wave form of Figure 19-8, determined by calculus methods, is also 6.32 A.

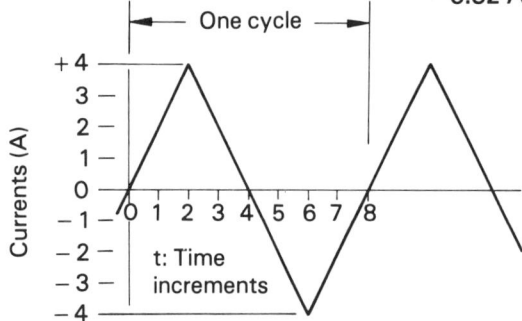

FIGURE 19-9 Periodic ac triangle wave form with one cycle divided into 8 time increments for Example 19-17.

Effective and Average Value of Current and Voltage

EXAMPLE 19-17 The triangular wave form of Figure 19-9 is displayed on the face of an oscilloscope. An rms ammeter is measuring the effective value of the current as 2.4 A. Determine if this reading is reasonable.

SOLUTION Apply Equation 19-13 to determine the approximate rms value of one cycle of the current wave form of Figure 19-9. Select values of i midway in each time increment.

$$I = \sqrt{\frac{1^2 + 3^2 + 3^2 + 1^2 + (-1)^2 + (-3)^2 + (-3)^2 + (-1)^2}{8}}$$

$$I = \sqrt{\frac{40}{8}} = \sqrt{5} = 2.2 \text{ A}$$

∴ An ammeter reading of 2.4 A is reasonable.

Observation Applying the methods of calculus results in an effective current of 2.3 A.

Computing the Effective Values of Current or Voltage

Table 19-3 pictures several wave forms common to electronic circuits. The effective value of voltage or current may be determined for a particular wave form, listed in Table 19-3, by substituting the peak value (maximum amplitude) of the wave form into Equation 19-14.

$$I = kI_p \text{ (A)} \quad \text{or} \quad E = kE_p \text{ (V)} \tag{19-14}$$

where I or E = effective value of current or voltage
I_p or E_p = peak (maximum) value of current or voltage
k = a constant selected from Table 19-3

Example 19-18 demonstrates the use of Equation 19-14 and Table 19-3 in determining the effective values of current and voltage.

EXAMPLE 19-18 Determine the effective value of current or voltage given the following periodic waves.
(a) Sine wave with a maximum amplitude of 20 V.
(b) Half-wave pulse with a peak amplitude of 8 A.
(c) Square wave with a maximum amplitude of 5 V.
(d) Triangular wave with a peak amplitude of 15 V.
(e) Rectangular pulse with a peak amplitude of 300 mA and equal *on* and *off* times.

TABLE 19-3   STEADY AND PERIODIC WAVE FORM CHARACTERISTICS

Wave-form Shape (i = instantaneous current)	Wave-form Classification*	$i = f(t)$	$I = kI_p$†	Notes (t = time)
	Positive amplitude Steady dc	No	$I = 1.0I_p$	
	Negative amplitude Steady dc	No	$I = 1.0I_p$	
	Periodic ac Square wave	Yes	$I = 1.0I_p$	Symmetrical wave
	Positive amplitude Periodic dc Rectangular pulse	Yes	$I = 0.707I_p$	ON t = OFF t
	Periodic ac Sine wave	Yes	$I = 0.707I_p$	Symmetrical wave
	Positive amplitude Periodic dc Half-wave pulse	Yes	$I = 0.50I_p$	ON t = OFF t
	Positive amplitude Periodic dc Full-wave pulse	Yes	$I = 0.707I_p$	
	Periodic ac Triangular wave	Yes	$I = 0.577I_p$	

* The term *steady* indicates that the amplitude of the wave is not a function of time [$i \neq f(t)$].
† The effective current (I) is related to peak current (I_p) by the equation $I = kI_p$, where k is a constant for a given wave shape.

SOLUTION Using the information in Table 19-3:
(a) Substitute $k = 0.707$ into Equation 19-14:

$$E_p = 20 \text{ V for the sine wave}$$
$$E = (0.707)(20)$$
$$\therefore E = 14.1 \text{ V}$$

(b) Substitute $k = 0.50$ into Equation 19-14:

$$I_p = 8 \text{ A for the half-wave pulse}$$
$$I = (0.50)(8)$$
$$\therefore I = 4.0 \text{ A}$$

(c) Substitute $k = 1.0$ into Equation 19–14:

$E_p = 5$ V for the square wave

$E = (1.0)(5)$

∴ $E = 5.0$ V

(d) Substitute $k = 0.577$ into Equation 19–14:

$E_p = 15$ V for the triangular wave

$E = (0.577)(15)$

∴ $E = 8.66$ V

(e) Substitute $k = 0.707$ into Equation 19–14:

$I_p = 0.300$ A for the rectangular pulse

$I = (0.707)(0.300)$

∴ $I = 0.212$ A

In summary, when determining the effective value (rms) of current or voltage, first determine if the wave form is like any of those listed in Table 19–3. If it is, use the appropriate formula to determine the effective value of current or voltage. If the wave form is not listed in Table 19–3, apply the techniques of Equation 19–13 to approximate the effective value of (rms) of current or voltage.

Average Value The average value of a periodic wave form is determined by computing the area under the curve for one cycle and then dividing this area by the period of the wave. The area above the time axis is assigned a positive sign, while the area below the time axis is assigned a negative sign. The area of the curve over one period of the wave ($T = 1/f$) is the algebraic sum of the areas.

A symmetrical wave form, that is, a wave form with equal areas above and below the reference or time axis, will have an average value equal to zero. Thus, periodic ac sine waves, cosine waves, square waves, and triangular waves will have a zero average value.

The average value of any (ac or dc) voltage or current may be measured with a dc meter with a D'Arsonval meter movement (used in a VOM). Thus, the average value over one cycle is the dc reading on a dc meter. Once again, symmetrical, periodic ac waves, like the sine and cosine waves pictured in Figures 19–1, 19–3, 19–4, and 19–5, have no dc level. However, steady dc and periodic dc pulses, such as those pictured in Table 19–4, all

TABLE 19-4 AVERAGE VALUE OF SELECTED STEADY DC AND PERIODIC DC PULSE WAVE FORMS

Wave-form Shape	Wave-form Classification	Average Value (dc)	Notes
	Positive amplitude Steady dc	$E_{AV} = 1E_p$	
	Periodic dc Rectangular pulse	$E_{AV} = 0.5E_p$	$t_1 = t_2$
	Periodic dc Half-wave pulse	$E_{AV} = 0.318E_p$	
	Periodic dc Full-wave pulse	$E_{AV} = 0.637E_p$	

have a dc level. The dc level is stated in terms of a multiplier and the peak (maximum) amplitude of the wave.

19–5 AVERAGE POWER

In the preceding section, we introduced the idea that the power delivered to the load depends on the effective value of the current and voltage, which, in turn, depends on the shape of the wave. Table 19-3 listed several steady and periodic wave forms and their characteristics. We recommend that you review this material at this time.

Power in a Resistive Load

When a periodic sinusoidal ac voltage of the general form $e = E_{mx} \sin(\omega t)$ is applied to a resistive load, a circuit current results in the form of $i = I_{mx} \sin(\omega t)$, which is in phase with the voltage having a phase angle equal to zero ($\theta = 0°$). Plotting the two waves represented by the instantaneous equations and then multiplying the instantaneous amplitudes results in a third wave, whose area represents the instantaneous power, as shown in Figure 19–10. Thus

$$p = ei$$

and

(1) $e = E_{mx} \sin(\omega t)$

(2) $i = I_{mx} \sin(\omega t)$

Substitute Equations (1) and (2) into $p = ei$:

$$p = [E_{mx} \sin(\omega t)][I_{mx} \sin(\omega t)]$$

$$p = E_{mx} I_{mx} \sin^2(\omega t)$$

521

Average Power

However, $\sin^2(\omega t) = \frac{1}{2}[1 - \cos(2\omega t)]$. Thus

$$p = \frac{E_{mx}I_{mx}}{2}[1 - \cos(2\omega t)]$$

Distributing,

$$p = \frac{E_{mx}I_{mx}}{2} - \frac{E_{mx}I_{mx}}{2}\cos(2\omega t)$$

Observation. The second term is a cosine wave with a frequency twice that of the voltage or current wave. Since the average value of a cosine wave is zero, the second term is zero and only the first term remains.

This term is called the *average power* and is expressed in a formula as

$$P = \frac{E_{mx}I_{mx}}{2} \quad \text{(W)} \tag{19-15}$$

EXAMPLE 19-19 Determine the power dissipated by a resistor that has a voltage of $e = 40 \sin(377t)$ and a current of $i = 0.6 \sin(377t)$.

SOLUTION Apply Equation 19-15:

$$P = \frac{E_{mx}I_{mx}}{2} = \frac{(40)(0.6)}{2} = 12 \text{ W}$$

∴ The power dissipated by the resistor is 12 W.

Power in a Reactive Load

When a sinusoidal ac voltage is applied to a reactive load (inductor or capacitor), the voltage is 90° out of phase with the current. As with a resistive load, when the instantaneous amplitudes of voltage and current are multiplied, the area under the resultant wave represents the instantaneous power of the reactor, as shown in Figure 19-11. The average power is the average of the power curve, which, for a reactor, is zero since there is equal area above and below the reference axis. Thus, reactive loads (inductors and capacitors) do not dissipate power. In any circuit, only resistance dissipates power.

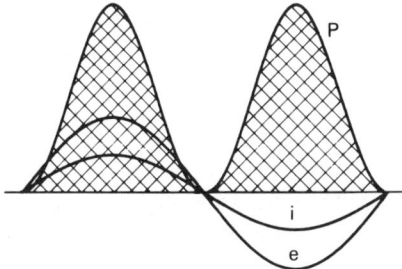

FIGURE 19-10
Instantaneous power (crosshatched) in a resistive load is always positive.

Effective Value of Sinusoidal Voltage and Current

It is the effective current and voltage (designated by the capital letters I and E) that are used in computing the average power in linear resistive loads. From Equation 19–15 ($P = E_{mx}I_{mx}/2$), we may derive an expression relating the amplitude of the ac sinusoidal wave form (I_{mx} and E_{mx}) to the effective value of current and voltage. Thus, factoring the right member of Equation 19–15 with the common denominator results in

$$P = \frac{E_{mx}}{\sqrt{2}} \times \frac{I_{mx}}{\sqrt{2}}$$

Replace $1/\sqrt{2}$ with 0.707:

$$P = (0.707 E_{mx})(0.707 I_{mx})$$

However, from Table 19–3, $I = 0.707 I_{mx}$; thus

$$I = 0.707 I_{mx} \quad (A) \tag{19–16}$$

$$E = 0.707 E_{mx} \quad (V) \tag{19–17}$$

and

$$P = EI = (0.707\ E_{mx})(0.707\ I_{mx}) = \frac{E_{mx}I_{mx}}{2} \quad (W) \tag{19–18}$$

The average power in a resistive circuit with an ac source is the product of the effective voltage (E) and the effective current (I). Table 19–5 is a summary of ac voltage and current notation.

EXAMPLE 19–20 Determine the effective values of current and voltage and the power dissipated by a load connected to an ac source with a voltage of $e = 24 \sin(400t + 90°)$ and a current of $i = 2 \sin(400t)$.

SOLUTION Use Equations 19–16 and 19–17 to determine the effective quantities:

$$I = 0.707 I_{mx} = 0.707(2) = 1.41 \text{ A}$$

$$E = 0.707 E_{mx} = 0.707(24) = 17.0 \text{ V}$$

∴ The effective current is 1.41 A and the effective voltage is 17.0 V.

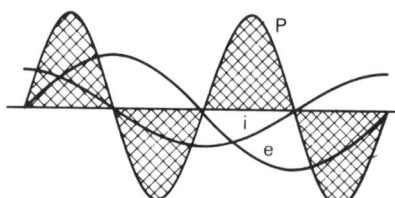

FIGURE 19–11
Instantaneous power in a reactive load is both positive and negative.

TABLE 19–5 AC VOLTAGE AND CURRENT NOTATION

AC Quantity	Notation	Example	Visual Representation	Comments
Peak-to-peak	E_{pp}, V_{pp} or I_{pp}	$E_{pp} = 20$ V	Peak-to-peak 20V	Twice maximum
Maximum or peak	E_{mx} or E_p I_{mx} or I_p	$E_p = 10$ V	Peak 10V	$\frac{1}{2}$ peak to peak
Effective or rms	E or I	$E = 7.07$ V	rms 7.07V	No subscript, $0.707 \times$ peak

Observation The power dissipated is zero watts because the load is an inductor, as indicated by the phase angle of 90°. Inductors, like capacitors, do not dissipate electric energy. Only resistance dissipates energy.

PROBLEMS

Section 19–1

19–1 Express the frequency, in hertz, of 120 cycles in 5 s.

19–2 Express the frequency, in hertz, of 18 rotations in 6 s.

19–3 Express the frequency, in hertz, of a wave with a period of 52 ms.

19–4 Express the frequency, in hertz, of a wave with a period of 14 μs.

19–5 Express the frequency, in hertz, of a wave with an angular velocity of 1225 rad/s.

19–6 Express the frequency, in hertz, of a wave with an angular velocity of 440 rad/s.

19–7 Determine (a) the period and (b) the angular velocity of a 7-kHz wave.

19–8 Determine (a) the frequency and (b) the angular velocity of a wave with a 27-ms period.

Section 19–2

19–9 Determine the angular displacement (in both radians and degrees) of a sinusoidal wave with an angular velocity of 250 rad/s when the wave is 5 ms into its cycle.

19–10 Determine the angular displacement (in radians and degrees) of a 1.6-kHz wave 150 μs into its cycle.

19–11 Determine the maximum amplitude of a sine wave of voltage having an instantaneous amplitude of 44 V at an angle of 1.6 rad.

19–12 Determine the instantaneous amplitude of a 60-Hz sine wave of voltage with a maximum amplitude of 92 V at a time of 14 ms.

19-13 Write the instantaneous equation for the voltage of a sine wave that has a period of 62.5 μs and a peak amplitude of 5.6 V.

19-14 From the voltage equation $e = 12 \sin(700t)$, determine (a) the frequency of the voltage, (b) the period of the voltage wave, and (c) the instantaneous amplitude of voltage 500 μs after the start of the cycle.

Section 19-3

19-15 An 800-Hz sine wave of current in a 100-Ω resistor has a peak amplitude of 8 mA. (a) Write the equation for voltage in terms of time, and (b) determine the instantaneous amplitude of current at 72° of the cycle.

19-16 Determine the inductive reactance, X_L, of a 15-mH inductor at (a) 620 Hz, (b) 10 kHz, and (c) 132 kHz.

19-17 Determine the inductive reactance, X_L, of a 10-H inductor at (a) 60 Hz, (b) 400 Hz, and (c) 1.8 kHz.

19-18 The voltage in a 0.5-H inductor is given by the equation $e = 18 \sin(377t + 20°)$ V. Write the equation for current in terms of time.

19-19 Determine the capacitive reactance, X_C, of a 330-pF capacitor at (a) 400 Hz, (b) 15 kHz, and (c) 250 kHz.

19-20 Determine the capacitive reactance, X_C, of a 0.22-μF capacitor at (a) 60 Hz, (b) 5 kHz, and (c) 2 MHz.

19-21 The current in a 0.1-μF capacitor is given by the equation $i = 50 \sin(12.5 \times 10^3 t)$ mA. Write the equation for voltage in terms of time.

19-22 The current in a circuit is represented by $i = 0.4 \sin(1280t - 37°)$ A. Determine the current at the origin of the wave ($t = 0$ s).

Section 19-4

19-23 Using the information of Table 19-3, determine the effective current or voltage of the following waves: (a) Steady dc wave with an amplitude of -8 A. (b) Periodic dc symmetrical rectangular pulse with an amplitude of 22.7 V. (c) Periodic dc full-wave pulse with a maximum amplitude of 600 mA.

19-24 From the information of Table 19-3, determine the effective current or voltage of the following waves: (a) Periodic dc half-wave pulse with a negative peak amplitude of -40 V. (b) Periodic dc rectangular pulse with an amplitude of 12 mA. (c) Steady dc wave with a negative amplitude of -180 μA.

19-25 Approximate the effective current of the periodic wave forms pictured in Figure 19-12 over one cycle by applying Equation 19-13.

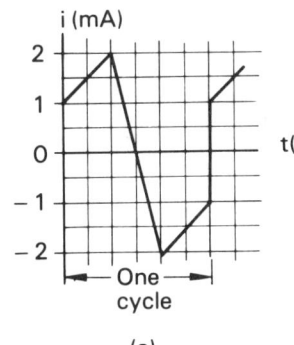
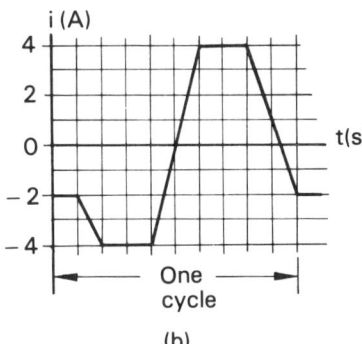

FIGURE 19–12
Wave forms for Problem 19–25.

(a) (b)

19–26 Determine the effective voltage of the periodic wave forms pictured in Figure 19–13 over one cycle by applying Equation 19–13.

Section 19–5 19–27 A 0.75-H inductor is connected to a 60-Hz, 18-V source.
(a) Determine the current (I) in the inductor.
(b) Determine the power dissipated by the inductor.
(c) With the voltage as reference, write the instantaneous sinusoidal equations of current and voltage.

19–28 A 4.7-μF capacitor is connected to a 60-Hz, 28-V source.
(a) Determine the current (I) in the capacitor.
(b) Determine the power dissipated by the capacitor.
(c) With the voltage as reference, write the instantaneous sinusoidal equations of current and voltage.

19–29 Determine the average value of (a) steady dc from a 12-V battery, and (b) a periodic dc half-wave pulse with a peak voltage of 17 V.

19–30 Determine the dc current of a periodic dc full-wave pulse with a peak current of 360 mA.

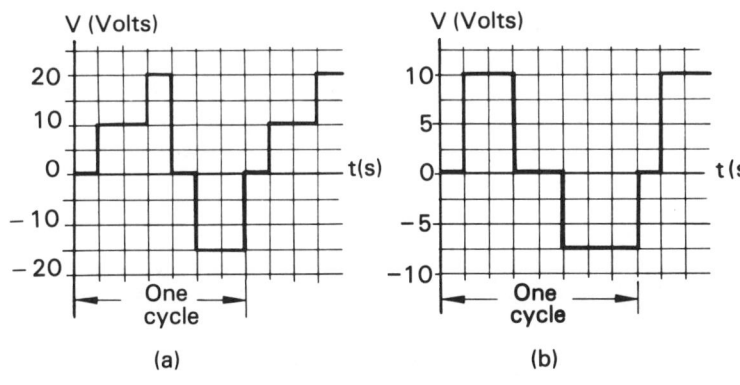

FIGURE 19–13
Wave forms for Problem 19–26.

(a) (b)

19-31 Determine the power dissipated by a resistive load having a voltage across it of $e = 165 \sin(812t)$ and a current of $i = 0.42 \sin(812t)$.

19-32 If the source voltage of an ac circuit is $e = 20 \sin(2513t)$, determine (a) the frequency of the source, (b) the effective value of voltage, and (c) the instantaneous amplitude of the voltage at $t = 1.6$ ms.

19-33 If the current in a series ac circuit is $i = 345 \sin(3770t)$ mA, determine (a) the frequency of the current, (b) the effective value of current, and (c) the instantaneous amplitude of the current at $t = 500$ μs.

19-34 The voltage of Problem 19-32 is applied to a 750 pF capacitor.
  (a) Determine the capacitive reactance of the capacitor.
  (b) Determine the current in the capacitor.
  (c) Write the instantaneous equation for the capacitor's current.

19-35 The current of Problem 19-33 is passed through a 10-H inductor.
  (a) Determine the inductive reactance of the inductor.
  (b) Determine the voltage across the inductor.
  (c) Write the instantaneous equation for the voltage across the inductor.

19-36 A 60-Hz sinusoidal current in a 330-Ω resistor dissipates 1.5 W. Determine (a) the peak current, (b) the effective current, and (c) the average current.

19-37 Determine the peak voltage of an ac source if the circuit current is 210 mA and the power dissipated by a resistive load is 6.3 W.

19-38 Determine the angular velocity of a voltage source that has an instantaneous amplitude of 70% of the maximum amplitude 180 μs after the start of the cycle.

20 Mathematics of Phasors

20-1 COMPLEX NUMBERS
20-2 TRANSFORMING COMPLEX NUMBER FORMS
20-3 ARITHMETIC OPERATIONS WITH COMPLEX NUMBERS
20-4 PHASORS
20-5 OPERATING WITH PHASOR QUANTITIES

CHAPTER PREVIEW Complex numbers are applied to alternating currents and voltages in a network to give quantity to them. Complex numbers are well suited for this application since they describe both the magnitude (amplitude) and the direction (phase) of ac quantities.

Stopping the ac periodic sinusoidal wave forms of voltage and current at selected times provides a means for diagramming the relationship of voltage and current in a network. These diagrams are called *phasor diagrams*. The currents and voltages noted in phasor diagrams are called phasor quantities or simply phasors.

In this chapter you will be introduced to the system of complex numbers, which is applied to current and voltage phasors so that the techniques of algebra may be used with the circuit rules, laws, and theorems.

PERFORMANCE OBJECTIVES Once you have read each section, worked each example with pencil, paper, and calculator, and answered each problem for every section, you will be able to:

☐ Write complex numbers in both rectangular and polar form.
☐ Transform complex number forms between rectangular and polar.
☐ Perform arithmetic operations with complex numbers.
☐ Derive a phasor from an instantaneous equation.
☐ Express a phasor in effective (rms) quantities of voltage and current.
☐ Know when an equation is to be operated on with scalar arithmetic and when it is to be operated on with complex arithmetic.
☐ Apply the rules of complex arithmetic to phasors.

20–1 COMPLEX NUMBERS Before phasors can be applied to the analysis of ac networks, an understanding of how to form complex numbers is needed. Once the forms are understood, the process of transforming complex numbers between forms must be mastered in order to carry out the arithmetic operations with complex numbers.

Rectangular Form A complex number is the coordinates of a point in a *complex plane*, as noted in Figure 20–1. The complex number may be written in either one of two forms: **rectangular** or **polar.**

The rectangular coordinate of the *complex plane* is written with a real number (r-coordinate) followed by an imaginary number (j-coordinate), as in

529
Complex Numbers

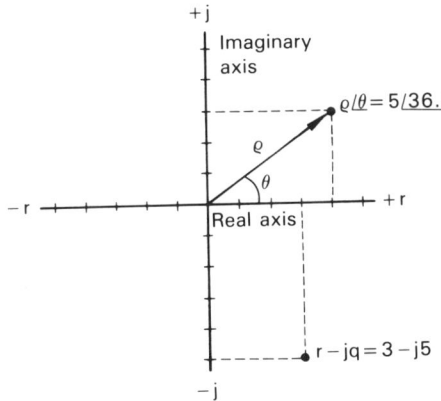

FIGURE 20-1
Complex plane. Coordinates of horizontal axis (r-axis) are real numbers; coordinates of vertical axis (j-axis) are imaginary numbers.

$$r + jq \qquad (20-1)$$

where $r + jq$ = *rectangular form* of a complex number
r = coordinate of the real axis of the complex plane
q = coordinate of the imaginary axis

EXAMPLE 20-1 List several coordinates of the complex plane written in the rectangular form of the complex number.

SOLUTION Use the structure of Equation 20-1: (a) $6 + j3$; (b) $3 - j5$; (c) $-4 - j6$.

Observation The point $3 - j5$ is noted in the complex plane of Figure 20-1.

EXAMPLE 20-2 Using the rectangular form of the complex number, record the coordinates of points A, B, F, H, I, and M of Figure 20-2. Assume each division to be one unit.

SOLUTION Use the structure of Equation 20-1:
Point A: $4 + j0$ Point H: $-2 - j2$
Point B: $2 + j3$ Point I: $0 - j2$
Point F: $-4 + j2$ Point M: $3 - j4$

Polar Form The polar coordinate of the complex plane is written with two numbers. The first number, called the *magnitude*, is the radial distance from the *pole* (location $0 + j0$) to the point. The second number, called the *argument*, is the direction (angle) from the positive real axis to the point. The argument is assigned a posi-

Mathematics of Phasors

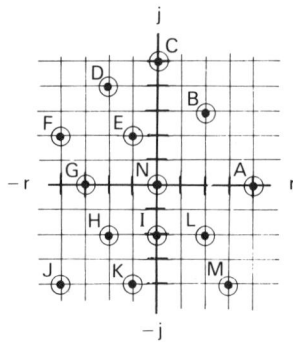

FIGURE 20-2
Points on a complex plane of Example 20-2.

tive sign when the movement is in a *ccw* (counterclockwise) direction and a negative sign when the movement is in a *cw* direction. Equation 20-2 indicates the form of the polar coordinate of the complex plane.

$$\rho \underline{/\theta} \qquad (20-2)$$

where $\rho \underline{/\theta}$ = *polar form* of a complex number
ρ = magnitude of the complex number
θ = direction of the complex number in degrees or radians

EXAMPLE 20-3 Locate the position of the points specified in the complex plane by the following polar coordinates: (a) $6\underline{/45°}$, (b) $3.2\underline{/-30°}$, and (c) $-5\underline{/60°}$.

SOLUTION (a) As noted in Figure 20-3(a).
(b) As noted in Figure 20-3(b).
(c) As noted in Figure 20-3(c).

Observation $-5\underline{/60°}$ is equal to $5\underline{/-120°}$ or $5\underline{/240°}$, as explained by Equation 20-3.

$$-\rho \underline{/\theta} = \rho \underline{/\theta \pm 180°} \qquad (20-3)$$

EXAMPLE 20-4 Express the following polar coordinates with positive magnitudes: (a) $-26\underline{/30°}$, (b) $-6\underline{/-40°}$,

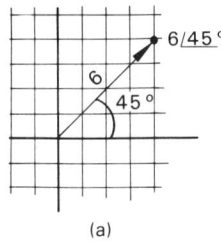
(a)

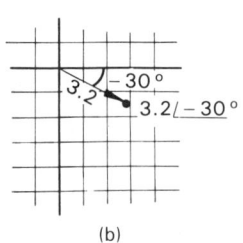
(b)

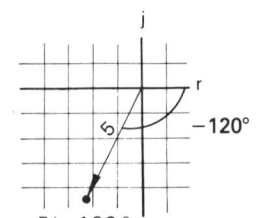
(c)

FIGURE 20-3
Location of polar coordinates of Example 20-3.

Transforming Complex Number Forms

(c) $-15\underline{/120°}$, and (d) $-10\underline{/-100°}$.

SOLUTION Use Equation 20–3:

(a) $26\underline{/-150°}$ (c) $15\underline{/-60°}$
(b) $6\underline{/140°}$ (d) $10\underline{/80°}$

Observation It is conventional to express the argument (angle) of the polar coordinate as a value between $+180°$ and $-180°$; thus, $-180° \leq \theta \leq 180°$.

20–2 TRANSFORMING COMPLEX NUMBER FORMS

Either rectangular coordinates or polar coordinates may be used to indicate a point on a complex plane. Figure 20–4 pictures this relationship. From this figure, we learn that the two systems must be equivalent. Thus

$$\rho\underline{/\theta} \Leftrightarrow r + jq \qquad (20\text{–}4)$$

where $\rho\underline{/\theta}$ = polar coordinate
    $r + jq$ = rectangular coordinate
    $\Leftrightarrow$ = indicates that one form may yield the other form

Rectangular to Polar Transformation

In working with current and voltage in ac circuits, it will be necessary to express these quantities using complex number forms. Depending on the application, these complex quantities may be expressed in either rectangular or polar forms. The process for transforming a complex number from the rectangular form to the polar form may be carried out in one of two ways.

The first and most expedient way is with the rectangular-to-polar calculator key $\boxed{\rightarrow P}$. The second method is defined by Rule 20–1.

RULE 20–1. RECTANGULAR TO POLAR TRANSFORMATION

To change a complex number in rectangular form to polar form $(r + jq \Rightarrow \rho\underline{/\theta})$:

1. $\rho = \sqrt{r^2 + q^2}$
2. $\theta = \text{Arctan}(q/r)$
   (a) If $(r + jq)$ is in quadrant II, then add $180°$ to θ.
   (b) If $(r + jq)$ is in quadrant III, then subtract $180°$ from θ.
3. Form the complex number in polar form: $\rho\underline{/\theta}$.

EXAMPLE 20-5 Express $-3 - j4$ in polar form. (a) Use the $\boxed{\to P}$ calculator key, and (b) use Rule 20-1.

SOLUTION (a) Use $\boxed{\to P}$ key:

$$\therefore \quad -3 - j4 \ \boxed{\to P} \ 5\underline{/-126.9°}$$

Observation Check your calculator owner's guide to determine how to use this powerful feature of your calculator.

(b) Use Rule 20-1:

Step 1: $\quad \rho = \sqrt{9 + 16} = 5$

Step 2: $\quad \theta = \text{Arctan}(-4/-3) = 53.13°$

Step 2(b): Since $-3 - j4$ is in quadrant III, then subtract 180° from θ. Thus

$$\theta = 53.13° - 180° = -126.9°$$

Step 3: $\quad \rho\underline{/\theta} = 5\underline{/-126.9°}$

$$\therefore \quad -3 - j4 \Rightarrow 5\underline{/-126.9°}$$

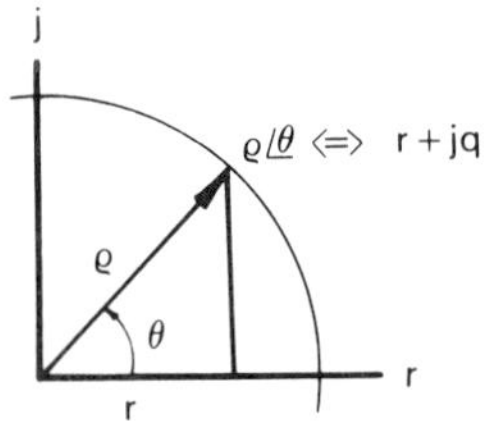

FIGURE 20-4 Relationship between rectangular and polar coordinates. The point on the plane may be expressed in the rectangular or polar form of a complex number.

EXAMPLE 20-6 Express $-7 + j5$ in polar form. (a) Use the $\boxed{\to P}$ calculator key, and (b) use Rule 20-1.

SOLUTION (a) Use $\boxed{\to P}$ key:

$$\therefore \quad -7 + j5 \ \boxed{\to P} \ 8.60\underline{/144.5°}$$

(b) Use Rule 20-1:

Step 1: $\quad \rho = \sqrt{49 + 25} = 8.60$

Step 2: $\quad \theta = \text{Arctan}(5/-7) = -35.54°$

Step 2(a): Since $-7 + j5$ is in quadrant II, then add 180° to θ. Thus

$$\theta = -35.54° + 180° = 144.5°$$

Step 3: $\quad \rho\underline{/\theta} = 8.60\underline{/144.5°}$

$$\therefore \quad -7 + j5 \Rightarrow 8.60\underline{/144.5°}$$

EXAMPLE 20-7 Express $9.3 + j18.9$ in polar form. (a) Use the $\boxed{\to P}$ calculator key and (b) use Rule 20-1.

SOLUTION (a) Use $\boxed{\to P}$ key:

$$\therefore \quad 9.3 + j18.9 \ \boxed{\to P} \ 21.1\underline{/63.8°}$$

(b) Use Rule 20-1:

533

Transforming Complex Number Forms

Step 1: $\rho = \sqrt{86.49 + 357.2} = 21.1$
Step 2: $\theta = \text{Arctan}(18.9/9.3) = 63.8°$
Step 3: $\rho\underline{/\theta} = 21.1\underline{/63.8°}$
∴ $9.3 + j18.9 \Rightarrow 21.1\underline{/63.8°}$

Observation   You are encouraged to use the $\boxed{\rightarrow P}$ calculator key rather than Rule 20–1 to transform from rectangular to polar coordinates.

Polar to Rectangular Transformation

Like the previous transformation of rectangular to polar coordinates, the process for transforming a complex number from polar form to rectangular form may be carried out in one of two ways.

The first and most expedient way is with the polar to rectangular calculator key $\boxed{\rightarrow R}$. The second method is defined by Rule 20–2.

RULE 20–2. POLAR TO RECTANGULAR TRANSFORMATION

To change a complex number in polar form to rectangular form ($\rho\underline{/\theta} \Rightarrow r + jq$):

1. $r = \rho \cos(\theta)$
2. $q = \rho \sin(\theta)$
3. Form the complex number in rectangular form: $r + jq$.

EXAMPLE 20–8   Express $5\underline{/-36.9°}$ in rectangular form. (a) Use the $\boxed{\rightarrow R}$ calculator key, and (b) use Rule 20–2.

SOLUTION   (a) Use $\boxed{\rightarrow R}$ key:

∴ $5\underline{/-36.9°}$ $\boxed{\rightarrow R}$ $4.0 - j3.0$

(b) Use Rule 20–2:

Step 1: $r = 5 \cos(-36.9°) = 4.0$
Step 2: $q = 5 \sin(-36.9°) = -3.0$
Step 3: $r + jq = 4 - j3$
∴ $5\underline{/-36.9°} \Rightarrow 4 - j3$

Observation   Rule 20–2 is only used if your calculator lacks the polar to rectangular key $\boxed{\rightarrow R}$. Consult your

534

Mathematics of Phasors

owner's guide for the use of this valuable feature of your calculator.

EXAMPLE 20–9 Express $24\underline{/128°}$ in rectangular form. (a) Use the $\boxed{\rightarrow R}$ calculator key, and (b) use Rule 20–2.

SOLUTION (a) Use $\boxed{\rightarrow R}$ key:

$\therefore \qquad 24\underline{/128°} \quad \boxed{\rightarrow R} \quad -14.8 + j18.9$

(b) Use Rule 20–2:

Step 1: $\qquad r = 24\cos(128°) = -14.8$

Step 2: $\qquad q = 24\sin(128°) = 18.9$

Step 3: $\qquad r + jq = -14.8 + j18.9$

$\therefore \qquad 24\underline{/128°} \Rightarrow -14.8 + j18.9$

20–3 ARITHMETIC OPERATIONS WITH COMPLEX NUMBERS

Complex numbers may be added, subtracted, multiplied, or divided by following a relatively simple rule for each operation. Unlike scalar numbers (e.g., 3, 18, and 23.6), complex numbers may be expressed in one of two forms. The selection of a particular complex number form depends upon the particular arithmetic operation to be done. As a general guideline, addition and subtraction must be performed with the complex numbers expressed in rectangular form. Multiplication and division are best performed with the complex numbers expressed in polar form.

Adding Complex Numbers

To add complex numbers, express each number in rectangular form and apply the steps of Rule 20–3.

RULE 20–3. ADDING COMPLEX NUMBERS

To add complex numbers, first express each number in rectangular form and then:

1. Add the real numbers.
2. Add the imaginary numbers.
3. Form the sum in rectangular form.

EXAMPLE 20–10 Add $-17 + j12$ and $10 - j20$.

SOLUTION Use Rule 20–3:

Arithmetic Operations with Complex Numbers

$$-17 + j12$$
$$\underline{10 - j20}$$
$$-7 - j8$$

∴ $(-17 + j12) + (10 - j20) = -7 - j8$

EXAMPLE 20–11 Add $11.3\underline{/-160°}$ and $5.4\underline{/35°}$. Express the answer in polar form.

SOLUTION Transform each to rectangular form:

$$11.3\underline{/-160°} \Rightarrow -10.6 - j3.86$$
$$5.4\underline{/35°} \Rightarrow 4.42 + j3.10$$

Add:

$$-10.6 - j3.86$$
$$\underline{4.4 + j3.10}$$
$$-6.2 - j0.76$$

Express in polar form:

$$-6.2 - j0.76 \Rightarrow 6.25\underline{/-173°}$$

∴ $(11.3\underline{/-160°}) + (5.4\underline{/35°}) = -6.2 - j0.76$
$= 6.25\underline{/-173°}$

Subtracting Complex Numbers To subtract complex numbers, express each number in rectangular form and apply the steps of Rule 20–4.

RULE 20–4. SUBTRACTING COMPLEX NUMBERS

To subtract complex numbers, first express each number in rectangular form and then:

1. Change the sign of both the real and the imaginary parts of the complex number to be subtracted.
2. Add as in Rule 20–3.

EXAMPLE 20–12 Subtract $-5 - j4$ from $9 - j2$. Express the difference in polar form.

SOLUTION Change the sign of $-5 - j4$:

$$-(-5 - j4) = 5 + j4$$

Add:

Mathematics of Phasors

$$9 - j2$$
$$\underline{5 + j4}$$
$$14 + j2$$

Express in polar form:

$$14 + j2 \Rightarrow 14.1\underline{/8.13°}$$

$$\therefore \quad (9 - j2) - (-5 - j4) = 14 + j2 = 14.1\underline{/8.13°}$$

EXAMPLE 20–13 Subtract $18.3\underline{/-38°}$ from $25.7\underline{/144°}$. Express the answer in polar form.

SOLUTION Transform each to rectangular form:

$$18.3\underline{/-38°} \Rightarrow 14.4 - j11.3$$
$$25.7\underline{/144°} \Rightarrow -20.8 + j15.1$$

Change the sign of $14.4 - j11.3$:

$$-(14.4 - j11.3) = -14.4 + j11.3$$

Add:

$$-20.8 + j15.1$$
$$\underline{-14.4 + j11.3}$$
$$-35.2 + j26.4$$

Express in polar form:

$$-35.2 + j26.4 \Rightarrow 44.0\underline{/143°}$$

$$\therefore \quad (25.7\underline{/144°}) - (18.3\underline{/-38°}) = -35.2 + j26.4$$
$$= 44.0\underline{/143°}$$

Multiplying Complex Numbers Although complex numbers may be multiplied in either the rectangular or polar form, the preferred method is to express each number in polar form and multiply by applying the steps of Rule 20–5.

RULE 20–5. MULTIPLYING COMPLEX NUMBERS

To multiply complex numbers in polar form:

1. Multiply the magnitudes.
2. Algebraically add the angles.
3. Form the product in polar form.

537

Arithmetic Operations with Complex Numbers

EXAMPLE 20-14 Multiply $5\underline{/20°}$ and $3\underline{/30°}$.

SOLUTION Use Rule 20-5.

Step 1: Multiply the magnitudes:

$$5 \times 3 = 15$$

Step 2: Add the angles:

$$20° + 30° = 50°$$

Step 3: Form the product in polar form:

$$15\underline{/50°}$$

$$\therefore \quad (5\underline{/20°})(3\underline{/30°}) = 15\underline{/50°}$$

EXAMPLE 20-15 Multiply $8.4 - j6.3$ by $12.8 + j15.2$.

SOLUTION Transform each to polar form:

$$8.4 - j6.3 \Rightarrow 10.5\underline{/-36.9°}$$

$$12.8 + j15.2 \Rightarrow 19.9\underline{/49.9°}$$

Multiplying using Rule 20-5 results in

$$(10.5\underline{/-36.9°})(19.9\underline{/49.9°}) = 209\underline{/13°}$$

$$\therefore \quad (8.4 - j6.3)(12.8 + j15.2) = 209\underline{/13°}$$

Dividing Complex Numbers

Although complex numbers may be divided in either the rectangular or polar form, the preferred method is to express each number in polar form and divide by applying the steps in Rule 20-6.

RULE 20-6. DIVIDING COMPLEX NUMBERS

To divide complex numbers in polar form:

1. Divide the magnitudes.
2. Subtract the angle of the denominator from the angle of the numerator.
3. Form the quotient in polar form.

EXAMPLE 20-16 Divide $15\underline{/30°}$ by $5\underline{/50°}$.

SOLUTION Use Rule 20-6.

Step 1: Divide the magnitudes:

$$15/5 = 3$$

Step 2: Subtract 50° from 30°:

$$30° - 50° = -20°$$

Step 3: Form the quotient:

$$3\underline{/-20°}$$

$$\therefore \quad (15\underline{/30°})/(5\underline{/50°}) = 3\underline{/-20°}$$

EXAMPLE 20–17 Divide $4.32 - j9.53$ by $7.07 - j5.84$.

SOLUTION Transform each to polar form:

$$4.32 - j9.53 \Rightarrow 10.5\underline{/-65.6°}$$

$$7.07 - j5.84 \Rightarrow 9.17\underline{/-39.6°}$$

Dividing using Rule 20–6 results in

$$\frac{10.5\underline{/-65.6°}}{9.17\underline{/-39.6°}} = 1.15\underline{/-26°}$$

$$\therefore \quad \frac{4.32 - j9.53}{7.07 - j5.84} = 1.15\underline{/-26°}$$

20–4 PHASORS

The phasor is a very useful tool in the analysis of ac networks. By representing sinusoidal quantities with phasors rather than with periodic equations, complex arithmetic rather than *graphic arithmetic* may be used in the solution of ac circuits. The following narrative examples contrast the expediency of applying phasors using complex arithmetic with the time-consuming application of instantaneous equations using graphical arithmetic.

EXAMPLE 20–18 Determine the instantaneous equation of the source voltage (e) in the circuit of Figure 20–5 by applying Kirchhoff's voltage law ($e = v_R + v_L$).

Observation Since the current is the same in each component of the series circuit, the current in the circuit of Figure 20–5 may serve as a reference to relate the voltage across the resistor to the voltage across the inductor. From Chapter 19, it was learned that the voltage across a resistor is *in phase* with the current. Thus, the instantaneous voltage equation of the resistor has zero phase angle in its periodic equation. From Figure 20–5,

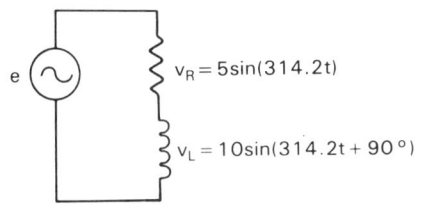

FIGURE 20-5
Circuit of Examples
20-18 and 20-19.

$$v_R = 5 \sin(314.2t + 0°)$$

or simply $v_R = 5 \sin(314.2t)$. The voltage across an inductor is *leading* the current by 90°; therefore, the instantaneous voltage equation of the inductor has a 90° phase angle in its periodic equation. From Figure 20-5,

$$v_L = 10 \sin(314.2t + 90°)$$

SOLUTION Since $e = v_R + v_L$ (from the circuit of Figure 20-5), the graphic addition of the instantaneous amplitudes of the sine-wave curve of v_R to the sine-wave curve of v_L will result in the sine-wave curve of e. Once the curve of e is known, the instantaneous equation of the curve may be written by first determining and then substituting the maximum voltage (E_{mx}), the frequency (f), and the phase angle (θ) into the general form.

$$e = E_{mx} \sin(\omega t \pm \theta)$$

Construct the graphs of the wave forms represented by the equations $v_R = 5 \sin(314.2t)$ and $v_L = 10 \sin(314.2t + 90°)$ by selecting values for t. The sum of v_R and v_L at the selected time (t) will equal the amplitude of e at that same time. The following are sample calculations for several selected times. The values, along with those listed in Table 20-1, were used to construct the curves of Figure 20-6.

TABLE 20-1 SUMMARY OF CALCULATIONS OF EXAMPLE 20-18

Time (ms)	v_R (V)	v_L (V)	$e = v_R + v_L$ (V)	Time (ms)	v_R (V)	v_L (V)	$e = v_R + v_L$ (V)
0	0.00	10.0	10.0	11	−1.55	−9.51	−11.1
1	1.55	9.51	11.1	12	−2.94	−8.09	−11.0
2	2.94	8.09	11.0	13	−4.05	−5.88	−9.9
3	4.05	5.88	9.9	14	−4.76	−3.09	−7.9
4	4.76	3.09	7.9	15	−5.00	0.00	−5.0
5	5.00	0.00	5.0	16	−4.76	3.09	−1.7
6	4.76	−3.09	1.7	17	−4.05	5.88	1.8
7	4.05	−5.88	−1.8	18	−2.94	8.09	5.2
8	2.94	−8.09	−5.2	19	−1.55	9.51	8.0
9	1.55	−9.51	−8.0	20	0.00	10.0	10.0
10	0.00	−10.0	−10.0				

540
Mathematics of Phasors

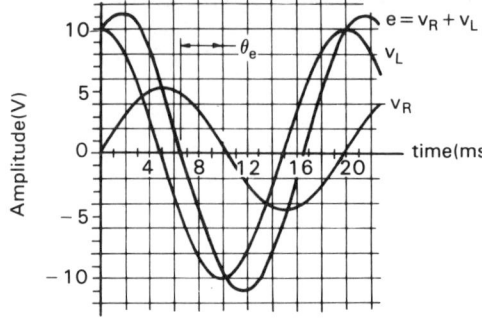

FIGURE 20-6
Sum of two out-of-phase sinusoidal voltage wave forms with the same frequency results in a sinusoidal wave form with the same frequency as the wave forms from which it was derived.

Set $t = 0$ ms:

$v_R = 5 \sin(314.2 \times 0) = 0$ V

$v_L = 10 \sin(314.2 \times 0 + 90°) = 10.0$ V

$e = v_R + v_L = 0 + 10 = 10$ V

∴ $e = 10$ V at $t = 0$ ms

Set $t = 2$ ms:

$v_R = 5 \sin(314.2 \times 2 \times 10^{-3}) = 2.94$ V

$v_L = 10 \sin\left(314.2 \times 2 \times 10^{-3} + \dfrac{\pi}{2}\right) = 8.09$ V

$e = v_R + v_L = 2.94 + 8.09 = 11.0$ V

∴ $e = 11.0$ V at $t = 2$ ms

Observation In the equation for v_L, $\pi/2$ rad has been substituted for 90° because ωt is in radians. Furthermore, the calculator has been formatted to operate in radians by depressing the *angular mode selection* key RAD or DRG .

Set $t = 6$ ms:

$v_R = 5 \sin(314.2 \times 6 \times 10^{-3}) = 4.76$ V

$v_L = 10 \sin\left(314.2 \times 6 \times 10^{-3} + \dfrac{\pi}{2}\right) = -3.09$ V

$e = v_R + v_L = 4.76 - 3.09 = 1.67$ V

∴ $e = 1.67$ V at $t = 6$ ms

From the wave form of e in Figure 20-6, determine the amplitude E_{mx}.

∴ $E_{mx} = 11.1$ V

Determine the frequency of the wave form e in Figure 20–6 by first determining the period and then computing the frequency. Thus

$$f = \frac{1}{T} = \frac{1}{20 \text{ ms}} = 50 \text{ Hz}$$

∴ $f = 50$ Hz

Compute the angular velocity of the curve. Thus

$$\omega = 2\pi f = 2\pi(50) = 314.2 \text{ rad/s}$$

∴ $\omega = 314.2$ rad/s

Using the *in phase* wave form of v_R as a reference, determine the phase angle (θ_e) of wave form e. From Figure 20–6, the time difference between the wave forms of v_R and e is approximately 3.5 ms. Convert this time displacement to radians and then to degrees. Thus

$$\theta_e = \alpha = \omega t = 314.2 \, \frac{\text{rad}}{\text{s}} \times 3.5 \times 10^{-3} \, \text{s}$$

$$\theta_e = 1.10 \text{ rad}$$

$$\theta_e = 1.10 \text{ rad} \times \frac{180°}{\pi \text{ rad}} = 63°$$

∴ The source voltage, e, leads the circuit current and the voltage across the resistance by 63°. Substitute into the general equation. Thus

$$e = E_{\text{mx}} \sin(\omega t \pm \theta)$$
$$e = 11.1 \sin(314.2t + 63°) \text{ V}$$

∴ $e = 11.1 \sin(314.2t + 63°)$ V is the instantaneous equation of the source voltage of the circuit of Figure 20–5 derived from the wave form of Figure 20–6.

Well, that was indeed time consuming! The following example will use phasors and complex arithmetic to determine the same instantaneous equation of the voltage source, e, of Figure 20–5.

EXAMPLE 20–19 From Example 20–18 and Figure 20–6, we have learned that the wave forms of the instantaneous voltages of a circuit give a continuous picture of the interrelation between the circuit voltages. The phasor diagram of the series circuit of Figure 20–5 will give a picture of the circuit conditions at a selected time. Since the circuit quantities *track* together (i.e., there is no net movement between the quantities), the phase angles of the sine waves become the angles between the phasors. Given sufficient information, the phasor may be derived from the instantaneous equation, and conversely. Thus

$$e = E_{mx} \sin(\omega t \pm \theta) \Leftrightarrow E_{mx}/\pm\theta \qquad (20\text{--}5)$$

where E_{mx} = peak amplitude of the wave
θ = phase angle between sine waves or, in the phasor, the angle between the phasors.

Write the phasors for the instantaneous voltage equation of the circuit of Figure 20–5. Add the phasors and write the instantaneous equation of the resultant.

SOLUTION Using Equation 20–5, derive the phasor for v_R and v_L from their instantaneous equation. Thus

$$v_R = 5 \sin(314.2t + 0°) \Rightarrow V_{Rmx} = 5/0° \text{ V}$$

$$v_L = 10 \sin(314.2t + 90°) \Rightarrow V_{Lmx} = 10/90° \text{ V}$$

Observation Figure 20–7 is the phasor diagram for the circuit of Figure 20–5.
Add the phasors to determine E_{mx}. Thus

$$E_{mx} = V_{Rmx} + V_{Lmx} = 5/0° + 10/90°$$

Transform the phasors from polar form to rectangular form in order to add:

$$5/0° \Rightarrow 5 + j0 \text{ V}$$
$$\underline{10/90° \Rightarrow 0 + j10 \text{ V}}$$
$$5 + j10 \text{ V}$$

∴ The sum of $5/0°$ V and $10/90°$ V is $5 + j10$ V. Transform from rectangular to polar form:

$$5 + j10 \Rightarrow 11.2/63.4° \text{ V}$$

Derive the instantaneous equation of e from its phasor. Thus

$$E_{mx} = 11.2/63.4° \Rightarrow e = 11.2 \sin(314.2t + 63.4°)$$

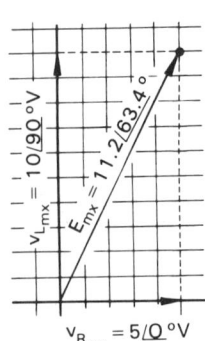

FIGURE 20–7
Phasor diagram for the circuit of Figure 20–5. The resultant of adding phasors V_{Rmx} and V_{Lmx} is shown as $E_{mx} = 11.2/63.4°$.

∴ $e = 11.2 \sin(314.2t + 63.4°)$ V is the instantaneous equation of the voltage source of Figure 20–5.

Observation The result of adding sine waves of the same frequency is another sine wave of the same frequency. This being the case, the time-consuming procedure of Example 20–17 is not necessary, since the simpler procedure of Example 20–18 yields the same results.

Expressing Phasors with Effective Value

When analyzing *ac circuits*, it is expeditious that the magnitude of the phasor representing current or voltage be expressed as an effective quantity. Thus, the phasor of a sinusoidal current or voltage is expressed as

$$I\underline{/\theta} \quad \text{and} \quad E\underline{/\theta}$$

where I and E represent the effective values of current and voltage and θ is the angle between the phasors. The following example will emphasize this concept.

EXAMPLE 20–20 Express the following voltage and current equations as phasors having effective values:
(a) $e = 162 \sin(800t - 30°)$ V
(b) $i = 372 \sin(200t + 15°)$ mA
(c) $i = 99.2 \sin(500t)$ mA

SOLUTION (a) Use Equation 20–5 to derive the phasor:

$$e = 162 \sin(800t - 30°) \Rightarrow 162\underline{/-30°} \text{ V (pk)}$$

Express the magnitude as an effective quantity:

$$E = 0.707 E_{mx} = 0.707(162) = 115 \text{ V}$$

Write the phasor:

∴ $E\underline{/\theta} = 115\underline{/-30°}$ V

(b) Use Equation 20–5 to derive the phasor:

$$i = 372 \sin(200t + 15°) \text{ mA} \Rightarrow 372\underline{/15°} \text{ (pk)}$$

Express the magnitude as an effective quantity:

$$I = 0.707 I_{mx} = 0.707(372 \times 10^{-3})$$
$$I = 263 \text{ mA}$$

Write the phasor:

$$\therefore \quad I\underline{/\theta} = 263\underline{/15°} \text{ mA}$$

(c) Use Equation 20-5 to derive the phasor:

$$i = 99.2 \sin(500t) \text{ mA} \Rightarrow 99.2\underline{/0°} \text{ mA (peak)}$$

Express the magnitude as an effective quantity:

$$I = 0.707 I_{\text{mx}} = 0.707(99.2 \times 10^{-3})$$
$$I = 70.1 \text{ mA}$$

Write the phasor:

$$\therefore \quad I\underline{/\theta} = 70.1\underline{/0°} \text{ mA}$$

Symbolizing Phasor Quantities

Before we proceed to the application of phasors, we need to consider the meaning of the letters used to symbolize current and voltage in equations. For example,

$$E = V_1 + V_2 + V_3$$

appears to be a familiar form of Kirchhoff's voltage law. The equation tells us that the source voltage may be determined by adding three voltage drops, V_1, V_2, and V_3. This, of course, is true. However, the equation does not indicate how to carry out the addition. Since we now have two number systems, scalar numbers used with dc circuits and complex numbers used with ac circuits, how do we know which system to use?

It has been traditional to use special notation for phasor quantities, such as using boldface letters to represent phasor quantities, as in **E** = **V**$_1$ + **V**$_2$ + **V**$_3$. A contemporary method, and the one we will be using, is to use lightface (not bold) letters to represent phasors and to use absolute notation when only the magnitude of the phasor is needed. The determination of what the symbol means depends upon your understanding of the context under which the equation is being applied. Thus, $E = V_1 + V_2 + V_3$ may be a scalar equation for a dc circuit that uses scalar quantities, or it may be a phasor equation for an ac circuit that uses phasor quantities. You must decide which it will be by understanding the characteristics of the circuit (ac or dc) in question.

20-5 OPERATING WITH PHASOR QUANTITIES

The rules of complex numbers may be used with phasors only when the frequency of all the wave forms from which the phasors are derived is the same. This is the case for the follow-

Operating with Phasor Quantities

ing examples, as well as for the analysis of ac networks. Wave forms with varying frequency are treated in Chapter 28.

EXAMPLE 20-21 Solve for V_2 in the following KVL equation.

$$35\underline{/0°} = 22\underline{/-28°} + V_2 + 8\underline{/65°} \text{ V}$$

SOLUTION Isolate V_2 in the left member of the equation:

$$V_2 = 35\underline{/0°} - 22\underline{/-28°} - 8\underline{/65°}$$

Change the terms with negative magnitudes to positive magnitudes by applying Equation 20-3 $(-\rho\underline{/\theta} = \rho\underline{/\theta \pm 180°})$:

$$V_2 = 35\underline{/0°} + 22\underline{/152°} + 8\underline{/-115°}$$

Add, using the rules of complex arithmetic:

$$\begin{array}{rl} 35\underline{/0°} \Rightarrow & 35.0 + j0 \\ 22\underline{/152°} \Rightarrow & -19.4 + j10.3 \\ 8\underline{/-115°} \Rightarrow & -3.38 - j7.25 \\ \hline V_2 = & 12.2 + j3.05 \end{array}$$

Express V_2 in polar form:

$$\therefore \quad V_2 = 12.6\underline{/14°} \text{ V}$$

EXAMPLE 20-22 Solve for I_1 in the following KCL equation.

$$0.58\underline{/0°} = I_1 + 0.37\underline{/70°} \text{ A}$$

SOLUTION Solve for I_1:

$$I_1 = 0.58\underline{/0°} - 0.37\underline{/70°}$$

Express $-0.37\underline{/70°}$ with a positive magnitude:

$$I_1 = 0.58\underline{/0°} + 0.37\underline{/-110°}$$

Add, using the rules of complex arithmetic:

$$\begin{array}{rl} 0.58\underline{/0°} \Rightarrow & 0.58 + j0 \\ 0.37\underline{/-110°} \Rightarrow & -0.127 - j0.348 \\ \hline I_1 = & 0.453 - j0.348 \end{array}$$

Express I_1 in polar form:

$$\therefore \quad I_1 = 0.571\underline{/-37.5°} \text{ A}$$

EXAMPLE 20-23 Determine the current (I) entering the node of a two-branch parallel ac circuit when the two branch currents are $I_1 = 32\underline{/-26°}$ mA and $I_2 = 18\underline{/51°}$ mA.

546

Mathematics of Phasors

SOLUTION Apply KCL:

$$I = I_1 + I_2$$

Observation Each of the letters (I, I_1, and I_2) represents a phasor quantity and must use the rules of complex arithmetic. Substitute:

$$I = 32 \times 10^{-3} \underline{/-26°} + 18 \times 10^{-3} \underline{/51°} \text{ A}$$

Transform the phasors to rectangular form and add:

$$\begin{aligned} 32 \times 10^{-3} \underline{/-26°} &\Rightarrow 28.8 - j14.0 \text{ mA} \\ 18 \times 10^{-3} \underline{/51°} &\Rightarrow \underline{11.3 + j14.0 \text{ mA}} \\ I &= 40.1 + j0 \quad \text{mA} \end{aligned}$$

Express I in polar form:

$$\therefore \quad I = 40.1 \underline{/0°} \text{ mA is the current entering the node of the two-branch circuit.}$$

EXAMPLE 20–24 Express the current i_2 in the block diagram of the circuit of Figure 20–8 as a sinusoidal equation.

SOLUTION Use KCL:

$$i = i_1 + i_2$$

Solve for i_2:

$$i_2 = i - i_1$$

Derive the phasors from the instantaneous equations:

$$i = 140 \times 10^{-3} \sin(377t - 30°) \Rightarrow 140\underline{/-30°} \text{ mA (peak)}$$

$$I = 0.707(140\underline{/-30°}) \text{ mA} = 98.98\underline{/-30°} \text{ mA}$$

$$i_1 = 30 \times 10^{-3} \sin(377t - 45°) \Rightarrow 30\underline{/-45°} \text{ mA (peak)}$$

$$I_1 = 0.707(30\underline{/-45°}) \text{ mA} = 21.21\underline{/-45°} \text{ mA}$$

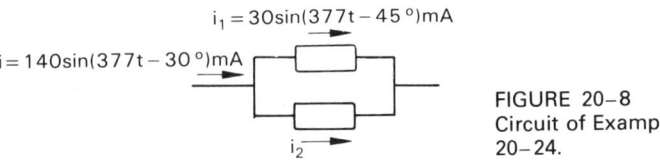

FIGURE 20–8
Circuit of Example 20–24.

State KCL in terms of the effective quantities:

$$I_2 = I - I_1$$

Substitute:

$$I_2 = 98.98 \times 10^{-3} \underline{/-30°} - 21.21 \times 10^{-3} \underline{/-45°}$$

Express $-21.21 \times 10^{-3}\underline{/-45°}$ with a positive magnitude and add:

$$I_2 = 98.98 \times 10^{-3}\underline{/-30°} + 21.21 \times 10^{-3}\underline{/135°}$$

$$\begin{aligned}
98.98 \times 10^{-3}\underline{/-30°} &\Rightarrow 85.7 - j49.5 \text{ mA} \\
21.21 \times 10^{-3}\underline{/135°} &\Rightarrow \underline{-15.0 + j15.0 \text{ mA}} \\
I_2 &= 70.7 - j34.5 \text{ mA}
\end{aligned}$$

Express I_2 in polar form:

$$I_2 = 78.7\underline{/-26°} \text{ mA}$$

Derive the sinusoidal instantaneous equation for i_2:

$$I_2 = 78.7\underline{/-26°} \text{ mA} \Rightarrow$$

$$i_2 = \sqrt{2}(78.7)\sin(377t - 26°) \text{ mA}$$

$$\therefore \quad i_2 = 111 \sin(377t - 26°) \text{ mA}$$

PROBLEMS

Section 20-1

20-1 From Figure 20-2, record the coordinates of the following points in the rectangular form of a complex number: (a) point C, (b) point E, (c) point J, and (d) point L.

20-2 From Figure 20-2, record the coordinates of the following points in the rectangular form of a complex number: (a) point D, (b) point G, (c) point K, and (d) point N.

20-3 Express the following polar coordinates with positive magnitudes and with arguments from $+180°$ to $-180°$:
(a) $-7\underline{/20°}$ (b) $-12\underline{/-40°}$
(c) $-3\underline{/-140°}$ (d) $-24\underline{/160°}$

20-4 Express the following polar coordinates with positive magnitudes and with arguments from $+180°$ to $-180°$:
(a) $-5\underline{/-95°}$ (b) $-10\underline{/-10°}$
(c) $-6\underline{/175°}$ (d) $-20\underline{/55°}$

Section 20-2

20-5 Express the following complex numbers in polar form:
(a) $6 + j10$ (b) $-4 + j4$
(c) $8 - j5.4$ (d) $-12.5 - j18.3$
(e) $0.34 + j0.15$ (f) $-620 + j890$

20-6 Express the following complex numbers in polar form:
(a) $174 - j105$ (b) $-20.2 + j153$
(c) $-0.866 - j0.5$ (d) $15.6 + j28.7$
(e) $-3.5 - j2.7$ (f) $-15.7 - j30.4$

20-7 Express the following complex numbers in rectangular form:
(a) $80/\underline{50°}$ (b) $15/\underline{30°}$
(c) $47/\underline{-75°}$ (d) $36.4/\underline{13.2°}$
(e) $-4.83/\underline{-19.6°}$ (f) $99.3/\underline{-83.8°}$

20-8 Express the following complex numbers in rectangular form:
(a) $26/\underline{-48°}$ (b) $-14/\underline{57°}$
(c) $483/\underline{-135°}$ (d) $24.6/\underline{156°}$
(e) $-12.8/\underline{-85°}$ (f) $0.924/\underline{162°}$

Section 20-3

20-9 Add the following complex numbers:
(a) $(3 + j6) + (6 - j4)$
(b) $(3 + j0) + (-8 - j6)$
(c) $(0.52 + j1.85) + (2.14 - j0.91)$
(d) $118/\underline{-42°} + 83/\underline{115°}$

20-10 Add the following complex numbers:
(a) $(-19 + j13) + (-12 - j9)$
(b) $(94 + j0) + (-28 - j46)$
(c) $43.8/\underline{8°} + 19.5/\underline{74°}$
(d) $0.83/\underline{-118°} + 0.47/\underline{21°}$

20-11 Subtract the following complex numbers:
(a) $(-5 + j8) - (9 + j7)$
(b) $(-27 - j16) - (40 - j3)$
(c) $(8.3/\underline{14°}) - (3.4/\underline{-25°})$
(d) $(0.73/\underline{-111°}) - (0.22/\underline{96°})$

20-12 Subtract the following complex numbers:
(a) $(4 + j9) - (3 - j5)$
(b) $(184 - j65) - (-93 - j107)$
(c) $(-61.2/\underline{87°}) - (92.3/\underline{-17°})$
(d) $(70.3/\underline{-152°}) - (4.23/\underline{31°})$

20-13 Multiply the following complex numbers:
(a) $(5/\underline{18°})(4/\underline{12°})$
(b) $(2/\underline{-43°})(6/\underline{-18°})$
(c) $(60.2/\underline{-28°})(14.7/\underline{-84°})$
(d) $(8.35 + j16.2)(-5.13 - j2.76)$

20-14 Multiply the following complex numbers:
(a) $(21/\underline{36°})(11/\underline{-58°})$ (b) $(88/\underline{-65°})(92/\underline{-18°})$
(c) $(0.63/\underline{8.8°})(1.3/\underline{68.4°})$ (d) $(7.3 - j4.4)(2.5 + j9.1)$

20-15 Divide the following complex numbers:

Problems

    (a) $(18\underline{/45°})/(3\underline{/15°})$ (b) $(55\underline{/-10°})/(5\underline{/-60°})$
    (c) $(38\underline{/116°})/(19\underline{/-47°})$ (d) $(5 - j6)/(7 + j3)$

20–16 Divide the following complex numbers:
    (a) $(52\underline{/52°})/(13\underline{/-28°})$
    (b) $(0.913\underline{/102°})/(0.491\underline{/58.8°})$
    (c) $(84.3\underline{/-27.6°})/(174\underline{/62.6°})$
    (d) $(-10 - j5)/(6 + j6)$

Section 20–4

20–17 Express the following voltage and current equations as phasors having effective values.
    (a) $i = 8.5 \times 10^{-3} \sin(825t - 18°)$ A
    (b) $i = 0.62 \sin(400t + 81°)$ A
    (c) $e = 164 \sin(377t)$ V
    (d) $e = 36 \sin(512t + 40°)$ V

20–18 Express the following voltage and current equations as phasors having effective values.
    (a) $i = 55 \sin(192t + 0°)$ mA
    (b) $e = 20 \sin(900t + 45°)$ V
    (c) $i = 2.3 \sin(500t - 74°)$ A
    (d) $e = 86 \sin(377t - 90°)$ V

20–19 Express the following voltage and current phasors as instantaneous equations of their sine waves. Let $\omega t = 512t$ radians.
    (a) $I = 85\underline{/-18°}$ A   (b) $I = 34\underline{/65°}$ mA
    (c) $E = 24\underline{/0°}$ V   (d) $E = (12.3 - j18.9)$ V

20–20 Express the following voltage and current phasors as instantaneous equations of their sine waves. Let $\omega t = 377t$ radians.
    (a) $E = 12\underline{/65°}$ V   (b) $I = 32.6\underline{/-44°}$ mA
    (c) $E = 117\underline{/-22°}$ V   (d) $I = (0.25 + j0.88)$ A

Section 20–5

20–21 Solve for E in the KVL equation, $E = 18\underline{/76°} + 6\underline{/-22°}$ V.

20–22 Solve for I in the KCL equation:
    $I = 0.83\underline{/57°} + 0.95\underline{/-12°}$ A.

20–23 Solve for I_2 in the KCL equation:
    $292\underline{/-47°} = 194\underline{/35°} + I_2$ mA.

20–24 Solve for V_1 in the KVL equation:
    $36\underline{/0°} = V_1 + 15.2\underline{/58.5°}$ V.

20–25 State the sinusoidal equation for the voltage v_1 in the block diagram of the circuit of Figure 20–9 when:
    $e = 200 \sin(377t - 34°)$ V and $v_2 = 97 \sin(377t - 10°)$ V

20–26 State the sinusoidal equation for the voltage e in the block diagram of the circuit of Figure 20–9 when:
    $v_1 = 5.3 \sin(512t)$ V and $v_2 = 9.6 \sin(512t - 48°)$ V

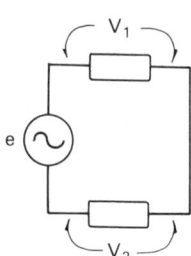

FIGURE 20–9
Circuit of Problems 20–25 and 20–26.

20-27 State the sinusoidal equation for the current i in the block diagram of the circuit of Figure 20-10 when: $i_1 = 5.28 \sin(800t - 78°)$ A and $i_2 = 7.32 \sin(800t + 8°)$ A

20-28 State the sinusoidal equation for the current i_2 in the block diagram of the circuit of Figure 20-10 when: $i = 219 \sin(200t + 65°)$ mA and $i_1 = 50.8 \sin(200t)$ mA

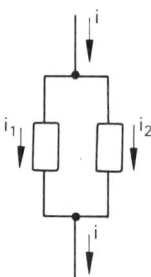

FIGURE 20-10
Circuit of Problems 20-27 and 20-28.

21 Series and Parallel AC Circuits

21-1 VOLTAGE PHASOR FOR SERIES AC CIRCUITS
21-2 IMPEDANCE OF SERIES AC CIRCUITS
21-3 SOLVING SERIES AC CIRCUITS
21-4 VOLTAGE DIVISION IN SERIES AC CIRCUITS
21-5 CURRENT PHASOR FOR PARALLEL AC CIRCUITS
21-6 ADMITTANCE OF PARALLEL AC CIRCUITS
21-7 SOLVING PARALLEL AC CIRCUITS
21-8 CURRENT DIVISION IN PARALLEL AC CIRCUITS
21-9 FORMING SERIES AND PARALLEL AC EQUIVALENT CIRCUITS

CHAPTER PREVIEW This chapter uses phasor voltage and phasor current diagrams to present the relationship between current, voltage, and the circuit components in an ac circuit. You will use impedance vector diagrams to aid in the solution of series ac circuits and admittance vector diagrams to aid in the solution of parallel ac circuits. Besides solving series and parallel circuits, the chapter also explores the concepts of equivalent circuits.

PERFORMANCE OBJECTIVES Once you have read each section, worked each example with pencil, paper, and calculator, and answered each problem for every section, you will be able to:

☐ Apply Kirchhoff's voltage law to the solution of series ac circuits.
☐ Construct an impedance vector diagram to aid in solving series ac circuits.
☐ Apply the voltage-divider equation to series ac circuits.
☐ Apply Kirchhoff's current law to the solution of parallel ac circuits.
☐ Construct an admittance vector diagram to aid in solving parallel ac circuits.
☐ Apply the current-divider equation to parallel ac circuits.
☐ Form either a series or a parallel equivalent circuit for an ac circuit.

21-1 VOLTAGE PHASOR FOR SERIES AC CIRCUITS From your work in Chapter 19, you learned that the ac voltage across a resistor is in phase with the current in the resistor, and the ac voltage across an inductor is *leading* the current in the inductor by 90°. You also learned that the ac voltage across a capacitor is *lagging* the current in the capacitor by 90°.

Because the phase relation of current and voltage plays an important role in series and parallel circuits, it should be memorized. To remind you that voltage leads current in an inductor and current leads voltage in a capacitor, use the following mnemonic phrase. ELI the ICE man. The key words are **ELI** and **ICE**. ELI means that in an inductor, L, voltage, E, is before (leads) current, I. ICE means that in a capacitor, C, current, I, is before (leads) voltage, E.

Voltage Phasor In an ac series circuit, as in any series circuit, the current is the same throughout the circuit. Thus, in drawing a phasor diagram showing the voltage drops in an ac series circuit, the current is selected as the reference phasor. Figure 21–1 pictures the generalized voltage phasor for a series RLC circuit. From this diagram, we see that the voltage phasor for the series circuit is con-

553 structed so that V_R, which is in phase with the circuit current, is along the positive R-axis; V_L, which leads the circuit current by 90°, is along the positive j-axis; and V_C, which lags the circuit current by 90°, is along the negative j-axis.

KVL for Series AC Circuits

Kirchhoff's voltage law (KVL), $E = V_1 + V_2 + V_3, \ldots$, is valid for ac series circuits, as demonstrated by Example 21-1. However, before working the example, we must once again remind you that each of the parameters in the ac circuit is a phasor quantity. Thus, each voltage is assigned a direction to its magnitude using the phasor diagram of Figure 21-1 and/or the mnemonic phrase, "ELI the ICE man."

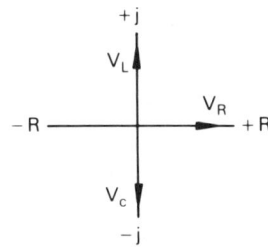

FIGURE 21-1
Generalized voltage phasor diagram for a series R-L-C circuit.

EXAMPLE 21-1 Verify Kirchhoff's voltage law for the circuit of Figure 21-2(a).

SOLUTION With current, $I = 5\underline{/0°}$ A, as the reference, then

$$|V_R| = IR = (5)(8) = 40 \text{ V}$$
$$|V_C| = IX_C = (5)(12) = 60 \text{ V}$$
$$|V_L| = IX_L = (5)(6) = 30 \text{ V}$$

Assign a direction to each voltage. Thus

$$V_R = 40\underline{/0°} \text{ V}, \quad V_C = 60\underline{/-90°} \text{ V},$$
$$V_L = 30\underline{/90°} \text{ V}$$

Substitute into KVL:

$$E = V_R + V_C + V_L$$
$$50\underline{/-36.87°} = 40\underline{/0°} + 60\underline{/-90°} + 30\underline{/90°}$$

To add, express each voltage phasor in the right member as a complex number in rectangular form:

$$V_R = 40\underline{/0°} \Rightarrow 40 + j0$$
$$V_C = 60\underline{/-90°} \Rightarrow 0 - j60$$
$$V_L = 30\underline{/90°} \Rightarrow 0 + j30$$
$$V_R + V_C + V_L = 40 - j30 \text{ V}$$

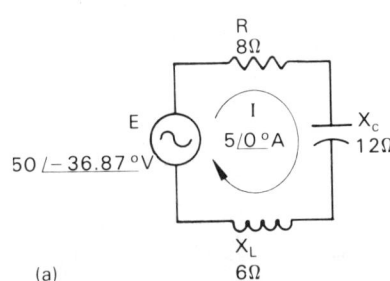

(a)

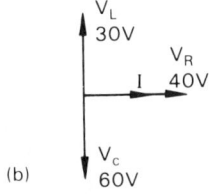

(b)

FIGURE 21-2
(a) Series ac circuit of Example 21-1.
(b) Voltage phasor diagram.

Express the sum, $40 - j30$ V, in polar form and compare it to the source, E, of $50\underline{/-36.87°}$ V:

$$40 - j30 \quad \boxed{\rightarrow P} \quad 50\underline{/-36.87°}$$

$$\therefore \quad E = V_R + V_C + V_L \Rightarrow 50\underline{/-36.87°} \text{ V}$$

EXAMPLE 21–2 As indicated by Figure 21–3(a), a 680-Ω metal-film resistor is connected in series with a 1-μF capacitor and a 250-mH inductor. The circuit is connected to a 400-Hz ac source. (a) Determine the voltage across each component when the measured current is 150 mA. (b) Determine the source voltage, E. (c) Construct the current-voltage phasor diagram showing the phase angle, θ. (d) Write the instantaneous equation for the source voltage and the circuit current.

SOLUTION (a) Determine the voltage across each component in the circuit when $\omega = 2\pi f = 2\pi 400 = 2513$ rad/s:

$$|V_R| = IR = (0.15)(680) = 102 \text{ V}$$

$$|V_C| = IX_C = I\frac{1}{\omega C} = \frac{0.15}{2513 \times 1 \times 10^{-6}}$$

$$|V_C| = 59.7 \text{ V}$$

$$|V_L| = IX_L = I(\omega L) = (0.15)(2513 \times 0.25)$$

$$|V_L| = 94.2 \text{ V}$$

Construct the voltage phasor diagram, Figure 21–3(b), and assign direction to each of the voltages.

$$\therefore \quad V_R = 102\underline{/0°} \text{ V}, \quad V_C = 59.7\underline{/-90°} \text{ V}$$
$$V_L = 94.2\underline{/90°} \text{ V}$$

(b) Determine the source voltage from KVL:

$$E = V_R + V_C + V_L$$

$$E = 102\underline{/0°} + 59.7\underline{/-90°} + 94.2\underline{/90°}$$

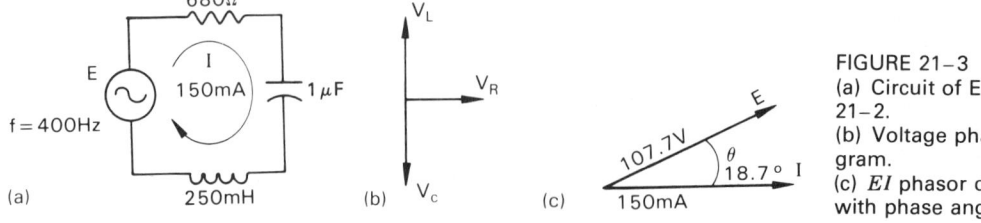

FIGURE 21–3
(a) Circuit of Example 21–2.
(b) Voltage phasor diagram.
(c) EI phasor diagram with phase angle noted.

To add, express each voltage phasor in the right member as a complex number in rectangular form:

$$V_R = 102\underline{/0°} \Rightarrow 102 + j0$$
$$V_C = 59.7\underline{/-90°} \Rightarrow 0 - j59.7$$
$$V_L = 94.2\underline{/90°} \Rightarrow \underline{0 + j94.2}$$
$$E = 102 + j34.5 \text{ V}$$

Express the source voltage, $102 + j34.5$ V, in polar form:

$$102 + j34.5 \boxed{\rightarrow P} \; 107.7\underline{/18.7°} \text{ V}$$

∴ The source voltage, E, is $107.7\underline{/18.7°}$ V.

(c) The EI phasor diagram is shown in Figure 21–3(c).

(d) The instantaneous voltage and current equations for $\omega = 2513$ rad/s are

$$e = \sqrt{2}\, E \sin(\omega t + \theta)$$
$$e = 152 \sin(2513t + 18.7°) \text{ V}$$
$$i = \sqrt{2}\, I \sin(\omega t)$$
$$i = 212 \sin(2513t) \text{ mA}$$

∴ $e = 152 \sin(2513t + 18.7°)$ V and $i = 212 \sin(2513t)$ mA.

Observation The maximum values of voltage and current are determined from the effective voltage and current by multiplying by $\sqrt{2}$, which is the reciprocal of 0.707. Thus, $1/0.707 = 1.414 = \sqrt{2}$. Recall from Equations 19–13 and 19–14, $I_{mx} = I/0.707$, which also equals $\sqrt{2}\, I$, and $E_{mx} = E/.707$, which also equals $\sqrt{2}\, E$.

21–2 IMPEDANCE OF SERIES AC CIRCUITS

When alternating current passes through a series RLC circuit, the current in the circuit is determined by the total opposition offered by the resistance of the resistor and the reactance of the inductor and capacitor. The opposition to the passage of alternating current through a circuit is called **impedance**. Thus, the word impedance may be used to describe the opposition offered by an ac circuit containing only resistance or only capacitance or only inductance. It may also be used to describe ac circuits containing any combination of resistance, capacitance, and inductance. Thus, the word *impedance* is a general label that is used to describe the opposition to the flow of alternating current by any combination of circuit components.

Computing Impedance Impedance is a *vector quantity* that may be written as a complex number in either the rectangular or polar form. The letter Z is used to represent impedance. Thus, $Z = 5\underline{/53.13°}\ \Omega = 3 + j4\ \Omega$ shows an impedance written in both the polar and rectangular form. Ohm's law may be stated in terms of impedance as Equation 21–1.

$$Z = E/I \quad (\Omega) \tag{21-1}$$

EXAMPLE 21–3 Determine the impedance of the circuit of Figure 21–2(a).

SOLUTION Use Equation 21–1 and substitute $I = 5\underline{/0°}$ A and $E = 50\underline{/-36.87°}$ V:

$$Z = \frac{E}{I} = \frac{50\underline{/-36.87°}}{5\underline{/0°}}$$

$$\therefore \quad Z = 10\underline{/-36.87°}\ \Omega$$

EXAMPLE 21–4 Determine the impedance of the circuit of Figure 21–3(a).

SOLUTION Use Equation 21–1 and substitute the values of E and I from Figure 21–3(c). Thus

$$Z = \frac{E}{I} = \frac{107.7\underline{/18.7°}}{0.15\underline{/0°}}$$

$$\therefore \quad Z = 718\underline{/18.7°}\ \Omega$$

Observation Notice that the angle of the circuit impedance is the phase angle of the circuit (the angle between the current and voltage in the EI phasor diagram).

Impedance Diagram In an ac series circuit consisting of a resistor, an inductor, and a capacitor, the following circuit conditions exist:

$$I = I_R = I_L = I_C$$

And from the voltage phasor diagram of Figure 21–1,

$$E = V_R + j|V_L| - j|V_C|$$

Dividing I of the first equation into each member of the second equation results in

$$\frac{E}{I} = \frac{V_R}{I} + \frac{j|V_L|}{I} + \frac{-j|V_C|}{I}$$

Impedance of Series AC Circuits

and

$$Z = R + (jX_L - jX_C) \quad (\Omega) \tag{21-2}$$

When the vector quantities of the right member of Equation 21-2 are graphed, the impedance vector diagram of Figure 21-4 is formed. The following guidelines may be used to aid in the construction of an impedance diagram.

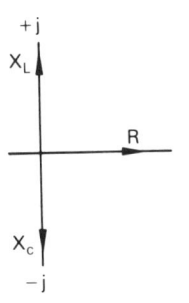

FIGURE 21-4
Impedance vector diagram has resistance plotted horizontally along the reference axis, inductive reactance plotted vertically up, and capacitive reactance plotted vertically down.

GUIDELINE 21-1. CONSTRUCTING AN IMPEDANCE DIAGRAM

When plotting resistance, inductive reactance, and capacitive reactance on a complex plane,

1. Plot resistance along the positive real axis.
2. Plot inductive reactance along the positive imaginary $(+j)$ axis.
3. Plot capacitive reactance along the negative imaginary $(-j)$ axis.

EXAMPLE 21-5 Construct an impedance vector diagram for the series-connected components of Figure 21-5(a).

SOLUTION From Figure 21-5(a),

$$R = 2 \text{ k}\Omega, \quad X_L = 1 \text{ k}\Omega, \quad X_C = 1.5 \text{ k}\Omega$$

∴ The impedance vector diagram is as shown in Figure 21-5(b).

Impedance of Series-Connected Components

The impedance of a series circuit is determined by applying Equation 21-2 to the circuit. Once the impedance is computed, it may be written in either the rectangular, $R \pm jX$, or the polar, $Z\underline{/\theta}$, form of the complex number. Notice that the rectangular form is written with the resistance first and then the reactance. A positive j $(+j)$ indicates an inductive circuit; a negative j $(-j)$ indicates a capacitive circuit.

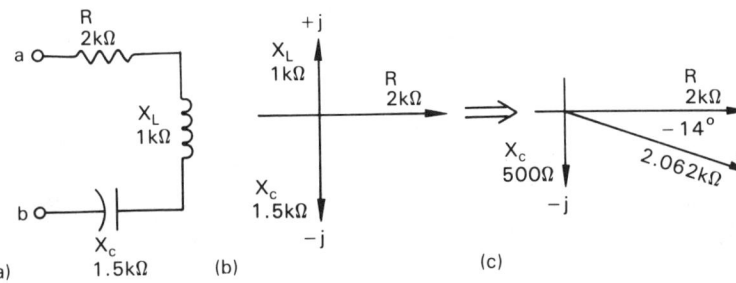

FIGURE 21-5
(a) Circuit of Example 21-5.
(b) Impedance vector diagram.
(c) Impedance vector diagram of Example 21-6.

EXAMPLE 21–6 (a) Express the impedance across terminals a-b of the circuit of Figure 21–5(a) in both the rectangular and polar form. (b) Construct the impedance vector diagram for both forms.

SOLUTION (a) Substitute into Equation 21–2:

$$Z = R + (jX_L - jX_C)$$

$$Z = 2 + (j1 - j1.5) \text{ k}\Omega$$

$$Z = 2 - j0.5 \text{ k}\Omega$$

∴ The impedance written in rectangular form is $(2 - j0.5)$ kΩ. Convert $(2 - j0.5)$ kΩ to polar form:

$$2 - j0.5 \text{ k}\Omega \boxed{\rightarrow P} \ 2.062\underline{/-14°} \text{ k}\Omega$$

∴ The impedance written in polar form is $2.062\underline{/-14°}$ kΩ.

(b) Figure 21–5(c) pictures both the rectangular and polar forms of the impedance.

21–3 SOLVING SERIES AC CIRCUITS

An ac series circuit may be made up of several resistors, capacitors, and inductors. When this is the case, form each type of component into an equivalent component, as was done in Chapter 7 for series-connected resistance, in Chapter 13 for series-connected capacitors, and in Chapter 17 for series-connected inductors. Thus

$$R_T = R_1 + R_2 + R_3 + \cdots + R_n \quad \Omega \quad (7\text{–}2)$$

$$L_T = L_1 + L_2 + L_3 + \cdots + L_n \quad \text{H} \quad (17\text{–}7)$$

$$C_T = \frac{1}{\dfrac{1}{C_1} + \dfrac{1}{C_2} + \dfrac{1}{C_3} + \cdots + \dfrac{1}{C_n}} \quad \text{F} \quad (13\text{–}21)$$

Of course, Ohm's law, Kirchhoff's laws, and all the rules for series dc circuits apply to series ac circuits, as demonstrated in the following examples.

EXAMPLE 21–7 Assume that an ac voltage source of $e = 170 \sin(377t - 14°)$ V is connected across terminals a-b of Figure 21–5(a).

Determine (a) the circuit current, I, given the impedance of $2.06\underline{/-14°}$ kΩ, and (b) the phase angle, θ, between the current and voltage phasor.

Solving Series AC Circuits

SOLUTION (a) Apply Ohm's law to determine current:

$$I = \frac{E}{Z}$$

$$|E| = 0.707 \, E_{mx} = (0.707)(170) = 120 \text{ V}$$

$$E = 120 \underline{/-14°} \text{ V}$$

$$I = \frac{120\underline{/-14°}}{2.06 \times 10^3 \underline{/-14°}} = 58.3\underline{/0°} \text{ mA}$$

∴ The circuit current is $58.3\underline{/0°}$ mA.

(b) Since I is at 0° when E is at $-14°$, the angle between E and I is $-14°$.

∴ $\underline{/\theta} = -14°$

EXAMPLE 21–8 The series circuit of Figure 21–6 is connected across a 400-Hz, 90-V ac source. (a) Determine the impedance of the circuit. (b) Determine the circuit current, I. (c) Construct the voltage-current phasor with voltage as the reference. (d) Construct the impedance vector diagram.

SOLUTION (a) Compute the equivalent components and then determine the impedance:

$$R_T = R_1 + R_2 = 560 + 1600 = 2.16 \text{ k}\Omega$$

$$L_T = L_1 = 5 \text{ H}$$

$$C_T = \frac{1}{\dfrac{1}{0.068 \times 10^{-6}} + \dfrac{1}{0.15 \times 10^{-6}}}$$

$$C_T = 0.0468 \text{ μF}$$

Determine the reactance where $f = 400$ Hz and $\omega = 2513$ rad/s:

$$X_L = \omega L = (2513)(5) = 12.6 \text{ k}\Omega$$

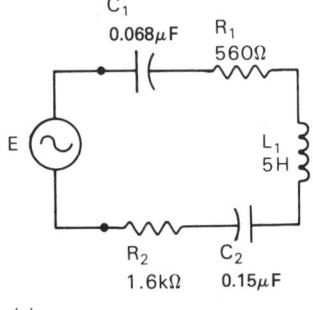

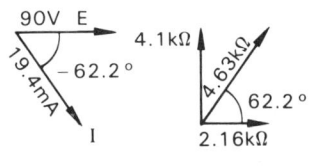

FIGURE 21–6
(a) Circuit of Example 21–8.
(b) EI phasor diagram with circuit phase angle, θ, indicated.
(c) Rectangular and polar impedance vector diagram.

$$X_C = \frac{1}{\omega C} = \frac{1}{(2513)(0.0468 \times 10^{-6})} = 8.5 \text{ k}\Omega$$

$$R = 2.16 \text{ k}\Omega$$

Substitute into Equation 21–2:

$$Z = R + (jX_L - jX_C)$$

$$Z = 2.16 \text{ k}\Omega + (j12.6 \text{ k}\Omega - j8.5 \text{ k}\Omega)$$

$$Z = 2.16 + j4.1 \text{ k}\Omega$$

Convert $2.16 + j4.1$ kΩ to polar form:

$$2.16 + j4.1 \text{ k}\Omega \;\boxed{\rightarrow P}\; 4.63 \underline{/62.2°} \text{ k}\Omega$$

$$\therefore \quad Z\underline{/\theta} = 4.63 \underline{/62.2°} \text{ k}\Omega$$

Observation The notation $2.16 + j4.1$ kΩ means 2.16 kΩ + $j4.1$ kΩ.

(b) Determine I:

$$I = \frac{E}{Z} = \frac{90\underline{/0°}}{4.63 \times 10^3 \underline{/62.2°}} = 19.4\underline{/-62.2°} \text{ mA}$$

$$\therefore \quad I = 19.4\underline{/-62.2°} \text{ mA when } E = 90\underline{/0°} \text{ V}.$$

(c) The voltage-current phasor with voltage at 0° is shown in Figure 21–6(b).

(d) The impedance vector diagram is pictured in Figure 21–6(c).

EXAMPLE 21–9 For the circuit of Figure 21–7, compute the circuit current and verify Kirchhoff's voltage law for a frequency of 5 kHz.

SOLUTION Determine the impedance of the circuit:

$$\omega = 2\pi f = 2\pi 5 \times 10^3 = 31.4 \times 10^3 \text{ rad/s}$$

$$X_L = \omega L = (31.4 \times 10^3)(470 \times 10^{-3})$$

$$X_L = 14.8 \text{ k}\Omega$$

$$X_C = \frac{1}{\omega C} = \frac{1}{(31.4 \times 10^3)(910 \times 10^{-12})}$$

$$X_C = 35.0 \text{ k}\Omega$$

$$R = 12 \text{ k}\Omega + 2.7 \text{ k}\Omega = 14.7 \text{ k}\Omega$$

$$Z = R + (jX_L - jX_C)$$

$$Z = 14.7 + (j14.8 - j35) \text{ k}\Omega$$

$$Z = 14.7 - j20.2 \text{ k}\Omega$$

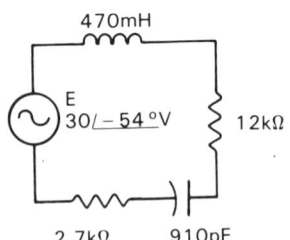

FIGURE 21–7
Circuit of Example 21–9.

Express Z in polar form:

$$Z = 14.7 - j20.2 \text{ k}\Omega \quad \boxed{\rightarrow P} \quad 25.0\underline{/-54°} \text{ k}\Omega$$

Apply Ohm's law:

$$I = \frac{E}{Z} = \frac{30\underline{/-54°}}{25 \times 10^3 \underline{/-54°}}$$

$$\therefore \quad I = 1.2\underline{/0°} \text{ mA}$$

Verify KVL:

$$E = V_R + V_L + V_C = IR + IX_L + IX_C$$

$30\underline{/-54°} = (1.2 \times 10^{-3}\underline{/0°})(14.7 \times 10^3 \underline{/0°}) +$
$\phantom{30\underline{/-54°} =} (1.2 \times 10^{-3}\underline{/0°})(14.8 \times 10^3 \underline{/90°}) +$
$\phantom{30\underline{/-54°} =} (1.2 \times 10^{-3}\underline{/0°})(35 \times 10^3 \underline{/-90°})$

Observation   The inductive reactance, 14.8 kΩ, has a +90° angle assigned, while the capacitive reactance of 35 kΩ has a −90° angle assigned. These assignments are made from the impedance vector diagram of Figure 21–4.

$$30\underline{/-54°} = 17.64\underline{/0°} + 17.76\underline{/90°} + 42\underline{/-90°} \text{ V}$$

Express the right member in rectangular form and add:

$$30\underline{/-54°} = 17.64 + (j17.76 - j42)$$

$$30\underline{/-54°} = 17.64 - j24.24$$

$$30\underline{/-54°} = 17.64 - j24.24 = 30.0\underline{/-54°}$$

$\therefore$ KVL is satisfied.

21–4 VOLTAGE DIVISION IN SERIES AC CIRCUITS

The voltage-divider equation of Chapter 10 (Equation 10–5) is restated here in terms of impedance.

$$V_n = \frac{EZ_n}{\Sigma Z} \text{ (V)} \tag{21-3}$$

where V_n = voltage drop across Z_n in volts
 E = source voltage in volts
 Z_n = selected impedance in ohms across which V_n is dropped
 ΣZ = total impedance of the series circuit in ohms

In applying the equation, the ac series circuit is blocked into two impedance blocks. One of these blocks contains all the components that the desired voltage is across. The other contains

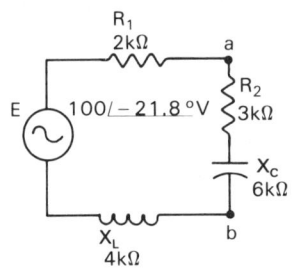
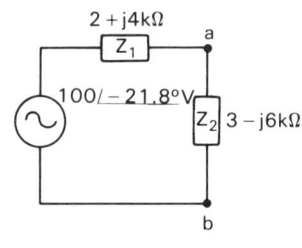

FIGURE 21-8
(a) Circuit of Example 21-10.
(b) Block diagram of circuit.

the remaining circuit components. This concept is demonstrated in the following examples.

EXAMPLE 21-10 Apply the voltage-divider equation, Equation 21-3, to the circuit of Figure 21-8(a) and determine V_{ab}.

SOLUTION Block the circuit into two impedance blocks, as pictured in Figure 21-8(b). Thus

$Z_1 = R_1 + L = 2 + j4$ kΩ →P $4.47\underline{/63.4°}$ kΩ
$Z_2 = R_2 + C = 3 - j6$ kΩ →P $6.71\underline{/-63.4°}$ kΩ
$\Sigma Z = Z_1 + Z_2 = 5 - j2$ kΩ →P $5.39\underline{/-21.8°}$ kΩ

Apply Equation 21-3:

$$V_{ab} = V_{Z2} = \frac{EZ_2}{\Sigma Z}$$

Substitute with all parameters in polar form:

$$V_{ab} = \frac{(100\underline{/-21.8°})(6.71 \times 10^3\underline{/-63.4°})}{5.39 \times 10^3\underline{/-21.8°}}$$

$$V_{ab} = 124.5\underline{/-63.4°} \text{ V}$$

∴ $V_{ab} = 124.5\underline{/-63.4°}$ V

EXAMPLE 21-11 Use the voltage-divider equation to determine V_{ac} and V_{bc} for the circuit of Figure 21-9.

SOLUTION Determine V_{ac} using Figure 21-9(b):

$$V_{ac} = \frac{EZ_2}{\Sigma Z}$$

where

$E = 15\underline{/0°}$
$Z_2 = 50 + j60$ →P $78.1\underline{/50.2°}$ Ω
$\Sigma Z = Z_1 + Z_2 = (0 - j20) + (50 + j60)$
$\Sigma Z = 50 + j40$ →P $64\underline{/38.7°}$ Ω

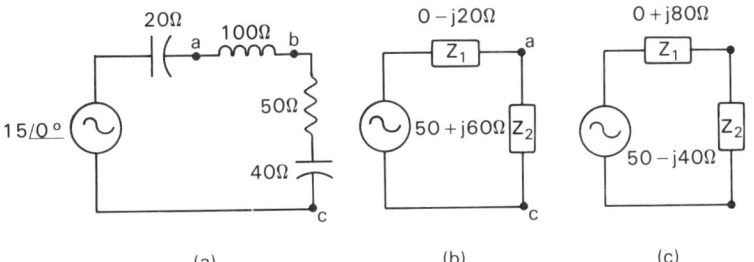

FIGURE 21-9
(a) Circuit of Example 21-11.
(b) Block diagram for determining V_{ac}.
(c) Block diagram for determining V_{bc}.

Substitute:

$$V_{ac} = \frac{(15\underline{/0°})(78.1\underline{/50.2°})}{64\underline{/38.7°}} = 18.3\underline{/11.5°} \text{ V}$$

∴ $V_{ac} = 18.3\underline{/11.5°}$ V

Determine V_{bc} using Figure 21-9(c):

$$V_{bc} = \frac{EZ_2}{\Sigma Z}$$

where

$E = 15\underline{/0°}$

$Z_2 = 50 - j40 \quad \boxed{\rightarrow P} \quad 64\underline{/-38.7°}$ Ω

$\Sigma Z = Z_1 + Z_2 = (0 + j80) + (50 - j40)$

$\Sigma Z = 50 + j40 \quad \boxed{\rightarrow P} \quad 64\underline{/38.7°}$ Ω

$$V_{bc} = \frac{(15\underline{/0°})(64\underline{/-38.7°})}{64\underline{/38.7°}} = 15\underline{/-77.4°} \text{ V}$$

∴ $V_{bc} = 15\underline{/-77.4°}$ V

21-5 CURRENT PHASOR FOR PARALLEL AC CIRCUITS

In an ac parallel circuit, as in any parallel circuit, the voltage is the same across all branches of the parallel circuit. Passing an alternating current through a parallel circuit having three branches, as pictured in Figure 21-10(a), results in the following circuit conditions:

☐ In the resistor, the current and applied voltage are in phase.
☐ In the inductor, the inductive current lags the applied voltage by 90°.
☐ In the capacitor, the capacitive current leads the applied voltage by 90°.

These concepts are summarized in Figure 21-10(b), the *current phasor diagram.*

Series and Parallel AC Circuits

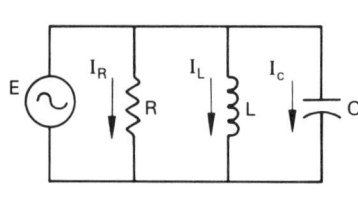 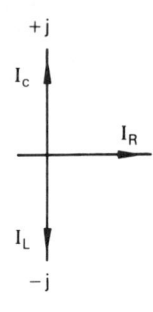

FIGURE 21–10
(a) Source voltage, E, is the same across each branch of the parallel circuit.
(b) General current phasor diagram for a parallel R-L-C circuit.

(a)   (b)

KCL for Parallel Circuits

Kirchhoff's current law (KCL), $I = I_1 + I_2 + I_3, \ldots$, is valid for ac parallel circuits. However, ac phasor quantities, such as current, are represented by complex numbers and must be operated on with complex arithmetic. Example 21–12 demonstrates KCL and the use of complex arithmetic.

EXAMPLE 21–12 Verify Kirchhoff's current law for the circuit of Figure 21–11(a).

SOLUTION With voltage, $E = 12\underline{/0°}$ V, as the reference, then

$$|I_R| = \frac{E}{R} = \frac{12}{3} = 4 \text{ A}$$

$$|I_L| = \frac{E}{X_L} = \frac{12}{3} = 4 \text{ A}$$

$$|I_C| = \frac{E}{X_C} = \frac{12}{12} = 1 \text{ A}$$

Assign an angle to each current using the current phasor of Figure 21–10(b). Thus

$$I_R = 4\underline{/0°} \text{ A}, \quad I_L = 4\underline{/-90°} \text{ A},$$
$$I_C = 1\underline{/90°} \text{ A}$$

Substitute into KCL:

$$I = I_R + I_L + I_C$$

$$5\underline{/-36.9°} = 4\underline{/0°} + 4\underline{/-90°} + 1\underline{/90°}$$

To add, express each current phasor in the right member as a complex number in rectangular form:

$$I_R = 4\underline{/0°} \Rightarrow 4 + j0$$
$$I_L = 4\underline{/-90°} \Rightarrow 0 - j4$$
$$I_C = 1\underline{/90°} \Rightarrow 0 + j1$$
$$I_R + I_L + I_C = 4 - j3 \text{ A}$$

Current Phasor for Parallel AC Circuits

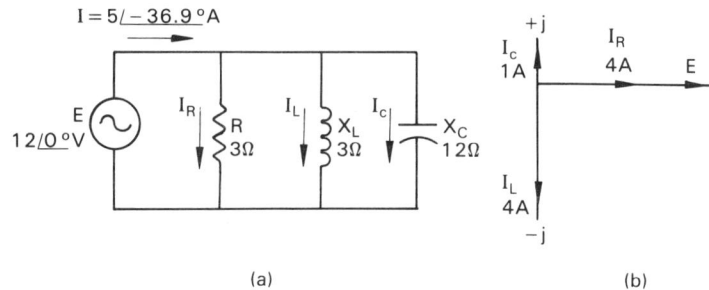

FIGURE 21-11
(a) Parallel ac circuit of Example 21-12.
(b) Current phasor diagram.

Express the sum, $4 - j3$ A in polar form and compare it to the circuit current, I, of $5/-36.9°$ A.

$$4 - j3 \quad \boxed{\rightarrow P} \quad 5/-36.9°$$

$$\therefore \quad I = I_R + I_L + I_C \Rightarrow 5/-36.9° = 5/-36.9° \text{ A}$$

EXAMPLE 21-13 As indicated by Figure 21-12(a), a 1.2-kΩ carbon-film resistor is connected in parallel with a 0.82-μF capacitor and a 750-mH inductor. The circuit is attached to a 400-Hz, 230-V alternator. (a) Determine the current in each component. (b) Determine the circuit current, I. (c) Construct the current-voltage phasor diagram. (d) Write the instantaneous equation for the source voltage and circuit current.

SOLUTION (a) Determine the current in each component when $\omega = 2\pi 400 = 2513$ rad/s.

$$|I_R| = \frac{E}{R} = \frac{230}{1.2 \times 10^3} = 192 \text{ mA}$$

$$|I_C| = \frac{E}{X_C} = E\omega C = (230)(2513)(0.82 \times 10^{-6})$$

$$|I_C| = 474 \text{ mA}$$

$$|I_L| = \frac{E}{X_L} = \frac{E}{\omega L} = \frac{230}{2513 \times 0.75} = 122 \text{ mA}$$

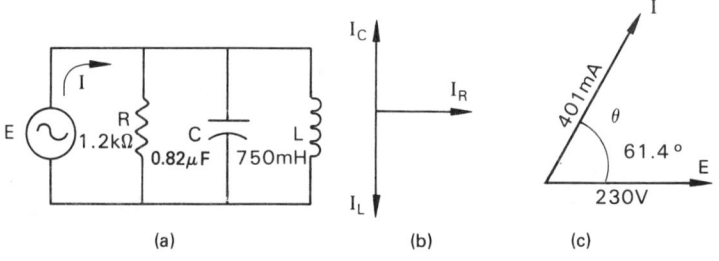

FIGURE 21-12
(a) Circuit of Example 21-13.
(b) Current phasor diagram.
(c) EI phasor diagram with phase angle, θ, noted.

Construct the current phasor diagram, Figure 21–12(b), and assign direction to each of the currents.

∴ $I_R = 192\underline{/0°}$ mA, $I_C = 474\underline{/90°}$ mA,
   $I_L = 122\underline{/-90°}$ mA

(b) Determine the circuit current from KCL:

$$I = I_R + I_C + I_L$$

$$I = (192\underline{/0°} + 474\underline{/90°} + 122\underline{/-90°}) \text{ mA}$$

To add, express each current phasor in the right member as a complex number in rectangular form:

$I_R = 192\underline{/0°}$ mA $\Rightarrow$ $192 + j0$ mA

$I_C = 474\underline{/90°}$ mA $\Rightarrow$ $0 + j474$ mA

$I_L = 122\underline{/-90°}$ mA $\Rightarrow$ $\underline{0 - j122 \text{ mA}}$

$I = 192 + j352$ mA

Express $192 + j352$ mA in polar form:

$192 + j352$ mA $\boxed{\rightarrow P}$ $401\underline{/61.4°}$ mA

∴ The circuit current, I, is $401\underline{/61.4°}$ mA.

(c) The EI phasor diagram is shown in Figure 21–12(c).

(d) The instantaneous voltage and current equations for $\omega = 2513$ rad/s are

$e = \sqrt{2}\, E \sin(\omega t)$
$e = 325 \sin(2513t)$ V
$i = \sqrt{2}\, I \sin(\omega t + \theta)$
$i = 567 \sin(2513t + 61.4°)$ mA

∴ $e = 325 \sin(2513t)$ V   and
   $i = 567 \sin(2513t + 61.4°)$ mA

21–6 ADMITTANCE OF PARALLEL AC CIRCUITS

Admittance is the vector quantity used to describe the ease of passage of alternating current. It is measured in siemens (S) and is noted by the letter Y. Like conductance (the reciprocal of resistance), which is used in the solution of dc parallel circuits, *admittance* (the reciprocal of impedance) is used in the solution of ac parallel circuits. Thus

$$Y = \frac{1}{Z} \quad \text{(S)} \tag{21-4}$$

Admittance of Parallel AC Circuits

$$Y = \frac{I}{E} \quad \text{(S)} \tag{21-5}$$

EXAMPLE 21-14 Express an impedance of $82.3 \underline{/-27.5°}\ \Omega$ as an admittance.

SOLUTION Use Equation 21-4:

$$Y = \frac{1}{Z} = \frac{1}{82.3\underline{/-27.5°}}$$

$$\therefore \quad Y = 12.2\underline{/27.5°}\ \text{mS}$$

EXAMPLE 21-15 The voltage source across a parallel circuit is $110\underline{/0°}$ V when the circuit current is $260\underline{/-20°}$ mA. Determine the admittance of the circuit.

SOLUTION Use Equation 21-5:

$$Y = \frac{I}{E} = \frac{0.260\underline{/-20°}}{110\underline{/0°}}$$

$$\therefore \quad Y = 2.36\underline{/-20°}\ \text{mS}$$

Admittance Diagram

You know that the reciprocal of resistance is conductance; thus

$$G = \frac{1}{R} \quad \text{(S)} \tag{21-6}$$

The reciprocal of reactance is called *susceptance* and is noted by the letter B. Susceptance is a measure of the ability of an inductor or a capacitor to pass alternating current. Thus

$$B_L = \frac{1}{X_L} \quad \text{(S)} \tag{21-7}$$

$$B_C = \frac{1}{X_C} \quad \text{(S)} \tag{21-8}$$

where B_L = inductive susceptance
B_C = capacitive susceptance

EXAMPLE 21-16 Express the following reactances as susceptance: (a) $X_C = 200\underline{/-90°}\ \Omega$, and (b) $X_L = 50\underline{/90°}\ \Omega$.

SOLUTION (a) Use Equation 21-8:

$$B_C = \frac{1}{X_C} = \frac{1}{200\underline{/-90°}} = 5\underline{/90°} \text{ mS}$$

$$\therefore \quad B_C = 5\underline{/90°} \text{ mS}$$

(b) Use Equation 21-7:

$$B_L = \frac{1}{X_L} = \frac{1}{50\underline{/90°}} = 20\underline{/-90°} \text{ mS}$$

$$\therefore \quad B_L = 20\underline{/-90°} \text{ mS}$$

Observation In an impedance diagram, X_L has an angle of $+90°$ $(+j)$; however, when X_L is reciprocated, the angle becomes $-90°$ $(-j)$. Thus

$$\frac{1}{X_L\underline{/90°}} = B_L\underline{/-90°} = -jB_L$$

and also

$$\frac{1}{X_C\underline{/-90°}} = B_C\underline{/90°} = +jB_C$$

In an ac parallel circuit consisting of a resistive branch, an inductive branch, and a capacitive branch, the following circuit conditions exist:

$$E = V_R = V_C = V_L$$

And from the current phasor diagram of Figure 21-10(b),

$$I = I_R + j|I_C| - j|I_L|$$

Dividing E of the first equation into each member of the second equation results in

$$\frac{I}{E} = \frac{I_R}{E} + \frac{j|I_C|}{E} + \frac{-j|I_L|}{E}$$

and

$$Y = G + (jB_C - jB_L) \quad \text{(S)} \tag{21-9}$$

When the vector quantities of the right member of Equation 21-9 are graphed, the admittance vector diagram of Figure

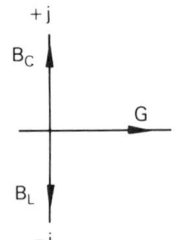

FIGURE 21-13
Admittance vector diagram for parallel circuits has the conductance plotted horizontally along the reference axis, the capacitive susceptance plotted vertically up, and the inductive susceptance plotted vertically down.

Admittance of Parallel AC Circuits

21–13 is formed. The following guidelines may be used to aid in constructing an admittance diagram.

GUIDELINE 21–2. CONSTRUCTING AN ADMITTANCE DIAGRAM

When plotting conductance (G), capacitive susceptance (B_C), and inductive susceptance (B_L) on a complex plane,

1. Plot conductance along the positive real axis.
2. Plot capacitive susceptance along the positive imaginary ($+j$) axis.
3. Plot inductive susceptance along the negative imaginary ($-j$) axis.

Admittance of Parallel-Connected Components

The admittance of an elementary parallel circuit, such as the one pictured in Figure 21–10(a), is determined by applying Equation 21–9 to the circuit. Once the admittance is computed, it then may be written in either the rectangular form, $G \pm jB$, or the polar form, $Y\underline{/\theta}$, of the complex number. The rectangular form is written with the conductance, G, first and then the susceptance, B. A positive j ($+j$) indicates a capacitive circuit; a negative j ($-j$) indicates an inductive circuit.

EXAMPLE 21–17 (a) Express the admittance across terminals a-b of the circuit of Figure 21–14(a) in both rectangular and polar form. (b) Construct the admittance vector diagram for both forms.

SOLUTION (a) Determine the conductance and susceptance:

$$G = \frac{1}{R} = \frac{1}{50} = 20 \text{ mS}$$

$$B_C = \frac{1}{X_C} = \frac{1}{80} = 12.5 \text{ mS}$$

$$B_L = \frac{1}{X_L} = \frac{1}{30} = 33.3 \text{ mS}$$

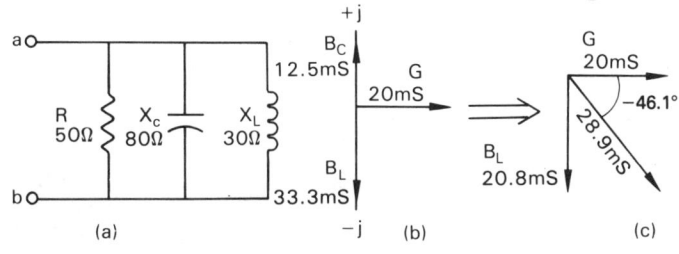

FIGURE 21 – 14
(a) Circuit of Examples 21–17 and 21–18.
(b) Admittance vector diagram of the circuit.
(c) Rectangular and polar admittance diagram.

Using the information of Figure 21–14(b), substitute into Equation 21–9:

$$Y = G + (jB_C - jB_L)$$
$$Y = 20 + (j12.5 - j33.3) \text{ mS}$$
$$Y = (20 - j20.8) \text{ mS}$$

∴ The admittance written in rectangular form is $(20 - j20.8)$ mS. Convert $(20 - j20.8)$ mS to polar form:

$$20 - j20.8 \text{ mS} \quad \boxed{\rightarrow P} \quad 28.9 \underline{/-46.1°} \text{ mS}$$

∴ The admittance written in polar form is $28.9\underline{/-46.1°}$ mS.

(b) Figure 21–14(c) pictures both the rectangular and the polar form of the admittance.

Observation The concepts associated with admittance are summarized in Table 21–1.

21–7 SOLVING PARALLEL AC CIRCUITS

An ac parallel circuit may be made up of two or more parallel branches, and each branch may have one or more components connected in series. In solving parallel ac circuits, the impedance of each branch is computed and then the admittance is determined. Once each branch admittance is known, they are added to obtain the admittance of the entire circuit. Thus

$$Y_T = Y_1 + Y_2 + Y_3 + \cdots + Y_n \quad \text{(S)} \quad (21–10)$$

The impedance of the entire circuit may be computed by taking the reciprocal of the admittance. Thus

$$Z_T = \frac{1}{Y_T} \quad (\Omega) \quad (21–11)$$

TABLE 21–1 ADMITTANCE SUMMARIZED

	Noted by	Equal to	Units	Polar Form	Rectangular Form
Conductance	G	$1/R$	S	$G\underline{/0°}$	$G + j0$
Susceptance (inductive)	B_L	$1/X_L$	S	$B_L\underline{/-90°}$	$0 - jB_L$
Susceptance (capacitive)	B_C	$1/X_C$	S	$B_C\underline{/90°}$	$0 + jB_C$
Admittance	Y	$1/Z$	S	$Y\underline{/\theta}$	$G \pm jB$

Solving Parallel AC Circuits

Of course, Ohm's law, Kirchhoff's laws, and all the rules for parallel dc circuits apply to parallel ac circuits, as demonstrated in the following examples.

EXAMPLE 21–18 Assume that an ac voltage source of $e = 165 \sin(377t)$ V is connected across terminals a-b of Figure 21–14(a). Determine (a) the current in each branch, (b) the source current, I, and (c) the phase angle between the current and voltage phasor.

SOLUTION (a) Since E is the same across each branch, then for $E = 117\underline{/0°}$ V,

$$I_R = \frac{E}{R} = \frac{117\underline{/0°}}{50\underline{/0°}} = 2.34\underline{/0°} \text{ A}$$

$$I_C = \frac{E}{X_C} = \frac{117\underline{/0°}}{80\underline{/-90°}} = 1.46\underline{/90°} \text{ A}$$

$$I_L = \frac{E}{X_L} = \frac{117\underline{/0°}}{30\underline{/90°}} = 3.90\underline{/-90°} \text{ A}$$

∴ $I_R = 2.34\underline{/0°}$ A, $I_C = 1.46\underline{/90°}$ A,

$I_L = 3.90\underline{/-90°}$ A

(b) Apply KCL, express each current in rectangular form, and add:

$$I = I_R + I_C + I_L$$

$$I = (2.34 + j0) + (0 + j1.46) + (0 - j3.90)$$

$$I = 2.34 - j2.44 \quad \boxed{\rightarrow P} \quad 3.38\underline{/-46.2°} \text{ A}$$

∴ The source current is $3.38\underline{/-46.2°}$ A.

(c) Since E is at 0° when I is at −46.2°, the angle between E and I is −46.2°.

∴ $\theta = -46.2°$

EXAMPLE 21–19 The parallel circuit of Figure 21–15 is connected across a 1-kHz, 15-V ac source. Determine (a) the impedance of each branch, (b) the admittance of each branch, (c) the current in each branch, (d) the impedance of the entire circuit, and (e) the source current.

SOLUTION (a) Compute the impedance of each branch. Start with branch 1:

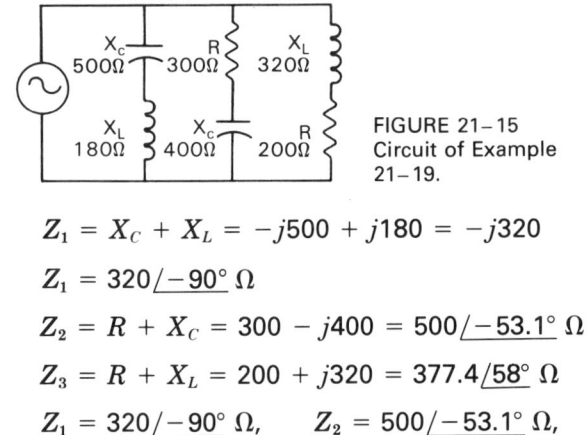

FIGURE 21–15
Circuit of Example 21–19.

$$Z_1 = X_C + X_L = -j500 + j180 = -j320$$

$$Z_1 = 320\underline{/-90°}\ \Omega$$

$$Z_2 = R + X_C = 300 - j400 = 500\underline{/-53.1°}\ \Omega$$

$$Z_3 = R + X_L = 200 + j320 = 377.4\underline{/58°}\ \Omega$$

∴ $Z_1 = 320\underline{/-90°}\ \Omega$, $Z_2 = 500\underline{/-53.1°}\ \Omega$, $Z_3 = 377.4\underline{/58°}\ \Omega$

(b) Compute the admittance of each branch:

$$Y_1 = \frac{1}{Z_1} = \frac{1}{320\underline{/-90°}} = 3.13\underline{/90°}\ \text{mS}$$

$$Y_2 = \frac{1}{Z_2} = \frac{1}{500\underline{/-53.1°}} = 2.0\underline{/53.1°}\ \text{mS}$$

$$Y_3 = \frac{1}{Z_3} = \frac{1}{377.4\underline{/58°}} = 2.65\underline{/-58°}\ \text{mS}$$

∴ $Y_1 = 3.13\underline{/90°}\ \text{mS}$, $Y_2 = 2.0\underline{/53.1°}\ \text{mS}$, $Y_3 = 2.65\underline{/-58°}\ \text{mS}$

(c) Compute the current of each branch:

$$I_1 = \frac{E}{Z_1} = EY_1 = (15\underline{/0°})(3.13 \times 10^{-3}\underline{/90°})$$

$$I_1 = 47\underline{/90°}\ \text{mA}$$

$$I_2 = \frac{E}{Z_2} = EY_2 = (15\underline{/0°})(2 \times 10^{-3}\underline{/53.1°})$$

$$I_2 = 30\underline{/53.1°}\ \text{mA}$$

$$I_3 = \frac{E}{Z_3} = EY_3 = (15\underline{/0°})(2.65 \times 10^{-3}\underline{/-58°})$$

$$I_3 = 39.8\underline{/-58°}\ \text{mA}$$

∴ $I_1 = 47\underline{/90°}\ \text{mA}$, $I_2 = 30\underline{/53.1°}\ \text{mA}$, $I_3 = 39.8\underline{/-58°}\ \text{mA}$

(d) Compute the impedance of the entire circuit:

Solving Parallel AC Circuits

$$Z_T = 1/Y_T \quad \text{and} \quad Y_T = Y_1 + Y_2 + Y_3$$

Express each branch admittance in rectangular form and add:

$Y_1 = 3.13 \underline{/90°}$ mS $\boxed{\rightarrow R}$ $\quad 0 + j3.13$ mS

$Y_2 = 2.0 \underline{/53.1°}$ mS $\boxed{\rightarrow R}$ $\quad 1.20 + j1.6$ mS

$Y_3 = 2.65 \underline{/-58°}$ mS $\boxed{\rightarrow R}$ $\quad \underline{1.40 - j2.25}$ mS

$\qquad\qquad\qquad\qquad\qquad Y_T = 2.60 + j2.48$ mS

Convert $Y_T = 2.60 + j2.48$ mS to polar form:

$Y_T = 2.60 + j2.48$ mS $\boxed{\rightarrow P}$ $3.59 \underline{/43.6°}$ mS

Compute Z_T:

$$Z_T = \frac{1}{Y_T} = \frac{1}{3.59 \times 10^{-3}\underline{/43.6°}} = 279\underline{/-43.6°} \ \Omega$$

$\therefore \quad Z_T = 279\underline{/-43.6°} \ \Omega$

(e) Compute the source current:

$I = E/Z_T = EY_T = (15\underline{/0°})(3.59 \times 10^{-3}\underline{/43.6°})$
$I = 53.9\underline{/43.6°}$ mA

$\therefore \quad I = 53.9\underline{/43.6°}$ mA

Check The source current, I, is also equal to $I_1 + I_2 + I_3$. Thus

$I = (47\underline{/90°} + 30\underline{/53.1°} + 39.8\underline{/-58°})$ mA

$47\underline{/90°}$ mA $\boxed{\rightarrow R}$ $\quad 0 + j47$ mA

$30\underline{/53.1°}$ mA $\boxed{\rightarrow R}$ $\quad 18 + j24$ mA

$39.8\underline{/-58°}$ mA $\boxed{\rightarrow R}$ $\quad \underline{21.1 - j33.8}$ mA

$I = 39.1 + j37.2$ mA $\boxed{\rightarrow P}$ $54\underline{/43.6°}$ mA

EXAMPLE 21–20 The parallel circuit of Figure 21–16(a) is connected across a current source of $i = 14.14 \sin(7540t)$ mA. Determine (a) the admittance of each branch (b) the impedance of the entire circuit, (c) the voltage across each branch and (d) the current in the 0.01-μF capacitor of branch 3.

SOLUTION (a) Compute the branch admittance:

$$Y_1 = \frac{1}{Z_1} = \frac{1}{R + jX_L}$$

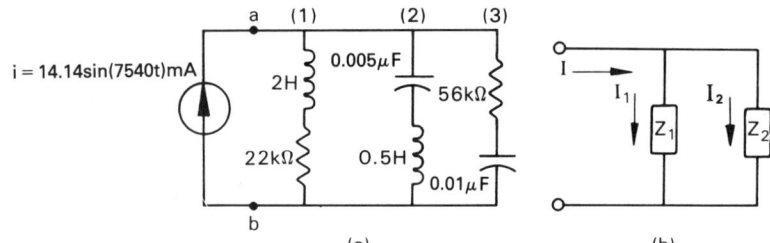

FIGURE 21-16
(a) Circuit of Examples 21-20 and 21-21.
(b) Circuit of Example 21-22.

Where

$$R = 22 \text{ k}\Omega$$

$$X_L = \omega L = 7540 \times 2 = 15.1 \text{ k}\Omega$$

$$Y_1 = \frac{1}{22 + j15.1 \text{ k}\Omega} = \frac{1}{26.7\underline{/34.5°} \text{ k}\Omega}$$

$$Y_1 = 37.5\underline{/-34.5°} \text{ }\mu\text{S}$$

$$Y_2 = \frac{1}{Z_2} = \frac{1}{(-jX_C + jX_L)}$$

Where

$$X_C = \frac{1}{\omega C} = \frac{1}{7540 \times 5 \times 10^{-9}} = 26.5 \text{ k}\Omega$$

$$X_L = \omega L = 7540 \times 0.5 = 3.77 \text{ k}\Omega$$

$$Y_2 = \frac{1}{-j26.5 \text{ k} + j3.77 \text{ k}\Omega} = \frac{1}{22.73\underline{/-90°} \text{ k}\Omega}$$

$$Y_2 = 44\underline{/90°} \text{ }\mu\text{S}$$

$$Y_3 = \frac{1}{Z_3} = \frac{1}{R - jX_C}$$

Where

$$R = 56 \text{ k}\Omega$$

$$X_C = \frac{1}{\omega C} = \frac{1}{7540 \times 1 \times 10^{-8}} = 13.3 \text{ k}\Omega$$

$$Y_3 = \frac{1}{56 - j13.3 \text{ k}\Omega} = \frac{1}{57.6\underline{/-13.4°} \text{ k}\Omega}$$

$$Y_3 = 17.4\underline{/13.4°} \text{ }\mu\text{S}$$

$$\therefore \quad Y_1 = 37.5\underline{/-34.5°} \text{ }\mu\text{S}, \quad Y_2 = 44\underline{/90°} \text{ }\mu\text{S},$$
$$Y_3 = 17.4\underline{/13.4°} \text{ }\mu\text{S}$$

(b) Compute the impedance of the entire circuit:

Current Division in Parallel AC Circuits

$$Z_T = \frac{1}{Y_T} \quad \text{and} \quad Y_T = Y_1 + Y_2 + Y_3$$

Express each branch admittance in rectangular form and add:

$Y_1 = 37.5\underline{/-34.5°}\ \mu S$ →R $\quad 30.9 - j21.2\ \mu S$

$Y_2 = 44\underline{/90°}\ \mu S$ →R $\quad 0 + j44\ \mu S$

$Y_3 = 17.4\underline{/13.4°}\ \mu S$ →R $\quad \underline{16.9 + j4.0\ \mu S}$

$\qquad\qquad\qquad\qquad\qquad Y_T = 47.8 + j26.8\ \mu S$

Convert $Y_T = 47.8 + j26.8\ \mu S$ to polar form:

$Y_T = 47.8 + j26.8\ \mu S$ →P $\ 54.8\underline{/29.3°}\ \mu S$

Compute Z_T:

$$Z_T = \frac{1}{Y_T} = \frac{1}{(54.8 \times 10^{-6}\underline{/29.3°})}$$

$Z_T = 18.2\underline{/-29.3°}\ k\Omega$

∴ $\quad Z_T = 18.2\underline{/-29.3°}\ k\Omega$

(c) Compute the voltage across the branches:

$V_1 = V_2 = V_3 = IZ_T$
$V_1 = (0.707)(14.14 \times 10^{-3}\underline{/0°})(18.2 \times 10^3\underline{/-29.3°})$

∴ $\quad V_1 = V_2 = V_3 = 182\underline{/-29.3°}\ V$

(d) Determine the current in the 0.01-μF capacitor of branch 3:

$$I_3 = I_{0.01\,\mu F} = \frac{V_3}{Z_3} = V_3 Y_3$$

$I_3 = (182\underline{/-29.3°})(17.4 \times 10^{-6}\underline{/13.4°})$

$I_3 = 3.17 \times 10^{-3}\underline{/-15.9°}\ A$

∴ The current in the 0.01-μF capacitor of branch 3 is $3.17\underline{/-15.9°}$ mA.

21-8 CURRENT DIVISION IN PARALLEL AC CIRCUITS

The current-divider equation of Chapter 10 (Equation 10-10) is restated here in terms of admittance.

$$I_n = \frac{I\,Y_n}{\Sigma\,Y} \quad (A) \qquad\qquad (21\text{-}12)$$

where I_n = current in the selected branch in amperes
$\qquad I$ = source current in amperes

Y_n = admittance of the selected branch in siemens
ΣY = total admittance of the parallel circuit in siemens

EXAMPLE 21–21 Using the circuit parameters determined in Example 21–20, apply the current-divider equation to the circuit of Figure 21–16(a) to determine the current in the 0.01-μF capacitor of branch 3.

SOLUTION Substitute into Equation 21–12 where (from Example 21–20) $I = (0.707)(14.14)\underline{/0°}$ mA, $Y_3 = 17.4\underline{/13.4°}$ μS, and $\Sigma Y = Y_T = 54.8\underline{/29.3°}$ μS. Thus

$$I_3 = I_{0.01\,\mu F} = \frac{IY_3}{\Sigma Y}$$

$$I_3 = \frac{(10 \times 10^{-3}\underline{/0°})(17.4 \times 10^{-6}\underline{/13.4°})}{54.8 \times 10^{-6}\underline{/29.3°}}$$

$$I_3 = 3.18\underline{/-15.9°} \text{ mA}$$

∴ The current in the 0.01-μF capacitor of branch 3 of Figure 21–16(a) is $3.18\underline{/-15.9°}$ mA.

EXAMPLE 21–22 Apply the current-divider equation to the circuit of Figure 21–16(b) to determine I_1 when $I = 2\underline{/30°}$ A, $Z_1 = 200 + j85$ Ω, and $Z_2 = 470 - j618$ Ω.

SOLUTION Use Equation 21–12:

$$I_1 = \frac{IY_1}{\Sigma Y}$$

where $\Sigma Y = Y_1 + Y_2$

$Y_1 = 1/Z_1 = 1/(200 + j85)$

$Y_1 = 1/(217\underline{/23°}) = 4.6\underline{/-23°}$ mS

$Y_1 = 4.6\underline{/-23°}$ mS $\boxed{\rightarrow R}$ $4.23 - j1.8$ mS

$Y_2 = \dfrac{1}{Z_2} = \dfrac{1}{470 - j618} = \dfrac{1}{776\underline{/-52.8°}}$

$Y_2 = 1.29\underline{/52.8°}$ mS

$Y_2 = 1.29\underline{/52.8°}$ mS $\boxed{\rightarrow R}$ $0.78 + j1.03$ mS

Add Y_1 and Y_2:

577

Forming Series and Parallel AC Equivalent Circuits

$$Y_1 = 4.23 - j1.8 \text{ mS}$$
$$Y_2 = 0.78 + j1.03 \text{ mS}$$
$$\Sigma Y = 5.01 - j0.77 \text{ mS} \quad \boxed{\rightarrow P}$$
$$\Sigma Y = 5.07 \underline{/-8.74°} \text{ mS}$$

Substitute into Equation 21–12:

$$I_1 = \frac{IY_1}{\Sigma Y} = \frac{(2\underline{/30°})(4.6 \times 10^{-3}\underline{/-23°})}{5.07 \times 10^{-3}\underline{/-8.74°}}$$

$$I_1 = 1.81\underline{/15.74°} \text{ A}$$

$$\therefore \quad I_1 = 1.81\underline{/15.74°} \text{ A}$$

21–9 FORMING SERIES AND PARALLEL AC EQUIVALENT CIRCUITS

For a given frequency, an ac series circuit may be simplified into either a series equivalent circuit or a parallel equivalent circuit. In either case, the equivalent circuit has the same source voltage, source current, frequency, and impedance as the original series circuit from which it was derived.

For a given frequency, an ac parallel circuit may also be simplified into either a series equivalent circuit or a parallel equivalent circuit having, in either case, the same source voltage, source current, frequency, and impedance as the original parallel circuit from which it was derived.

Series Equivalent Circuit

The series equivalent circuit of either a series or parallel ac circuit is formed by connecting a resistance in series with a reactance. The values of the components making up the series equivalent circuit are derived from the rectangular form of the **impedance** of the series or parallel circuit. That is, $Z_T = R_{eq} + jX_{eq}$, where R_{eq} is the resistive component of the series equivalent circuit and X_{eq} is the reactive component of the series equivalent circuit.

EXAMPLE 21–23 Determine the components of a series equivalent circuit for the series circuit of Figure 21–17(a).

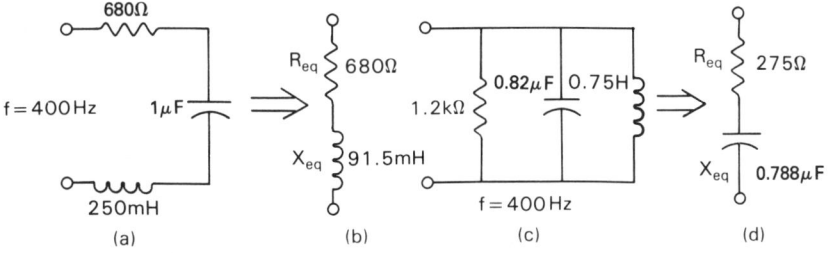

FIGURE 21–17
(a) Series circuit, and (b) its series equivalent circuit at a frequency of 400 Hz. (c) Parallel circuit, and (d) its series equivalent circuit at a frequency of 400 Hz.

SOLUTION Determine the impedance of the circuit:

$$Z_T = R + (jX_L - jX_C)$$

$$R = 680 \; \Omega$$

$$X_L = \omega L = (2\pi 400)(0.25) = 628 \; \Omega$$

$$X_C = \frac{1}{\omega C} = \frac{1}{2\pi 400 \times 1 \times 10^{-6}} = 398 \; \Omega$$

$$Z_T = 680 + (j628 - j398) = 680 + j230$$

$$Z_T = 680 + j230 \, \Omega$$

∴ The series equivalent circuit consists of a 680-Ω resistance in series with 230 Ω of inductive reactance.

Observation The value of the inductance of the equivalent inductor at 400 Hz is equal to X_L/ω or $230/(2\pi 400) = 91.5$ mH. Figure 21–17(b) pictures the series equivalent circuit.

EXAMPLE 21–24 Determine the components of a series equivalent circuit for the parallel circuit of Figure 21–17(c).

SOLUTION From Figure 21–17(c), determine the impedance of the circuit at 400 Hz:

$$Y_1 = \frac{1}{R} = \frac{1}{1.2 \times 10^3} = 0.833 \text{ mS}$$

$$Y_2 = \frac{1}{-jX_C} = j(2\pi 400 \times 0.82 \times 10^{-6})$$

$$Y_2 = j2.06 \text{ mS}$$

$$Y_3 = \frac{1}{jX_L} = \frac{1}{j(2\pi 400 \times 0.75)} = -j0.531 \text{ mS}$$

$$Y_T = Y_1 + Y_2 + Y_3$$

$$Y_T = 0.833 + j2.06 - j0.531 \text{ mS}$$

$$Y_T = 0.833 + j1.53 \text{ mS} \;\boxed{\rightarrow P}\; 1.74\underline{/61.4°} \text{ mS}$$

$$Z_T = \frac{1}{Y_T} = \frac{1}{1.74 \times 10^{-3}\underline{/61.4°}} = 575\underline{/-61.4°} \; \Omega$$

$$Z_T = 575\underline{/-61.4°} \;\boxed{\rightarrow R}\; 275 - j505 \; \Omega$$

∴ The series equivalent circuit consists of a 275-Ω resistance in series with 505Ω of capacitive reactance.

*Forming
Series and Parallel AC
Equivalent Circuits*

Observation The value of the equivalent capacitor, at 400 Hz, is equal to $1/\omega X_C$ or $1/(2\pi 400 \times 505) = 0.789\ \mu F$. Figure 21–17(d) pictures the series equivalent circuit.

Parallel Equivalent Circuit The parallel equivalent circuit of either a series or parallel ac circuit is formed by connecting a resistance in shunt with a reactance. The value of the components making up the parallel equivalent circuit are derived from the rectangular form of the **admittance** of the series or parallel circuit. That is, $Y_T = G_{eq} + jB_{eq}$. The parallel equivalent circuit is formed by taking the reciprocal of

☐ G_{eq}, to get the equivalent parallel resistance R_p.
☐ B_{eq}, to get the equivalent parallel reactance X_p.

The value of the reactive component is then computed, if the source frequency is known.

EXAMPLE 21–25 Determine the components of a parallel equivalent circuit for the series circuit of Figure 21–18(a).

SOLUTION Determine the admittance of the circuit:

$$Y_T = \frac{1}{Z_T} \quad \text{and} \quad Z_T = R + (jX_L - jX_C)$$

$$R = 680\ \Omega$$

$$X_L = \omega L = (2\pi 400)(0.25) = 628\ \Omega$$

$$X_C = \frac{1}{\omega C} = \frac{1}{2\pi 400 \times 1 \times 10^{-6}} = 398\ \Omega$$

$$Z_T = 680 + (j628 - j398) = 680 + j230$$

$$Z_T = 680 + j230 \ \boxed{\rightarrow P}\ 718\underline{/18.7°}\ \Omega$$

$$Y_T = \frac{1}{Z_T} = \frac{1}{718\underline{/18.7°}} = 1.39\underline{/-18.7°}\ \text{mS}$$

$$Y_T = 1.32 - j0.446\ \text{mS}$$

FIGURE 21–18
(a) Series circuit, and
(b) its parallel equivalent circuit at a frequency of 400 Hz.
(c) Parallel circuit, and
(d) its parallel equivalent circuit at a frequency of 400 Hz.

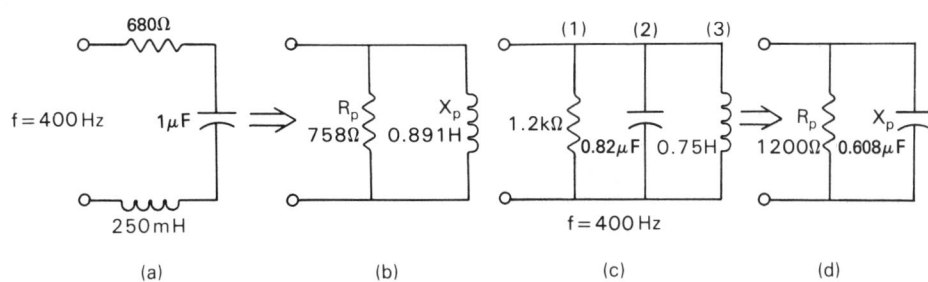

Compute the parallel equivalent resistance, R_p, and the parallel equivalent reactance, X_p, from the admittance, $Y_T = 1.32 - j0.446$ mS:

$$R_p = \frac{1}{G_{eq}} = \frac{1}{1.32 \times 10^{-3}} = 758 \text{ } \Omega$$

$$X_p = \frac{1}{B_{eq}} = \frac{1}{-j0.446 \times 10^{-3}} = j2.24 \text{ k}\Omega$$

∴ The parallel equivalent circuit consists of a 758-Ω resistance in parallel with 2.24 kΩ of inductive reactance.

Observation The value of the equivalent inductor at 400 Hz is equal to X_L/ω or $(2.24 \times 10^3)/(2\pi 400) = 0.891$ H. Figure 21–18(b) pictures the parallel equivalent circuit.

EXAMPLE 21–26 Determine the components of a parallel equivalent circuit for the parallel circuit of Figure 21–18(c).

SOLUTION From Figure 21–18(c), determine the admittance of the circuit at 400 Hz:

$$Y_1 = \frac{1}{R} = \frac{1}{1.2 \times 10^3} = 0.833 \text{ mS}$$

$$Y_2 = \frac{1}{-jX_C} = j(2\pi 400 \times 0.82 \times 10^{-6})$$

$$Y_2 = j2.06 \text{ mS}$$

$$Y_3 = \frac{1}{jX_L} = \frac{1}{j(2\pi 400 \times 0.75)} = -j0.531 \text{ mS}$$

$$Y_T = Y_1 + Y_2 + Y_3$$

$$Y_T = 0.833 + j2.06 - j0.531 \text{ mS}$$

$$Y_T = 0.833 + j1.53 \text{ mS}$$

Compute the parallel equivalent resistance, R_p, and the parallel equivalent reactance, X_p, from the admittance, $Y_T = 0.833 + j1.53$ mS:

$$R_p = \frac{1}{G_{eq}} = \frac{1}{0.833 \times 10^{-3}} = 1200 \text{ } \Omega$$

$$X_p = \frac{1}{B_{eq}} = \frac{1}{j1.53 \times 10^{-3}} = -j654 \text{ } \Omega$$

Forming Series and Parallel AC Equivalent Circuits

∴ The parallel equivalent circuit consists of a 1200-Ω resistance in parallel with 654 Ω of capacitive reactance.

Observation The value of the equivalent capacitor at 400 Hz is equal to $1/(\omega X_C)$ or $1/(2\pi 400 \times 654) = 0.608\ \mu$F.

Summary When forming an equivalent circuit for either a series or a parallel ac circuit, you have two options for the equivalent circuit. The equivalent circuit may be a series equivalent circuit or a parallel equivalent circuit. The choice of which it will be is up to you since each equivalent circuit form is identical to the other in value for a given frequency. Thus, for the same series circuit of Figures 21–17(a) and 21–18(a), the resulting equivalent circuits, Figures 21–17(b) and 21–18(b), are identical in value but different in form.

EXAMPLE 21–27 Show that the series equivalent circuit of Figure 21–17(b) is equal in value to the parallel equivalent circuit of Figure 21–18(b) at 400 Hz.

SOLUTION Compute the impedance of each circuit and compare. For Figure 21–17(b),

$$Z = R_{eq} + jX_{eq}$$

where

$$R_{eq} = 680\ \Omega$$
$$X_{eq} = \omega L = (2\pi 400)(91.5 \times 10^{-3}) = 230\ \Omega$$
$$Z = 680 + j230\ \Omega \quad \boxed{\rightarrow P}\quad 718\underline{/18.7°}\ \Omega$$

For Figure 21–18(b),

$$Z = \frac{1}{Y_T} \quad \text{and} \quad Y_T = \frac{1}{R_p} - j\frac{1}{X_p}$$

where

$$\frac{1}{R_p} = \frac{1}{758} = 1.32\ \text{mS}$$

$$\frac{1}{X_p} = \frac{1}{\omega L} = \frac{1}{(2\pi 400)(0.891)} = 0.447\ \text{mS}$$

$$Y_T = 1.32 - j0.447\ \text{mS} \quad \boxed{\rightarrow P}$$
$$Y_T = 1.39\underline{/-18.7°}\ \text{mS}$$

$$Z = \frac{1}{Y_T} = \frac{1}{1.39 \times 10^{-3}/-18.7°}$$
$$Z = 719/18.7° \; \Omega$$

∴ The impedance of the series equivalent circuit, $718/18.7° \; \Omega$, equals the impedance of the parallel equivalent circuit $719/18.7° \; \Omega$.

The concept that a given ac circuit may be represented by either a series equivalent circuit or a parallel equivalent circuit is apparent when the ac circuit in question is sealed in a box and only the source voltage and current are known.

EXAMPLE 21–28 For the ac circuit contained in the box of Figure 21–19(a), determine (a) the value of the components in the series equivalent circuit and (b) the value of the components in the parallel equivalent circuit for a frequency of 1000 Hz when $E = 20/0°$ V and $I = 80/30°$ mA.

SOLUTION (a) For the series equivalent circuit, compute the circuit impedance:

$$Z = \frac{E}{I} = \frac{20/0°}{80 \times 10^{-3}/30°} = 250/-30° \; \Omega$$

Form the series equivalent circuit from the rectangular form of the impedance $(Z = R_{eq} + jX_{eq})$:

$$Z = 250/-30° \;\; \boxed{\rightarrow R} \;\; 217 - j125 \; \Omega$$

$R_{eq} = 217 \; \Omega$
$X_{eq} = j125 \; \Omega$

Determine the value for the equivalent capacitor:

$$C_{eq} = \frac{1}{\omega X_{eq}} = \frac{1}{2\pi 1000 \times 125} = 1.27 \; \mu F$$

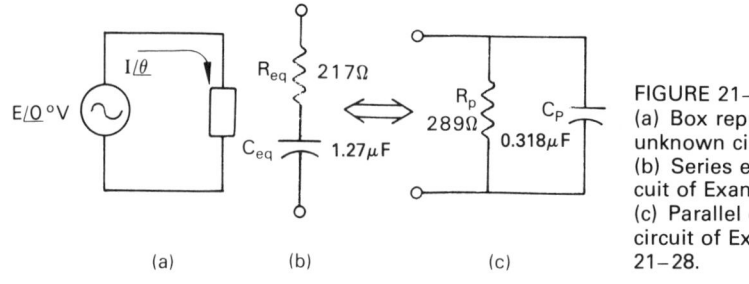

FIGURE 21–19 (a) Box represents an unknown circuit. (b) Series equivalent circuit of Example 21–28. (c) Parallel equivalent circuit of Example 21–28.

∴ The series equivalent circuit is made up of a resistance of 217 Ω in series with a capacitor of 1.27 μF when the frequency is 1000 Hz, as shown in Figure 21–19(b).

(b) For the parallel equivalent circuit, compute the circuit admittance:

$$Y = \frac{I}{E} = \frac{80 \times 10^{-3}\underline{/30°}}{20\underline{/0°}} = 4\underline{/30°} \text{ mS}$$

Form the parallel equivalent circuit from the rectangular form of the admittance ($Y = G_{eq} + jB_{eq}$):

$$Y = 4\underline{/30°} \text{ mS} \boxed{\rightarrow R} \; 3.46 + j2 \text{ mS}$$

$$G_{eq} = 3.46 \text{ mS}$$

$$B_{eq} = j2 \text{ mS}$$

Determine the value for the equivalent resistance, R_p, and the equivalent capacitor, C_p:

$$R_p = \frac{1}{G_{eq}} = \frac{1}{3.46 \times 10^{-3}} = 289 \text{ Ω}$$

$$X_{C_p} = \frac{1}{B_{eq}} = \frac{1}{j2 \times 10^{-3}} = -j500 \text{ Ω}$$

and

$$C_p = \frac{1}{\omega X_{C_p}} = \frac{1}{2\pi 1000 \times 500} = 0.318 \; \mu\text{F}$$

∴ The parallel equivalent circuit is made up of 289 Ω of resistance in parallel with a capacitor of 0.318 μF when the frequency is 1000 Hz, as shown in Figure 21–19(c).

PROBLEMS

Section 21–1

21–1 Verify KVL for a series ac circuit having a source voltage of $70\underline{/-55.32°}$ V. The current in the circuit ($I = 181\underline{/0°}$ mA) passes through a 220-Ω resistor connected in series with a 0.05-μF capacitor having a capacitive reactance of 318 Ω.

21–2 Verify KVL for a series ac circuit having a source voltage of $16.8\underline{/38.2°}$ V. The current in the circuit ($I = 4.0\underline{/0°}$ mA) passes through a 3.3-kΩ resistor and a 470-mH inductor having an inductive reactance of 2.6 kΩ.

21–3 From the information of Problem 21–1, (a) determine the frequency of the voltage source and (b) write the instantaneous periodic equation for the source voltage.

Series and Parallel AC Circuits

21-4 From the information of Problem 21-2, (a) determine the frequency of the voltage source and (b) write the instantaneous periodic equation for the circuit current.

21-5 A series circuit consisting of a 0.33-μF capacitor and a carbon-film resistor color coded white-brown-brown-gold is connected to a 2.4-kHz ac voltage source. (a) Determine the voltage across each component when the measured current is 24/0° mA. (b) Determine the peak source voltage. (c) Determine the circuit phase angle.

21-6 A 400-Hz ac current passing through a series circuit consisting of an 8.2-mH inductor, a 10-μF capacitor, and a 27-Ω resistor is 2/0° A. Determine (a) the voltage across each circuit component and (b) the source voltage E in polar form. Construct (c) the voltage phasor diagram and (d) the phasor diagram of the source voltage and circuit current (label the phase angle).

Section 21-2

21-7 Determine the impedance of a series ac circuit having 300/25° V as the source voltage and $I = 10.5/0°$ mA as the circuit current.

21-8 Determine the impedance of a series ac circuit having 78/-64° V as the source voltage and $I = 2.7/0°$ mA as the circuit current.

21-9 A resistor of 47 kΩ and a 0.015-μF capacitor are connected in series. Determine the impedance in polar and rectangular forms at angular velocities (ω) of (a) 377 rad/s, (b) 2513 rad/s, and (c) 6283 rad/s.

21-10 A resistor of 180 Ω and a 500-μH choke (inductor) are connected in series. Determine the impedance in polar and rectangular forms at angular velocities (ω) of (a) 37.7 krad/s, (b) 314 krad/s, and (c) 7.54 Mrad/s.

21-11 For the circuit of Figure 21-20(a), $E = 18/0°$ V at a frequency of 100 kHz. Determine the impedance of the circuit and construct both the rectangular and polar impedance vector diagrams.

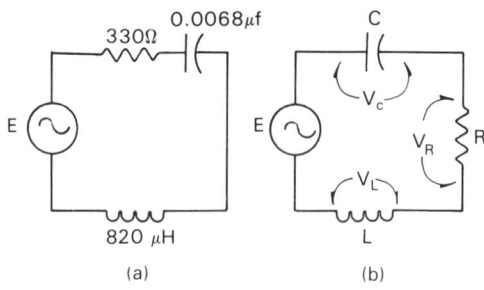

FIGURE 21-20
(a) Circuit of Problems 21-11 and 21-12.
(b) Circuit of Problems 21-19 and 21-20.

Section 21-3

21-12 For the circuit of Figure 21-20(a), $E = 56\underline{/0°}$ V at a frequency of 60 kHz. Determine the impedance of the circuit and construct both the rectangular and polar impedance vector diagrams.

21-13 Assume that an ac current source of 9.9-A is connected to terminals a-b of Figure 21-5(a). Determine the voltage across each component in the circuit.

21-14 Assume that an ac voltage source of $E = 36.76\underline{/-14°}$ V is connected to terminals a-b of Figure 21-5(a). Determine the circuit current, I.

21-15 For the series ac circuit of Figure 21-6(a), $e = 21.2 \sin(3000t)$ V. Determine (a) the impedance of the circuit, (b) the circuit current, I, and (c) the phase angle, θ.

21-16 For the series ac circuit of Figure 21-6(a), $e = 60 \sin(2200t + 30°)$ V. Determine (a) the impedance of the circuit, (b) the circuit current, I, and (c) the phase angle, θ.

21-17 Assume the source voltage of Figure 21-7 is changed to $100\underline{/0°}$ V and $\omega = 29.5$ krad/s. (a) Determine the impedance of the circuit. (b) Verify Kirchhoff's voltage law.

21-18 Assume the source voltage of Figure 21-7 is changed to $e = 16.5 \sin(37.7 \times 10^3\, t - 30°)$ V. Determine (a) the impedance of the circuit, (b) the circuit current, I, and (c) the voltage across the 470-mH inductor, V_L.

21-19 For the circuit of Figure 21-20(b), R is 2.2 kΩ and the source, E, is operating at a frequency of 15 kHz. If $V_C = 10$ V, $V_L = 18$ V, and $V_R = 22$ V, then determine:
(a) Source voltage (E).
(b) Circuit current (I).
(c) Circuit phase angle (θ).
(d) Value of the inductor (L).
(e) Value of the capacitor (C).
(f) Impedance of the circuit ($Z\underline{/\theta}$).
(g) Write the instantaneous periodic equation for the source voltage (assume the current phasor is at zero degrees).

21-20 For the circuit of Figure 21-20(b), R is 10 kΩ and the source, E, is operating at a frequency of 710 kHz. If $V_C = 800$ mV, $V_L = 300$ mV, and $V_R = 200$ mV, then determine:
(a) Source voltage (E).
(b) Circuit current (I).
(c) Circuit phase angle (θ).
(d) Value of the inductance (L).
(e) Value of the capacitor (C).

(f) Impedance of the circuit (Z/θ).

(g) Write the instantaneous periodic equation for the circuit current when the current phasor is at zero degrees.

Section 21–4

21–21 Using the voltage-divider equation, determine V_{ac} in the circuit of Figure 21–21(a).

21–22 Using the voltage-divider equation, determine V_{bc} in the circuit of Figure 21–21(a).

21–23 Determine V_{bc} in the circuit of Figure 21–21(b) when the source voltage is $e = 18 \sin(6.28 \times 10^3 t + 51°)$ V.

21–24 Determine V_{ab} in the circuit of Figure 21–21(b) when the source voltage is $e = 7.07 \sin(5027t + 22.8°)$ V.

Section 21–5

21–25 Verify KCL for a two-branch parallel circuit having a source current of $1.24/-76°$ A and a source voltage of $24.6/0°$ V. The first branch is an 82-Ω resistor and the second branch is an inductor with a reactance of 20.5 Ω.

21–26 Verify KCL for a three-branch parallel circuit having a source current of $45.3/22°$ mA and a source voltage of $67.2/0°$ V. The first branch is a 1.6-kΩ resistor, the second branch is a capacitor with a reactance of 772 Ω, and the third branch is an inductor with a reactance of 960 Ω.

21–27 For the inductor in Problem 21–25, determine the inductance if the source current is $i = 1.75 \sin(377t - 80°)$ A.

21–28 For the capacitor in Problem 21–26, determine the capacitance if the source current is $i = 64.1 \times 10^{-3} \sin(1257t + 22°)$ A.

Section 21–6

21–29 Determine the admittance of a parallel ac circuit having $120/0°$ V as the source voltage and $I = 10.5/48°$ mA as the circuit current.

21–30 Determine the admittance of a parallel ac circuit having $6.3/0°$ V as the source voltage and $I = 42/-18°$ mA as the circuit current.

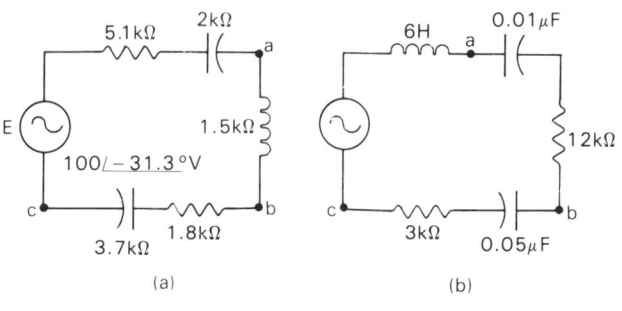

FIGURE 21–21
(a) Circuit of Problems 21–21 and 21–22.
(b) Circuit of Problems 21–23 and 21–24.

587
Problems

21-31 Determine the circuit admittance and the circuit impedance for a two-branch parallel circuit where branch 1 is $750 - j430 \, \Omega$ and branch 2 is $470 - j800 \, \Omega$.

21-32 Determine the circuit admittance and the circuit impedance for a two-branch parallel circuit where branch 1 is $51 - j18 \, \Omega$ and branch 2 is $100 + j132 \, \Omega$.

21-33 For the circuit of Figure 21-22(a), $E = 280 \underline{/0°}$ mV at a frequency of 1.2 MHz. Determine the admittance of the circuit and construct the admittance vector diagram.

21-34 For the circuit of Figure 21-22(a), $E = 52 \underline{/0°}$ mV at a frequency of 680 kHz. Determine the admittance of the circuit and construct the admittance vector diagram.

Section 21-7

21-35 Assume that a source of $e = 14.14 \sin(512t)$ V is connected across terminals a-b of Figure 21-14(a). Determine (a) the current in each branch, (b) the source current, I, and (c) the admittance of the entire circuit.

21-36 Assume that a source of $e = 70.7 \sin(1257t)$ V is connected across terminals a-b of Figure 21-14(a). Determine (a) the current in each branch, (b) the source current, I, and (c) the impedance of the entire circuit.

21-37 Assume the parallel circuit of Figure 21-15 is connected across a 1-kHz, 100-V ac source. Determine (a) the impedance of each branch, (b) the admittance of each branch, (c) the current in each branch, and (d) the source current.

21-38 Assume the parallel circuit of Figure 21-15 is connected across a 60-Hz, 200-V ac source. Determine (a) the impedance of each branch, (b) the admittance of each branch, (c) the current in each branch, and (d) the source current.

21-39 Assume the parallel circuit of Figure 21-16(a) is connected across an 800-Hz, 300-V ac source. Determine (a) the impedance of each branch and (b) the admittance of the entire circuit.

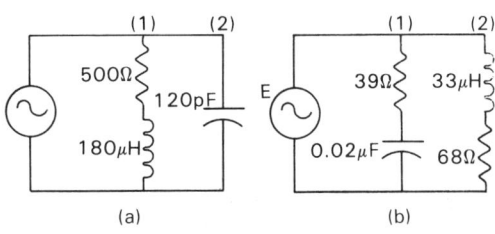

FIGURE 21-22
(a) Circuit of Problems 21-33 and 21-34.
(b) Circuit of Problems 21-41 and 21-42.

21-40 Assume the parallel circuit of Figure 21-16(a) is connected across a 1400-Hz, 440-V ac source. Determine (a) the current in the second branch, (b) the voltage across the 56-kΩ resistor, and (c) the impedance of the entire network.

21-41 For the circuit of Figure 21-22(b), the current in branch 1 is $i = 133 \sin(1.38 \times 10^6 t + 42.9°)$ mA. Determine (a) the source voltage, E, and (b) the impedance of the circuit.

21-42 For the circuit of Figure 21-22(b), the voltage across the 68-Ω resistor in branch 2 is $e = 28.3 \sin(1.89 \times 10^6 t - 42.5°)$ V. Determine (a) the current in branch 1 and (b) the souce voltage, E.

Section 21-8

21-43 Using the current-divider equation, determine the current in the capacitor of branch 2 of Figure 21-15 when the current out of the voltage source is $100/43.7°$ mA.

21-44 Using the current-divider equation, determine the current in the 200-Ω resistor of branch 3 of Figure 21-15 when the current out of the voltage source is $0.32/43.7°$ A.

21-45 Using the current-divider equation, determine I_1 in branch 1 of the circuit of Figure 21-22(a) when the current out of the voltage source is $i = 115 \sin(5 \times 10^6 t - 15°)$ mA.

21-46 Using the current-divider equation, determine I_2 in branch 2 of the circuit of Figure 21-22(b) when the current out of the voltage source is $i = 740 \times 10^{-3} \sin(0.95 \times 10^6 t - 41°)$ A.

Section 21-9

In each of the following problems, determine the values of the components of (a) the series equivalent circuit and (b) the parallel equivalent circuit of the ac circuit in the box of Figure 21-19(a), given the following information.

21-47 $E = 28/0°$ V, $I = 1.8/64°$ A, $f = 60$ Hz
21-48 $E = 100/0°$ V, $I = 84/31°$ mA, $f = 400$ Hz
21-49 $E = 10/0°$ V, $I = 500/-55°$ μA, $f = 200$ kHz
21-50 $E = 380/0°$ mV, $I = 30/25°$ μA, $f = 1.0$ MHz

22 Series-Parallel AC Networks

22-1 SOLVING SERIES–PARALLEL AC NETWORKS

CHAPTER PREVIEW This chapter is a continuation of the previous chapter. Like Chapter 9, this chapter will use the concepts of blocking the series-parallel network into series blocks and parallel blocks. The series blocks are solved with the previously stated rules, laws, and techniques of series ac circuit analysis. The parallel blocks are solved with the rules, laws, and techniques of parallel ac circuit analysis. This chapter will provide you with an understanding of the solution of series-parallel ac networks by demonstrating the methods and techniques used for the solution of single-source series-parallel networks.

PERFORMANCE OBJECTIVES Once you have read each section, worked each example with pencil, paper, and calculator, and answered each problem for every section, you will be able to:

☐ Solve single-source series-parallel ac networks.
☐ Form an equivalent circuit of a series-parallel ac network.

22-1 SOLVING SERIES-PARALLEL AC NETWORKS

The technique used in Chapter 9 to solve dc series-parallel networks will be used here to solve ac networks. These techniques include blocking the network into series blocks and parallel blocks and sectioning the network. The following examples will demonstrate the techniques used to solve for various ac circuit parameters in series-parallel networks.

EXAMPLE 22-1 Determine the impedance across terminals $a-b$ of Figure 22-1(a).

SOLUTION Combine $Z_3 \parallel Z_2$ into an equivalent impedance; then add it to Z_1. Thus

$$Z_3 = R_2 + j\omega L_2$$
$$Z_3 = 10 \times 10^3 + j100 \times 10^3 \times 100 \times 10^{-3}$$
$$Z_3 = 10 + j10 \text{ k}\Omega \boxed{\rightarrow P} \ 14.14\underline{/45°} \text{ k}\Omega$$

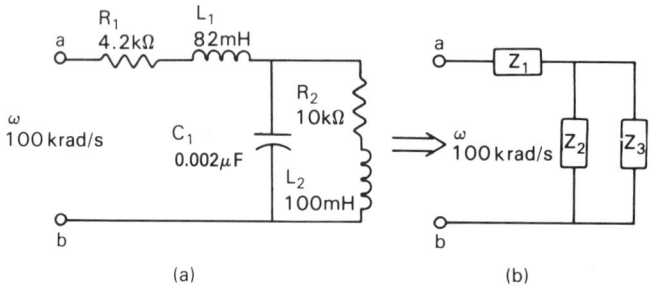

FIGURE 22-1
(a) Network of Example 22-1.
(b) Blocked network.

Solving Series-Parallel AC Networks

$$Y_3 = \frac{1}{Z_3} = \frac{1}{14.14\underline{/45°}\text{ k}\Omega} = 70.7\underline{/-45°}\text{ }\mu\text{S}$$

$$Y_3 = 70.7\underline{/-45°}\text{ }\mu\text{S} \boxed{\rightarrow\text{R}} \ 50 - j50 \text{ }\mu\text{S}$$

and

$$Z_2 = 0 - j\frac{1}{\omega C_1} = 0 - j\frac{1}{100 \times 10^3 \times 2 \times 10^{-9}}$$

$$Z_2 = 0 - j5 \text{ k}\Omega \boxed{\rightarrow\text{P}} \ 5\underline{/-90°}\text{ k}\Omega$$

$$Y_2 = 1/Z_2 = 1/5\underline{/-90°}\text{ k}\Omega = 200\underline{/90°}\text{ }\mu\text{S}$$

$$Y_2 = 200\underline{/90°}\text{ }\mu\text{S} \boxed{\rightarrow\text{R}} \ 0 + j200 \text{ }\mu\text{S}$$

Determine $Z_3 \| Z_2$:

$$Z_{3\|2} = \frac{1}{Y_3 + Y_2} = \frac{1}{(50 - j50) + (0 + j200) \text{ }\mu\text{S}}$$

$$Z_{3\|2} = \frac{1}{50 + j150 \text{ }\mu\text{S}} = \frac{1}{158\underline{/71.6°}\text{ }\mu\text{S}}$$

$$Z_{3\|2} = 6.32\underline{/-71.6°}\text{ k}\Omega \boxed{\rightarrow\text{R}} \ 1.99 - j6.0 \text{ k}\Omega$$

Add $Z_{3\|2}$ and Z_1 to determine the impedance of the network:

$$Z_1 = R_1 + j\omega L_1 = 4.2 + j8.2 \text{ k}\Omega$$

$$Z_T = Z_{3\|2} + Z_1$$

$$Z_T = (1.99 - j6.0) + (4.2 + j8.2) \text{ k}\Omega$$

$$\therefore \quad Z_T = 6.19 + j2.2 \text{ k}\Omega \boxed{\rightarrow\text{P}} \ 6.57\underline{/19.6°}\text{ k}\Omega$$

EXAMPLE 22-2 Determine the voltage across Z_2 of Figure 22-2(b) when $E = 100\underline{/0°}$ V.

SOLUTION First, determine $Z_{3\|2}$; then apply the voltage-divider equation to determine V_{Z2}:

$$Z_3 = R_3 - jX_{C2}$$

$$Z_3 = 180 - j90 \text{ }\Omega \boxed{\rightarrow\text{P}} \ 201\underline{/-26.6°}\text{ }\Omega$$

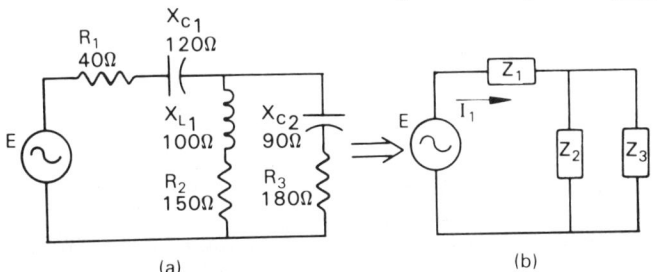

FIGURE 22-2
(a) Network of Example 22-2.
(b) Blocked network.

Series-Parallel AC Networks

$$Y_3 = \frac{1}{Z_3} = 4.98\underline{/26.6°} \text{ mS} \quad \boxed{\rightarrow R}$$

$$Y_3 = 4.45 + j2.23 \text{ mS}$$

and

$$Z_2 = R_2 + jX_{L1}$$

$$Z_2 = 150 + j100 \text{ }\Omega \quad \boxed{\rightarrow P} \quad 180\underline{/33.7°} \text{ }\Omega$$

$$Y_2 = \frac{1}{Z_2} = 5.55\underline{/-33.7°} \text{ mS} \quad \boxed{\rightarrow R}$$

$$Y_2 = 4.62 - j3.08 \text{ mS}$$

Determine $Z_3 \parallel Z_2$:

$$Z_{3\parallel 2} = \frac{1}{Y_3 + Y_2}$$

$$Z_{3\parallel 2} = \frac{1}{(4.45 + j2.23) + (4.62 - j3.08) \text{ mS}}$$

$$Z_{3\parallel 2} = \frac{1}{9.07 - j0.85 \text{ mS}} = \frac{1}{9.11\underline{/-5.35°} \text{ mS}}$$

$$Z_{3\parallel 2} = 110\underline{/5.35°} \text{ }\Omega \quad \boxed{\rightarrow R} \quad 110 + j10.3 \text{ }\Omega$$

Apply the voltage-divider equation:

$$V_{Z2} = V_{Z3} = V_{Z3\parallel 2} = \frac{EZ_{3\parallel 2}}{\Sigma Z}$$

where $E = 100\underline{/0°}$ V, $Z_{3\parallel 2} = 110\underline{/5.35°}$ $\Omega = 110 + j10.3$ Ω, and $\Sigma Z = Z_1 + Z_{3\parallel 2}$. Determine ΣZ and substitute:

$$\Sigma Z = (40 - j120) + (110 + j10.3) \text{ }\Omega$$

$$\Sigma Z = 150 - j109.7 \text{ }\Omega \quad \boxed{\rightarrow P} \quad 186\underline{/-36.2°} \text{ }\Omega$$

$$V_{Z2} = \frac{EZ_{3\parallel 2}}{\Sigma Z} = \frac{(100\underline{/0°})(110\underline{/5.35°})}{186\underline{/-36.2°}}$$

$$V_{Z2} = 59.1\underline{/41.6°} \text{ V}$$

$\therefore \quad V_{Z2} = 59.1\underline{/41.6°}$ V

EXAMPLE 22–3 Determine the current in R_2 of Z_4 (Figure 22–3) when $I = 10\underline{/21°}$ mA.

SOLUTION Start by determining the equivalent impedance of $Z_{2,3,4}$ of Figure 22–3(b); then compute the voltage across the equivalent $Z_{2,3,4}$. Finally, compute the current in branch Z_4, which is the current in

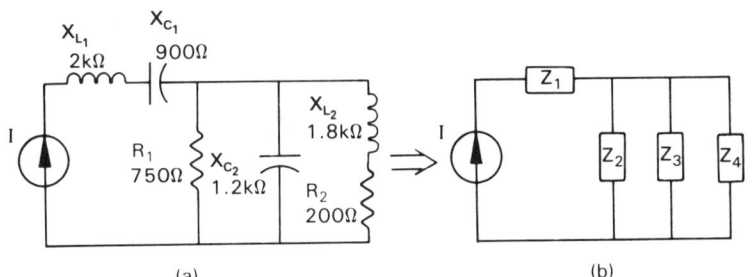

FIGURE 22-3
(a) Network of Example 22-3.
(b) Blocked network.

R_2. Thus, the equivalent impedance $Z_{2,3,4} = 1/(Y_2 + Y_3 + Y_4)$:

$$Z_2 = 750 + j0 \ \Omega \ \boxed{\rightarrow P} \ 750\underline{/0°} \ \Omega$$

$$Y_2 = \frac{1}{Z_2} = \frac{1}{750\underline{/0°}} = 1.33\underline{/0°} \text{ mS} \ \boxed{\rightarrow R}$$

$$Y_2 = 1.33 + j0 \text{ mS}$$

$$Z_3 = 0 - j1200 \ \Omega \ \boxed{\rightarrow P} \ 1200\underline{/-90°} \ \Omega$$

$$Y_3 = \frac{1}{Z_3} = \frac{1}{1200\underline{/-90°} \ \Omega}$$

$$Y_3 = 0.833\underline{/90°} \text{ mS} \ \boxed{\rightarrow R} \ 0 + j0.833 \text{ mS}$$

$$Z_4 = 200 + j1800 \ \Omega \ \boxed{\rightarrow P} \ 1811\underline{/83.7°} \ \Omega$$

$$Y_4 = \frac{1}{Z_4} = \frac{1}{1811\underline{/83.7°} \ \Omega}$$

$$Y_4 = 0.552\underline{/-83.7°} \text{ mS} \ \boxed{\rightarrow R}$$

$$Y_4 = 0.061 - j0.549 \text{ mS}$$

Determine the equivalent impedance $Z_{2,3,4}$:

$$Z_{2,3,4} = \frac{1}{Y_2 + Y_3 + Y_4}$$

where

$$Y_{2+3+4} = (1.33 + j0 \text{ mS}) + (0 + j0.833 \text{ mS}) + (0.061 - j0.549 \text{ mS})$$

$$Y_{2+3+4} = 1.39 + j0.284 \text{ mS} \ \boxed{\rightarrow P}$$

$$Y_{2+3+4} = 1.42\underline{/11.5°} \text{ mS}$$

$$Z_{2,3,4} = \frac{1}{Y_{2+3+4}} = \frac{1}{1.42 \times 10^{-3}\underline{/11.5°}}$$

$$Z_{2,3,4} = 704\underline{/-11.5°} \ \Omega$$

$$V_{2,3,4} = IZ_{2,3,4}$$

$$V_{2,3,4} = (10 \times 10^{-3}\underline{/21°})(704\underline{/-11.5°})$$

$$V_{2,3,4} = 7.04\underline{/9.5°} \text{ V}$$

$$I_4 = \frac{V_{2,3,4}}{Z_4} = \frac{7.04\underline{/9.5°}}{1811\underline{/83.7°}} = 3.89\underline{/-74.2°} \text{ mA}$$

∴ The current in R_2 is $3.89\underline{/-74.2°}$ mA.

EXAMPLE 22-4 Determine the voltage across R_2 of Figure 22-4(a) when $E = 100\underline{/0°}$ V.

SOLUTION Start by determining the equivalent impedance $Z_{4,3}$; then apply the voltage-divider equation to Z_2 in series with $Z_{4,3}$. Thus

$$Z_{4,3} = \frac{1}{Y_4 + Y_3}$$

where

$$Y_4 = \frac{1}{Z_4} = \frac{1}{0 - j400} = \frac{1}{400\underline{/-90°}}$$

$$Y_4 = 2.5\underline{/90°} \text{ mS}$$

$$Y_4 = 2.5\underline{/90°} \text{ mS } \boxed{\rightarrow R} = 0 + j2.5 \text{ mS}$$

$$Y_3 = \frac{1}{Z_3} = \frac{1}{600 + j300} = \frac{1}{671\underline{/26.6°}}$$

$$Y_3 = 1.49\underline{/-26.6°} \text{ mS}$$

$$Y_3 = 1.49\underline{/-26.6°} \text{ mS } \boxed{\rightarrow R}$$

$$Y_3 = 1.33 - j0.667 \text{ mS}$$

Determine $Z_{4,3}$:

$$Z_{4,3} = \frac{1}{Y_4 + Y_3}$$

$$Z_{4,3} = \frac{1}{(0 + j2.5) + (1.33 - j0.667) \text{ mS}}$$

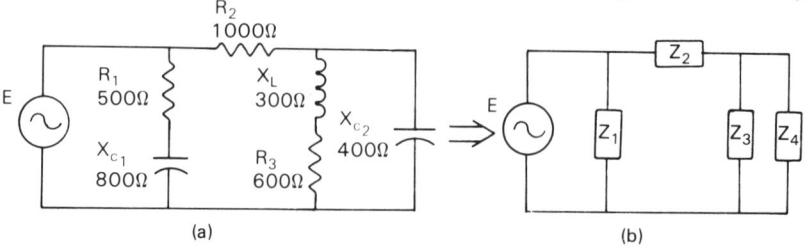

FIGURE 22-4 (a) Network of Example 22-4. (b) Blocked network.

Solving Series-Parallel AC Networks

$$Z_{4,3} = \frac{1}{1.33 + j1.82 \text{ mS}} = \frac{1}{2.25\underline{/54°} \text{ mS}}$$

$$Z_{4,3} = 444\underline{/-54°} \ \Omega$$

$$Z_{4,3} = 444\underline{/-54°} \ \boxed{\rightarrow R} \ 261 - j359 \ \Omega$$

Apply the voltage-divider equation and determine V_{Z2} which equals V_2:

$$V_2 = \frac{EZ_2}{\Sigma Z}$$

where

$$E = 100\underline{/0°} \text{ V}$$

$$Z_2 = 1000\underline{/0°} \ \Omega$$

$$\Sigma Z = Z_2 + Z_{4,3}$$

$$\Sigma Z = (1000 + j0) + (261 - j359)$$

$$\Sigma Z = 1261 - j359 \ \Omega$$

$$\Sigma Z = 1261 - j359 \ \boxed{\rightarrow P} \ 1311\underline{/-15.9°} \ \Omega$$

and

$$V_2 = \frac{(100\underline{/0°})(1000\underline{/0°})}{1311\underline{/-15.9°}} = 76.3\underline{/15.9°} \text{ V}$$

∴ The voltage across $R_2 = 76.3\underline{/15.9°}$ V.

EXAMPLE 22–5 Determine the current in L_2 of impedance Z_4 (Figure 22–5) when I is $0.2\underline{/-35.3°}$ A.

SOLUTION Start by forming the equivalent impedance of $Z_5 \parallel Z_4$; then add this equivalent to Z_3. This result is used to form the equivalent impedance of $Z_2 \parallel Z_{3+5\parallel4}$. Once $Z_{2\parallel(3+4\parallel5)}$ is known, then the voltage across this equivalent impedance in turn gives the voltage across Z_2 since $V_2 = V_{3+5\parallel4} = V_{2\parallel(3+5\parallel4)}$. Computing I_{Z2} and applying KCL to the node formed by the junction of Z_1, Z_2, and Z_3 will yield I_{Z3}. Finally, applying the current-divider equation to the current entering the node formed by the junction of Z_3, Z_4, and Z_5 results in the current in Z_4 and, consequently, I_{L2}. Thus

$$Z_{5\parallel4} = \frac{1}{Y_5 + Y_4}$$

Where $1/\omega C_2 = 323 \ \Omega$ and $\omega L_2 = 310 \ \Omega$ and

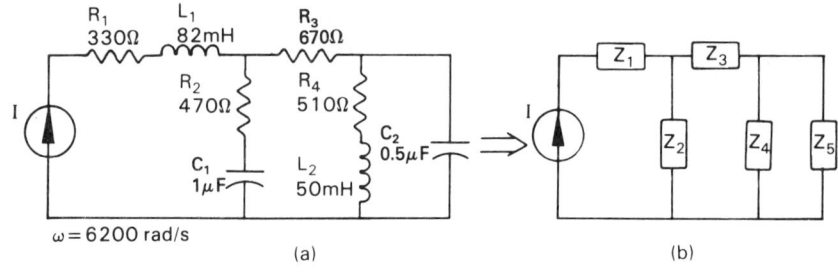

FIGURE 22-5 (a) Network of Example 22-5. (b) Blocked network.

$$Y_5 = \frac{1}{Z_5} = \frac{1}{0 - j323} = \frac{1}{323\underline{/-90°}}$$

$$Y_5 = 3.10\underline{/90°} \text{ mS}$$

$$Y_5 = 3.10\underline{/90°} \text{ mS} \quad \boxed{\rightarrow R} \quad 0 + j3.10 \text{ mS}$$

and

$$Y_4 = \frac{1}{Z_4} = \frac{1}{510 + j310} = \frac{1}{597\underline{/31.3°}}$$

$$Y_4 = 1.68\underline{/-31.3°} \text{ mS}$$

$$Y_4 = 1.68\underline{/-31.3°} \text{ mS} \quad \boxed{\rightarrow R}$$

$$Y_4 = 1.44 - j0.873 \text{ mS}$$

Determine $Z_{5\|4}$:

$$Z_{5\|4} = \frac{1}{Y_5 + Y_4} = \frac{1}{1.44 + j2.23 \text{ mS}}$$

$$Z_{5\|4} = \frac{1}{2.65\underline{/57.1°} \text{ mS}}$$

$$Z_{5\|4} = 377\underline{/-57.1°} \ \Omega \quad \boxed{\rightarrow R} \quad 205 - j317 \ \Omega$$

Add $Z_{5\|4}$ to Z_3:

$$Z_{3+5\|4} = (670 + j0) + (205 - j317)$$

$$Z_{3+5\|4} = 875 - j317$$

$$Z_{3+5\|4} = 875 - j317 \ \Omega \quad \boxed{\rightarrow P} \quad 931\underline{/-19.9°} \ \Omega$$

Form the equivalent impedance of Z_2 parallel to $Z_{3+5\|4}$:

$$Z_2 \parallel Z_{3+5\|4} = \frac{1}{Y_2 + Y_{3+5\|4}}$$

Where $1/\omega C_1 = 161 \ \Omega$

$$Y_2 = \frac{1}{Z_2} = \frac{1}{470 - j161} = \frac{1}{497\underline{/-18.9°}}$$

$$Y_2 = 2.01\underline{/18.9°} \text{ mS}$$

$$Y_2 = 2.01\underline{/18.9°} \text{ mS} \quad \boxed{\rightarrow R} \quad 1.90 + j0.651 \text{ mS}$$

and

$$Y_{3+5\|4} = \frac{1}{Z_{3+5\|4}} = \frac{1}{931\underline{/-19.9°}} = 1.07\underline{/19.9°} \text{ mS}$$

$$Y_{3+5\|4} = 1.07\underline{/19.9°} \text{ mS} \quad \boxed{\rightarrow R}$$

$$Y_{3+5\|4} = 1.01 + j0.364 \text{ mS}$$

Determine $Z_{2\|(3+5\|4)}$:

$$Z_{2\|(3+5\|4)} = \frac{1}{Y_2 + Y_{3+5\|4}} = \frac{1}{2.91 + j1.02 \text{ mS}}$$

$$Z_{2\|(3+5\|4)} = \frac{1}{3.08\underline{/19.3°} \text{ mS}} = 325\underline{/-19.3°} \text{ }\Omega$$

Compute the voltage across the equivalent impedance $Z_{2\|(3+5\|4)}$. Thus

$$V_{2\|(3+5\|4)} = IZ_{2\|(3+5\|4)}$$

$$V_{2\|(3+5\|4)} = (0.2\underline{/-35.3°})(325\underline{/-19.3°})$$

$$V_{2\|(3+5\|4)} = 65.0\underline{/-54.6°} \text{ V}$$

Determine I_2 from

$$I_2 = \frac{V_2}{Z_2}$$

where

$$V_2 = V_{2\|(3+5\|4)} = 65.0\underline{/-54.6°} \text{ V}$$

$$Z_2 = 497\underline{/-18.9°}$$

$$I_{Z2} = \frac{65.0\underline{/-54.6°}}{497\underline{/-18.9°}} = 131\underline{/-35.7°} \text{ mA}$$

Apply KCL to determine I_{Z3}:

$$I_{Z3} = I - I_{Z2} = 0.2\underline{/-35.3°} - 0.131\underline{/-35.7°}$$

$$I_{Z3} = (0.163 - j0.116) - (0.106 - j0.076)$$

$$I_{Z3} = (0.057 - j0.040) = 70\underline{/-35.1°} \text{ mA}$$

Apply the current-divider equation to determine I_{Z4}. Thus

$$I_{Z4} = \frac{I_{Z3}Y_4}{\Sigma Y}$$

where

$$I_{Z3} = 70\underline{/-35.1°}\ \text{mA}$$

$$Y_4 = 1.68\underline{/-31.3°}\ \text{mS}$$

$$\Sigma Y = Y_4 + Y_5 = 2.65\underline{/57.1°}\ \text{mS}$$

and

$$I_{Z4} = \frac{(70\underline{/-35.1°})(1.68 \times 10^{-3}\underline{/-31.3°})}{2.65 \times 10^{-3}\underline{/-57.1°}}$$

$$I_{Z4} = 44.4\underline{/-123.5°}\ \text{mA}$$

∴ The current in L_2 of impedance Z_4 is $44.4\underline{/-123.5°}$ mA.

EXAMPLE 22–6 Determine the components in both the series and parallel equivalent circuit of the network of Figure 22–1(a) for $\omega = 100$ krad/s.

SOLUTION From Example 22–1, the total impedance of the network is

$$Z_T = 6.19 + j2.2\ \text{k}\Omega\ \boxed{\rightarrow P}\ 6.57\underline{/19.6°}\ \text{k}\Omega$$

The series equivalent circuit consists of a resistor of 6.19 kΩ and an inductor with an inductive reactance of 2.2 kΩ. The inductance of the inductor is $L = X_L/\omega$ or $L = (2.2 \times 10^3)/(100 \times 10^3) = 22$ mH. Figure 22–6(b) pictures the series equivalent circuit.

Determine the parallel equivalent circuit.

$$Y_T = \frac{1}{Z_T} = \frac{1}{6.57 \times 10^3\underline{/19.6°}}$$

$$Y_T = 0.152\underline{/-19.6°}\ \text{mS}$$

$$Y_T = 152\underline{/-19.6°}\ \mu\text{S}\ \boxed{\rightarrow R}\ 143 - j51\ \mu\text{S}$$

$$R_p = \frac{1}{G_{eq}} = \frac{1}{143 \times 10^{-6}} = 6.99\ \text{k}\Omega$$

$$X_p = \frac{1}{B_L} = \frac{1}{51 \times 10^{-6}} = 19.6\ \text{k}\Omega$$

∴ The parallel equivalent circuit consists of a 6.99-kΩ resistor shunted by an inductor with an inductive reactance of 19.6 kΩ. The inductance

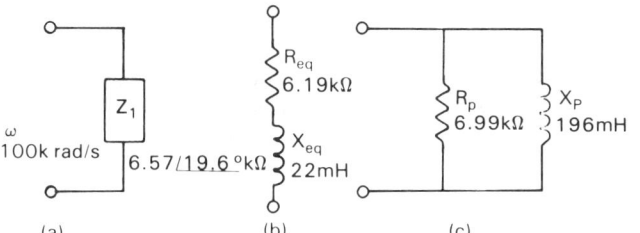

FIGURE 22-6
(a) The total impedance, Z_T, of the network of Figure 22-1(a) is represented by an impedance block.
(b) Series equivalent circuit with component values noted.
(c) Parallel equivalent circuit with component values noted.

of the inductor is $L = X_L/\omega$ or $L = (19.6 \times 10^3)/(100 \times 10^3) = 196$ mH. Figure 22-6(c) pictures the parallel equivalent circuit.

PROBLEMS

Section 22-1

22-1 Assume that a current source, $i = 21.2 \sin(100 \times 10^3 t - 19.6°)$ mA, is connected to terminals a-b of Figure 22-1(a). Determine (a) the current in L_2 of Z_3 and (b) the voltage across C_1 of Z_2.

22-2 Assume that a voltage source, $e = 311 \sin(100 \times 10^3 t)$ V, is connected across terminals a-b of Figure 22-1(a). Determine (a) the current in R_1 of Z_1 and (b) the voltage across R_2 of Z_3.

22-3 The angular velocity, ω, in the network of Figure 22-1(a) is changed from 100 to 75 krad/s. Determine (a) the impedance across terminals a-b and (b) the values of the components of the parallel equivalent circuit.

22-4 The angular velocity, ω, in the network of Figure 22-1(a) is changed from 100 to 125 krad/s. Determine (a) the admittance across terminals a-b and (b) the values of the components of the series equivalent circuit.

22-5 For the network of Figure 22-2(a), when $E = 50/\underline{0°}$ V, determine (a) the current in R_2 and (b) the voltage across R_1.

22-6 For the network of Figure 22-2(a), determine (a) the voltage across X_{L1} and (b) the voltage across R_3 when $E = 220/\underline{0°}$ V.

22-7 For the network of Figure 22-2(a), determine the values of the components of the series equivalent circuit when $\omega = 377$ rad/s.

22-8 For the network of Figure 22-2(a), determine the value of the components of the parallel equivalent circuit when $\omega = 377$ rad/s.

22-9 For the network of Figure 22-3(a), determine the voltage across R_2 when $I = 27/\underline{21°}$ mA.

22-10 For the network of Figure 22-3(a), determine the voltage across C_2 when $I = 50/\underline{21°}$ mA.

22-11 Determine the impedance of the network of Figure 22-3(a).
22-12 Determine the current in L_2 of the network of Figure 22-3(a) when $I = 50\underline{/-54.3°}$ mA.
22-13 Determine the impedance of the network of Figure 22-4(a).
22-14 Determine the voltage across R_1 of Figure 22-4(a) when $E = 120\underline{/0°}$ V.
22-15 For the network of Figure 22-4(a), determine the current in R_3 when $E = 60\underline{/0°}$ V.
22-16 For the network of Figure 22-4(a), determine the voltage across X_{C2} when $E = 28\underline{/0°}$ V.
22-17 Determine (a) the impedance of the network of Figure 22-5 when $\omega = 6200$ rad/s and (b) the value of the components of the parallel equivalent circuit.
22-18 Determine (a) the impedance of the network of Figure 22-5 when $\omega = 5600$ rad/s and (b) the values of the components of the parallel equivalent circuit.
22-19 Determine the voltage across R_2 in the network of Figure 22-5 when the source current is $i = 396 \sin(7500t - 30°)$ mA.
22-20 Determine the current in C_1 of the network of Figure 22-5 when the source current is $i = 226 \sin(6800t - 20°)$ mA.

23 Power in Alternating Current Circuits

23-1 INTRODUCTION TO AC POWER PARAMETERS
23-2 AVERAGE POWER
23-3 REACTIVE POWER
23-4 POWER IN AC NETWORKS
23-5 MAXIMUM POWER TRANSFER

CHAPTER PREVIEW

This chapter is an extension of the discussion of power dissipation started in Chapter 6 and continued in Chapter 19. In this chapter, we will consider how power is determined in ac circuits containing both resistance and reactance. As always, resistance is the only component of the circuit impedance that dissipates average power. However, the discussion of power will also consider how the reactive component of impedance affects the power factor of the circuit, as well as the transfer of power from source to load. Besides average power, reactive and apparent power will be studied in this chapter.

PERFORMANCE OBJECTIVES

Once you have read each section, worked each example with pencil, paper, and calculator, and answered each problem of every section, you will be able to:

☐ Understand the relationship among apparent power, average power, and reactive power.
☐ Use the power factor to determine reactive and average power from apparent power.
☐ Apply the power triangle to ac circuits to determine the total power.
☐ Understand the meaning of leading and lagging power factors.
☐ Determine the value of capacitance needed to correct an ac power line for low power factor.
☐ Determine the conditions of the load impedance for maximum power transfer from an ac voltage source with internal reactive impedance.

23–1 INTRODUCTION TO AC POWER PARAMETERS

The following discussion explores the parameters governing the power dissipation in an ac circuit. The parameters include *average power*, P (also known as true power, actual power, or real power), *apparent power*, P_a, *reactive power*, P_q, and the *power factor*, $\cos(\theta)$.

Apparent Power

We will start our discussion of power in an ac circuit by defining *apparent power*. The apparent power in a circuit, P_a, is the product of the voltage applied to a circuit, E, and the circuit current, I. Apparent power is measured in units of volt-amperes (VA). Thus

$$P_a = |E| |I| \quad \text{(VA)} \tag{23-1}$$

With this definition in mind, look once again at Equation 19–15. Here, the average power (also known as true power) is

Introduction To AC Power Parameters

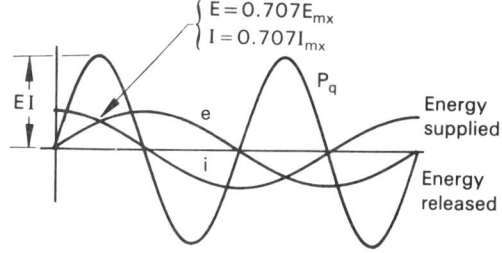

FIGURE 23-1
The product of EI of the voltage and current waves is the reactive power, P_q, when waves e and i are out of phase by 90°.

defined as $P = EI$ (W). By comparing these two equations, we must conclude that the apparent power, P_a, equals the average power, P, in a resistive circuit. This is indeed the case.

Reactive Power

Shifting our attention from a resistive circuit to a reactive circuit, we recall that the power curve of Figure 19-8 (reproduced here as Figure 23-1) has an average power of zero watts. This indicates that the perfect reactive component (no resistance) is not a consumer of power. However, at any instant of time, an inductor or capacitor is either taking energy from or returning energy to the source. The product of the effective source voltage, E, and the effective current, I, of a purely reactive circuit results in the peak value of the reactive power curve of Figure 23-1. The peak reactive power is simply called the *reactive power*, P_q. However, when this definition for reactive power, P_q, is compared to that for apparent power, P_a, we must conclude that the reactive power equals the apparent power in a purely reactive circuit.

The Power Triangle

To summarize, the apparent power equals the average power in a resistive circuit. Thus, $P_a = P$ in a resistive circuit. In a purely reactive circuit (where $\theta = 90°$), the apparent power equals the reactive power. Thus, $P_a = P_q$ in a reactive circuit. It may be learned from these statements that the apparent power may be composed entirely of average power or entirely of reactive power, or it may be a combination of average power and reactive power. The relationship between these three powers, average (P), reactive (P_q), and apparent (P_a), is diagrammed in the *power triangle*.

The power triangle, as pictured in Figure 23-2, is a *scalar* presentation used to determine the relationship among the three previously defined powers. Angle θ in the power triangle is the current-voltage phase angle of the circuit. By applying the trigonometric functions to the power triangle (which is a right triangle), the following relationships are determined. Thus, applying the cosine function results in

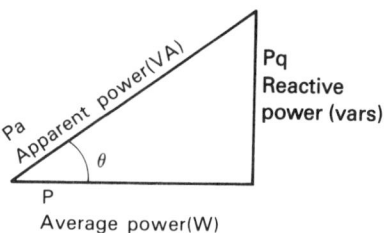

FIGURE 23-2 Power triangle picturing the relationship among apparent power, reactive power, and average power.

$$\cos(\theta) = \frac{\text{average power}}{\text{apparent power}} = \frac{P}{P_a} \qquad (23\text{-}2)$$

Solving for average power and substituting EI for P_a,

$$P = P_a \cos(\theta) = EI \cos(\theta) \quad (\text{W}) \qquad (23\text{-}3)$$

Applying the sine function results in

$$\sin(\theta) = \frac{\text{reactive power}}{\text{apparent power}} = \frac{P_q}{P_a} \qquad (23\text{-}4)$$

Solving for reactive power and substituting EI for P_a,

$$P_q = P_a \sin(\theta) = EI \sin(\theta) \quad (\text{var}) \qquad (23\text{-}5)$$

Power Factor In Equation 23-2, $\cos(\theta)$ is called the *power factor* of the circuit. The power factor, as noted, is defined as the ratio between the average power and the apparent power. From the impedance vector diagram, the power factor is also equal to the ratio of the circuit resistance to the magnitude of the circuit impedance. Thus

$$\cos(\theta) = \frac{R}{|Z|} \qquad (23\text{-}6)$$

In purely resistive ac circuits, the impedance equals the resistance ($Z = R$), the phase angle is zero ($\theta = 0°$), and the power factor is 1 [$\cos(0°) = 1$]. In purely reactive ac circuits, the impedance equals the reactance ($Z = X$), the phase angle is 90° ($\theta = 90°$), and the power factor is 0 [$\cos(90°) = 0$]. Thus, the power factor is a positive, unitless number that can be expressed as a percentage or a decimal fraction having a range between 1 and 0. Determining and using the power factor are demonstrated in the following examples.

EXAMPLE 23-1 Determine the power factor for each of the following circuits: (a) $P = 10$ W and $P_a = 15$ W, (b) $R = 27\ \Omega$ and $|Z| = 97\ \Omega$, and (c) $Z = 210 + j575\ \Omega$.

SOLUTION (a) $\cos(\theta) = P/P_a = 10/15 = 0.667$.

∴ The power factor is 0.667.
(b) $\cos(\theta) = R/|Z| = 27/97 = 0.278$.

∴ The power factor is 0.278.
(c) Convert $210 + j575$ to polar form:

$$210 + j575 \boxed{\rightarrow P} \;612\underline{/70°}$$

$$\cos(\theta) = \cos(70°) = 0.342$$

∴ The power factor is 0.342.

EXAMPLE 23–2 Determine the average power dissipated in an ac circuit having a phase angle of 22° and an apparent power of 200 VA.

SOLUTION Use Equation 23–3:

$$P = P_a \cos(\theta) = 200 \cos(22°) = 185 \text{ W}$$

∴ The average power is 185 W.

23–2 AVERAGE POWER

The power dissipated by the resistive part of the load may be computed from one of several equations. The equations for computing average power in an ac circuit include the power factor, as noted in the following:

$$P = |E| |I| \cos(\theta) \quad \text{(W)} \qquad (23\text{–}7)$$

$$P = |I|^2 |Z| \cos(\theta) \quad \text{(W)} \qquad (23\text{–}8)$$

$$P = \frac{|E|^2}{|Z|} \cos(\theta) \quad \text{(W)} \qquad (23\text{–}9)$$

EXAMPLE 23–3 Determine the average power dissipated in an ac circuit having an impedance of $Z = 30\underline{/60°}\;\Omega$ and a source voltage of 220 V.

SOLUTION Substitute into Equation 23–9:

$$P = \frac{|E|^2}{|Z|} \cos(\theta) = \frac{220^2}{30} \cos(60°) = 807 \text{ W}$$

Leading and Lagging Power Factors

The words *leading* and *lagging* are often used with the power factor. The relationship of the current through the load to the voltage as the reference phasor determines the *lead* or *lag* of the circuit. Because current leads voltage in a capacitive load, a capacitive circuit has a leading power factor. Since current lags voltage in an inductive load, an inductive circuit has a lagging power factor.

EXAMPLE 23–4 A certain ac circuit has a lagging power factor of 0.4, a load impedance of 180 Ω, and a circuit current $i = 14.14 \sin(377t)$ A. Determine (a) the power dissipation in the resistive part of the load, and (b) the components in the series equivalent circuit.

SOLUTION (a) Use Equation 23–8:

$$P = |I|^2 |Z| \cos(\theta)$$

$$P = [(14.14)(0.707)]^2 (180)(0.4)$$

∴ $P = 7.2$ kW

(b) Determine the series equivalent components:

$$|\theta| = \arccos(\text{power factor})$$

$$|\theta| = \arccos(0.4) = 66.4°$$

$$|\theta| = 66.4°$$

Observation The circuit phase angle, θ, is assigned a positive sign because the load has a lagging power factor, indicating an inductive load impedance. Thus

$$Z = 180 \underline{/66.4°} \;\Omega \quad \boxed{\rightarrow R} \quad 72.1 + j165 \;\Omega$$

$$R_{eq} = 72.1 \;\Omega \quad \text{and} \quad X_{eq} = 165 \;\Omega$$

Compute the equivalent inductance:

$$L = \frac{X_L}{\omega} = \frac{165}{377} = 0.438 \text{ H}$$

∴ The equivalent series circuit consists of a resistance of 72.1 Ω and an inductor having an inductance of 438 mH.

Power Measurement

The power dissipated by the resistance in an ac load is measured with a *wattmeter*. The circuit connections to measure power with a wattmeter are shown in Figure 23–3(a).

As shown, the wattmeter has four terminals: two terminals for the *current* coil and two terminals for the *potential* coil. Like an ammeter, the current coil is connected in series with the circuit and, like a voltmeter, the potential coil is connected across the circuit. When connecting the *electrodynamometer* wattmeter (a two-coil meter movement) into the circuit, the top of the potential coil must be connected to the same lead as the current coil; otherwise, the meter will be deflected down scale.

607

Reactive Power

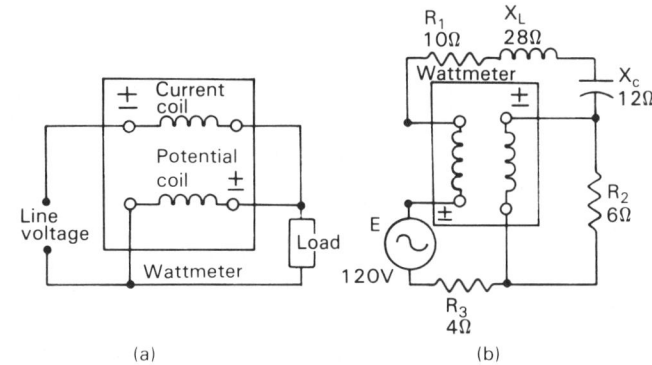

FIGURE 23-3
(a) Wattmeter connections.
(b) Wattmeter connected to measure power delivered to R_2.

EXAMPLE 23-5 Determine the wattmeter reading of Figure 23-3(b).

SOLUTION Determine the current in R_2:

$$|I| = |I_2| = \frac{|E|}{|Z|} = \frac{120}{20 + j16}$$

$$|I_2| = \frac{120}{25.6} = 4.69 \text{ A}$$

Compute the power dissipated by R_2:

$$P_2 = I^2 R_2 = (4.69)^2(6) = 132 \text{ W}$$

∴ The wattmeter indicates 132 W.

Observation Because the phase angle, θ, between the current in R_2 and the voltage across R_2 is zero, the power factor of the resistor is 1; $\cos(0°) = 1$.

23-3 REACTIVE POWER

Reactive power (also referred to as *wattless* power) is present in a load whenever the apparent power is greater than the average power. When this is the case, the circuit power factor is less than 100% (1.0). Suppose that a wattmeter connected to a 117-V, 60-Hz line indicates 600 W of power while an ammeter indicates 10 A. Calculating the apparent power results in 1170 VA ($P_a = EI = 117 \times 10 = 1170$ VA). Obviously, there is also some reactive power present in the load. This is verified by the power factor of 51% [$\cos(\theta) = P/P_a = 600/1170 = 0.51$].

Reactive Power in Power Lines

Reactive power at the power line frequency of 60 Hz is due to the coils in industrial ac inductive motors and transformers. It is the inductance of the coils that causes the difference between the average power and the apparent power. From Equation 23–5, $P_q = P_a \sin(\theta)$, it was learned that the reactive power is found by multiplying the apparent power by the sine of the circuit phase angle. The symbol used in this text for reactive power is P_q, where the q stands for quadrature (90°). The unit of reactive power is the var (volt-ampere reactive).

EXAMPLE 23–6 Determine the reactive power in a circuit with a 51% power factor when the apparent power is 1170 VA.

SOLUTION Determine the circuit phase angle and then substitute into Equation 23–5. Thus

$$\cos(\theta) = 0.51$$
$$\theta = \arccos(0.51) = 59.3°$$
$$P_q = P_a \sin(\theta) = 1170 \sin(59.3°)$$
$$P_q = 1006 \text{ vars}$$

∴ The reactive power is 1006 vars.

When the power factor of a power line is less than 100% (1.0), the line is carrying reactive current along with real current. It is the reactive current that reduces the capacity of the line for carrying real power-developing current.

EXAMPLE 23–7 For the listed power factors, determine the amount of real current that contributes to useful work and the amount of reactive current that contributes no useful work in a power line when an ammeter indicates 60 A: (a) 1.0, (b) 0.85, (c) 0.5, and (d) 0.3.

Observation From the current phasor diagram of Chapter 21, reactive current is 90° out of phase with real current. Thus, expressing the current in rectangular form will give the answer to this problem.

SOLUTION (a) Express the 60-A current in polar form when $\cos(\theta) = 1.0$:

$$\theta = \arccos(1) = 0°$$
$$I = 60\underline{/0°} \quad \boxed{\rightarrow R} \quad 60 + j0 \text{ A}$$

∴ The resistive current is 60 A and the reactive current is 0 A with a power factor of 1.0.

(b) Express the 60-A current in polar form when $\cos(\theta) = 0.85$:

$\theta = \arccos(0.85) = 31.8°$

$I = 60\underline{/31.8°}\ \boxed{\rightarrow R}\ 51 + j31.6\ A$

∴ The resistive current is 51 A and the reactive current is 31.6 A.

(c) Express the 60-A current in polar form when $\cos(\theta) = 0.5$:

$\theta = \arccos(0.5) = 60°$

$I = 60\underline{/60°}\ \boxed{\rightarrow R}\ 30 + j52\ A$

∴ The resistive current is 30 A and the reactive current is 52 A.

(d) Express the 60-A current in polar form when $\cos(\theta) = 0.3$:

$\theta = \arccos(0.3) = 72.5°$

$I = 60\underline{/72.5°}\ \boxed{\rightarrow R}\ 18 + j57\ A$

∴ The resistive current is 18 A and the reactive current is 57 A.

The next example uses the parameters of the previous example to demonstrate how the power factor may be used to indicate ineffective power transmission in a power line. Keep in mind that the presence of reactive current in a power line does no useful work and that the size of the power company's generators, transformers, bus bars, and wires must be made larger to accommodate the *wattless* reactive current. Even though reactive power does not register on the consumer kilowatt-hour meter (Figure 6-17), the reactive current, along with the resistive current, does indicate on an ammeter.

EXAMPLE 23-8 Using the values of the power factors and source current, $I = 60$ A, from the previous example, determine the apparent power, average power, and reactive power delivered to a load when the source is $220\underline{/0°}$ V.

SOLUTION (a) For $\cos(\theta) = 1.0$, $\theta = 0°$. Thus

$P_a = EI = (220)(60) = 13.2$ kVA

From the power triangle and Equation 23-3,

$P = P_a \cos(\theta) = (13.2 \times 10^3)(1)$

$P = 13.2$ kW

And from the power triangle and Equation 23-5,

$$P_q = P_a \sin(\theta) = (13.2 \times 10^3) \sin(0°)$$

$$P_q = 0 \text{ vars}$$

∴ $P_a = 13.2$ kVA, $P = 13.2$ kW,

$P_q = 0$ vars

Observation When the power factor is 1.0 (100%), the power is being transmitted very effectively from the power company's generators through the power lines to the consumer's load. When the power factor is 1.0, 100% of the apparent power is doing useful work in the load.

(b) For $\cos(\theta) = 0.85$, $\theta = 31.8°$. Thus

$$P_a = EI = (220)(60) = 13.2 \text{ kVA}$$

From the power triangle and Equation 23-3,

$$P = P_a \cos(\theta) = (13.2 \times 10^3)(0.85)$$

$$P = 11.22 \text{ kW}$$

And from the power triangle and Equation 23-5,

$$P_q = P_a \sin(\theta) = (13.2 \times 10^3) \sin(31.8°)$$

$$P_q = 6.96 \text{ kvars}$$

∴ $P_a = 13.2$ kVA, $P = 11.22$ kW,

$P_q = 6.96$ kvars

(c) For $\cos(\theta) = 0.5$, $\theta = 60°$. Thus

$$P_a = EI = (220)(60) = 13.2 \text{ kVA}$$

From the power triangle and Equation 23-3,

$$P = P_a \cos(\theta) = (13.2 \times 10^3)(0.5)$$

$$P = 6.6 \text{ kW}$$

And from the power triangle and Equation 23-5,

$$P_q = P_a \sin(\theta) = (13.2 \times 10^3) \sin(60°)$$

$$P_q = 11.43 \text{ kvars}$$

∴ $P_a = 13.2$ kVA, $P = 6.6$ kW,

$P_q = 11.43$ kvars

(d) For $\cos(\theta) = 0.3$, $\theta = 72.5°$. Thus

$$P_a = EI = (220)(60) = 13.2 \text{ kVA}$$

From the power triangle and Equation 23-3,

$$P = P_a \cos(\theta) = (13.2 \times 10^3)(0.3)$$

$$P = 3.96 \text{ kW}$$

And from the power triangle and Equation 23-5,

$$P_q = P_a \sin(\theta) = (13.2 \times 10^3) \sin(72.5°)$$

$$P_q = 12.6 \text{ kvars}$$

$\therefore \quad P_a = 13.2 \text{ kVA}, \quad P = 3.96 \text{ kW},$

$P_q = 12.6 \text{ kvars}$

Observation Notice how ineffective the generation and distribution system is in delivering average power to the load when the power factor is low. When the power factor is 0.3, only 30% of the apparent power is doing useful work in the load. This example demonstrates that reactive power is not desirable in a power line since the current capacity of the line is being used to deliver *wattless* power.

Correcting for Low Power Factor

Industrial users of electrical energy, unlike residential users, have a unique problem because most of their electric load is inductive. Besides supplying the industrial user with electric current that does useful work, the power companies also may supply the reactive current, called *magnetizing* current, to develop the magnetic fields needed to operate the large reactive motors used by industry. Because supplying reactive current increases the power company's cost (since large generators, transformers, and wires are needed), the industrial consumer is charged a power-factor penalty by the utility company for the kilovars supplied. Recall that reactive power is *wattless* power and, as such, does not register on the kilowatt-hour meter.

Rather than pay the power company, many industrial consumers supply their own kilovars by placing a capacitor across the power line in parallel with the inductive load. Figure 23-4(b) pictures such a connection.

In Figure 23-4(a), the power company supplies both the power-producing current, 26 A, and the reactive current, 30.4 A. In Figure 23-4(b), the line current has been reduced because

612

Power in Alternating Current Circuits

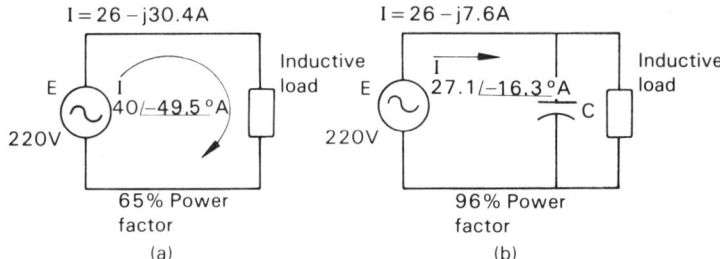

FIGURE 23-4
(a) Inductive load with both the real and reactive current (magnetizing current) supplied by the power line.
(b) The addition of an appropriate shunt capacitor to the inductive load reduces the line current by removing the need for most of the reactive current.

most of the reactive (magnetizing) current is no longer supplied by the power line; instead, the reactive current passes between the capacitor and the inductive load. Thus, when an inductive load is shunted by an appropriately sized capacitor, the ac reactive current will flow back and forth between the inductor and capacitor with little or no additional reactive current taken from the power line. Because the power transmission system is relieved of supplying the magnetization current to the inductive loads, it can carry additional power-producing current.

EXAMPLE 23-9 Determine the capacitor size needed to *correct* a 220-V, 60-Hz power line with a power factor of 0.65 to a power factor of 0.95. The current in the uncorrected line is 40 A and the load is reactive.

SOLUTION Determine both the reactive power, P_q, and the average power, P, when the power factor is 0.65. Thus

$$P_a = EI = (220)(40) = 8.8 \text{ kVA}$$

From Figure 23-5(a),

$$P_q = P_a \sin(\theta)$$

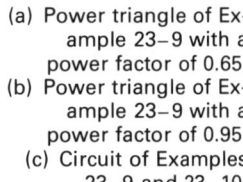

FIGURE 23-5
(a) Power triangle of Example 23-9 with a power factor of 0.65.
(b) Power triangle of Example 23-9 with a power factor of 0.95.
(c) Circuit of Examples 23-9 and 23-10.

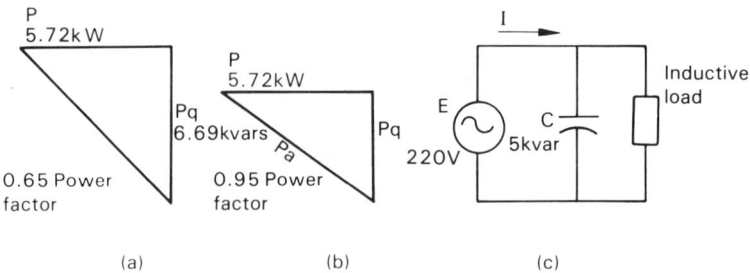

where $\theta = \arccos(0.65) = 49.5°$,

$$P_q = 8.8 \times 10^3 \sin(49.5°) = 6.69 \text{ kvars}$$

and

$$P = P_a \cos(\theta) = (8.8 \times 10^3)(0.65) = 5.72 \text{ kW}$$

Observation The average power to the load ($P = 5.72$ kW) will not change when the power factor is changed. However, both the apparent power and the reactive power will change when the power factor is corrected from 0.65 to 0.95. This is due to the lowering of the line current, I, from 40 A at $\cos(\theta) = 0.65$ to a lower value at $\cos(\theta) = 0.95$. With this in mind, determine the reactive power when the power factor is 0.95. Thus, from the power triangle of Figure 23–5(b), and when $\theta = \arccos 0.95 = 18.2°$,

$$\tan(\theta) = \frac{P_q}{P}$$

$$P_q = P \tan(\theta) = 5.72 \times 10^3 \tan (18.2°)$$

$$P_q = 1.88 \text{ kvars}$$

Observation The difference in the reactive power of 6.69 kvars at 0.65 power factor and the reactive power of 1.88 kvars at 0.95 power factor is the reactive power that must be *supplied* by the capacitor, C, connected in parallel with the inductive load, as pictured in Figure 23–5(c). Thus

$$P_q \text{ (capacitor)} = (6.69 - 1.88) \text{ kvars}$$

$$P_q = 4.81 \text{ kvars}$$

∴ A capacitor, C, with a rating of 4.81 kvars is needed to raise the power factor from 0.65 to 0.95.

Observation Capacitors used for correcting the power factor are rated in kvars rather than microfarads. Thus, a standard industrial capacitor for this application would be rated as 5 kvars at 240 V.

EXAMPLE 23–10 Use the values of Example 23–9 to determine the line current, I, when the 5-kvar capacitor is shunted across the inductive load of Figure 23–5(c).

SOLUTION Use the 0.95 power factor power triangle of Figure 23–5(b) to compute the apparent power from which the line current, I, is computed. Apply the Pythagorean theorem to the power triangle and determine P_a. Thus, where $P = 5.72$ kW and $P_q = (6.69 - 5.0)$ kvars = 1.69 kvars,

$$P_a^2 = P^2 + P_q^2 \qquad (23-10)$$

$$P_a = \sqrt{P^2 + P_q^2}$$

$$P_a = \sqrt{(5.72 \times 10^3)^2 + (1.69 \times 10^3)^2}$$

$$P_a = 5.96 \text{ kVA}$$

Determine the line current $|I|$ for a 0.95 power factor:

$$P_a = |E|\,|I| \quad \text{and} \quad |I| = \frac{P_a}{|E|}$$

$$|I| = \frac{5.96 \times 10^3}{220} = 27.1 \text{ A}$$

∴ The current, I, in the line is 27.1 A.

Observation With the addition of the 5-kvar rated capacitor across the reactive load, the line current dropped from 40 A (power factor 0.65) to 27.1 A (power factor 0.95).

23–4 POWER IN AC NETWORKS

When determining the total watts, vars, and volt-amperes of an ac network, the average powers are added and the reactive powers are added. However, the total apparent power (volt-amperes) is computed from the total watts and vars of the circuit using the Pythagorean theorem.

$$P_a^2 = P^2 + P_q^2 \qquad (23-10)$$

As you know from your previous work with power in dc networks, power is not dependent on circuit configuration or the direction of current. To facilitate determining the total power of an ac network, an algorithm similar to that used with vector and phasor quantities will be used to combine the scalar powers.

Figure 23–6 pictures two power triangles that have their position determined by leading or lagging power factors. That is, capacitive reactive power is pictured as positive, while inductive reactive power is pictured as negative. In this case, voltage has been selected as the reference quantity.

Power in AC Networks

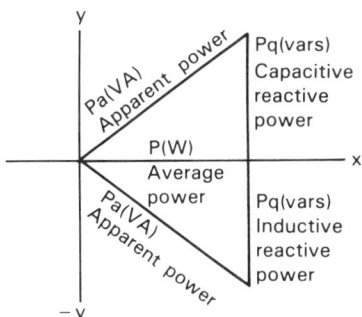

FIGURE 23-6
With voltage as the reference, capacitive reactive power is plotted in the direction of the +y axis; inductive reactive power in the direction of the −y axis; the average power is plotted in the direction of the x axis.

The algorithm is applied to an ac network by stating the apparent power as a polar number and the average and reactive powers as rectangular numbers. Thus,

$$P_a \underline{/\theta} \;\; \boxed{\rightarrow R} \;\; (P, P_q) \tag{23-11}$$

The following examples use these concepts to determine the power in ac networks.

EXAMPLE 23-11 For the blocked loads in the network of Figure 23-7(a), determine the total watts, vars, volt-amperes, and the network power factor, as well as the network current, I. Load A is 500 W and 200 vars inductive; load B is 500 vars capacitive; load C is 2000 VA with a 0.8 lagging power factor.

SOLUTION Express each load in rectangular form with the reactive power assigned a direction, as noted in Figure 23-6. Thus,
Load A: (500 W, −200 vars)
Load B: (0 W, 500 vars)
Load C: 2000$\underline{/-36.9°}$ VA $\boxed{\rightarrow R}$ (1600 W, −1200 vars)

Observation 2000$\underline{/\theta}$ VA results in 2000$\underline{/-36.9°}$ VA from the lagging power factor of 0.8, and $\theta =$

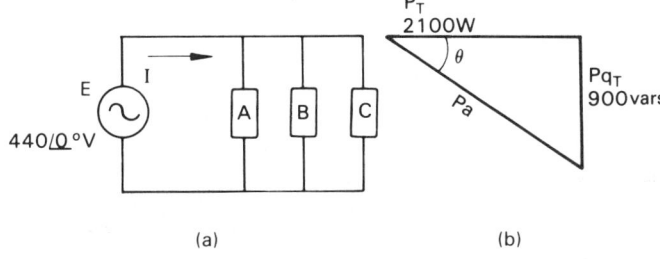

FIGURE 23-7
(a) Blocked system diagram for Example 23-11.
(b) Power triangle of the system.

arccos(0.8) = 36.9°. A minus sign is assigned to the phase angle because a lagging power factor results from an inductive load: thus, 2000/−36.9° VA. Use the rectangular power forms to algebraically add the watts and the vars. Thus

$$P_T = 500 \text{ W} + 0 \text{ W} + 1600 \text{ W} = 2100 \text{ W}$$

$$P_{qT} = -200 \text{ vars} + 500 \text{ vars} - 1200 \text{ vars}$$

$$P_{qT} = -900 \text{ vars}$$

Apply the Pythagorean theorem, Equation 23–10, to determine the apparent power:

$$P_{aT} = \sqrt{P_T^2 + P_{qT}^2} = \sqrt{2100^2 + (-900)^2}$$

$$P_{aT} = 2285 \text{ VA}$$

Construct the network power triangle, as shown in Figure 23–7(b), and determine the network power factor. Thus

$$\tan(\theta) = \frac{P_{qT}}{P_T}$$

$$\theta = \arctan\left(\frac{P_{qT}}{P_T}\right) = \arctan\left(\frac{900}{2100}\right)$$

$$\theta = 23.2°$$

Power factor = $\cos(\theta) = \cos(23.2°) = 0.92$ (lagging). Determine the network current, I:

$$P_a = |E| \, |I| \quad \text{and} \quad |I| = \frac{P_a}{|E|}$$

$$|I| = \frac{2285}{440} = 5.2 \text{ A}$$

and

$$I = 5.2/-23.2° \text{ A}$$

∴ $P_T = 2100 \text{ W}, P_{qT} = 900 \text{ vars}$ (ind)

$P_{aT} = 2285 \text{ VA},$

$\cos(\theta) = 0.92$ lagging, and $I = 5.2/-23.2°$ A.

EXAMPLE 23–12 For the network of Figure 23–8(a), $Z_A = 500 + j0 \text{ }\Omega$, $Z_B = 330 + j600 \text{ }\Omega$, and $Z_C = 0 - j300 \text{ }\Omega$. (a) For each branch, determine the average, reactive, and apparent powers. (b) De-

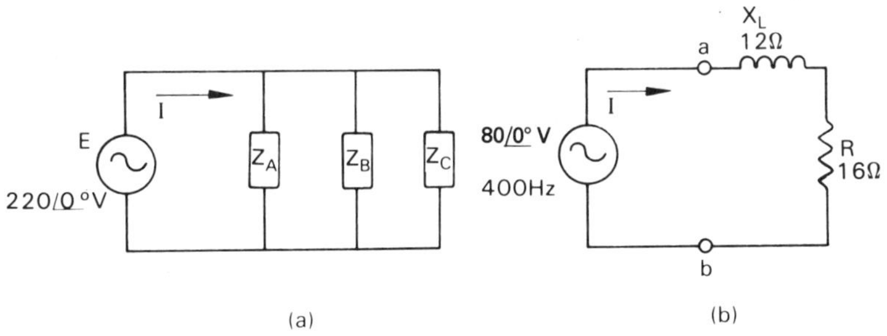

FIGURE 23-8
(a) Network of Example 23-12.
(b) Network of Example 23-13.

termine the entire network's average power, reactive power, and apparent power, as well as the network's power factor. (c) Determine the source current I.

SOLUTION (a) For Z_A, where $500 + j0 \; \Omega = 500\underline{/0°} \; \Omega$:

$$I_A = \frac{E}{Z_A} = \frac{220\underline{/0°}}{500\underline{/0°}} = 0.440 \text{ A}$$

$$P = EI \cos(\theta) = (220)(0.440) \cos(0°)$$

$$P = 96.8 \text{ W}$$

$$P_q = EI \sin(\theta) = (220)(0.440) \sin(0°)$$

$$P_q = 0 \text{ vars}$$

$$P_a = EI = (220)(0.440) = 96.8 \text{ VA}$$

∴ For branch A, $P = 96.8$ W, $P_q = 0$ vars, and $P_a = 96.8$ VA.

For Z_B, where $330 + j600 \; \Omega = 685\underline{/61.2°} \; \Omega$:

$$I_B = \frac{E}{Z_B} = \frac{220\underline{/0°}}{685\underline{/61.2°}} = 0.321\underline{/-61.2°} \text{ A}$$

$$P = EI \cos(\theta) = (220)(0.321) \cos(61.2°)$$

$$P = 34.0 \text{ W}$$

$$P_q = EI \sin(\theta) = (220)(0.321) \sin(61.2°)$$

$$P_q = 61.9 \text{ vars (ind)}$$

$$P_a = EI = (220)(0.321) = 70.6 \text{ VA}$$

∴ For branch B, $P = 34.0$ W, $P_q = 61.9$ vars (ind), and $P_a = 70.6$ VA.

For Z_C, where $0 - j300 \; \Omega = 300\underline{/-90°} \; \Omega$:

$$I_C = \frac{E}{Z_C} = \frac{220\underline{/0°}}{300\underline{/-90°}} = 0.733\underline{/90°} \text{ A}$$

$$P = EI \cos(\theta) = (220)(0.733) \cos(90°)$$
$$P = 0 \text{ W}$$
$$P_q = EI \sin(\theta) = (220)(0.733) \sin(90°)$$
$$P_q = 161 \text{ vars(cap)}$$
$$P_a = EI = (220)(0.733) = 161 \text{ VA}$$

∴ For branch C, $P = 0$ W, $P_q = 161$ vars (cap), and $P_a = 161$ VA.

(b) Determine the network's average, reactive, and apparent powers. Start by adding the average power of each branch. Thus

$$P_T = 96.8 \text{ W} + 34.0 \text{ W} + 0 \text{ W} = 130.8 \text{ W}$$

Assign polarity to each branch reactive power, using Figure 23–6, and add. Thus

$$P_{qT} = 0 \text{ vars} - 61.9 \text{ vars} + 161 \text{ vars}$$
$$P_{qT} = 99.1 \text{ vars (cap)}$$

Apply Equation 23–10 to determine the apparent power:

$$P_a = \sqrt{P_T^2 + P_{qT}^2} = \sqrt{130.8^2 + 99.1^2}$$
$$P_a = 164.1 \text{ VA}$$

Compute the network power factor, using Equation 23–2:

$$\cos(\theta) = \frac{P}{P_a} = \frac{130.8}{164.1} = 0.80 \quad \text{(leading)}$$

Observation The power factor is leading because the reactive power is capacitive and capacitive loads have leading power factors.

∴ The network has an average power of 130.8 W, a capacitive reactive power of 99.1 vars, an apparent power of 164.1 VA, and a leading power factor of 0.80.

(c) The source current, I, may be determined from the network's apparent power. Thus

$$P_a = |E| |I| \quad \text{and} \quad |I| = \frac{P_a}{|E|}$$

$$|I| = \frac{164.1}{220} = 0.746 \text{ A}$$

Observation The source current, I, has a positive phase angle since the source voltage is the reference phasor and the load is capacitive. I leads E in a capacitive circuit.

EXAMPLE 23–13 For the network of Figure 23–8(b), (a) determine the power factor of the load connected across terminals ab, and then (b) specify the capacitance of a capacitor, in microfarads, needed to correct the power factor to 0.9. The capacitor will be connected between terminals ab.

SOLUTION (a) Compute the impedance and determine the circuit phase angle:

$$Z = 16 + j12 \quad \boxed{\rightarrow P} \quad 20\underline{/36.9°}$$

$$\cos(\theta) = \cos(36.9°) = 0.80 \text{ (lagging)}$$

(b) Since the average power will be the same with or without the capacitor connected between terminals ab, determine the average power from the network of Figure 23–8(b), which has a 0.8 lagging power factor. Thus

$$I = \frac{E}{Z} = \frac{80\underline{/0°}}{20\underline{/36.9°}} = 4\underline{/-36.9°} \text{ A}$$

$$P = I^2 R = (4^2)(16) = 256 \text{ W}$$

Also, determine the reactive power:

$$P_q = I^2 X_L = (4^2)(12) = 192 \text{ vars (ind)}$$

Now, determine the apparent power of the network with the unknown capacitor connected between terminals ab. The power factor is 0.9.

$$P = P_a \cos(\theta) \quad \text{and} \quad P_a = \frac{P}{\cos(\theta)}$$

$$P_a = \frac{256}{0.9} = 284 \text{ VA}$$

Convert to rectangular form, using Equation 23–11:

$P_a/\theta \;\boxed{\rightarrow R}\; (P, P_q)$, where
$\theta = \arccos(0.9) = 25.8°$

$284/25.8° \;\boxed{\rightarrow R}\; 256$ W, 124 vars (ind)

Determine the reactive power of the capacitor by computing the difference between the reactive power at a 0.8 power factor and the reactive power at a 0.9 power factor.

$$P_{q(\text{cap})} = P_{q0.8} - P_{q0.9} = 192 - 124$$

$$P_{q(\text{cap})} = 68 \text{ vars}$$

Determine the capacity of the 68-var capacitor:

$$P_q = \frac{E^2}{X_C} \quad \text{and} \quad X_C = \frac{E^2}{P_q}$$

$$X_C = \frac{80^2}{68} = 94 \;\Omega$$

and

$$X_C = \frac{1}{\omega C} \text{ where } \omega = 2\pi 400 = 2513 \text{ rad/s}$$

$$C = \frac{1}{\omega X_C} = \frac{1}{2513 \times 94} = 4.23 \times 10^{-6}$$

$\therefore \quad C = 4.23 \;\mu\text{F}$

23–5 MAXIMUM POWER TRANSFER

In Section 11–5, you learned that in a dc network maximum power transfer occurs from a source when the load resistance is equal to the internal source resistance ($R_L = R_{\text{int}}$). Figure 23–9(a) pictures a dc source of 12 V with an internal resistance of 100 Ω matched for maximum power transfer.

An ac voltage source, with a series equivalent source impedance, is matched for maximum power transfer when the load resistance is equal to the internal source resistance and the power factor of the entire circuit is 1.0. That is, the reactive power in the circuit must be zero vars. This, of course, only occurs when the impedance of the entire network has no reactive component ($Z = R + j0$).

Figure 23–9(b) pictures an ac equivalent voltage source that has an internal capacitance of 500 pF in series with an internal resistance of 72 Ω. This source is matched for maximum power transfer to a 72-Ω resistive load by the addition of an inductor. The value of the inductor is selected so that the reactance of the inductor, when added to the reactance of the capacitor, is 0 Ω.

Maximum Power Transfer

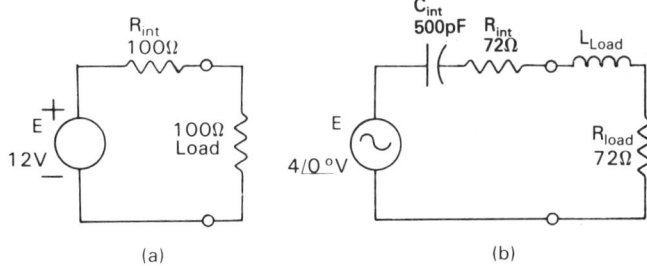

FIGURE 23-9
(a) dc voltage source matched for maximum power transfer.
(b) ac voltage source matched for maximum power transfer.

EXAMPLE 23-14 Compute the value of inductance needed to match the 72-Ω resistive load to the 2-MHz ac source pictured in Figure 23-9(b).

SOLUTION Compute the capacitive reactance and then determine the inductor value. Thus

$$X_C = \frac{1}{2\pi fC} = \frac{1}{2\pi 2 \times 10^6 \times 500 \times 10^{-12}}$$

$$X_C = 159 \ \Omega$$

Let $X_L = X_C = 159 \ \Omega$:

$$X_L = 2\pi fL \quad \text{and} \quad L = \frac{X_L}{2\pi f}$$

$$L = \frac{159}{2\pi 2 \times 10^6} = 12.65 \ \mu\text{H}$$

∴ The load is matched to the ac source when the inductor L_{load} of Figure 23-9(b) is 12.65 μH.

EXAMPLE 23-15 Determine the maximum power transferred to the 72-Ω load of Figure 23-9(b) when the inductor $L_{\text{load}} = 12.65 \ \mu\text{H}$ and the operating frequency is 2 MHz.

SOLUTION Compute the impedance of the network.

$$X_C = \frac{1}{\omega C} = \frac{1}{2\pi 2 \times 10^6 \times 500 \times 10^{-12}} = 159 \ \Omega$$

$$X_L = \omega L = 2\pi 2 \times 10^6 \times 12.65 \times 10^{-6}$$

$$X_L = 159 \ \Omega$$

$$Z = R + jX = 144 + j159 - j159$$
$$Z = 144 + j0 \;\Omega = 144\underline{/0°}\;\Omega$$

Because the power factor is 1.0, the load is matched for maximum power transfer ($Z = R + j0$). Thus, the power developed in the load is half the power developed in the network:

$$P_{\text{load}} = \tfrac{1}{2}\frac{E^2}{Z}\cos(\theta) = \tfrac{1}{2}\frac{4^2}{144}\cos(0°)$$

$$\therefore \quad P_{\text{load}} = 55.6 \text{ mW}$$

Observation To achieve maximum power transfer when the source impedance has a reactive component, select a load impedance with a reactive component equal but opposite in sign to that of the reactive component of the source impedance. Thus, an inductive source impedance would need a capacitive load impedance selected so the net circuit reactance is 0 Ω.

As was previously stated in Section 11–5, the major technical problem in circuits and systems operating at radio frequencies is matching the output impedance to the input impedance to obtain maximum transfer of RF power from one circuit into another. Thus, maximum power is transferred from one electronic circuit to the next when the resistive component of the series equivalent load impedance is equal to the resistive component of the series equivalent source impedance, and the reactance of the series equivalent load impedance is equal in magnitude but opposite in sign to the reactance of the series equivalent source impedance.

EXAMPLE 23–16 An ac RF oscillator operating at 10 MHz has a source impedance of $47 + j18\;\Omega$. (a) Compute the component values of the *series* equivalent load for maximum power from the oscillator, and (b) compute the component values of the *parallel* equivalent load for maximum power from the oscillator.

SOLUTION (a) The series load for maximum power is $47 - j18\;\Omega$. Thus

$$R_{\text{eq}} = 47\;\Omega \quad \text{and} \quad X_{C\text{eq}} = 18\;\Omega$$

$$C_{eq} = 1/(\omega X_C) = 1/(2\pi 10 \times 10^6 \times 18)$$

$$C_{eq} = 884 \text{ pF}$$

∴ A 47-Ω resistance and an 884-pF capacitance, connected in series to the oscillator, will produce maximum power in the load resistance.

(b) Change the series load impedance to admittance and determine the parallel load components. Thus

$$Z_L = 47 - j18 \; \Omega = 50.3\underline{/-21°} \; \Omega$$

$$Y_L = \frac{1}{Z_L} = \frac{1}{50.3\underline{/-21°}} = 19.9\underline{/21°} \text{ mS}$$

$$Y_L = 19.9\underline{/21°} \text{ mS} \quad \boxed{\rightarrow R} \quad 18.6 + j7.1 \text{ mS}$$

$$R_p = \frac{1}{G_L} = \frac{1}{18.6 \times 10^{-3}} = 53.8 \; \Omega$$

$$X_{Cp} = \frac{1}{B_C} = \frac{1}{7.1 \times 10^{-3}} = 140.8 \; \Omega$$

$$C_p = \frac{1}{\omega X_{Cp}} = \frac{1}{2\pi 10 \times 10^6 \times 140.8} = 113 \text{ pF}$$

∴ A 53.8-Ω resistance and a 113-pF capacitance connected in parallel to the oscillator will produce maximum power in the load resistance.

PROBLEMS

Section 23–1

23–1 A 440-V, 60-Hz source is connected in series with an ammeter to a reactive load. Determine (a) the apparent power, (b) the reactive power, and (c) the average power in the load when the ammeter indicates 92 A and the circuit phase angle is 36°.

23–2 Repeat Problem 23–1 for an ammeter reading of 110 A and a circuit phase angle of 30°.

23–3 A load consisting of two parallel impedances, $Z_A = 20\underline{/-28°} \; \Omega$ and $Z_B = 12\underline{/32°} \; \Omega$, is connected to an ac 220-V, 60-Hz source. Determine (a) the power factor of the load, (b) the apparent power of the load, and (c) the power in watts dissipated by the load.

23–4 Repeat Problem 23–3 for a load consisting of two parallel impedances, $Z_A = 6 - j10 \; \Omega$ and $Z_B = 14 + j8 \; \Omega$, connected to an ac 117-V, 60-Hz source.

Section 23-2

23-5 Determine the component values of a series equivalent circuit for an ac network having a power dissipation of 60 W and a leading power factor of 0.74. The source voltage is $e = 168 \sin(377t)$ V.

23-6 Determine the component values of a series equivalent circuit for an ac network having a power dissipation of 15 W and a lagging power factor of 60%. The source current is $i = 14.14 \sin(377t - 53°)$ A.

23-7 Determine the apparent power and the reactive power of the network of Problem 23-6.

23-8 Determine the apparent power and the reactive power of the network of 23-5.

23-9 For the circuit of Figure 23-10(a), determine (a) the power factor of the load, and (b) the parallel equivalent circuit of the load.

23-10 For the circuit of Figure 23-10(b), determine (a) the power factor of the load, and (b) the parallel equivalent circuit of the load.

Section 23-3

23-11 An induction motor, which develops 5 horsepower of mechanical power, is connected to a 220-V, 60-Hz power line. Determine the total current in the power line when the power factor is 0.82 and the motor is 86% efficient.

23-12 For the values of Problem 23-11, determine the reactive current in the line and compute the kvar rating of a shunt capacitor needed to correct the power factor to 1.0.

23-13 An induction motor is connected across a 440-V, 60-Hz line. The motor, with a 0.70 lagging power factor, is connected in parallel with a capacitive load with a 0.80 leading power factor. Determine the power in watts dissipated by the combined load when the current in the induction motor is 55 A and the current in the capacitive load is 40 A.

23-14 For the values of Problem 23-13, determine (a) the total current in the power line, (b) the reactive current in the power line, and (c) the resistive current (watt-producing) in the power line.

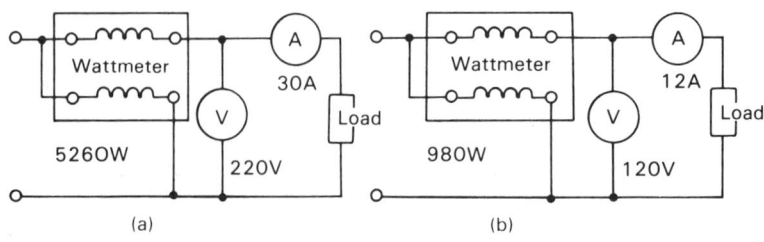

FIGURE 23-10
(a) Circuit of Problem 23-9.
(b) Circuit of Problem 23-10.

23-15 For the values of Problem 23-13, determine the power factor for the combined load.

23-16 For the values of Problem 23-13, determine the component values for both the series and parallel equivalent circuits of the combined load.

23-17 Two industrial plants take their power from the same power line. The power taken by plant A is 86 kW with a power factor of 0.82 lagging. If the total power delivered by the line is 154 kW at a power factor of 0.70 lagging, determine the power factor of plant B.

23-18 Determine the capacitor rating, in kvars, needed to raise the lagging power factor of a 440-V, 60-Hz power line from 0.74 to 0.88 when the current in the uncorrected power line is 50 A.

23-19 An inductive load is connected to a 220-V, 60-Hz power line. The inductive load takes a current of 18 A from the line, and a resistive load connected in parallel with the reactive load takes 12 A. Determine the power factor of the inductive load when an ammeter in the power line reads 28.3 A and the line power factor is 0.92.

23-20 Using the values of Problem 23-19, determine the value, in microfarads, of the capacitor that must be connected across the line to bring the power factor to 1.0.

Section 23-4

23-21 For the loads pictured in Figure 23-7(a), determine (a) the circuit current, I, and (b) the power factor of the combined loads when load A is 15 kVA with a lagging power factor of 0.78, load B is 10 kW with a 1.0 power factor, and load C is 8 kVA with a leading power factor of 0.88.

23-22 For the values of Problem 23-21, determine the capacitor rating, in microfarads, needed to correct the 440-V, 400-Hz line to a power factor of 1.0.

23-23 For the loads pictured in Figure 23-8(a), $Z_A = 22.3 + j8.5 \, \Omega$, $Z_B = 15 + j0 \, \Omega$, and $Z_C = 10 - j6 \, \Omega$.
(a) Determine the average, reactive, and apparent power for each branch in the network.
(b) Determine the combined average power, reactive power, and apparent power.
(c) Determine the power factor of the combined loads.

23-24 For the block diagram of Figure 23-11(a), determine the total average, reactive, and apparent power in the network.

23-25 For the block diagram of Figure 23-11(b), determine the circuit current, I, and the combined power factor of the network.

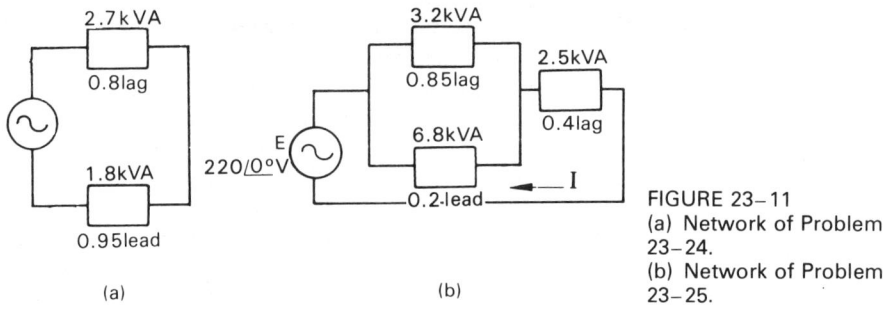

FIGURE 23-11
(a) Network of Problem 23-24.
(b) Network of Problem 23-25.

Section 23-5
23-26 A 144-MHz, $10\underline{/0°}$-V source has an internal impedance of $70 - j15.5$ Ω. Determine (a) the load impedance for maximum power transfer, (b) the component values of the series equivalent load for maximum transfer, and (c) the power developed in the load when matched to the source for maximum power transfer.

23-27 An ac RF source is operating at 27 MHz and has a source impedance of $265 - j140$ Ω. (a) Compute the component values of the series equivalent load for maximum power to the load, and (b) compute the component values of the parallel equivalent load for maximum power to the load.

24 Transformers

24–1 TRANSFORMER BEHAVIOR
24–2 TRANSFORMER VOLTAGE AND CURRENT RATIO
24–3 TURNS RATIO, IMPEDANCE RATIO, AND REFLECTED IMPEDANCE
24–4 TRANSFORMER CONFIGURATIONS
24–5 TRANSFORMER OPERATING CONSTRAINTS

CHAPTER PREVIEW

Over the years, the study of the transformer has captured the imagination and focused the creative genius of such noted men as Michael Faraday, who in 1831 first demonstrated the *induction coil* (transformer), John Fleming, the inventor of the vacuum diode, who wrote about transformers, Nikola Tesla, who did extensive experimentation with wireless power transmission, and Charles Steinmetz, who researched the relation between hysteresis and flux density. The modern-day transformer is basically the same device used by Faraday to demonstrate electromagnetic induction. Its principal function remains the same, that of transforming voltage, current, and impedance from one level to another. In this chapter, you will learn about the ferromagnetic (iron-core) transformer and the transformation of ac voltage, current, and impedance.

Because of the apparent simplicity of the construction of the *iron-core* transformer (copper wire, iron, and insulation), as well as its longevity (no moving parts), you may fail to appreciate the quality of design, which allows us to treat the modern-day ferromagnetic transformer as a linear device and to summarize its properties with three simple equations.

PERFORMANCE OBJECTIVES

Once you have read each section, worked each example with pencil, paper, and calculator, and answered each problem for every section, you will be able to:

☐ Understand the principle of operation of the transformer.
☐ Apply a transformer to a circuit to match impedance.
☐ Understand the limitation of transformer ratings.

24–1 TRANSFORMER BEHAVIOR

The elementary transformer consists of two coils in close proximity to one another so that the electromagnetic field created by an alternating current in one coil links the turns of the other coil. The coil connected to the ac source is the *primary* of the transformer; the coil connected to the load is the *secondary* of the transformer. This concept is pictured in Figure 24–1(a). The changing amplitude of the current in the primary coil, i_p, produces a changing magnetic flux, ϕ_p. All, or a part, of the primary flux may link the secondary coil. The *mutual flux* is the part of the primary flux that is common to both the primary and secondary coils. It is the mutual flux that accounts for the transfer of electric energy between the source and load.

Mutual Flux

The mutual flux, ϕ_m, is responsible for two phenomena in the transformer.

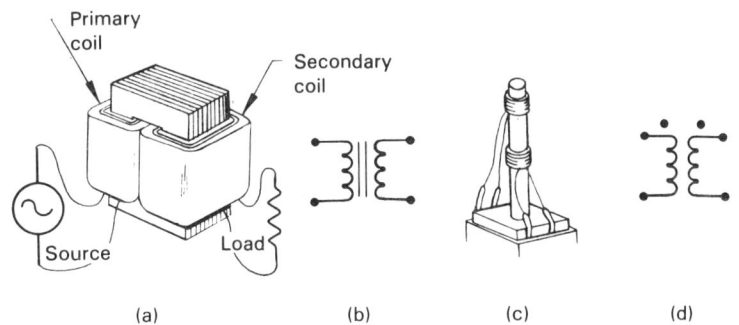

FIGURE 24-1
(a) Iron-core transformer
(b) Schematic symbol of iron-core transformer
(c) Air-core transformer
(d) Schematic symbol of an air-core (nonmagnetic core) transformer. Note *dot* convention labeling the end of the primary and secondary coil having the same ralative polarity.

☐ First, it induces a voltage in the primary coil that is nearly equal to the applied voltage but opposite in polarity. It is this opposing voltage or counter electromotive force (cemf) in the primary coil that limits the primary current to the relatively small *magnetizing* current, i_m, when the secondary is open. Without the self-inductance of the primary winding, the primary current would be limited only by the resistance of the copper wire in the primary coil and would, of course, be very large.

☐ Second, the mutual flux links the secondary coil and induces a voltage in the winding of the secondary coil.

Mutual Inductance The two functions of the mutual flux provide the transformer with its unique properties. Thus, by *mutual induction*, the alternating current in the primary coil creates a mutual flux that links both the primary and the secondary coils, generating a voltage in each winding. In the second case, a varying current in one winding creates a changing voltage in a second winding with no physical connection between the two coils, only the *mutual inductance*. The mutual inductance, M, is equal to the square root of the product of the primary and secondary coil inductances L_p and L_s, times the coefficient of coupling, k. Thus

$$M = k\sqrt{L_p L_s} \quad (H) \tag{24-1}$$

Coefficient of Coupling The degree to which the mutual flux links the secondary coil is defined by the coefficient of coupling, k. Thus

$$k = \frac{\phi_m}{\phi_p} \tag{24-2}$$

where ϕ_m is the component of the primary flux, ϕ_p, that links the secondary coil. Since the mutual flux can never exceed the value of the primary flux, the coefficient of coupling, k, ranges between 0 and 1.

When the coils are separated by air or material with permeabilities equal to air, only a small fraction (<0.1) of the mutual flux links the secondary coil. Thus, radio-frequency, air-core transformers are characterized by low coefficients of coupling. However, when the coils are wound on the same iron core, the mutual flux is concentrated in the magnetic core and the coefficient of coupling approaches unity ($k \approx 1$). For our discussion of *ferromagnetic* transformers (iron-core transformers), we will assume that $k = 1$ and $\phi_p = \phi_m = \phi_s$.

24–2 TRANSFORMER VOLTAGE AND CURRENT RATIO

When the coefficient of coupling is equal to one ($k = 1$), then $\phi_m = \phi_p = \phi_s$, and the voltage induced in the primary coil, e_p, by the mutual flux, as well as the voltage induced in the secondary coil, e_s, by the mutual flux, may be determined by Faraday's law. Thus

$$e_p = \frac{N_p \, \Delta\phi_m}{\Delta t} \quad \text{and} \quad e_s = \frac{N_s \, \Delta\phi_m}{\Delta t}$$

Since we are assuming that both the primary and the secondary inductances are linear for variation in frequency, temperature, and source voltage, the expression of Faraday's law for the generation of instantaneous voltage may be applied to the generated effective (rms) voltage, as well. Thus,

$$E_p = \frac{N_p \, \Delta\phi_m}{\Delta t} \quad \text{and} \quad E_s = \frac{N_s \, \Delta\phi_m}{\Delta t}$$

Dividing the secondary voltage equation into the primary equation to form a proportion results in,

$$\frac{E_p}{E_s} = \frac{N_p \, \Delta\phi_m / \Delta t}{N_s \, \Delta\phi_m / \Delta t}$$

Factoring produces:

$$\frac{E_p}{E_s} = \frac{N_p}{N_s} \tag{24-3}$$

Equation 24–3 indicates that voltage is *transformed* by simply changing the turns ratio (N_p/N_s) of the transformer.

Current Ratio Assuming an *ideal transformer*, that is, a transformer in which the transformation of energy from the source to the load occurs through the transformer because of perfect magnetic coupling and no loss of energy in the form of heat, then the average power developed by the source in the transformer primary is equal to the average power dissipated in the load. Thus

Transformer Voltage and Current Ratio

$$P_{\text{source}} = P_{\text{load}} \quad \text{and} \quad E_p I_p \cos(\theta) = E_s I_s \cos(\theta) \quad \text{(W)}$$

The power factor, $\cos(\theta)$, of the iron-core transformer is unity since all of the current in the primary and secondary develops power (watts), and none of the current is used to create wattless reactive power (vars). Thus, $\cos(\theta) = 1$ and $P_{\text{source}} = P_{\text{load}}$ may be stated in terms of voltage and current as $E_p I_p = E_s I_s$. Dividing $E_p I_p = E_s I_s$ by $E_s I_p$ results in

$$\frac{E_p I_p}{E_s I_p} = \frac{E_s I_s}{E_s I_p}$$

and factoring,

$$\frac{E_p}{E_s} = \frac{I_s}{I_p}$$

Since $E_p/E_s = N_p/N_s$ from Equation 24–3, then

$$\frac{I_s}{I_p} = \frac{N_p}{N_s} \qquad (24\text{--}4)$$

Like Equation 24–3, Equation 24–4 also indicates that current may be transformed by simply changing the turns ratio. Because of the limitless number of possible combinations of primary and secondary turns, the number of voltages and currents that the transformer may produce is literally without limit.

EXAMPLE 24–1 For the iron-core transformer pictured in Figure 24–2(a), determine (a) the effective voltage induced in the primary coil, E_p, and the effective voltage induced in the secondary coil, E_s, by ϕ_m, and (b) the effective current in the secondary, I_s, and the effective current in the primary, I_p, when a 100-Ω load is connected to the secondary, as noted in Figure 24–2(b).

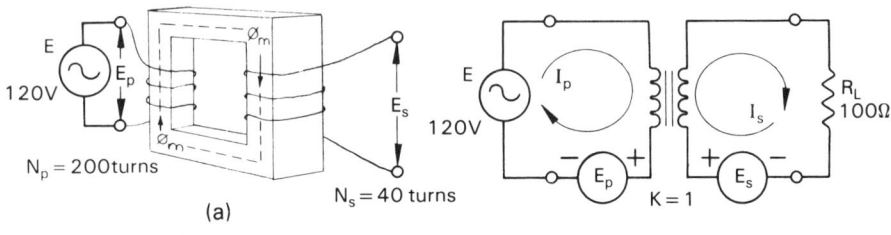

FIGURE 24–2
(a) Iron-core transformer for Example 24–1.
(b) Induced primary voltage, E_p, shown as being in series with the primary and the induced secondary voltage, E_s, shown as being in series with the secondary.

SOLUTION (a) The voltage induced in the primary coil by the mutual flux is equal to the source voltage, E, when it is assumed that there are no *core* or *copper* losses (eddy currents, hysteresis, and wire resistance) or leakage flux. Thus

$$E_p = E = 120 \text{ V}$$

Apply Equation 24-3 to determine E_s:

$$E_s = \frac{E_p N_s}{N_p} = \frac{120 \times 40}{200} = 24 \text{ V}$$

$\therefore \quad E_p = 120 \text{ V} \quad \text{and} \quad E_s = 24 \text{ V}$

(b) For the circuit of Figure 24-2(b), $E_s = 24$ V. By Ohm's law

$$I_s = \frac{E_s}{R} = \frac{24}{100} = 0.24 \text{ A}$$

Apply Equation 24-4 to determine I_p:

$$I_p = \frac{I_s N_s}{N_p} = \frac{0.24 \times 40}{200} = 0.048 \text{ A}$$

$\therefore \quad I_p = 48 \text{ mA} \quad \text{and} \quad I_s = 240 \text{ mA}$

Observation These results, $E_p = 120$ V, $I_p = 48$ mA, and $E_s = 24$ V, $I_s = 240$ mA, may be checked by comparing primary power to secondary power. Thus, $120 \times 0.048 = 24 \times 0.24 = 5.76$ W.

24-3 TURNS RATIO, IMPEDANCE RATIO, AND REFLECTED IMPEDANCE

The ratio of N_p/N_s is defined as the transformer *turns ratio* and it will be represented by the letter n in this text. Thus

$$n = \frac{N_p}{N_s} \tag{24-5}$$

The value of n for a given transformer indicates whether the transformer is a step-down transformer ($n > 1$) or a step-up transformer ($n < 1$).

EXAMPLE 24-2 A certain iron-core transformer has 280 turns in the primary coil and 70 turns in the secondary coil. (a) Determine the turns ratio and state whether the transformer is a step-up or step-down transformer, and (b) determine the current

Turns Ratio, Impedance Ratio, and Reflected Impedance

in a 1.2-kΩ resistive load attached to the secondary winding when the ac source voltage is 117 V.

SOLUTION (a) Apply Equation 24–5 to determine the turns ratio:

$$n = \frac{N_p}{N_s} = \frac{280}{70} = 4$$

∴ The transformer is a step-down transformer ($n > 1$) with a turns ratio of 4 to 1.

(b) Determine the current in the load by first determining the secondary voltage and then applying Ohm's law. Thus

$$\frac{E_p}{E_s} = \frac{N_p}{N_s} = n \quad \text{and} \quad E_s = \frac{E_p}{n}$$

$$E_s = \frac{117}{4} = 29.25 \text{ V}$$

Solve for I_L:

$$I_L = \frac{E_s}{R_L} = \frac{29.25}{1.2 \times 10^3} = 24.4 \text{ mA}$$

∴ The load current is 24.4 mA.

Impedance Ratio   In our discussion of voltage and current ratios, it was noted that the values of current and voltage are simply transformed by adjusting the turns ratio, n. The primary impedance, Z_p, and the secondary impedance, Z_s, of the transformer may be *matched* by altering the transformer turns ratio.

From Figure 24–2(b), the primary impedance may be defined as $Z_p = E_p/I_p$, and the secondary impedance as $Z_s = E_s/I_s$. The impedance ratio is derived from the voltage and current ratio by multiplying these two ratios. Thus, from Equations 24–3 and 24–4,

$$n = \frac{E_p}{E_s} \quad \text{and} \quad n = \frac{I_s}{I_p}$$

Multiplying results in

$$(n)(n) = \left(\frac{E_p}{E_s}\right)\left(\frac{I_s}{I_p}\right) \quad \text{and} \quad n^2 = \left(\frac{E_p}{I_p}\right)\left(\frac{I_s}{E_s}\right)$$

and, $E_p/I_p = Z_p$, while $I_s/E_s = 1/Z_s$. Thus

$$n^2 = \frac{Z_p}{Z_s} \qquad (24\text{–}6)$$

and

$$Z_p = n^2 Z_s \quad (\Omega) \tag{24-7}$$

Equation 24–6 indicates that the primary and secondary impedances are transformed or matched by adjusting the square of the turns ratio.

Reflected Impedance From Equation 24–7, we learn that the primary impedance is dependent on the product of the turns ratio squared and the secondary impedance. To fully understand the meaning of this equation, you must first have an understanding of the makeup of the secondary and primary impedance of the ideal transformer. Under load, the secondary circuit appears as pictured in Figure 24–3(a), where $Z_s = E_s/I_s$. In the ideal transformer, the attached load is the only impedance seen by the secondary current. Thus, in the ideal transformer, $Z_s = Z_L$. A well-designed, modern-day, iron-core transformer will, in fact, have a very small secondary impedance consisting mainly of an inductive reactance and a resistive element resulting from the resistance of the wire of the secondary winding. When compared to the load impedance, the secondary impedance may be neglected and $Z_s \approx Z_L$, and Equation 24–7 becomes

$$Z_p = n^2 Z_L \quad (\Omega) \tag{24-8}$$

Equation 24–8 indicates that, with the ideal transformer, the source is looking into a *reflected impedance*, which is equal to $n^2 Z_L$. For the modern-day iron-core transformer, this is a very good approximation of the primary impedance as seen by the source. Thus, in a well-designed iron-core transformer, the primary impedance is a direct function of the load on the second-

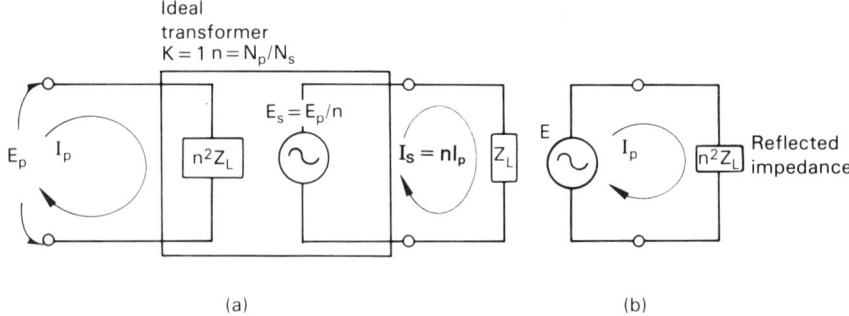

FIGURE 24–3
(a) An ideal transformer.
(b) Source E "sees" a primary impedance equal to the reflected impedance ($Z_p = n^2 Z_L$).

635

Turns Ratio, Impedance Ratio, and Reflected Impedance

ary, and the source "*sees*" an impedance equal to n^2Z_L, as pictured in Figure 24–3(b).

The use of iron-core transformers to match loads to source impedance is very desirable since little loss in power is experienced in the process. The following examples demonstrate the use of an iron-core transformer to match impedance.

EXAMPLE 24–3 The preamplifier circuit of Figure 24–4 has been designed for minimum distortion to operate into a resistive load of 10 kΩ ± 10%. The preamplifier is to be used to *drive* a power amplifier that has a resistive equivalent impedance of 800 Ω. Specify the turns ratio of the transformer needed to match the impedance.

SOLUTION Use Equation 24–8 and solve for n:

$$Z_p = n^2 Z_L \quad \text{and} \quad n = \sqrt{Z_p/Z_L}$$

Substitute:

$$n = \sqrt{\frac{10 \times 10^3}{800}} = 3.54$$

∴ The transformer is a step-down transformer with a turns ratio of 3.54 to 1.

EXAMPLE 24–4 The audio system pictured in Figure 24–5 is a background music system for a large medical office complex. The central amplifier, rated at 80 W continuous output, is designed to provide a 70-V constant voltage into the audio distribution line. As noted in the diagram, three different audio power levels are specified: 0.625 W for speakers located in waiting rooms, 0.5 W for hallway and corridor speakers, 1.25 W for speakers located in the inner courtyard. Determine the turns ratio of the transformers for each of the three power levels and compute the reflected

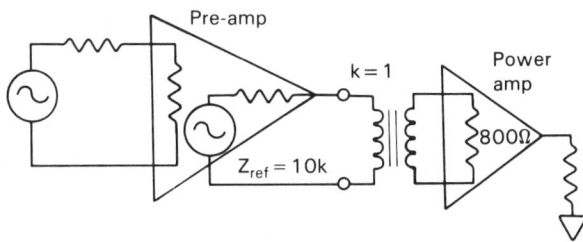

FIGURE 24–4
Circuit of Example 24–3.

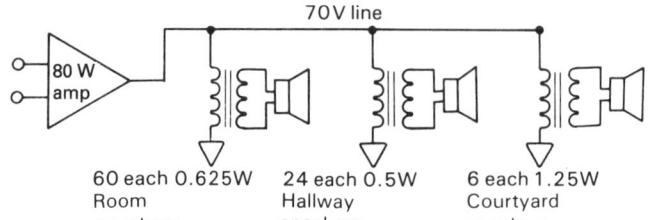

FIGURE 24-5
Audio distribution system for Example 24-4.

impedance of each. Each of the 90 speakers in the system is rated at 4 Ω.

SOLUTION For the 0.625-W level, where $P = V^2/R$,

$$V_{\text{speaker}} = \sqrt{PR} = \sqrt{0.625 \times 4} = 1.58 \text{ V}$$

Determine the turns ratio, n:

$$n = \frac{V_p}{V_s} = \frac{70}{1.58} = 44.3$$

Compute the reflected impedance:

$$Z_{\text{ref}} = Z_p = n^2 Z_L = (44.3^2)4 = 7.85 \text{ k}\Omega$$

∴ The turns ratio is 44.3 to 1, step-down, and the reflected impedance is 7.85 kΩ.

For the 0.5-W level, where $P = V^2/R$,

$$V_{\text{speaker}} = \sqrt{PR} = \sqrt{0.5 \times 4} = 1.41 \text{ V}$$

Determine the turns ratio, n:

$$n = \frac{V_p}{V_s} = \frac{70}{1.41} = 50$$

Compute the reflected impedance:

$$Z_{\text{ref}} = Z_p = n^2 Z_L = (50^2)4 = 10 \text{ k}\Omega$$

∴ The turns ratio is 50 to 1, step-down, and the reflected impedance is 10 kΩ.

For the 1.25-W level, where $P = V^2/R$

$$V_{\text{speaker}} = \sqrt{PR} = \sqrt{1.25 \times 4} = 2.24 \text{ V}$$

Determine the turns ratio, n:

$$n = \frac{V_p}{V_s} = \frac{70}{2.24} = 31.3$$

Compute the reflected impedance:

$$Z_{ref} = Z_p = n^2 Z_L = (31.3^2)4 = 3.92 \text{ k}\Omega$$

∴ The turns ratio is 31.3 to 1, step-down, and the reflected impedance is 3.92 kΩ.

Observation Notice how the low value of the speaker impedance is transformed to a higher value, thus limiting the power to each speaker and preventing any one speaker from taking more than its designed share of power.

24–4 TRANSFORMER CONFIGURATIONS

Autotransformer

In this section we will consider three transformer configurations. These include the autotransformer, the center-tapped primary transformer, and the multiple secondary transformer.

In iron-core variable autotransformer pictured in Figure 24–6 is commonly used to adjust 120-V single-phase line voltage from 0 to 140 V. In the autotransformer, the primary winding is connected *series-aiding* with the secondary winding. Because of this connection, an autotransformer of a given size has a higher volt-ampere rating than a comparable two-winding transformer. However, the autotransformer does not have any isolation between the primary and the secondary circuit. To obtain isolation and to ensure safe operation, the variable autotransformer is connected to the power line through an isolation transformer ($n = 1$). Like the transformer with two windings, the induced secondary voltage of the autotransformer, E_s, is related to the impressed primary voltage, E_p, by the turns ratio, N_p/N_s, where $E_s = E_p N_s / N_p$.

Center-Tapped Primary

The center-tapped transformer pictured in Figure 24–7 finds application in transformer-coupled audio circuits where a source is connected to each side of the center tap. The windings are

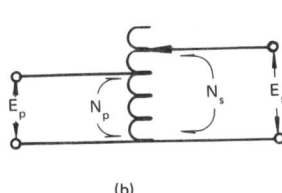

(a)   (b)

FIGURE 24–6 Variable autotransformer.
    (a) Typical laboratory bench mount autotransformer.
    (b) Schematic symbol with primary and secondary windings noted.

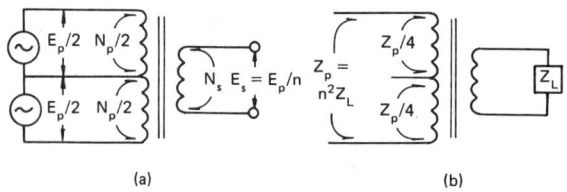

(a)   (b)

FIGURE 24–7   A center-tapped transformer (a) where $E_p/E_s = N_p/N_s$ and (b) where the reflected impedance in each half of the primary winding is equal to $Z_p/4$.

phased so that the secondary voltage is equal to the primary voltage divided by the turns ratio. Thus for Figure 24–7 (a)

$$\frac{E_p}{E_s} = \frac{N_p}{N_s} = n \quad \text{and} \quad E_s = \frac{E_p}{n} \tag{24-9}$$

Assuming an equal number of turns on either side of the tap in the primary, then the reflected impedance into each half of the primary $Z_{1/2}$ may be determined from the following.

From Equations 24–5 and 24–8

$$\frac{Z_p}{Z_L} = \left(\frac{N_p}{N_s}\right)^2 = n^2$$

$$\frac{(Z_{1/2})}{Z_L} = \left(\frac{N_p/2}{N_s}\right)^2 = \left(\frac{N_p}{2N_s}\right)^2 = \frac{1}{4}\left(\frac{N_p}{N_s}\right)^2 = \frac{n^2}{4}$$

$$Z_{1/2} = \frac{n^2 Z_L}{4}$$

Since $Z_p = n^2 Z_L$, then

$$Z_{1/2} = \frac{Z_p}{4} = \frac{n^2 Z_L}{4} \tag{24-10}$$

Multiple Secondary   The secondary of a transformer may consist of two or more separate windings, which may support individual loads as pictured in Figure 24–8(a). The separate secondary windings may be connected series-aiding to form a tapped secondary as shown in Figure 24–8(b). From the circuit in Figure 24–8(a),

$$n_{s1} = \frac{N_p}{N_{s1}} = \frac{E_p}{E_{s1}}, \quad n_{s2} = \frac{N_p}{N_{s2}} = \frac{E_p}{E_s}, \quad \frac{N_{s1}}{N_{s2}} = \frac{E_{s1}}{E_{s2}} \tag{24-11}$$

To determine the reflected impedance of the multiple secondary transformer, recall that in the ideal transformer the power into the primary equals the power dissipated by the load(s).

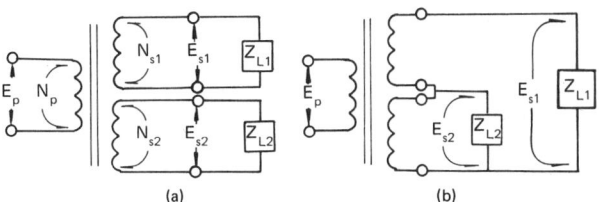

FIGURE 24-8 Multiple secondary transformer with (a) separate winding sourcing separate loads and with (b) the winding connected series-aiding to form a tapped secondary.

$$P_{\text{primary}} = P_{L1} + P_{L2}$$

$$\frac{E_p^2}{Z_p} = \frac{E_{s1}^2}{Z_{L1}} + \frac{E_{s2}^2}{Z_{L2}}$$

Solving for Z_p,

$$Z_p = \frac{E_p^2 Z_{L1} Z_{L2}}{E_{s1}^2 Z_{L2} + E_{S2}^2 Z_{L1}}$$

Since $E_{s1} = E_p/n_1$ and $E_{s2} = E_p/n_2$, then

$$Z_p = \frac{E_p^2 Z_{L1} Z_{L2}}{(E_p/n_1)^2 Z_{L2} + (E_p/n_2)^2 Z_{L1}}$$

Factor E_p^2:

$$Z_p = \frac{Z_{L1} Z_{L2}}{(Z_{L1}/n_2^2) + (Z_{L2}/n_1^2)} \qquad (24\text{--}12)$$

You might notice that the structure of Equation 24–12 indicates that the load impedances attached to the multiple secondaries are reflected into the primary as parallel loads. Furthermore, Equations 24–12 may be applied to the tapped secondary of Figure 24–8(b).

24–5 TRANSFORMER OPERATING CONSTRAINTS

Small iron-core transformers used in electronic circuits are affected by several parameters that will influence the performance of the transformer. These include voltage, frequency, temperature, and volt-ampere rating.

The discussion in this section will center on the small-sized, 60-Hz *power transformer* used to provide power to an electronic circuit. This type of transformer, as well as any transformer, must be operated according to the manufacturer's specifications. If the specifications are not followed, and the power transformer is connected to a higher than rated line voltage or to a frequency lower than specified, or to a load in excess of the volt-ampere rating, or operated at a temperature above the specified ambient, the performance of the transformer will be adversely affected.

Voltage As was previously noted, the equations of the earlier sections were formulated under the assumption that the flux density in the ferromagnetic core of the transformer is operating below the saturation level of the BH curve and along the linear portion of the curve. For the BH curve of transformer steel, pictured in Figure 16–7, the linear region is from $0.2\ T$ to $0.8\ T$.

When the power transformer is initially designed, the designer fixes the magnetizing current, I_M, so that the core flux density, B_c, is below the saturation flux density, B_m. By operating the core in the linear range of the BH curve, the magnetizing current and the mutual flux will both be sinusoidal. This is very important to the normal operation of the power transformer.

Once the secondary winding of the power transformer is loaded, the current in the secondary coil sets up a flux in the transformer core, which, by Lenz's law, opposes the sinusoidal flux in the core. To overcome the opposing flux created by the load current, the primary current increases to meet the demand of the load. Thus, the mutual flux is maintained in the core at the designed value. Therefore, in a core in which the mutual flux density remains linear, an increase in the flux from the load component of the primary current nullifies the increase in flux created by the load current in the secondary coil.

When the source voltage is increased beyond the design maximum, the permissible magnetizing current, I_m, will cause B_c in the core to exceed the upper limit, B_m. The core will be set to operate in the nonlinear region of the BH curve, resulting in excessive magnetizing current, I_m. Thus, the power factor will be considerably less than 1, as indicated by the current phasor of Figure 24–9. The efficiency of transformation will be dramatically decreased, and harmonic distortion will be present in the load, owing to the nonlinear operation of the core.

The designer has placed a constraint on the effective voltage that may be applied to the primary winding of the power transformer to ensure a certain flux density in the core. This value of voltage must not be exceeded. Equation 24–13 is presented here to give you an insight into the interdependency among the five fundamental transformer design parameters, which are voltage, frequency, turns, core area, and flux density. Thus

$$E = 0.707\omega NAB_m \quad \text{or} \quad E = 4.44fN\phi_m \quad \text{(V)} \tag{24–13}$$

641

Transformer Operating Constraints

Because the operating frequency, f, is fixed by the power company at 60 Hz, and the number of turns, N, and cross-sectional area, A, are fixed by the designer, then from Equation 24–13 flux density in the core is a direct function of the source voltage, E. Therefore, to prevent the designed value of flux density from being exceeded in a transformer, the specified voltage *must* not be exceeded!

Frequency

From Equation 24–13, we see that an increase in the operating frequency of the source attached to the primary results in a decrease in the flux density. The power loss in the transformer decreases with an increase in frequency. Thus, a 60-Hz transformer may be used at 400-Hz without damage, but a 400-Hz transformer cannot be used at 60 Hz. For a given size of transformer, increasing the line frequency from 60 to 400 Hz results in at least a two-fold increase in the transformer VA rating. Because of this, 400-Hz transformers are made much smaller in size for a given power rating than those operating at 60 Hz.

The increase in power rating at frequencies above 60 Hz is a result of a more efficient design that is not possible at 60 Hz, owing to the constraint to maintain the magnetizing current at a value resulting in a flux density below the saturation level. Because of this constraint at 60 Hz, more turns and less current are used to achieve the magnetomotive force ($\mathcal{F} = NI$) needed to magnetize the core, resulting in higher power losses than at 400 Hz. At 400 Hz, an optimum balance between turns and magnetizing current may be achieved, thereby lessening the losses and increasing the VA rating of the transformer.

Temperature

A well-designed, small, 60-Hz power transformer will heat up during operation, as this is its nature. In most power transformers, the core may be as hot as 105°C, yet in others it may be as much as 130°C.

A slight reduction in the transformer's physical size can be achieved by allowing the core to operate at a higher tempera-

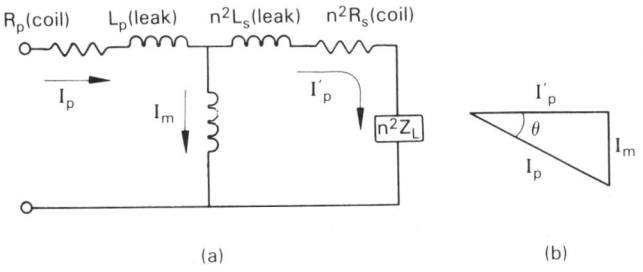

FIGURE 24–9
(a) The low frequency transformer equivalent T circuit without the core loss shown.
(b) The primary current phasor with the primary current, I_p, load current, I'_p, and magnetizing current, I_m, shown.
$I_p = I'_p - jI_m$

ture. However, size reduction is much more readily achieved by increasing the operating frequency.

The small-sized 60-Hz power transformer may lose 20% to 40% of its source power to heat losses resulting from eddy currents, hysteresis, leakage flux (unlinked flux in primary and secondary), and wire resistance. Because of these heat losses, the transformer will have an elevated operating temperature and must be provided with adequate ventilation.

Volt-Ampere Rating Power transformers are rated in volt-amperes available to the load at a specified frequency. Because the load attached to the secondary winding of the transformer may be reactive or resistive, using apparent power to rate the transformer provides a uniform specification for all load conditions.

EXAMPLE 24-5 A 60-Hz power transformer, rated at 800 VA, 117/60 V delivers power from a 117-V line to a 5-Ω totally reactive load. Determine the current in the load, as well as the reactive and apparent power.

SOLUTION The secondary is rated at 60 V. Thus

$$|I_L| = \frac{E_s}{Z_L} = \frac{60}{5} = 12 \text{ A}$$

I_L is all reactive current. Thus

$$P_q = E_s I_L = (60)(12) = 720 \text{ vars}$$

Compute the apparent power:

$$P_a = E_s I_L = (60)(12) = 720 \text{ VA}$$

∴ The current in the load is 12 A and the 720 vars of reactive power equals the 720 VA of apparent power.

Observation Since no average power is dissipated in this case, a transformer rating in units of watts would be meaningless.

Summary The study of the transformer will be an ongoing subject in your career in electronics. This chapter has introduced you to the transformer and, as need arises, you will do additional study in such areas as the loosely coupled transformer, tuned transformer, and pulse transformer. To assist you in your ongoing study, many books and articles have been written on the subject of the transformer.

PROBLEMS

Section 24-1

24-1 The coefficient of coupling in a loosely coupled RF transformer with a primary inductance of 180 μH is 0.5. Determine the mutual inductance when the secondary coil is 30 μH.

24-2 For the transformer of Problem 24-1, determine the mutual inductance when the secondary coil is 1 mH.

24-3 Determine the flux created in the primary coil of a transformer having a coefficient of coupling of 0.92 when the flux in the core, ϕ_m, is 0.8 mWb.

24-4 A power transformer with a 10 cm² cross section has a flux density of 0.5 T. Determine the coefficient of coupling of the transformer when the magnetomotive force of the primary coil produces a flux of 1.2 mWb in the primary coil.

Section 24-2

24-5 Determine the current in the 800-turn primary winding of a 60-Hz power transformer connected to a 220-V line when a 28-Ω resistive load is connected to the 20-turn secondary.

24-6 Repeat Problem 24-5 for a 10-Ω resistive load connected to a 40-turn secondary.

24-7 A 16-Ω resistive load is connected across the secondary of an iron-core transformer with a 5 to 1 turns ratio. The transformer has a primary current of 4 A. Determine both the voltage across the load and the voltage of the source.

24-8 Determine the number of turns in the primary coil of an iron-core transformer when a resistive load of 8 Ω is attached to the 120-turn secondary. The voltage across the load is 24 V when the current in the primary coil is 200 mA.

Section 24-3

24-9 Determine the reflected impedance in an iron-core audio transformer having a turns ratio of 1 to 4 when the load across the secondary is 9.6 kΩ.

24-10 Determine the turns ratio of an audio transformer used to match a 2.7-kΩ source resistance to an 8-Ω speaker resistance so that maximum power is developed in the speaker.

24-11 For a low-impedance capacitor microphone to operate with minimum distortion over a wide audio-frequency range, it must operate into a 250-Ω load. An audio transformer is used to match the 4.7-kΩ input resistance of the microphone preamplifier to the microphone. Determine the turns ratio of the transformer.

24–12 Determine the reflected impedance for the transformer of Problem 24–7.

24–13 The audio system pictured in Figure 24–5 is to be expanded to include 10 additional 8-Ω speakers. Six speakers are to have a 0.625-W power level while the remaining four speakers are to operate at 1.25 W. (a) Determine the turns ratio of the transformer for each of the two power levels, and (b) compute the reflected impedance of each.

24–14 Determine the power taken from a 25-V audio line by a 4-Ω speaker attached to the line through a 4 to 1 step-down transformer.

Section 24–4

24–15 Determine the secondary voltage of the 120-V (E_p) autotransformer pictured in Figure 24–6 when $N_p/N_s = 1:0.4$.

24–16 Determine the turns ratio (n) for the adjustable autotransformer pictured in Figure 24–6 when $E_p = 240$ V and $E_s = 80$ V.

24–17 For the center-tapped transformer of Figure 24–7 (a), determine the voltage across each side of the primary ($E_p/2$) when the turns ratio is 18 and the secondary voltage is 30 V.

24–18 Assuming equal voltage across each half of the center-tapped primary of Figure 24–7 (a), determine the secondary voltage when $E_p/2$ is 2 V and $n = 0.25$

24–19 Determine the impedance reflected into each half of the center-tapped transformer of Figure 24–7 (b) when $E_p = 25$ V, $E_s = 0.781$ V, and $Z_L = 16$ Ω.

24–20 For the center-tapped transformer of Figure 24–7 (b), determine the load impedance, Z_L, when $Z_p/4 = 600$ Ω and $n = 0.06$.

24–21 For the multiple secondary transformer of Figure 24–8 (a), determine the reflected impedance, Z_p, due to the two secondary loads when $N_p = 60$, $N_{s1} = 12$, $N_{s2} = 20$, $Z_{L1} = 50\,\underline{/32°}$ Ω and $Z_{L2} = 30\,\underline{/-68°}$ Ω.

24–22 For the transformer of Figure 24–8 (a), determine the reflected impedance, Z_p, due to the two secondary loads when $N_p = 45$, $N_{s1} = 9$, $N_{s2} = 5$, $Z_{L1} = 4\,\underline{/12°}$ Ω, and $Z_{L2} = 8\,\underline{/18°}$ Ω.

Section 24–5

24–23 The core of a 117-V, 60-Hz *ideal* transformer has a cross section of 12 cm². Determine the number of turns in the primary coil needed to create a flux density of 0.7 T in the core.

24–24 Determine the voltage induced in a 900-turn coil attached to a 60-Hz source when the flux in the core is 500 μWb.

24-25 A 60-Hz power transformer rated at 1 kVA, 117/24 V delivers power from a 117-V line to a reactive load with a 0.9 lagging power factor. (a) Determine the reactive current in the load when the secondary has an apparent power of 1 kVA. (b) Determine the average power in the load for an apparent power of 800 VA when $k = 1$. (c) Determine the power (watts) producing component of the primary current when the secondary has an apparent power of 500 VA.

24-26 An iron-core transformer ($k = 1$) is attached to a 48-V, 60-Hz source. The primary of the transformer has 210 turns. Determine (a) the induced secondary voltage when $n = 0.6$, (b) the flux in the core, ϕ_m, and (c) the current in a 1-kΩ load with a unity power factor.

25 Decibels

25-1 DECIBEL DEFINITION AND NOTATION
25-2 APPROXIMATING DECIBEL GAIN AND LOSS
25-3 COMPUTING ABSOLUTE POWER
25-4 VOLTAGE AND CURRENT RATIOS AND DECIBELS
25-5 APPLICATION OF DECIBELS

CHAPTER PREVIEW

The human ear hears in a logarithmic manner. Thus, an acoustic source of energy must be increased by a factor of 10 in order for the ear to interpret a sound as being twice as loud. Because the ear responds to a change in acoustic energy in a logarithmic manner, the decibel (a logarithmic expression of two power levels) has been adopted as the unit for measuring the response of the ear. Over the years, the decibel (dB) has been used to indicate the performance of all types of electronic circuits and devices.

Because logarithmic equations will be used in the examples, we recommend that you review common logarithms in a mathematics text. The [log], [10^x], and/or [INV] [LOG] strokes of the calculator will be used in the solution of the examples in this chapter.

PERFORMANCE OBJECTIVES

Once you have read each section, worked each example with pencil, paper, and calculator, and answered each problem for every section, you will be able to:

☐ Express gain or loss in decibels.
☐ Determine decibel gain or loss without formal calculations by approximating.
☐ Understand the difference between relative power gain and absolute power gain.
☐ Express decibel gain or loss from current or voltage ratios.
☐ Correct a dBm reading made across a resistance other than 600 Ω.

25–1 DECIBEL DEFINITION AND NOTATION

The decibel (dB) is the unit used to express the relationship between the logarithmic ratio of two power levels. Stated mathematically,

$$N_{dB} = 10 \log \left(\frac{P_{out}}{P_{in}} \right) \text{ (dB)} \qquad (25-1)$$

where N_{dB} = gain or loss in decibels* (dB)
P_{out} = output power (W)
P_{in} = input power (W)

Amplification and Attenuation

The following examples will introduce you to the concepts of amplification and attenuation as they relate to relative power gain and loss in decibels.

---

* The decibel is one-tenth of a bel (B). The name bel was selected to honor the inventor of the telephone, Alexander Graham Bell.

Decibel Definition and Notation

EXAMPLE 25-1 An audio preamplifier has a measured output power of 500 mW when a power of 2 mW is applied to the input. Determine the relationship between the two powers in decibels.

SOLUTION Substitute into Equation 25-1:

$$N_{dB} = 10 \log \left(\frac{P_{out}}{P_{in}}\right)$$

$$N_{dB} = 10 \log \left(\frac{500 \times 10^{-3}}{2 \times 10^{-3}}\right)$$

$$\therefore \quad N_{dB} = +24 \text{ dB}$$

In Example 25-1, the ratio of P_{out}/P_{in} was greater than 1. This, in turn, resulted in a positive logarithm and a positive decibel calculation. If the ratio of P_{out}/P_{in} was less than 1 (for example, 0.5), the logarithm would have been negative (-0.30) as would the decibel calculation (-3.0 dB). A positive decibel indicates a *gain*; a negative decibel indicates a *loss*.

The term *gain* is associated with *amplification*, whereas *loss* is associated with *attenuation*. In a system flow diagram, amplification is indicated by the triangle symbol pictured in Figure 25-1(a) and attenuation by the square symbol pictured in Figure 25-1(b).

EXAMPLE 25-2 Determine the loss in decibels of the attenuator network pictured in Figure 25-2 when the output is 5 mW and the input is 500 mW.

SOLUTION Substitute into Equation 25-1:

$$N_{dB} = 10 \log \left(\frac{P_{out}}{P_{in}}\right)$$

$$N_{dB} = 10 \log \left(\frac{5 \times 10^{-3}}{500 \times 10^{-3}}\right)$$

$$\therefore \quad N_{dB} = -20 \text{ dB}$$

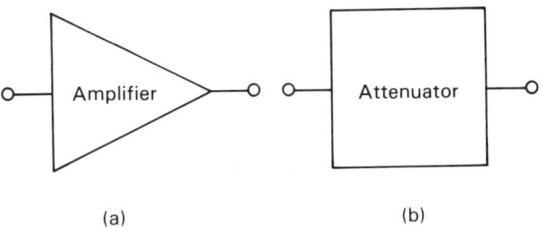

FIGURE 25-1 Symbols indicating (a) amplification—gain in power, and (b) attenuation—loss in power.

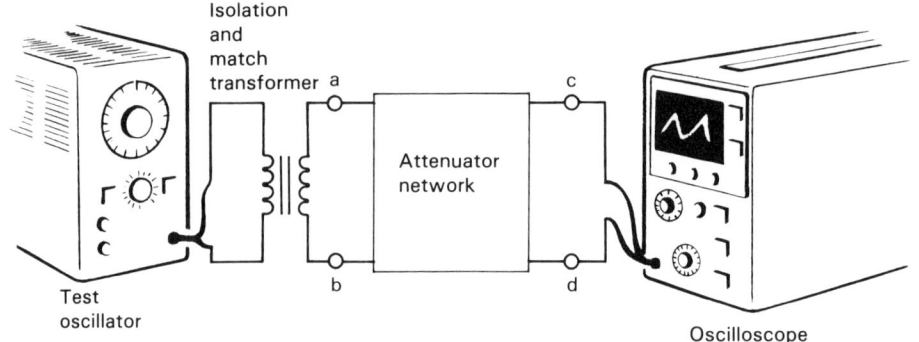

FIGURE 25-2 Circuit for Example 25-2.

25-2 APPROXI-MATING DECIBEL GAIN AND LOSS

A practical approach to decibel calculation is through the use of a curve, such as pictured in Figure 25-3. The use of these curves is demonstrated in the following examples.

Approximating with a Decibel Curve

EXAMPLE 25-3 Use Figure 25-3(b) to determine the current ratio required for -40 dB.

SOLUTION Enter the horizontal axis at -40 dB, project vertically upward, intersect the current curve, and project horizontally to the left. Read 1×10^{-2}. Thus, the current ratio is $1:0.01$.

∴ A loss of 40 dB is represented by a current ratio of 1/100.

EXAMPLE 25-4 Using Figure 25-3(a), express the output to input voltage ratio of 200/1 as a power gain in decibels.

SOLUTION Enter the vertical axis at 2×10^2. Project horizontally to the right, intersecting the voltage curve, then project vertically down. Read $+46$ dB.

∴ A voltage ratio of 200/1 is approximately a 46-dB power gain.

Approximating without a Calculator

The power gain or loss in decibels may be approximated to the nearest whole decibel without using a calculator by knowing the information in Table 25-1. This technique is used in the following series of examples. Remember that decibel gain and loss are added algebraically.

EXAMPLE 25-5 Approximate the output power from a 32-dB amplifier when the input power is 5 mW.

SOLUTION Using the information of Table 25-1, think of the amplifier as being two 10-dB amplifiers and four

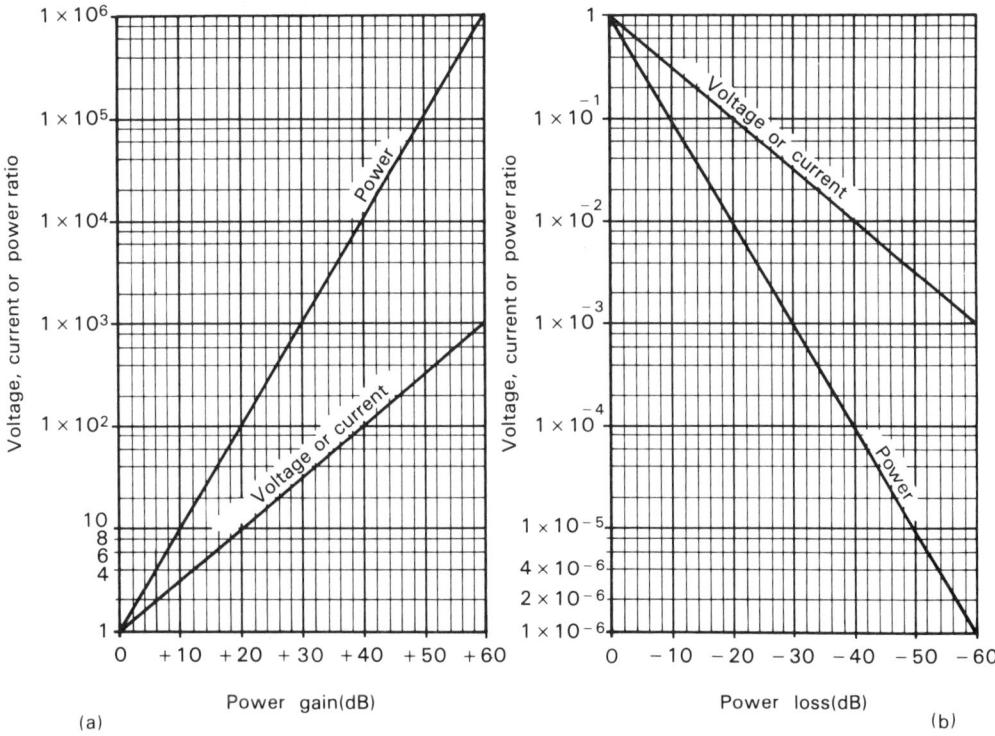

FIGURE 25-3 Curves used to approximate (a) power gain (dB), and (b) power loss (dB).

3-dB amplifiers. This concept is pictured in Figure 25-4. Increase the power by 10 for each 10-dB amplifier and double the power for each 3-dB amplifier. Thus

$$P_\text{out} = 5 \text{ mW} \times 10 \times 10 \times 2 \times 2 \times 2 \times 2$$
$$P_\text{out} = 500 \text{ mW} \times 2 \times 2 \times 2 \times 2$$
$$P_\text{out} = 1000 \text{ mW} \times 8$$
$$\therefore \quad P_\text{out} = 8 \text{ W}$$

TABLE 25-1 RELATIVE POWER GAIN AND LOSS IN DECIBELS COMPARED TO RELATIVE CHANGE IN POWER LEVEL

Gain	Change	Loss	Change
+10 dB	10 times greater	−10 dB	1/10 as great
+3 dB	2 times greater	−3 dB	1/2 as great

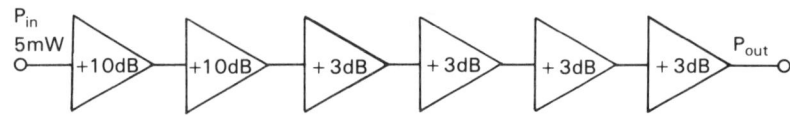

FIGURE 25-4
System flow diagram for Example 25-5. A +32 dB amplifier is divided into two +10 dB amplifiers and four +3 dB amplifiers so the output power may be approximated.

The procedure of Example 25-5 is summarized as Figure 25-5. Notice how the relationships of decibel and power ratio are used for this approximation.

EXAMPLE 25-6 Approximate the input power to the symmetrical π, 24-dB attenuator pictured in Figure 25-6.

SOLUTION Think of the 24-dB attenuator as being three -10-dB attenuators and two $+3$-dB amplifiers. That is, $(-10 \text{ dB}) + (-10 \text{ dB}) + (-10 \text{ dB}) + (3 \text{ dB}) + (3 \text{ dB}) = -24 \text{ dB}$. Draw the system flow diagram, Figure 25-7(a), and work from the output power to the input power.

From Figure 25-7(b), top right to top left,

$P_{in} = 200 \text{ mW} \times \tfrac{1}{2} \times \tfrac{1}{2} \times 10 \times 10 \times 10$

$P_{in} = 100 \text{ mW} \times \tfrac{1}{2} \times 1000 = 50 \text{ mW} \times 1000$

$\therefore \quad P_{in} = 50 \text{ W}$

EXAMPLE 25-7 Approximate the power gain in decibels to increase 500 μW to 20 mW.

SOLUTION Begin by multiplying P_{in} by 10 (+10 dB):

	Comments
500 μW $\times$ 10 = 5 mW	+10 dB
5 mW $\times$ 10 = 50 mW	50 mW > 20 mW; try $\times$ 2
5 mW $\times$ 2 = 10 mW	+3 dB
10 mW $\times$ 2 = 20 mW	+3 dB

$\therefore$ Power gain equals 10 dB + 3 dB + 3 dB = 16 dB.

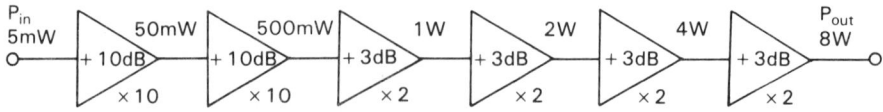

FIGURE 25-5 Summary of Example 25-5.

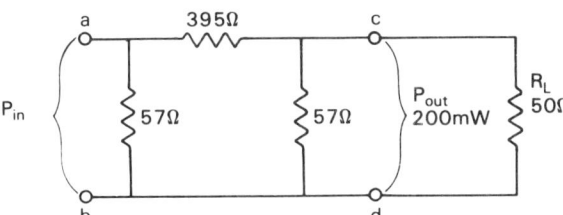

FIGURE 25-6 The symmetrical π attenuator of Example 25-6. This attenuator matches a 50 Ω source to the 50 Ω load with a fixed attenuation of −24 dB.

25-3 COMPUTING ABSOLUTE POWER

The statement "+10 dB" is meaningless unless the reference power level is given: +10 dB above what power level? Thus, the power gain in decibels has meaning only when related to a power level; this, then, is a *relative power gain*.

Gain and Loss

In contrast to relative power gain, *absolute power* has a defined reference power level. Several absolute power systems have been created to express power gain in decibels. Of these various systems, the one referenced to 1 mW (1 mW = 0 dBm) has found wide use in the elctronics industry.

Decibel Milliwatt (dBm)

A third letter is added to the decibel unit designation to indicate an absolute power, as in dBm. The m in dBm indicates a reference level of 1 mW.

Equation 25-2 is used to compute absolute power in dBm.

$$N_{dBm} = 10 \log \left(\frac{P}{1 \text{ mW}}\right) \text{ (dBm)} \tag{25-2}$$

EXAMPLE 25-8 Determine the absolute power represented by +15 dBm.

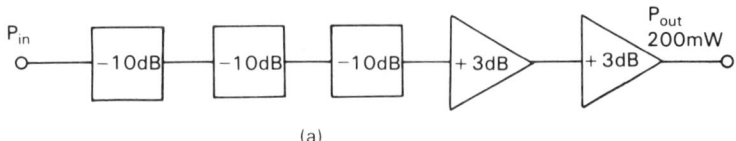

(a)

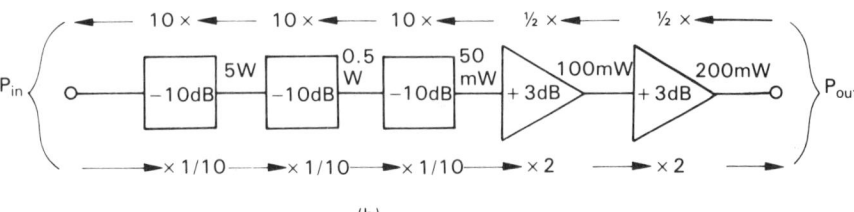

(b)

FIGURE 25-7 (a) System flow diagram for approximating the P_{in} of the −24 dB attenuator of Example 25-6.
(b) Move from P_{out} to P_{in} (right to left top), using the indicated multipliers.

SOLUTION Substitute into Equation 25-2 and solve for P:

$$15 = 10 \log \left(\frac{P}{1 \text{ mW}}\right)$$

$$1.5 = \log \left(\frac{P}{1 \text{ mW}}\right)$$

Take the antilog of each member:

$$10^{1.5} = \frac{P}{1 \text{ mW}}$$

$$P = (1 \times 10^{-3})(10^{1.5})$$

$$\therefore \quad P = 31.6 \text{ mW}$$

EXAMPLE 25-9 Using the approximations of Table 25-1, determine the power represented by +23 dBm.

SOLUTION Assume the +23 dBm is made up of two +10-dB amplifiers and one +3-dB amplifier. The input power is, of course, 1 mW. Increase the power by 10 for each 10-dB amplifier and double the power for each 3-dB amplifier. Thus

$$P = 1 \text{ mW} \times 10 \times 10 \times 2$$

$$\therefore \quad P = 200 \text{ mW}$$

Measuring dBm Although power in dBm does not depend upon the resistance of the circuit under consideration, measurement with instruments having dBm scales does depend upon the circuit resistance. These instruments are referenced to 1 mW across a 600-Ω load. Figure 25-8 pictures a dBm scale with the range selector and several typical range settings noted.

When making decibel measurements, the *reading* is the algebraic sum of the indication of the meter pointer and the range

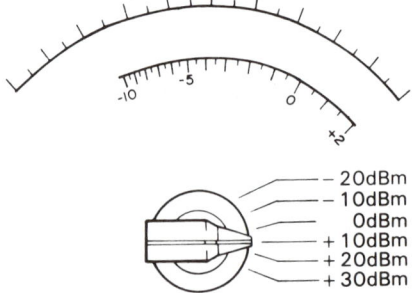

FIGURE 25-8
A dB scale and range switch. The actual reading, in dBm, is the algebraic sum of the meter indication and the range setting.

655

Computing Absolute Power

setting. That is, a meter indication of -5 dB with a range setting of $+20$ dBm is a measurement of $+15$ dBm ($+20$ dBm $-$ 5 dB).

If the decibel measurement is made across a load other than 600 Ω, a correction of N_{dB} is algebraically added to the reading. This correction then places the decibel reading back onto the 1-mW reference. The amount of correction may be determined by reading an *impedance correction graph* provided by the equipment manufacturer.

EXAMPLE 25–10 Determine the actual power in a 100-Ω load when a decibel meter is indicating -7 dB and the range switch is set to $+30$ dBm. Use Figure 25–9 to correct the decibel reading to dBm (1 mW, 600 Ω).

SOLUTION Read the correction for 100 Ω from Figure 25–9 as $+7.5$ dB. Thus, meter reading + range setting + correction = N_{dBm}.

$$-7 \text{ dB} + 30 \text{ dBm} + 7.5 \text{ dB} = 30.5 \text{ dBm}$$

Solve for absolute power:

$$N_{dBm} = 10 \log \left(\frac{P}{1 \text{ mW}}\right)$$

$$30.5 = 10 \log \left(\frac{P}{1 \text{ mW}}\right)$$

$$10^{3.05} = \frac{P}{1 \text{ mW}}$$

$$\therefore \quad P = 1.12 \text{ W}$$

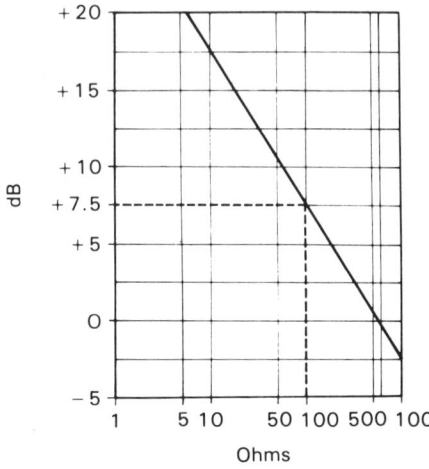

FIGURE 25–9 *Impedance correction graph* used to convert dB reading to dBm of Example 25–10.

Creating an Absolute Decibel System

A decibel system may be created at anytime to meet a particular need. When this is done, the only requirement is that the reference power be stated so that the absolute power may be determined. This concept is demonstrated in Example 25–11.

EXAMPLE 25–11 Create a decibel system referenced to 1 fW (femtowatt); then determine the absolute power for +100 dBf.

SOLUTION Use f to indicate femto. Thus

$$0 \text{ dBf} = 1 \text{ fW} = 1 \times 10^{-15} \text{ W}$$

Determine the absolute power for +100 dBf:

$$N_{dBf} = 10 \log \left(\frac{P}{1 \text{ fW}}\right)$$

$$100 = 10 \log \left(\frac{P}{1 \text{ fW}}\right)$$

$$10^{10} = \frac{P}{1 \text{ fW}}$$

$$P = 1 \times 10^{-15} \times 10^{10} = 1 \times 10^{-5} \text{ W}$$

$$\therefore \quad P = 10 \text{ }\mu\text{W}$$

25–4 VOLTAGE AND CURRENT RATIOS AND DECIBELS

Voltage and current ratios may be used to compute power gain and loss in decibels when working with amplifiers and attenuators. The procedure is to determine the input and output voltage or current; then the gain or loss of the circuit is calculated using V_{out}/V_{in} or I_{out}/I_{in}. Finally, the power gain or loss of the amplifier or attenuator is expressed in decibels, using either Equation 25–3 or 25–4.

$$N_{dB} = 20 \log \left(\frac{V_{out}}{V_{in}}\right) + 10 \log \left(\frac{R_{in}}{R_{out}}\right) \text{ (dB)} \quad (25\text{–}3)$$

$$N_{dB} = 20 \log \left(\frac{I_{out}}{I_{in}}\right) + 10 \log \left(\frac{R_{out}}{R_{in}}\right) \text{ (dB)} \quad (25\text{–}4)$$

EXAMPLE 25–12 Determine the gain in decibels of an audio amplifier that has an input voltage of 2.00 V across a resistance of 600 Ω and an output voltage of 70 V across 120 Ω.

SOLUTION Use Equation 25–3 and substitute:

$$N_{dB} = 20 \log \left(\frac{70}{2}\right) + 10 \log \left(\frac{600}{120}\right)$$

$N_{dB} = 30.88 + 6.99$

∴ $N_{dB} = 37.87$ dB

EXAMPLE 25–13 Determine the loss in decibels of the attenuator pad pictured in Figure 25-10.

SOLUTION First, determine V_{in}, V_{out}, R_{in}, and R_{out}; then compute the decibel loss:

$$V_{in} = \frac{ER_{ab}}{R_{source} + R_{ab}}$$

Where R_{ab} is the equivalent resistance viewed between points ab looking toward the load, determine R_{ab}:

$$R_{ab} = 260 + \frac{1}{\frac{1}{87} + \frac{1}{75}} = 300 \ \Omega$$

Substitute and determine V_{in}:

$$V_{in} = \frac{10(300)}{300 + 300} = 5.00 \text{ V}$$

Determine V_{out}:

$$V_{out} = V_{cd} = \frac{ER_{cd}}{R_{source} + R_1 + R_{cd}}$$

$$V_{out} = \frac{10(40)}{300 + 260 + 40}$$

$$V_{out} = 0.667 \text{ V}$$

Determine R_{in}:

$$R_{in} = R_{ab} = 300 \ \Omega$$

Determine R_{out}:

$$R_{out} = R_{load} = 75 \ \Omega$$

Use Equation 25–3 and substitute $V_{out} = 0.667$ V, $V_{in} = 5.00$ V, $R_{in} = 300 \ \Omega$, and $R_{out} = 75 \ \Omega$:

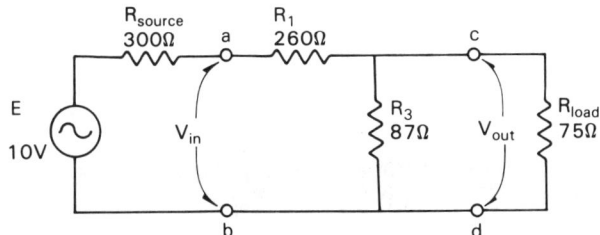

FIGURE 25–10 Minimum loss attenuator pad (R_1 and R_3) used to match a 300 Ω source resistance to a 75 Ω load used in Example 25–13.

$$N_{dB} = 20 \log \left(\frac{V_{out}}{V_{in}}\right) + 10 \log \left(\frac{R_{in}}{R_{out}}\right)$$

$$N_{dB} = 20 \log \left(\frac{0.667}{5.00}\right) + 10 \log \left(\frac{300}{75}\right)$$

$$N_{dB} = -17.50 + 6.02$$

$$\therefore N_{dB} = -11.48 \text{ dB}$$

Decibel Gain and Loss with Equal Resistance

Equations 25–3 and 25–4 may be simplified to Equations 25–5 and 25–6 when the input and output resistance are equal.

$$N_{dB} = 20 \log \left(\frac{V_2}{V_1}\right) \text{ (dB)} \tag{25-5}$$

$$N_{dB} = 20 \log \left(\frac{I_2}{I_1}\right) \text{ (dB)} \tag{25-6}$$

Where V_1 and V_2 are measured across the same or equal resistances, I_1 and I_2 pass through the same or equal resistances.

Notice that each equation is missing the term containing the resistance. With equal resistances, the ratio of R_1 and R_2 is 1. The log of 1 is equal to 0 [log(1) = 0]. Furthermore, the subscript notation has been changed from *out* and *in* to 1 and 2. This was done because both measurements may be made at the same point in the circuit; thus, *out* and *in* would have no meaning. Example 25–14 will serve to clarify this point.

EXAMPLE 25–14 Determine the power gain or loss in decibels of an amplifier when the output current through an 8-Ω load is 1.0 A at 1 kHz but drops to 300 mA at 80 Hz.

SOLUTION Select 1.0 A at 1 kHz as the reference (I_1), and calculate the power gain or loss of the amplifier at 80 Hz using Equation 25–6.

NOTE: Equation 25–6 is used because each current is passing through the same 8-Ω resistance.

$$N_{dB} = 20 \log \left(\frac{I_2}{I_1}\right)$$

$$N_{dB} = 20 \log \left(\frac{0.30}{1.0}\right)$$

$$\therefore N_{dB} = -10.5 \text{ dB}$$

Observation The power of the signal at 80 Hz is 10.5 dB below the power at 1.0 kHz.

25–5 APPLICATION OF DECIBELS

The following series of examples will provide you with an insight into the use of decibels to determine circuit parameters.

EXAMPLE 25–15 The preamplifier of Figure 25–11 is to be *interfaced* (connected) to a +40-dB *power amplifier*. Determine the power developed in the 16-Ω load for each of the two noted preamplifier voltages. The preamplifier has an open circuit output voltage that is adjustable between 200 mV and 2.2 V.

SOLUTION Determine P_{in} across R_{in} of 1.0 kΩ when $E = 200$ mV.

$$P_{in} = \frac{V_{in}^2}{R_{in}} = \frac{\left(\frac{ER_{in}}{R_{in} + R_{int}}\right)^2}{R_{in}}$$

$$P_{in} = \frac{\left[\frac{(200 \times 10^{-3})(1 \times 10^3)}{11 \times 10^3}\right]^2}{1 \times 10^3}$$

$$P_{in} = 331 \text{ nW}$$

Determine P_{out} across R_{load} of 16 Ω when $P_{in} = 331$ nW:

$$N_{dB} = 10 \log \left(\frac{P_{out}}{P_{in}}\right)$$

$$40 = 10 \log \left(\frac{P_{out}}{331 \times 10^{-9}}\right)$$

$$4 = \log \left(\frac{P_{out}}{331 \times 10^{-9}}\right)$$

Take the antilog of each member:

$$10^4 = \frac{P_{out}}{331 \times 10^{-9}}$$

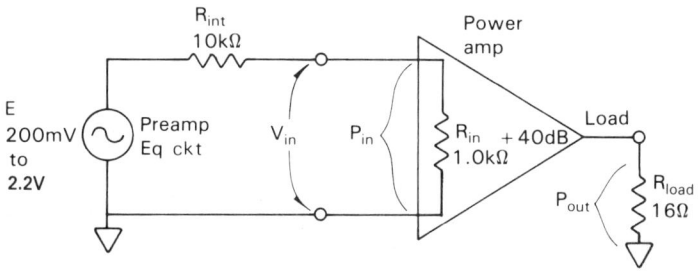

FIGURE 25–11
Circuit for Example 25–15. $P_{in} = V_{in}^2/R_{in}$.

$$P_{\text{out}} = (331 \times 10^{-9})(10^4) = 3.31 \text{ mW}$$

Determine P_{in} across R_{in} of 1.0 kΩ when $E = 2.2$ V:

$$P_{\text{in}} = \frac{V_{\text{in}}^2}{R_{\text{in}}} = \frac{\left(\frac{ER_{\text{in}}}{R_{\text{in}} + R_{\text{int}}}\right)^2}{R_{\text{in}}}$$

$$P_{\text{in}} = \frac{\left[\frac{(2.2)(1 \times 10^3)}{11 \times 10^3}\right]^2}{1 \times 10^3}$$

$$P_{\text{in}} = 40 \ \mu\text{W}$$

Determine P_{out} across R_{load} of 16 Ω when $P_{\text{in}} = 40 \ \mu$W:

$$N_{\text{dB}} = 10 \log \left(\frac{P_{\text{out}}}{P_{\text{in}}}\right)$$

$$40 = 10 \log \left(\frac{P_{\text{out}}}{40 \times 10^{-6}}\right)$$

$$4 = \log \left(\frac{P_{\text{out}}}{40 \times 10^{-6}}\right)$$

Take the antilog of each member:

$$10^4 = \frac{P_{\text{out}}}{40 \times 10^{-6}}$$

$$P_{\text{out}} = (40 \times 10^{-6})(10^4) = 400 \text{ mW}$$

∴ The range of power developed in the 16-Ω load for the range of preamplifier voltage is 3.31 mW to 400 mW depending on the preamplifier voltage setting.

The power amplifier of Example 25–15 was designed to deliver up to 50 W of power to a 16-Ω load. It is apparent that the preamplifier was not capable of *driving* the power amplifier to its full 50-W output. The following example determines the input voltage (V_{in}) needed to *drive* the power amplifier to its rated power.

EXAMPLE 25–16 Determine the input voltage (V_{in}) needed to develop 50 W in the 16-Ω load of Figure 25–11.

SOLUTION Solve for P_{in}:

$$N_{dB} = 10 \log \left(\frac{P_{out}}{P_{in}}\right)$$

$$40 = 10 \log \left(\frac{50}{P_{in}}\right)$$

$$4 = \log \left(\frac{50}{P_{in}}\right)$$

Take the antilog of each member:

$$10^4 = \frac{50}{P_{in}}$$

$$P_{in} = \frac{50}{10^4} = 5 \text{ mW}$$

Determine V_{in}:

$$P_{in} = \frac{V_{in}^2}{R_{in}}$$

$$V_{in} = \sqrt{R_{in} P_{in}} = \sqrt{(1 \times 10^3)(5 \times 10^{-3})}$$

$$V_{in} = 2.24 \text{ V}$$

∴ The input voltage across the input terminals of the power amplifier must be 2.24 V for the power amplifier to develop 50 W in the 16-Ω load.

From the previous examples, it has been learned that the preamplifier of Figure 25–11, as now constructed, will not drive the power amplifier to full power. This is because the preamplifier has too high an internal resistance (10 kΩ) compared to the amplifier input resistance (1.0 kΩ). Also, the preamplifier is too limited in open-circuit voltage (E). The following examples explore the solution to the shortcomings of the preamplifier.

EXAMPLE 25–17 The internal resistance (10 kΩ) of the preamplifier of Figure 25–11 is to be matched for maximum power transfer to the input resistance ($R_{in} = 1.0$ kΩ) of the power amplifier with an audio transformer. This is shown in Figure 25–12. (a) Determine the turns ratio for maximum power transfer. Assume 100% coupling. (b) Determine the loss in decibels from the source, E, to the load, R_{in}, when the load and source are matched for maximum power transfer.

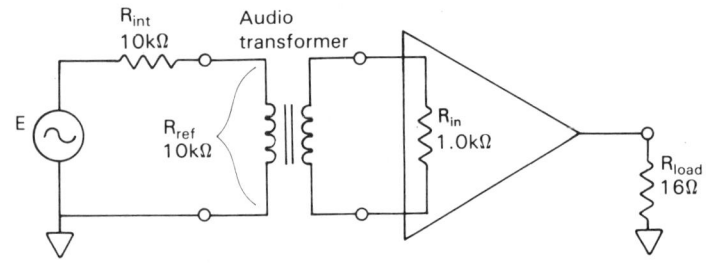

FIGURE 25-12
Circuit for Example 25-17. The transformer load (R_{in}) of 1.0 kΩ is *transformed* into a reflected resistance (R_{ref}) of 10 kΩ for maximum power transfer.

SOLUTION (*a*) Use $T^2 = R_{ref}/R_{in}$ to compute the turns ratio. Let the reflected resistance, R_{ref}, equal the preamplifier internal resistance, R_{int}, for maximum power transfer ($R_{ref} = R_{int} = 10$ kΩ).

$$T^2 = \frac{R_{ref}}{R_{in}}$$

$$T = \sqrt{\frac{R_{ref}}{R_{in}}} = \sqrt{\frac{10 \times 10^3}{1 \times 10^3}}$$

$$T = 3.16$$

∴ A turns ratio of 3.16:1 will cause the R_{in} of 1.0 kΩ to be reflected into the primary as 10 kΩ. Maximum power transfer from the source, E, to the load, R_{in}, is achieved because the source resistance ($R_{int} = 10$ kΩ) is equal to the transformed load resistance ($R_{ref} = 10$ kΩ).

SOLUTION (*b*) Under maximum power transfer, only half of the available source power is transferred to the load; thus, if $P_{source} = 1.0$ W, then $P_{load} = 0.5$ W.
Determine the decibel loss for maximum power transfer:

$$N_{dB} = 10 \log \left(\frac{P_{load}}{P_{source}}\right) = 10 \log \left(\frac{0.5}{1}\right)$$

∴ $N_{dB} = -3.0$ dB

NOTE: Maximum power transfer is characterized by a 3.0-dB loss.

In Example 25-17, the transformer was assumed to be *perfect*. However, audio transformers are typically 60% to 80% efficient. The following example explores this fact.

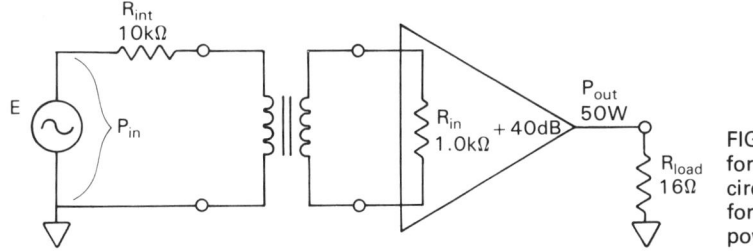

FIGURE 25–13 Circuit for Example 25–19. The circuit includes a transformer for maximum power transfer.

EXAMPLE 25–18 Determine the power loss in decibels of a transformer that is 63% efficient. That is, 1-W input yields 0.63-W output.

SOLUTION $N_{dB} = 10 \log \left(\dfrac{P_{out}}{P_{in}} \right) = 10 \log \left(\dfrac{0.63}{1.0} \right)$

$\therefore \quad N_{dB} = -2.0 \text{ dB}$

The amplifying system of Figure 25–11 has been redrawn as Figure 25–13. This figure will be used to compute the preamplifier open-circuit voltage (E) needed to *drive* the power amplifier to 50 W. The system flow diagram of Figure 25–14 will also be used, and it includes the losses of maximum power transfer (-3 dB), the transformer efficiency (-2 dB), and the gain of the power amplifier ($+40$ dB). The following example will outline the procedure for determining the preamplifier voltage (E).

EXAMPLE 25–19 Determine the preamplifier voltage (E) of Figure 25–13 needed to *drive* the power amplifier to 50 W.

SOLUTION Using the system diagram of Figure 25–14, algebraically combine the gains and losses of the system of Figure 25–14 to get the total gain in decibels (N_{dB}).

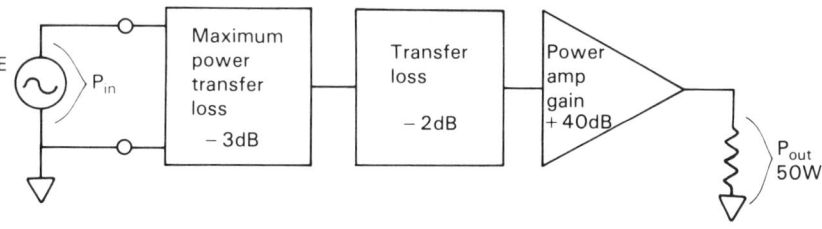

FIGURE 25–14
System flow diagram for Example 25–19. The overall system gain or loss in decibels is found by algebraically adding the gains and losses. The total power gain of this system is $+35$ dB.

$$N_{dB} = -3 - 2 + 40 = +35 \text{ dB}$$

Determine P_{in}:

$$N_{dB} = 10 \log \left(\frac{P_{out}}{P_{in}}\right)$$

$$35 = 10 \log \left(\frac{50}{P_{in}}\right)$$

$$3.5 = \log \left(\frac{50}{P_{in}}\right)$$

Take the antilog of each member:

$$10^{3.5} = \frac{50}{P_{in}}$$

$$P_{in} = \frac{50}{10^{3.5}} = 15.8 \text{ mW}$$

Determine the preamplifier open-circuit voltage (E). Use Figure 25–15:

$$P_{in} = \frac{E^2}{R_{int} + R_{ref}}$$

$$E = \sqrt{(P_{in})(R_{int} + R_{ref})}$$

$$E = \sqrt{(15.8 \times 10^{-3})(20 \times 10^3)}$$

$$\therefore E = 17.78 \text{ V}$$

PROBLEMS

Section 25–1

25–1 Determine the power gain, in decibels, of an amplifier that produces an output power of 100 W with an input power of 250 mW.

25–2 Determine the gain, in decibels, of an audio amplifier that develops 12 W in the output load from an input of 28 mW.

25–3 Determine the loss, in decibels, of an attenuator network when the output power is 180 μW and input power is 18 mW.

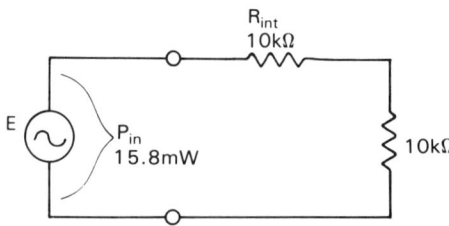

FIGURE 25–15
Circuit for Example 25–19 used to determine E when P_{in} is known.

Problems

25-4 Determine the loss, in decibels, of a balanced symmetrical H attenuator when the output power is 80 μW and the input power is 2 mW.

25-5 An audio equalizer has an insertion loss of 7.5 dB when placed in an audio line. Determine the output power when the input power is 320 mW.

25-6 A *high-pass* filter has an attenuation of 18 dB at 15 kHz. Determine the input power to the filter at 15 kHz when the output power is 450 mW.

Section 25-2

25-7 The separation ratio between input A and input B of a stereo phonograph cartridge is 20:1. Using the curve of Figure 25-3(a), express this voltage ratio, in decibels, as a power ratio.

25-8 The rejection ratio of an RF filter is 1:12. Using the curve of Figure 25-3(b), express this voltage ratio, in decibels, as a power loss.

25-9 Without the aid of a calculator, approximate the output power of a 20-dB amplifier when the input power is 8 mW.

25-10 Without the aid of a calculator, approximate the input power to a 16-dB amplifier when the output power is 12 W.

25-11 Without the aid of a calculator, approximate the input power to a −6-dB T-pad attenuator when the output power is 15 mW.

25-12 Without the aid of a calculator, approximate the output power from a −36-dB low-pass filter when the input power is 24 W.

Section 25-3

25-13 Express 280 mW in dBm.

25-14 State the power represented by +16 dBm.

25-15 State the power represented by −9 dBm.

25-16 A certain amplifier has an output of 600 mV across 600 Ω. Express this signal level in dBm.

25-17 Determine the actual power in a 10-Ω load when a decibel meter is indicating +2 dB and the range switch is set to −10 dBm. Use Figure 25-9 to correct the reading.

25-18 Determine the current in a 600-Ω load connected to an audio mixer if the power to the load is +16 dBm.

25-19 Determine the actual power in a 600-Ω load of an audio system with the following components when the input power is +8 dBm. The system consists of a +30-dB preamplifier attached to a −22-dB attenuator, which is connected through a +45-dB amplifier to a −12-dB equalizer. The equalizer is terminated in a 600-Ω load.

25-20 Determine the voltage across a 600-Ω load in an audio system that has a −3-dBm input signal to a +18-dB preamplifier. The preamplifier is connected through a matching transformer of −4 dB to an amplifier of +32 dB. The output of the amplifier is connected through a −9-dB mixer to the 600-Ω load.

25-21 Develop a decibel system referenced to 1 nW (nanowatt) and determine the absolute power for +40 dBn.

25-22 Develop a decibel system referenced to 1 μW (microwatt) and determine the absolute power for +18 dBμ.

Section 25-4

25-23 A linear IC amplifier has an input resistance of 200 Ω and an output load of 4.7 kΩ. A 2-mV signal, which is measured across the input terminals, results in 72 mV across the load. Determine the decibel gain of the amplifier.

25-24 For the −24-dB attenuator pictured in Figure 25-6, the voltage V_{ab} is 15 V. Determine the output voltage, V_{cd}, across the 50-Ω load.

25-25 Determine the decibel loss of the network pictured within the square of Figure 25-16.

25-26 Determine the voltage across an 8-Ω load connected to the output of a +38-dB amplifier. The microphone connected to the input of the amplifier develops 10 mV across 600 Ω.

25-27 The input to a 25-m, 300-Ω *twin lead* cable is 280 mV. The output is 140 mV when the cable is terminated with a 300-Ω resistive load. What is the loss in the cable expressed in dB/100 m?

25-28 The input to a preamplifier is 2 mV and the output is 6.4 V. Determine the power gain of the preamplifier if each voltage is developed across the same amount of resistance.

Section 25-5

25-29 A line-matching transformer, having an efficiency of 78%, is used to match a microphone to an amplifier. Assuming maximum power transfer, state the decibel

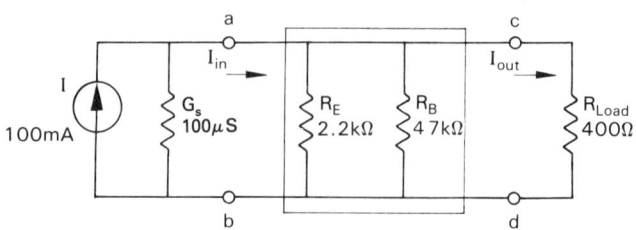

FIGURE 25-16
Network of Problem 25-25.

25-30 Determine the decibel attenuation of the *T* network within the box pictured in Figure 25-17.

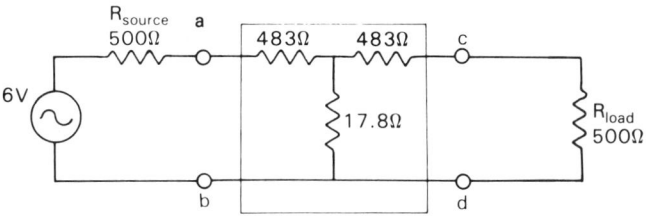

FIGURE 25-17
Network of Problem 25-30.

26 Resonance

26-1 INTRODUCTION TO RESONANCE
26-2 SERIES-RESONANT CIRCUIT
26-3 PARALLEL-RESONANT CIRCUIT
26-4 IMPEDANCE OF THE UNLOADED PARALLEL-RESONANT CIRCUIT
26-5 LOADING THE PARALLEL-RESONANT CIRCUIT

CHAPTER PREVIEW   In this chapter, the characteristics and operating conditions of an important ac circuit condition will be studied. This circuit condition, which occurs at a particular frequency, is called *resonance;* the frequency at which it occurs is called the *resonant* frequency.

Communication circuits are designed to operate within a narrow band of frequencies centered on the resonant frequency. These narrow bands of radio frequencies, called *channels*, have a specific *bandwidth*.

For a radio receiver to respond to a transmitted radio wave, it must be able to select only one channel of frequencies at a time, and it must be sensitive enough to distinguish a particular band of transmitted frequencies from the *noise* that occurs at radio frequencies. Thus, two important properties of a resonant circuit are its *selectivity* and *sensitivity*. In this chapter, these characteristics will be discussed in relation to the two types of resonant circuits, *series* resonance and *parallel* resonance.

PERFORMANCE OBJECTIVES   Once you have read each section, worked each example with pencil, paper, and calculator, and answered each problem for every section, you will be able to:

☐ Determine the quality factor for both series- and parallel-resonant circuits.
☐ Determine the bandwidth of series- and parallel-resonant circuits.
☐ Determine the equivalent impedance of series- and parallel-resonant circuits.
☐ Differentiate between the characteristics of series- and parallel-resonant circuits.

26–1 INTRODUCTION TO RESONANCE   Resonance occurs in an ac *RLC* circuit when certain operating conditions are met. For the ideal series *RLC* circuit, resonance occurs when $X_L = X_C$. Figure 26–1 pictures a series *RLC* circuit along with its impedance vector diagram and voltage phasor.

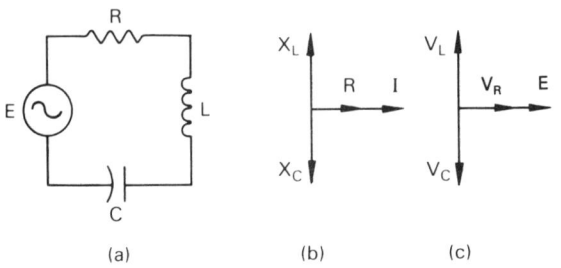

FIGURE 26–1
(a) Series RLC circuit.
(b) Impedance vector diagram of ideal RLC circuit at resonance.
(c) Voltage phasor diagram of ideal series RLC circuit at resonance.

Introduction to Resonance

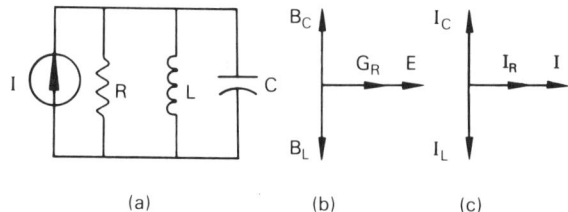

FIGURE 26-2 (a) Parallel RLC circuit.
(b) Admittance vector diagram of ideal parallel RLC circuit at resonance.
(c) Current phasor diagram of ideal parallel RLC circuit at resonance.

For the ideal parallel RLC circuit, resonance occurs when $B_C = B_L$. Figure 26-2 pictures a parallel RLC circuit along with its admittance vector diagram and current phasor.

Series-Resonant Characteristics

From Figure 26-1(b), it is learned that $X_L = X_C$ at the resonant frequency. Also, from this same impedance vector diagram, it is learned that the impedance at the resonant frequency is resistive and equal to R ($Z = R + jX_L - jX_C = R + 0 = R$). Furthermore, from Figure 26-1(c), it is learned that the source voltage, E, is in phase with the circuit current, I; thus the circuit phase angle, θ, is zero ($\theta = 0°$). Also, from this voltage phasor diagram, it is learned that at resonance $|V_L| = |V_C|$ and $E = V_R$. In summary, the series RLC circuit has the following characteristics when it is resonant.

$$X_L = X_C, \quad Z = R, \quad \theta = 0°, \quad |V_L| = |V_C|, \quad E = V_R$$

Parallel-Resonant Characteristics

From Figure 26-2(b), it is learned that $B_C = B_L$ at the resonant frequency. Also, from the same admittance vector diagram, it is learned that the admittance at the resonant frequency is conductive (resistive) and equal to G_R ($Y = G_R + jB_C - jB_L = G_R + 0 = G_R$). Furthermore, from Figure 26-2(c), it is learned that the source current, I, is in phase with the circuit voltage, E; thus the circuit phase angle, θ, is zero ($\theta = 0°$). Also, from this current phasor diagram, it is learned that at resonance $|I_L| = |I_C|$ and $I = I_R$. In summary, the parallel RLC circuit has the following characteristics when it is resonant.

$$B_L = B_C, \quad Y = G_R, \quad \theta = 0°, \quad |I_L| = |I_C|, \quad I = I_R$$

Observation: Once again we see the duality between the series and parallel circuits. This duality is very apparent when the series equivalent circuit parameters of the series-resonant circuit are compared to the series equivalent circuit parameters of the parallel-resonant circuit.

Series and Parallel Equivalent Circuit Parameters

In each case of resonance, the RLC circuit is resistive at the resonant frequency. In the case of the series-resonant circuit of Figure 26–1, the resistance at the resonant frequency is quite small ($10 < R < 100$). Since Z is at its minimum at resonance in the series RLC circuit, the current, I, is at its maximum, being limited only by R. Figure 26–3(a) pictures Z as a function of frequency. At the resonant frequency, the series RLC circuit is resistive. Below the resonant frequency, the reactance of the capacitor is greater than the reactance of the inductor. Therefore, the circuit impedance is capacitive ($Z/-\theta$). Above the resonant frequency, the reactance of the inductor is greater than the reactance of the capacitor. Therefore, the circuit impedance is inductive ($Z/+\theta$).

Contrasting the impedance characteristics of the series RLC circuit, shown in Figure 26–3(a), to those of the parallel RLC circuit, pictured in Figure 26–3(b), it is seen that the impedance at resonance in the parallel circuit is large ($10\ k\Omega < R < 1\ M\Omega$) and is at its maximum value. Because of the large impedance at the resonant frequency, the source current is at its minimum value. Above the resonant frequency, the current in the inductor of Figure 26–2 decreases due to the increase in inductive reactance, and the current in the capacitor increases due to the decrease in capacitive reactance. The net source current is a combination of resistive and capacitive currents, indicating that the circuit impedance is capacitive ($Z/-\theta$). Below the resonant frequency, the current in the inductor of Figure 26–2 increases due to the decrease in inductive reactance, and the current in the capacitor decreases due to the increase in capacitive reactance. The net source current is a combination of resistive and inductive currents, indicating that the circuit impedance is inductive ($Z/+\theta$).

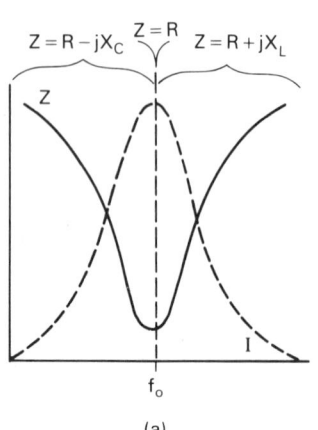

(a)

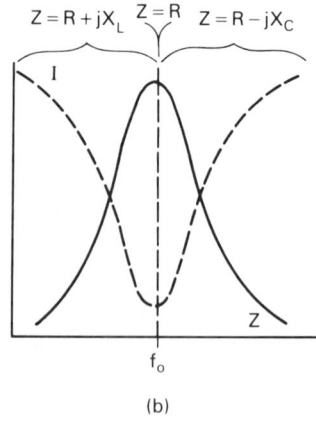

(b)

FIGURE 26–3
(a) Relative equivalent circuit impedance and current as a function of frequency in series RLC circuit of Figure 26–1.
(b) Relative equivalent circuit impedance and source current as a function of frequency in parallel RLC circuit of Figure 26–2. Note: f_o indicates resonant frequency.

673

Series-Resonant Circuit

Comparing the impedance of the series RLC to that of the parallel RLC circuit of Figure 26–3, you will notice that the impedance *increases* above and below the resonant frequency in the series RLC circuit. In the parallel RLC circuit, you will notice that the impedance *decreases* above and below the resonant frequency.

As a final observation, if the characteristics of the series RLC circuit at the resonant frequency are considered to be those of a resonant circuit, the characteristics of the parallel RLC circuit at the resonant frequency must not be resonant. Because of the unresonantlike behavior of the parallel circuit at the resonant frequency, it is often called an *antiresonant* circuit. The remaining sections in this chapter will further investigate the properties and characteristics of the series- and parallel-resonant circuit.

26–2 SERIES-RESONANT CIRCUIT

In the chapter preview, it was mentioned that a radio receiver must have both selectivity and sensitivity in order to *tune* a channel of frequencies. Sensitivity is the ability of a radio receiver to respond to weak input signals. We will open our discussion of the series-resonant circuit by looking at how sensitivity may be affected by the ratio between the values of L and C in the series-resonant RLC circuit.

L/C Ratio and Sensitivity

The circuit of Figure 26–4(a) pictures an air-core transformer with a tuned secondary. This circuit is used to match the antenna, which is attached to the primary coil, into the tuning section (secondary) of a radio receiver and to provide a voltage gain of the signal at the resonant frequency. Notice in the circuit of Figure 26–4(b) that the induced secondary voltage, E_s, is shown connected in series with the RLC components. The secondary of the transformer is a series-resonant circuit. For purposes of discussion, R is held constant at 10 Ω and E_s at 250 µV. In Example 26–1, L and C will be varied to demonstrate how the selected values of L and C affect the output voltage, V_o, and the sensitivity of reception.

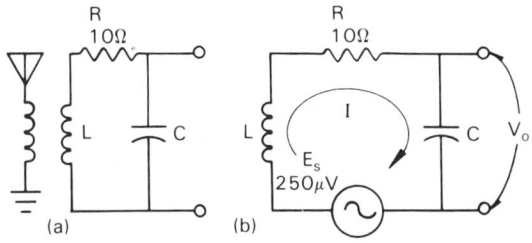

FIGURE 26–4
(a) Schematic of RLC circuit of Example 26–1.
(b) Equivalent circuit.

EXAMPLE 26-1 Each of the following sets of values of L and C will cause the circuit of Figure 26-4(b) to be resonant at 1.0 MHz. Determine both the voltage across C (V_o) and the voltage gain (A_v) for each set of LC values: (a) $L = 400$ μH, $C = 63.33$ pF; (b) $L = 2000$ μH, $C = 12.65$ pF; (c) $L = 4000$ μH, $C = 6.333$ pF. $E_s = 250$ μV in each case.

Observation Remember, at the resonant frequency of 1.0 MHz, $X_L = X_C$ and $Z = R + j0$. Thus, $R = 10\ \Omega$ is the only impedance to the current in the circuit.

SOLUTION Since $R = 10\ \Omega$ is the net impedance for each set of values, the circuit current is constant and equal to

$$I = \frac{E_s}{R} = \frac{250 \times 10^{-6}}{10} = 25\ \mu A$$

(a) Where $L = 400$ μH and $C = 63.33$ pF, determine V_o and A_v at 1.0 MHz.

$$V_o = IX_C = IX_L$$
$$V_o = (25 \times 10^{-6})(2\pi \times 1 \times 10^6 \times 400 \times 10^{-6})$$
$$V_o = 62.8\ mV$$
$$A_v = \frac{V_o}{E_s} = \frac{62.8 \times 10^{-3}}{250 \times 10^{-6}} = 251$$

∴ The output voltage of 62.8 mV is 250 times greater than the 250-μV source voltage.

(b) Where $L = 2000$ μH and $C = 12.65$ pF, determine V_o and A_v at 1.0 MHz.

$$V_o = IX_C = IX_L$$
$$V_o = (25 \times 10^{-6})(2\pi \times 1 \times 10^6 \times 2000 \times 10^{-6})$$
$$V_o = 314\ mV$$
$$A_v = \frac{V_o}{E_s} = \frac{314 \times 10^{-3}}{250 \times 10^{-6}} = 1256$$

∴ The output voltage of 314 mV is 1250 times greater than the 250-μV source.

(c) Where $L = 4000$ μH and $C = 6.333$ pF, determine V_o and A_v at 1.0 MHz.

$$V_o = IX_C = IX_L$$

Series-Resonant Circuit

$$V_o = (25 \times 10^{-6})(2\pi \times 1 \times 10^6 \times 4000 \times 10^{-6})$$

$$V_o = 628 \text{ mV}$$

$$A_v = \frac{V_o}{E_s} = \frac{628 \times 10^{-3}}{250 \times 10^{-6}} = 2512$$

∴ The output voltage of 628 mV is 2500 times greater than the 250-μV source voltage.

Observation Because the product of the values of L and C results in the same number ($L \times C = 2.53 \times 10^{-14}$), the resonant frequency is the same for each set of LC values. However, the ratio of the values of L and C changes with each set of values. From Figure 26–5, which summarizes Example 26–1, we learn that as the L/C ratio increases the sensitivity of the circuit increases.

Effective Resistance and Selectivity

For a receiver to tune just one channel of frequencies and reject all the others, it must have *selectivity*. The *effective* resistance of the resonant circuit, along with the L/C ratio, controls both the selectivity and the sensitivity of the resonant circuit.

The effective resistance, R_{eff}, of the resonant circuit stems mainly from the inductor, since capacitors used in resonant circuits have very little heat (I^2R) loss even at frequencies greater than 100 MHz. Assuming the inductor to be the sole source of resistance in the resonant circuits, the following properties of

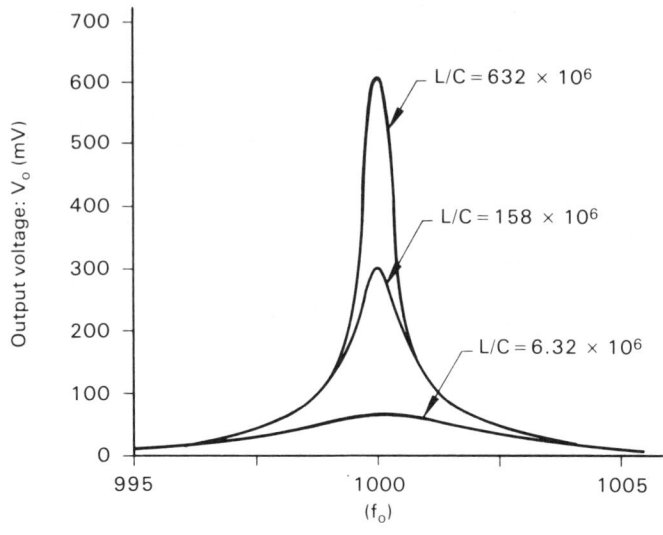

FIGURE 26–5 Resonance curve of voltage output versus frequency of Example 26–1.

676

Resonance

the inductor's core and windings may be identified as the source of the effective or heat-producing resistance.

☐ Eddy currents in the core represent major I^2R losses for iron-core inductors used above audio frequencies (>20 kHz). Because of these losses, inductors used at radio frequencies have cores made of ferrite materials. These nonmetallic, ceramic materials have very low eddy current loss.

☐ Ohmic or dc resistance of the copper wire in the inductor coil contributes to the effective resistance.

☐ Skin effect is a direct function of frequency and is the result of magnetic induction in the wire of the inductor coil. At radio frequencies, the current in the coil winding is mainly carried in the surface or *skin* of the wire. Because of the nonuniform distribution of current in the conductor, a greater I^2R loss is experienced in the coil winding as the operating frequency is increased. Figure 26-6 shows the skin effect resistance as a function of frequency. Skin effect may be minimized by using a ribbon conductor, a hollow conductor, or a specially constructed multistranded conductor called *litz* wire. Skin effect resistance accounts for most of the effective resistance (power-producing resistance) in air-core inductors operating at radio frequencies.

Bandwidth (BW) In a channel of frequencies, as pictured in Figure 26-7, all the amplitudes of voltage or current that are above the 70.7% or -3-dB level (half-power point) carry sufficient energy to be useful. The ideal resonant response curve would have very steep vertical sides, as pictured by the ideal response curve of

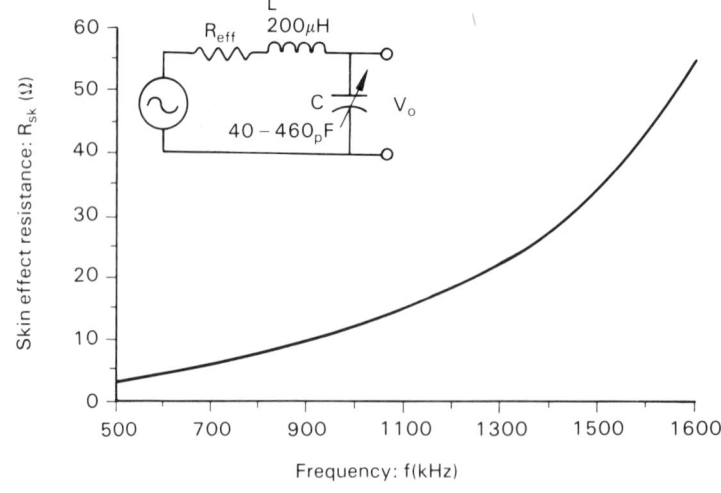

FIGURE 26-6
Inductor under test is a 200 μH, single-layer, air-core solenoid with a dc resistance of <1 Ω. Skin effect resistance increases with frequency and accounts for most of the effective resistance in the resonant circuit at radio frequencies.

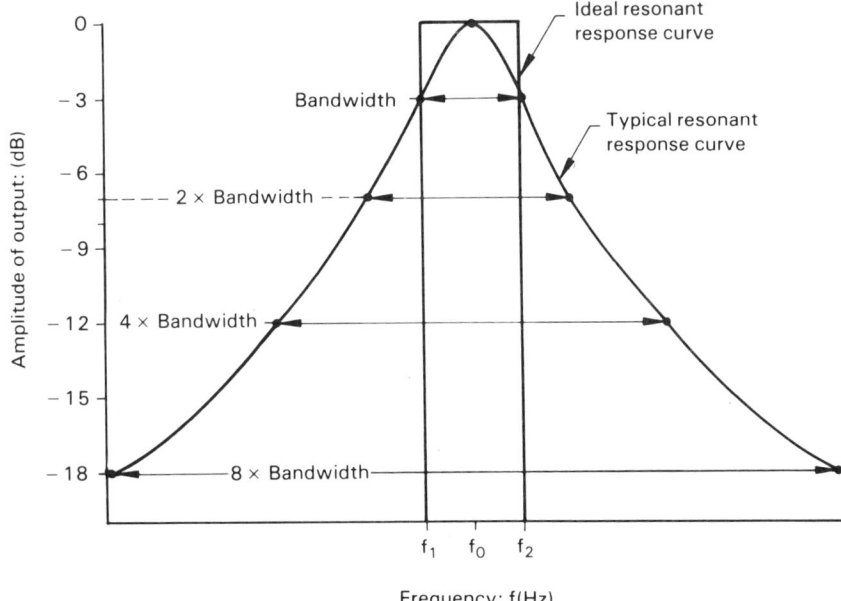

FIGURE 26-7 Typical resonant response curve (bell-shaped) has its bandwidth defined as BW = f₂ − f₁ at the −3 dB (half power level) of the output signal. The ideal response curve has a uniform bandwidth.

Figure 26-7. This curve would be very selective since it would pass only the selected band of frequencies. The width of the channel at 70.7% of the maximum amplitude (−3 dB) is called the *bandwidth*. In the ideal curve, the bandwidth would only be wide enough to pass the desired frequencies and reject all others. The typical resonant response curve, pictured in Figure 26-7, is far from ideal as its curve is *bell shaped*. The bandwidth of the practical resonant response curve, as pictured in Figure 26-7, is defined as

$$BW = f_2 - f_1 \quad \text{(Hz)} \tag{26-1}$$

As pictured in Figure 26-7, the bandwidth is determined at the −3-dB points (also called half-power points) that correspond to 0.707 of the signal at the resonant frequency. The impedance at the half-power frequencies is 1.414 times the impedance at the resonant frequency, which is $Z = R_{\text{eff}}$. Thus, from the impedance triangle of Figure 26-8, we learn that the net difference in reactance equals R_{eff} at the half-power freuency (−3 dB). Thus, at f_1 of Figure 26-7, X_C has increased from its resonant value by an amount equal to $\frac{1}{2} R_{\text{eff}}$, and X_L has decreased from its resonant value by an amount equal to $\frac{1}{2} R_{\text{eff}}$

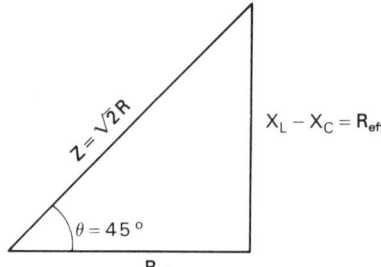

FIGURE 26-8
Impedance triangle for RLC circuit at f_1 and f_2 of Figure 26-7.

so that at f_1, $X_{L_1} - X_{C_1} = R_{\text{eff}}$. At f_2 of Figure 26-7, X_C has decreased from its resonant value by an amount equal to $\frac{1}{2}R_{\text{eff}}$, and X_L has increased from its resonant value by an amount equal to $\frac{1}{2}R_{\text{eff}}$, so that at f_2, $X_{L_2} - X_{C_2} = R_{\text{eff}}$. Therefore, the absolute difference of X_{L_2} and X_{L_1} is equal to R_{eff} at the -3-dB frequencies of Figure 26-7. This is also true for the absolute difference between X_{C_2} and X_{C_1}. Thus

$$|X_{L_2}| - |X_{L_1}| = R_{\text{eff}}$$
$$2\pi f_2 L - 2\pi f_1 L = R_{\text{eff}}$$
$$2\pi L(f_2 - f_1) = R_{\text{eff}}$$

However, $f_2 - f_1$ is defined as bandwidth in Equation 26-1. Thus

$$2\pi L(\text{BW}) = R_{\text{eff}}$$

$$\text{BW} = \frac{R_{\text{eff}}}{2\pi L} \quad (\text{Hz}) \tag{26-2}$$

From Equation 26-2, we learn that the bandwidth of the practical resonant response curve of Figure 26-7 depends directly on the effective resistance, R_{eff}, of the resonant circuit. Example 26-2 explores the relationship between the bandwidth and the effective resistance.

EXAMPLE 26-2 Investigate the influence that R_{eff} has on the bandwidth (selectivity) of the resonant curve of the circuit pictured in Figure 26-6. Determine the skin effect resistance from the curve pictured in Figure 26-6 for resonant frequencies of (a) 650 kHz, (b) 1100 kHz, (c) 1450 kHz, and (d) 1600 kHz. The dc resistance of the coil is 0.8 Ω.

SOLUTION (a) Use Equation 26-2 to determine the bandwidth for a resonant frequency of 650 kHz.

$$\text{BW} = \frac{R_\text{eff}}{2\pi L} \quad \text{where } R_\text{eff} = R_\text{sk} + R_\text{dc}$$
$$R_\text{eff} = 5 + 0.8$$
$$R_\text{eff} = 5.8 \; \Omega$$
$$\text{BW} = \frac{5.8}{2\pi \times 200 \times 10^{-6}} = 4.62 \text{ kHz}$$

(b) Use Equation 26-2 to determine the bandwidth for a resonant frequency of 1100 kHz.

$$\text{BW} = \frac{R_\text{eff}}{2\pi L} \quad \text{where } R_\text{eff} = R_\text{sk} + R_\text{dc}$$
$$R_\text{eff} = 15 + 0.8$$
$$R_\text{eff} = 15.8 \; \Omega$$
$$\text{BW} = \frac{15.8}{2\pi \times 200 \times 10^{-6}} = 12.6 \text{ kHz}$$

(c) Use Equation 26-2 to determine the bandwidth for a resonant frequency of 1450 kHz.

$$\text{BW} = \frac{R_\text{eff}}{2\pi L} \quad \text{where } R_\text{eff} = R_\text{sk} + R_\text{dc}$$
$$R_\text{eff} = 30 + 0.8$$
$$R_\text{eff} = 30.8 \; \Omega$$
$$\text{BW} = \frac{30.8}{2\pi \times 200 \times 10^{-6}} = 24.5 \text{ kHz}$$

(d) Use Equation 26-2 to determine the bandwidth for a resonant frequency of 1600 kHz.

$$\text{BW} = \frac{R_\text{eff}}{2\pi L} \quad \text{where } R_\text{eff} = R_\text{sk} + R_\text{dc}$$
$$R_\text{eff} = 55 + 0.8$$
$$R_\text{eff} = 55.8 \; \Omega$$
$$\text{BW} = \frac{55.8}{2\pi \times 200 \times 10^{-6}} = 44.4 \text{ kHz}$$

∴ R_eff affects the bandwidth directly. A tenfold increase in R_eff produces a tenfold increase in the bandwidth.

Observation Figure 26-9 summarizes the resonant curves of Example 26-2. The curves have been *stacked*, one on top of the other, so that the bandwidths may be compared. The vertical position of the

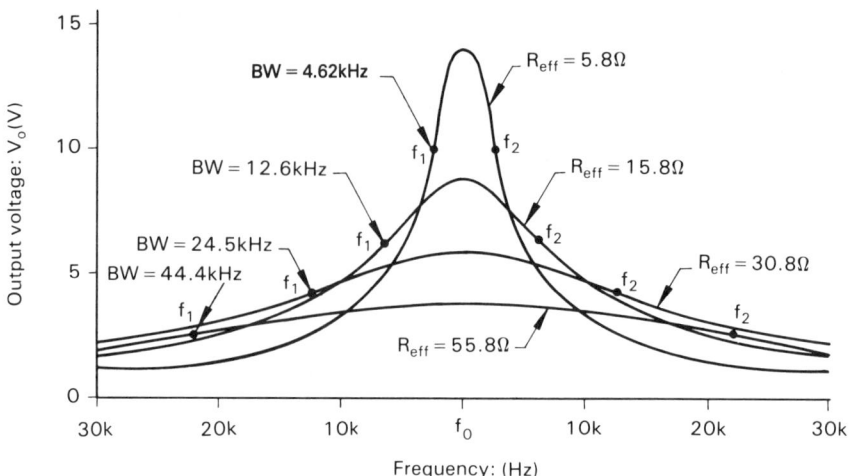

FIGURE 26-9 Summation of Example 26-2, showing effect on selectivity and sensitivity of resonant response curve due to effective resistance, R_{eff}.

curve has been determined by the output voltage, V_o, across C at the resonant frequency, using the circuit of Figure 26-6 when $E = 100$ mV.

Quality Factor (Q) Thus far we have investigated how the L/C ratio and R_{eff} relate to sensitivity and selectivity in a resonant circuit. We have learned that increasing the L/C ratio and/or decreasing the effective resistance in the resonant circuit produces a greater degree of selectivity (narrower bandwidth) and sensitivity (greater output across the capacitor). These two important resonant circuit parameters (L/C and R_{eff}) are related to the resonant response curve through the *quality factor*, Q, of the resonant circuit.

The Q of a resonant circuit describes both the ability of the resonant circuit to produce a large output at the resonant frequency and to tune a selective band of frequencies. It is the single factor that relates both the effects of the L/C ratio and the effective resistance to the shape of the resonant response curve. The *Q-factor* at the resonant frequency, Q_o, is given by the following equation:

$$Q_o = \frac{1}{R_{eff}} \sqrt{\frac{L}{C}} \quad \text{(no units)} \qquad (26\text{--}3)$$

This equation relates what we have previously experienced to the Q of the circuit. Thus, lower effective resistance produces higher quality factors, which, in turn, indicate more sensitivity

Series-Resonant Circuit

and selectivity. Furthermore, higher L/C ratios produce higher quality factors, which indicate an improvement in both sensitivity and selectivity.

The circuit Q is related to its bandwidth (circuit selectivity) by the following equation:

$$\text{BW} = \frac{f_o}{Q_o} \quad \text{(Hz)} \tag{26-4}$$

where BW = bandwidth at the -3-dB points (0.707 of the maximum signal level)
f_o = resonant frequency
Q_o = quality factor at f_o

The circuit Q is also related to the output voltage (circuit sensitivity) across the reactive components (usually taken across the capacitor) by the following equation:

$$V_o = V_C = V_L = Q_o E \quad \text{(V)} \tag{26-5}$$

where V_o = output across either the inductor or capacitor at the resonant frequency
Q_o = quality factor
E = source voltage

Equation 26-5 is sometimes referred to as the Q-gain equation since it relates the input and output voltages in the series-resonant circuit.

From earlier discussions, it was learned that the inductor is responsible for most, if not all, of the effective resistance in the resonant circuit. Because this is the case, the circuit quality factor is usually related to the inductive reactance by the following equation:

$$Q_o = \frac{X_L}{R_{\text{eff}}} \quad \text{(no units)} \tag{26-6}$$

The following examples will apply the quality factor to determine quantities of the series-resonant circuit.

EXAMPLE 26–3 Determine the output voltage, V_o, and the bandwidth, BW, of the series RLC circuit of Figure 26–10(a), which has a resonant frequency of 1 MHz.

SOLUTION Determine the Q of the circuit, using Equation 26–3.

$$Q_o = \frac{1}{R_{\text{eff}}}\sqrt{\frac{L}{C}} = \frac{1}{12}\sqrt{\frac{200 \times 10^{-6}}{127 \times 10^{-12}}}$$

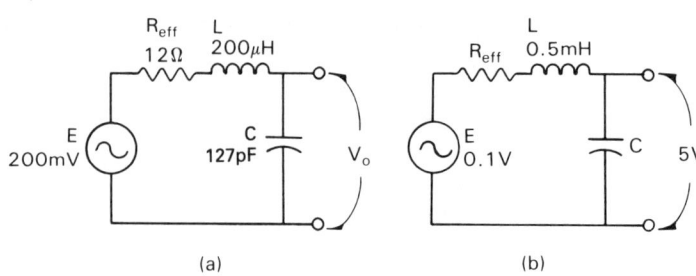

FIGURE 26-10
(a) Circuit of Example 26-3.
(b) Circuit of Example 26-4.

$$Q_o = 104.6$$

Compute V_o, the voltage across C:

$$V_o = EQ_o = (0.200)(104.6) = 20.9 \text{ V}$$

Determine the bandwidth, using Equation 26-4.

$$\text{BW} = \frac{f_o}{Q_o} = \frac{1 \times 10^6}{104.6} = 9.56 \text{ kHz}$$

∴ The output voltage, V_o, is 20.9 V and the bandwidth is 9.56 kHz.

EXAMPLE 26-4 Determine the value of the effective resistance in the circuit of Figure 26-10(b) if the circuit is resonant at 200 kHz.

SOLUTION Determine the effective resistance, R_{eff}, of the circuit by first determining Q_o from Equation 26-5, and then using Equation 26-6. Thus

$$V_o = EQ_o \quad \text{and} \quad Q_o = \frac{V_o}{E}$$

$$Q_o = \frac{5}{0.1} = 50$$

From Equation 26-6:

$$Q_o = \frac{X_L}{R_{\text{eff}}} \quad \text{and} \quad R_{\text{eff}} = \frac{X_L}{Q_o}$$

$$R_{\text{eff}} = \frac{2\pi \times 200 \times 10^3 \times 0.5 \times 10^{-3}}{50} = 12.6 \text{ }\Omega$$

∴ The effective resistance is 12.6 Ω.

EXAMPLE 26-5 The coil of Figure 26-10(b) is an air-core inductor with a dc resistance of 2 Ω. Determine the skin effect resistance for the conditions determined in Example 26-4.

Series-Resonant Circuit

SOLUTION From Example 26-4, $R_{eff} = 12.6\ \Omega$. Thus

$$R_{eff} = R_{dc} + R_{sk} \quad \text{and} \quad R_{sk} = R_{eff} - R_{dc}$$

$$R_{sk} = 12.6 - 2 = 10.6\ \Omega$$

∴ The skin effect resistance at 200 kHz is 10.6 Ω.

Observation This calculation was made on the assumption that the capacitor has no I^2R losses.

EXAMPLE 26-6 Determine the value of the capacitor in the circuit of Figure 26-10(b) in order for the circuit to be resonant at a frequency of 600 kHz.

SOLUTION Since $X_L = X_C$, then

$$X_L = 2\pi fL = 2\pi \times 600 \times 10^3 \times 0.5 \times 10^{-3}$$

$$X_L = X_C = 1885\ \Omega$$

And $X_C = 1/(2\pi fC)$; thus

$$C = \frac{1}{2\pi f X_C} = \frac{1}{2\pi \times 600 \times 10^3 \times 1885}$$

∴ $C = 141$ pF

Resonant Frequency (f_o) The resonant frequency of the series-resonant circuit depends upon the values of L and C in the circuit. As stated in the introduction, resonance occurs when $X_C = X_L$. The following equation, which relates L and C to the resonant frequency, may be derived from the statement $X_L = X_C$. Thus

$$X_L = X_C$$

$$2\pi fL = \frac{1}{2\pi fC}$$

$$4\pi^2 f^2 LC = 1$$

$$f^2 = \frac{1}{4\pi^2 LC}$$

$$f_o = \frac{1}{2\pi\sqrt{LC}} \quad \text{(Hz)} \tag{26-7}$$

From Equation 26-7, it is learned that the resonant frequency, f_o, varies inversely with the value of the inductor, L, and the capacitor, C.

EXAMPLE 26-7 A series circuit consisting of an 80-μH inductor is connected to a 20-pF capacitor. (a) Determine the resonant frequency of this series LC circuit. (b) If the effective resistance is 8 Ω, determine both the output voltage across the capacitor when the source voltage is 2.4 V and the bandwidth.

SOLUTION (a) Use Equation 26–7 to determine the resonant frequency.

$$f_o = \frac{1}{2\pi\sqrt{LC}}$$

$$f_o = \frac{1}{2\pi\sqrt{80 \times 10^{-6} \times 20 \times 10^{-12}}}$$

∴ $f_o = 3.98$ MHz

(b) Determine Q_o, using Equation 26–3.

$$Q_o = \frac{1}{R_{\text{eff}}}\sqrt{\frac{L}{C}} = \frac{1}{8}\sqrt{\frac{80 \times 10^{-6}}{20 \times 10^{-12}}} = 250$$

Compute the output voltage, V_o.

$$V_o = Q_o E = (250)(2.4) = 600 \text{ V}$$

Use Equation 26–4 to determine bandwidth.

$$\text{BW} = \frac{f_o}{Q_o} = \frac{3.98 \times 10^6}{250} = 15.9 \text{ kHz}$$

∴ The voltage across the capacitor is 600 V and the bandwidth is 15.9 kHz.

Observation Be very careful when working around high-Q series-resonant circuits, as the voltages can be very high. When specifying the voltage rating of the LC components in a series-resonant circuit, you must take into account the voltage gain afforded by the Q of the circuit.

Defining the Quality Factor (Q) In the discussion of power factor correction in Section 23–3, it was pointed out that, at or near unity power factor, reactive currents pass back and forth between the inductor and capacitor with little or no reactive current being taken from the power line. Thus, the reactive power, P_q, is exchanged between the L and C circuit components.

This same phenomenon also occurs in the resonant RLC circuits at the resonant frequency. At resonance, the reactive power of the inductor is equal to the reactive power of the capacitor, and the source supplies only the average power to the resistance of the circuit. Q-factor, Q_o, of the series-resonant cir-

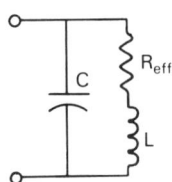

FIGURE 26-11
Parallel LC circuit is called a tank circuit.

cuit is defined as the ratio between the reactive power, P_q, of either the inductor or the capacitor, and the average power dissipated by the effective resistance of the RLC circuit at the resonant frequency. Thus

$$Q_o = \frac{P_q}{P} = \frac{I^2 X_L}{I^2 R_{eff}} = \frac{X_L}{R_{eff}} \quad \text{(no units)} \tag{26-8}$$

From Equation 26-8, it is learned that circuit sensitivity and selectivity, which are both described by the Q-factor, depend on how well the energy stored in the capacitor is transferred to the inductor, and vice versa. If the losses in the exchange of energy are nearly zero, the quality factor, Q_o, is very high, indicating a high circuit selectivity and sensitivity. Conversely, if the losses in the exchange of energy are high, the quality factor is low, indicating a low circuit selectivity and sensitivity.

26-3 PARALLEL-RESONANT CIRCUIT

The parallel LC circuit pictured in Figure 26-11 is called a *tank circuit* and is resonant at some frequency that depends on the value of the inductor, L, the capacitor, C, and the effective resistance of the inductor, R_{eff}. These circuit parameters are related to the parallel-resonant frequency, f_{ar}, by Equation 26-9. Thus

$$f_{ar} = \frac{1}{2\pi L} \sqrt{\frac{L}{C} - R_{eff}^2} \quad \text{(Hz)} \tag{26-9}$$

Equation 26-9 simplifies to Equation 26-10 when the L/C ratio is much greater than R_{eff}^2, which is the usual case in a well-designed resonant circuit. Thus

$$f_{ar} = \frac{1}{2\pi L} \sqrt{\frac{L}{C}} = \frac{1}{2\pi \sqrt{LC}} \quad \text{(Hz)} \tag{26-10}$$

For Equation 26-9, when the L/C ratio is much greater than R_{eff}^2, the Q-factor, Q_o, is greater than 10 ($Q_o > 10$) and $f_{ar} = f_o$. Thus

$$f_o = \frac{1}{2\pi \sqrt{LC}} \quad \text{(Hz)} \tag{26-11}$$

We see then that the same equation form is used to determine the resonant frequency in both the series- and parallel-resonant circuits.

Bandwidth of the Parallel-Resonant Circuit

The Q-factor, Q_o, is determined from the effective resistance of the resonant tank circuit, which, for all practical purposes, is the resistance resulting from the coil (ohmic, skin effect, and eddy current resistance). Because the coil is the source of nearly all $I^2 R$ losses in the parallel-resonant circuit, the same equation used to determine Q_o in the series resonant circuit may be applied to the Q-factor of the resonant tank circuit. Thus

$$Q_o = \frac{X_L}{R_{\text{eff}}} \quad \text{(no units)} \tag{26-12}$$

Like the series-resonant circuit, the parallel-resonant circuit has both a high sensitivity and selectivity when Q_o is large, and both a low sensitivity and selectivity when Q_o is small. The bandwidth of the parallel-resonant circuit is computed from the same equation form as the series-resonant circuit. Thus

$$\text{BW} = \frac{f_o}{Q_o} \quad \text{(Hz)} \tag{26-13}$$

EXAMPLE 26-8 For the circuit of Figure 26-11, determine the resonant frequency, f_o, the Q-factor, Q_o, and the bandwidth, BW, when $C = 100$ pF, $L = 330$ μH, and $R_{\text{eff}} = 40$ Ω.

SOLUTION Assume that $Q_o > 10$ and use Equation 26-11 to determine the resonant frequency.

$$f_o = \frac{1}{2\pi\sqrt{LC}} = \frac{1}{2\pi\sqrt{330 \times 10^{-6} \times 100 \times 10^{-12}}}$$

∴ $f_o = 876$ kHz

Determine the quality factor.

$$Q_o = \frac{X_L}{R_{\text{eff}}} = \frac{2\pi \times 876 \times 10^3 \times 330 \times 10^{-6}}{40}$$

∴ $Q_o = 45.4$

Determine the bandwidth.

$$\text{BW} = \frac{f_o}{Q_o} = \frac{876 \times 10^3}{45.4}$$

∴ BW = 19.3 kHz

Observation Had the original assumption that $Q_o > 10$ not been correct, the entire problem would be reworked starting with Equation 26-9 instead of Equation 26-11.

26-4 IMPEDANCE OF THE UNLOADED PARALLEL-RESONANT CIRCUIT

The parallel circuit of Figure 26-11 has an equivalent circuit impedance at the resonant frequency that is all resistive. The value of the impedance (resistance) at resonance depends on the Q of the coil, Q_o, which in turn results from the equation

Impedance of the Unloaded Parallel-Resonant Circuit

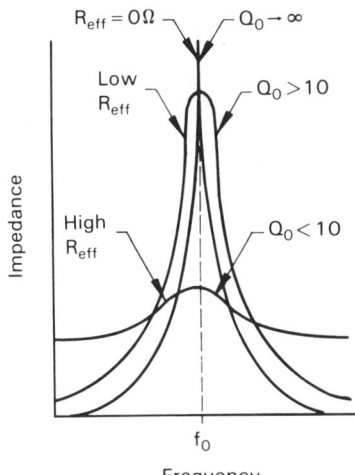

FIGURE 26-12
Impedance of a parallel resonant circuit is maximum at the resonant frequency. The value of the impedance depends on the effective resistance of the coil, R_{eff}.

$Q_o = X_L/R_{eff}$. Figure 26-12 pictures several curves of impedance as a function of frequency. From this figure, we may see that the circuit impedance is at its maximum value at the resonant frequency, and the impedance is a high value when Q_o is very large and a low value when Q_o is small. Thus, the impedance (equivalent resistance) of the resonant tank circuit is a direct function of the Q of the coil, Q_o. Because the capacitor is assumed to be perfect, Q_o is also the quality factor of the entire parallel-resonant circuit pictured in Figure 26-11.

Forming the Equivalent Circuit of the Parallel-Resonant Circuit

In the general two-branch parallel-resonant tank circuit of Figure 26-13(a), the impedance of the circuit (resistance) at the resonant frequency may be found by first forming the RL branch into its parallel equivalent circuit. As noted in Figure 26-13(b), this parallel equivalent circuit consists of a resistance, R_p, in parallel with an inductor, L_p, and the original circuit capacitor, C.

The technique for forming the parallel equivalent circuit was demonstrated in Section 21-9. The following steps, taken from that section, must be applied to the two-branch resonant circuit of Figure 26-13(a) in order to form the equivalent parallel-resonant circuit of Figure 26-13(b). Thus for the inductive branch of Figure 26-13(a),

$$R_{eff} + jX_L \quad \boxed{\rightarrow P} \quad Z/\underline{\theta}$$

and

$$Y = \frac{1}{Z/\underline{\theta}} \quad \boxed{\rightarrow R} \quad G - jB_L$$

from which

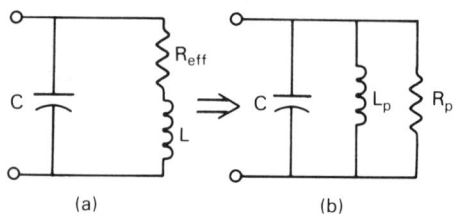

FIGURE 26–13 (a) Two-branch parallel resonant circuit with effective resistance of the inductor shown in series with the inductor.
(b) Equivalent parallel resonant circuit shown with L_p and R_p in place of the series RL branch.

$$R_p = \frac{1}{G} \quad \text{and} \quad X_{L_p} = \frac{1}{-jB_L}$$

With f_o known,

$$L_p = \frac{X_{L_p}}{2\pi f_o}$$

The preceding steps are used in the next example to form the equivalent circuit of the parallel-resonant circuit of Figure 26–14(a).

EXAMPLE 26–9 Determine the impedance of the parallel circuit pictured in Figure 26–14(a) at its resonant frequency of 1 MHz.

SOLUTION Convert the series RL branch of Figure 26–14(a) into the parallel equivalent circuit pictured in Figure 26–14(b) by applying the steps previously outlined. Thus

$$X_L = 2\pi f L = 2\pi \times 1 \times 10^6 \times 100 \times 10^{-6}$$

$$X_L = 628 \ \Omega$$

$$R_{\text{eff}} = 25 \ \Omega$$

From Figure 26–14(a), determine the branch impedance.

$$R + jX_L = 25 + j628 \ \boxed{\rightarrow P} \ 628.5 \underline{/87.72°} \ \Omega$$

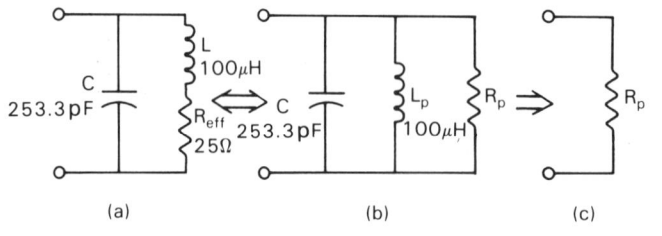

FIGURE 26–14
(a) Parallel circuit for Example 26–9.
(b) Equivalent circuit at resonance.
(c) Equivalent impedance at resonance.

Impedance of the Unloaded Parallel-Resonant Circuit

Determine the branch admittance.

$$Y = \frac{1}{Z\underline{/\theta}} = \frac{1}{628.5\underline{/87.72°}} = 1.59\underline{/-87.72°} \text{ mS}$$

Convert Y to rectangular form.

$$Y = 1.59\underline{/-87.72°} \text{ mS } \boxed{\rightarrow R} \;\; 63.3 - j1590 \; \mu S$$

Form the equivalent parallel circuit components.

$$R_p = \frac{1}{G} = \frac{1}{63.3 \times 10^{-6}} = 15.8 \text{ k}\Omega$$

$$X_{L_p} = \frac{1}{-jB_L} = \frac{1}{-j1590 \times 10^{-6}} = 629 \; \Omega$$

Compute the value of the equivalent inductor, L_p.

$$L_p = \frac{X_{L_p}}{2\pi f_o} = \frac{629}{2\pi \times 1 \times 10^6} = 100 \; \mu H$$

∴ The impedance and the equivalent circuit of the parallel circuit of Figure 26–14(a) are both a resistance of 15.8 kΩ, as pictured in Figure 26–14(c). Thus, $Z = R_p = 15.8$ kΩ at $f_o = 1$ MHz.

Observation From the example, it is learned that the equivalent inductance, L_p, is assumed to be equal to the LR branch inductance, L, when $Q_o > 10$. Thus, $L_p = L = 100 \; \mu H$ when $Q_o > 10$.

From Example 26–9, it is learned that the parallel-resonant circuit is equivalent to the resistance derived from converting the series effective resistance, R_{eff}, of the inductor into its parallel equivalent, R_p.

A less time-consuming approximation of the impedance of the parallel-resonant circuit may be made by applying one of the following equations to the two-branch parallel-resonant circuit. The only constraint on using these equations to approximate R_p is that Q_o be greater than 10.

$$R_p = Q_o X_L \quad (\Omega) \tag{26–14}$$

Since $Q_o = X_L/R_{\text{eff}}$,

$$R_p = \frac{X_L^2}{R_{\text{eff}}} \quad (\Omega) \tag{26–15}$$

However, $X_L = X_C$ at f_o. Thus

690
Resonance

$$R_p = \frac{X_L X_C}{R_{\text{eff}}} = \frac{2\pi fL \frac{1}{2\pi fC}}{R_{\text{eff}}} = \frac{L}{CR_{\text{eff}}} \quad (\Omega) \qquad (26\text{-}16)$$

EXAMPLE 26-10 Apply Equations 26-14 through 26-16 to the parallel-resonant circuit of Figure 26-14(a) to determine the resistance of the equivalent circuit, Figure 26-14(c), at the resonant frequency of 1.0 MHz. At resonance, $Q_o = 25.1$ and $X_L = 628.3 \, \Omega$.

SOLUTION From Equation 26-14,

$$R_p = Q_o X_L = (25.1)(628.3)$$

$$\therefore \quad R_p = 15.8 \text{ k}\Omega$$

From Equation 26-15,

$$R_p = \frac{X_L^2}{R_{\text{eff}}} = \frac{628.3^2}{25}$$

$$\therefore \quad R_p = 15.8 \text{ k}\Omega$$

From Equation 26-16,

$$R_p = \frac{L}{CR_{\text{eff}}} = \frac{100 \times 10^{-6}}{253.3 \times 10^{-12} \times 25}$$

$$\therefore \quad R_p = 15.8 \text{ k}\Omega$$

Observation From Example 26-9, R_p was computed to be 15.8 kΩ. From Example 26-10, R_p was also computed to be 15.8 kΩ. Thus, very good results are obtained using Equation 26-14, 26-15, or 26-16 when $Q_o > 10$.

26-5 LOADING THE PARALLEL-RESONANT CIRCUIT

Because of the large value of equivalent resistance exhibited by the *high-Q* parallel-resonant circuit, the parallel-resonant circuit is not *driven* by a voltage source. The voltage source, having low internal resistance, will *load* the parallel-resonant circuit, causing it to lose its resonant properties. Instead, the parallel-resonant circuit is connected to a current source, as pictured in Figure 26-15. The current source has a high internal impedance. Devices such as bipolar transistors and field-effect transistors are current-source devices.

The current, I, entering the circuit tank of Figure 26-16, provides current to R_p to overcome the heat losses (I^2R) in the resistive component. At resonance, the current in the capacitor,

FIGURE 26–15 (a) Current source is used to *drive* parallel resonant circuits. (b) Equivalent circuit resistance, R_p, is in parallel with the source internal resistance, R_{int}.

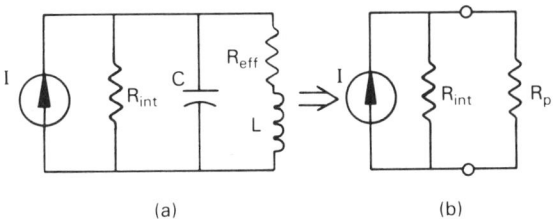

I_C, is equal to the current in the inductor, I_L, and is Q_o times greater than the current entering the tank circuit. Thus

$$I_C = I_L = Q_o I \quad (A) \tag{26–17}$$

From Equation 26–17, it is learned that there is a *Q-gain* in current in the parallel-resonant circuit. At resonance, equal reactive currents *circulate* back and forth between the L and C components in the parallel-resonant circuit. The current entering the resonant tank is, for all practical purposes, all resistive since it carries only energy to overcome losses in the effective resistance.

Loading the Tank Circuit

When the two-branch resonant circuit is connected to a current source, as pictured in Figure 26–15(a), the internal source resistance loads the resonant circuit causing the circuit Q-factor, Q_o, to decrease to a new value, Q_o'. Since Q_o is decreased in value, the bandwidth of the resonant circuit, BW, widens to a new value, BW'. The impedance of the loaded parallel-resonant circuit, as shown in Figure 26–15(b), is the parallel combination of R_{int} and R_p. Thus

$$R_p' = \frac{R_{int} R_p}{R_{int} + R_p} \quad (\Omega) \tag{26–18}$$

and

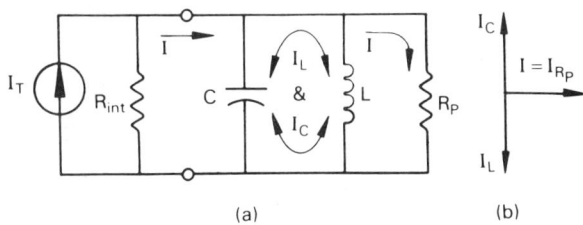

FIGURE 26–16 (a) Equivalent circuit of parallel resonant circuit connected to a current source.
(b) Current phasor for the resonant circuit. At resonance, $I_C = I_L$ which results in $I_C + I_L = 0$ and I, the current entering the tank circuit, equaling I_{R_p} ($I = I_{R_p}$).

$$R'_p = Q'_o X_L \quad (\Omega) \tag{26-19}$$

Solve for Q'_o:

$$Q'_o = \frac{R'_p}{X_L} \quad \text{(no units)} \tag{26-20}$$

and

$$BW' = \frac{f_o}{Q'_o} \quad \text{(Hz)} \tag{26-21}$$

The following examples demonstrate the concepts discussed in this section.

EXAMPLE 26–11 The resonant tank circuit ($Q_o > 10$) of Figure 26–15(a) is made up of an 84.4-pF capacitor and a 300-μH air-core inductor. The bandwidth of this two-branch circuit was measured before attaching it to the current source, and it was found to have a bandwidth of 8 kHz. Determine the bandwidth of the resonant circuit once the circuit is attached to the current source if the internal resistance is 100 kΩ.

SOLUTION First determine the Q of the unloaded circuit from the bandwidth equation, BW = f_o/Q_o.

$$Q_o = \frac{f_o}{BW}$$

Solve for f_o and substitute:

$$f_o = \frac{1}{2\pi\sqrt{LC}}$$

$$f_o = \frac{1}{2\pi\sqrt{300 \times 10^{-6} \times 84.4 \times 10^{-12}}}$$

$$f_o = 1.00 \text{ MHz}$$

$$Q_o = \frac{f_o}{BW} = \frac{1.00 \times 10^6}{8 \times 10^3} = 125$$

Determine the impedance of the tank circuit, R_p, of Figure 26–15(b).

$$R_p = Q_o X_L$$

$$R_p = (125)(2\pi \times 300 \times 10^{-6} \times 1 \times 10^6)$$

$$R_p = 236 \text{ k}\Omega$$

Using Equation 26–18, determine the impedance of the circuit with R_{int} connected.

Loading the Parallel-Resonant Circuit

$$R'_p = \frac{R_{int} R_p}{R_{int} + R_p}$$

$$R'_p = \frac{(100 \times 10^3)(236 \times 10^3)}{336 \times 10^3} = 70.2 \text{ k}\Omega$$

Solve for Q'_o, using Equation 26–20.

$$Q'_o = \frac{R'_p}{X_L} = \frac{70.2 \times 10^3}{2\pi \times 1 \times 10^6 \times 300 \times 10^{-6}} = 37.2$$

Determine BW', using Equation 26–21.

$$\text{BW}' = \frac{f_o}{Q'_o} = \frac{1 \times 10^6}{37.2} = 26.9 \text{ kHz}$$

∴ The loaded resonance circuit has a bandwidth of 27 kHz.

Observation With the connection of the current source to the resonant circuit, the Q of the resonant circuit diminished and the bandwidth was made wider. If the bandwidth of the resonant circuit is to remain relatively unaffected by the loading from the current source or other circuits placed in parallel with the resonant tank circuit, then R_{int} must be much greater than R_p.

EXAMPLE 26–12 The parallel-resonant circuit of Figure 26–17(a) is tested and found to have a resonant frequency of 10.7 MHz and a Q-factor of 80. The resonant tank circuit is connected to a current source. Determine the minimum internal resistance of the source if the circuit bandwidth is limited to 250 kHz.

SOLUTION From Figure 26–17(c) and (d), it is noted that R'_p is the result of R_p shunted by R_{int}. Thus, from Equation 26–18, solve for R_{int}.

$$R'_p = \frac{R_p R_{int}}{R_p + R_{int}}$$

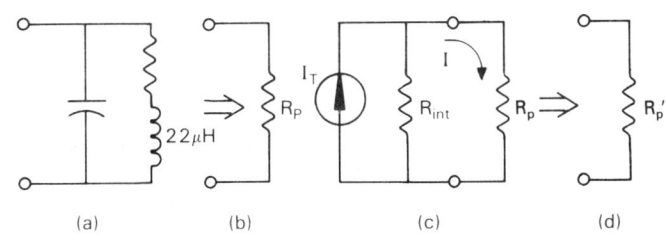

FIGURE 26–17 Circuit of Example 26–12.
(a) Unloaded tank circuit.
(b) Equivalent circuit at f_0.
(c) Resonant circuit is loaded by the internal resistance of the current source.
(d) Equivalent loaded circuit at f_0.

$$R_{\text{int}} = \frac{R_p R_p'}{R_p - R_p'}$$

Determine R_p and R_p' and substitute.

$$R_p = Q_o X_L$$
$$R_p = (80)(2\pi \times 10.7 \times 10^6 \times 22 \times 10^{-6})$$
$$R_p = 118 \text{ k}\Omega$$

$$R_p' = X_L Q_o' \quad \text{and} \quad Q_o' = \frac{f_o}{\text{BW}'}$$
where BW' = 250 kHz

$$Q_o' = \frac{10.7 \times 10^6}{250 \times 10^3} = 42.8$$

Compute R_p':

$$R_p' = Q_o' X_L$$
$$R_p' = (42.8)(2\pi \times 10.7 \times 10^6 \times 22 \times 10^{-6})$$
$$R_p' = 63.3 \text{ k}\Omega$$

Compute R_{int}:

$$R_{\text{int}} = \frac{R_p R_p'}{R_p - R_p'}$$
$$R_{\text{int}} = \frac{(118 \times 10^3)(63.3 \times 10^3)}{54.7 \times 10^3} = 137 \text{ k}\Omega$$

∴ The internal resistance of the current source must be at least 137 kΩ if the circuit bandwidth is no greater than 250 kHz.

EXAMPLE 26–13 Using the parameters of the previous problem, determine the current, I, in R_p of Figure 26–17(c) when the voltage across the resonant circuit is 8.0 V. Also, determine the reactive current circulating between the inductor and capacitor at resonance.

SOLUTION From Ohm's law:

$$I = \frac{V_p}{R_p} = \frac{8}{118 \times 10^3} = 67.8 \text{ }\mu\text{A}$$

∴ The resistive current in the tank circuit is 67.8 μA.

Determine the reactive currents, I_C and I_L, using Equation 26–17.

$$I_C = I_L = Q_o I = 80(67.8 \times 10^{-6}) = 5.4 \text{ mA}$$

∴ The reactive current of 5.4 mA is exchanged between C and L in the resonant tank circuit.

PROBLEMS

Section 26–2

26–1 A 22-mH inductor is connected in series with a 5000-pF capacitor. Determine the resonant frequency and the L/C ratio.

26–2 A 4.7-mH inductor is connected in series with a 0.022-μF capacitor. Determine the resonant frequency and the L/C ratio.

26–3 The voltage across the capacitor in a series-resonant circuit is 40 V at the resonant frequency of 600 kHz. Determine the voltage and frequencies (f_1 and f_2) at the half-power level of the output signal when the bandwidth is 12 kHz.

26–4 Using Figure 26–7 as a model, determine the voltage and frequencies at -12 dB for the circuit of Problem 26–3.

26–5 A series-resonant circuit, made up of a 63.3-pF capacitor and a 400-μH air-core inductor, has a measured bandwidth of 10 kHz. Determine the skin effect resistance when the dc (ohmic) resistance in the circuit is 5 Ω.

26–6 A series circuit consists of a 1.2-mH inductor and a 0.005-μF capacitor. Determine (a) the resonant frequency, and (b) the effective resistance, R_{eff}, of the resonant circuit when the bandwidth is 2 kHz.

26–7 Determine the output voltage, V_o, and the bandwidth, BW, of the RLC circuit of Figure 26–10(a) at resonance when E is changed to 8 V and R_{eff} to 15 Ω.

26–8 Repeat Problem 26–7 for the circuit of Figure 26–10(a) when E is changed to 5 V and R_{eff} to 22 Ω.

26–9 Determine (a) the value of the effective resistance, R_{eff}, in the circuit of Figure 26–10(b) when the circuit is resonant at 100 kHz, and (b) the value of the capacitance, C.

26–10 For the circuit of Figure 26–10(b), determine (a) the value of the capacitance, C, at the resonant frequency of 250 kHz, and (b) the value of the effective resistance, R_{eff}.

Section 26–3

26–11 Determine the resonant frequency of a 50-μH inductor and a 47-pF capacitor.

26–12 Determine the resonant frequency of a 10-mH inductor and a 1000-pF capacitor.

26–13 Determine the bandwidth of the resonant tank circuit of Problem 26–11 when $R_{eff} = 80$ Ω.

696

Resonance

26–14 Determine the bandwidth of the resonant tank circuit of Problem 26–12 when $R_{\text{eff}} = 20\ \Omega$.

26–15 A 180-μH inductor in shunt with a capacitor is resonant at 2.4 MHz. Determine the value of the capacitor shunted across the inductor.

26–16 Determine the effective resistance of the resonant tank circuit of Problem 26–15 when the measured bandwidth is 40 kHz.

Section 26–4

26–17 The circuit of Figure 26–13(a) is resonant at 500 kHz and has a bandwidth of 8 kHz. Approximate the equivalent impedance, R_p, of the network at resonance when $L = 220\ \mu$H.

26–18 For the circuit values given in Problem 26–17, determine (a) the effective resistance, (b) the value of the parallel capacitor, and (c) the -3-dB frequencies (f_1 and f_2) of the resonant response curve.

26–19 For the circuit of Figure 26–13(a), $C = 0.01\ \mu$F, $L = 100$ mH, and $R_{\text{eff}} = 60\ \Omega$. Approximate the equivalent impedance, R_p.

26–20 For the circuit values given in Problem 26–19, determine (a) the resonant frequency, (b) the bandwidth of the circuit, and (c) the -3-dB frequencies (f_1 and f_2) of the resonant response curve.

Section 26–5

26–21 For the circuit of Figure 26–18, determine the reactive current in the resonant tank circuit when $I_T = 600\ \mu$A and $R_{\text{int}} = 100$ kΩ.

26–22 For the circuit of Figure 26–18, determine the reactive current in the resonant tank circuit when $I_T = 1$ mA and $R_{\text{int}} = 40$ kΩ.

26–23 Using the circuit parameters of Problem 26–21, determine the bandwidth of the *loaded* resonant circuit.

26–24 Using the circuit parameters of Problem 26–22, determine the bandwidth of the *loaded* resonant circuit.

26–25 A resonant tank circuit consisting of a 1-mH air-core inductor shunted by a 330-pF capacitor was measured and

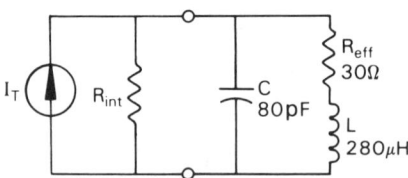

FIGURE 26–18
Circuit of Problems 26–21 and 26–22.

Problems

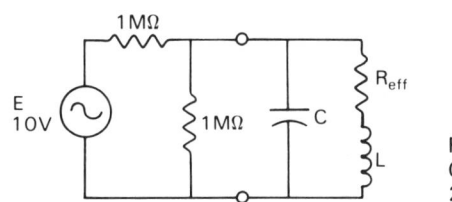

FIGURE 26–19
Circuit of Problem 26–27.

found to have a quality factor of 100. Determine the minimum internal resistance of a 1-mA current source that is connected to the parallel-resonant tank circuit if the circuit bandwidth is limited to 10 kHz.

26–26 Using the values determined in Problem 26–25, determine the voltage across the 330-pF capacitor with the 1-mA current source connected.

26–27 For the circuit of Figure 26–19, $C = 50$ pF. Determine the value of R_{eff} and L to produce a circuit bandwidth of 18 kHz at a resonant frequency of 1.6 MHz.

27 An Introduction to Three-Phase Power Systems

27–1 CHARACTERISTICS OF THREE-PHASE SYSTEM
27–2 THREE-PHASE POWER GENERATION
27–3 THREE-PHASE POWER TRANSMISSION AND DISTRIBUTION
27–4 CONFIGURATIONS AND PARAMETERS OF THE THREE-PHASE LOAD
27–5 POWER AND POWER FACTOR OF THE THREE-PHASE LOAD

CHAPTER PREVIEW

This chapter is a continuation of the discussion of power in Chapter 23. In this chapter, we will consider how three-phase power is generated, transmitted, and distributed to industrial, commercial, and residential users. The terms and terminology common to the power industry will be introduced along with the typical transformer and load configurations. Once again, power (average, reactive, and apparent) will be studied along with power factor.

PERFORMANCE OBJECTIVES

Once you have read each section, worked each example with pencil, paper, and calculator, and answered each problem of every section, you will be able to:

☐ Diagram Δ- and Y-connected loads.
☐ Compute line and phase voltages and currents.
☐ Understand the commonly used terminology of the power industry.
☐ Comprehend the fundamentals of power generation, power transmission, and line protection.
☐ Recognize many electrical symbols used in distribution and single-line schematics.

27–1 CHARACTERISTICS OF THREE-PHASE SYSTEM

The elementary generator pictured in Chapter 19 [Figure 19–1(a)] produces a single alternating voltage per cycle. This type of generated voltage is called *single-phase* voltage. When three coils, spaced 120° apart, are rotated in the elementary generator of Figure 27–1(a), three alternating voltages per cycle are produced [Figure 27–1(b)]. This type of generated voltage is called *three-phase voltage*, also noted as 3ϕ voltage.

Three-phase power systems are very important to the industrialized world because electrical power is generated, trans-

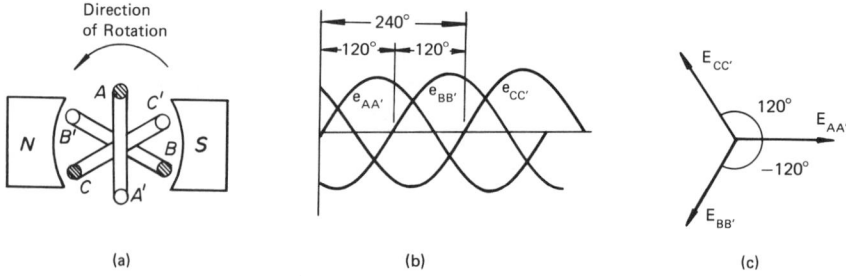

FIGURE 27–1
Three-phase voltage. (a) Three-phase generation with positive rotation noted (ccw rotation). (b) Three-phase periodic sine wave. (c) Phasor voltage diagram of three-phase voltage.

Characteristics of Three-Phase System

mitted, and distributed using three-phase technology. Three-phase power systems have many advantages over single-phase power systems. These advantages include:

- ☐ **Constant power to a balanced load.** Unlike single-phase power, which *pulsates*, three-phase power is constant, resulting in the development of an even torque in three-phase motors. Because of even torque, three-phase motors have little or no mechanical vibration. Figure 27–2(a) and (b) compares three-phase and single-phase load power. In Figure 27–2(a), the instantaneous power (p_t) is determined by adding the three instantaneous load powers at a given instant of time. Here we see that for every instant of time the instantaneous power is a constant. However, the single-phase power to a purely resistive load is zero twice in each power cycle. As shown in Figure 27–2(b), the power goes through zero four times per cycle when the load is reactive (power factor < 1). Not only is the power zero four times in the single-phase reactive power cycle, but the load even returns power to the source during the negative-power periods (p_{neg}).
- ☐ **A rotating magnetic field.** With the coil arrangement of the three-phase generator, the genrated voltages produce *rotating* magnetic fields that have constant flux densities. The rotating magnetic field makes the starting of three-phase motors possible.
- ☐ **Smaller size, less weight and cost.** Because the wiring supplying power to the load in a 3ϕ system is 25% smaller (by weight) than that needed for the same voltage and power in a single-phase system, a savings in wiring cost is realized with 3ϕ power systems. Too, a 3ϕ motor is usually considerably smaller in weight and size when compared to a single-phase motor with the same *nameplate* power rating. Again, because

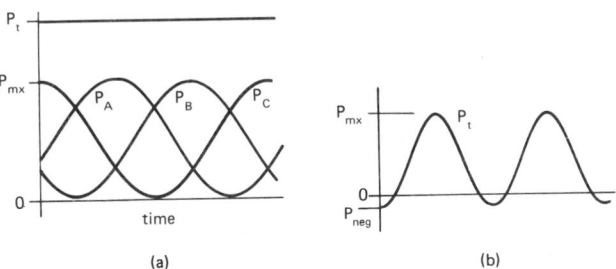

FIGURE 27–2
Instantaneous power. (a) Balanced three-phase load. (b) Single-phase load < 1 power factor. *Note:* A balanced three-phase load has the same power and power factor in each phase of the three loads.

of a savings in manufacturing cost, 3ϕ motors with the same power rating as single-phase motors are less expensive.

☐ **Smaller filter components in dc power supplies.** When 3ϕ power is rectified and filtered for dc power, much smaller values of capacitors and inductors may be used than are used with single-phase power supplies. This reduction in size results in a considerable saving in cost and space.

27–2 THREE-PHASE POWER GENERATION

Hydroelectric Generation

In the *energy storage dam,* such as Hoover Dam on the Colorado River, a very large volume of water ($>28 \times 10^6$ acre-feet) is stored behind the dam. This stored water represents a huge reservoir of potential energy, which is converted to electrical energy when the water from the giant reservoir behind the dam is fed into the turbines in the power house. Figure 27–3(a) pictures an energy storage dam.

As shown in Figure 27–3(b), the *run-of-river dams* (typically 80 feet high) use giant *draft tubes* to direct the water on the upriver side into the turbine in the power house.

In each of the two types of hydroelectric generating facilities, the turbine is turned by the force of the falling water. The turbine, connected to the generator by a shaft, spins a powerful dc electromagnet inside the coils of wire of the three-phase generator.

Figure 27–4 pictures the *rotor* and *stator,* the two main parts of the three-phase synchronous generator (commonly called an

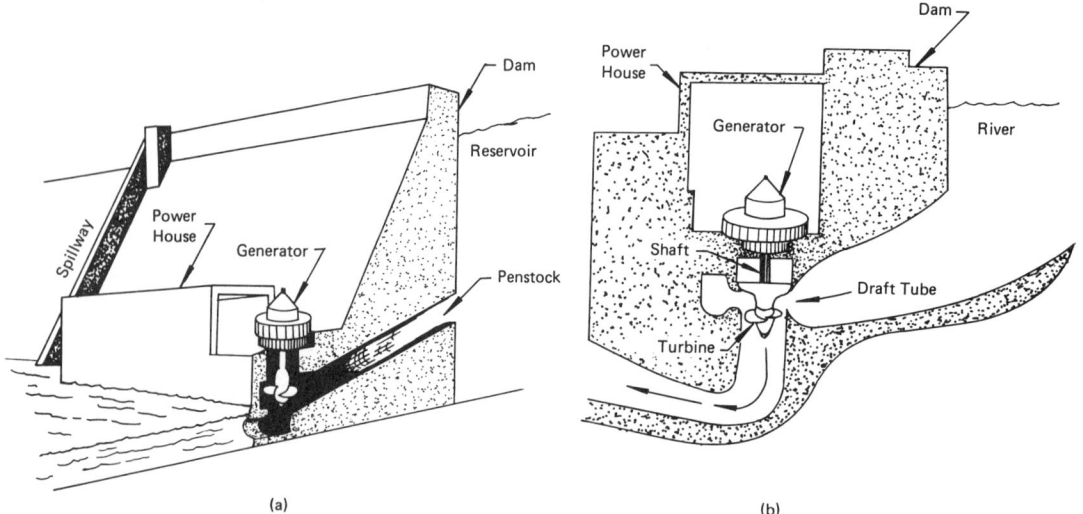

FIGURE 27–3
Hydroelectric generation. (a) Energy storage dam. (b) Run-of-river dam.

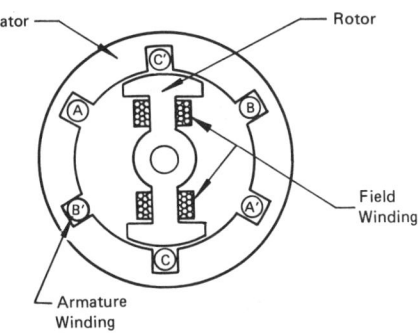

FIGURE 27-4
Elementary two-pole three-phase synchronous generator (also called an alternator).

alternator). The armature windings or coils are placed in grooves in the cylindrically shaped, laminated, low-loss steel structure making up the stator, while the field windings or coils are wound on the rotor. The field coil is *excited* by a dc source that produces a strong dc magnetic field. The rotor, with the field coil attached, is turned at a constant speed past the stationary armature coils. Voltage is induced in each stator coil, producing a three-phase sinusoidal voltage. Referring back to Figure 27-1(b), you will see that each voltage wave is noted with a subscript corresponding to the stator coils of Figure 27-4.

The frequency of the generated voltage depends on the number of pairs of poles in the rotor and the rotational speed of the rotor. Equation 27-1 relates these parameters.

$$f = \frac{SP}{60} \tag{27-1}$$

where f = frequency in hertz (Hz)
S = speed of rotation in revolutions per minute (rev/min)
P = number of pairs of poles
60 = constant to convert minutes to seconds

EXAMPLE 27-1 Because of the characteristics of the turbine used in hydroelectric generation, the turbine must be turned at a low speed. Determine the number of poles in the rotor of a 60-Hz generator if the turbine turns 80 rev/min.

SOLUTION Since the rotor is directly connected to the turbine by the turbine shaft, the alternator rotor is turning at 80 rev/min. Thus, from Equation 27-1,

$$P = \frac{60f}{S} = \frac{60(60)}{80} = 45 \text{ pairs}$$

∴ The rotor consists of 45 pairs of poles, or 90 poles.

As noted in the previous example, large numbers of poles are used in hydroelectric generators to produce the desired 60-Hz line frequency. This is due to the low rotation speed of hydro-turbines. The poles of the rotor pictured in Figure 27–5(a) are called *salient* poles. Salient pole design is used in the rotors of hydroelectric generators (80 to 180 rev/min). However, the salient pole rotor is not used in steam-turbine-driven electric generators. At the high speeds (1800 to 3600 rev/min) needed for optimum performance of this type of turbogenerator, the salient pole rotors experience destructive forces on the outer protruding parts of the structure. Instead, generators driven by high-speed turbines (1800 to 3600 rpm) use *cylindrical* two- or four-pole rotors as pictured in Figure 27–5(b).

Thermoelectric Generation

Thermoelectric generating facilities are classified by the fuel used to heat water in a boiler to produce steam for the turbogenerator. Typical fuels include coal, oil, neutral gas, and uranium (nuclear energy). Of these fuels, coal generates about half of the United States' electricity and nuclear energy about one-sixth.

Figure 27–6 diagrams a coal-fired boiler-turbogeneration facility. Some of the nation's most efficient coal-fueled generating facilities are also the biggest in the world. One such facility produces 1300 MW of electric power by burning about 3.5 million tons of coal a year.

In the thermoelectric generation facility, the temperature and pressure of the steam from the boiler determine the output capacity of the electric generator. The higher the velocity of the steam passing through the turbine, the greater is the force. To get the most efficiency out of the energy contained in the steam, a number of successive turbine stages are used. Each stage is connected to the same shaft, which is connected to the electric generator. Steam at a very high pressure (1500 to 3700 psi) enters the high-pressure end of the turbine where the turbine blades are mounted in small-diameter wheels. As the pressure drops in the medium-pressure section of the turbine, the blades are mounted in proportionally larger wheels. Finally, at the low-

FIGURE 27–5
Rotor design. (a) Salient two-pole rotor. (b) Cylindrical two-pole rotor.

(a)   (b)

Three-Phase Power Transmission and Distribution

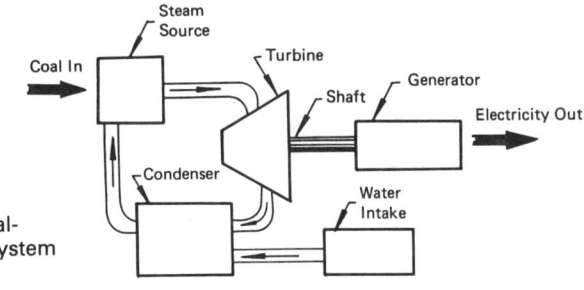

FIGURE 27-6
Thermal electric coal-fueled generating system diagram.

pressure end of the turbine, the blades reach their maximum diameter. The turbine turns at a constant speed.

Alternate and Renewable Energy Sources

The future of electrical energy generation may partially lie in one of several renewable generation technologies, especially those based on photovoltaic and solar-thermal principles. The reasons utilities are attracted to these technologies include limited capital risk, short start-up time, and environmental safety. Rather than investing huge amounts of capital and time in the construction of traditional fossil fuel, hydroelectric, or nuclear generating facilities, an established utility can quickly add to its base geneating capacity in smaller, less expensive increments.

Utilities are now required by law (the Public Utilities Regulatory Policy Act) to interconnect with and to buy power from producers of less than 80-MW capacity. At this time, power produced by renewable solar generating technologies is not economically competitive with the traditional methods of power generation. However, advances in these technologies promise to make them economically viable by the mid 1990s.

27-3 THREE-PHASE POWER TRANSMISSION AND DISTRIBUTION

Introduction

The three coils of the three-phase alternator may be connected in several ways to produce power. In Figure 27-7(a), the three armature coils are brought out as three separate single-phase circuits. This type of connection is very expensive since the transmission line connected to the coils would have six conductors.

To lower the number of conductors in the transmission line to three, the end of each coil is connected together, as shown in Figure 27-7(b), to form a *Y-connected generator*. The common connection point of the three coils is called the *neutral*, and the three conductors connected to A, B, and C are called *lines*. In the Y-connected generator, the voltage from line to line (V_{AB}, V_{BC}, V_{CA}) is called the *line voltage*, while the voltage across the coils from line to neutral ($V_{AA'}$, $V_{BB'}$, $V_{CC'}$) is called the *phase voltage*.

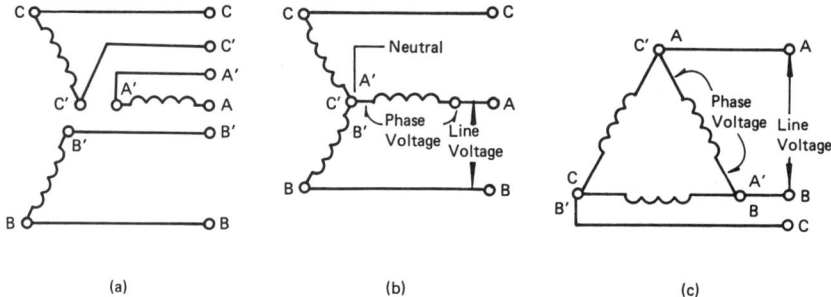

FIGURE 27-7
Three-phase generator connections. (a) The single-phase connection of the generator requires two conductors per phase (six conductors total) in the transmission line. (b) With a balanced load, the Y-connected alternator requires one conductor per phase (three conductors total) in the transmission line. (c) The Δ-connected alternator requires one conductor per phase (three conductors total) in the transmission line.

By connecting the beginning of one coil to the end of another in sequential order (coil A to coil B, etc.), a *delta-connected generator* (Δ-connected) is formed. The Δ-connected generator of Figure 27-7(c) has a phase sequence of ABC; that is, the voltage of coil AA' reaches its positive peak value first, followed by coil BB' and then coil CC'. In the single-phase connection of Figure 27-7(a), the phase sequence is of no consequence, but in the three-phase Δ- and Y-connections the phase sequence determines the direction of rotation of three-phase motors.

In the Δ-connected generator, the line and phase voltages are equal. Thus, for the Δ-connected generator of Figure 27-7(c), the line voltage (V_{AB}, V_{BC}, V_{CA}) is equal to the phase voltage ($V_{AA'}$, $V_{BB'}$, $V_{CC'}$).

To summarize, the line and phase voltages of a Y-connected generator (also called a *star connection*) are not equal. However, the line and phase voltages of a Δ-connected generator are equal. Also, as indicated in Figure 27-7(b) and (c), a three-phase balanced circuit requires three lines (one for each phase in the three-phase circuit).

Transmission Line The electric power from generating facilities is transported to receiving facilities over extra-high-voltage (EHV) transmission lines. Figure 27-8 pictures several types of transmission line towers that are used to support the overhead EHV transmission lines. Transmission line voltages have historically increased to progressively higher values due to economic pressures. Higher transmission voltage results in both lower line losses per unit of transmitted power and higher volume of transmitted power per unit width of power line right-of-way.

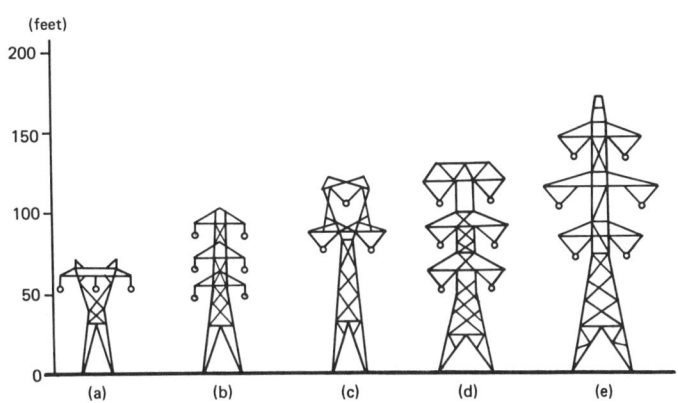

FIGURE 27-8
Several types of transmission line towers. (a) Single-circuit, *flat*, 230-kV tower. (b) Double-circuit, *stack*, 230-kV tower. (c) Single-circuit, *delta*, 500-kV tower. (d) Double-circuit, *stack*, 345-kV tower. (e) Double-circuit, *stack*, 500-kV tower.

Table 27-1 summarizes the growth in EHV power transmission. Notice the dramatic increase in power capacity that results from higher line voltage. The way of the future appears to be in ultrahigh-voltage (UHV) circuits. One 1.2-MV UHV circuit has the capacity of seven EHV 500-kV circuits. Besides moving huge amounts of power, an ultrahigh-voltage transmission line uses proportionately less land for right-of-way. It is projected that one 200-ft-wide right-of-way for a single-circuit, 8000-MW, 1.2-MV transmission line could replace seven single-circuit 1200-MW, 500-kV transmission lines for a saving of 850 ft in right-of-way width. In addition to savings in land, UHV transmission systems have an approximate 2% increase in efficiency

TABLE 27-1 EHV–UHV POWER TRANSMISSION SYSTEMS

Line Voltage (kV ac)	*Capacity/3ϕ Circuit (MW)*	*Right-of-Way Width (ft)*	*Put in Service*
220	100	80	1920s
287	120	120	1930s
345	320	140	1950s
500	1200	150	1960s
800 (dc) (±400 kV dc)	1400	135	1970s
1200 (experimental)	8000	200	1980s

due to the lowering of the $I^2 R$ losses, a significant savings in power.

As the transmission line voltages increased, the configuration of the line changed from a single conductor per phase, constructed as shown in Figure 27–9(a), to two or more conductors spaced a few inches apart, as noted in Figure 27–9(b). When two or more conductors are used per phase to make up the line, the line is called a *bundle-conductor line*. Figure 27–9(c) pictures several designs for a bundle-conductor line. These designs are used to lower line losses due to the *corona effect* associated with EHV-UHV conductors. The corona effect is an electric discharge caused by the breaking down of the air surrounding the power line. By reducing the corona effect, power loss is lowered and radio-frequency interference (RFI) is diminished.

Intertie The interconnection of several large utilities to share power and improve system reliability has been an important trend in modern power systems. An example of this is in the 850-mile-long Pacific intertie that interconnects the hydroelectric generation facilities of northwest United States with the thermoelectric facilities in the Southwest.

Without the intertie, thermal plants would need to keep some generating units on *spinning reserve* (burning fuel to turn unloaded generators) so that they may be switched on and off to match the peak demands of the system. This is a very costly and wasteful solution to the problem of load management. However, with access to the intertie, less expensive power is taken from the hydro plants to meet the peak demand periods.

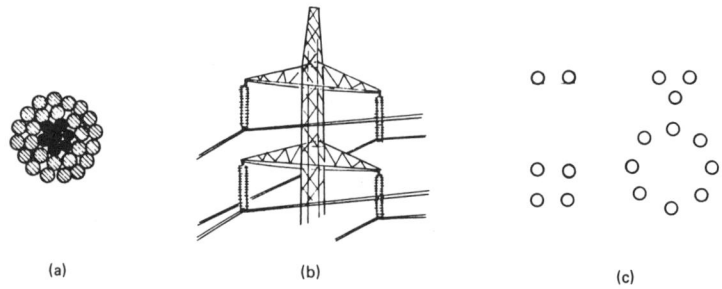

FIGURE 27–9
(a) Cross section of a transmission line conductor designed with aluminum strands wrapped on a steel core designated ACSR (aluminum conductor steel reinforced). (b) A 230-kV line with two closely spaced conductors per phase forming a *bundle-conductor* line. (c) Cross-sectional configuration of bundle-conductor lines used to reduce corona effects associated with EHV and UHV transmission lines.

Unlike hydro plants that can quickly meet the increase in demand by increasing their power output by simply running more water through their turbines, thermal generation cannot be throttled up and down as simply or cost effectively. Thermoelectric generating systems operate most efficiently when they generate at a uniform level.

Figure 27-10 shows how power demand is met by a combination of hydro generation for the peaks and thermal generation for the base load. At night, during off-peak times, one of two options is available through the intertie. Because thermal generation capacity exceeds the demand, the excess power may be sent over the intertie so that hydro generation may be reduced and water saved in the reservoirs of the energy-saving dams. Or if water is plentiful (spring and early summer), thermal generation is cut back and inexpensive hydropower is used to to meet the night off-peak power demand.

To manage their load and level out the summer demand curve, some utilities have incentive programs to encourage customers to turn off air conditioners and water heaters during the hours of peak power demand.

Power Transformer

The power transformer plays a major role in many parts of the three-phase power system. It is used to interface the synchronous generator at the generating site to the transmission line. In this application, the transformer is used to lower the transmission line loss (I^2R) and raise the capacity by stepping up the moderate voltage (13.8 kV) of the alternator to high (69 to 138 kV), extra-high (220 to 700 kV), or ultra-high voltge (1200 kV).

At the receiving station side of the transmission line, as shown in Figure 27-11, the voltage is stepped down and sent on

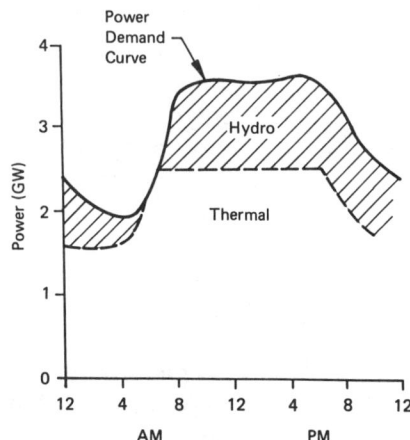

FIGURE 27-10
A power demand curve showing the mix of thermal-generated power and hydro-generated power.

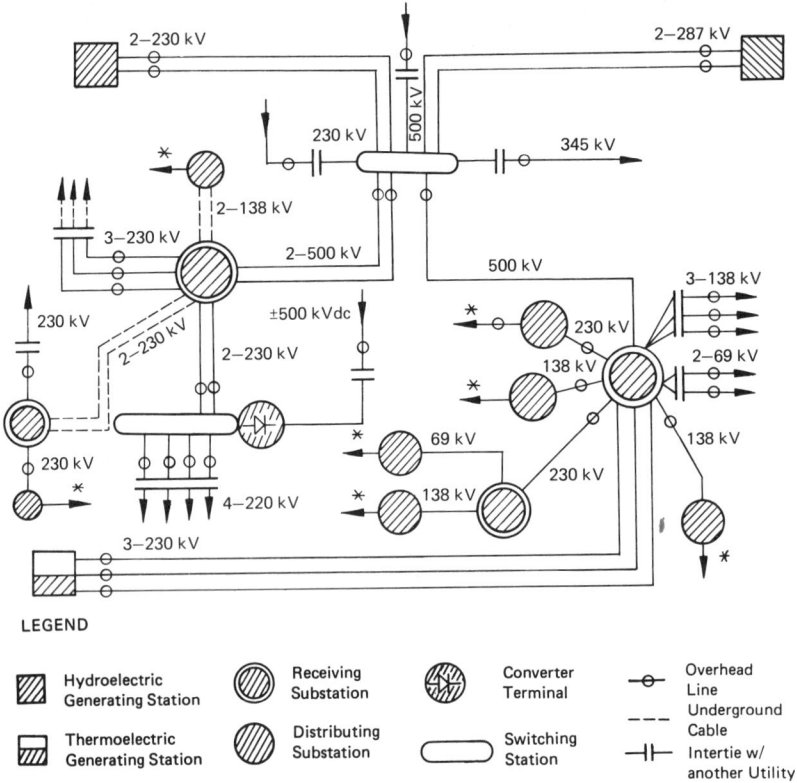

FIGURE 27-11
Typical three-phase power system diagram. *Any number of lines to consumers—34.5-kV industrial users, 4.16-kV commercial/residential users.

to the distributing stations, where once again the voltage is stepped down and distributed to the consumer.

The three-phase transformer may be a single integrated unit or it may be constructed from individual single-phase transformers connected to form the desired delta (Δ) or wye (Y) three-phase transformer bank. Figure 27–12 pictures five different configurations of the Δ- and Y-connection used to make up three-phase transformers.

In the Δ-Δ connection pictured in Figure 27–12, both primary and secondary windings are connected in delta. Additional connections pictured are Δ-Y, Y-Δ, Y-Y, and *open delta* (V-V). In the Y-connection, the primary and/or secondary winding may be grounded at the neutral to form a grounded wye connection. Also pictured in Figure 27–12 is an open-delta or V-connection. The power available in the V-connection is 57.7% of the power of the closed-delta connection.

Three-Phase Power Transmission and Distribution

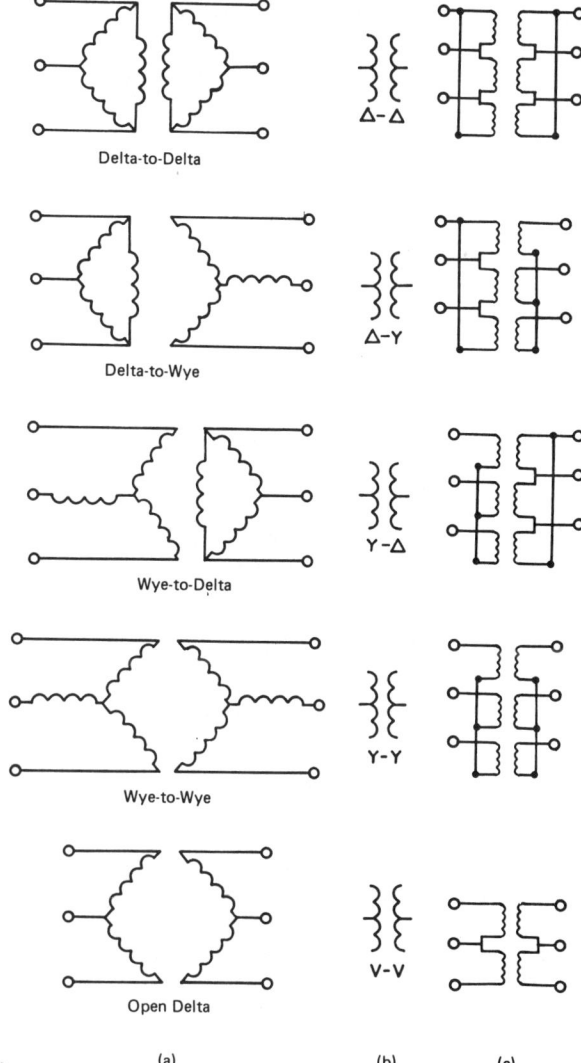

FIGURE 27-12
Three-phase transformer connections.

The V-connection is used in areas of a city where future growth is expected. By initially installing only two transformers, the utility company saves the cost of the third transformer. Once the demand for power increases, the third transformer is installed. This brings the capacity of the system up from 57.7% to 100%. The V-connection is used to advantage when one of the three transformers in a Δ-Δ bank is damaged or is removed from service for maintenance. The remaining two transformers may be V-connected, thereby restoring some power to operate vital services, such as hospitals and traffic signals.

Power System Protection

The purpose of the protection system is to continually monitor the power system parameters (current, voltage, frequency, etc.). When a fault (an out-of-tolerance parameter) is detected, a control signal is generated within the protection system, which is used to isolate the faulted section of the power system.

The technique used in electric system protection is to divide the system into overlapping zones. The object of system protection is to quickly identify a faulted component in the system and to disconnect the zone containing that component, thereby preventing interruption of power to the entire system and/or damage to the operating equipment.

Faults in three-phase systems may occur in any part of the system, including the generators, transformers, transmission lines, underground cables, switching stations, and receiving stations. A fault condition includes such events as two power lines coming in contact with one another, the insulation within a generator or transformer breaking down, or one or more phases in the power system shorting to ground.

Figure 27–13 is a single-line schematic of part of a power system with some of the protection zones noted. Notice how the zones are overlapped to give redundancy to the protection scheme. Special *protection relays* are used to measure and detect fault conditions. When a fault is detected, the protective relay closes, which in turn signals the zone *circuit breakers* to open and disconnect the faulted section of the system.

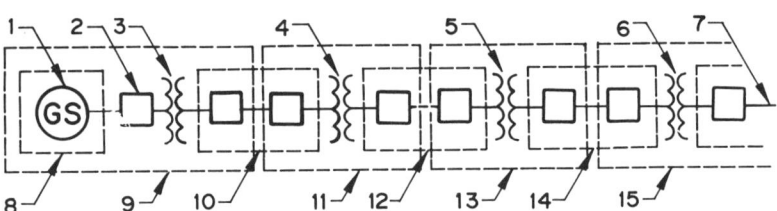

DEVICE
1. Synchronous Generator
2. Circuit Breaker
3. Generator to Line Step-up Transformer 13.8–500 kV
4. Receiving Station Step-down Transformer 500–138 kV
5. Distribution Station Step-down Transformer 138–34.5 kV
6. Industrial Sub-station Step-down Transformer 34.5–4.16 kV
7. To in Plant Step-down Transformer 4160–480 V

PROTECTION ZONE
8. Generator Zone
9. Transformer-Generator Zone
10. Transmission Line Zone
11. Receiving Station Zone
12. Cable Zone
13. Distribution Station Zone
14. Distribution Line Zone
15. Industrial Plant Sub-station Zone

FIGURE 27–13
Generalized zone protection system.

Some types of protective relays have provisions for adjusting the time of operation between the detection of a fault and the signaling of the fault. This time-delay feature permits the overlapping of the zones in the protection system. Without time delay, the detection of a fault would cause all the system circuit breakers to be simultaneously activated. But with time delay, relays may be set up in sequential order so that only the faulted zone is disconnected while the power system in the overlap zone remains operational. The power system in the overlap zone is only disconnected if the relays in the faulted zone fail to act.

27–4 CONFIGURATIONS AND PARAMETERS OF THE THREE-PHASE LOAD

Introduction

In this section we will consider three load configurations that are common to three-phase systems. They are the Δ-connected load, the four-wire Y-connected load, and the three-wire Y-connected load. Each of these load connections may either be balanced or unbalanced and each may be sourced from a Δ- or Y-connected transformer secondary. The analysis of these three-phase load configurations is made using the concepts previously developed for single-phase circuits, along with a new concept of Y- to Δ-transformation.

As was previously mentioned, the load may be balanced or unbalanced. The three-phase induction motor pictured in Figure 27–14 has equal phase impedances and is considered a balanced three-phase load. However, the Δ- or Y-connected three-phase load may be made up of single phase loads as shown in Figure 27–15. These loads might consist of lamps (incandescent and fluorescent), single-phase motors, heating elements, and household appliances. Since each leg of the load does not require the same power, the line currents are not equal and the load is unbalanced.

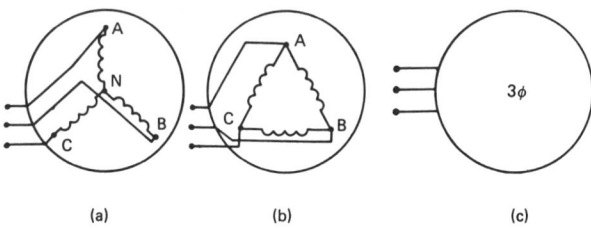

FIGURE 27–14
Three-phase motor connection. (a) Y-connected three-phase motor. (b) Δ-connected three-phase motor. (c) Schematic symbol for three-phase motor.

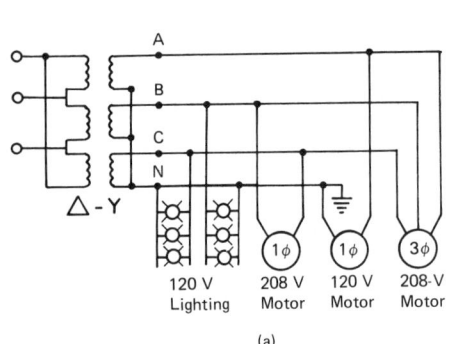
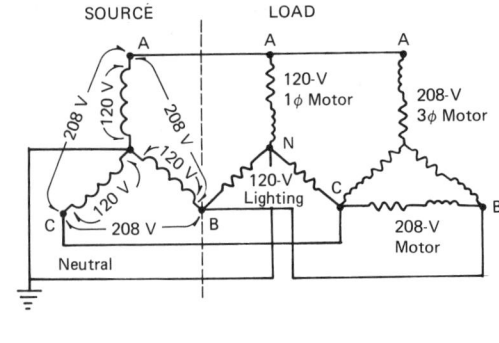

FIGURE 27–15
Unbalance load. (a) A delta to wye step-down transformer providing 208-V three-phase (line to line to line), 208-V (line to line), and 120-V single-phase (line to neutral). (b) Schematic of load circuit.

Delta-Connected Load

The Δ-connected load may be balanced or unbalanced, and it may be sourced from either a Δ- or Y-connected transformer secondary. The following facts about the Δ-connected load will be applied in the examples that follow.

☐ For either the balanced or unbalanced Δ-connected load, the voltage across each leg of the load is the load phase voltage. The load phase voltage is equal to its respective source line voltage.

$$E_{line} = V_{phase} \tag{27-2}$$

☐ In the balanced Δ-connected load, the magnitude of the line current is $\sqrt{3}$ times larger than the phase current in each leg of the load.

$$|I_{line}| = \sqrt{3}\,|I_{phase}| \tag{27-3}$$

☐ In the balanced Δ-connected load, the phase current leads its respective line current by 30°.
☐ In either the balanced or unbalanced Δ-connected load, the sum of the line currents is zero.

$$I_A + I_B + I_C = 0 \tag{27-4}$$

☐ In the unbalanced Δ-connected load, the line currents are unequal.

$$I_A \neq I_B \neq I_C \tag{27-5}$$

EXAMPLE 27–2 For the balanced Δ-connected load of Figure 27–16,

$$Z_{AB} = Z_{BC} = Z_{CA} = 10 + j24 = 26\underline{/67.38°}\ \Omega$$

714

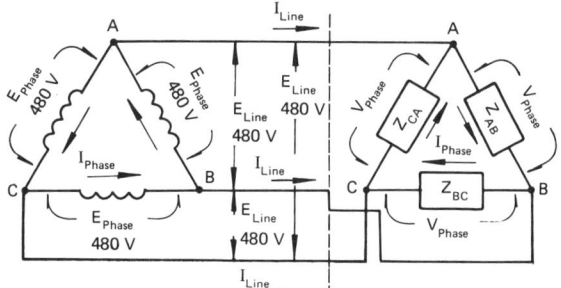

FIGURE 27-16
Circuit for Example 27-2, where $Z_{AB} = Z_{BC} = Z_{CA}$.

Determine (a) the line and phase voltage, (b) the line and phase current, and (c) that the sum of the line currents is zero.

SOLUTION (a) The magnitude of the phase voltage of the load is equal to the line voltage of the source. From Equation 27-2:

$$|E_{line}| = |V_{phase}| = 480 \text{ V}$$

Observation From Figure 27-1(c), we learn that each of the phase voltages is 120° apart. Thus, for the load phase voltage:

$$V_{AB} = 480 \underline{/0°} \text{ V}$$

$$V_{BC} = 480 \underline{/-120°} \text{ V}$$

$$V_{CA} = 480 \underline{/120°} \text{ V}$$

Since $|E_{line}| = |V_{phase}|$, the line voltages are therefore:

$$E_{AB} = 480 \underline{/0°} \text{ V}$$

$$E_{BC} = 480 \underline{/-120°} \text{ V}$$

$$E_{CA} = 480 \underline{/120°} \text{ V}$$

SOLUTION (b) The phase currents are the currents in each leg of the Δ-connected load. By Ohm's law, the phase currents in the load of Figure 27-16 are therefore:

$$I_{AB} = \frac{V_{AB}}{Z_{AB}} = \frac{480 \underline{/0°}}{26 \underline{/67.38°}} = 18.46 \underline{/-67.38°} \text{ A}$$

$$I_{BC} = \frac{V_{BC}}{Z_{BC}} = \frac{480 \underline{/-120°}}{26 \underline{/67.38°}} = 18.46 \underline{/-187.4°} \text{ A}$$

$$I_{CA} = \frac{V_{CA}}{Z_{CA}} = \frac{480 \underline{/120°}}{26 \underline{/67.38°}} = 18.46 \underline{/52.62°} \text{ A}$$

Observation By inspecting the nodes A, B, and C of the load, we learn that each line current entering the load is the resultant of two phase currents. Also recall that currents entering a node are assumed positive, and currents leaving are assumed negative. By KCL, the line currents are as follows:

Solving for I_A (current in line A) at node A:

$$I_A + I_{CA} - I_{AB} = 0$$

$$I_A = I_{AB} - I_{CA}$$

$$I_A = 18.46\,\underline{/-67.38°} - 18.46\,\underline{/52.62°}$$

$$I_A = (7.100 - j17.04) - (11.21 + j14.67)$$

$$I_A = -4.110 - j31.71 = 31.98\,\underline{/-97.39°}\text{ A}$$

∴ The current in line A, I_A, is $31.98\,\underline{/-97.39°}$ A.

Solving for I_B (current in line B) at node B:

$$I_B + I_{AB} - I_{BC} = 0$$

$$I_B = I_{BC} - I_{AB}$$

$$I_B = 18.46\,\underline{/-187.4°} - 18.46\,\underline{/-67.38°}$$

$$I_B = (-18.31 + j2.378) - (7.100 - j17.04)$$

$$I_B = -25.41 + j19.42 = 31.98\,\underline{/142.6°}\text{ A}$$

∴ The current in line B, I_B, is $31.98\,\underline{/142.6°}$ A.

Solving for I_C (current in line C) at node C:

$$I_C + I_{BC} - I_{CA} = 0$$

$$I_C = I_{CA} - I_{BC}$$

$$I_C = 18.46\,\underline{/52.62°} - 18.46\,\underline{/-187.4°}$$

$$I_C = (11.21 + j14.67) - (-18.31 + j2.378)$$

$$I_C = 29.52 + j12.29 = 31.98\,\underline{/22.60°}\text{ A}$$

∴ The current in line C, I_C, is $31.98\,\underline{/22.60°}$ A.

Observation The following facts for the Δ-connected load have been verified.

1. The magnitudes of the line and phase currents in the **balanced** Δ-connected load are related by the $\sqrt{3}$. That is, $|I_{\text{line}}| = \sqrt{3}|I_{\text{phase}}|$ or, for this example, $31.98 = \sqrt{3}\,18.46$ A.
2. For the **balanced** Δ-connected load, the phase current leads its respective line current by 30°.

Configurations and Parameters of the Three-Phase Load

For this example, the phase current I_{AB} $\underline{/-67.38°}$ leads its respective line I_A $\underline{/-97.39°}$ by 30°.

3. In the **balanced** Δ-connected load, the magnitudes of the line currents are equal, and each is displaced from the other by 120°. The magnitudes of the phase currents are also equal, and each phase current is displaced from the other by 120°.

SOLUTION (c) Add the line currents to check the current calculations.

$$I_A + I_B + I_C = 0$$

$$(-4.110 - j31.71) + (-25.41 + j19.42) +$$

$$(29.52 + j12.29) = 0.0 + j0.0 \text{ A}$$

∴ All the current calculations are correct.

In the previous example, the characteristics and parameters of the balanced Δ-connected load were noted. In the next example, the characteristics and parameters of the unbalanced Δ-connected load will be investigated.

EXAMPLE 27–3 Determine the line and phase currents in the unbalanced Δ-connected load of Figure 27–17, where $f = 60$ Hz and $\omega = 377$ rad/s.

SOLUTION Since the line and phase voltages are equal in a Δ-connected circuit, then by Ohm's law the phase currents are:

$$I_{AB} = \frac{V_{AB}}{Z_{AB}}, \quad \text{where } Z_{AB} = 50 - j120.6 \text{ }\Omega$$

$$I_{AB} = \frac{240\underline{/0°}}{130.6\underline{/-67.48°}} = 1.838\underline{/67.48°} \text{ A}$$

∴ $I_{AB} = 1.838\underline{/67.48°}$ A

$$I_{BC} = \frac{V_{BC}}{Z_{BC}}, \quad \text{where } Z_{BC} = 20 + j0 \text{ }\Omega$$

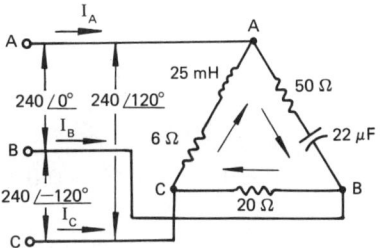

FIGURE 27–17
Circuit for Example 27–3.

$$I_{BC} = \frac{240/-120°}{20/0°} = 12/-120°$$

$\therefore \quad I_{BC} = 12/-120°$ A

$$I_{CA} = \frac{V_{CA}}{Z_{CA}}, \quad \text{where } Z_{CA} = 6 + j9.425 \text{ A}$$

$$I_{CA} = \frac{240/120°}{11.17/57.52°} = 21.49/62.48°$$

$\therefore \quad I_{CA} = 21.49/62.48°$ A

Apply KCL to determine the line current, I_A.

$I_A + I_{CA} - I_{AB} = 0$

$I_A = I_{AB} - I_{CA}$

$I_A = 1.838/67.48° - 21.49/62.48°$

$I_A = (0.704 + j1.698) - (9.930 + j19.06)$

$I_A = -9.226 - j17.36 = 19.65/-118°$

$\therefore \quad I_A = 19.66/-118°$ A

Apply KCL to determine the line current I_B.

$I_B + I_{AB} - I_{BC} = 0$

$I_B = I_{BC} - I_{AB}$

$I_B = 12/-120° - 1.838/67.48°$

$I_B = (-6.000 - j10.39) - (0.704 + j1.698)$

$I_B = -6.704 - j12.09 = 13.82/-119°$

$\therefore \quad I_B = 13.82/-119°$ A

Apply KCL to determine the line current I_C.

$I_C + I_{BC} - I_{CA} = 0$

$I_C = I_{CA} - I_{BC}$

$I_C = 21.49/62.48° - 12/-120°$

$I_C = (9.930 + j19.06) - (-6.000 - j10.39)$

$I_C = 15.93 + j29.45 = 33.48/61.59°$

$\therefore \quad I_C = 33.48/61.59°$ A

Check $\quad I_A + I_B + I_C = 0$

$(-9.226 - j17.36) + (-6.704 - j12.09) +$

$(15.93 + j29.45) = 0.0 + j0.0$

Observation
1. In the **unbalanced** Δ-connected load, both the line currents and the phase currents are unequal.
2. In any balanced or unbalanced **three-wire** connected load, the sum of the line currents is zero.

Four-wire Y-connected Load

The four-wire Y-connected load may be balanced or unbalanced, and it may be sourced from either a Δ- or Y-connected transformer secondary. The following facts about the Y-connected load will be applied in the examples that follow.

☐ In both the balanced or unbalanced four-wire Y-connected loads, the phase voltage across each arm of the load (line to neutral) is equal to its respective source phase voltage.

$$E_{\text{phase}} = V_{\text{phase}} \qquad (27-6)$$

☐ In the balanced four-wire Y-connected load, the currents of the load are equal, which results in zero current in the neutral.

$$I_N = I_A + I_B + I_C = 0\,\text{A} \qquad (27-7)$$

☐ In the unbalanced four-wire Y-connected load, the unbalanced loads cause an unequal phase current in each arm of the load, resulting in a current in the neutral.

$$I_N = I_A + I_B + I_C \qquad (27-8)$$

☐ In both the balanced and unbalanced four-wire Y-connected loads, the respective line and phase currents are equal in a given arm of the load.

$$I_{\text{line}} = I_{\text{phase}} \qquad (27-9)$$

☐ For either the balanced or unbalanced four-wire Y-connected load, the line voltage is $\sqrt{3}$ times the phase voltage.

$$|E_{\text{line}}| = \sqrt{3}\,|V_{\text{phase}}| \qquad (27-10)$$

☐ In either balanced or unbalanced four-wire Y-connected load, the line voltage leads its respective phase voltage by 30°.

EXAMPLE 27–4 For the balanced four-wire Y-connected load of Figure 27–18, $Z_A = Z_B = Z_C = 4.00 + j3.00 = 5.00\underline{/36.87°}\,\Omega$. Determine (a) the line and phase voltages, (b) the line and phase currents, and (c) the neutral current.

SOLUTION (a) Since the magnitude of the line voltage of the source is 208 V, the magnitude of the phase voltage is:

$$|E_{line}| = \sqrt{3}\,|V_{phase}|$$

$$|V_{phase}| = \frac{|E_{line}|}{\sqrt{3}} = \frac{208}{\sqrt{3}} = 120\text{ V}$$

Observation From Figure 27–1(c), we learn that the phase voltages are displaced by 120° from one another.

∴ $V_{AN} = 120\underline{/0°}$ V

∴ $V_{BN} = 120\underline{/-120°}$ V

∴ $V_{CN} = 120\underline{/120°}$ V

Observation Each line voltage of the four-wire Y-connected load of Figure 27–18 is the resultant of two of the phase voltages.

For the line voltage V_{AB}, where $V_{NB} = -V_{BN}$;

$$V_{AB} = V_{AN} + V_{NB} = V_{AN} - V_{BN}$$

$$V_{AB} = 120\underline{/0°} - 120\underline{/-120°}$$

$$V_{AB} = (120 + j0) - (-60.00 - j103.9)$$

$$V_{AB} = 180 + j103.9 = 208\underline{/30°}$$

∴ The line voltage V_{AB} is $208\underline{/30°}$ V.

For the line voltage V_{BC}, where $V_{NC} = -V_{CN}$:

$$V_{BC} = V_{BN} + V_{NC} = V_{BN} - V_{CN}$$

$$V_{BC} = 120\underline{/-120°} - 120\underline{/120°}$$

$$V_{BC} = (-60.00 - j103.9) - (-60.00 + j103.9)$$

$$V_{BC} = 0 - j207.8 = 208\underline{/-90°}$$

∴ The line voltage V_{BC} is $208\underline{/-90°}$ V.

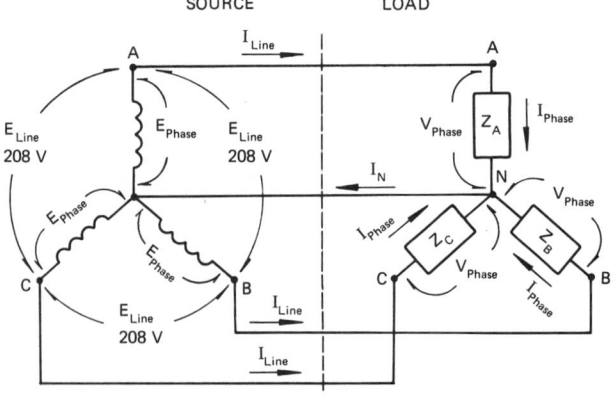

FIGURE 27–18
Circuit for Example 27–4, where $Z_A = Z_B = Z_C$.

Configurations and Parameters of the Three-Phase Load

For the line voltage V_{CA}, where $V_{NA} = -V_{AN}$:

$$V_{CA} = V_{CN} + V_{NA} = V_{CN} - V_{AN}$$

$$V_{CA} = 120\underline{/120°} - 120\underline{/0°}$$

$$V_{CA} = (-60.00 + j103.9) - (120 + j0)$$

$$V_{CA} = -180.0 + j103.9 = 208\underline{/150°}$$

∴ The line voltage V_{CA} is $208\underline{/150°}$ V.

Observation The following facts for the four-wire Y-connected load have been verified.

1. The magnitudes of the line and phase voltages across the four-wire Y-connected load are related by $\sqrt{3}$. That is, $|E_{line}| = \sqrt{3}\,|E_{phase}|$. For this example, $208 = \sqrt{3} \times 120$ V.
2. For the four-wire Y-connected load, the line voltage leads its respective phase voltage by 30°. For this example, the line voltage $V_{AB}\underline{/30°}$ leads its respective phase voltage $V_{AN}\underline{/0°}$ by 30°.
3. The magnitudes of the line voltages across the four-wire Y-connected load are equal, and each is displaced from the other by 120°. The magnitudes of the phase voltages are also equal, and each phase voltage is displaced from the other by 120°.

SOLUTION (b) In the four-wire Y-connected load, the respective line and phase currents are equal in a given arm of the load.

For load Z_A, where $I_{line} = I_{phase} = I_A$:

$$I_A = \frac{V_{AN}}{Z_A} = \frac{120\underline{/0°}}{5\underline{/36.87°}} = 24.0\underline{/-36.87°}$$

∴ The line and phase current in load Z_A is $24.0\underline{/-36.87°}$ A.

For load Z_B, where $I_{line} = I_{phase} = I_B$:

$$I_B = \frac{V_{BN}}{Z_B} = \frac{120\underline{/-120°}}{5\underline{/36.87°}} = 24.0\underline{/-156.9°}$$

∴ The line and phase current in load Z_B is $24.0\underline{/-156.9°}$ A.

For load Z_C, where $I_{line} = I_{phase} = I_C$:

$$I_C = \frac{V_{CN}}{Z_C} = \frac{120\underline{/120°}}{5\underline{/36.87°}} = 24.0\underline{/83.13°}$$

∴ The line and phase current in load Z_C is 24.0/83.13° A.

Observation Since the loads are balanced, the magnitudes of the line and phase currents are all equal.

SOLUTION (c) Because the load is balanced, the neutral current, I_N, is zero.

$$I_N = I_A + I_B + I_C = 0$$
$$I_N = 24.0/\underline{-36.87°} + 24.0/\underline{-156.9°} + 24.0/\underline{83.13°}$$
$$I_N = (19.20 - j14.40) + (-22.08 - j9.416)$$
$$+ (2.871 + j23.83) = 0.0 + j0.0$$

∴ $I_N = 0.0 + j0.0$ A

Observation Notice that the four-wire Y-connected load has two different values of voltage. The 120/208-V four-wire Y-connected 3ϕ transformer secondary finds wide use in commercial establishments and light manufacturing plants. The 208-V line-to-line-to-line 3ϕ voltage is used for three-phase induction motors. The 120-V, 1ϕ line-to-neutral voltage is used for single-phase lighting, appliances, and motors. The 208-V, 1ϕ line-to-line voltage is used for single-phase heaters and motors.

In the previous example, the characteristics and parameters of the balanced four-wire Y-connected load were noted. In the next example, the characteristics and parameters of the unbalanced four-wire Y-connected load will be investigated.

EXAMPLE 27–5 Determine (a) the line and phase voltages and (b) the neutral current in the four-wire unbalanced Y-connected load of Figure 27–19, where $f = 60$ Hz and $\omega = 377$ rad/s.

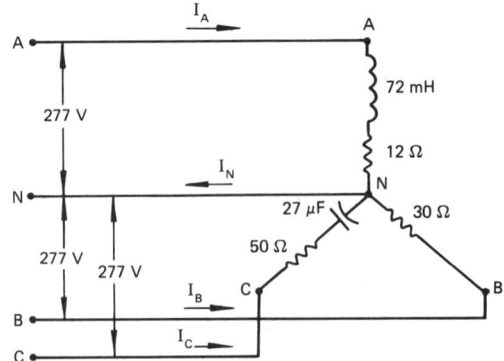

FIGURE 27–19
Circuit for Example 27–5.

Configurations and Parameters of the Three-Phase Load

SOLUTION (a) Since the magnitude of the phase voltage is 277 V and the phase voltages are displaced by 120° from one another, then:

$$V_{AN} = 277\underline{/0°} \text{ V}$$
$$V_{BN} = 277\underline{/-120°}$$
$$V_{CN} = 277\underline{/120°}$$

From Equation 27–10:

$$|E_{\text{line}}| = \sqrt{3}\,|V_{\text{phase}}|$$
$$|E_{\text{line}}| = \sqrt{3} \times 277 = 480 \text{ V}$$

Observation For the four-wire Y-connected load, the line voltage leads its respective phase voltage by 30°. Thus, the line voltages are:

$$E_{AB} = V_{AN} + V_{NB} = V_{AN} - V_{BN} = 480\underline{/30°} \text{ V}$$
$$E_{BC} = V_{BN} + V_{NC} = V_{BN} - V_{CN} = 480\underline{/-90°} \text{ V}$$
$$E_{CA} = V_{CN} + V_{NA} = V_{CN} - V_{AN} = 480\underline{/150°} \text{ V}$$

SOLUTION (b) By Ohm's law the load phase currents are:

$$I_A = \frac{V_{AN}}{Z_A}, \quad \text{where } Z_A = 12 + j27.14 \text{ }\Omega$$

$$I_A = \frac{277\underline{/0°}}{29.67\underline{/66.15°}} = 9.336\underline{/-66.15°}$$

$$\therefore \quad I_A = 9.336\underline{/-66.15°} \text{ A}$$

$$I_B = \frac{V_{BN}}{Z_B}, \quad \text{where } Z_B = 30 + j0 \text{ }\Omega$$

$$I_B = \frac{277\underline{/-120°}}{30\underline{/0°}} = 9.233\underline{/-120°}$$

$$\therefore \quad I_B = 9.233\underline{/-120°} \text{ A}$$

$$I_C = \frac{V_{CN}}{Z_C}, \quad \text{where } Z_C = 50 - j98.24 \text{ }\Omega$$

$$I_C = \frac{277\underline{/120°}}{110.2\underline{/-63.03°}} = 2.514\underline{/183.0°}$$

$$\therefore \quad I_C = 2.514\underline{/183.0°} \text{ A}$$

Solving for the neutral current:

$$I_N = I_A + I_B + I_C$$
$$I_N = 9.336\underline{/-66.15°} + 9.233\underline{/-120°} + 2.514\underline{/183.0°}$$

$$I_N = (3.775 - j8.539) + (-4.617 - j7.996)$$
$$+ (-2.511 - j0.132)$$
$$\therefore \quad I_N = -3.353 - j16.67 = 17.0\underline{/-101.4°} \text{ A}$$

Three-wire Y-connected Load

The three-wire Y-connected load may be balanced or unbalanced, and it may be sourced from either a Δ- or Y-connected transformer secondary. The following facts about the Y-connected load will be applied in the examples that follow.

☐ In both the balanced and unbalanced three-wire Y-connected load, the respective line and phase currents are equal in a given arm of the load.

$$I_A = I_{AN}, \quad I_B = I_{BN}, \quad I_C = I_{CN} \quad (27\text{-}11)$$

☐ In the balanced or unbalanced three-wire Y-connected load, the sum of the line currents is zero.

$$I_A + I_B + I_C = 0 \quad (27\text{-}12)$$

☐ For the balanced three-wire Y-connected load, the line voltage is $\sqrt{3}$ times the phase voltage.

$$|E_{\text{line}}| = \sqrt{3}\,|V_{\text{phase}}| \quad (27\text{-}13)$$

☐ In the balanced three-wire Y-connected load, the line voltage leads its respective phase voltage by 30°.
☐ In the unbalanced three-wire Y-connected load, the line currents are unequal.

$$I_A \neq I_B \neq I_C \quad (27\text{-}14)$$

☐ For the unbalanced three-wire Y-connected load, the phase voltage across each arm of the load is both unknown and unequal.

$$V_{AN} \neq V_{BN} \neq V_{CN} \quad (27\text{-}15)$$

☐ The wye-to-delta (Y-to-Δ) transformation, as noted in Figure 27-20, is used in the solution of the load parameters of an unbalanced three-wire Y-connected load.

$$\begin{bmatrix} Z_{AB} = \dfrac{Z}{Z_C} \\[6pt] Z_{BC} = \dfrac{Z}{Z_A} \\[6pt] Z_{CA} = \dfrac{Z}{Z_B} \end{bmatrix} \quad (27\text{-}16)$$

where $Z = Z_A Z_B + Z_B Z_C + Z_C Z_A$.

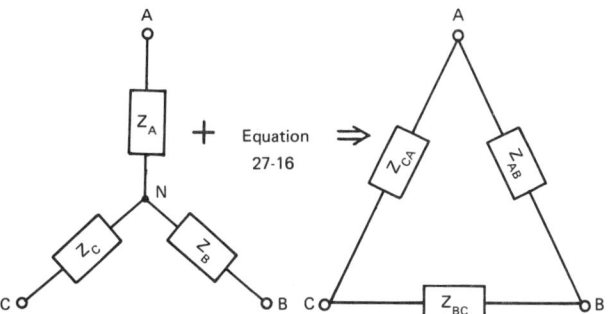

FIGURE 27-20
The Y-connected load may be transformed into an equivalent Δ-connected load using Equation 27-16.

EXAMPLE 27-6 For the balanced three-wire Y-connected load of Figure 27-21, determine (a) the line and phase voltages and (b) the line and phase currents when $Z_A = Z_B = Z_C = 1.50 + j1.13 = 1.88\underline{/37.0°}$ Ω.

SOLUTION (a) Since the magnitude of the line voltage is 480 V (Figure 27-21), the magnitude of the phase voltage is:

$$|E_{line}| = \sqrt{3}\,|V_{phase}|$$

$$|V_{phase}| = \frac{|E_{line}|}{\sqrt{3}} = \frac{480}{\sqrt{3}} = 277 \text{ V}$$

Since the load phase voltages are displaced by 120° from one another, then:

$$V_{AN} = 277\underline{/0°} \text{ V}$$

$$V_{BN} = 277\underline{/-120°} \text{ V}$$

$$V_{CN} = 277\underline{/120°} \text{ V}$$

The magnitude of the line voltage from Figure 27-21 is 480 V. The line voltage across the balanced three-wire Y-connected load is displaced 30° from its respective phase voltage. Then:

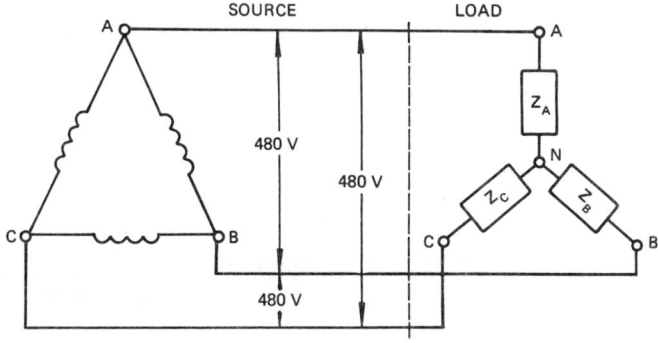

FIGURE 27-21
Circuit for Example 27-6, where $Z_A = Z_B = Z_C$.

$V_{AB} = 480\underline{/30°}$ V

$V_{BC} = 480\underline{/-90°}$ V

$V_{CA} = 480\underline{/150°}$ V

Observation The following facts for the **balanced** three-wire Y-connected load have been verified.

1. The magnitude of the line voltage is $\sqrt{3}$ times the phase voltage.
2. The line voltage leads its respective phase voltage by 30°.
3. The magnitudes of the line voltages are equal, and each line voltage is displaced 120° from the other. The magnitudes of the phase voltages are also equal, and each phase voltage is displaced 120° from the other.

SOLUTION (b) Since the respective line and phase currents are equal in a given arm of the load, then, for load Z_A where $I_{\text{line}} = I_{\text{phase}} = I_A$:

$$I_A = \frac{V_{AN}}{Z_A} = \frac{277\underline{/0°}}{1.88\underline{/37.0°}} = 147.3\underline{/-37.0°}$$

∴ The line and phase current in load Z_A is $147.3\underline{/-37.0°}$ A.

For load Z_B, where $I_{\text{line}} = I_{\text{phase}} = I_B$:

$$I_B = \frac{V_{BN}}{Z_B} = \frac{277\underline{/-120°}}{1.88\underline{/37.0°}} = 147.3\underline{/-157°}$$

∴ The line and phase current in load Z_B is $147.3\underline{/-157°}$ A.

For load Z_C, where $I_{\text{line}} = I_{\text{phase}} = I_C$:

$$I_C = \frac{V_{CN}}{Z_C} = \frac{277\underline{/120°}}{1.88\underline{/37.0°}} = 147.3\underline{/83.0°}$$

∴ The line and phase current in load Z_C is $147.3\underline{/83.0°}$ A.

Observation Because the loads are balanced, the magnitudes of the line and phase currents are all equal.

Check In any balanced or unbalanced **three-wire** connected load, the sum of the line currents is zero.

$$I_A + I_B + I_C = 0$$

$$(117.6 - j88.65) + (-135.6 - j57.55)$$
$$+ (17.95 + j146.2) = 0.0 + j0.0$$

727

Configurations and Parameters of the Three-Phase Load

In the previous example, the characteristics and parameters of the balanced three-wire Y-connected load were noted. In the next example, the characteristics and parameters of the unbalanced three-wire Y-connected load will be investigated using the wye-to-delta transformation of the load.

EXAMPLE 27-7 Determine (a) the line and phase currents and (b) the line and phase voltages in the unbalanced three-wire load of Figure 27–22 when $f = 60$ Hz and $\omega = 377$ rad/s.

Observation Since the loads are unequal, the phase voltages (V_{AN}, V_{BN}, V_{CN}) are also unequal and unknown. They are not $\sqrt{3}$ times $|E_{line}|$. To determine the phase voltage across each leg of the unbalanced three-wire Y-connected load, the phase currents in each arm must be known. The key to the solution lies in the transformation of the Y-connected load to an equivalent Δ-connected load from which the line currents may be calculated. Once the line currents of the equivalent Δ-connected load are known, the line currents of the original Y-connected load are also known. The solution begins by transforming the load using Equations 27–16 along with the configuration shown in Figure 27–20. From Figure 27–22:

$$Z_A = 1.2 + j3.016 = 3.246 \underline{/68.3°} \ \Omega$$

$$Z_B = 0.62 + j5.655 = 5.689 \underline{/83.74°} \ \Omega$$

$$Z_C = 0.3 + j1.508 = 1.538 \underline{/78.75°} \ \Omega$$

FIGURE 27-22
Circuit for Example 27-7.

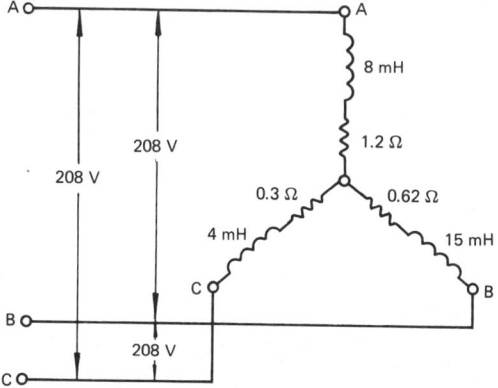

Solve for Z in Equation 27–16 where:

$$Z = Z_A Z_B + Z_B Z_C + Z_C Z_A$$

$$Z = 18.47\underline{/152°} + 8.750\underline{/162.5°} + 4.992\underline{/147.1°}$$

$$Z = (-16.31 + j8.671) + (-8.345 + j2.631) + (-4.191 + j2.712)$$

$$Z = -28.85 + j14.01 = 32.07\underline{/154.1°}\ \Omega$$

Substitute into Equation 27–16:

$$Z_{BC} = \frac{Z}{Z_A} = \frac{32.07\underline{/154.1°}}{3.246\underline{/68.3°}}$$

$$\therefore\quad Z_{AB} = 20.85\underline{/75.35°}\ \Omega$$

$$Z_{BC} = \frac{Z}{A_A} = \frac{32.07\underline{/154.1°}}{3.246\underline{/68.3°}}$$

$$\therefore\quad Z_{BC} = 9.880\underline{/85.80°}\ \Omega$$

$$Z_{CA} = \frac{Z}{Z_B} = \frac{32.07\underline{/154.1°}}{5.689\underline{/83.74°}}$$

$$\therefore\quad Z_{CA} = 5.637\underline{/70.36°}\ \Omega$$

Observation   Figure 27–23 summarizes the wye-to-delta transformation.

Calculate the phase currents in the Δ-connected load of Figure 27–23 and then determine the line currents.

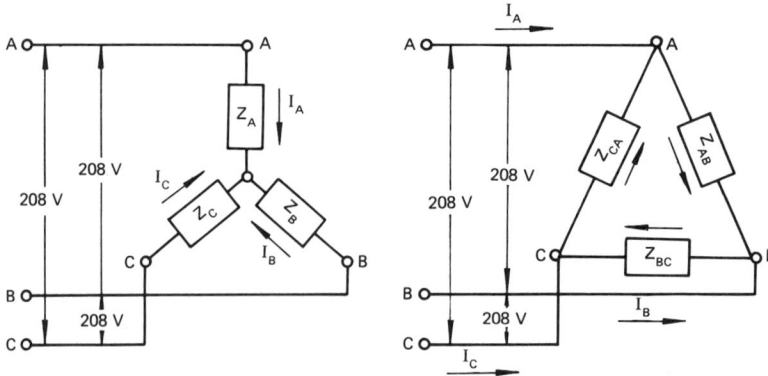

FIGURE 27–23
Equivalent loads of Example 27–7 are formed by transforming the Y-connected load to a Δ-connected load.

$Z_A = 3.246\ \underline{/68.3}$ ⟹ $Z_{AB} = 20.85\ \underline{/75.35}$

$Z_B = 5.689\ \underline{/83.74}$ ⟹ $Z_{BC} = 9.880\ \underline{/85.80}$

$Z_C = 1.538\ \underline{/78.75}$ ⟹ $Z_{CA} = 5.637\ \underline{/70.36}$

Configurations and Parameters of the Three-Phase Load

$$I_{AB} = \frac{V_{AB}}{Z_{AB}} = \frac{208\underline{/0°}}{20.85\underline{/75.35°}}$$

$\therefore \quad I_{AB} = 9.976\underline{/-75.35°}$ A

$$I_{BC} = \frac{V_{BC}}{Z_{BC}} = \frac{208\underline{/-120°}}{9.880\underline{/85.80°}}$$

$\therefore \quad I_{BC} = 21.05\underline{/-205.8°}$ A

$$I_{CA} = \frac{V_{CA}}{Z_{CA}} = \frac{208\underline{/120°}}{5.637\underline{/70.36°}}$$

$\therefore \quad I_{CA} = 36.90\underline{/49.64°}$ A

By KCL the line currents are:

$I_A + I_{CA} - I_{AB} = 0$

$I_A = I_{AB} - I_{CA}$

$I_A = 9.976\underline{/-75.35°} - 36.90\underline{/49.64°}$

$I_A = (2.523 - j9.652) - (23.90 + j28.12)$

$\therefore \quad I_A = -21.38 - j37.77 = 43.40\underline{/-119.5°}$ A

$I_B + I_{AB} - I_{BC} = 0$

$I_B = I_{BC} - I_{AB}$

$I_B = 21.05\underline{/-205.8°} - 9.976\underline{/-75.35°}$

$I_B = (-18.95 + j9.162) - (2.523 - j9.652)$

$\therefore \quad I_B = -21.47 + j18.81 = 28.54\underline{/138.8°}$ A

$I_C + I_{BC} - I_{CA} = 0$

$I_C = I_{CA} - I_{BC}$

$I_C = 36.90\underline{/49.64°} - 21.05\underline{/-205.8°}$

$I_C = (23.90 + j28.12) - (-18.95 + j9.162)$

$\therefore \quad I_C = 42.85 + j18.96 = 46.86\underline{/23.87°}$ A

SOLUTION (a) The line currents (I_A, I_B, and I_C) in the original Y-connected load are the same as the line currents in the load formed by transforming the Y-connected load to a Δ-connected load. Since the line and phase currents in a Y-connected load are equal, the phase currents in Figure 27–22 are:

$\therefore \quad I_A = -21.38 - j37.77 = 43.40\underline{/-119.5°}$ A

$\therefore \quad I_B = -21.47 + j18.81 = 28.54\underline{/138.8°}$ A

$\therefore \quad I_C = 42.85 + j18.96 = 46.86\underline{/23.87°}$ A

Check $I_A + I_B + I_C = 0$

$(-21.38 - j37.77) + (-21.47 + j18.81) + (42.85 + j18.96) = 0.0 + j0.0$

SOLUTION (b) The phase voltages are:

$V_{AN} = I_A Z_A = (43.40\underline{/-119.5°})(3.246\underline{/68.3°})$

$\therefore \quad V_{AN} = 140.9\underline{/-51.20°}$ V

$V_{BN} = I_B Z_B = (28.54\underline{/138.8°})(5.689\underline{/83.74°})$

$\therefore \quad V_{BN} = 162.4\underline{/222.5°}$ V

$V_{CN} = I_C Z_C = (46.86\underline{/23.87°})(1.538\underline{/78.75°})$

$\therefore \quad V_{CN} = 72.07\underline{/102.6°}$ V

Observation Each line voltage of the three-wire Y-connected load of Figure 27–22 is the resultant of two of the phase voltages.
For the line voltage V_{AB}, where $V_{NB} = -V_{BN}$:

$V_{AB} = V_{AN} + V_{NB} = V_{AN} - V_{BN}$

$V_{AB} = 140.9\underline{/-51.20°} - 162.4\underline{/222.5°}$

$V_{AB} = (88.29 - j109.8) - (-119.7 - j109.7)$

$\therefore \quad V_{AB} = 208.0 - j0.1 = 208\underline{/0.0°}$ V

For the line voltage V_{BC}, where $V_{NC} = -V_{CN}$:

$V_{BC} = V_{BN} + V_{NC} = V_{BN} - V_{CN}$

$V_{BC} = 162.4\underline{/222.5°} - 72.07\underline{/102.6°}$

$V_{BC} = (-119.7 - j109.7) - (-15.72 + j70.33)$

$\therefore \quad V_{BC} = -104.0 - j180.0 = 208\underline{/-120°}$ V

For the line voltage V_{CA}, where $V_{NA} = -V_{AN}$:

$V_{CA} = V_{CN} + V_{NA} = V_{CN} - V_{AN}$

$V_{CA} = 72.07\underline{/102.6°} - 140.9\underline{/-51.20°}$

$V_{CA} = (-15.72 + j70.33) - (88.29 - j109.8)$

$\therefore \quad V_{CA} = -104.0 + j180.1 = 208\underline{/120°}$ V

27–5 POWER AND POWER FACTOR OF THE THREE-PHASE LOAD

Before starting this section, it is recommended that you reread Section 23–1 as a review of the parameters that govern power dissipation.

In general, the total average power (watts) supplied to any three-phase load (balanced or unbalanced, Δ- or Y-connected,

Power in Any Three-Phase Load three- or four-wire) may be computed by adding the power dissipated in each of the three phases of the load.

$$P = P_{phaseA} + P_{phaseB} + P_{phaseC} \qquad (27\text{--}17)$$

where P = total average power (watts) supplied to the three-phase load by the source
P_{phaseA} = average power (watts) dissipated by the resistive component in phase A of the load
P_{phaseB} = average power (watts) dissipated by the resistive component in phase B of the load
P_{phaseC} = average power (watts) dissipated by the resistive component in phase C of the load

The average power of each phase of the load may be computed using any of the forms of the power equation, where angle ϕ is the symbol used to designate the phase angle between the phase voltage (V_{phase}) and the phase current (I_{phase}) in each of the three phases of the load.

$$P_{phase} = |V_{phase}| |I_{phase}| \cos(\phi) \qquad (27\text{--}18)$$

$$P_{phase} = |I_{phase}|^2 |Z| \cos(\phi) \qquad (27\text{--}19)$$

$$P_{phase} = \frac{|V_{phase}|^2}{|Z|} \cos(\phi) \qquad (27\text{--}20)$$

where P = power dissipated in a given phase (A, B, C) of the three-phase load
V_{phase} = phase voltage across a given phase of the three-phase load
I_{phase} = phase current in a given phase of the three-phase load
$\cos(\phi)$ = power factor for a given phase of the three-phase load
ϕ = phase angle of a given phase of the load and is the angle associated with the impedance of the given phase; ϕ is also the angle between the phase voltage and the phase current
Z = impedance of a given phase of the three-phase load

EXAMPLE 27-8 (*a*) Determine the average power is in each phase of a balanced Δ-connected three-phase load with a line voltage of 240 V when $Z_{AB} = Z_{BC} = Z_{CA} = 10 + j24 = 26.0\underline{/67.38°}$ Ω. (*b*) Determine the average power supplied to the entire three-phase load.

SOLUTION (*a*) Since the load is balanced and $|E_{line}| = |V_{phase}|$ in a Δ-connected load,

$$P_{phaseA} = P_{phaseB} = P_{phaseC} = \frac{|V_{phase}|^2}{|Z|} \cos(\phi)$$

$$P_{phaseA} = P_{phaseB} = P_{phaseC} = \frac{240^2}{26.0} \cos(67.38°)$$

$$\therefore \quad P_{phaseA} = P_{phaseB} = P_{phaseC} = 852 \text{ W}$$

(b) Use Equation 27–17 to determine the power supplied to the entire three-phase load.

$$P = P_{phaseA} + P_{phaseB} + P_{phaseC}$$

$$\therefore \quad P = 852 + 852 + 852 = 2.56 \text{ kW}$$

EXAMPLE 27–9 (a) Determine the average power in each phase of an unbalanced three-wire Y-connected load. (b) Determine the power to the entire load when:

$$I_A = 43.40\underline{/-119.5°} \text{ A and } V_{AN} = 140.9\underline{/-51.20°} \text{ V}$$

$$I_B = 28.55\underline{/138.8°} \text{ A and } V_{BN} = 162.4\underline{/222.5°} \text{ V}$$

$$I_C = 46.86\underline{/23.87°} \text{ A and } V_{CN} = 72.07\underline{/102.6°} \text{ V}$$

SOLUTION (a) Determine the phase angle for each of the phases by computing the angle between the phase voltage and the phase current.

$$|\phi_A| = -119.5° - (-51.20°) = 68.3°$$

$$|\phi_B| = 138.8° - (222.5°) = 83.7°$$

$$|\phi_C| = 23.87° - (102.6°) = 78.7°$$

Compute the power of each phase

$$P_{phaseA} = |V_{phaseA}||I_{phaseA}| \cos(\phi_A)$$

$$\therefore \quad P_{phaseA} = (140.9)(43.40) \cos(68.3°) = 2.261 \text{ kW}$$

$$P_{phaseB} = |V_{phaseB}||I_{phaseB}| \cos(\phi_B)$$

$$\therefore \quad P_{phaseB} = (162.4)(28.55) \cos(83.7°) = 508.8 \text{ W}$$

$$P_{phaseC} = |V_{phaseC}||I_{phaseC}| \cos(\phi_C)$$

$$\therefore \quad P_{phaseC} = (72.07)(46.86) \cos(78.7°) = 661.7 \text{ W}$$

SOLUTION (b) Use Equation 27–17 to determine the total power:

$$P = P_{phaseA} + P_{phaseB} + P_{phaseC}$$

$$\therefore \quad P = 2261 + 508.8 + 661.7 = 3.432 \text{ kW}$$

Power in Balanced Three-Phase Loads

The power in a balanced three-phase load connected delta or wye (three- or four-wire) is simply three times the power in each phase.

$$P = 3|V_{phase}||I_{phase}|\cos(\phi) \qquad (27\text{--}21)$$

The power in a balanced three-phase load in terms of the line voltage and line current is computed using the following equation: For the balanced Δ-connected load, where $|E_{line}| = |V_{phase}|$ and $|I_{line}| = \sqrt{3}\,|I_{phase}|$,

$$P = 3|E_{line}|\frac{|I_{line}|}{\sqrt{3}}\cos(\phi)$$

Rationalize the denominator

$$P = 3|E_{line}|\frac{|I_{line}|}{\sqrt{3}} \times \frac{\sqrt{3}}{\sqrt{3}}\cos(\phi) = \frac{3}{3}|E_{line}||I_{line}|\sqrt{3}\cos(\phi)$$

$$P = \sqrt{3}\,|E_{line}||I_{line}|\cos(\phi) \qquad (27\text{--}22)$$

For the balanced Y-connected load (three- or four-wire), where $|I_{line}| = |I_{phase}|$ and $|E_{line}| = \sqrt{3}\,|V_{phase}|$,

$$P = 3\frac{|E_{line}|}{\sqrt{3}}|I_{line}|\cos(\phi)$$

Rationalize the denominator:

$$P = 3\frac{|E_{line}|}{\sqrt{3}}|I_{line}| \times \frac{\sqrt{3}}{\sqrt{3}}\cos(\phi)$$

$$P = \frac{3}{3}|E_{line}||I_{line}|\sqrt{3}\cos(\phi)$$

$$P = \sqrt{3}\,|E_{line}||I_{line}|\cos(\phi) \qquad (27\text{--}22)$$

EXAMPLE 27–10 Use Equations 27–21 and 27–22 to determine the power dissipated in the balanced three-phase load in the following circuits:
(a) Four-wire Y-connected: $V_{phase} = 2400$ V, $I_{line} = 20.4$ A, and $\phi = 36°$.
(b) Δ-connected: $E_{line} = 480$ V, $I_{phase} = 14.7$ A, and $\phi = 24.5°$.
(c) Three-wire Y-connected: $V_{phase} = 277$ V, $I_{phase} = 21.7$ A, and $\phi = 0°$.

SOLUTION (a) Since $|I_{line}| = |I_{phase}|$ in the Y-connected load, then

$$|I_{line}| = |I_{phase}| = 20.4 \text{ A}$$

$$P = 3|V_{phase}||I_{phase}|\cos(\phi)$$

$$\therefore \quad P = 3(2400)(20.4)\cos(36°) = 119 \text{ kW}$$

Alternate solution, where $|E_{line}| = \sqrt{3}|V_{phase}|$ in a balanced Y-connected load:

$$|E_{line}| = \sqrt{3}(2400) = 4.16 \text{ kV}$$
$$P = \sqrt{3}|E_{line}||I_{line}| \cos(\phi)$$
$$\therefore P = \sqrt{3}(4160)(20.4) \cos(36°) = 119 \text{ kW}$$

SOLUTION (b) Since $|V_{phase}| = |E_{line}|$ in the Δ-connected load,

$$|V_{phase}| = |E_{line}| = 480 \text{ V}$$
$$P = 3|V_{phase}||I_{phase}| \cos(\phi)$$
$$\therefore P = 3(480)(14.7) \cos(24.5°) = 19.3 \text{ kW}$$

Alternate solution, where $|I_{line}| = \sqrt{3}|I_{phase}|$ in the balanced Δ-connected load:

$$|I_{line}| = \sqrt{3}(14.7) = 25.5 \text{ A}$$
$$P = \sqrt{3}|E_{line}||I_{line}| \cos(\phi)$$
$$\therefore \sqrt{3}(480)(25.5) \cos(24.5°) = 19.3 \text{ kW}$$

SOLUTION (c) $P = 3|V_{phase}||I_{phase}| \cos(\phi)$

$$\therefore P = 3(277)(21.7) \cos(0°) = 18.0 \text{ kW}$$

Alternate solution, where $|I_{line}| = |I_{phase}|$ and $|E_{line}| = \sqrt{3}|V_{phase}|$ in the balanced Y-connected load:

$$|I_{line}| = |I_{phase}| = 21.7 \text{ A}$$
$$|E_{line}| = \sqrt{3}(277) = 480 \text{ V}$$
$$P = \sqrt{3}|E_{line}||I_{line}| \cos(\phi)$$
$$\therefore P = \sqrt{3}(480)(21.7) \cos(0°) = 18.0 \text{ kW}$$

Reactive Power in the Three-Phase Load

The reactive power (vars) supplied to any three-phase load (balanced or unbalanced, Δ- or Y-connected, three or four wire) may be computed by combining the reactive powers in each of the three phases of the load.

$$Q = Q_{phaseA} + Q_{phaseB} + Q_{phaseC} \qquad (27\text{--}23)$$

where Q = combined reactive power (vars) supplied to the three-phase load
Q_{phaseA} = reactive power in phase A of the load
Q_{phaseB} = reactive power in phase B of the load
Q_{phaseC} = reactive power in phase C of the load

The reactive power of each phase of the load may be com-

Power and Power Factor of the Three-Phase Load

puted using any of the following equations, where ϕ is the symbol used to designate the phase angle between the phase voltage (V_{phase}) and the phase current (I_{phase}) in each of the three phases of the load.

$$Q_{\text{phase}} = |V_{\text{phase}}||I_{\text{phase}}| \sin(\phi) \qquad (27\text{-}24)$$

$$Q_{\text{phase}} = |I_{\text{phase}}|^2 |Z| \sin(\phi) \qquad (27\text{-}25)$$

$$Q_{\text{phase}} = \frac{|V_{\text{phase}}|^2}{|Z|} \sin(\phi) \qquad (27\text{-}26)$$

EXAMPLE 27-11 (a) Determine the reactive power in each phase of a balanced Y-connected three-phase load with a line voltage of 208 V and a load impedance of $Z_A = Z_B = Z_C = 12.0 \underline{/24°}\ \Omega$. (b) Determine the average power supplied to the entire three-phase load.

SOLUTION (a) Since the load is balanced,

$$Q_{\text{phase}A} = Q_{\text{phase}B} = Q_{\text{phase}C} = \frac{|V_{\text{phase}}|^2}{|Z|} \sin(\phi)$$

$$V_{\text{phase}} = \frac{E_{\text{phase}}}{\sqrt{3}} = \frac{208}{\sqrt{3}} = 120\ \text{V}$$

$$\therefore \quad Q_{\text{phase}A} = Q_{\text{phase}B} = Q_{\text{phase}C} = \frac{120^2}{12} \sin(24°)$$

$$= 488\ \text{vars} \quad (\text{inductive})$$

Observation Since the load impedance is inductive, the reactive power is also inductive.

SOLUTION (b) The combined reactive power from Equation 27-23 is:

$$Q = Q_{\text{phase}A} + Q_{\text{phase}B} + Q_{\text{phase}C}$$

$$\therefore \quad Q = 488 + 488 + 488 = 1.46\ \text{kvars} \quad (\text{inductive})$$

Apparent Power in the Three-Phase Load

The apparent power of the three-phase load may be determined from the average power and the reactive power of the load.

$$(P, Q) \boxed{\rightarrow P}\ S\underline{/\theta} \qquad (27\text{-}27)$$

where S = apparent power in volt-amperes (VA) of the entire three-phase load
P = average power in watts (W) of the entire three-phase load
Q = reactive power in vars of the entire three-phase load
θ = phase angle of the entire three-phase load

EXAMPLE 27–12 Determine the average power (P), reactive power (Q), apparent power (S), and the power factor ($\cos \theta$) of an unbalanced, four-wire, Y-connected three-phase load that is attached to a 480-V, 60-Hz, Δ-connected transformer secondary. The load phase impedance is:

$$Z_A = 5.82\underline{/27.0°}\ \Omega, \qquad Z_B = 2.40\underline{/38.0°}\ \Omega,$$
$$Z_C = 3.85\underline{/47.0°}\ \Omega$$

SOLUTION Determine the phase voltage (V_{phase}) of the four-wire Y-connected load.

$$|V_{\text{phase}}| = \frac{|E_{\text{line}}|}{\sqrt{3}} = \frac{480}{\sqrt{3}} = 277\ \text{V}$$

Compute the average power in each phase of the load.

$$P_{\text{phase}A} = \frac{|V_{\text{phase}A}|^2}{|Z_A|} \cos(\phi_A), \quad \text{where}\ \phi_A = 27.0°$$

$$\therefore\quad P_{\text{phase}A} = \frac{277^2}{5.82}\cos(27.0°) = 11.7\ \text{kW}$$

$$P_{\text{phase}B} = \frac{|V_{\text{phase}B}|^2}{Z_B} \cos(\phi_B), \quad \text{where}\ \phi_B = 38.0°$$

$$\therefore\quad P_{\text{phase}B} = \frac{277^2}{2.40}\cos(38.0°) = 25.2\ \text{kW}$$

$$P_{\text{phase}C} = \frac{|V_{\text{phase}C}|^2}{Z_C} \cos(\phi_C), \quad \text{where}\ \phi_C = 47.0°$$

$$\therefore\quad P_{\text{phase}C} = \frac{277^2}{3.85}\cos(47.0°) = 13.6\ \text{kW}$$

Determine the average power for the entire load.

$$P = P_{\text{phase}A} + P_{\text{phase}B} + P_{\text{phase}C}$$
$$\therefore\quad P = 11.7 \times 10^3 + 25.2 \times 10^3 + 13.6 \times 10^3$$
$$= 50.5\ \text{kW}$$

Compute the reactive power in each phase and then indicate whether the reactive power is inductive or capacitive.

$$Q_{\text{phase}A} = \frac{|V_{\text{phase}A}|^2}{Z_A} \sin(\phi_A), \quad \text{where}\ \phi_A = 27.0°$$

$$\therefore\quad Q_{\text{phase}A} = \frac{277^2}{5.82}\sin(27.0°) = 5.99\ \text{kvar}\quad (\text{ind})$$

Power and Power Factor of the Three-Phase Load

$$Q_{phaseB} = \frac{|V_{phaseB}|^2}{Z_B} \sin(\phi_B), \quad \text{where } \phi_B = 38.0°$$

$$\therefore \quad Q_{phaseB} = \frac{277^2}{2.40} \sin(38.0°) = 19.7 \text{ kvar} \quad (\text{ind})$$

$$Q_{phaseC} = \frac{|V_{phaseC}|^2}{Z_C} \sin(\phi_C), \quad \text{where } \phi_C = 47.0°$$

$$\therefore \quad Q_{phaseC} = \frac{277^2}{3.85} \sin(47.0°) = 14.6 \text{ kvar} \quad (\text{ind})$$

Determine the reactive power for the entire load.

$$Q = Q_{phaseA} + Q_{phaseB} + Q_{phaseC}$$

$$\therefore \quad Q = 5.99 \times 10^3 + 19.7 \times 10^3 + 14.6 \times 10^3$$

$$= 40.29 \text{ kvar} \quad (\text{inductive})$$

Determine the apparent power and the load phase angle; use Equation 27–27.

$$(P, Q) \boxed{\rightarrow P} \; S\underline{/\theta}$$

$$(50.5 \times 10^3, 40.3 \times 10^3) \boxed{\rightarrow P} \; 64.6\underline{/38.6°} \text{ kVA}$$

$$\therefore \quad S = 64.6 \text{ kVA} \quad (\text{inductive}) \quad \text{and} \quad \theta = 38.6°$$

Determine the load power factor.

$$\therefore \quad \cos(\theta) = \cos(38.6°) = 0.782 \quad (\text{lagging})$$

Observation Since the load is inductive, the power factor is lagging.

Both the average power of the three-phase load and the reactive power of the three-phase load may be expressed in terms of the load phase angle, θ, and the load apparent power, S. Thus from the power triangle of Figure 27–24:

$$P = S \cos(\theta) \tag{27-28}$$

$$Q = S \sin(\theta) \tag{27-29}$$

FIGURE 27–24
The power triangle picturing the relationship among the apparent, reactive, and average power.

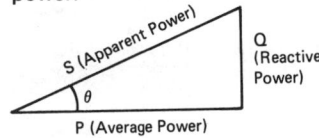

EXAMPLE 27–13 Determine the average and reactive power in a 200-kVA three-phase load with an 88% leading power factor.

SOLUTION First determine the circuit phase angle from the power factor.

$$\theta = \text{arc cos } (0.88) = 28.36°$$

Use Equation 27–28 to solve for the average power.

$$\therefore \quad P = S \cos (\theta) = 200 \times 10^3 \cos (28.36°) = 176 \text{ kW}$$

Use Equation 27–29 to solve for the reactive power.

$$\therefore \quad Q = S \sin (\theta) = 200 \times 10^3 \sin (28.36°)$$
$$= 95 \text{ kvars} \quad \text{(capacitive)}$$

Observation Since the power factor is leading, both the load and the reactive power are capacitive.

Power Factor Correction in Three-Phase Systems

The concept of correcting a single-phase power factor by adding an appropriate *power capacitor* was introduced in Section 23–3. These same concepts may be applied to three-phase loads.

The power capacitors that are used with three-phase power systems are assembled as a single unit with three equal capacitors. These capacitor assemblies may be connected to the main power bus through a circuit breaker as noted in Figure 27–25(a) or across the line feeding a three-phase induction motor as shown in Figure 27–25(b). Power capacitors are rated in kilovars rather than microfarads as noted in Table 27–2.

EXAMPLE 27–14 Select a three-phase power capacitor from Table 27–2 to correct an industrial plant's lagging

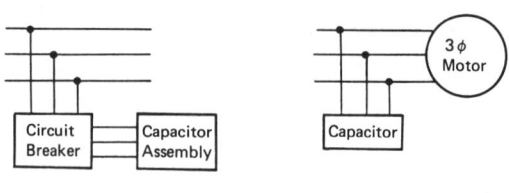

FIGURE 27–25
Three-phase power capacitor (a) connected across the main bus and (b) connected across an induction motor. The capacitor raises the power factor of the line and lowers the cost of operation.

Power and Power Factor of the Three-Phase Load

TABLE 27–2 SELECTED VALUES OF 240-, 480-, AND 600-V, THREE-PHASE, 60-Hz POWER CAPACITORS

Capacitive Rating (kvar)			
6	25	60	120
8	30	70	150
10	35	80	200
15	40	90	250
20	50	100	300

Note: (a) 240-V capacitors provide 0.75 of their rating when operated at 208 V.
(b) The weight of three-phase power capacitor assemblies ranges from 15 to 500 pounds.

power factor from 72% to 95%. The plant has a balanced three-phase Δ-connected load of 120 kW, 480 V, 60 Hz.

SOLUTION First determine the reactive power in the load for a power factor of 72%.

$$P = S \cos(\theta)$$

$$S = \frac{P}{\cos(\theta)}$$

$$S = \frac{120 \times 10^3}{0.72} = 167 \text{ kVA}$$

Solve for the reactive power when $\theta = \arccos(0.72) = 43.9°$.

$$Q = S \sin(\theta)$$

$$Q = 167 \times 10^3 \sin(43.9°)$$

$$= 116 \text{ kvar} \quad \text{(inductive)}$$

∴ The reactive power is 116 kvars (inductive) for a lagging power factor of 72%.

Next determine the reactive power in the load for a power factor of 95%.

$$P = S \cos(\theta)$$

$$S = \frac{P}{\cos(\theta)}$$

$$S = \frac{120 \times 10^3}{0.95} = 126 \text{ kVA}$$

Solve for the reactive power when $\theta = \arccos = (0.95) = 18.2°$.

$$Q = S \sin(\theta)$$
$$Q = 126 \times 10^3 \sin(18.2°)$$
$$= 39.4 \text{ kvar} \quad \text{(inductive)}$$

∴ The reactive power is 39.4 kvars (inductive) for a lagging power factor of 95%.

Finally, determine the difference between the two reactive powers. This difference is equal to the value of the three-phase power capacitor.

$$\Delta Q = 116 \text{ kvar} - 39.4 \text{ kar} = 76.6 \text{ kar}$$

∴ Select an 80-kvar, 480-V power capacitor from the values listed in Table 27–2.

Increasing a low power factor to 85% or more decreases the reactive current in both the power transformers and the conductors in the distribution system. With reactive current reduced, more power-producing current may pass through the conductors and transformers to power additional equipment. Thus, by raising the power factor, more motors, lights, or heaters may be added to an existing electrical system without the expense of rewiring or installing larger transformers.

The next example uses values of the previous example to demonstrate how additional power-producing current can be accommodated in a distribution system by simply raising the power factor of the system.

EXAMPLE 27–15 Determine the amount of additional power available to the balanced Δ-connected load in a 167-kVA, 480-V electrical distribution system when the lagging power factor of the system is corrected from 72% to 95%.

SOLUTION Determine the current in the 167-kVA distribution system at 72% lagging power factor.

$$S = |E||I|$$
$$|I| = \frac{S}{|E|} = \frac{167 \times 10^3}{480} = 348 \text{ A}$$
$$\theta = \arccos(0.72) = 43.9°$$

Observation Assign a minus sign to the phase angle to indicate an inductive current.

$$I \underline{/\theta} = 348 \underline{/-43.9°} \text{ A}$$

and

$$348 \underline{/-43.9°} \boxed{\rightarrow R} \ 251 - j241 \text{ A}$$

∴ The system can handle 348 A of current. When the power factor is 72%, the mix of current is 251 A of resistive power-producing current (watts) and 241 A of wattless inductive current (vars).

With a 348-A capacity and a 95% lagging power factor, determine the new mix of current.

$$|I| = 348 \text{ A}$$

$$\theta = \text{arc cos } (0.95) = 18.2°$$

$$I \underline{/\theta} = 348 \underline{/-18.2°} \boxed{\rightarrow R} \ 331 - j109 \text{ A}$$

∴ When the power factor is 95%, the mix of current is 331 A of resistive power-producing current (watts) and 109 A of wattless inductive current (vars).

The power available to the load in each case is the resistive current in the load times the 480 V across the load. At 72% power factor

$$P = 251 \times 480 = 120 \text{ kW}$$

At 95% power factor

$$P = 331 \times 480 = 159 \text{ kW}$$

∴ The increase in available power to the load is the difference between the two powers or 159 kW − 120 kW = 39 kW. Thus a 33% increase in load power is realized by raising the power factor from 72% to 95%.

PROBLEMS

Sections 27–1 through 27–4

27–1 Determine the number of pairs of poles in the rotor of a 60-Hz generator that turns 180 rev/min.

27–2 Determine the number of pairs of poles in the rotor of a 400-Hz generator that turns 4000 rev/min.

27–3 Determine the speed (rev/min) of a 50-pole rotor in a 60-Hz generator.

27–4 Determine the frequency produced by an alternator with a cylindrical four-pole rotor that turns 1500 rev/min.

27–5 The impedance of each phase of a balanced Δ-connected load is $10 \underline{/0°} \ \Omega$. Determine both the phase and line currents when the magnitude of the 60-Hz, three-phase line voltage is 240 V.

27–6 Repeat Problem 27–5 for a line voltage of 480 V.

27-7 The impedance of an unbalanced Δ-connected load is $Z_{AB} = 7.3\,\underline{/24°}\ \Omega$, $Z_{BC} = 10\,\underline{/0°}\ \Omega$, and $Z_{CA} = 8.2\,\underline{/32°}\ \Omega$. Determine the line currents when the magnitude of the 60-Hz, three-phase line voltage is 480 V.

27-8 Repeat Problem 27-7 for a 60-Hz line voltage of 600 V.

27-9 A balanced Δ-connected load is attached to a Y-connected 60-Hz source. The load impedance ($Z_{AB} = Z_{BC} = Z_{CA}$) is $3.6 + j2.0\ \Omega$. The source has a phase voltage of 277 V. Determine the magnitude of the line current.

27-10 Repeat Problem 27-9 for a 60-Hz source phase voltage of 120 V.

27-11 An unbalanced four-wire, Y-connected load is made up of $Z_A = 12.4\,\underline{/22°}\ \Omega$, $Z_B = 6.8\,\underline{/38°}\ \Omega$, and $Z_C = 15\,\underline{/0°}\ \Omega$. The 60-Hz line voltage is 208 V. Determine each line current and the neutral current.

27-12 Repeat Problem 27-11 for a 60-Hz line voltage of 480 V.

27-13 An unbalanced three-wire Y-connected load is made up of $Z_A = 18\,\underline{/0°}\ \Omega$, $Z_B = 20\,\underline{/-12°}\ \Omega$, and $Z_C = 16\,\underline{/34°}\ \Omega$. The 60-Hz line voltage is 208 V. Determine the line currents.

27-14 Repeat Problem 27-13 for a 60-Hz line voltage of 480 V.

27-15 A balanced three-wire Y-connected load attached to a 60-Hz, 208-V line has three equal impedances ($Z_A = Z_B = Z_C$), each consisting of a resistance of 4.8 Ω connected in series with an inductance of 7.43 mH. Determine the line and phase currents.

27-16 Repeat Problem 27-15 for a 60-Hz line voltage of 480 V.

27-17 An unbalanced four-wire, Y-connected load attached to a 60-Hz, 4160-V line has three impedances. Z_A consists of 8.52 Ω of resistance, and Z_B consists of an inductance of 7.50 mH in series with 5 Ω of resistance. Z_C consists of an inductance of 4.80 mH in series with 3 Ω of resistance. Determine each line and phase current. Also determine the neutral current.

27-18 Repeat Problem 27-17 for a 60-Hz, 8320-V line.

Section 27-5

27-19 A 4160-V, three-phase, Y-connected distribution line supplies the following loads:
90 kVA, 0.72 lagging power factor
130 kVA, 0.88 lagging power factor
150 kVA, 0.82 leading power factor
110 kVA, 1.0 power factor

Determine (a) the total average power in kilowatts, (b) the total reactive power in kvars, (c) the total apparent power in kVA, and (d) the system power factor.

27–20 A 4800-V, three-phase, Δ-connected distribution line supplies the following loads:
375 kVA, 0.88 lagging power factor
150 kVA, 0.92 lagging power factor
500 kVA, 0.78 lagging power factor
Determine (a) the total average power in kilowatts, (b) the total reactive power in kvars, (c) the total apparent power in kVA, and (d) the system power factor.

27–21 Determine the capacity (kvar) of a power capacitor needed to increase the power factor of a 60-Hz, 600-V, 50-A line from 76% lagging to 96% lagging.

27–22 Determine the capacity (kvar) of a power capacitor needed to increase the power factor of a 60-Hz, 240-V, 80-A line from 78% lagging to 92% lagging.

27–23 A 208-V, three-phase motor with a 0.84 lagging power factor requires 28.33 kVA for its operation. From the information given in Table 27–2, select an appropriate capacitor to correct the line to an approximate unity power factor.

27–24 A 208-V, three-phase motor with a 0.78 lagging power factor requires 16.5 kVA for its operation. From the information given in Table 27–2, select an appropriate capacitor to correct the line to an approximate unity power factor.

27–25 A 480-V, three-phase, 2-kvar power capacitor is Δ-connected across the load of Figure 27–19. Assuming a 60-Hz source, determine the power factor of the load without the power capacitor connected and then determine the power factor of the load with the power capacitor connected.

27–26 Repeat Problem 27–25 for a 1-kvar power capacitor and a line frequency of 400 Hz.

IV Frequency Domain and Filters

This section introduces you to several important concepts. Included are complex wave forms, frequency domain, filters, and both harmonic and intermodulation distortion. With a knowledge of these concepts, you will be ready to begin your study of communication circuits and systems.

28 Complex Wave Forms and the Frequency Domain

28-1 SUPERPOSITION OF SINUSOIDS
28-2 FOURIER THEOREM
28-3 FREQUENCY DOMAIN VERSUS TIME DOMAIN
28-4 SPECTRUM ANALYSIS
28-5 HARMONIC DISTORTION
28-6 INTERMODULATION DISTORTION
28-7 PRACTICAL APPLICATIONS

CHAPTER PREVIEW

The sine wave can be considered the basic building block of all electronic signals. By itself, the sine wave is used to teach ac principles. However, in reality, an extremely small proportion of useful information or data is transmitted by a pure sine wave. One can compare a single sine wave and its usefulness to a single note in a musical composition. Any musical composition consists of many different notes arranged so that they interrelate with each other in terms of time, amplitude, and frequency.

In this chapter, we will show that it is the relationship between sine waves, that is, their frequency, amplitude, and phase, that carries useful information in an electronic circuit. Furthermore, it will be shown that all repetitive wave forms, such as square waves, triangular waves, and pulses, can be broken down into a set of individual sine waves, each having a different frequency, amplitude, and phase. It will also be shown that, theoretically, by adding appropriate combinations of sine waves any repetitive wave form may be created.

PERFORMANCE OBJECTIVES

Once you have read each section, worked each example with pencil, paper, and calculator, and answered each problem for every section, you will be able to:

☐ Understand how complex wave forms are made by summing sine waves.
☐ Understand the relationship between the frequency domain and the time domain.
☐ Understand spectrum analysis.
☐ Understand harmonic and intermodulation distortion.

28–1 SUPERPOSITION OF SINUSOIDS

Consider the case of a single loudspeaker reproducing a pure tone, say 1 kHz. Basically, the cone just vibrates back and forth at a rate of 1000 times per second. Obviously, a loudspeaker is capable of reproducing more than one tone at a time. Otherwise, music could not be heard on the radio. It is the principle of superposition that provides an understanding of how more than one tone may be reproduced. If two or more sine waves are applied to a loudspeaker at the same time, the loudspeaker will have a motion that is a replica of the sum of the two sine waves. Figure 28–1 pictures two sine waves, a and b, and their sum, c. A speaker that is driven by signal c, as pictured in Figure 28–1, will sound like two different sine waves, a and b.

As the amplitude of two signals, a and b, is changed, so will the resultant, c. Similarly, if the frequencies are changed, the resultant will change. Furthermore, if the phase between two

Superposition of Sinusoids

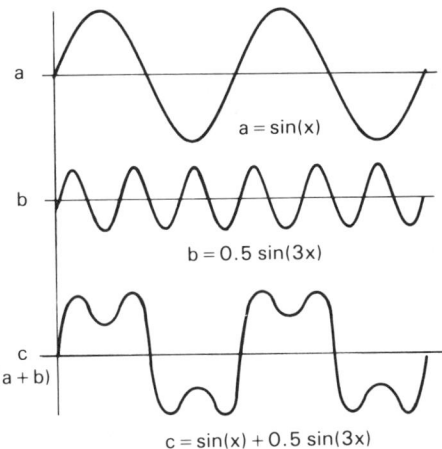

FIGURE 28-1
Wave form c is the sum of sine waves a and b.

signals is changed, the resultant will change. For example, take the case of a signal, a, with a frequency f, and a second signal, b, with a frequency $3f$. When they are in phase, the resultant signal will appear as previously shown in Figure 28-1. However, when the phase of signal b is changed by 180° (π), the resultant of $a + b'$ will also change, as noted in Figure 28-2. This same technique may be applied to three, four, or any number of sine waves. It is this principle of the superposition of sine waves that allows us to take complex wave forms and break them down into a set of sine waves of varying frequency, amplitude, and phase. Conversely, the linear addition of a number of sine waves will yield a single complex wave form. An example of the

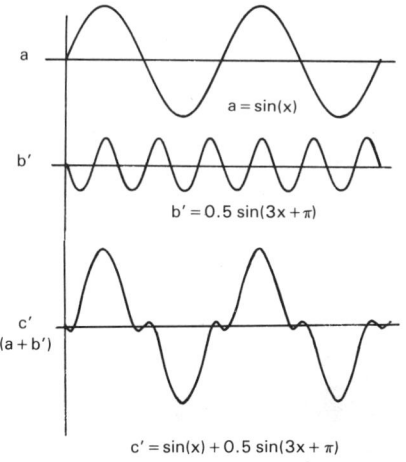

FIGURE 28-2
Resultant wave form c' is different from wave form c in Figure 28-1 because the sine wave b has changed phase as seen in wave b'.

Complex Wave Forms and the Frequency Domain

addition of a set of sine waves occurs when you listen to a loudspeaker. A single-cone loudspeaker, which is ideally a rigid cone, must reproduce all the frequencies of a symphony orchestra or a rock band. It does this not by vibrating at all the various frequencies at the same time but, rather, by vibrating at the one complex wave form that is the sum of all the frequencies generated.

If the electrical signal coming out of the amplifier feeding the loudspeaker were monitored using an oscilloscope, a wave form such as the one pictured in Figure 28–3 might be seen. Notice that even though the wave form is complex it is only a *single* wave form. It is this single wave form that is followed by the loudspeaker in accurately reproducing a sound that is the total sound of the orchestra or band.

It is this very important concept, that simple wave shapes can be added to form a complex wave shape and that a complex wave shape can be broken down into simple wave shapes, that is the subject of the next section.

28–2 FOURIER THEOREM

The Fourier theorem, derived by the French mathematician, states that any repetitive wave form is made up of or can be represented by a series of sine waves of appropriate frequency, amplitude, and phase. By applying calculus, a theoretical determination can be made of the characteristics (amplitude, frequency, and phase) of the sine waves needed to reproduce any repetitive wave shape. However, except for the very simple repetitive wave shapes, calculations of the frequency, amplitude, and phase of each sine wave become very tedious and complex. Mathematical tables exist for some of the simple wave forms but, in practice, complex wave forms are not theoretically analyzed.

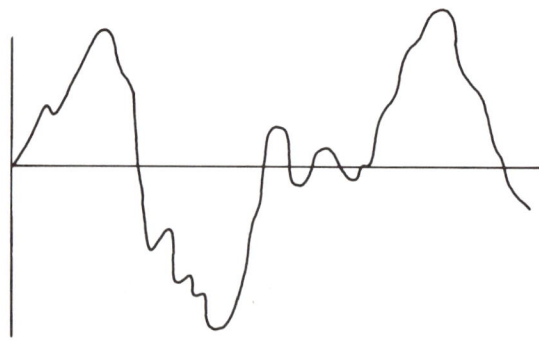

FIGURE 28–3
Example of a complex wave form.

Spectrum Analysis

At this point, you may ask why all this is important. The importance is not in the ability to calculate all the details of the wave shape but, rather, to understand that each repetitive wave shape is made up of a unique set of sine waves. Once you accept this fact, then you will be able to understand that the shape of the resultant wave form may be determined by knowing which sine waves are present in the formation of the complex wave form.

The examples of adding waves, pictured in Figures 28–1 and 28–2, have shown you that phase is important. An amplifier may have the ability to reproduce any single frequency accurately. But if the proper phase relationship does not exist for all frequencies, reproduction of a complex wave will not be good. Note that the resultant wave shapes in Figures 28–1 and 28–2 are significantly different because of the different phase relationships between the individual sine waves.

28–3 FREQUENCY DOMAIN VERSUS TIME DOMAIN

Both the frequency domain and time domain are important for a thorough understanding of electronics. For time-domain measurements, the most commonly used electronic instrument is the *oscilloscope*. The oscilloscope measures the amplitude of a signal in relation to time. On the other hand, the *spectrum analyzer* is the most commonly used instrument for obtaining frequency-domain information. The spectrum analyzer presents a display of the frequencies present in a wave form and the relative amplitudes of each of the frequencies.

It is important for the electronics technician to understand when to use the oscilloscope and when to use the spectrum analyzer. An oscilloscope is more useful when determining the timing relationship in a digital circuit or when checking the shape of a particular wave form, such as a square wave, a sawtooth wave, or a sine wave. However, it is not useful in determining what frequencies are present in a recording of Beethoven's Fifth Symphony played by an orchestra. This information, however, is easily obtained with a spectrum analyzer.

Frequency domain is especially useful in the analysis of complex wave forms that would not be readily analyzed in the time domain. Many seemingly complex wave forms in the time domain have very simple frequency domain structures, and vice versa.

28–4 SPECTRUM ANALYSIS

The term frequency spectrum refers to the frequency components that exist in a continuous range or spectrum. Spectrum analysis is, therefore, the study of the frequency components in

Complex Wave Forms and the Frequency Domain

a complex wave form or signal. When the components of a particular electrical signal are known, operations can be performed on that signal by adding, removing, or changing the various components that make up the signal.

Consider the wave form shown in Figure 28–4(a). This wave form is basically a 1000-Hz sine wave with some additional frequency or frequencies added to it. For our hypothetical circuit to work properly, we must have a pure 1000-Hz signal without any extraneous signal added to it. If the extra signal can be determined, other electronic techniques, such as filtering, may be used to remove the unwanted signal. However, by just looking at the signal in Figure 28–4(a), it is very difficult, if not impossible, to determine the individual signals making up the displayed resultant.

The application of the spectrum analyzer to the signal results in a direct display of the frequency components and their amplitudes on the screen of the spectrum analyzer. For purposes of discussion, assume that the complex signal of Figure 28–4(a) consists of $f_1 = 1000$ Hz and $f_2 = 10\,000$ Hz. Furthermore, the amplitude of f_2 is one-tenth that of f_1. Then the spectrum analyzer would have an output on its screen as shown in Figure 28–4(b).

If the complex wave forms of Figures 28–1 and 28–2 were injected into a spectrum analyzer, the outputs would be identical because the spectrum analyzer cannot differentiate phase. The frequency spectrum for both signals is pictured in Figure 28–5.

Frequency Response

The term *frequency response* is commonly used when talking about hi-fi amplifiers, speakers, and other related equipment. Frequency response is a plot of the resulting amplitudes of different frequencies within the spectral range of the component under test when input frequencies of the same amplitude are injected into the device. A *flat* frequency response means that the amplitudes of the output frequencies are all identical when equal-amplitude input frequencies are used. A flat frequency

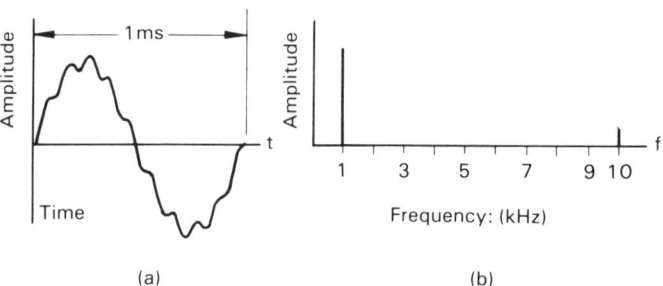

FIGURE 28–4
(a) 1000 Hz sine wave with additional frequency added to it, as viewed on a time base oscilloscope.

(b) Frequency spectrum of the signal, as viewed on a spectrum analyzer.

Harmonic Distortion

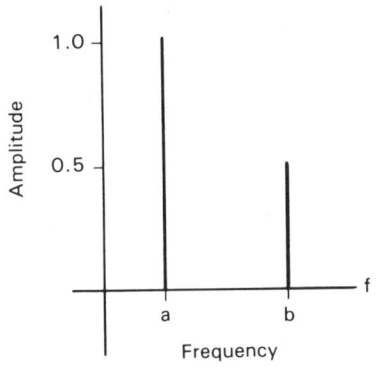

FIGURE 28-5
Frequency spectrum of the complex wave form of Figures 28-1 and 28-2.

response is necessary for true reproduction of an input signal since a complex signal made up of identical harmonic frequencies but different amplitude frequencies will sound different.

If a specification states that the frequency response is 20 to 20 000 Hz ± 2 dB, then given any input with identical amplitudes and a frequency between 20 and 20 000 Hz, the output amplitudes will be within ±2 dB of each other. All the frequency responses shown in Figure 28-6 satisfy this particular specification. In fact, there are an infinite number of possibilities that would fit this particular criterion.

28-5 HARMONIC DISTORTION

Harmonic distortion is present to some degree in almost all electronic devices. A device has harmonic distortion if its transfer function (the relationship between the output signal and the original input signal) is not perfectly linear. By definition, a device has harmonic distortion if the output of the device

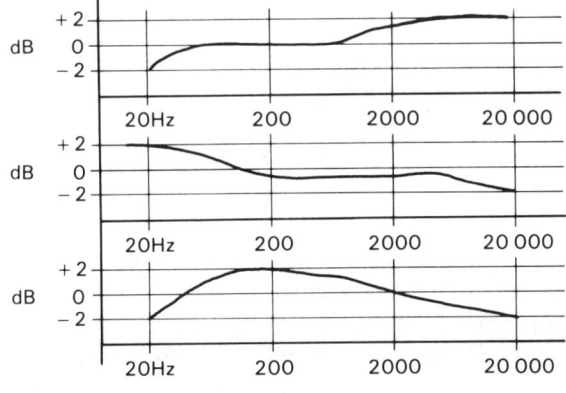

FIGURE 28-6
Selected frequency response plots which satisfy the specification 20 Hz to 20 000 Hz ± 2 dB.

contains frequency components that are harmonics or integral multiples of the frequency of the input. For instance, if a 1-kHz signal is injected into a hi-fi amplifier, ideally the output will contain only an amplified 1-kHz signal. However, the perfect amplifier has yet to be created. Thus, with a 1-kHz signal into the amplifier, the output would also contain varying amounts of 2-kHz, 3-kHz, and so on, signals. In measuring the harmonic content of the output with the spectrum analyzer, the fewer harmonics observed the better, since hi-fi amplifiers are designed to have an output signal identical to their input signal.

Mathematically, harmonic distortion is measured as a percentage of the harmonics present, as related to the original or fundamental signal. If f_1 is the fundamental, then $f_2 = 2f_1$ is the second harmonic, and $f_3 = 3f_1$ is the third harmonic, and so on. The equation for percentage of harmonic distortion (% HD) is stated in terms of the amplitude of the fundamental, A_1, and the amplitude of the harmonics, A_2, A_3, and so on. Thus

$$\% \text{ HD} = \sqrt{\frac{A_2^2 + A_3^2 + \cdots + A_n^2}{A_1^2}} \times 100 \qquad (28\text{--}1)$$

EXAMPLE 28–1 Determine the percent of harmonic distortion of a device when the output contains a second harmonic of 200 mV, a third harmonic of 70 mV, and a fourth harmonic of 10 mV. The fundamental measures 5 V.

SOLUTION Using Equation 28–1, substitute $A_1 = 5$ V, $A_2 = 0.2$ V, $A_3 = 0.07$ V, and $A_4 = 0.01$ V. Thus

$$\% \text{ HD} = \sqrt{\frac{0.2^2 + 0.07^2 + 0.01^2}{5^2}} \times 100 = 4.24$$

∴ The percent of harmonic distortion is 4.24%.

The presence of harmonics in a wave is not necessarily undesirable. Consider the case of different musical instruments playing the same note. Each instrument sounds different from the other because each generates harmonics of different amplitude from those of the other. When the piano and the violin, each playing the same note, are compared using the spectrum analyzer (see Figure 28–7), the difference in the harmonics is apparent.

Electronic music is created by synthesizing, or adding together, sine waves of various frequencies to create the desired tone. This tone can then be turned on or off with different rise and fall times to re-create an infinite number of possible sounds. By re-creating the harmonics and the appropriate rise and fall

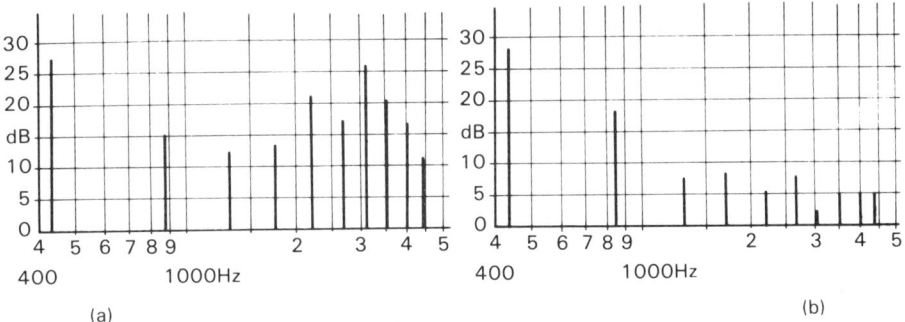

FIGURE 28-7 (a) Spectrum of 440 Hz note on violin. (b) Spectrum of 440 Hz note on piano.

times (attack and decay times), the sound of various instruments is closely approximated through electronic synthesis.

28-6 INTERMODULATION DISTORTION

In the previous section, harmonic distortion was discussed. It was learned that harmonic distortion results when undesirable harmonics of a signal are generated within a device, a circuit, or a system. Mathematically, if f_1 is the input signal, then harmonic distortion occurs when signals equal to $2f_1$, $3f_1$, $4f_1$, and so on, are generated.

Unlike harmonic distortion, intermodulation distortion (IM distortion) occurs when two or more signals generate frequencies that are the sum and difference of the signals. IM distortion may be stated mathematically, when f_1 and f_2 are present at the same time in a device, a circuit, or a system, as

$$f_{IM} = \pm mf_1 \pm nf_2 \quad (Hz) \tag{28-2}$$

where m and $n = 0, 1, 2, 3, 4$, and so on.

Since m and n can assume any whole-number value, there are infinite combinations of $\pm mf_1 \pm nf_2$, resulting in an infinite number of frequencies. In well-designed linear devices, as m and n assume large values, the amplitude of the resulting signals is very small and can be ignored. In fact, the perfect linear device only changes the amplitude of the input signal and not the frequency content. The perfect linear device does not generate new frequencies in its output signal.

EXAMPLE 28-2 Two frequencies, $f_1 = 35$ MHz and $f_2 = 40$ MHz, are fed into an amplifier. Determine the possible intermodulation distortion products that might result when m and n assume values from 0 to 3.

SOLUTION Use Equation 28-2 where $f_1 = 35$ MHz and $f_2 = 40$ MHz; $m = 0, 1, 2, 3$, and $n = 0, 1, 2, 3$.

m	n	$f_{IM} = mf_1 + nf_2$ (MHz)	$f_{IM} = mf_1 - nf_2$ (MHz)	$f_{IM} = -mf_1 + nf_2$ (MHz)
0	0	0	0	0
0	1	40	N/A	40
0	2	80	N/A	80
0	3	120	N/A	120
1	0	35	35	N/A
1	1	75	N/A	5
1	2	115	N/A	45
1	3	155	N/A	85
2	0	70	70	N/A
2	1	110	30	N/A
2	2	150	N/A	10
2	3	190	N/A	50
3	3	225	N/A	15

Observation The letters N/A mean *not applicable*.

From this example, notice how many different frequencies exist as a result of intermodulation distortion. When each wave, represented by a frequency in the example, is added to the others, the resulting wave shape will be quite different from the original.

By using a piano, it may be shown that intermodulation distortion is more detrimental than harmonic distortion to the output of a linear device. By playing notes that are multiples of each other, such as an octave, the sound is quite pleasant. However, by playing random notes that are not multiples of each other, such as notes right next to each other, the sound heard is a dissonant sound.

When intermodulation distortion occurs, many frequencies are present that are not even multiples of each other, as was demonstrated in the previous example. It is very important that intermodulation distortion be minimized since most electronic systems process signals having more than just one frequency. Unlike the field of music where it is very important for musical instruments to generate harmonics, the field of electronics does not want devices creating new harmonics.

Using IM Distortion Although linear amplifiers are designed to eliminate intermodulation distortion, there are some devices designed specifically to create intermodulation distortion. One such device is the *mixer*, which is used extensively in the radio communications field. The mixer is also known as an *up converter* or a *down converter*.

By using a mixer, a signal at one frequency may be shifted up or down by mixing it with another frequency. This process is pictured in Figure 28-8. In the example pictured, a low frequency of 1 kHz is fed into the mixer along with the 100-kHz frequency from the local oscillator. Since the mixer has been specifically designed to emphasize the sum of the two input frequencies (the intermediate frequency, I, and the local oscillator frequency, L), the output signal has a frequency of 101 kHz. As you might suspect, other combinations of frequencies exist, but these are filtered out, leaving only the radio frequency signal, R, in the output.

In closing our discussion of intermodulation distortion, you should realize that determining the products of intermodulation distortion is difficult because of the large number of possible combinations that might be present in the output signal. Contrast this with harmonic distortion where we only have to know the original frequency to have a knowledge of the harmonics since they are integer multiples of the input frequency.

28-7 PRACTICAL APPLICATIONS

An understanding of the frequency-domain spectral analysis and the composition of complex waves leads to a wealth of applications in both theoretical and practical electronics.

Square-Wave Testing

In addition to the sine-wave frequency response plot made with hi-fi amplifiers, most are tested with square waves to check their behavior when amplifying complex wave forms. The use of a square wave saves a great deal of time because a square wave is composed of an infinite number of sine waves whose frequencies are odd multiples of the fundamental frequency. In essence, many frequency measurements are made simultaneously with the square wave. Several of the output responses that might be viewed on an oscilloscope when a square wave is put into an amplifier are pictured in Figure 28-9. When a square wave is not perfectly reproduced, the cause may be a combination of both amplitude and phase distortion. Remember that the

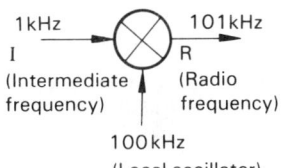

FIGURE 28-8
Symbol for a mixer showing various input and output signals.

Complex Wave Forms and the Frequency Domain

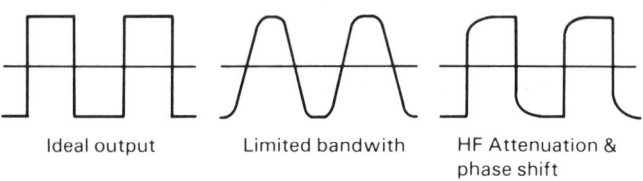
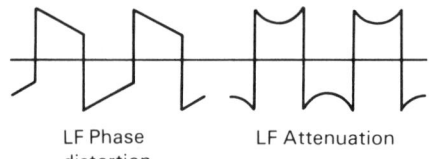

FIGURE 28-9 Output responses to a square wave input.

phase of the harmonics of a complex wave must be unchanged in order for a perfect reproduction to take place.

Filter Applications Filters, which will be discussed in more detail in the next chapter, are used to remove unwanted frequencies. The knowledge of complex wave forms and spectral analysis allows you to determine what frequencies are present in a given signal and, hence, allows you to design or select a filter to remove those frequencies that are unwanted.

Radio-Frequency Communications All forms of RF communications, including radio, television, radar, and satellite systems, require a detailed knowledge and control of the frequencies being transmitted and received.

Music Synthesis The synthesis of music through the use of electronics can be achieved when the required frequencies needed to create the desired sounds are known. Electronic organs use this fact to imitate the sounds of different musical instruments.

Distortion Analysis Knowing the mathematics behind harmonic and intermodulation distortion gives you an understanding of what to look for when analyzing electronic circuits or devices for distortion. Some of the most critical design considerations involve the presence of undesirable or *spurious signals*.

PROBLEMS

Sections 28-1 and 28-2

28-1 Using your own words, state the Fourier theorem.

28-2 Sketch the result of adding two signals having equal amplitude and frequency but different phase. Use phase differences of (a) 90° and (b) 180°.

28-3 A three-way speaker system has a perfect frequency response when tested with a sine-wave generator. However, because the physical location of the speakers in the

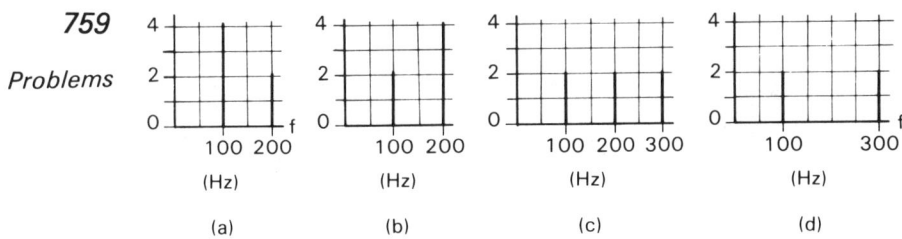

FIGURE 28-10 Spectral diagrams for Problems 28-4 through 28-7.

cabinet is not appropriate, phase distortion has occurred. In your own words, describe the result of this phase distortion on the complex wave forms that are present in music.

Sections 28-3 and 28-4

28-4 Sketch two cycles of the complex wave form (in the time domain) whose spectrum diagram is shown in Figure 28-10(a). Assume that the frequency components are in-phase sine waves.

28-5 Repeat Problem 28-4 for the information of Figure 28-10(b).

28-6 Repeat Problem 28-4 for the information of Figure 28-10(c).

28-7 Repeat Problem 28-4 for the information of Figure 28-10(d).

28-8 Sketch the complex wave form resulting when a 60-Hz sine wave signal with an amplitude of 10 is combined with a 600-Hz signal with an amplitude of 1.

28-9 Using the parameters of Problem 28-8, draw the spectrum diagram of the two signals in Problem 28-8. Use semilog paper.

Section 28-5

28-10 Determine the third harmonic of a 350-kHz frequency.

28-11 A device has only odd harmonic distortion; state the first three frequencies that are present in addition to the fundamental frequency of 125 MHz.

28-12 Given the frequency information in the spectrum diagram of Figure 28-11, determine the percent of harmonic distortion of the fundamental.

28-13 Determine the percent of harmonic distortion of a device when the output contains a second harmonic of 150 mV, a fourth harmonic of 100 mV, and a sixth harmonic of 80 mV. The fundamental measures 6 V.

Section 28-6

28-14 Two signals having frequencies of 2000 and 3000 Hz are injected into an amplifier. Using Equation 28-2, deter-

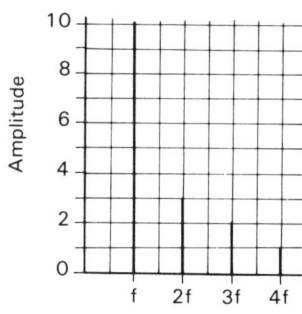

FIGURE 28-11
Frequency spectrum of signal in Problem 28-12.

mine all possible output frequencies from $m, n = 0$ to $m, n = 3$.

28-15 A 200- and a 300-Hz signal are put into an amplifier. Determine the possible 2 by 2 intermodulation products. *Note:* A 2 by 2 *IM* product means that $f_{IM} = \pm 2f_1 \pm 2f_2$.

28-16 Determine the 3 by 3 intermodulation products of the two frequencies 125 and 480 MHz. See the note in Problem 28-15.

28-17 A microwave mixer, as pictured in Figure 28-8, is designed to convert a 4-GHz input signal down to a 160-MHz output signal by producing the difference in the output. Determine the two possible local oscillator frequency settings.

29 Introduction to Filters

29-1 LOW-PASS FILTERS
29-2 HIGH-PASS FILTERS
29-3 BODE PLOT
29-4 BANDPASS FILTERS
29-5 BAND-REJECT FILTERS
29-6 TUNABLE FILTERS
29-7 FILTER SPECIFICATIONS AND TERMINOLOGY

CHAPTER PREVIEW You may not be aware of it, but you already have some understanding of the function filters play in electronic circuits. For, you see, filters used in electronic circuits serve the same general purpose as filters used in a vacuum cleaner, forced-air furnace, or aquarium, that is, to select what is to pass through the filter and what is to be blocked by the filter. As an example, the filter in a furnace or a vacuum cleaner lets air through but stops small particles. In electronic circuits, filters are designed to select certain frequencies and block others.

In the following sections, different types of filters will be introduced along with their application. Once you have completed this chapter, you will have a knowledge of the basic terminology used with filters as well as an introductory overview understanding of filters.

PERFORMANCE OBJECTIVES Once you have read each section, worked each example with pencil, paper, and calculator, and answered each problem of every section, you will be able to:

☐ Understand the basic functions of the different types of filters.
☐ Have a basic knowledge of filter specifications.
☐ Construct and analyze several simple low-pass and high-pass filters.
☐ Draw Bode plots of single and multistage filters.

29–1 LOW-PASS FILTERS

A low-pass filter is an electrical network that allows low frequencies to pass but attenuates the high frequencies. The *ideal* low-pass filter has a frequency characteristic curve as pictured in Figure 29–1(a). Figure 29–1(b) pictures the frequency characteristic curve of the *practical* low-pass filter.

LR Low-Pass Filter

To identify the simplest form of low-pass filter, think of a device that passes dc or low frequencies and suppresses high frequencies. As you have previously learned, an inductor meets

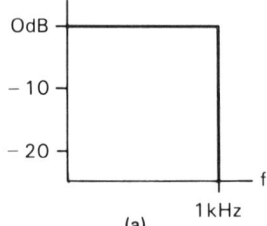 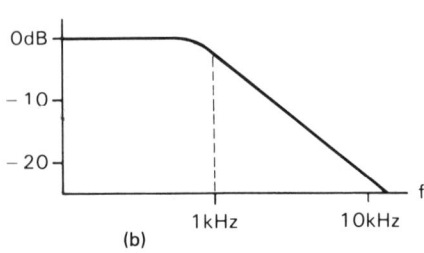

FIGURE 29–1 (a) Characteristic curve of an ideal 1 kHz low-pass filter. (b) Characteristic curve of a practical 1 kHz low-pass filter.

Low-Pass Filters

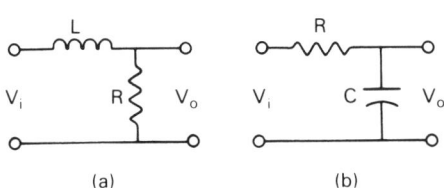

FIGURE 29-2
Two elementary low-pass filter circuits.

this criterion since its impedance increases with frequency. The circuit of a simple low-pass filter is pictured in Figure 29-2(a).

The output voltage, V_o, is determined by applying the following equation to the circuit:

$$V_o = \frac{V_i R}{R + jX_L} \quad (V) \tag{29-1}$$

where V_o = output voltage, V_i = input voltage, and $X_L = \omega L$. From Equation 29-1, $V_o = V_i$ when $f = 0$ Hz (dc) and $X_L = 0\ \Omega$. V_o decreases toward zero as $X_L = 2\pi f L$ increases.

Cutoff Frequency

By definition, the *cutoff* frequency, f_c, of a filter is that frequency at which the output power has dropped to one-half of the input power. In terms of decibels, it is the frequency that corresponds to the -3-dB point in the frequency characteristic curve. This condition occurs when $X_L = R$ or when $2\pi f_c L = R$. Rearranging and solving for the cutoff frequency, f_c, results in

$$f_c = \frac{R}{2\pi L} \quad (Hz) \tag{29-2}$$

EXAMPLE 29-1 Determine the cutoff frequency of the circuit of Figure 29-2(a) when $L = 10$ mH and $R = 50\ \Omega$.

SOLUTION Use Equation 29-2.

$$f_c = \frac{R}{2\pi L} = \frac{50}{2\pi 10 \times 10^{-3}} = 796\ \text{Hz}$$

∴ This filter will pass frequencies below 796 Hz with relatively little attenuation.

Observation At 796 Hz, there will be 3 dB of attenuation, and the attenuation increases as the frequency increases, as noted in Figure 29-1(b).

RC Low-Pass Filter

Another simple circuit configuration of a low-pass filter is shown in Figure 29-2(b). In this circuit, the capacitor appears as an open circuit at low frequencies, resulting in V_i equaling V_o. As the frequency of the input source rises, the impedance of C

decreases, and V_o decreases toward zero. The output voltage, V_o, may be determined by applying the following equation:

$$V_o = \frac{V_i X_C}{R + jX_C} \quad (V) \tag{29-3}$$

where V_o = the output voltage, V_i = the input voltage, and $X_C = 1/(2\pi f C)$.

The cutoff frequency occurs when $X_C = R$ or when $1/(2\pi f C) = R$. Rearranging and solving for the cutoff frequency, f_c, results in

$$f_c = \frac{1}{2\pi RC} \quad (Hz) \tag{29-4}$$

EXAMPLE 29-2 Determine the cutoff frequency of the circuit of Figure 29-2(b) when $C = 0.01\ \mu F$ and $R = 100\Omega$.

SOLUTION Use Equation 29-4.

$$f_c = \frac{1}{2\pi RC} = \frac{1}{2\pi 100 \times 0.01 \times 10^{-6}}$$

$$\therefore \quad f_c = 159\ kHz$$

Observation As noted in this example, a low-pass filter does not mean that the cutoff frequency is a low frequency. It simply means that it passes all the frequencies below the cutoff frequency.

DC Power-Supply Filter

The low-pass filter has many applications, including the filter for the dc power supply. Other than batteries, dc power is derived from a rectified ac source in which all frequency components greater than zero hertz (dc) are filtered out.

A *block* diagram of a dc power supply is shown in Figure 29-3. The primary source, noted as an ac generator, is from the 60-Hz power line. The full-wave rectifier inverts the 60-Hz sine wave into a 120-Hz pulsating dc wave form, as indicated in Figure 29-3. When the signal is passed through the low-pass filter, all frequencies 120 Hz and greater are attenuated to an insignifi-

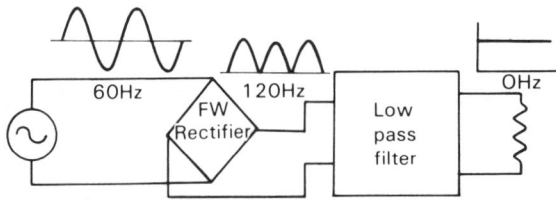

FIGURE 29-3. Block diagram of a dc power supply with frequency domain wave forms noted.

FIGURE 29-4
Frequency spectrum of the signals present in a dc power supply. (a) from the 60 Hz ac line, (b) at the input of the low-pass filter, and (c) at the load of Figure 29-3.

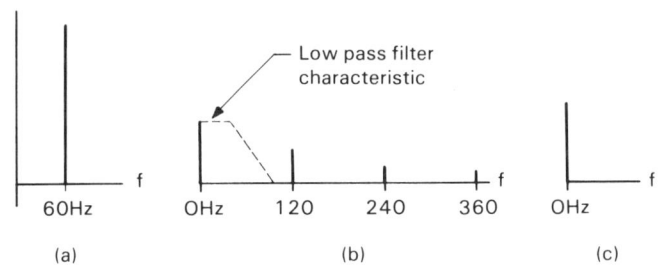

cant value, resulting in only the dc component of the wave present in the load. Figure 29-4 pictures the frequency spectrum present in the power supply of Figure 29-3. In the frequency spectrum at the output of the rectifier, Figure 29-4(b), it may be seen that the signal contains a dc component equal to the average value of the ac signal plus harmonics of the basic signal frequency.

In manufacturing power supplies, the cost of production rises with an increase in the amount of filtering. Inexpensive power supplies usually have a measurable amount of the residual 120-Hz frequency present in the output, as pictured in Figure 29-5. This component in the output of the dc power supply is called *ripple*.

29-2 HIGH-PASS FILTERS

A high-pass filter, as pictured in Figure 29-6, is an electrical network that allows high frequencies to pass but attenuates the low frequencies. Look carefully at Figures 29-6(a) and (b) to see how the components are interchanged for a high-pass filter versus the low-pass filter of Figure 29-2.

The cutoff frequency described in the previous section is determined in the same manner using the same equations. The only difference is in the definition of cutoff frequency. In the high-pass filter, the cutoff frequency is that frequency above which the signal will pass the filter unaltered. For your conve-

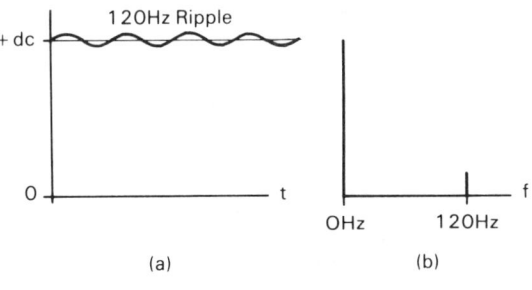

FIGURE 29-5
(a) Output wave form of a dc power supply containing ripple.
(b) Frequency spectrum of the wave form in Figure 29-5(a).

Introduction to Filters

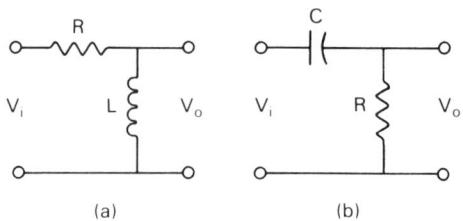

FIGURE 29–6
Two elementary high-pass filter circuits.

nience, Equations 29–2 and 29–4 are restated. Thus, for the RL high-pass filter,

$$f_c = \frac{R}{2\pi L} \quad \text{(Hz)} \tag{29-2}$$

and for the RC high-pass filter,

$$f_c = \frac{1}{2\pi RC} \quad \text{(Hz)} \tag{29-4}$$

The characteristic curves of both ideal and practical high-pass filters are pictured in Figure 29–7.

EXAMPLE 29–3 Determine the cutoff frequency of the circuit of Figure 29–6(b) when $R = 47$ kΩ and $C = 200$ pF.

SOLUTION Use Equation 29–4.

$$f_c = \frac{1}{2\pi RC} = \frac{1}{2\pi 47 \times 10^3 \times 200 \times 10^{-12}}$$

$$\therefore \quad f_c = 16.93 \text{ kHz}$$

Observation Frequencies above 16.93 kHz will be passed relatively unattenuated, whereas frequencies below 16.93 kHz will be attenuated by an increasingly greater impedance as the frequency decreases.

LC High-Pass Filter To understand how the LC high-pass filter of Figure 29–8 behaves, consider what occurs at the frequency extremes. At dc

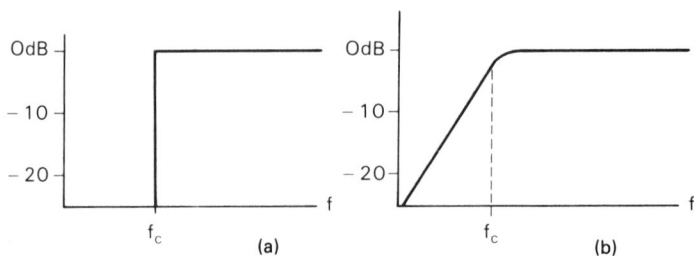

FIGURE 29–7
(a) Characteristic curve of an ideal high-pass filter.
(b) Characteristic curve of a practical high-pass filter.

High-Pass Filters

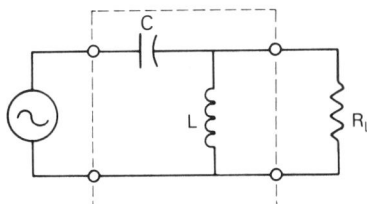

FIGURE 29-8 An LC high-pass filter, shown within dashed box, with a resistive load attached.

(zero hertz) the capacitor acts as an open circuit while the inductor acts as a short circuit. Conversely, at very high frequencies, the capacitor acts as a short circuit while the inductor acts as an open circuit. Thus, the circuit behaves as a high-pass filter. The cutoff frequency is that frequency that results when $X_C = X_L$. Thus

$$X_C = X_L \quad \text{and} \quad \frac{1}{2\pi f_c C} = 2\pi f_c L$$

Multiply both members by $2\pi f_c C$:

$$1 = 4\pi^2 f_c^2 LC$$

Solve for f_c by taking the square root of each member:

$$f_c^2 = \frac{1}{4\pi^2 LC}$$

$$f_c = \frac{1}{2\pi \sqrt{LC}} \quad \text{(Hz)} \tag{29-5}$$

EXAMPLE 29-4 Determine the cutoff frequency of the circuit of Figure 29-8 when $L = 10$ mH and $C = 20$ µF.

SOLUTION Use Equation 29-5:

$$f_c = \frac{1}{2\pi \sqrt{LC}} = \frac{1}{2\pi \sqrt{10 \times 10^{-3} \times 20 \times 10^{-6}}}$$

$$\therefore \quad f_c = 356 \text{ Hz}$$

Observation You may develop the analogy for the LC low-pass filter by simply interchanging the L and C components in Figure 29-8 and then applying Equation 29-5.

DC-Blocking Capacitor The most complete yet simple high-pass filter is the *dc-blocking capacitor*. The purpose of this device is to allow ac signals to pass without attenuation while blocking dc voltage and current. The blocking capacitor is used in electronic active circuits (amplifiers, oscillators, etc.), where different sections of the cir-

EXAMPLE 29-5 The ac output signal from the amplifier, as pictured in Figure 29-9, has a dc voltage level called a dc *offset*. When this signal, containing a dc component, is *fed* directly into a loudspeaker, it causes the speaker cone to be physically displaced from its normal center position, thereby limiting the mechanical motion of the speaker cone. However, by feeding the signal through a capacitor, which acts as a high-pass filter, the dc level is blocked, and the speaker will have use of its full range of mechanical travel to reproduce the amplified sound.

EXAMPLE 29-6 In the circuit of Figure 29-10, the voltage across R_1 is a 2-V peak-to-peak (2-V p-p) 10-kHz sine wave with an average voltage (dc voltage) of +10 V. Because the signal generator, e_s, cannot have an offset dc voltage or current applied to it, a dc-blocking capacitor is placed in series with the generator to prevent dc from flowing through it.

Observation A quick calculation will show that the 100-μF capacitor is virtually a short circuit at the 10-kHz frequency (X_C = 0.16 Ω).

29-3 BODE PLOT

The frequency response of filters can be easily plotted using the *Bode plot*, a technique named for the scientist who invented it. The Bode plot of the simple RL low-pass filter pictured in Figure 29-1(b) is determined by applying the voltage-divider equation. Thus

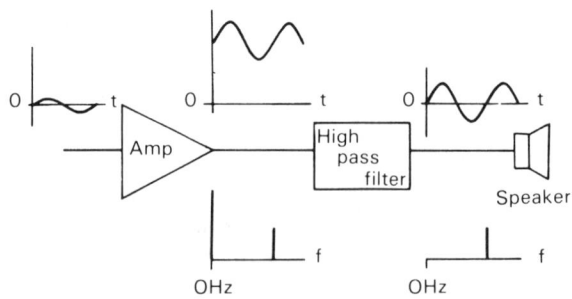

FIGURE 29-9
A high-pass filter removes the dc component from the output of the audio amplifier.

Bode Plot

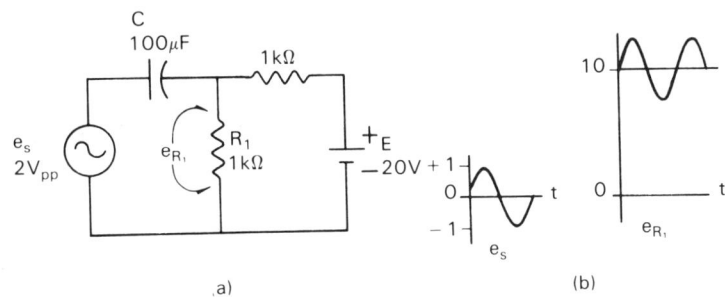

FIGURE 29-10
(a) A capacitor is used to block the dc component from entering the signal generator.
(b) Wave forms of the signals e_s and e_{R_1}.

$$V_o = \frac{V_i R}{R + j\omega L} \quad \text{(V)} \tag{29-6}$$

From Equation 29-6, when $\omega = 0$ rad/s, then

$$\frac{V_o}{V_i} = \frac{R}{R + j0} = \frac{R}{R} = 1$$

With $\omega = 0$ rad/s as a reference point, an increase in ω will result in a decrease in the ratio of V_o/V_i. Figure 29-1(b) is the Bode plot of the amplitude of the ratio of V_o to V_i as a function of frequency ($\omega = 2\pi f$). Because of the linear nature of the curve, a very accurate plot can be made without having to calculate numerous points. This, then, is the power of the Bode plot.

Constructing the Bode Plot

To properly draw the Bode plot of a simple RC or RL filter, you must use *semilog* graph paper. As pictured in Figure 29-11, the semilog paper has linear graduations in the vertical or y-axis and logarithmic graduations in the horizontal or x-axis. The linear axis is labeled in units of decibels, while the log axis is labeled in either units of angular velocity, radians per second (rad/s), or frequency (hertz).

Because of its logarithmic nature, the horizontal axis can only be changed by factors of 10^n, where n is an integer number.

To construct the curve of the simple low-pass filter, as pictured in Figure 29-11, draw a straight line at the 0-dB level from the lowest frequency to the cutoff frequency, f_c. As in previous cases, the cutoff frequency is calculated using the following formulas. For the RL filters,

$$f_c = \frac{R}{2\pi L} \quad \text{(Hz)} \quad (29\text{-}2) \quad \text{or} \quad \omega_c = \frac{R}{L} \quad \text{(Hz)} \tag{29-7}$$

And for the RC filter,

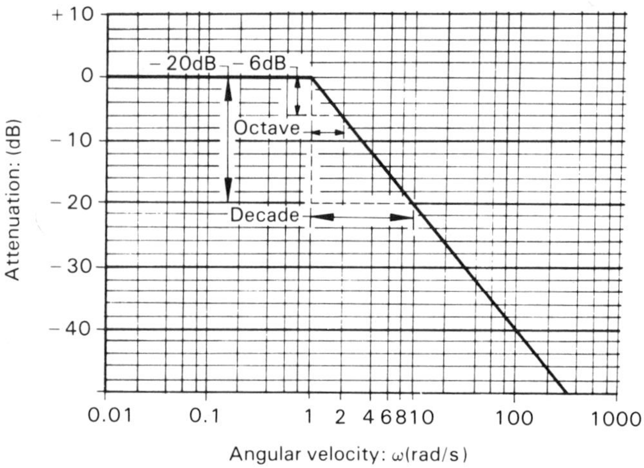

Introduction to Filters

FIGURE 29-11 Example of semi-log graph paper used for the Bode plot of a single-stage low-pass filter.

$$f_c = \frac{1}{2\pi RC} \quad \text{(Hz)} \quad (29\text{–}4) \quad \text{or} \quad \omega_c = \frac{1}{RC} \quad \text{(Hz)} \quad (29\text{–}8)$$

Observation: The cutoff frequency is also known as the *break frequency* because the amplitude *"breaks"* into a different response at that point.

The remainder of the plot above the cutoff frequency, f_c, is constructed by allowing the slope of the curve to diminish at the rate of -6 dB/octave, which is also equal to -20 dB/decade, as pictured in Figure 29-11. An *octave* is a doubling of frequency (or angular velocity); a *decade* is a tenfold increase in frequency (or angular velocity).

EXAMPLE 29-7 Draw the Bode plot of an RC low-pass filter when $R = 3.2$ kΩ and $C = 0.01$ μF.

SOLUTION Use Equation 29-4 to determine the break frequency.

$$f_c = \frac{1}{2\pi RC} = \frac{1}{2\pi 3.2 \times 10^3 \times 0.01 \times 10^{-6}}$$

$$f_c = 5.0 \text{ kHz}$$

Construct the Bode plot, as pictured in Figure 29-12.

∴ The break frequency is 5 kHz and the curve drops off at the rate of -20 dB/decade.

Observation The Bode plot is an approximation of the actual curve of a filter. The approximation is worst at f_c, where the curve has an error in amplitude of 3 dB. To construct a more accurate response

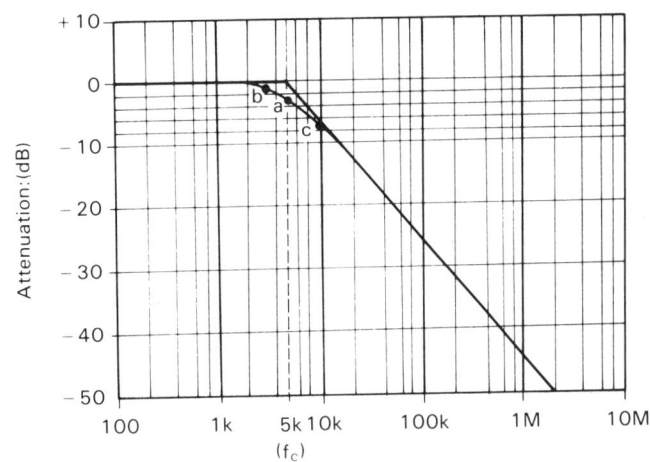

FIGURE 29-12
Bode plot of the filter in Example 29-7.

curve in this region, determine a point 3 dB below the actual break frequency, f_c (point a of Figure 29-12). Sketch the response curve that passes through point a and two other points, b and c, which are determined by locating these points 1 dB below the original straight lines at frequencies equal to $f_c/2$ (point b) and $2f_c$ (point c). With three points known, a fairly accurate response curve of the filter can now be made.

Multiple-Pole Filters

Up to this point, we have been investigating the parameters of simple filters that have a −6-dB/octave or −20-dB/decade attenuation rate. In most applications, a more rapid attenuation is desired. The rates of attenuation usually considered are multiples of the simple −6 dB/octave, such as −12 dB/octave or −18 dB/octave. The majority of practical filters are in the −6-dB/octave to −60-dB/octave range.

As pictured in Figure 29-13, the Bode plot can be used to determine the characteristic of filters that have attenuation rates that are multiples of the simple −6-dB/octave filter. The characteristics of these filters are determined by drawing the attenuation line with a slope of the prescribed multiple of $n(-6$ dB/octave) or $n(-20$ dB/decade). However, in doing this, the error at the breakpoint frequency is also multiplied by the factor n. Therefore, a filter with 18-dB/octave attenuation rate would have an error of three times 3 dB or 9 dB at the breakpoint frequency. Similarly, at points $2f_c$ and $f_c/2$, there is an error of three times 1 dB or 3 dB.

Introduction to Filters

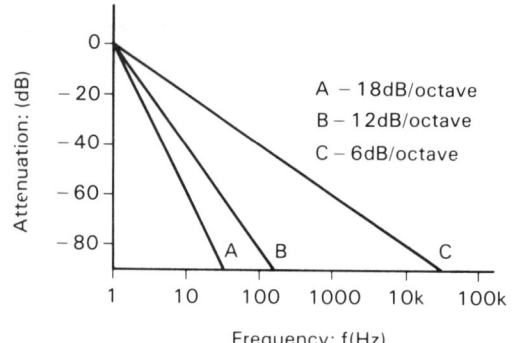

FIGURE 29-13
Bode plots of filters with attenuation rates of −6, −12, and −18 dB/octave.

The rate of attenuation of a filter is designated by the number of *stages* or *poles*. Hence, a two-stage or two-pole filter has an attenuation rate of two times the basic rate, or −12 dB/octave (−40 dB/decade). A three-stage or three-pole filter would have a −18 dB/octave or −60 dB/decade attenuation rate.

29-4 BANDPASS FILTERS

A bandpass filter is one that allows a certain range or *band* of frequencies to pass through it relatively unattenuated, as shown in Figure 29-14. Depending upon the design and purpose of the filter, the bandwidth may be very narrow or very wide. In radio receivers, bandpass filters are used to select out the specific frequencies being transmitted by a particular broadcast station. The bandpass filter in this application must be narrow enough to reject other nearby stations close in frequency to the one tuned to, yet wide enough to allow all the useful information to be received.

Another application of a bandpass filter is in the *three-way* hi-fi speaker system. A bandpass filter allows the mid-frequencies, typically 500 Hz to 5 kHz, to drive the midrange speaker while a low-pass filter diverts the frequencies below

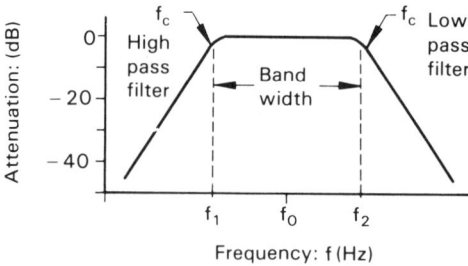

FIGURE 29-14
Bode plot of a bandpass filter.

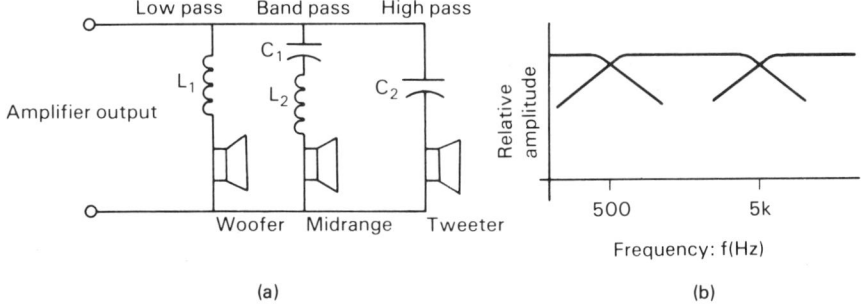

FIGURE 29-15 (a) Crossover network of a 3-way speaker system. (b) Bode plot of the characteristics of the 3-way crossover network.

500 Hz to the *woofer* and a high-pass filter diverts the frequencies above 5 kHz to the *tweeter*. A simple, three-way, −6-dB/octabe *crossover* network is shown in Figure 29-15(a).

A bandpass filter can be thought of in two ways. One is to consider it as a resonant circuit, as in Chapter 26. The other way is to think of it as a low-pass filter in series with a high-pass filter where the low-pass cutoff frequency is above that of the high-pass filter, as pictured in Figure 29-14.

EXAMPLE 29-8 A three-way speaker system needs a bandpass filter for its midrange speaker for proper operation. Determine the values of C_1 and L_2 of the bandpass filter in Figure 29-15(a). Assume the midrange speaker is equivalent to an 8-Ω resistor and that its optimum operating range is 500 Hz to 5000 kHz.

SOLUTION Consider the L and the C in two separate parts, where the RL portion is the low-pass filter and the RC portion is the high-pass filter. For a cutoff frequency of 500 Hz, high pass,

$$f_c = \frac{1}{2\pi RC} \quad \text{and} \quad C = \frac{1}{2\pi R f_c}$$

$$C = \frac{1}{2\pi 8 \times 500} = 39.8 \ \mu F$$

For a cut-off frequency of 5000 Hz, low pass,

$$f_c = \frac{R}{2\pi L} \quad \text{and} \quad L = \frac{R}{2\pi f_c}$$

$$L = \frac{8}{2\pi 5000} = 254.6 \ \mu H$$

$$\therefore \quad C_1 = 40 \ \mu\text{F} \quad \text{and} \quad L_2 = 250 \ \mu\text{H}$$

Observation To make sure that the capacitor contributes an insignificant impedance at 5000 Hz, solve for X_C.

$$X_C = \frac{1}{2\pi f_c} = \frac{1}{2\pi 5000 \times 40 \times 10^{-6}} = 0.8 \ \Omega$$

Test the inductor at 500 Hz to see that it contributes an insignificant impedance. Solve for X_L.

$$X_L = 2\pi f L = 2\pi 500 \times 250 \times 10^{-6} = 0.8 \ \Omega$$

Thus, $X_C = 0.8 \ \Omega$ at 5 kHz and $X_L = 0.8 \ \Omega$ at 500 Hz.

To summarize the previous example, at frequencies lower than 500 Hz, the capacitor has an increasingly greater impedance. At 500 Hz, the capacitor impedance is equal to the resistance of the speaker, hence the 3-dB point. At 500 Hz, the inductor contributes less than 10% of the total impedance. In the frequency range between 500 Hz and 5 kHz, both the capacitor and the inductor have impedances less than 8 Ω, and the majority of the power is applied to the speaker. At 5 kHz, the capacitor contributes less than 10% of the impedance and the inductor is equal to 8 Ω and, again, only half the available power reaches the speaker. At frequencies greater than 5 kHz, the inductor has increasing impedance and prevents the power from reaching the speaker.

29–5 BAND-REJECT FILTERS

A band-reject filter behaves in a manner exactly opposite to that of a bandpass filter. Instead of passing a selected band of frequencies, as in the bandpass filter, the band-reject filter attenuates a selected band of frequencies. Like the bandpass filter, the band-reject filter may be formed by combining a low-pass and high-pass filter in parallel, as shown in Figure 29–16(a). However, unlike the bandpass filter, the band-reject filter is designed so that the cutoff frequency of the low-pass filter is below that of the high-pass filter, as pictured in Figure 29–16(b). At low frequencies, the signal passes through the low-pass filter but is attenuated by the high-pass filter. In the *reject* range, both filters block the signal. At higher frequencies, the signal is passed by the high-pass filter and blocked by the low-pass filter.

Notch Filter Band-reject filters, also known as *notch* filters when the rejection band is narrow, are used to eliminate specific undesirable signals. For example, a radio communication system may experience interference and distortion in a receiver from a nearby

Tunable Filters

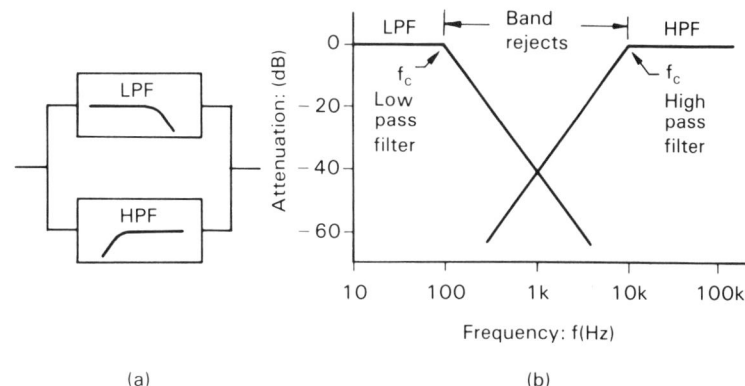

(a) (b)

FIGURE 29–16 (a) A band-reject filter may be formed from a low-pass filter (LPF) paralleled with a high-pass filter (HPF).
(b) Characteristics of a band-reject filter with a reject band of 100 Hz to 10 kHz and 20 dB/decade attenuation rates.

radio transmitter. Although the interfering transmitter is not operating on the same frequency, its close proximity and its strong signal overpower the receiver, preventing it from receiving the desired signal. By placing a band-rejection filter in the antenna system, the unwanted signal is attenuated and proper operation of the receiver is restored.

Because band-rejection filters are usually narrow-band devices used to eliminate specific frequencies, they are used in conjunction with combinations of low-pass, high-pass, and bandpass filters when a wide range of frequencies is to be eliminated.

Band-Reject Filter Characteristics The characteristics of a band-reject filter are shown in Figure 29–17. The specifications for the *reject bandwidth* can be made at various attenuations. For instance, in Figure 29–17, the 3- and 10-dB rejection bandwidths are noted. Also, notice that the rejection at the center frequency, f_0, is also indicated. When specifying the desired characteristic of a band-reject filter, a minimum rejection value at the center frequency is called out. The value specified would depend on the need of a specific system.

29–6 TUNABLE FILTERS

Up to this point, we have investigated filters with fixed characteristics. However, many of the most useful filters are tunable, either by mechanical or electrical methods.

Mechanically Tunable Filters Mechanically tunable filters usually use a capacitor which consists of plates that intermesh in varying amounts, depending

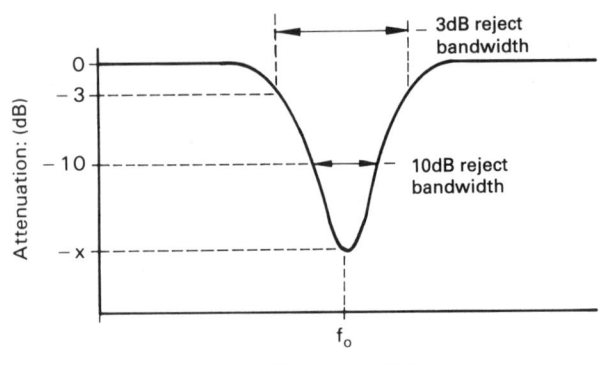

FIGURE 29-17
Band-reject filter parameters.

upon the setting. Most radios with mechanical tuning controls will have this type of capacitor somewhere in their circuits.

Electronically Tunable Filters

Electronically tunable circuits are becoming prevalent in the more sophisticated radio receivers. Many FM tuners found in the home or car hi-fi system are now *digitally* or electronically tunable. By pushing the appropriate button, a particular station is selected through the application of the proper voltage to a *varactor* or voltage-dependent capacitor. A varactor is a solid-state device whose capacitance depends on the voltage across it. Contrast electronic tuning to the mechanical push-button tuning where the capacitance in a filter (resonant circuit) is selected by *physically* changing the position of the variable plate capacitor. In each case, the basic filter principles remain the same, regardless of the tuning techniques involved.

29-7 FILTER SPECIFICATIONS AND TERMINOLOGY

Through the use of computers, the design of filters has become a fairly routine function. Aside from designing simple filters, the electronic circuit or system engineers will more often than not purchase filters from a company that specializes in filters. When purchasing from a vendor, it is important for the engineer to correctly specify the characteristics that are desired. To specify a filter, one must determine:

☐ The type of filter (i.e., low-pass, high-pass, etc.).
☐ The frequency characteristics, such as:
    The cutoff frequency for the low- or high-pass filter.
    The low and high cutoff frequencies for the bandpass and band-reject filter.

- The center frequency and the bandwidth of the bandpass and band-reject filter.
- ☐ The rate of attenuation defined by any of the following methods:
  - Number of poles or stages (i.e., 2-pole or 3-stage, etc.)
  - Decibel/octave (i.e., 12 dB/octave or 30 dB/octave, etc.)
  - Decibel/decade (i.e., 20 dB/decade or 60 dB/decade, etc.)
- ☐ Other specifications which are general to any passive device, including:
  - Insertion loss: the inherent signal power lost in the filter when a signal passes through it.
  - Input and output impedance of the filter network.
  - Power rating, to ensure that the filter does not get damaged by too powerful a signal.

Summary Since this chapter has been designed as an introduction to filters, we have only examined simple filters that have a flat amplitude response over their *passband*. In the detailed study of filter design, covered in undergraduate and graduate courses, there are many different ways to control the passband response, the phase response (which was not mentioned in this chapter), the attenuation rates, and the impedance match. As your experience grows, you will hear and read about filter terminology such as Butterworth, Chebyshev, constant k, m-derived, elliptical, and other mathematical terms that describe the functions and equations of filters. As a user of filters, this chapter has given you enough information to be able to understand the terms used in the purchase of a filter.

PROBLEMS

Section 29–1

29–1 A low-pass filter consists of the following components connected in series: $R = 10$ kΩ and $C = 0.01$ μF. Draw the schematic of this simple RC filter and determine the cutoff frequency.

29–2 An RL low-pass filter consists of an inductor of 15 mH and a resistor of 500 Ω. Draw the schematic and determine the cutoff frequency.

29–3 An 8-Ω speaker actually has a dc resistance of 7 Ω and an inductance of 0.25 mH. Determine the cutoff frequency of this speaker and determine if it would be better used as a woofer (low-frequency) or tweeter (high frequency) without knowing any of its other parameters.

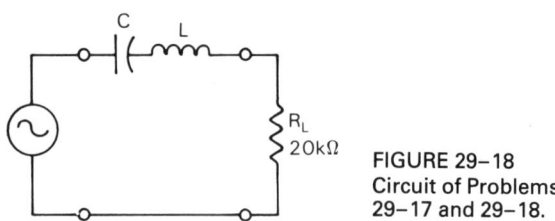

FIGURE 29-18
Circuit of Problems
29-17 and 29-18.

29-4 Determine the value of inductance needed in series with the speaker of the previous problem if you want it to start *rolling off* at 500 Hz.

Section 29-2   29-5 Draw the schematic and determine the cutoff frequency of an RC high-pass filter when $R = 50$ kΩ and $C = 1000$ pF.

29-6 For a particular 4-Ω tweeter to work properly, it needs a series capacitance to attenuate frequencies below 7 kHz. Determine the value of the capacitor needed to attenuate the signal by 3 dB at 7 kHz.

29-7 In Problem 29-6, assume that you cannot use a capacitor. Determine how an inductor could achieve the same results. Draw a schematic showing the value of the inductor and the circuit configuration.

29-8 You own an old turntable that has a lot of *rumble* (low-frequency noise) below 20 Hz. To attenuate the low frequencies, you decide to design a high-pass filter that has a 3-dB cutoff point at 50 Hz. The input to the amplifier has a resistance of 50 kΩ. Determine the value of the series capacitance needed in this application.

Note: Use semilog graph paper for the following problems.

Section 29-3   29-9 Draw the Bode plot of the filter in Problem 29-1.
29-10 Draw the Bode plot of the filter in Problem 29-2.
29-11 Draw the frequency response of the system in Problem 29-4.
29-12 Draw the Bode plot of the filter in Problem 29-5.

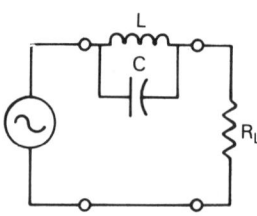

FIGURE 29-19
Circuit of Problems
29-21 and 29-22.

Problems 779

29-13 Draw the Bode plot of the filter in Problem 29-8.

29-14 Draw the Bode plot of a two-stage low-pass filter with $f_c = 10$ kHz.

29-15 Draw the Bode plot of a three-stage high-pass filter with $f_c = 250$ Hz.

29-16 Draw the resultant of a two-stage low-pass filter with $f_c = 20$ kHz connected in series with a two-stage high-pass filter with $f_c = 40$ Hz.

Section 29-4

29-17 Determine the values of L and C in Figure 29-18 required to achieve a bandpass filter with -3-dB frequencies of 10 and 200 kHz when R_L is 20 kΩ.

29-18 For Figure 29-18, $R_L = 8\ \Omega$; determine the values of L and C in order to obtain a bandpass of 700 Hz to 3 kHz.

29-19 Draw the Bode plot of a two-stage bandpass filter with breakpoints of 20 and 50 kHz.

29-20 Draw the Bode plot of a three-stage bandpass filter with breakpoints of 90 and 250 MHz.

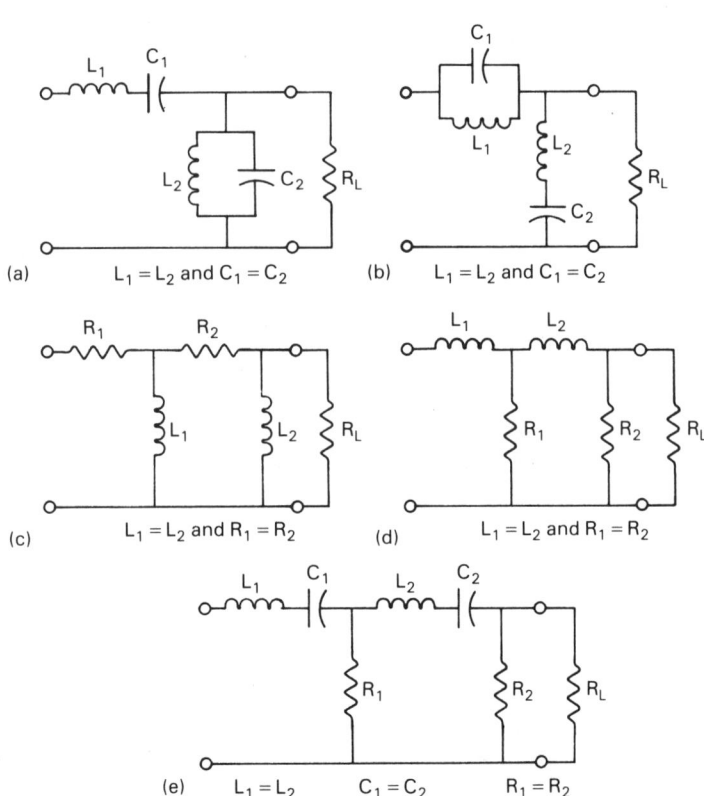

FIGURE 29-20
Filter configurations for Problem 29-23.

Section 29-5 29-21 Draw the very low frequency and the very high frequency equivalent circuits of the circuit in Figure 29-19.

29-22 For Figure 29-19, determine the value of L needed for a center-reject frequency (f_0) of 455 kHz when $C = 100$ pF.

29-23 Identify the types of filters shown in Figure 29-20.

APPENDIX

Answers To Selected Problems

CHAPTER 1

1-1 (a) Battery (b) Regulated Power Supply
1-3 Overload
1-5 Electric Circuit
1-7 See Figure A1.
1-9 See Figure A2.
1-11 See Figure A3.
1-13 See Figure A4.
1-15 See Figure A5.

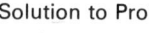

FIGURE A1
Solution to Problem 1-7

CHAPTER 2

2-1 (a) Mechanical (b) Electrical (c) Heat (d) Light
2-3 The law of conservation of energy indicates energy cannot be created or destroyed, so when an amount of energy in one form *"disappears,"* it must *"reappear"* in another form in an equivalent amount.
2-5 (a) Work (b) Energy
2-7 *Contact Force*—an object is moved or held in place by physical contact.
Field Force—an object is moved or held in place by an invisible force acting at a distance.
2-9 Run a comb through your hair on a dry day. The electric charge on the plastic will attract a small bit of paper.
2-11 (a) Repel (b) Attract
2-13 See Figure A6. Because the L shell has 8 electrons, this substance is very stable and is unlikely to form a compound with another element.
2-15 Neutral atom—number of protons is equal to the number of electrons.
Negative ion—more electrons than a neutral atom.
Positive ion—fewer electrons than a neutral atom.
2-17 *Ionic bonding* occurs when one atom transfers the electrons in its valence shell to another atom to form ions, each with 8 electrons in the outer shell, but with opposite charges. In contrast, *covalent bonding* results when electrons of two or more atoms are shared.
2-19 The *lattice* is made up of the atoms that are held in place by bonding and the *crystal* is formed by lattices, thus forming solid metal.

FIGURE A2
Solution to Problem 1-9

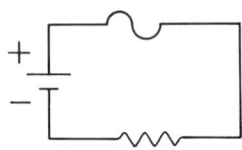

FIGURE A3
Solution to Problem 1-11

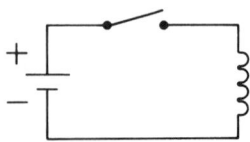

FIGURE A4
Solution to Problem 1-13

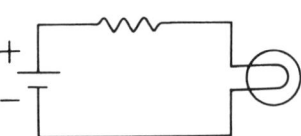

FIGURE A5
Solution to Problem 1-15

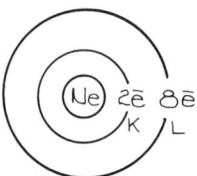

FIGURE A6
Solution to Problem 2-13

2-21 A coulomb equals the electron charge of 6.242 × 10¹⁸ electrons.
2-23 Voltage
2-25 Work
2-27 Generator, solar cell, thermocouple, and crystal cartridge are devices that raise the energy level of electric charge.
2-29 Energy is transmitted from a place of higher energy (the source) to a place of lower energy (the load). Because the load is lower in energy level than the source, energy will not move from the load to the source.
2-31 Using the word *drift* is appropriate because the movement of the electron is slow.
2-33 Electric charge per time
2-35 An electric circuit has a voltage source, metallic conductors, and a load. When the circuit is complete, there is electric current in the circuit which does work. When the circuit is incomplete, there is no current in the circuit.
2-37 Exposure to lasers; ultraviolet, microwave, and x-ray radiation; chemicals; hand and power tools; static and dynamic electricity; and high, intermediate, and low levels of voltage are sources of safety hazards.
2-39 (a) Unplug the equipment from the power source.
(b) Know the equipment by studying the operation manual.
(c) Know the circuit by checking the schematic diagram carefully.
(d) Pay attention to references to safety hazards mentioned by the manufacturer.
(e) Make sure a circuit is OFF by measuring before working on it.
(f) Keep yourself insulated from grounded objects when working on an energized circuit.
(g) Use only one hand when making measurements so you don't become a completed path for current to travel through.

CHAPTER 3

3-1 (a) Numerical (b) Unit
3-3 International System of Units
3-5 (a) Length: meter, m (b) Electric current: ampere, A
(c) Plane angle: radian, rad (d) Time: second, s
(e) Mass: kilogram, kg
3-7 (1) f (2) i (3) g (4) c (5) h (6) b
(7) a and d (8) g and h (9) e (10) d
3-9 (a) 0.230 × 10³ (b) 1.750 × 10³ (c) 47.300 × 10³
(d) 2.220 000 × 10⁶ (e) 8.240 × 10³
(f) 560.000 × 10³ or 0.560 000 × 10⁶
3-11 (a) 0.052 (b) 0.000 001 25 (c) 0.000 000 520
(d) 0.000 000 000 043 (e) 0.0025 (f) 0.000 820
3-13 (a) 43 kΩ (b) 800 ns (c) 4.5 mC (d) 9.43 kW
(e) 1.8 MΩ (f) 33 mV (g) 250 μA (h) 750 mm

3-15 (a) 9.24 kV (b) 78.3 mC (c) 510 kΩ or 0.51 MΩ
     (d) 285 mA (e) 3 μs   (f) 82.3 kW
3-17 64.8 s
3-19 200 mg
3-21 8.33 V
3-23 9.44 lb
3-25 19.29 mi/gal
3-27 22 mA

CHAPTER 4
Part I

4-1 Energy bands
4-3 It takes a discreet amount of energy to move an electron from the valence band to the conduction band.
4-5 4.54 MeV
4-7 (a) *Insulators* have a wide energy gap
    (b) *Semiconductors* have a narrow energy gap
    (c) *Conductors* have no energy gap
4-9 The separate molecules of monomers are linked to each other over and over again, forming a structure of several repeating parts. The linking of thousands and thousands of monomers forms polymers.
4-11 *ATFT teflon extruded flexible tubing:* shrink tubing; nonflammable; moistureproof; chemically inert.
     Teflon wire MIL-W-16878 D: Single conductor, stranded silver-plated copper.
4-13 Mica or asbestos
4-15 (a) Paper: 1200 volts/mil   (b) Styrene: 600 volts/mil
     (c) Steatite: 300 volts/mil (d) Bakelite: 300 volts/mil
     (e) Teflon: 1000 volts/mil   (f) Mica: 5000 volts/mil
4-17 1 kV
4-19 Covalent
4-21 Holes, electrons

4-23

Material	Relative Conductivity	Conductivity @ 20°C
(a) Silver	100	6.10×10^7
(b) Copper	95	58×10^6
(c) Gold	67	40.9×10^6
(d) Aluminum	58	35.4×10^6
(e) Tungsten	30	18.3×10^6
(f) Tin-lead	10	6.1×10^6
(g) Nichrome	1.6	976×10^3
(h) Carbon	4×10^{-2}	24.4×10^3
(i) Germanium	3×10^{-6}	1.83
(j) Silicon	3×10^{-9}	1.83×10^{-3}
(k) PVC	1×10^{-16}	61×10^{-12}
(l) Mica	1×10^{-18}	610×10^{-15}
(m) Teflon	1×10^{-21}	610×10^{-18}

4-25 78.1×10^{-9} Ω·m

785

4-27 2.04 Ω
4-29 980 μΩ
4-31 8.38 × 10¹⁸ Ω
4-33 In silicon and mica, covalent bonds are broken as the temperature rises, resulting in more charge carriers and, therefore, more free electrons and holes. In copper and aluminum, the atoms (positive ions) making up the crystalline lattice vibrate due to an increase in thermal energy which causes the moving atoms to present an obstacle to the otherwise free electrons.
4-35 10.3 Ω @ 1200°C
4-37 162 mΩ @ 85°C
4-39 391 μΩ @ 75°C

Part II
4-41 1.04 Ω
4-43 83.5 Ω @ 2000°C
4-45 196 Ω @ 900°C
4-47 (a) 10 Ω/1000 ft   (b) 100 Ω/1000 ft   (c) 1000 Ω/1000 ft
4-49 (a) 2.5 Ω/500 ft   (b) 12.5 Ω/500 ft   (c) 250 Ω/500 ft
4-51 (a) 7.71 Ω/300 ft   (b) 1600 cmil   (c) 10 mils
4-53 12.47 Ω/8.2 ft @ 55°C
4-55 4372 ft
4-57 No. 12 wire

CHAPTER 5
5-1 3 mΩ
5-3 50 MΩ
5-5 125 S
5-7 2.5 mS
5-9 10 Ω
5-11 17 A
5-13 12.1 V
5-15 Resistance
5-17 Current, while passing in either direction through the device, will experience the same electrical resistance.
5-19 Thermal
5-21 300 Ω
5-23 180 677 Ω
5-25 Connect two of the three terminals to make a rheostat. One terminal must be fixed and the other one variable.
5-27 250 Ω
5-29 (a) ≈100 Ω   (b) ≈2.5 kΩ
5-31 (a) ≈ −10°C   (b) ≈85°C
5-33 ≈2.7 kΩ
5-35 ≈6 kΩ
5-37 ≈2.8 fc
5-39 (a) 91 Ω   (b) 7.5 Ω   (c) 3.83 kΩ   (d) 1.2 Ω
     (e) 68 kΩ   (f) 0.47 Ω   (g) 53.6 Ω   (h) 0.2 Ω
5-41 (a) 180 kΩ ±10%   (b) 2.05 Ω ±1%   (c) 47 kΩ ±20%
     (d) 33 Ω ±5%   (e) 49.3 kΩ ±0.5%   (f) 2.2 kΩ ±2%

5-43 (a) Green Brown Brown Gold
(b) Gray Red Orange Silver
(c) Red Yellow Yellow Red
5-45 $10^{9/24} = 2.37$
5-47 110 Ω
5-49 108 Ω

CHAPTER 6

6-1 1.4 kW
6-3 833 W
6-5 3.38 kW
6-7 ≈150 mW
6-9 715 kΩ, 1 W, 500 V, ±1%, 7153
6-11 3.4 W at 175°C
6-13 (a) 1.75 kW (b) 85% (c) $6.86
6-15 (a) 4.4 kW (b) 38% (c) $59.14
6-17 (a) 1.55 kW (b) 64.7%
6-19 1.2 W
6-21 See Figure A7.
6-23 ≈300 ms
6-25 AGC, 1A

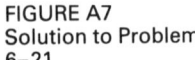

FIGURE A7
Solution to Problem
6-21

CHAPTER 7

7-1 65 Ω
7-3 570 Ω
7-5 $I_2 = I_3 = 2$ mA
7-7 40 Ω
7-9 130 V
7-11 3 A
7-13 (a) 12 mA (b) 6 V (c) 24 V (d) 90 V (e) 360 mW
7-15 (a) 8 mA (b) 60 V
7-17 (a) 211.6 V (b) 8.4 V
7-19 (a) Red Violet Brown (b) $R_3 = 358$ Ω
(c) 55.6 mA (d) 834 mW
7-21 (a) See Figure A8. (b) 3 A
7-23 45 V + (−9 V) + (−27 V) + (−9 V) = 0
7-25 (a) 800 Ω (b) 15.8 kΩ (c) 2.47 kΩ

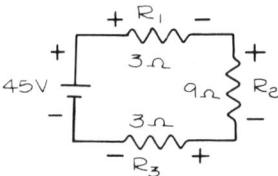

FIGURE A8
Solution to Problem
7-21(a)

7-27 (a) 5 V   (b) 10 V   (c) 8 kΩ
7-29 (a) 20 V   (b) 25 V   (c) 40 V   (d) 85 V
7-31 (a) Top—negative; bottom—positive   (b) 2.25 V
7-33 −9 V
7-35 5.19 mA, cw
7-37 (a) 1 mA   (b) 80 mW
7-39 2.55 mA, cw
7-41 (a) 45 V   (b) −16.1 V   (c) 23.3 V   (d) −45 V
7-43 (a) 58.5 V   (b) 47.2 V   (c) 129 V   (d) 106 V
7-45 (a) −33 V   (b) −9 V   (c) −30 V
7-47 (a) 353 mV   (b) −1.77 V   (c) −2.65 V
7-49 $R_1 = 333\ \Omega$ $R_2 = 667\ \Omega$
7-51 (a) 1.25 V   (b) 781 mW   (c) 83%
7-53 (a) 12 V   (b) 0 V   (c) 0 V   (d) 12 V   (e) −12 V
7-55 (a) 10.4 mA   (b) 18.5 mA   (c) 24 mA   (d) 0 A

CHAPTER 8

8-1 (a) 75 mS   (b) 13.3 Ω
8-3 (a) 500 μS   (b) 2 kΩ
8-5 30 Ω
8-7 (a) 22 V   (b) 22 V
8-9 2 A
8-11 2 A
8-13 (a) 714 mA   (b) 14.4 Ω   (c) 100 V   (d) 125 W
8-15 (a) 200 mA   (b) 397 mA   (c) 1.25 W
8-17 (a) 36.9 mA   (b) 24.6 mA   (c) 18.5 mA
8-19 6.25 Ω
8-21 880 mA + (−130 mA) + (−410 mA) + (−340 mA) = 0
8-23 137 V
8-25 (a) 11.7 mA   (b) 2.35 mA   (c) 3.91 mA
8-27 (a) 10 mA   (b) 5.13 mA   (c) 9.56 mA   (d) I = 24.7 mA
8-29 0 A
8-31 −5 mA
8-33 8 mA
8-35 (a) −34.8 V   (b) 5.12 mA   (c) 2.9 mA
     (d) 36 mA + (−28 mA) + (−5.12 mA) + (−2.9 mA) = 0
8-37 (a) 3.33 kΩ   (b) 8.4 mA   (c) 28 V
8-39 (a) 180 mA   (b) 70.6 V   (c) 392 Ω   (d) 12.7 W
8-41 750 Ω or less
8-43 (a) 0 W   (b) 232 mW for R_{int} = 14.5 kΩ
8-45 I_L = 2.74 mA; I_{int} = 268 μA
8-47 (a) 12.3 V   (b) 12.3 kΩ

CHAPTER 9

9-1 See Figure A9 where R_A = 50 Ω, R_B = 50 Ω, E_T = 10 V, and R_{eq} = 100 Ω
9-3 See Figure A9 where R_A = 5.6 kΩ, R_B = 4.27 kΩ, E_T = 18 V, R_{eq} = 9.87 kΩ and I = 1.82 mA

788

9-5 See Figure A10; $V_{ab} = 12.1$ V
9-7 $R_{eq} = 1.86$ kΩ, $E_T = 22$ V, and $I = 11.8$ mA
9-9 $R_{eq} = 80$ Ω and $V_{ba} = -16$ V
9-11 $R_{eq} = 84$ Ω and $I = 119$ mA
9-13 1.48 mA
9-15 (a) 643 μA (b) 8.57 V
9-17 (a) 17.85 V (b) 11.3 mA
9-19 5.45 kΩ
9-21 22.3 mA
9-23 -61 V
9-25 68.8 mA
9-27 -1.32 V
9-29 **434 mW**
9-31 2.92 W
9-33 (a) 4.04 V (b) 10.6 kΩ (c) 1.54 mW
9-35 No. 7 wire
9-37 262 ft
9-39 120 Ω resistor is open
9-41 1.2 kΩ $\pm 5\%$, $\frac{1}{2}$ W

CHAPTER 10

10-1 (a) 16 V (b) 4 V
10-3 (a) 37.5 kΩ (b) 12.5 kΩ
10-5 (a) $R_1 = 816$ Ω $R_2 = 184$ Ω
      (b) $R_1 = 825$ Ω $\pm 1\%$ $R_2 = 182$ Ω $\pm 1\%$
      (c) $P_1 = 1.98$ W $P_2 = 431$ mW
10-7 (a) 15.7 V (b) $P_{4.7k} = 24.4$ mW $P_{2.2k} = 11.4$ mW
10-9 (a) Orange Blue Brown Gold
      (b) 2 W—provides a derating by a factor of 2
10-11 (a) 60.7 mA (b) 24.3 mA (c) 40.4 mA
10-13 $R_1 = 200$ Ω $R_2 = 600$ Ω
10-15 (a) 526 mΩ (b) 101 mΩ (c) 50 mΩ
10-17 629 mΩ
10-19 (a) 45.7 V (b) -1.08%
10-21 129 kΩ
10-23 50 V
10-25 (a) $R_1 = 1.6$ kΩ $R_B = 2.67$ kΩ
       (b) $R_1 = 1.6$ kΩ $\pm 5\%$ $R_B = 2.7$ kΩ $\pm 5\%$
       (c) $P_1 = \frac{1}{4}$ W $P_B = \frac{1}{4}$ W

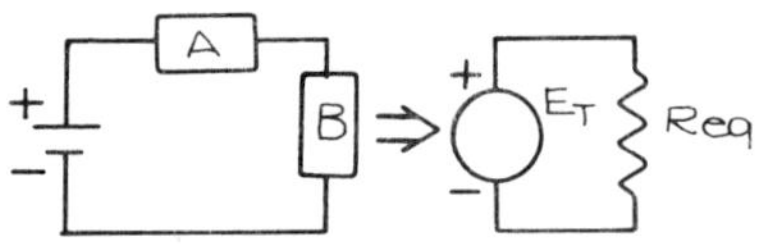

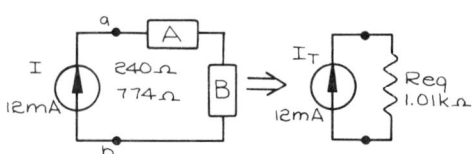

FIGURE A9
Diagram of Problems
9-1 and 9-3

FIGURE A10
Solution to Problem 9-5

10-27 (a) 1 kΩ (b) 0.38%
10-29 20 mA
10-31 13.0%

CHAPTER 11

11-1   12.14 V
11-3   12.5 V
11-5   0.5 V
11-7   1.22 V
11-9   −1.54 V
11-11 $E_{Th} = 12.5$ V $R_{Th} = 25\ \Omega$
11-13   (a) 21.3 mA   (b) 17.0 mA   (c) 10.6 mA
11-15   100 mV
11-17   50 mA
11-19   6.9 mA
11-21 $I_N = 500$ mA $R_N = 25\ \Omega$
11-23   8.57 mA
11-25 $E_{Th} = 7.15$ V $R_{Th} = 1.36$ kΩ
11-27   143 Ω
11-29   17.3 kΩ
11-31   12.14 V
11-33   −6.67 V
11-35   11.8 V
11-37   −5.88 V

CHAPTER 12

12-1 $E_1 - V_1 - E_2 - V_2 + E_3 = 0$
12-3 $15\ kI_1 - 10\ kI_2 = 10$
        $10\ kI_1 - 30\ kI_2 = 15$
        $-5\ kI_1 - 20\ kI_2 = 5$
12-5   4
12-7 $V_1 = 35$ V, $V_2 = -10$ V, $V_3 = 15$ V, $V_4 = 5$ V
12-9   3
12-11   Node a: $200 = 3\ V_a - V_b$ Node b: $0 = 3\ V_a - 8\ V_b$
12-13   2
12-15   4.62 V
12-17   (a) 2   (b) 1
12-19   −0.97 V
12-21   0 V
12-23   **$V_4 = 14.91$ V**
12-25   3.4 mA (down)
12-27   489.3 V
12-29   3.01 A

CHAPTER 13

13-1   86.5 fN
13-3   1.15 pN
13-5   125 kV/m

13-7   1.54 fN
13-9   10 mC/m^2
13-11 150 μC
13-13 2 μF
13-15 12.2 J
13-17 4.2 mJ
13-19 15.3 pF
13-21 67.8 V
13-23 0.0266 μF @ 250 V
13-25 (a) 65 μF   (b) 3.25 mC   (c) 81.3 mJ
13-27 (a) $V_{500\,pF}$ = 200 V   (b) $V_{330\,pF}$ = 90 V   (c) $V_{270\,pF}$ = 110 V

CHAPTER 14
14-1   9.4 μs
14-3   50 μs
14-5   6 kΩ
14-7   80 ms
14-9 v_R = −20 V, v_C = 20 V, current ccw
14-11 At t = 0$^+$ v_1 = 100 V v_2 = 0 V v_C = 0 V
       At t = ∞ v_1 = 20 V v_2 = 80 V v_C = 80 V
14-13 See Figure A11.
14-15 3.11 kΩ
14-17 See Figure A12.
14-19 (a) 1.9 V   (b) 7.87 V   (c) 12.6 V   (d) 18.4 V   (e) 20 V
14-21 41.5 kΩ

CHAPTER 15
15-1   (a) *SIMILARITIES:* Both have north and south poles and attract iron and other magnetic materials.
       (b) *DIFFERENCES:* Man-made magnets are made of alloys of iron, whereas natural magnets are made of iron oxide. Man-made magnets are stronger than natural magnets.

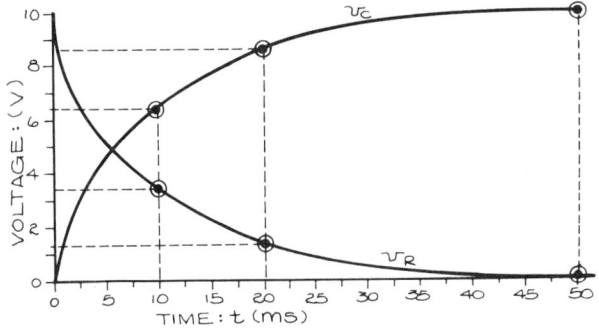

FIGURE A11
Solution to Problem
14-13

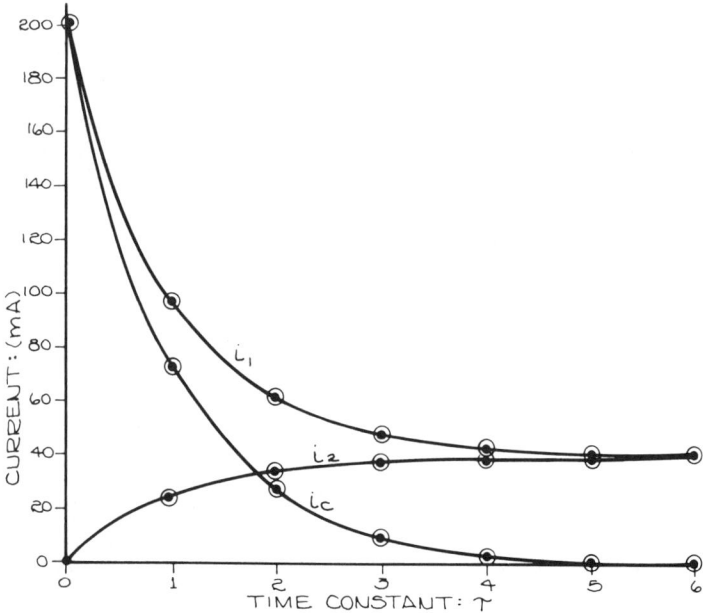

FIGURE A12
Solution to Problem
14–17

15–3 Paper, glass, plastic, copper, air, etc.
15–5 (a) Magnetic lines of force move from the north pole to the south pole.
(b) Field force of a magnet diminishes by the square of the distance from the pole.
(c) Unlike poles attract each other.
(d) Lines of force are only a visual representation of the field force and cannot be seen.
(e) True
(f) Magnetic lines form a complete continuous loop.
15–7 90 μWb
15–9 40 T
15–11 240 A
15–13 300 A/m
15–15 506 mA
15–17 1
15–19 5 mWb/(A · m)
15–21 419 μT
15–23 (a) B_r/B_m Ferrite = 0.90 B_r/B_m Fe alloy = 0.50
(b) Ferrite
15–25 $\mu = 342 \ \mu\text{Wb/A} \cdot \text{m}, \ \mu_r = 272$

CHAPTER 16

- 16–1 cw
- 16–3 To the left
- 16–5 0.269 N
- 16–7 814 mA
- 16–9 (a) 600 A (b) 1592 A/m (c) 2 mT (d) 1.23 μWb
- 16–11 488 mA/Wb
- 16–13 3.27 A
- 16–15 89.7 mA
- 16–17 1.6 A
- 16–19 (a) 1.24 T (b) 16.8 N
- 16–21 117 mA
- 16–23 (a) 52.8 μWb (b) 1.97 kA/m (c) 230 mT

CHAPTER 17

- 17–1 30 mWb
- 17–3 13.3 kV
- 17–5 Up
- 17–7 Top pole is north; bottom pole is south
- 17–9 150 V
- 17–11 5 H
- 17–13 90.5 μH
- 17–15 67 turns
- 17–17 (a) 101 mH (b) $W_1 = 18.8$ μJ, $W_2 = 48.8$ μJ, $W_3 = 58.8$ μJ
- 17–19 (a) 625 mH (b) 176 mJ

CHAPTER 18

- 18–1 (a) 60 ms (b) 18.5 μs (c) 37.5 ms (d) 357 ns
- 18–3 500 mH
- 18–5 /1.4 μs
- 18–7 (a) $i_1(0^+) = 100$ mA $i_1(\infty) = 0$ A
         (b) $i_2(0^+) = 400$ mA $i_2(\infty) = 0$ A
         (c) $v_{R1}(0^+) = 25$ V $v_{R1}(\infty) = 0$ V
         (d) $v_{R2}(0^+) = 20$ V $v_{R2}(\infty) = 0$ V
         (e) $v_{R3}(0^+) = 20$ V $v_{R3}(\infty) = 0$ V
- 18–9 400 μs
- 18–11 At $t = 0^-$ $i = 480$ mA   At $t = 0^+$ $i = 480$ mA
- 18–13 See Figure A13.
- 18–15 (a) $v_{L1}(0^-) = 0$ V $v_{L1}(0^+) = -20$ V
         (b) $v_{L2}(0^-) = 0$ V $v_{L2}(0^+) = 0$ V
         (c) $v_{R1}(0^-) = 20$ V $v_{R1}(0^+) = 20$ V

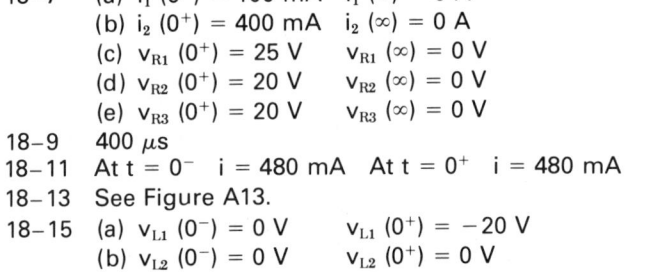

FIGURE A13
Solution to Problem 18–13

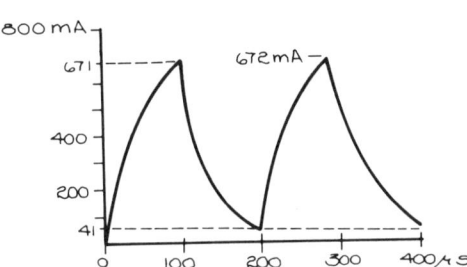

FIGURE A14
Solution to Problem
18–21

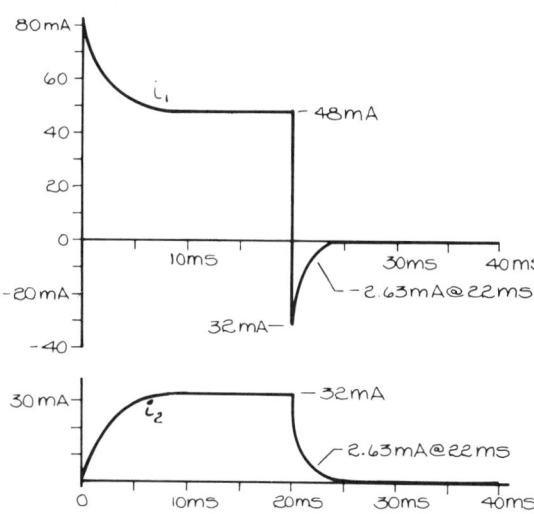

FIGURE A15
Solution to Problem
18–23

(d) $v_{R2}(0^-) = 20$ V $v_{R2}(0^+) = 20$ V
(e) $i_1(0^-) = 100$ mA $i_1(0^+) = 100$ mA
(f) $i_2(0^-) = 20$ mA $i_2(0^+) = 20$ mA

18–17 291 mA
18–19 16.9 ms
18–21 See Figure A14.
18–23 See Figure A15.
18–25 See Figure A16.
18–27 500 Ω

CHAPTER 19
19–1 24 Hz
19–3 19.2 Hz
19–5 195 Hz
19–7 (a) 143 μs   (b) 44 krad/s
19–9 (a) 1.25 radians   (b) 71.6°
19–11 44 V
19–13 $e = 5.6 \sin(101 \times 10^3 t)$ V
19–15 (a) $e = 0.8 \sin(5027t)$ V   (b) 7.61 mA
19–17 (a) 3.77 kΩ   (b) 25.1 kΩ   (c) 113 kΩ
19–19 (a) 1.21 MΩ   (b) 32.2 kΩ   (c) 1.93 kΩ
19–21 $e = 40 \sin(12.5 \times 10^3 t - \pi/2)$ V
19–23 (a) −8 A   (b) 16 V   (c) 424 mA
19–25 (a) 1.37 mA   (b) 3.0 A   *Note:* Mid-interval values used.
19–27 (a) 63.7 mA   (b) 0 W
       (c) $e = 25.5 \sin(377t)$ V $i = 90.1 \sin(377t - 90°)$ mA
19–29 (a) 12 V   (b) 5.41 V
19–31 34.7 W

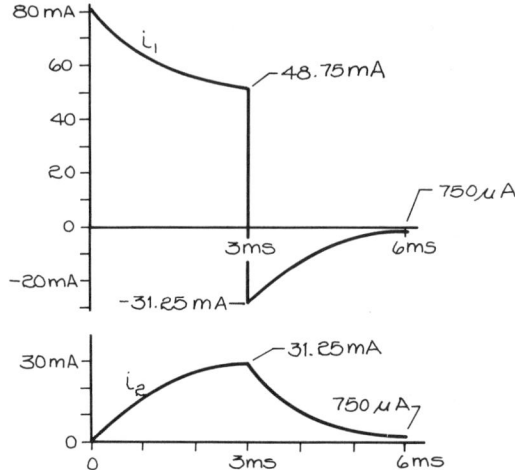

FIGURE A16
Solution to Problem
18–25

19–33 (a) 600 Hz (b) 244 mA (c) 328 mA
19–35 (a) 37.7 kΩ (b) 9.2 kV (c) e = 13 sin(3770t + 90°) kV
19–37 42.4 V

CHAPTER 20

20–1 (a) 0 + j5 (b) −1 + j2 (c) −4 − j4 (d) 2 − j2
20–3 (a) 7/−160° (b) 12/140° (c) 3/40° (d) 24/−20°
20–5 (a) 11.7/59° (b) 5.66/135° (c) 9.65/−34°
  (d) 22.2/−124° (e) 0.372/23.8° (f) 1085/125°
20–7 (a) 51.4 + j61.3 (b) 13 + j7.5 (c) 12.2 − j45.4
  (d) 35.4 + j8.31 (e) −4.55 + j1.62 (f) 10.7 − j98.7
20–9 (a) 9 + j2 (b) −5 − j6
  (c) 2.66 + j0.94 (d) 52.6 − j3.8
20–11 (a) −14 + j1 (b) −67 − j13
  (c) 4.97 + j3.45 (d) −0.239 − j0.901
20–13 (a) 20/30° (b) 12/−61° (c) 885/−112° (d) 106/−89.3°
20–15 (a) 6/30° (b) 11/50° (c) 2/163° (d) 1.02/−73.4°
20–17 (a) 6.01/−18° mA (b) 438/81° mA
  (c) 116/0° V (d) 25.5/40° V
20–19 (a) 120 sin(512t − 18°) A
  (b) 48.1 × 10⁻³ sin(512t + 65°) A
  (c) 34 sin(512t) V
  (d) 32 sin(512t − 56.9°) V
20–21 18.2/57.1° V
20–23 327/−83° mA
20–25 118 sin(377t − 53.5°) V
20–27 9.32 sin(800t − 26.4°) A

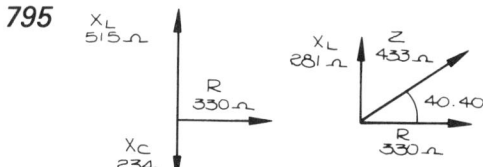

FIGURE A17
Solution to Problem
21–11

CHAPTER 21
21–1 $39.82 - j57.56$ V $= 70\underline{/-55.32°}$ V
21–3   (a) 10 kHz   (b) $e = 99 \sin(62.9 \times 10^3 t - 55.32°)$ V
21–5   (a) $V_C = 4.82$ V $V_R = 21.8$ V   (b) 31.6 V   (c) 12.5°
21–7 $28.6\underline{/25°}$ kΩ
21–9   (a) $47 - j177$ k$\Omega = 183\underline{/-75.1°}$ kΩ
     (b) $47 - j26.5$ k$\Omega = 54\underline{/-29.4°}$ kΩ
     (c) $47 - j10.6$ k$\Omega = 48.2\underline{/-12.7°}$ kΩ
21–11 $330 + j281$ $\Omega = 433\underline{/40.4°}$ Ω See Figure A17
21–13 $V_R = 19.8$ kV $V_L = 9.9$ kV $V_C = 14.85$ kV
21–15   (a) $2.16 + j7.88$ k$\Omega = 8.17\underline{/74.7°}$ kΩ
     (b) $1.84\underline{/-74.7°}$ mA   (c) 74.7°
21–17   (a) $14.7 - j23.4$ k$\Omega = 27.6\underline{/-57.9°}$ kΩ
     (b) $100 + j0 \Rightarrow 100\underline{/0°}$ V
21–19   (a) $23.4\underline{/20°}$ V   (b) 10 mA   (c) 20°   (d) 19.1 mH
     (e) 0.0106 μF   (f) $2.34\underline{/20°}$ kΩ
     (g) $e = 33.1 \sin(94.25 \times 10^3 t + 20°)$ V
21–21 $35.15\underline{/-50.7°}$ V
21–23 $2.33\underline{/-46.7°}$ V
21–25 $0.3 - j1.2 = 1.24\underline{/-76°}$ A
21–27   54.4 mH
21–29 $87.5\underline{/48°}$ μS
21–31 $Y = 2.17\underline{/44.1°}$ mS $Z = 461\underline{/-44.1°}$ Ω
21–33 $239 + j256$ μS $= 350\underline{/47°}$ μS
21–35   (a) $I_R = 200\underline{/0°}$ mA $I_C = 125\underline{/90°}$ mA $I_L = 333\underline{/-90°}$ mA
     (b) $289\underline{/-46.1°}$ mA   (c) $28.9\underline{/-46.1°}$ mS
21–37   (a) $Z_1 = 320\underline{/-90°}$ Ω $Z_2 = 500\underline{/-53.1°}$ Ω $Z_3 = 377\underline{/58°}$ Ω
     (b) $Y_1 = 3.13\underline{/90°}$ mS $Y_2 = 2.0\underline{/53.1°}$ mS $Y_3 = 2.65\underline{/-58°}$ mS
     (c) $I_1 = 313\underline{/90°}$ mA $I_2 = 200\underline{/53.1°}$ mA $I_3 = 265\underline{/-58°}$ mA
     (d) $359\underline{/43.6°}$ mA
21–39   (a) $Z_1 = 24.2\underline{/24.6°}$ kΩ $Z_2 = 37.3\underline{/-90°}$ kΩ $Z_3 = 59.4\underline{/-19.6°}$ kΩ
     (b) $55.5\underline{/15.9°}$ μS
21–41   (a) $5\underline{/0°}$ V   (b) $40.5\underline{/-14.1°}$ Ω
21–43 $55.7\underline{/53.1°}$ mA
21–45 $148\underline{/-48°}$ mA
21–47   (a) $R_{eq} = 6.82$ Ω; $C_{eq} = 189$ μF
     (b) $R_p = 35.5$ Ω; $C_p = 153$ μF
21–49   (a) $R_{eq} = 11.5$ kΩ; $L_{eq} = 13$ mH
     (b) $R_p = 34.9$ kΩ; $L_p = 19.4$ mH

CHAPTER 22

- 22-1   (a) $6.71\underline{/-136°}$ mA   (b) $94.9\underline{/-91°}$ V
- 22-3   (a) $8.68\underline{/-6°}$ kΩ (b) $R_p = 8.74$ kΩ $C_p = 160$ pF
- 22-5   (a) $165\underline{/7.9°}$ mA   (b) $10.8\underline{/36.3°}$ V
- 22-7 $R_{eq} = 149\ \Omega$; $C_{eq} = 24.2\ \mu F$
- 22-9 $2.1\underline{/-74.2°}$ V
- 22-11 $1.18\underline{/54.3°}$ kΩ
- 22-13 $586\underline{/-40.4°}\ \Omega$
- 22-15 $30.1\underline{/-64.7°}$ mA
- 22-17   (a) $753\underline{/32.1°}\ \Omega$ (b) $R_p = 885\ \Omega$, $L_p = 228$ mH
- 22-19 $83.9\underline{/-31.6°}$ V

CHAPTER 23

- 23-1   (a) 40.48 kVA   (b) 23.8 kvars   (c) 32.75 kW
- 23-3   (a) 0.984   (b) 5.65 kVA   (c) 5.56 kW
- 23-5 $R_{eq} = 129\ \Omega$; $C_{eq} = 22.7\ \mu F$
- 23-7   (a) 25 VA   (b) 20 vars
- 23-9   (a) 0.797   (b) $R_p = 9.26\ \Omega$; $X_p = 12.2\ \Omega$
- 23-11   24 A
- 23-13   31.02 kW
- 23-15   0.977
- 23-17   0.57
- 23-19   0.78
- 23-21   (a) 66.6 A   (b) 0.98
- 23-23   (a) Load A $P_a = 2028$ VA $P = 1895$ W
                          $P_q = 722.5$ vars (ind)
              Load B $P_a = 3227$ VA $P = 3227$ W
                          $P_q = 0$ vars
              Load C $P_a = 4151$ VA $P = 3558$ W
                          $P_q = 2138$ vars (cap)
     (b) $P_a = 8795$ VA $P = 8680$ W $P_q = 1416$ vars (cap)
     (c) 0.987
- 23-25 $I = 26.12$ A $\cos(\theta) = 0.884$
- 23-27   (a) $R_{eq} = 265\ \Omega$; $L_{eq} = 825$ nH
         (b) $R_p = 339\ \Omega$; $L_p = 3.78\ \mu H$

CHAPTER 24

- 24-1 $3.67\ \mu H$
- 24-3 $870\ \mu Wb$
- 24-5   4.91 mA
- 24-7 $E_{source} = 1.6$ kV $E_{load} = 320$ V
- 24-9 $600\ \Omega$
- 24-11   0.23
- 24-13   (a) $n_{0.625\ W} = 31.3$; $n_{1.25\ W} = 22.1$
         (b) $Z_{0.625\ W} = 7.84$ kΩ; $Z_{1.25\ W} = 3.91$ kΩ
- 24-15 $E_s = 48$ V
- 24-17 $E_p/2 = 270$ V
- 24-19 $Z_{1/2} = Z_p/4 = 4.1$ kΩ
- 24-21 $Z_p = 274\underline{/-55.5°}\ \Omega$
- 24-23   523
- 24-25   (a) 18.16 A   (b) 720 W   (c) 3.85 A

CHAPTER 25

25–1	26 dB
25–3	−20 dB
25–5	**56.9 mW**
25–7	26 dB
25–9	800 mW
25–11	60 mW
25–13	24.5 dBm
25–15	125 μW
25–17	8.9 mW
25–19	79.4 W
25–21	10 μW
25–23	17.4 dB
25–25	−0.76 dB
25–27	−24 dB/100 m
25–29	−4.08 dB

CHAPTER 26

26–1	$f_o = 15.2$ kHz; L/C $= 4.4 \times 10^6$
26–3	28.28 V @ 594 kHz and 28.28 V @ 606 kHz
26–5	20.1 Ω
26–7	$V_o = 670$ V BW $= 11.9$ kHz
26–9	(a) 6.28 Ω (b) 5065 pF
26–11	3.28 MHz
26–13	254 kHz
26–15	24.4 pF
26–17	43.2 kΩ
26–19	167 kΩ
26–21	17.3 mA
26–23	36.8 kHz
26–25	66.7 kΩ
26–27	$R_{eff} = 14.46$ Ω L $= 198$ μH

CHAPTER 27

27–1	20 pairs	27–3	144 rev/s
27–5	$I_A = 41.6 \underline{/-30°}$ A	27–7	$I_A = 103 \underline{/-55.7°}$ A
	$I_B = 41.6 \underline{/-150°}$ A		$I_B = 85.4 \underline{/-170°}$ A
	$I_C = 41.6 \underline{/90°}$ A		$I_C = 103 \underline{/75.4°}$ A
27–9	$|I_A| = |I_B| = |I_C| = |I_{line}| = 202$ A		
27–11	$I_A = I_{phaseA} = 9.68 \underline{/-22°}$ A		
	$I_B = I_{phaseB} = 17.7 \underline{/-158°}$ A		
	$I_C = I_{phaseC} = 8 \underline{/120°}$ A		
	$I_N = 11.8 \underline{/-164°}$ A		
27–13	$I_A = 4.99 \underline{/-40.5°}$ A	27–15	$I_A = I_{phaseA} = 21.6 \underline{/-30.3°}$ A
	$I_B = 7.56 \underline{/-145°}$ A		$I_B = I_{phaseB} = 21.6 \underline{/-150.3°}$ A
	$I_C = 7.93 \underline{/72.3°}$ A		$I_C = I_{phaseC} = 21.6 \underline{/89.7°}$ A
27–17	$I_A = I_{phaseA} = 282 \underline{/0°}$ A	27–19	(a) P $= 412.2$ kW
	$I_B = I_{phaseB} = 417 \underline{/-150°}$ A		(b) Q $= 38.4$ kvars (ind)
	$I_C = I_{phaseC} = 686 \underline{/88.9°}$ A		(c) S $= 414$ kVA
	$I_N = 482 \underline{/97.9°}$ A		(d) cos(5.32°) $= 0.996$ (lag)

798   27-21   Capacity = 11.1 kvars   27-23   Select 20 kvars since 208-V operation lowers rating by 0.75.

27-25   cos (θ) = 0.914 lagging without 2-kvar capacitor
cos (θ) = 0.998 leading with 2-kvar capacitor

CHAPTER 28   28-1   Any repetitive waveform is made up of a series of sine waves of appropriate frequency, amplitude, and phase. By applying calculus, a theoretical determination can be made of the characteristics of the sine waves needed to reproduce any repetitive wave shape.

28-3   The shape of the sound wave will not be a true reproduction of the electric wave shape because the sound wave is altered by phase distortion.

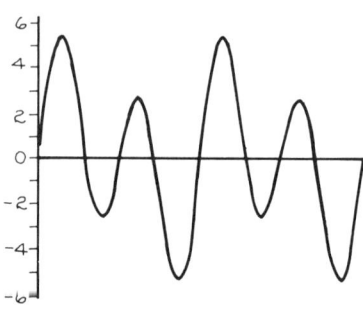

FIGURE A18
Solution to Problem
28-5

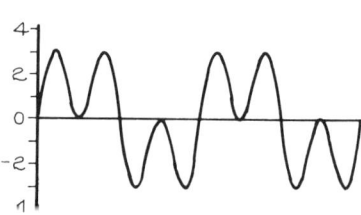

FIGURE A19
Solution to Problem
28-7

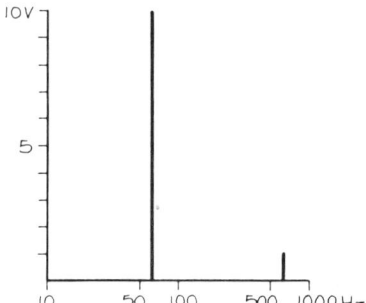

FIGURE A20
Solution to Problem
28-9

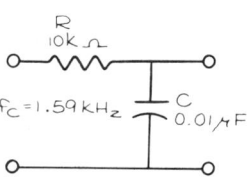

FIGURE A21
Solution to Problem
29-1

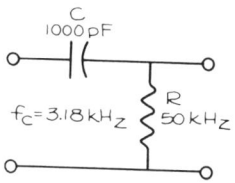

FIGURE A22
Solution to Problem
29–5

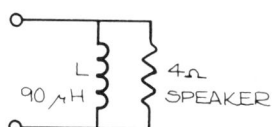

FIGURE A23
Solution to Problem
29–7

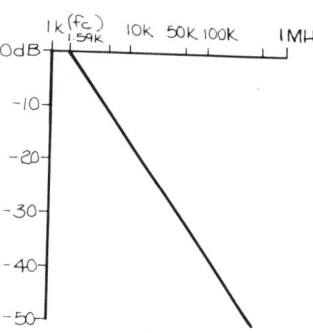

FIGURE A24
Solution to Problem
29–9

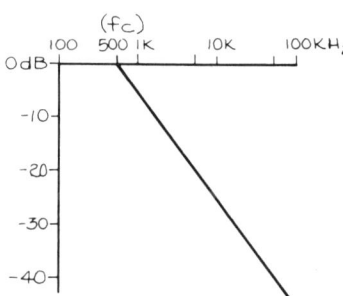

FIGURE A25
Solution to Problem
29–11

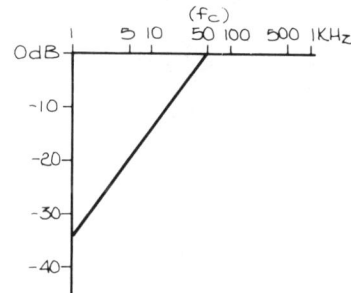

FIGURE A26
Solution to Problem
29–13

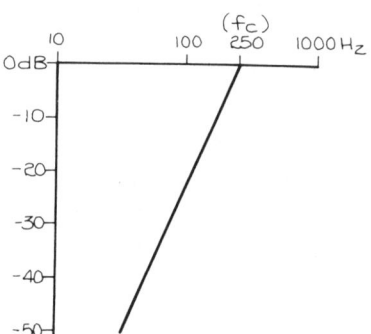

FIGURE A27
Solution to Problem
29–15

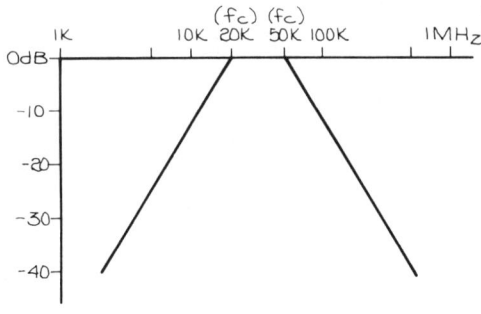

FIGURE A28
Solution to Problem
29–19

	28-5	See Figure A18.
	28-7	See Figure A19.
	28-9	See Figure A20.
	28-11	3rd = 375 MHz 5th = 625 MHz 7th = 875 MHz
	28-13	3.29%
	28-15	1 kHz and 200 Hz
	28-17	3.84 GHz and 4.16 GHz

CHAPTER 29

29-1	See Figure A21.
29-3	4.46 kH; Woofer
29-5	See Figure A22.
29-7	See Figure A23.
29-9	See Figure A24.
29-11	See Figure A25.
29-13	See Figure A26.
29-15	See Figure A27.
29-17	C = 796 pF; L = 15.9 mH
29-19	See Figure A28.
29-21	See Figure A29.
29-23	(a) Band-Pass (b) Band Reject (c) High-Pass (d) Low-Pass (e) Band-Pass

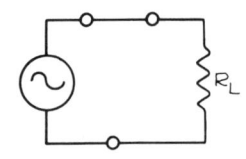

FIGURE A29
Solution to Problem 29-21

INDEX

Actual power. *See* Average power
Admittance, 579
    definition of, 566
    determining, 570–75
    of parallel ac circuits, 566–70
    of parallel-connected components, 569–70
    table summarizing, 570
    vector diagram construction, 567–69
Alternating current
    definition of, 24
    sine wave properties of, 500
    terminology of, 500–502
    voltge and current notation, 523
Alternating current circuits
    power in, 601–23
Alternating current networks
    determining total power in, 614–20
Alternator. *See* Synchronous generator
Aluminum
    absolute value of inferred zero resistance temperature, 80
    conductive properties of, 64
    resistivity of, 65, 79
    temperature coefficient of, 69
Aluminum electrolytic capacitors, 400
Ambient temperature, definition of, 120
American Wire Gauge, 81–85
    applications with formulas, 85–89
    definition of, 89
    equation for approximate cross section of any gauge number, 84
    equation for approximate resistance per 1000 ft for any gauge in, 84–85
    table of gauge numbers, diameters, cross sections, 82
Ammeter, 92, 93, 211, 242, 609
    symbol used for, 37
Ampere, definition of, 22–23, 32–33, 36, 49, 446
Amplification, 648, 649
Amplifiers, 137, 138, 269, 282, 293, 309, 767, 768
Amplitude, definition of, 147
Analog adder circuits, operation of, 332–34
Angular displacement
    equation for, 503
    equation relating angular velocity to, 502
Angular velocity, 501–502, 503
    definition of, 501
    equation relating frequency to, 502
Apparent power, 126, 602
    computing total, 614
    definition of, 602
    in three-phase load, 735
    relationship to average power, 602–603
    relationship to reactive power, 603
Approximations, 278–82
Argument, definition of, 529
Asbestos, 58
Atoms
    definition of, 15
    electrically neutral, 15, 16
    in metals, 18
    structure of, 15–19
Attenuation, 648, 649
    rate of, 771, 772, 777
Average power, 519–23, 602

Average power, in ac circuits
    equations for computing, 605
    relationship to apparent power, 602–603
Average power
    average value, 520–21
    definition of, 513–14, 147, 521
    effective value of sinusoidal voltage and current, 522-23
    equations for computing, 513–14, 126–28, 133, 521
    power in a reactive load, 522
    power in a resistive load, 520–21
    power in three-phase balanced load, 731-38
AWG. *See* American Wire Gauge
Axial leads, 100
    definition of, 120

Backlash, 105
Bandpass filters, 772-74
    applications of, 772-73
    definition of, 772
    description of, 773
Band-reject filters, 774-75
    behavior of, 774
    characteristics of, 775
    notch filter type, 774-75
Bandwidth, definition of, 677
Barium titanate, 59, 60, 108
Bilateral, definition of, 120
Bipolar transistors, 202, 690
Bleeder resistors
    definition of, 283
    selection of, 276–78
Bode plot, 768–71
    constructing, 769–71
    uses of, 768, 769, 771
Bonding of atoms
    covalent, 17, 18
    ionic, 17
    metallic, 17, 18–19
Branch, definition of, 210
Break frequency, 771
    definition of, 770
Bulk-property resistors
    characteristics of, 101

Calculators. *See* Electronic calculators
Candela, 32
Capacitance, 379–401, 466
    definition of, 386
    equations for, 386, 388, 392–93
    function of interface plate area, plate spacing and permitivity, 388–92
    method of increasing, 390–92
    sine wave phase relation of, 508–511
    stray. *See* Stray capacitance
    summary of ac effects on, 512
    use of, 386
Capacitance, of parallel plate capacitors, 387–90
    with dielectric, 391–92
Capacitance tolerance, definition of, 401
Capacitive reactance
    definition of, 509
Capacitors, 279, 293, 380, 763, 767, 774, 775
    applications of, 398, 406
    in automobile ignition systems, 492
    characteristics of, 398–410
    charge and discharge of, 406–407
    charging RC circuit, 415–17
    classification of, 398–400
    connecting in parallel and series, 396–98
    dc-blocking, 767-68
    discharging RC circuit, 417–18
    effect of leakage current, 422
    energy stored by, 387
    opposing a change in voltage, 509
    power in, 522, 523
    specifications for, 400-401
    under steady-state conditions, 484
    used for correcting power factor, 611, 613
    varactor, 776
    voltage across, 409–410
    voltage division between two series, 413–14
Carbon, 57
    conductive properties of, 64
Carbon-composition resistors, 98, 193, 262, 278
    characteristics of, 100–101, 103, 117
    computing the rated continuous working voltage for, 132–33
    power rating of, 128–29, 133–35

Carbon-film resistors, 278, 557
    characteristics of, 101, 103, 117
    power rating of, 128, 129
    rated continuous working voltage, 132
Ceramic capacitors, 399
Ceramic insulators, 58–59
Cermet-film resistors
    characteristics of, 101, 102, 103
    power derating curve for, 129–31
    typical working voltages for, 132
Charge carrier. *See* Electric current
Chemical energy, 21
Chlorine
    bonding with sodium, 17
Circuit breakers, 7, 141, 147
Circuit control. *See* Electric circuit control and protective devices
Circuits. *See* Electric circuits
Circular mil, definition of, 78, 89
Coercive force, 438, 439
Complex numbers, arithmetic operations with, 534–38
    adding, 534–35
    dividing, 537–38
    multiplying, 536–37
    subtracting, 535–36
Complex numbers, forms of
    polar, 528, 529–31
    rectangular, 528–29
Complex numbers, transformation of, 531–34
    polar to rectangular, 533-34
    rectangular to polar, 531-33
Complex wave forms
    creation of, 748–50
    harmonic distortion of. *See* Harmonic distortion
    intermodulation distortion. *See* Intermodulation distortion
    spectrum analysis of, 751–52
Computers, 126
Computers, timing circuits of, 420
    effects of stray capacitance, 423–24
    electrolytic capacitors in, 422-23
Conductance, 566, 567
    definition of, 63, 73, 94
    equations for, 94
Conductance band, 56, 57, 60, 62, 63
Conductive path, 7, 9
    definition of, 8

Conductivity
    definition of, 73
    of selective materials, 63, 64
Conductors, 56
    definition of, 73
    energy-band diagram of, 57
    equation for any one of six parameters, 73
    resistivity of selected, 65
    round metallic electric. *See* Round conductors
    temperature coefficient of selected, 69
Constantan, 80, 81
    absolute value of inferred zero resistance temperature, 80
    resistivity of, 65, 79
    temperature coefficient of, 68–69
Constant-current diode, 202
Constant-current sources, 202, 203, 204, 205
    construction of, 207–210
Conventional current direction
    rules for, 156, 157, 165–67, 179, 356
Copper, 7, 8, 18, 19, 23, 56
    absolute value of inferred zero resistance temperatures, 80
    conductive properties of, 63, 64
    resistivity of, 65, 68, 79
    temperature coefficient of, 69
    wire resistance per 1000 ft at 20 degrees C, 82, 83–84
Coulomb, 20, 21, 33
    definition of, 19, 36, 49
Coulomb's law, 380–81, 383
    definition of, 381
    equations for, 381, 382
Counterelectromotive force, 471
    definition of, 471
Critical resistance
    definition of, 138
    for several power series, 138
Crystals, 18, 19, 21
Current. *See* Electric current
Current dividers, 263-72
    applications, 269–72, 597–98
Current dividers, current divider principle, 264–69, 331
    equations for, 265, 575
Current dividers
    relation of currents to conductances, 266-67

relation of currents to resistance, 268
Current limited, 269, 270
    definition of, 283
Current phasor
    for parallel ac circuits, 563–66
Cutoff frequency
    as break frequency, 770
    definition of, 763
    equations for, 763, 764

Dc offset, definition of, 768
Dc power-supply filters, 764–65
Dc voltage source
    symbol used for, 37
Decade, definition of, 770
Decibels
    computing absolute power, 653–56
    computing power gain and loss with
        voltage and current ratios, 656–58
    creating an absolute decibel system, 656
    notation, 649
    relative power gain in, 653
Decibels, Decibel milliwatt unit designation, 653–54
    measuring, 654–55
Decibels, definition of, 648
Decibels, gain and loss in, 648, 649
    approximating, 650–52
Decimal notation
    compared to other notations, 39
    converting from engineering notation to, 41–42
    definition of, 40
    expressing in engineering notation, 39–40
Degrees Celsius, 35
    definition of, 49
Derating, definition of, 147
Dielectric constants
    definition of, 392
    values of selected materials, 392, 393
Dielectrics
    applications of, 391–92
    definition of, 390, 392
Dielectrics, permittivity of, 390
    effects of, 391, 392
Dielectric strengths, 394–95
    definition of, 395
    values of selected materials, 395

Diodes, 68, 96, 97, 293
    resistance of, 96, 97
Dipole, definition of, 394
Direct current, definition of, 23
Domain, definition of, 437
Doping, definition of, 62
Dry cell battery, 21, 22, 24

Eddy currents, definition of, 474
Efficiency, definition of, 147
Efficiency, equations for computing
    of a cascaded system, 137–38
    of an energy-converting device, 135–36
Electric ac generators, 12, 13, 14, 21, 24, 383
    definition of, 500
    three-phase, 702–704
Electric charge, 12, 14, 19, 21, 33, 34, 36
    measurement of. *See* Coulomb, definition of
    potential energy difference of the. *See* Potential difference
    two states of, 15
    work done on. *See* Electromotive force
Electric circuit control and protective devices, 141–47
    circuit breakers, 7, 141, 147
    definition of, 9
    fuses. *See* Fuses
    switches. *See* Switches
Electric circuits
    complete, 23
    components of, 7
    conventional current direction, 156, 157, 165–67, 179
    definition of, 7
    diagram of. *See* Schematic diagrams
    elementary, 21–22
    important concepts relating to, 23
    overload, 7
    troubleshooting, 171, 260
Electric current, 21–23, 34, 36–37
    alternating. *See* Alternating current
    ampere of, 22–23, 32–33
    conventional. *See* Conventional current direction
    definition of, 12, 22, 26, 36
    direct. *See* Direct current
    effective value of, 522
    flow of charge in, 21
    flow in electric circuits, 23–24

instantaneous, 514, 515, 518
  notation, 523
Electric current sources
  application of current-divider principle in circuits with, 269–71
Electric current sources, characteristics of, 202–210
  constant current. *See* Constant-current sources
  ideal, 202, 206, 272
  internal shunt resistance of. *See* Internal shunt resistance
  practical, 202, 203, 205
Electric current sources, types of connections
  parallel aiding, 197, 198
  parallel opposing, 197, 198
Electric energy, 7, 9, 12, 13, 14, 19, 20, 21
  computing cost of, 139–41
Electric field, 14–15
  definition of, 14, 26
Electric field strength, 380, 383–84
  between plates of parallel plate capacitor, 387
  equation for, 383
  expressing, 384
Electric flux density, 380
  definition of, 384
  dependent on permittivity and field strength, 391
  equations for, 385
Electricity, 7, 13–28
  cost of. *See* Electric energy, computing cost of
  definition of, 9, 12, 26
  properties of, 12, 19
Electric materials
  categories of, 52
  selected properties of, 55–90
Electric motors, 135, 136–37, 428, 455
Electric resistance, 33, 37
  definition of, 38
Electric shock, 24–25, 209
  body responses to current, 25
Electrodynamometer wattmeter, 606
Electrolytic capacitors, 422, 423
Electromagnetic induction, 466–67
  definition of, 466, 471
Electromagnetism, 444–47
  force on current carrying conductors, 444–46
  torque on a current loop, 446–47
Electromagnets, 450
  application of right-hand rule, 447
  formation of, 447
Electromotive force, 33, 37, 432
  definition of, 21, 26, 38
  measurement of, 21, 33, 38
  sources of, 21, 23, 24
Electronic calculators
  functions and operations used in text, 6, 30, 415, 531, 533, 540, 648
  use of, 282–83, 532, 533–35
Electronic devices
  active, 252, 282
  current operative, 263
  solid-state, 202, 203, 252
Electronic music, creation of, 754, 758
Electronics
  definition of, 23
  safety procedures in, 25
Electronic symbols, 8
Electrons, 15
  distribution of, 16
  shells of, 16
  spin up and spin down of, 436
  valence, 16, 17
Electrostatic force, 17, 18
  definition of, 16
Electrostatics, 380–85
Emf. *See* Electromotive force
Energy, 12, 13, 33
  average rate of delivering. *See* Average power
  in a closed system, 13
  cost of. *See* Electric energy, cost of
  definition of, 13, 20, 26, 36
  law of conservation of, 14, 135
Energy bands
  definition of, 56
  diagrams of, 56–57
  electron movement within, 56–57
Energy gap, 56, 57
Engineering notation, 30, 38–40
  compared to other notations, 39
  converting to decimal notation, 41–42
  definition of, 38, 49
  expressing decimal numbers in, 39–40
Equivalent circuits
  Norton's. *See* Norton's equivalent circuit
  series. *See* Equivalent series circuits

805

Thevenin's. *See* Thevenin's equivalent circuit
Equivalent series circuits, definition of, 179
Equivalent series circuits, for parallel circuits
    forming, 198–202
Equivalent series circuits, for series circuits
    forming, 167–70
Equivalent series circuits, for series parallel networks
    forming, 218–26

Fault
    definition of, 712
Faraday's law, 468, 469, 470, 471
    equation for, 468
    used to determine transformer voltage, 630
Ferromagnetic materials
    definition of, 437
    domains of, 437
    hysteresis loop of, 439
    magnetization of, 437–38, 452
Ferromagnetic transformers, 632
    coefficient of coupling, 630
    flux density in, 640
    longevity of, 628
    operating constraints of, *See* Transformers, operating constraints of
    power factor of, 631
    primary impedance of, 634–35
    secondary impedance of, 634
    used to match impedance, 635–37
    See also Transformers
Field effect transistors, 202, 690
Film resistors, 98, 100, 101, 103
    characteristics of, 101, 103
    power rating of, 128, 129
    types of, 101
Filters, 761–77
    applications of, 752, 758
    attenuation rate of, 771
    specifications and terminology, 776–77
Fixed linear resistors, 98, 100–103
    categories of, 100–103
    characteristics of, 100–103
    dc and audio frequency applications, 102
    industrial/consumer types, 103, 112, 113
    labeling systems used. *See* Resistance, designations of

    military type, 103, 115–17
    noise of, 101, 102
    specifying values and tolerances of. *See* Resistance, designations of
    voltage ratings of, 102
Flux linkage, 467, 470, 471
Force, 33, 34, 36
    contact, 14
    definition of, 35
    field, 14,
    See also Gravitational force
Fourier theorem, 750–51
    definition of, 750
Free electrons, 19, 22, 54–55, 60, 61
    definition of, 19, 23, 26
    randomness of, 19, 22
Frequency
    definition of, 501
    in elementary generators, 501
    equation relating angular velocity to, 502
Frequency domain
    measurements of, 751
Frequency response, 752–53
    definition of, 752
    flat, 752–53
    plotting for filters, 768
Full scale, definition of, 283
Fuses, 7, 8, 9, 141, 144–46
    current and voltage ratings of, 145
    fusing characteristics of, 145, 146
    fusing time of, 145, 146
    glass-cartridge, 144, 145–46
    ratings governing application of, 144–45

Generation
    alternate forms of, 705
    hydroelectric, 702
    thermoelectric, 704
Generator
    delta-connected, 714–19
    wye-connected, 719–30
Germanium
    electron structure for, 16
    recommended maximum operating temperature for devices of, 68
Germanium, intrinsic
    conductive properties of, 64
    resistivity of, 65
Glass-cartridge fuses, 144, 145–46

Glass insulators, 58, 60
Gold
    absolute value of inferred zero resistance temperatures, 80
    conductive properties of, 63, 64
    resistivity of, 65, 79
    temperature coefficient of, 69
Gravitational force, 14, 35
Gravity, 14
Ground, definition of, 179
Ground state, definition of, 56

Harmonic distortion, 753–54, 755
    definition of, 753–54
    equation for percentage of, 754
Harmonics
    presence in a wave, 754–55
Heat energy, 9, 12, 14, 20, 21
Heat sinks, use of, 68
Hertz, definition of, 501
High-pass filters
    cutoff frequency of, 765–67
    dc-blocking capacitor application, 767
    definition of, 765
    inductive capacitive, 766–67
Hole, definition of, 63, 73
Horsepower
    converting to watts, 136
    definition of, 136
Hydrogen
    bonding with oxygen, 17–18
Hysteresis, definition of, 439, 474
Hysteresis loop, 439

IM distortion. *See* Intermodulation distortion
Impedance, 566, 577
    computing, 556
    definition of, 555–56
    of series ac circuits, 555–58
    of series-connected components, 555–58
    vector diagram construction, 556–57
Induced voltage. *See* Counterelectromotive force
Inductance, 465–75
    of a coil, 472–74
    equations for, 471
    sine wave phase relation of, 506–508
    summary of ac effects on, 512
    total, 475

Induction, types of
    electromagnetic, 466–67, 471
    mutual, 467, 470, 471
    self, 471
Inductive reactance
    definition of, 506
    equation for, 506
Inductors, 279, 293, 466
    charge and discharge of, 480–81
    connected in series and in parallel, 475
    core and coil energy losses, 474
    definition of, 472
    energy stored by, 474
    equations describing wave forms of current and voltage of, 507–508
    filtering out unwanted frequencies, 507
    opposing a change in current, 507
    power in, 522, 523
    under steady-state conditions, 484
    types of, 472
Inferred zero resistance point
    definition of, 79, 89
    equation for, 80
Instantaneous equations
    for voltage and current, 511–13, 538
Instantaneous voltage
    equation relating to angular displacement, 503
    equation relating time to, 503
Insulating materials. *See* Insulators
Insulators, 56
    breakdown of, 60–62
    carbon-based, 58, 59
    characteristic of, 62
    conductive properties of, 63, 64
    definition of, 57, 73
    energy-band diagram of, 57
    material characteristics of, 57–58
    resistivity of selected, 65
    silicon-based, 58–59
    temperature coefficient of, 68
    thermal runaway of, 67–68
    wall thickness of, 61
Integrated circuits, 68, 252
Intermodulation distortion, 755–57
    contrasted to harmonic distortion, 756, 757
    creating with a mixer, 756–57
    definition of, 755

Internal shunt resistance
    determining, 206–207
    relation to constant current, 202–205
International System of Units. *See* SI system
Intertie
    definition of, 708
Ions, 18, 19
    definition of, 26
Iron
    electron spin in atom of, 437

Joule, 20, 21, 33, 36, 138
    definition of, 13

KCL. *See* Kirchhoff's current law
Kelvin, 32, 35, 49
Kilogram, definition of, 32
Kilowatthour meter, definition of, 138, 139
Kirchhoff's current law, 193–96, 228, 298, 300, 361, 558, 561
Kirchhoff's current law, application of, 597
    definition of, 193, 210
    in node analysis, 344, 361–62, 364
    to parallel ac circuits, 564–66
    to parallel circuits, 194–96, 264
    various forms of, 196, 197
Kirchhoff's voltage law, 158–61, 197, 228, 261, 271, 294–95, 296, 297, 481, 558, 560, 561, 571
Kirchhoff's voltage law, application of
    determining instantaneous equation of source voltage, 538
    in loop analysis, 344, 349–50, 360
    in RC series circuit, 418–19
    in RL circuits, 484–86
    to series ac circuits, 553–55
    to the series circuit, 158–60, 163–65
Kirchhoff's voltage law
    definition of, 158, 179
    various forms of, 160–61
KVL. *See* Kirchhoff's voltage law

Lamps, 127, 135, 144, 155–56
Lattices, 18–19, 56
    definition of, 18
Lead
    conductive properties of, 64

Lead polarization, 98
Leakage current, definition of, 60
Length, 32
    definition of, 34
Lenz's law, 466, 468–71
    definition of, 468–69
Light energy, 9, 12, 14, 20, 21
Linear resistance, 96, 97
    definition of, 120
Litz wire, definition of, 676
Load, 8, 23
    definition of, 7, 9
Load three-phase, 713–30
    delta-connected, 714–19
    wye-connected four-wire, 719–24
    wye-connected three-wire, 724–30
Loaded, definition of, 283
Lodestones
    definition of, 428
    flux density of, 432
Loop, definition of, 344
Loop analysis, 343, 344–45
    definition of, 344
    reasons for selecting instead of node analysis, 365–73
Loop equations
    assuming loop current direction, 348, 361
    establishing the number of, 345–47, 360
    establishing voltage polarities of elements within loop, 348–49, 360–61
    selection of loops, 348, 360
    solving by techniques of simultaneous equations, 351–61
    writing, 349–51
Loudspeakers, 135, 136, 138, 269–70, 428, 455, 748, 750
Low-pass filters, 762–65, 772
    dc power supply application, 764–65
    definition of, 762
Low-pass filters, inductive resistive (LR), 762–63
    determining output voltage of, 763
    solving for cutoff frequency, 763, 764
Low-pass filters, resistive capacitive (RC), 763–64
    determining output voltage of, 764
    solving for cutoff frequency, 764, 765

Magnetic circuits, 450–58
Magnetic circuits, air gap in, 455–57
    effects of, 455–56
    force resulting from, 456–57
Magnetic circuits
    definition of, 432
    devices employing, 450
    solving symmetrical parallel, 457–58
Magnetic circuits, series
    solving linear, 450–52
    solving nonlinear using BH magnetization curves, 451, 452–54
Magnetic field, characteristics of, 429–31
    magnetic induction, 430
    magnetic lines of force, 429–30
Magnetic field strength, 433–34
    equation for, 433–34
Magnetic flux
    definition of, 431, 450
    determining direction by right-hand rule, 444
Magnetic flux density, 431–32
    affected by permeability, 435
    definition of, 431
    determining, 435–36
    equation for, 431
    proportional to magnetic field strength, 434
Magnetic materials, properties of
    due to spin of electrons, 436
    ferromagnetic. *See* Ferromagnetic materials
    high retentivity, 430
    hysteresis, 438–39
    lower reluctance, 429–30, 431
Magnetic Ohm's law, 450
Magnetic quantities, 431–36
Magnetic units, 459
    common quantities, symbols, and units of the SI, CGS, and English systems, 460
Magnetism, 427–39
    atomic theory of, 436–37
    residual, 438, 439
Magnetite, 428
Magnetizing current, 611, 629
Magneto, 14
Magnetomotive force, 432–33, 435
    compared to electromotive force, 434
    definition of, 432

    equations for, 432, 450
    increased by air gap in circuit, 455, 456
Magnets
    artificial, 428–29, 430, 437
    natural, 428
    permanent. *See* Magnets, artificial
Magnitude, definition of, 529
Mass, 14, 32, 34–35
    definition of, 34
Maximum power transfer
    from ac voltage source, 620–23
Maximum power transfer theorem, 318–23
    definition of, 318
    use of Norton's equivalent circuit, 322–23, 330–31
    use of Thevenin's equivalent circuit, 318, 321, 328–29
Measurement, systems of, 30–32
    conversion between, 48–49
    English, 30, 31, 33, 35, 36, 44, 48, 49
    SI metric. *See* SI system
Mechanical energy, 12, 13, 14, 21
Metal-film resistors, 554
    characteristics of, 101, 103, 117
    power rating of, 129
    typical working voltages for, 132
Metallic wire. *See* Round conductors
Metal-oxide resistors
    characteristics of, 101
Metal-oxide varistors. *See* Varistors
Metals
    temperature coefficient of, 68, 69, 70
Meter, 31
    definition of, 32
Metric system. *See* SI system
Mica, 56, 56, 58
    breakdown voltage of, 60
    capacitors, 399, 310
    conductive properties of, 64
    as dielectric, 391–92, 393, 395
    resistivity of, 65
Millman's theorem, 323–26
Millman's theorem, application of, 324–26
    to demonstrate the operation of an analog adder circuit, 332–34
    guidelines for forming Thevenin's equivalent circuit, 323
Millman's theorem
    definition of, 323–24
    equations for, 324

MIL-spec, definition of, 120
Mixers, 756–57
MMF. *See* Magnetomotive force
Molecule, 17
    definition of, 18
Monomers, 58
Multipole-filters, 771–72
    determining characteristics of, 771
    rate of attenuation of, 772
Mutual induction, 470
    definition of, 467, 471

Network, definition of, 218, 245
Network theorems, 289–338
    application of, 326–38
Neutrons, 16
Newton, 33, 35, 36
Nichrome, 66, 101, 103
    absolute value of inferred zero resistance temperatures, 80
    conductive properties of, 63, 64
    resistivity of, 65, 79
    temperature coefficient of, 69
Node, definition of, 188, 210
Node analysis, 343, 344, 361
    reasons for selecting instead of loop analysis, 365–66
Node equations
    definition of major node, 364
    determining node voltages, 363, 364
    establishing the number of, 362–63
    selecting the reference node, 363
    writing and solving, 363–65
Nonlinear resistors, 107–112
    types of, 108
Norton's equivalent circuit, calculation of
    separating problem into two parts, 317
    using superposition theorem in a multiple-source circuit, 318
Norton's equivalent circuit
    compared to Thevenin's equivalent circuit, 314–15
    conversion to Thevenin's equivalent circuit, 315, 316–17
    forming of, 312–15
Norton's theorem, 311–18, 420
    compared to Thevenin's theorem, 311, 312
    definition of, 311

Notch filters, 774–75
Nucleus, 15, 16

Octave, definition of, 770
Ohm, definition of, 49
Ohmmeter, 107
    definition of, 87, 106
Ohm's law, application of, 97, 506, 559, 561
    computing capacitive circuit conditions, 510, 511
    computing inductive circuit conditions, 508
    determining voltage drop, 254–55
    to parallel circuits, 190–93, 203–205, 206, 207, 264
    to series circuits, 154–56, 158
    to series parallel networks, 237–38
Ohm's law
    equation for, 95, 152, 157, 194, 508
Open circuits, 177, 178–79
    characteristics of, 179
    definition of, 177, 179
    voltage of, 206, 210, 272, 274
Order of magnitude, definition of, 279, 283
Oscilloscope, 750, 757
    function of, 751
Overload, 7
    definition of, 9
Overload protection. *See* Electric circuit control and overload devices
Oxygen
    bonding with hydrogen, 17–18

Paper capacitors, 399
Parallel ac circuits, admittance of, 566–70
    of parallel-connected components, 569–70
    vector diagram construction, 567–69
Parallel ac circuits
    applying Kirchhoff's current law to, 564–66
    current division in, 575–77
    current phasor for, 563–66
    resonant characteristics of. *See* Parallel-resonant circuits
    solving, 570–75
Parallel ac equivalent circuits
    forming, 577, 579–83

Parallel circuits, 187–211
    application of ten-to-one rule in, 279
    computing equivalent resistance of, 199–202
    current of, 189
    current sources in, 196–98
    definition of, 188, 211
    determining amount of current in, 264
    resultant current direction, 198
    rules and laws governing, 189–90, 193, 194, 198
    simplifying to an equivalent series circuit, 199–202
    total conductance of, 189, 201
    voltage in, 189
Parallel plate capacitors, 387–92, 393
    determining capacitance of, 388–90
    use of dielectrics with, 391–92, 393–94
Parallel-resonant circuits
    bandwidth of, 686, 691, 692–93
    characteristics of, 671
    forming the parallel equivalent circuit, 687–90
    resonant frequency of, 685
Parallel-resonant circuits, impedance of
    loaded, 691–93
    unloaded, 686–90
Parallel-resonant circuits, factor of, 685–86
    determining, 685–86
    relationship to impedance, 685–86, 686–87, 689
Parallel-resonant circuits, tank circuit, 685, 687
    loading, 691–95
Period, of induced voltage wave form
    definition of, 501
    equation expressing relationship to frequency, 501
Permeability, 434–36
    equation for, 435
    of free space, 435
    of material in magnetic path, 434–36
    relative, 435, 436, 437
Permittivity, 380, 388
    definition of, 381–82
Phase angle, definition of, 505
Phasor diagrams, 542
    definition of, 528
Phasors, application of, 544–47
    compared to application of instantaneous equations, 538–43
Phasors
    expressing as effective values, 543–44
    symbolizing, 544
    using rules of complex numbers with, 544–47
Photoconductive cells. *See* Photoresistors
Photoresistors, 92, 108
    definition of, 111
    surface illumination of, 110–111
    uses of, 112
Photo-voltaic cell, 21
Piezoelectric effect, definition of, 21
Plastic film capacitors, 400
Plastics, 58
    insulating materials for wire and cable, 59
Polarity, definition of, 179
Polarization, definition of, 394
Polymerization, definition of, 58
Polymers. *See* Plastics
Polyvinylchloride
    breakdown voltage of, 60
    conductive properties of, 63, 64
    resistivity of, 65
    uses of, 59
Porcelain insulators, 59, 60
Potential difference, 19, 21, 382–83
    in a complete or closed circuit, 23
    definition of, 20, 26, 38, 382
    equation for, 382
    measurement of, 20, 21, 33, 38, 382
Potential energy
    difference of the electric charge. *See* Potential difference
    levels of, 16
Potentiometers, 92, 103
    carbon-composition, 104, 105–106, 107
    characteristics of, 104–105, 106
    definition of, 104, 121
    as a rheostat, 104, 105
    taper of, 106, 107
    as a voltage divider, 104, 262–63
    wire-wound, 105
Power, 33, 34, 514
    apparent. *See* Apparent power
    average rate of. *See* Average power
    definition of, 36, 127
    equations for. *See* Average power, computing
    in three-phase loads, 730–38

811

reactive. *See* Reactive power
real. *See* Average power
true. *See* Average power
Power dissipation, in ac circuits
    measurement of, 606–607
    parameters governing, 602–605
Power dissipation, definition of, 128
Power factors
    definition of, 604
    determining and using, 604–605
    in three-phase load, 738–41
    leading and lagging, 605–606
    used to indicate ineffective power line transmission, 609–10
Power lines, ac
    correcting for low power factor, 611–14
    reactive power in, 607–11
Powers of ten notation
    compared to other notations, 39
Power triangle, 603–604
Protons, 16
Pulse waves
    approximating effective (rms) quantity from, 516
Pulse waves, steady dc and periodic dc
    average value of, 520, 521
PVC. *See* Polyvinylchloride

Quadrature, 608

Radial leads, 100
    definition of, 121
Radian, 32
Radio and television receivers
    bandpass filter application in, 772
    maximum power transfer to, 321
    notch filter application in, 774–75
Radio and television transmitters, 126, 775
Radio frequencies
    problem obtaining power transfer, 622
Radio-frequency communications, 757, 758
RC circuits, 405–24
    applications, 420–24, 490
    charging, 415–17, 420
    as computer timing circuits, 420, 422–24
    current in, 418–19
    determining charge and discharge waveforms, 414–15, 489–90

determining exact voltage during transient period, 415–17
    determining transient time of, 408–409
    discharging, 417–18, 420
    instantaneous values of, 415–20
    procedure for solving, 489–90
    transient behavior of, 408
    transient period of, 415
    voltage division in, 413–15
RC circuits, conditions of
    after transient state, 409, 412
    prior to transient state, 409–12
RC circuits, time constant of
    definition of, 407
    equation for, 407, 408
    function of, 407, 409
Reactance, 567, 577, 579, 602
Reactive loads. *See* Inductors; Capacitors
Reactive power, 126, 602
    in ac power lines, 607–11
    in three-phase load, 734–35
    determining presence in load, 607
    relation to apparent power, 603
Real power. *See* Average power
Relays, operation of, 457
Reluctance, 450
    definition of, 429
    determining, 452
    equation for, 451
    increased by air gap in circuit, 455
Remanence, 438, 439
Resistance, 66, 466, 567, 602
    critical. *See* Critical resistance
    definition of, 64, 92, 93, 96
Resistance, designations of, 103, 112–120
    color code, 103, 113, 117–20
    effects of temperature on, 67–68
    MIL-spec, 103, 115–17
    preferred number series, 113–15
Resistance, equations for
    circuit elements, 93
    factors affecting, 64–65, 78
    as a function of temperature, 69–73
    linear. *See* Linear resistance
    nonlinear, 96, 97
    Ohm's law, 95
    of round conductors. *See* Round conductors, resistance of
    selected materials, 64, 71, 72, 73
    sine wave phase relation of, 505–506

summary of ac effects on, 512
Resistivity, 65, 66, 67
    definition of, 65, 73
    equation for, 68
    of round conductors, 79
    of selected materials, 65
    temperature coefficient of, 68–69
Resistors, 92, 96, 97
    accuracy of, 279
    application of ten-to-one rule to, 279–82
    bilateral characteristics of, 98
    choosing wattage rating for, 135
    definition of, 98
    power derating curve, 129–31
    power rating for, 128–29
    purpose of, 98, 261
    resistance of, 96, 97
    series dropping, 261
    specifying values and tolerances of. *See* Resistance, designations of
    voltage dependent. *See* Varistors
Resistors, linear, 98–107, 293
    fixed. *See* Fixed linear resistors
    power considerations in, 129–34
    specifications of, 128
    thermal stability of, 98–100
    variable. *See* Variable linear resistors
Resonance, 669–95
    parallel-resonant characteristics, 671
    series and parallel equivalent circuit parameters, 672–73
    series-resonant characteristics, 670, 671
Resonant circuits, properties of
    selectivity, 670, 675, 680, 681, 685, 686
    sensitivity, 670, 673, 675, 680, 681, 685, 686
Rheostat, 105
    definition of, 104, 121
Right-hand rule, 444, 469
Ripple, definition of, 765
RL circuits, 479–92
    applications, 490–92
    conditions before and after transient state, 484–86
    procedure for solving, 489–90
RL circuits, instantaneous values of, 486–90
    charge equation for current, 487–88, 489–90
    discharge equation for current, 486–87, 489–90
    solving for voltage across resistance, 488–89
RL circuits, time constant of, 481–84
    computing, 482–84
    definition of, 481–82
    equation for, 482
    relation of inductance and resistance to, 482
    used to generate high-voltage spikes, 490–92
Rms value. *See* Current, effective value of
Root mean square
    definition of, 513–14
    see also Current, effective value of
Round conductors, 78–89
    applications of formulas and AWG, 85–89
Round conductors, resistance of, 78, 79
    effect of temperature on, 79–81
    equation for, 78
    resistivity of, 78–79

Safety, in electronics
    hazards to, 24–25
    procedures for, 25
Scalar numbers, 544
Scalar presentation, 603
Schematic diagram, 8
    definition of, 8, 9, 37
    symbols used in, 37
Scientific method, 30
Second, definition of, 32
Selectivity, in resonant circuits
    definition of, 675
Self-inductance, 471, 472
Self-induction, definition of, 471
Semiconductors, 56
    amount of electric current in, 63
    as charge carriers, 62–63
    conductive properties of, 63, 64
    energy-band diagram of, 57
    explanation of term, 62
    extrinsic, 62
    intrinsic silicon, 62, 63, 64
    resistivity of selected, 65
    temperature coefficient of, 68
    thermal runaway of, 67–68
Sensitivity, in resonant circuits
    definition of, 673

Series ac circuits
    application of Kirchhoff's voltage law to, 553–55
    resonant characteristics of. *See* Series-resonant circuits
    solving, 558–61
    voltage division in, 561–63
    voltage phasor for, 552–53
Series ac circuits, impedance of, 555–58
    series-connected components, 557–58
    vector diagram construction, 556–57
Series ac equivalent circuits
    forming, 577–78, 581–83
Series circuits, 151–79
    amount of current in, 152, 157
    application of ten-to-one rule in, 279
    behavior of, 152
    conventional current direction, 156, 157, 165–67, 179
    definition of, 152, 179
    determining series internal resistance, 176–77
    equivalent. *See* Equivalent series circuits
    polarizing components of, 156–57
    total resistance of, 152, 157
    use of resistors in, 261–63
    voltage rise and drop in, 153, 157
Series circuits, with multiple voltage sources, 161–67
    application of KVL to, 163–65
    conventional current direction in, 165–67
    polarity of resultant voltage, 162–63
    types of connections, 161–62
Series circuits, rules and laws governing, 152–53, 156, 157, 158, 162, 165
    application of, 154–56, 157, 158–61, 162–65, 165–67
Series-parallel ac networks
    forming equivalent circuit of, 598–99
    solving, 589–98
Series-parallel networks, 217–45, 361
    definition of, 245
    forming an equivalent series circuit for, 218–26
    solving, 226–32
Series-parallel networks, applications with, 236–45
    determining one branch resistance, 236–37
    determining total resistance of two parallel-connected resistances, 236
    effects of an open circuit and short circuit, 239–41
    resistance of conductors, 243–44
Series-parallel networks, power dissipation in, 232–36
    equation for total power dissipated, 232
Series-resonant circuits
    bandwidth of, 676-80
    characteristics of, 670, 671
    effect of L/C ratio on sensitivity, 673–75
    resonant frequency of, 683–84
    voltages of, 684
Series-resonant circuits, effective resistance of, 675–76
    inductor properties as source of, 676
    relationship between bandwidth and, 678–80
Series-resonant circuits, quality factor of, 680–83
    defining the, 684–85
    Q-gain equation, 681
    related to inductive reactance, 681
Shells, of electron, 16
Short circuits, 177–79, 206
    characteristics of, 179
    definition of, 179
    explanation of, 177–78
Shunt, definition of, 211
Silicates, 58
Silicon, 56, 57
    recommended maximum operating temperature for devices of, 68
Silicon, intrinsic, 63, 64, 66
    bonding of atoms, 62, 63
    resistivity of, 65
Silicon dioxide, 58
Silver
    absolute value of inferred zero resistance temperatures, 80
    conductive properties of, 63, 64
    resistivity of, 65, 68, 79
    temperature coefficient of, 69
Sine waves, 514, 748
    approximating effective (rms) quantity from, 514–15
    average value of, 519–20
    characteristics of, 748, 749, 750

equations describing wave forms of resistor current and voltage, 506
linear addition of, 749–50
in phase, 505
phase relationships between individual, 748–49, 751
phase relation of capacitance, 508–511
phase relation of inductance, 506–508
phase relation of resistance, 505–506
Sine waves, produced in electric generators, 502–505
    time and displacement functions, 503–505
Sine waves
    superposition of, 748–49
SI system, 31–38
    base units of, 32–33, 38
    combining prefixed units, 44
    conversion between English and, 48–49
    derived units of, 33, 38
    prefixes and symbols for multiples and submultiples of, 42
    procedure for forming prefixed units in, 42–44
    selected identities in, 45
    supplementary units of, 32
    units with common definition and dimensional origin, 47
    use of unit analysis for conversion within, 44–48
Skin effect resistance, 676
Slug, 31, 35
Sodium
    bonding with chlorine, 17
Solenoids, 447–50
    action in relays, 457
    applications, 449–50
    definition of, 447
    determining inductance of, 472–74
    as electromagnetic devices, 449–50
    flux density in, 447–49
    flux leakage, 447
    toroid type, 448–49, 472
Solid-state devices, 252
    constant-current, 202, 203
Source resistance, definition of, 283, 318
Spectrum analysis, 751–52
    application of spectrum analyzer, 752
    definition of, 752
Sectrum analyzer, function of, 751

Splice, definition of, 89
Square waves, definition of, 757
Square-wave testing, 757–58
Standard conductor, definition of, 78
Steady dc waves
    characteristics of, 518
    determining the effective (rms) quantity from, 518
Steady-state conditions, 412
Steatite, 59, 60
Steinmetz, Charles, 628
Step current. *See* Step voltage, definition of
Stereo systems, 126
Stray capacitance
    definition of, 401
    effects of, 423–24
Superposition, principle of, 748
Superposition theorem, 290–301
Superposition theorem, application of, 293–301
    calculation of Thevenin's and Norton's equivalent circuits, 317–18, 328
Superposition theorem, definition of, 290–291
Superposition theorem, use concepts
    circuit conditions to be satisfied for, 293
    notation of polarity of each circuit voltage and current, 293
    setting sources to zero and replacing by internal resistance, 293
Surge current, definition of, 70
Susceptance, definition of, 567
Switches, 7, 8, 9, 141–44
    contact arrangements for, 148
    failure of, 142–43
Switches, types of
    DIP (dual in-line package), 142, 143–44
    knife, 141
    rotary, 142, 143
    toggle, 142, 143
Switches, uses of, 141, 143
Switching point, definition of, 109
Synchronous generator, 702–703

Tantalum electrolytic capacitors, 400, 423
Teflon
    breakdown voltage of, 60
    characteristics of, 59
    conductive properties of, 64

dielectric constant value of, 393
resistivity of, 65
Temperature, 32, 34, 49, 65, 73
    definition of, 35
    effects on conductor resistance, 67–68
    effect on wire (round conductor) resistance, 79–81
Temperature coefficient, 68–69, 70, 73
    definition of, 68, 73
    of selected metals and alloys, 69
Ten-to-one rule, 279
Tesla, Nikola, 628
Thermistors, 92
    characteristics of, 108–109
    definition of, 108
    relationship between temperature and resistance for, 108–109
    types of, 108–109
    uses of, 110
Thermocouple, 21
Thevenin's equivalent circuit, calculation of
    separating problem into two parts, 317
    using superposition theorem in a multiple-source circuit, 317, 328
Thevenin's equivalent circuit
    compared to Norton's equivalent circuit, 314–15
    conversion to Norton's equivalent circuit, 315–16, 335
Thevenin's equivalent circuit, forming of, 302–10, 323–29
    by applying Millman's theorem. *See* Millman's theorem
    to determine voltage across load and power dissipation, 307–309
    guidelines for, 306
Thevenin's equivalent circuit
    network conditions necessary to be a Thevenin's equivalent circuit, 309
Thevenin's theorem, 301–10
    additional concepts for understanding, 302
Thevenin's theorem, application of, 302–10
    forming Thevenin's equivalent circuit. *See* Thevenin's equivalent circuit, forming of
    to RC circuits, 419–20
Thevenin's theorem
    compared to Norton's theorem, 311, 312
    definition of, 301–302

point of reference for, 310
Time domain, measurement of, 751
Tin, conductive properties of, 64
Tolerance, definition of, 121
Toroid, definition of, 448
Transfer function, definition of, 753
Transformers, 450, 627–95, 709–11
    assuming ideal, 630, 634
    autotransformer, 637
    behavior of, 628–30
    center-tapped primary, 637
    coefficient of coupling in, 629–31
    current ratio for, 630–32
    determining voltage of, 630
    ferromagnetic (iron core). *See* Ferromagnetic transformers
    impedance ratio for, 633–34
    inductive coil, 628
    multiple secondary, 638
    mutual flux in, 628–29
    mutual inductance in, 629
    principle function of, 628
    reflected impedance of, 634–37
    step-down and step-up, 632–33
    three-phase, 709–11
    turns ratio for, 632–33, 634
Transformers, operating constraints of, 639–42
    frequency, 641
    temperature, 641–42
    voltage, 640
    volt-ampere rating, 642
Transformers, power
    application of, 709
    achieving size reduction of, 641
    causes of heat losses, 642
    effect of nonlinear operation of core, 640
    not exceeding specified voltage, 640
    sinusoidal magnetizing current and mutual flux in, 640
    temperature of, 641
    volt-ampere rating of, 642
Transient time, definition of, 408
Transistors, 62, 68, 202, 203, 252, 282
Transmission line, 706–8
Trimmer capacitors, 400
Trimmers, 103, 105, 106
True power. *See* Average power
Tunable filters
    electronically, 776

mechanically, 775–76
Tungsten, 18, 20, 81
    absolute value of inferred zero resistance temperatures, 80
    cold and hot resistance of, 69–70, 72–73
    conductive properties of, 64
    resistivity of, 65, 68, 69, 79
    temperature coefficient of, 69

Unit analysis technique, application of, 44–45
    for conversion between systems of units, 44, 48–49
    for conversion within SI system, 44–48
Unloaded, definition of, 283

Vacuum diode, 628
Valence band, 56, 57, 60, 62, 63
Valence electrons, 16
    definition of, 17, 26
    elevation to conduction band, 56–57, 62
    free motion of. *See* Free electrons
    sharing or transferring. *See* Bonding of atoms
Var, 608
Varactor, definition of, 776
Variable capacitors, 400
Variable linear resistors, 98, 103–107
    characteristics of, 103–105, 107
    used for volume control, 103, 104, 105–6, 107
Variable linear resistors, types of potentiometers. *See* Potentiometers
    trimmers, 103, 105, 106
    variable wire-wound, 103–104, 106
Varistors, 92, 108
    definition of, 110
    uses of, 110
Volt, 33, 38
    definition of, 20, 21, 49, 382
Voltage, 19–21
    across capacitors, 409–10
    approximating the effective (rms) quantity for, 126, 514
    computing the effective (rms) value for, 517–19
    definition of, 21, 26, 38
    induced. *See* Counterelectromotive force
    instantaneous. *See* Instantaneous voltage
    maximum continuous rms. *See* Voltage, working
    measurements of, 171–74
    notation, 170–71, 523
    polarity of, 173–74, 297
    reference symbols, 171–74
Voltage, in series circuits
    voltage divider principle, 255–60
    polarity of resultant voltage, 162–63
    types of source connections. *See* Voltage sources, types of connections
Voltage dividers, 252–63
    analysis of, 254–60
    applications, 260–63
    attaching load resistance to, 273–75
    designed loaded for two or more voltage levels, 276
    design options, 253–54
    determining load voltage with equation for, 272–73
    determining resistances in, 259–60
    importance of understanding, 252
Voltage dividers, voltage divider principle, 255–60
    applications of, 294, 295–96, 592, 595, 768
    equations for, 257, 488, 561
Voltage drop
    in a series circuit, 255
Voltage phasor
    for series ac circuits, 552–53
Voltage sources
    characteristics of, 174–79, 272
Voltage sources, ac
    applied to resistive load, 506
    maximum power transfer from, 620–23
Voltage sources, loaded
    definition of, 283
    determining load voltage, 272–75
Voltage sources, polarity of, 156–57, 162–63
    in loop equations, 348–49
Voltage sources, types of connections
    series aiding, 161, 162, 163, 165, 166, 167
    series opposing, 161–62, 165, 167
Voltage sources, unloaded, 272
    definition of, 283
Voltage, working, 131–133, 395
    computing the rated continuous working voltage, 131–133
    definition of, 131, 401

Voltmeter, 92, 93, 275–76
    symbol used for, 37

Watt, 36, 37
    definition of, 33, 49, 136
Wattage
    benefits for resistors with excess, 135
Wattless power. *See* Reactive power
Wattmeter, description of, 606
Wave forms, 514
    approximating effective (rms) quantities from, 514–18
    average value of, 519–20, 521
    characteristics of steady and periodic,
    computing effective (rms) value from, 517–19

Wave forms, periodic
    computing average value of, 519
    pulse. *See* Pulse waves
    sine ac waves. *See* Sine waves
    steady dc waves. *See* Steady dc waves
    symmetrical periodic ac, 520
Wire. *See* Round conductors
Wire-wound resistors, 98, 100
    characteristics of, 101–102, 103
Wire-wound resistors, types of
    power, 102, 103, 128, 129
    precision, 102, 103, 279
Work, 12–13, 33, 36
    definition of, 13
Wye-to-delta transformation, 724

Zone refining, 62